MEDICAL PHYSICS AND BIOMEDICAL ENGINEERING

Medical Science Series

MEDICAL PHYSICS AND BIOMEDICAL ENGINEERING

B H Brown, R H Smallwood, D C Barber,
P V Lawford and D R Hose

Department of Medical Physics and Clinical Engineering,
University of Sheffield and Central Sheffield University Hospitals,
Sheffield, UK

Institute of Physics Publishing
Bristol and Philadelphia

British Library Cataloguing-in-Publication Data

A catalogue record for this book is available from the British Library.

ISBN 0 7503 0367 0 (hbk)
ISBN 0 7503 0368 9 (pbk)

Library of Congress Cataloging-in-Publication Data are available

Published by Institute of Physics Publishing, wholly owned by The Institute of Physics, London

Institute of Physics Publishing, Dirac House, Temple Back, Bristol BS1 6BE, UK

US Office: Institute of Physics Publishing, The Public Ledger Building, Suite 1035, 150 South Independence Mall West, Philadelphia, PA 19106, USA

Typeset in LaTeX using the IOP Bookmaker Macros

Printed in the UK by Bookcraft Ltd, Bath

The *Medical Science Series* is the official book series of the International Federation for Medical and Biological Engineering (IFMBE) and the International Organization for Medical Physics (IOMP).

IFMBE

The IFMBE was established in 1959 to provide medical and biological engineering with an international presence. The Federation has a long history of encouraging and promoting international cooperation and collaboration in the use of technology for improving the health and life quality of man.

The IFMBE is an organization that is mostly an affiliation of national societies. Transnational organizations can also obtain membership. At present there are 42 national members, and one transnational member with a total membership in excess of 15 000. An observer category is provided to give personal status to groups or organizations considering formal affiliation.

Objectives

- To reflect the interests and initiatives of the affiliated organizations.
- To generate and disseminate information of interest to the medical and biological engineering community and international organizations.
- To provide an international forum for the exchange of ideas and concepts.
- To encourage and foster research and application of medical and biological engineering knowledge and techniques in support of life quality and cost-effective health care.
- To stimulate international cooperation and collaboration on medical and biological engineering matters.
- To encourage educational programmes which develop scientific and technical expertise in medical and biological engineering.

Activities

The IFMBE has published the journal *Medical and Biological Engineering and Computing* for over 34 years. A new journal *Cellular Engineering* was established in 1996 in order to stimulate this emerging field in biomedical engineering. In *IFMBE News* members are kept informed of the developments in the Federation. *Clinical Engineering Update* is a publication of our division of Clinical Engineering. The Federation also has a division for Technology Assessment in Health Care.

Every three years, the IFMBE holds a World Congress on Medical Physics and Biomedical Engineering, organized in cooperation with the IOMP and the IUPESM. In addition, annual, milestone, regional conferences are organized in different regions of the world, such as the Asia Pacific, Baltic, Mediterranean, African and South American regions.

The administrative council of the IFMBE meets once or twice a year and is the steering body for the IFMBE. The council is subject to the rulings of the General Assembly which meets every three years.

For further information on the activities of the IFMBE, please contact Jos A E Spaan, Professor of Medical Physics, Academic Medical Centre, University of Amsterdam, PO Box 22660, Meibergdreef 9, 1105 AZ, Amsterdam, The Netherlands. Tel: 31 (0) 20 566 5200. Fax: 31 (0) 20 691 7233. E-mail: IFMBE@amc.uva.nl. WWW: http://vub.vub.ac.be/~ifmbe.

IOMP

The IOMP was founded in 1963. The membership includes 64 national societies, two international organizations and 12 000 individuals. Membership of IOMP consists of individual members of the Adhering National Organizations. Two other forms of membership are available, namely Affiliated Regional Organization and Corporate Members. The IOMP is administered by a Council, which consists of delegates from each of the Adhering National Organization; regular meetings of Council are held every three years at the International

Conference on Medical Physics (ICMP). The Officers of the Council are the President, the Vice-President and the Secretary-General. IOMP committees include: developing countries, education and training; nominating; and publications.

Objectives

• To organize international cooperation in medical physics in all its aspects, especially in developing countries.
• To encourage and advise on the formation of national organizations of medical physics in those countries which lack such organizations.

Activities

Official publications of the IOMP are *Physiological Measurement*, *Physics in Medicine and Biology* and the *Medical Science Series*, all published by Institute of Physics Publishing. The IOMP publishes a bulletin *Medical Physics World* twice a year.

Two Council meetings and one General Assembly are held every three years at the ICMP. The most recent ICMPs were held in Kyoto, Japan (1991), Rio de Janeiro, Brazil (1994) and Nice, France (1997). The next conference is scheduled for Chicago, USA (2000). These conferences are normally held in collaboration with the IFMBE to form the World Congress on Medical Physics and Biomedical Engineering. The IOMP also sponsors occasional international conferences, workshops and courses.

For further information contact: Hans Svensson, PhD, DSc, Professor, Radiation Physics Department, University Hospital, 90185 Umeå, Sweden. Tel: (46) 90 785 3891. Fax: (46) 90 785 1588. E-mail: Hans.Svensson@radfys.umu.se.

CONTENTS

PREFACE

This book is based upon *Medical Physics and Physiological Measurement* which we wrote in 1981. That book had grown in turn out of a booklet which had been used in the Sheffield Department of Medical Physics and Clinical Engineering for the training of our technical staff. The intention behind our writing had been to give practical information which would enable the reader to carry out a very wide range of physiological measurement and treatment techniques which are often grouped under the umbrella titles of medical physics, clinical engineering and physiological measurement. However, it was more fulfilling to treat a subject in a little depth rather than at a purely practical level so we included much of the background physics, electronics, anatomy and physiology relevant to the various procedures. Our hope was that the book would serve as an introductory text to graduates in physics and engineering as well as serving the needs of our technical staff.

Whilst this new book is based upon the earlier text, it has a much wider intended readership. We have still included much of the practical information for technical staff but, in addition, a considerably greater depth of material is included for graduate students of both medical physics and biomedical engineering. At Sheffield we offer this material in both physics and engineering courses at Bachelor's and Master's degree levels. At the postgraduate level the target reader is a new graduate in physics or engineering who is starting postgraduate studies in the application of these disciplines to healthcare. The book is intended as a broad introductory text that will place the uses of physics and engineering in their medical, social and historical context. Much of the text is descriptive, so that these parts should be accessible to medical students with an interest in the technological aspects of medicine. The applications of physics and engineering in medicine have continued to expand both in number and complexity since 1981 and we have tried to increase our coverage accordingly.

The expansion in intended readership and subject coverage gave us a problem in terms of the size of the book. As a result we decided to omit some of the introductory material from the earlier book. We no longer include the basic electronics, and some of the anatomy and physiology, as well as the basic statistics, have been removed. It seemed to us that there are now many other texts available to students in these areas, so we have simply included the relevant references.

The range of topics we cover is very wide and we could not hope to write with authority on all of them. We have picked brains as required, but we have also expanded the number of authors to five. Rod and I very much thank Rod Hose, Pat Lawford and David Barber who have joined us as co-authors of the new book.

We have received help from many people, many of whom were acknowledged in the preface to the original book (see page xxiii). Now added to that list are John Conway, Lisa Williams, Adrian Wilson, Christine Segasby, John Fenner and Tony Trowbridge. Tony died in 1997, but he was a source of inspiration and we have used some of his lecture material in Chapter 13. However, we start with a recognition of the encouragement given by Professor Martin Black. Our thanks must also go to all our colleagues who tolerated our hours given to the book but lost to them. Sheffield has for many years enjoyed joint University and Hospital activities in medical physics and biomedical engineering. The result of this is a large group of professionals with a collective knowledge of the subject that is probably unique. We could not have written this book in a narrow environment.

We record our thanks to Kathryn Cantley at Institute of Physics Publishing for her long-term persistence and enthusiasm. We must also thank our respective wives and husband for the endless hours lost to them. As before, we place the initial blame at the feet of Professor Harold Miller who, during his years as Professor of Medical Physics at Sheffield and in his retirement until his death in 1996, encouraged an enthusiasm for the subject without which this book would never have been written.

Brian Brown and Rod Smallwood

Sheffield, 1998

PREFACE TO '*MEDICAL PHYSICS AND PHYSIOLOGICAL MEASUREMENT*'

This book grew from a booklet which is used in the Sheffield Department of Medical Physics and Clinical Engineering for the training of our technical staff. The intention behind our writing has been to give practical information which will enable the reader to carry out the very wide range of physiological measurement and treatment techniques which are often grouped under the umbrella title of medical physics and physiological measurement. However, it is more fulfilling to treat a subject in depth rather than at a purely practical level and we have therefore included much of the background physics, electronics, anatomy and physiology which is necessary for the student who wishes to know why a particular procedure is carried out. The book which has resulted is large but we hope it will be useful to graduates in physics or engineering (as well as technicians) who wish to be introduced to the application of their science to medicine. It may also be interesting to many medical graduates.

There are very few hospitals or academic departments which cover all the subjects about which we have written. In the United Kingdom, the Zuckermann Report of 1967 envisaged large departments of 'physical sciences applied to medicine'. However, largely because of the intractable personnel problems involved in bringing together many established departments, this report has not been widely adopted, but many people have accepted the arguments which advocate closer collaboration in scientific and training matters between departments such as Medical Physics, Nuclear Medicine, Clinical Engineering, Audiology, ECG, Respiratory Function and Neurophysiology. We are convinced that these topics have much in common and can benefit from close association. This is one of the reasons for our enthusiasm to write this book. However, the coverage is very wide so that a person with several years' experience in one of the topics should not expect to learn very much about their own topic in our book—hopefully, they should find the other topics interesting.

Much of the background introductory material is covered in the first seven chapters. The remaining chapters cover the greater part of the sections to be found in most larger departments of Medical Physics and Clinical Engineering and in associated hospital departments of Physiological Measurement. Practical experiments are given at the end of most of the chapters to help both individual students and their supervisors. It is our intention that a reader should follow the book in sequence, even if they omit some sections, but we accept the reality that readers will take chapters in isolation and we have therefore made extensive cross-references to associated material.

The range of topics is so wide that we could not hope to write with authority on all of them. We considered using several authors but eventually decided to capitalize on our good fortune and utilize the wide experience available to us in the Sheffield University and Area Health Authority (Teaching) Department of Medical Physics and Clinical Engineering. We are both very much in debt to our colleagues, who have supplied us with information and made helpful comments on our many drafts. Writing this book has been enjoyable to both of us and we have learnt much whilst researching the chapters outside our personal competence. Having said that, we nonetheless accept responsibility for the errors which must certainly still exist and we would encourage our readers to let us know of any they find.

Our acknowledgments must start with Professor M M Black who encouraged us to put pen to paper and Miss Cecile Clarke, who has spent too many hours typing diligently and with good humour whilst looking after a busy office. The following list is not comprehensive but contains those to whom we owe particular debts: Harry Wood, David Barber, Susan Sherriff, Carl Morgan, Ian Blair, Vincent Sellars, Islwyn Pryce, John Stevens, Walt O'Dowd, Neil Kenyon, Graham Harston, Keith Bomford, Alan Robinson, Trevor Jenkins, Chris Franks, Jacques Hermans and Wendy Makin of our department, and also Dr John Jarratt of the Department of Neurology and Miss Judith Connell of the Department of Communication. A list of the books which we have used and from which we have profited greatly is given in the Bibliography. We also thank the Royal Hallamshire Hospital and Northern General Hospital Departments of Medical Illustration for some of the diagrams.

Finishing our acknowledgments is as easy as beginning them. We must thank our respective wives for the endless hours lost to them whilst we wrote, but the initial blame we lay at the feet of Professor Harold Miller who, during his years as Professor of Medical Physics in Sheffield until his retirement in 1975, and indeed since that time, gave both of us the enthusiasm for our subject without which our lives would be much less interesting.

Brian Brown and Rod Smallwood

Sheffield, 1981

NOTES TO READERS

Medical physics and biomedical engineering covers a very wide range of subjects, not all of which are included in this book. However, we have attempted to cover the main subject areas such that the material is suitable for physical science and engineering students at both graduate and postgraduate levels who have an interest in following a career either in healthcare or in related research.

Our intention has been to present both the scientific basis and the practical application of each subject area. For example, Chapter 3 covers the physics of hearing and Chapter 15 covers the practical application of this in audiology. The book thus falls broadly into two parts with the break following Chapter 14. Our intention has been that the material should be followed in the order of the chapters as this gives a broad view of the subject. In many cases one chapter builds upon techniques that have been introduced in earlier chapters. However, we appreciate that students may wish to study selected subjects and in this case will just read the chapters covering the introductory science and then the application of specific subjects. Cross-referencing has been used to show where earlier material may be needed to understand a particular section.

The previous book was intended mainly for technical staff and as a broad introductory text for graduates. However, we have now added material at a higher level, appropriate for postgraduates and for those entering a research programme in medical physics and biomedical engineering. Some sections of the book do assume a degree level background in the mathematics needed in physics and engineering. The introduction to each chapter describes the level of material to be presented and readers should use this in deciding which sections are appropriate to their own background.

As the book has been used as part of Sheffield University courses in medical physics and biomedical engineering, we have included problems at the end of each chapter. The intention of the short questions is that readers can test their understanding of the main principles of each chapter. Longer questions are also given, but answers are only given to about half of them. Both the short and longer questions should be useful to students as a means of testing their reading and to teachers involved in setting examinations.

The text is now aimed at providing the material for taught courses. Nonetheless we hope we have not lost sight of our intention simply to describe a fascinating subject area to the reader.

ACKNOWLEDGMENTS

We would like to thank the following for the use of their material in this book: the authors of all figures not originated by ourselves, Butterworth–Heinemann Publishers, Chemical Rubber Company Press, Churchill Livingstone, Cochlear Ltd, John Wiley & Sons, Inc., Macmillian Press, Marcel Dekker, Inc., Springer-Verlag GmbH & Co. KG, The MIT Press.

CHAPTER 1

BIOMECHANICS

1.1. INTRODUCTION AND OBJECTIVES

In this chapter we will investigate some of the biomechanical systems in the human body. We shall see how even relatively simple mechanical models can be used to develop an insight into the performance of the system. Some of the questions that we shall address are listed below.

- What sorts of loads are supported by the human body?
- How strong are our bones?
- What are the engineering characteristics of our tissues?
- How efficient is the design of the skeleton, and what are the limits of the loads that we can apply to it?
- What models can we use to describe the process of locomotion? What can we do with these models?
- What are the limits on the performance of the body?
- Why can a frog jump so high?

The material in this chapter is suitable for undergraduates, graduates and the more general reader.

1.2. PROPERTIES OF MATERIALS

1.2.1. Stress/strain relationships: the constitutive equation

If we take a rod of some material and subject it to a load along its axis we expect that it will change in length. We might draw a load/displacement curve based on experimental data, as shown in figure 1.1.

We could construct a curve like this for any rod, but it is obvious that its shape depends on the geometry of the rod as much as on any properties of the material from which it is made. We could, however, chop the rod up into smaller elements and, apart from difficulties close to the ends, we might reasonably assume that each element of the same dimensions carries the same amount of load and extends by the same amount. We might then describe the displacement in terms of extension per unit length, which we will call *strain* (ε), and the load in terms of load per unit area, which we will call *stress* (σ). We can then redraw the load/displacement curve as a stress/strain curve, and this should be independent of the dimensions of the bar. In practice we might have to take some care in the design of a test specimen in order to eliminate end effects.

The shape of the stress/strain curve illustrated in figure 1.2 is typical of many engineering materials, and particularly of metals and alloys. In the context of biomechanics it is also characteristic of bone, which is studied in more detail in section 1.2.2. There is a linear portion between the origin O and the point Y. In this

1

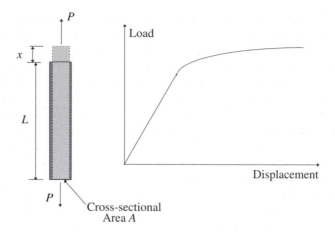

Figure 1.1. *Load/displacement curve: uniaxial tension.*

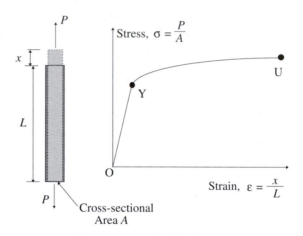

Figure 1.2. *Stress/strain curve: uniaxial tension.*

region the stress is proportional to the strain. The constant of proportionality, E, is called *Young's modulus*,

$$\sigma = E\varepsilon.$$

The linearity of the equivalent portion of the load/displacement curve is known as Hooke's law.

For many materials a bar loaded to any point on the portion OY of the stress/strain curve and then unloaded will return to its original unstressed length. It will follow the same line during unloading as it did during loading. This property of the material is known as *elasticity*. In this context it is not necessary for the curve to be linear: the important characteristic is the similarity of the loading and unloading processes. A material that exhibits this property and has a straight portion OY is referred to as *linear elastic* in this region. All other combinations of linear/nonlinear and elastic/inelastic are possible.

The linear relationship between stress and strain holds only up to the point Y. After this point the relationship is nonlinear, and often the slope of the curve drops off very quickly after this point. This means that the material starts to feel 'soft', and extends a great deal for little extra load. Typically the point Y represents a critical stress in the material. After this point the unloading curve will no longer be the same as the loading curve, and upon unloading from a point beyond Y the material will be seen to exhibit a permanent distortion. For this reason Y is often referred to as the *yield point* (and the stress there as the yield stress), although in principle there is no fundamental reason why the limit of proportionality should coincide with the limit of elasticity. The portion of the curve beyond the yield point is referred to as the *plastic region*.

The bar finally fractures at the point U. The stress there is referred to as the (uniaxial) *ultimate tensile stress* (UTS). Often the strain at the point U is very much greater than that at Y, whereas the ultimate tensile stress is only a little greater (perhaps by up to 50%) than the yield stress. Although the material does not actually fail at the yield stress, the bar has suffered a permanent strain and might be regarded as being damaged. Very few engineering structures are designed to operate normally above the yield stress, although they might well be designed to move into this region under extraordinary conditions. A good example of post-yield design is the 'crumple zone' of an automobile, designed to absorb the energy of a crash. The area under the load/displacement curve, or the volume integral of the area under the stress/strain curve, is a measure of the energy required to achieve a particular deformation. On inspection of the shape of the curve it is obvious that a great deal of energy can be absorbed in the plastic region.

Materials like rubber, when stretched to high strains, tend to follow very different loading and unloading curves. A typical example of a uniaxial test of a rubber specimen is illustrated in figure 1.3. This phenomenon is known as *hysteresis*, and the area between the loading and unloading curves is a measure of the energy lost during the process. Over a period of time the rubber tends to creep back to its original length, but the capacity of the system as a shock absorber is apparent.

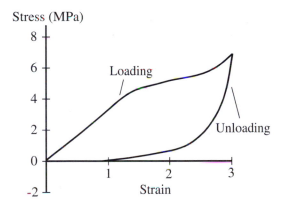

Figure 1.3. *Typical experimental uniaxial stress/strain curve for rubber.*

We might consider that the uniaxial stress/strain curve describes the behaviour of our material quite adequately. In fact there are many questions that remain unanswered by a test of this type. These fall primarily into three categories: one associated with the nature and orientation of loads; one associated with time; and one associated with our definitions of stress and strain. Some of the questions that we should ask and need to answer, particularly in the context of biomechanics, are summarized below: key words that are associated with the questions are listed in italics. We shall visit many of these topics as we discuss the properties of bone and tissue and explore some of the models used to describe them. For further information the reader is referred to the works listed in the bibliography.

- Our curve represents the response to tensile loads. Is there any difference under compressive loads? Are there any other types of load?

Compression, Bending, Shear, Torsion.

- The material is loaded along one particular axis. What happens if we load it along a different axis? What happens if we load it along two or three axes simultaneously?
  Homogeneity, Isotropy, Constitutive equations.

- We observe that most materials under tensile load contract in the transverse directions, implying that the cross-sectional area reduces. Can we use measures of this contraction to learn more about the material?
  Poisson's ratio, Constitutive equations.

- What happens if the rod is loaded more quickly or more slowly? Does the shape of the stress/strain curve change substantially?
  Rate dependence, Viscoelasticity.

- What happens if a load is maintained at a constant value for a long period of time? Does the rod continue to stretch? Conversely, what happens if a constant extension is maintained? Does the load diminish or does it hold constant?
  Creep, Relaxation, Viscoelasticity.

- What happens if a load is applied and removed repeatedly? Does the shape of the stress/strain curve change?
  Cyclic loads, Fatigue, Endurance, Conditioning.

- When calculating increments of strain from increments of displacement should we always divide by the original length of the bar, or should we recognize that it has already stretched and divide by its extended length? Similarly, should we divide the load by the original area of the bar or by its deformed area prior to application of the current increment of load?
  Logarithmic strain, True stress, Hyperelasticity.

The concepts of homogeneity and isotropy are of particular importance to us when we begin a study of biological materials. A *homogeneous* material is one that is the same at all points in space. Most biological materials are made up of several different materials, and if we look at them under a microscope we can see that they are not the same at all points. For example, if we look at one point in a piece of tissue we might find collagen, elastin or cellular material; the material is inhomogeneous. Nevertheless, we might find some uniformity in the behaviour of a piece of the material of a length scale of a few orders of magnitude greater than the scale of the local inhomogeneity. In this sense we might be able to construct characteristic curves for a 'composite material' of the individual components in the appropriate proportions. Composite materials can take on desirable properties of each of their constituents, or can use some of the constituents to mitigate undesirable properties of others. The most common example is the use of stiff and/or strong fibres in a softer matrix. The fibres can have enormous strength or stiffness, but tend to be brittle and easily damaged. Cracks propagate very quickly in such materials. When they are embedded in an elastic matrix, the resulting composite does not have quite the strength and stiffness of the individual fibres, but it is much less susceptible to damage. Glass, aramid, carbon fibres and epoxy matrices are widely used in the aerospace industries to produce stiff, strong and light structures. The body uses similar principles in the construction of bone and tissue.

An *isotropic* material is one that exhibits the same properties in all directions at a given point in space. Many composite materials are deliberately designed to be anisotropic. A composite consisting of glass fibres aligned in one direction in an epoxy matrix will be stiff and strong in the direction of the fibres, but its properties in the transverse direction will be governed almost entirely by those of the matrix material. For such a material the strength and stiffness obviously depend on the orientation of the applied loads relative to the orientation of the fibres. The same is true of bone and of tissue. In principle, the body will tend to orientate its fibres so that they coincide with the load paths within the structures. For example, a long bone will have fibres orientated along the axis and a pressurized tube will have fibres running around the circumference. There is even a remodelling process in living bone in which fibres can realign when load paths change.

Despite the problems outlined above, simple uniaxial stress/strain tests do provide a sound basis for comparison of mechanical properties of materials. Typical stress/strain curves can be constructed to describe the mechanical performance of many biomaterials. In this chapter we shall consider in more detail two very different components of the human body: bones and soft tissue. Uniaxial tests on bone exhibit a linear load/displacement relationship described by Hooke's law. The load/displacement relationship for soft tissues is usually nonlinear, and in fact the gradient of the stress/strain curve is sometimes represented as a linear function of the stress.

1.2.2. Bone

Bone is a composite material, containing both organic and inorganic components. The organic components, about one-third of the bone mass, include the cells, osteoblasts, osteocytes and osteoid. The inorganic components are hydroxyapatites (mineral salts), primarily calcium phosphates.

- The osteoid contains collagen, a fibrous protein found in all connective tissues. It is a low elastic modulus material ($E \approx 1.2$ GPa) that serves as a matrix and carrier for the harder and stiffer mineral material. The collagen provides much of the tensile strength (but not stiffness) of the bone. Deproteinized bone is hard, brittle and weak in tension, like a piece of chalk.
- The mineral salts give the bone its hardness and its compressive stiffness and strength. The stiffness of the salt crystals is about 165 GPa, approaching that of steel. Demineralized bone is soft, rubbery and ductile.

The skeleton is composed of *cortical (compact)* and *cancellous (spongy)* bone, the distinction being made based on the porosity or density of the bone material. The division is arbitrary, but is often taken to be around 30% porosity (see figure 1.4).

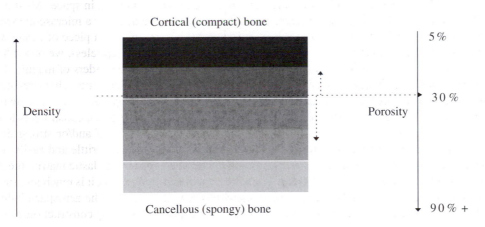

Figure 1.4. *Density and porosity of bone.*

Cortical bone is found where the stresses are high and cancellous bone where the stresses are lower (because the loads are more distributed), but high distributed stiffness is required. The aircraft designer uses honeycomb cores in situations that are similar to those where cancellous bone is found.

Cortical bone is hard and has a stress/strain relationship similar to many engineering materials that are in common use. It is anisotropic, and the properties that are measured for a bone specimen depend on the orientation of the load relative to the orientation of the collagen fibres. Furthermore, partly because of its composite structure, its properties in tension, in compression and shear are rather different. In principle, bone is strongest in compression, weaker in tension and weakest in shear. The strength and stiffness of bone also vary with the age and sex of the subject, the strain rate and whether it is wet or dry. Dry bone is typically slightly stiffer (higher Young's modulus) but more brittle (lower strain to failure) than wet bone. A typical uniaxial tensile test result for a wet human femur is illustrated in figure 1.5. Some of the mechanical properties of the femur are summarized in table 1.1, based primarily on a similar table in Fung (1993).

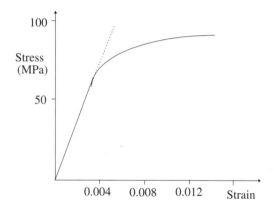

Figure 1.5. *Uniaxial stress/strain curve for cortical bone.*

Table 1.1. *Mechanical properties of bone (values quoted by Fung (1993)).*

| Bone | Tension | | | Compression | | | Shear | | Poisson's ratio |
	σ (MPa)	ε (%)	E (GPa)	σ (MPa)	ε (%)	E (GPa)	σ (MPa)	ε (%)	ν
Femur	124	1.41	17.6	170	1.85		54	3.2	0.4

For comparison, a typical structural steel has a strength of perhaps 700 MPa and a stiffness of 200 GPa. There is more variation in the strength of steel than in its stiffness. *Cortical bone is approximately one-tenth as stiff and one-fifth as strong as steel.* Other properties, tabulated by Cochran (1982), include the yield strength (80 MPa, 0.2% strain) and the fatigue strength (30 MPa at 10^8 cycles).

Living bone has a unique feature that distinguishes it from any other engineering material. It remodels itself in response to the stresses acting upon it. The re-modelling process includes both a change in the volume of the bone and an orientating of the fibres to an optimal direction to resist the stresses imposed. This observation was first made by Julius Wolff in the late 19th Century, and is accordingly called Wolff's law. Although many other workers in the field have confirmed this observation, the mechanisms by which it occurs are not yet fully understood.

Experiments have shown the effects of screws and screw holes on the energy-storing capacity of rabbit bones. A screw inserted in the femur causes an immediate 70% decrease in its load capacity. This is consistent with the stress concentration factor of three associated with a hole in a plate. After eight weeks the stress-raising effects have disappeared completely due to local remodelling of the bone. Similar re-modelling processes occur in humans when plates are screwed to the bones of broken limbs.

1.2.3. Tissue

Tissue is the fabric of the human body. There are four basic types of tissue, and each has many subtypes and variations. The four types are:

- epithelial (*covering*) tissue;
- connective (*support*) tissue;
- muscle (*movement*) tissue;
- nervous (*control*) tissue.

In this chapter we will be concerned primarily with connective tissues such as tendons and ligaments. Tendons are usually arranged as ropes or sheets of dense connective tissue, and serve to connect muscles to bones or to other muscles. Ligaments serve a similar purpose, but attach bone to bone at joints. In the context of this chapter we are using the term tissue to describe soft tissue in particular. In a wider sense bones themselves can be considered as a form of connective tissue, and cartilage can be considered as an intermediate stage with properties somewhere between those of soft tissue and bone.

Like bone, soft tissue is a composite material with many individual components. It is made up of cells intimately mixed with intracellular materials. The intracellular material consists of fibres of collagen, elastin, reticulin and a gel material called ground substance. The proportions of the materials depend on the type of tissue. Dense connective tissues generally contain relatively little of the ground substance and loose connective tissues contain rather more. The most important component of soft tissue with respect to the mechanical properties is usually the collagen fibre. The properties of the tissue are governed not only by the amount of collagen fibre in it, but also by the orientation of the fibres. In some tissues, particularly those that transmit a uniaxial tension, the fibres are parallel to each other and to the applied load. Tendons and ligaments are often arranged in this way, although the fibres might appear irregular and wavy in the relaxed condition. In other tissues the collagen fibres are curved, and often spiral, giving rise to complex material behaviour.

The behaviour of tissues under load is very complex, and there is still no satisfactory first-principles explanation of the experimental data. Nevertheless, the properties can be measured and constitutive equations can be developed that fit experimental observation. The stress/strain curves of many collagenous tissues, including tendon, skin, resting skeletal muscle and the scleral wall of the globe of the eye, exhibit a stress/strain curve in which the gradient of the curve is a linear function of the applied stress (figure 1.6).

1.2.4. Viscoelasticity

The tissue model considered in the previous section is based on the assumption that the stress/strain curve is independent of the rate of loading. Although this is true over a wide range of loading for some tissue types, including the skeletal muscles of the heart, it is not true for others. When the stresses and strains are dependent upon time, and upon rate of loading, the material is described as *viscoelastic*. Some of the models that have been proposed to describe viscoelastic behaviour are discussed and analysed by Fung (1993). There follows a brief review of the basic building blocks of these viscoelastic models. The nomenclature adopted is that of Fung. The models that we shall consider are all based on the assumption that a rod of viscoelastic material behaves as a set of linear springs and viscous dampers in some combination.

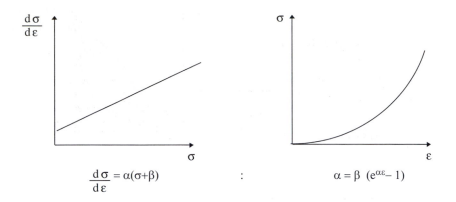

$$\frac{d\sigma}{d\varepsilon} = \alpha(\sigma+\beta) \qquad : \qquad \alpha = \beta\ (e^{\alpha\varepsilon}-1)$$

Figure 1.6. *Typical stress/strain curves for some tissues.*

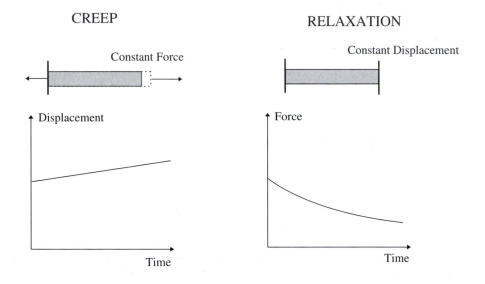

Figure 1.7. *Typical creep and relaxation curves.*

Creep and relaxation

Viscoelastic materials are characterized by their capacity to *creep under constant loads and to relax under constant displacements* (figure 1.7).

Springs and dashpots

A linear spring responds instantaneously to an applied load, producing a displacement proportional to the load (figure 1.8).

The displacement of the spring is determined by the applied load. If the load is a function of time, $F = F(t)$, then the displacement is proportional to the load and the rate of change of displacement is

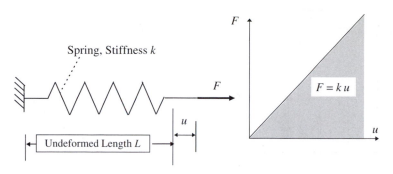

Figure 1.8. *Load/displacement characteristics of a spring.*

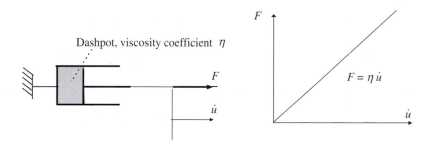

Figure 1.9. *Load/velocity characteristics of a dashpot.*

proportional to the rate of change of load,

$$u_{spring} = \frac{F}{k} \qquad \dot{u}_{spring} = \frac{\dot{F}}{k}.$$

A dashpot produces a velocity that is proportional to the load applied to it at any instant (figure 1.9).

For the dashpot the velocity is proportional to the applied load and the displacement is found by integration,

$$\dot{u}_{dashpot} = \frac{F}{\eta} \qquad u_{dashpot} = \int \frac{F}{\eta}\,dt.$$

Note that the displacement of the dashpot will increase forever under a constant load.

Models of viscoelasticity

Three models that have been used to represent the behaviour of viscoelastic materials are illustrated in figure 1.10.

The *Maxwell model* consists of a spring and dashpot in series. When a force is applied the velocity is given by

$$\dot{u} = \dot{u}_{spring} + \dot{u}_{dashpot}$$

$$\dot{u} = \frac{\dot{F}}{k} + \frac{F}{\eta}.$$

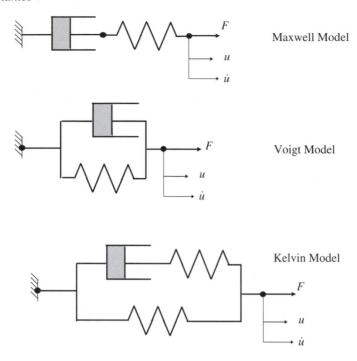

Maxwell Model

Voigt Model

Kelvin Model

Figure 1.10. *Three 'building-block' models of viscoelasticity.*

The displacement at any point in time will be calculated by integration of this differential equation.

The *Voigt model* consists of a spring and dashpot in parallel. When a force is applied, the displacement of the spring and dashpot is the same. The total force must be that applied, and so the governing equation is

$$F_{\text{dashpot}} + F_{\text{spring}} = F$$
$$\eta \dot{u} + ku = F.$$

The *Kelvin* model consists of a Maxwell element in parallel with a spring. The displacement of the Maxwell element and that of the spring must be the same, and the total force applied to the Maxwell element and the spring is known. It can be shown that the governing equation for the Kelvin model is

$$E_{\text{R}}\left(\tau_\sigma \dot{u} + u\right) = \tau_\varepsilon \dot{F} + F$$

where

$$E_{\text{R}} = k_2 \qquad \tau_\sigma = \frac{\eta_1}{k_2}\left(1 + \frac{k_2}{k_1}\right) \qquad \tau_\varepsilon = \frac{\eta_1}{k_1}.$$

In this equation the subscript 1 applies to the spring and dashpot of the Maxwell element and the subscript 2 applies to the parallel spring. τ_ε is referred to as the relaxation time for constant strain and τ_σ is referred to as the relaxation time for constant stress.

These equations are quite general, and might be solved for any applied loading defined as a function of time. It is instructive to follow Fung in the investigation of the response of a system represented by each of these models to a unit load applied suddenly at time $t = 0$, and then held constant. The unit step function $\mathbf{1}(t)$ is defined as illustrated in figure 1.11.

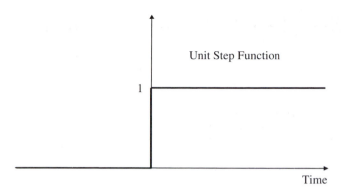

Figure 1.11. *Unit step function.*

For the Maxwell solid the solution is

$$u = \left(\frac{1}{k} + \frac{1}{\eta}t\right)\mathbf{1}(t).$$

Note that this equation satisfies the initial condition that the displacement is $1/k$ as soon as the load is applied.

For the Voigt solid the solution is

$$u = \frac{1}{k}\left(1 - e^{-(k/\eta)t}\right)\mathbf{1}(t).$$

In this case the initial condition is that the displacement is zero at time zero, because the spring cannot respond to the load without applying a velocity to the dashpot. Once again the solution is chosen to satisfy the initial conditions.

For the Kelvin solid the solution is

$$u = \frac{1}{E_R}\left(1 - \left(1 - \frac{\tau_\varepsilon}{\tau_\sigma}\right)e^{-t/\tau_\sigma}\right)\mathbf{1}(t).$$

It is left for the reader to think about the initial conditions that are appropriate for the Kelvin model and to demonstrate that the above solution satisfies them.

The solution for a load held constant for a period of time and then removed can be found simply by adding a negative and phase-shifted solution to that shown above. The response curves for each of the models are shown in figure 1.12. These represent the behaviour of the models under constant load. They are sometimes called creep functions. Similar curves showing force against time, sometimes called relaxation functions, can be constructed to represent their behaviour under constant displacement. For the Maxwell model the force relaxes exponentially and is asymptotic to zero. For the Voigt model a force of infinite magnitude but infinitesimal duration (an impulse) is required to obtain the displacement, and thereafter the force is constant. For the Kelvin model an initial force is required to displace the spring elements by the required amount, and the force subsequently relaxes as the Maxwell element relaxes. In this case the force is asymptotic to that generated in the parallel spring.

The value of these models is in trying to understand the observed performance of viscoelastic materials. Most soft biological tissues exhibit viscoelastic properties. The forms of creep and relaxation curves for the materials can give a strong indication as to which model is most appropriate, or of how to build a composite model from these basic building blocks. Kelvin showed the inadequacy of the simpler models in accounting

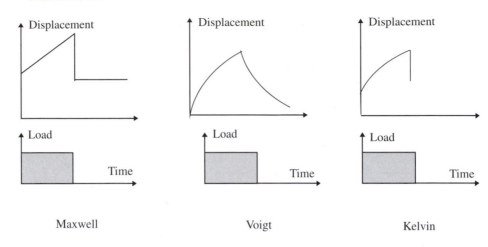

Figure 1.12. *Creep functions for Maxwell, Kelvin and Voigt models of viscoelasticity.*

for the rate of dissipation of energy in some materials under cyclic loading. The Kelvin model is sometimes called the standard linear model because it is the simplest model that contains force and displacement and their first derivatives.

More general models can be developed using different combinations of these simpler models. Each of these system models is *passive* in that it responds to an externally applied force. Further *active* (load-generating) elements are introduced to represent the behaviour of muscles. An investigation of the characteristics of muscles is beyond the scope of this chapter.

1.3. THE PRINCIPLES OF EQUILIBRIUM

1.3.1.   *Forces, moments and couples*

Before we begin the discussion of the principles of equilibrium it is important that we have a clear grasp of the notions of force and moment (figure 1.13).

A *force* is defined by its *magnitude, position* and *direction*. The SI unit of force is the newton, defined as the force required to accelerate a body of mass 1 kg through 1 m s^{-2}, and clearly this is a measure of the magnitude. In two dimensions any force can be resolved into components along two mutually perpendicular axes.

The *moment* of a force about a point describes the tendency of the force to turn the body about that point. Just like a force, a moment has position and direction, and can be represented as a vector (in fact the moment can be written as the cross-product of a position vector with a force vector). The magnitude of the moment is the force times the perpendicular distance to the force,

$$M = |F|d.$$

The SI unit for a moment is the newton metre (N m). A force of a given magnitude has a larger moment when it is further away from a point—hence the principle of levers. If we stand at some point on an object and a force is applied somewhere else on the object, then in general we will feel both a force and a moment. To put it another way, any force applied through a point can be interpreted at any other point as a force plus a moment applied there.

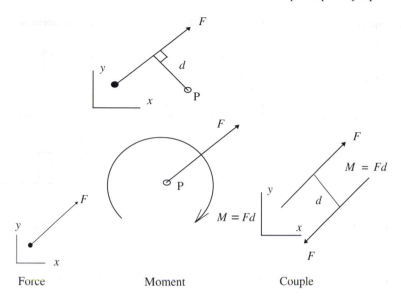

Figure 1.13. *Force, moment and couple.*

A *couple* is a special type of moment, created by two forces of equal magnitude acting in opposite directions, but separated by a distance. The magnitude of the couple is independent of the position of the point about which moments are taken, and no net force acts in any direction. (Check this by taking moments about different points and resolving along two axes.) Sometimes a couple is described as a pure bending moment.

1.3.2. *Equations of static equilibrium*

When a set of forces is applied to any structure, two processes occur:

- the body deforms, generating a system of internal stresses that distribute the loads throughout it, and
- the body moves.

If the forces are maintained at a constant level, and assuming that the material is not viscoelastic and does not creep, then the body will achieve a deformed configuration in which it is in a state of static equilibrium.

By definition: *A body is in static equilibrium when it is at rest relative to a given frame of reference.*

When the applied forces change only slowly with time the accelerations are often neglected, and the equations of static equilibrium are used for the analysis of the system. In practice, many structural analyses in biomechanics are performed based on the assumption of static equilibrium.

Consider a two-dimensional body of arbitrary shape subjected to a series of forces as illustrated in figure 1.14.

The body has three potential rigid-body movements:

- it can translate along the x-axis;
- it can translate along the y-axis;
- it can rotate about an axis normal to the plane, passing through the frame of reference (the z-axis).

Any other motion of the body can be resolved into some combination of these three components. By definition, however, if the body is in static equilibrium then it is at rest relative to its frame of reference. Thus the resultant

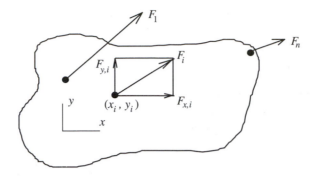

Figure 1.14. *Two-dimensional body of arbitrary shape subjected to an arbitrary combination of forces.*

load acting on the body and tending to cause each of the motions described must be zero. There are therefore three equations of static equilibrium for the two-dimensional body.

Resolving along the x-axis:

$$\sum_{i=1}^{n} F_{x,i} = 0.$$

Resolving along the y-axis:

$$\sum_{i=1}^{n} F_{y,i} = 0.$$

Note that these two equations are concerned only with the magnitude and direction of the force, and its position on the body is not taken into account.

Taking moments about the origin of the frame of reference:

$$\sum_{i=1}^{n} \left(F_{y,i} x_i - F_{x,i} y_i \right) = 0.$$

The fact that we have three equations in two dimensions is important when we come to idealize physical systems. We can only accommodate three unknowns. For example, when we analyse the biomechanics of the elbow and the forearm (figure 1.15), we have the biceps force and the magnitude and direction of the elbow reaction. We cannot include another muscle because we would need additional equations to solve the system.

The equations of static equilibrium of a three-dimensional system are readily derived using the same procedure. In this case there is one additional rigid-body translation, along the z-axis, and two additional rigid-body rotations, about the x- and y-axes, respectively. There are therefore six equations of static equilibrium in three dimensions.

By definition: *A system is statically determinate if the distribution of load throughout it can be determined by the equations of static equilibrium alone.*

Note that the position and orientation of the frame of reference are arbitrary. Generally the analyst will choose any convenient reference frame that helps to simplify the resulting equations.

1.3.3.   Structural idealizations

All real structures including those making up the human body are three-dimensional. The analysis of many structures can be simplified greatly by taking idealizations in which the three-dimensional geometry is

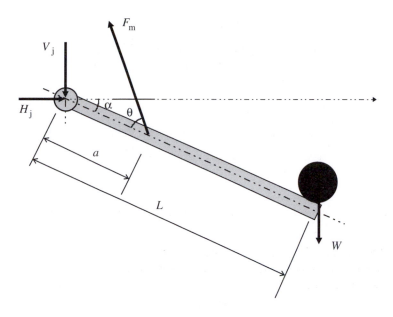

Figure 1.15. *A model of the elbow and forearm. The weight W is supported by the force of the muscle F_m and results in forces on the elbow joint V_j and H_j.*

Table 1.2. *One-dimensional structural idealizations.*

Element type	Loads	Example
Beams	Tension Compression Bending Torsion	Bones
Bars or rods	Tension Compression	
Wires or cables	Tension	Muscles, ligaments, tendons

represented by lumped properties in one or two dimensions. One-dimensional line elements are used commonly in biomechanics to represent structures such as bones and muscles. The following labels are commonly applied to these elements, depending on the loads that they carry (table 1.2).

1.3.4. Applications in biomechanics

Biomechanics of the elbow and forearm

A simple model of the elbow and forearm (figure 1.15) can be used to gain an insight into the magnitudes of the forces in this system.

Taking moments about the joint:

$$F_m a \sin\theta = WL \cos\alpha$$

$$F_m = W \frac{L}{a} \frac{\cos\alpha}{\sin\theta}.$$

Resolving vertically:

$$V_j = W - F_m \sin(\alpha + \theta) = W\left(1 - \frac{L}{a} \frac{\cos\alpha}{\sin\theta} \sin(\alpha + \theta)\right).$$

Resolving horizontally:

$$H_j = F_m \cos(\alpha + \theta) = W \frac{L}{a} \frac{\cos\alpha}{\sin\theta} \cos(\alpha + \theta).$$

The resultant force on the joint is

$$R = \sqrt{H_j^2 + V_j^2}.$$

For the particular case in which the muscle lies in a vertical plane, the angle $\theta = \pi/2 - \alpha$ and

$$F_m = W \frac{L}{a}.$$

For a typical person the ratio L/a might be approximately eight, and the force in the muscle is therefore eight times the weight that is lifted. The design of the forearm appears to be rather inefficient with respect to the process of lifting. Certainly the force on the muscle could be greatly reduced if the point of attachment were moved further away from the joint. However, there are considerable benefits in terms of the possible range of movement and the speed of hand movement in having an 'inboard' attachment point.

We made a number of assumptions in order to make the above calculations. It is worth listing these as they may be unreasonable assumptions in some circumstances.

- We only considered one muscle group and one beam.
- We assumed a simple geometry with a point attachment of the muscle to the bone at a known angle. In reality of course the point of muscle attachment is distributed.
- We assumed the joint to be frictionless.
- We assumed that the muscle only applies a force along its axis.
- We assumed that the weight of the forearm is negligible. This is not actually a reasonable assumption. Estimate the weight of the forearm for yourself.
- We assumed that the system is static and that dynamic forces can be ignored. Obviously this would be an unreasonable assumption if the movements were rapid.

1.4.   STRESS ANALYSIS

The loads in the members of statically determinate structures can be calculated using the methods described in section 1.3. The next step in the analysis is to decide whether the structure can sustain the applied loads.

1.4.1.   Tension and compression

When the member is subjected to a simple uniaxial tension or compression, the stress is just the load divided by the cross-sectional area of the member at the point of interest. Whether a tensile stress is sustainable can often be deduced directly from the stress/strain curve. A typical stress/strain curve for cortical bone was presented in figure 1.5.

Compressive stresses are a little more difficult because there is the prospect of a *structural instability*. This problem is considered in more detail in section 1.5. In principle, long slender members are likely to be subject to structural instabilities when loaded in compression and short compact members are not. Both types of member are represented in the human skeleton. Provided the member is stable the compressive stress/strain curve will indicate whether the stress is sustainable.

1.4.2. Bending

Many structures must sustain bending moments as well as purely tensile and compressive loads. We shall see that a bending moment causes both tension and compression, distributed across a section. A typical example is the femur, in which the offset of the load applied at the hip relative to the line of the bone creates a bending moment as illustrated in figure 1.16. W is the weight of the body. One-third of the weight is in the legs themselves, and each femur head therefore transmits one-third of the body weight.

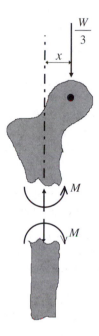

Figure 1.16. *Moment on a femur: two-leg stance.*

This figure illustrates an important technique in the analysis of structures. The equations of static equilibrium apply not only to a structure as a whole, but also to any part of it. We can take an arbitrary cut and apply the necessary forces there to maintain the equilibrium of the resulting two portions of the structure. The forces at the cut must actually be applied by an internal system of stresses within the body. For equilibrium of the head of the femur, the internal bending moment at the cut illustrated in figure 1.16 must be equal to $Wx/3$, and the internal vertical force must be $W/3$.

Engineer's theory of bending

The most common method of analysis of beams subjected to bending moments is the *engineer's theory of bending*. When a beam is subjected to a uniform bending moment it deflects until the internal system of stresses is in equilibrium with the externally applied moment. Two fundamental assumptions are made in the development of the theory.

- It is assumed that every planar cross-section that is initially normal to the axis of the beam remains so as the beam deflects. This assumption is often described as 'plane sections remain plane'.
- It is assumed that the stress at each point in the cross-section is proportional to the strain there, and that the constant of proportionality is Young's modulus, E. The material is assumed to be linearly elastic, homogeneous and isotropic.

The engineer's theory of bending appeals to the principle of equilibrium, supplemented by the assumption that plane sections remain plane, to derive expressions for the curvature of the beam and for the distribution of stress at all points on a cross-section. The theory is developed below for a beam of rectangular cross-section, but it is readily generalized to cater for an arbitrary cross-section.

Assume that a beam of rectangular cross-section is subjected to a uniform bending moment (figure 1.17).

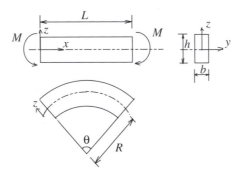

Figure 1.17. *Beam of rectangular cross-section subjected to a uniform bending moment.*

A reference axis is arbitrarily defined at some point in the cross section, and the radius of curvature of this axis after bending is R. Measuring the distance z from this axis as illustrated in figure 1.17, the length of the fibre of the beam at z from the reference axis is

$$l = (R + z)\theta.$$

If the reference axis is chosen as one that is unchanged in length under the action of the bending moment, referred to as the neutral axis, then

$$l_{z=0} = R\theta = L \qquad \theta = \frac{L}{R}.$$

The strain in the x direction in the fibre at a distance z from the neutral axis is

$$\varepsilon_x = \frac{(l_z - L)}{L} = \frac{(R+z)L/R - L}{L} = \frac{z}{R}.$$

If the only loads on the member act parallel to the x-axis, then the stress at a distance z from the neutral axis is

$$\sigma_x = E\varepsilon = E\frac{z}{R}.$$

Hence the stress in a fibre of a beam under a pure bending moment is proportional to the distance of the fibre from the neutral axis, and this is a fundamental expression of the engineer's theory of bending.

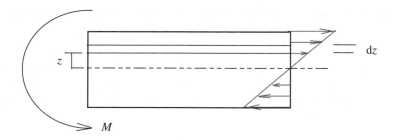

Figure 1.18. *Internal forces on beam.*

Consider now the forces that are acting on the beam (figure 1.18). The internal force F_x acting on an element of thickness dz at a distance z from the neutral axis is given by the stress at that point times the area over which the stress acts,

$$F_x = \sigma_x b\, dz = E \frac{z}{R} b\, dz.$$

The moment about the neutral axis of the force on this elemental strip is simply the force in the strip multiplied by the distance of the strip from the neutral axis, or $M_y = F_x z$. Integrating over the depth of the beam,

$$M_y = \int_{-h/2}^{h/2} E \frac{z}{R} bz\, dz = \frac{E}{R} \int_{-h/2}^{h/2} z^2 b\, dz = \frac{E}{R}\left[\frac{bz^3}{3}\right]_{-h/2}^{h/2}$$

$$= \frac{E}{R}\frac{bh^3}{12} = \frac{E}{R} I.$$

The term

$$\int_{-h/2}^{h/2} z^2 b\, dz \left[= \int_{\text{area}} z^2\, dA\right] = \frac{bh^3}{12} = I$$

is the *second moment of area* of the rectangular cross-section.

The equations for stress and moment both feature the term in E/R and the relationships are commonly written as

$$\frac{\sigma_x}{z} = \frac{E}{R} = \frac{M_y}{I}.$$

Although these equations have been developed specifically for a beam of rectangular cross-section, in fact they hold for any beam having a plane of symmetry parallel to either of the y- or z-axes.

Second moments of area

The application of the engineer's theory of bending to an arbitrary cross-section (see figure 1.19) requires calculation of the second moments of area.

By definition:

$$I_{yy} = \int_{\text{area}} z^2\, dA \qquad I_{zz} = \int_{\text{area}} y^2\, dA \qquad I_{yz} = \int_{\text{area}} yz\, dA.$$

$\uparrow$
product moment of area

The cross-sections of many practical structures exhibit at least one plane of geometrical symmetry, and for such sections it is readily shown that $I_{yz} = 0$.

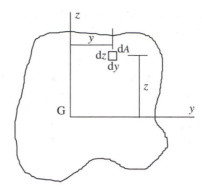

Figure 1.19. *Arbitrary cross-section.*

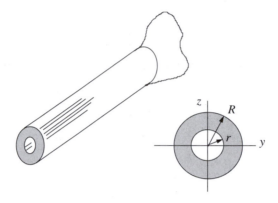

Figure 1.20. *Hollow circular cross-section of a long bone.*

The second moment of area of a thick-walled circular cylinder

Many of the bones in the body, including the femur, can be idealized as thick-walled cylinders as illustrated in figure 1.20.

The student is invited to calculate the second moment of area of this cross-section. It is important, in terms of ease of calculation, to choose an element of area in a cylindrical coordinate system ($dA = r \, d\theta \, dr$). The solution is

$$I_{\text{long bone}} = \frac{\pi (R^4 - r^4)}{4}.$$

Bending stresses in beams

The engineer's theory of bending gives the following expression for the bending stress at any point in a beam cross-section:

$$\sigma = \frac{Mz}{I}.$$

The second moment of area of an element is proportional to z^2, and so it might be anticipated that the optimal design of the cross-section to support a bending moment is one in which the elemental areas are separated as far as possible. Hence for a given cross-sectional area (and therefore a given weight of beam) the stress in the beam is inversely proportional to the separation of the areas that make up the beam. Hollow cylindrical bones are hence a good design.

This suggests that the efficiency of the bone in sustaining a bending moment increases as the radius increases, and the thickness is decreased in proportion. The question must arise as to whether there is any limit on the radius to thickness ratio. In practice the primary danger of a high radius to thickness ratio is the possibility of a local structural instability (section 1.5). If you stand on a soft-drinks can, the walls buckle into a characteristic diamond pattern and the load that can be supported is substantially less than that implied by the stress/strain curve alone. The critical load is determined by the bending stiffness of the walls of the can.

Deflections of beams in bending

The deflections of a beam under a pure bending moment can be calculated by a manipulation of the results from the engineer's theory of bending. It was shown that the radius of curvature of the neutral axis of the beam is related to the bending moment as follows:

$$\frac{E}{R} = \frac{M}{I}.$$

It can be shown that the curvature at any point on a line in the (x, z) coordinate system is given by

$$-\frac{1}{R} = \frac{\mathrm{d}^2 w/\mathrm{d}x^2}{(1 + (\mathrm{d}w/\mathrm{d}x)^2)^{3/2}} \approx \frac{\mathrm{d}^2 w}{\mathrm{d}x^2}.$$

The negative sign arises from the definition of a positive radius of curvature. The above approximation is valid for small displacements.

Then

$$\frac{\mathrm{d}^2 w}{\mathrm{d}x^2} = -\frac{M}{EI}.$$

This equation is strictly valid only for the case of a beam subjected to a uniform bending moment (i.e. one that does not vary along its length). However, in practice the variation from this solution caused by the development of shear strains due to a non-uniform bending moment is usually small, and for compact cross-sections the equation is used in unmodified form to treat all cases of bending of beams. The calculation of a displacement from this equation requires three steps. Firstly, it is necessary to calculate the bending moment as a function of x. This can be achieved by taking cuts normal to the axis of the beam at each position and applying the equations of static equilibrium. The second step is to integrate the equation twice, once to give the slope and once more for the displacement. This introduces two constants of integration; the final step is to evaluate these constants from the boundary conditions of the beam.

1.4.3. *Shear stresses and torsion*

Shear stresses can arise when tractile forces are applied to the edges of a sheet of material as illustrated in figure 1.21.

Shear stresses represent a form of biaxial loading on a two-dimensional structure. It can be shown that for any combination of loads there is always an orientation in which the shear is zero and the direct stresses (tension and/or compression) reach maximum and minimum values. Conversely any combination of tensile and compressive (other than hydrostatic) loads always produces a shear stress at another orientation. In the uniaxial test specimen the maximum shear occurs along lines orientated at 45° to the axis of the specimen.

Pure two-dimensional shear stresses are in fact most easily produced by loading a thin-walled cylindrical beam in torsion, as shown in figure 1.22.

Figure 1.21. *Shear stresses on an elemental area.*

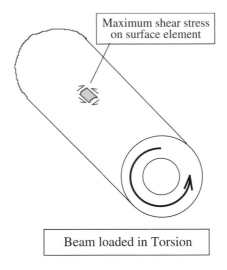

Figure 1.22. *External load conditions causing shear stresses.*

The most common torsional failure in biomechanics is a fracture of the tibia, often caused by a sudden arrest of rotational motion of the body when a foot is planted down. This injury occurs when playing games like soccer or squash, when the participant tries to change direction quickly. The fracture occurs due to a combination of compressive, bending and torsional loads. The torsional fracture is characterized by spiral cracks around the axis of the bone.

Combined stresses and theories of failure

We have looked at a number of loading mechanisms that give rise to stresses in the skeleton. In practice the loads are usually applied in combination, and the total stress system might be a combination of direct and shear stresses. Can we predict what stresses can cause failure? There are many theories of failure and these can be applied to well defined structures. In practice in biomechanics it is rare that the loads or the geometry are known to any great precision, and subtleties in the application of failure criteria are rarely required: they are replaced by large margins of safety.

1.5.   STRUCTURAL INSTABILITY

1.5.1.   *Definition of structural instability*

The elastic stress analysis of structures that has been discussed so far is based on the assumption that the applied forces produce small deflections that do not change the original geometry sufficiently for effects associated with the change of geometry to be important. In many practical structures these assumptions are valid only up to a critical value of the external loading. Consider the case of a circular rod that is laterally constrained at both ends, and subjected to a compressive loading (figure 1.23).

Figure 1.23. *Buckling of a thin rod.*

If the rod is displaced from the perfectly straight line by even a very small amount (have you ever seen a perfectly straight bone?) then the compressive force P produces a moment at all except the end points of the column. As the compressive force is increased, so the lateral displacement of the rod will increase. Using a theory based on small displacements it can be shown that there is a critical value of the force beyond which the rod can no longer be held in equilibrium under the compressive force and induced moments. When the load reaches this value, the theory shows that the lateral displacement increases without bound and the rod collapses. This phenomenon is known as buckling. It is associated with the *stiffness* rather than the strength of the rod, and can occur at stress levels which are far less than the yield point of the material. Because the load carrying capacity of the column is dictated by the lateral displacement, and this is caused by bending, it is the bending stiffness of the column that will be important in the determination of the critical load.

There are two fundamental approaches to the calculation of buckling loads for a rod.

- The *Euler theory* is based on the engineer's theory of bending and seeks an exact solution of a differential equation describing the lateral displacement of a rod. The main limitation of this approach is that the differential equation is likely to be intractable for all but the simplest of structures.
- An alternative, and more generally applicable, approach is to estimate a buckled shape for a structural component and then to apply the *principle of stationary potential energy* to find the magnitude of the applied loads that will keep it in equilibrium in this geometric configuration. This is a very powerful technique, and can be used to provide an *approximation* of the critical loads for many types of structure. Unfortunately the critical load will always be overestimated using this approach, but it can be shown that the calculated critical loads can be remarkably accurate even when only gross approximations of the buckled shape are made.

1.5.2.   *Where instability occurs*

Buckling is always associated with a compressive stress in a structural member, and whenever a light or thin component is subjected to a compressive load, the possibility of buckling should be considered. It should be noted that a *pure shear* load on a plate or shell can be resolved into a tensile and a *compressive component*, and so the structure might buckle under this load condition.

In the context of biomechanics, buckling is most likely to occur:

- in long, slender columns (such as the long bones);
- in thin shells (such as the orbital floor).

1.5.3. Buckling of columns: Euler theory

Much of the early work on the buckling of columns was developed by Euler in the mid-18th Century. Euler methods are attractive for the solution of simple columns with simple restraint conditions because the solutions are closed-form and accurate. When the geometry of the columns or the nature of the restraints are more complex, then alternative approximate methods might be easier to use, and yield a solution of sufficient accuracy. This section presents analysis of columns using traditional Euler methods. Throughout this section the lateral deflection of a column will be denoted by the variable y, because this is used most commonly in textbooks.

Long bone

Consider the thin rod illustrated in figure 1.23, with a lateral deflection indicated by the broken line. If the column is in equilibrium, the bending moment, M, at a distance x along the beam is

$$M = Py.$$

By the engineer's theory of bending:

$$\frac{d^2 y}{dx^2} = \frac{-M}{EI} = \frac{-Py}{EI}.$$

Defining a variable $\lambda^2 = (P/EI)$ and re-arranging:

$$\frac{d^2 y}{dx^2} + \lambda^2 y = 0.$$

This linear, homogeneous second-order differential equation has the standard solution:

$$y = C \sin \lambda x + D \cos \lambda x.$$

The constants C and D can be found from the boundary conditions at the ends of the column. Substituting the boundary condition $y = 0$ at $x = 0$ gives $D = 0$. The second boundary condition, $y = 0$ at $x = L$, gives

$$C \sin \lambda L = 0.$$

The first and obvious solution to this equation is $C = 0$. This means that the lateral displacement is zero at all points along the column, which is therefore perfectly straight. The second solution is that $\sin \lambda L = 0$, or

$$\lambda L = n\pi \qquad n = 1, 2, \ldots, \infty.$$

For this solution the constant C is indeterminate, and this means that the magnitude of the lateral displacement of the column is indeterminate. At the value of load that corresponds to this solution, the column is just held in equilibrium with an arbitrarily large (or small) displacement. The value of this critical load can be calculated by substituting for λ,

$$\sqrt{\frac{P_{cr}}{EI}} L = n\pi$$

from which

$$P_{cr} = \frac{n^2 \pi^2 EI}{L^2}.$$

Although the magnitude of the displacement is unknown, the shape is known and is determined by the number of half sine-waves over the length of the beam. The lowest value of the critical load is that when $n = 1$,

$$P_{cr} = \frac{\pi^2 EI}{L^2}$$

and this is a well-known formula for the critical load of a column.

It is interesting to illustrate the results of this analysis graphically (see figure 1.24).

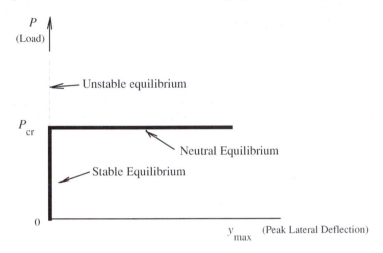

Figure 1.24. *Euler load/displacement curve for a slender column.*

If the load is less than the critical load, then the only solution to the differential equation is $C = 0$, and the column is stable. At the critical load, two solutions become possible. Either the column remains straight ($C = 0$) and the load can continue to increase, or the lateral displacement become indeterminate and no more load can be sustained. The first solution corresponds to a situation of unstable equilibrium and any perturbation (even one caused by the initial imperfection of the column) will cause collapse. Although this curve can be modified using more sophisticated theories based on finite displacements, and it can be demonstrated that loads higher than the critical load can be sustained, the associated lateral displacements are high and design in this region is normally impractical. The Euler instability load represents a realistic upper bound to the load that the column can carry in compression. The graph can be re-drawn with the axial deflection on the x-axis, and then it is apparent that there is a sudden change in axial stiffness of the column at the critical load.

1.5.4. *Compressive failure of the long bones*

It has been demonstrated that the stability of a uniform column is proportional to its flexural stiffness (EI) and inversely proportional to the square of its length. It is likely, therefore, that the most vulnerable bones in the skeleton will be the long bones in the leg. Since the governing property of the cross-section is the second moment of area, as for bending stiffness, it is apparent that a hollow cylindrical cross-section might be most efficient in resisting buckling. A compressive load applied to the head of the femur is illustrated in figure 1.25.

In a two-leg stance approximately one-third of the body weight is applied to the head of each femur. The remaining weight is distributed in the legs, and this makes the analysis more difficult. For simplicity (and conservatively) we might assume that one-half of the body weight is applied to the top of the femur. We have the usual problem in biomechanics in deciding which parts of the structure to include in our model. We might calculate separately the buckling loads of the femur and of the tibia, but this assumes that each provides a lateral support to the other at the knee. It might be more appropriate to take the femur/tibia structure as a continuous member, assuming that there is continuity of bending moment across the knee. Taking the skeleton of the leg as a whole, the length is approximately 1 m.

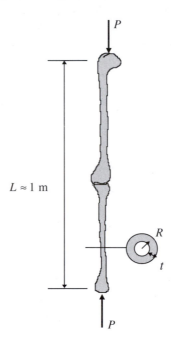

Figure 1.25. *Compressive buckling of the skeleton of the leg.*

The buckling strength of a non-uniform column will always be higher than that of a uniform column of its minimum dimensions. The buckling stress is simply the buckling load divided by the cross-sectional area.

$$P_{cr} = \frac{\pi^2 E I}{L^2} \qquad \sigma_{cr} = \frac{\pi^2 E I}{A L^2}.$$

The buckling stress can be calculated and compared with the ultimate compressive stress to determine whether failure will be by buckling or by simple overloading. In either case the result will be the same—a broken leg—but the appearance of the fracture might be different, and the design steps required to prevent a recurrence will certainly be so.

1.6.   MECHANICAL WORK AND ENERGY

1.6.1.   *Work, potential energy, kinetic energy and strain energy*

We have seen that we can gain an insight into the forces and stresses involved in many biomechanical systems by using the equations of static equilibrium alone. However, if we are to study problems that involve dynamic actions such as running and jumping, we shall need to develop additional techniques. One of the most powerful tools available to us is the principle of conservation of energy.

Energy is a scalar quantity with SI units of joules (J). Joules are derived units equivalent to newton metres (N m). We should not confuse energy with moment or torque, which are vector quantities whose magnitudes also happen to be measured in newton metres. The concept of work is similar to that of energy. We usually associate energy with a body and work with a force. In this chapter we shall concern ourselves only with mechanical energy, and not with other forms such as heat energy.

The *work done* by a constant force acting on a body is equal to the force multiplied by the distance moved in the direction of the force. If the force and the distance moved are taken to be vectors, then the work done is their dot product. By definition, therefore, a force that produces no displacement does no work.

In a *conservative* system all of the work that is done on the body is recoverable from it. In some sense it is 'stored' within the body as a form of energy.

Potential energy is associated with the potential of a body to do work under the action of a force (see figure 1.26). In order for a body to have potential energy it must be subjected to a force (or a field). Potential energy can perhaps best be understood by considering the work done by the force acting on the body. The work done by the force is equal to the potential energy lost by the body.

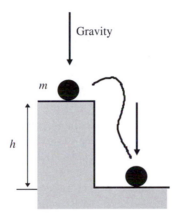

Figure 1.26. *Potential energy is a function of position and applied force.*

The weight of a mass in gravity is mg. In order to lift the mass (slowly) from one stair to another we would have to do work on the body. This work could, however, be recovered if we allowed the body to fall back to the lower step. The work that would be done by the gravitational force during the process of falling would be the force times the distance moved, or *work* $= mgh$. We might therefore consider that the body has more potential to do work, or potential energy, when it is on the top step relative to when it is on the bottom step. Obviously we can choose any position as the reference position, and potential energy can only be a relative quantity.

Kinetic energy is associated with the motion of a body. Newton's second law tells us that a body subjected to a constant force will accelerate at a constant rate. The work done by the force on the body is converted into kinetic energy of the body. The kinetic energy K of a body of mass m travelling at velocity v is

$$K = \tfrac{1}{2}mv^2.$$

Strain energy is associated with the deformation of a body. If the deformation is elastic then all of the strain energy is recoverable if the load is removed. The strain energy in a bar can be defined in terms of the area under the load/displacement curve (see figure 1.27).

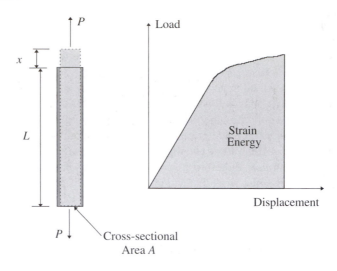

Figure 1.27. *Strain energy in a uniaxial test.*

Once again the energy of the body can be calculated from the work done by the force acting on it. In this case the force is not constant but is applied gradually, so that a particular value of the force is associated only with an increment of displacement.

Using the above definitions, potential energy and kinetic energy are terms that we might use in studying the dynamics of rigid particles. Deformable bodies have strain energy as well as potential and kinetic energy. When we apply the principles of conservation of energy we tacitly assume that all of the work done to a body is converted into a form of energy that can be recovered. In some situations this is not true, and the energy is converted into a form that we cannot recover. The most common examples include plastic deformations and frictional forces. When a bar is bent beyond the elastic limit it suffers permanent deformation, and we cannot recover the energy associated with the plastic deformation: indeed, if we wished to restore the bar to its original condition we would have to do more work on it. Similarly when a body is pulled along a rough surface, or through a fluid, we must do work to overcome the frictional forces which we can never recover. Systems in which the energy is not recoverable are called *non-conservative*. Sometimes, particularly in assessment of problems in fluid dynamics, we might modify our equations to cater for an estimated irrecoverable energy loss.

We should not always regard non-conservative systems in a negative light. If, for example, we need to absorb the energy of an impact, it might be most appropriate to use a dashpot system that dissipates the energy using the viscous characteristics of the fluid. Cars are designed with crush zones that absorb energy in gross plastic deformation—unfortunately they can only be used once!

1.6.2.   Applications of the principle of conservation of energy

Gravitational forces on the body: falling and jumping

The reader will be familiar with the calculation of the velocities of rigid bodies falling under the action of gravity (see figure 1.28). The principle of conservation of energy dictates that the sum of the potential and kinetic energy is constant. In this case the strain energy is negligible.

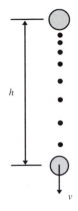

Consider a body of mass m dropped from a height h
under the action of gravity.
Change in potential energy, $dU = -mgh$.
Change in kinetic energy, $dK = \frac{1}{2}mv^2$.
For a conservative system, $dU + dK = 0$.

$$v = \sqrt{2gh}$$

Figure 1.28. *Body falling under gravity.*

We can extend the principle to estimate the forces applied to the body during falling and jumping actions. Consider, for example, the case of a person dropping from a wall onto level ground. There is a three-stage process.

- There is a period of free-fall during which potential energy of the body is converted into kinetic energy.
- The feet hit the ground and the reaction force from the ground does work to arrest the fall.
- The energy stored as elastic strain energy in the bones and muscles during the arresting process will be released causing a recoil effect.

The first two stages of the process are indicated in figure 1.29. Two possibilities are illustrated: in the first the knees are locked and the body is arrested rather abruptly, and in the second the knees are flexed to give a gentler landing.

We can apply the principle of conservation of energy from the time at which the fall commences to that at which it is arrested and the body is again stationary. At both of these times the kinetic energy is zero. The work done on the body by the reaction to the ground must therefore be equal to the potential energy that the body has lost in the falling process. We have an immediate problem in that we do not know how the reaction force changes with time during the arresting process (F_1 and F_2 are functions of time).

The second problem, landing on a bended knee, is addressed in section 1.9. The method adopted could produce unreliable results for the first problem because of the difficulty in estimating the dynamic compression of the skeleton. An alternative approach would be to take the legs as a spring, with the compressive force being proportional to the displacement, and to equate the strain energy in the legs at the end of the arresting process with the potential energy lost by the body.

1.7.   KINEMATICS AND KINETICS

The degrees of freedom that govern the motion of a system can often be determined from the geometry alone. The branch of mechanics associated with the study of such systems is called kinematics. Kinematics plays an important part in our understanding of the function of the human skeleton. The study of the motions and forces generated is called kinetics.

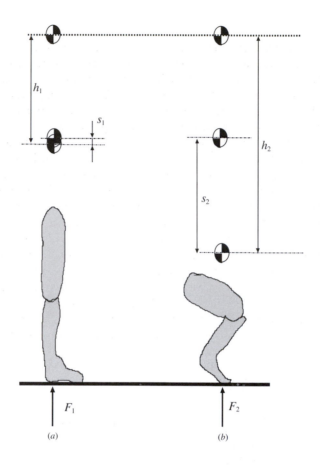

Figure 1.29. *Arrest of a fall: (a) knees locked; (b) knees bent.*

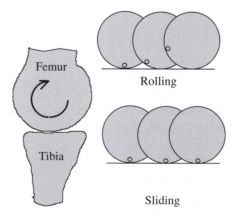

Figure 1.30. *Rolling and sliding action of the knee.*

1.7.1. Kinematics of the knee

The knee is a very complex joint that features many degrees of freedom in three dimensions (see figure 1.30). Its primary purpose, however, is to allow a flexion in the sagittal plane, and it is most instructive to develop a simple two-dimensional kinematic model of the knee in this plane. The femoral condyle rotates relative to the tibial plateau to give the leg the freedom to bend at the knee. We can quickly show, however, that the degree of rotation (approximately 150°) cannot be achieved by a rolling action alone: the femur would simply roll right off the tibia. Similarly, if the action were sliding alone the geometry of the bones would limit the degree of flexure to approximately 130°. In practice the femoral condyle must roll and slide on the tibial plateau.

It is illuminating to consider how one might go about the task of designing a system to achieve the required two-dimensional motion of the knee. One possibility is to make an engaging mechanism, like a standard door hinge, in which effectively one structure is embedded inside the other. On inspection of the anatomy of the femur and the tibia it is apparent that this is not the type of mechanism that is featured in the knee. The two bones do not overlap to any substantial degree, and certainly neither 'fits inside' the other. Clearly then the kinematics of the knee must involve straps (or ligaments) that attach the two bodies together. Assuming that the point of attachment of the straps on the bone cannot move once made, and that the straps cannot stretch, one strap (or two attached to the same point on one of the members) could be used only to provide a centre for the pure rolling action. We have seen that this cannot describe the motion of the knee. The simplest arrangement of rigid links that can provide the degree of rotation required is the four-bar link. When we study the anatomy of the knee, we see that there is indeed a four-bar link mechanism, formed by the tibial plateau, the femoral condyle and the two cruciate ligaments (see figure 1.31).

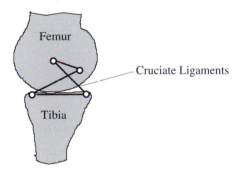

Figure 1.31. *The knee as a four-bar linkage.*

An understanding of the mechanics of this system can be gained by constructing a simple mechanical model using taut strings to represent the ligaments and a flat bar to represent the face of the tibia. As the tibia is rotated a line can be drawn illustrating its current position, and it is found that the resulting set of lines trace out the shape of the head of the femur. If the lengths of the strings or the mounting points are changed slightly the effect on the shape of the head of the femur can be seen. Consider the implications for the orthopaedic surgeon who is changing the geometry of a knee joint.

1.7.2. Walking and running

The study of the forces on the skeleton during the processes of walking and running is called gait analysis. The analysis of gait provides useful information that can assist in clinical diagnosis of skeletal and muscular abnormalities, and in the assessment of treatment. It has even been used to help to design a running track on which faster speeds can be attained. The following discussion is based on Chapter 8, 'Mechanics of Locomotion', of McMahon (1984), and this source should be studied for more comprehensive information.

For the purposes of definition, walking constitutes an action in which there is always at least one foot in contact with the ground. In the process of running there is an interval of time in which no part of the body is in contact with the ground. It is interesting to investigate the changes of the kinetic energy and the potential energy of the body during the processes of running and walking. It is conventional in the analysis of gait to separate the kinetic energy into a vertical and a horizontal component. In the simplest possible model we might consider only the motion of the centre of gravity of the body, as illustrated in figure 1.32.

- The kinetic energy associated with the forward motion of the centre of mass is $K_x = \frac{1}{2}mv_x^2$.
- The kinetic energy associated with the vertical motion of the centre of mass is $K_y = \frac{1}{2}mv_y^2$.
- The potential energy associated with the vertical position of the centre of mass is $U_y = mgh$.

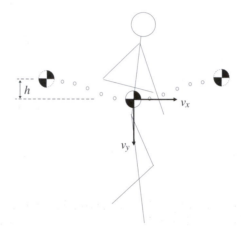

Figure 1.32. *Motion of the centre of gravity in running or walking.*

A record of the external reactions on the feet from the ground can be obtained from force-plate experiments. From these records the three components of energy can be derived, and each varies as a sine wave or as a trapezoidal wave. It turns out that:

- In the walking process the changes in the forward kinetic energy are of the same amplitude as and 180° out of phase with those in the vertical potential energy, and the vertical kinetic energy is small. Hence the total of the three energies is approximately constant.
- In the running process the vertical components of the energies move into phase with the forward kinetic energy. The total energy has the form of a trapezoidal wave, and varies substantially over the gait cycle.

The difference in the energy profiles during running and walking suggests that different models might be appropriate. For walking, the system appears simply to cycle between states of high potential energy and high kinetic energy, with relatively little loss in the process (in fact at a normal walking speed of 5 km hr^{-1} about 65% of the vertical potential energy is recovered in forward kinetic energy). In running there is effectively no recovery of energy in this manner, and alternative mechanisms such as muscle action must be responsible for the energy cycle. This gives us a strong clue as to how we might model the two systems: an appropriate walking model might be based on the conservation of energy, whereas a running model might feature viscoelastic and dynamically active muscle.

Ballistic model of walking

We have seen that there is an efficient mechanism of conversion of potential to kinetic energy in the walking process. This suggests that the action of walking might be described in terms of some sort of compound pendulum. One attempt to describe the process based on this conception is the ballistic model of walking, illustrated in figure 1.33.

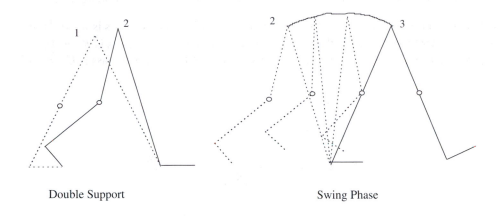

Double Support   Swing Phase

Figure 1.33. *Ballistic model of walking (based on McMahon (1984)).*

In this model it is assumed that the support leg remains rigid during the double support and swing phase periods, with the required rotation occurring at the ankle to keep the foot flat on the ground. The swing leg hinges at the knee. The swing phase commences when the toe leaves the ground and ends when the heel strikes the ground as illustrated above. During this period the foot remains at right angles to the swing leg. There follows another period of double support (not shown). It is assumed that all muscular action occurs during the double-support period and that thereafter the legs swing under the action of gravity alone (hence the term ballistic model). The mass of the body might be assumed to act at the hip joint, whilst the mass in the legs might be distributed in a realistic way to give appropriate centroids and mass moments of inertia.

The model is a three-degrees-of-freedom system, comprising the angle of the stiff leg and the two angles of the swing leg. All other parameters are determined by the geometry of the system. The equations of motion for the system can be written down. The initial conditions at the start of the swing phase are determined by the muscular action and are thus unknown. However, choosing an arbitrary set of initial conditions, the system can be traced (probably numerically) until the point at which the heel strikes, signalling the end of the swing phase. The initial conditions can be adjusted until the swing leg comes to full extension at the same moment at which the heel strikes the ground. Other constraints are that the toe must not stub the ground during the swing phase and that the velocity must not get so high that the stiff leg leaves the ground due to the centrifugal force of the body moving in an arch. The model can be used to find a correspondence between step length and time taken for the swing. Although the solution of the equations for any particular set of initial conditions is determinate, there is more than one set of initial conditions that corresponds to each step length, and therefore there are a range of times that the step can take. Typical results from the model, showing the range of swing time against subject height for a step length equal to the length of the leg, are presented by McMahon. Experimental data for a number of subjects are also shown, and there is a good correspondence with the results of the theoretical model.

A model of running

The components of energy that we considered for the walking model are demonstrably not conserved when we run. In fact, their sum has a minimum in the middle of the support phase and then reaches a maximum as we take off and fly through the air. This suggests that perhaps there is some form of spring mechanism becoming active when we land. This spring absorbs strain energy as our downwards progress is arrested, and then recoils propelling us back into the air for the next stride. McMahon analyses the model illustrated in figure 1.34. This idealization is strictly valid only during the period in which the foot is in contact with the track. We can analyse this system using standard methods for a two-degrees-of-freedom system. To gain a preliminary insight into the performance of the system we might first of all investigate its damped resonant frequencies, assuming that the foot remains permanently in contact with the track. The period during which the foot actually remains in contact with the track would be expected to be about one-half of the period of the resonant cycle. We can plot foot contact time against track stiffness. The surprising result of this exercise is that there is a range of track stiffnesses for which the foot contact time is less than it would be for a hard track. There is experimental evidence that running speed is inversely proportional to track contact time, and so the above analysis suggests that a track could be built that would improve times for running events. A further benefit of such a track would be that its compliance would reduce the peak forces on the foot relative to those associated with a hard surface, thus leading to a reduction in injuries. The first track constructed to test this hypothesis, a 220 yard indoor track built at Harvard University in 1977, was a real success. Training injuries were reduced to less than one-half of those previously sustained on a cinder track, and times were reduced by about 2%.

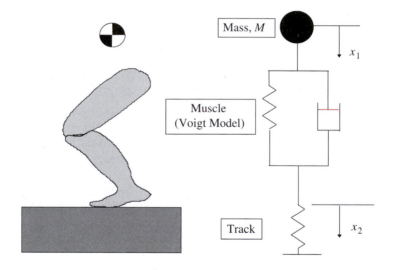

Figure 1.34. *Model of running.*

Experimental gait analysis

Experimental techniques used to measure gait include:

- visual systems, perhaps including a video camera;
- position, strain and force measurement systems;
- energy measurement systems.

Some of the techniques for defining and measuring normal gait are described in section 20.3.2. All of the techniques are of interest to the physicist working in medical applications of physics and engineering.

1.8. DIMENSIONAL ANALYSIS: THE SCALING PROCESS IN BIOMECHANICS

When we look around at the performance of other creatures we are often surprised at the feats that they can achieve. A flea can jump many times its own height, whilst we can only manage about our own (unless we employ an external device such as a pole). An ant can carry enormous weights relative to its own body weight and yet we cannot. We can often explain these phenomena using simple arguments based on our understanding of biomechanics.

1.8.1. Geometric similarity and animal performance

The simplest scaling process that we can imagine is that based on *geometric similarity*. If objects scale according to this principle then the only difference between them is their size. All lengths are scaled by the same amount, and we should not be able to tell the difference between them geometrically. If a characteristic length in the object is L, then the volume of the object is proportional to L^3.

Using the principle of geometric similarity together with Newton's second law, and assuming that the stress that can be sustained by a muscle is independent of its size, Hill reasoned in 1949 (see McMahon) that the absolute running and jumping performance of geometrically similar animals should be independent of their size.

There are three fundamental parameters in structural mechanics:

$$M \qquad L \qquad T$$
$$\text{mass} \qquad \text{length} \qquad \text{time}$$

All other parameters such as velocity, acceleration, force, stress, etc can be derived from them.

The *stress* developed in muscle is constant, independent of muscle size,

$$\sigma \propto L^0.$$

The *force* developed by the muscle is the stress multiplied by the cross-sectional area, and so is proportional to the square of the length,

$$F = \sigma A \propto L^2.$$

Newton's second law tells us that force is mass multiplied by acceleration. Assuming that the density of the tissue is constant for all animals, the force can be written as

$$F = ma \propto L^3 L T^{-2} \equiv L^2 \left(L T^{-1} \right)^2.$$

Comparing these two expressions for the force, it is apparent that the group (LT^{-1}) must be independent of the body size. The velocity of the body has these dimensions, and so must be independent of the body size. Hence the maximum running speed of geometrically similar animals is independent of their size! Hill supported his argument by the observation that the maximum running speeds of whippets, greyhounds and horses are similar despite the wide variation in size. In fact, the maximum running speeds of animals as diverse as the rat, the elephant, man and the antelope is constant to within a factor of about three, despite a weight ratio of up to 10 000.

The argument can readily be extended to assess the maximum height to which an animal might be able to jump. The principle of conservation of energy tells us that the maximum height will be proportional to the square of the take-off velocity. Since the velocity is independent of size, so will be the maximum jumping height. This is true to within a factor of about two for a wide range of animals.

1.8.2. Elastic similarity

When we apply the principle of geometric similarity we assume that the body has a single characteristic length, L, and that all dimensions scale with L. We can conceive alternative rules featuring more than one characteristic dimension. For example, we might assume that a long bone is characterized by its length and by its diameter, but that the proportion of length to diameter is not constant in a scaling process. A more general scaling process could be based on the assumption that the length and diameter scales are based on the relationship

$$L \propto d^q, \quad \text{where } q \text{ is an arbitrary constant.}$$

One rule that we might construct is based on the assumption that the structure will scale so that the stability of the column is independent of the animal size. In section 1.4.2 we developed a method of calculating the deflection of a beam. The deflection at the tip of a cantilever beam under self-weight can be shown to be proportional to $AL^4 = I$, where L is the length of the beam, A its cross-sectional area and I its second moment of area. For a beam of circular cross section the area is proportional to d^2 and the second moment of area is proportional to d^4,

$$\frac{\delta_{\text{tip}}}{L} \propto \frac{1}{L} \frac{L^4 d^2}{d^4} = \frac{L^3}{d^2}.$$

Hence if the deflection of the tip is to be proportional to the length of the beam, the relationship between d and L must be $d \propto L^{3/2}$.

This scaling rule applies to other elastic systems. The critical length of column with respect to buckling under self-weight is one example. This process of scaling is called elastic similarity.

The weight, W, of the cylindrical body is proportional to the square of the diameter multiplied by the length,

$$W \propto d^2 L = d^{q+2} = L^{(q+2)/q} \quad \text{or} \quad L \propto W^{q/(q+2)} \quad d \propto W^{1/(q+2)}.$$

For *geometric similarity*:

$$L \propto W^{1/3} \quad d \propto W^{1/3}.$$

For *elastic similarity*:

$$L \propto W^{1/4} \quad d \propto W^{3/8}.$$

Evidence from the animal kingdom suggests that *elastic similarity is the governing process of scaling in nature*.

1.9. PROBLEMS

1.9.1. Short questions

a   Is the average pressure under the foot when standing higher or lower than normal systolic arterial pressure?

b   Is bone weakest when subjected to tension, compression or shear stress?

c   What is the 'yield point' of a material?

d   Is most tissue homogeneous or inhomogeneous in terms of its mechanical properties?

e   Categorize bone and soft tissue as typically having linear or nonlinear mechanical properties.

f   What is a viscoelastic material?

g   If a spring is considered as being equivalent to an electrical resistance where *force* $\equiv$ *voltage* and *displacement* $\equiv$ *current*, then what electrical component is equivalent to a dashpot?

h   What three basic models can be used to describe 'creep'?
i   Approximately what fraction of body weight is applied to the head of each femur when standing?
j   Is energy conserved during running?
k   Why can a frog jump so high?
l   What is the characteristic property of an elastic material?
m   What are the approximate magnitudes of the ultimate tensile stress and of the ultimate tensile strain of cortical bone?
n   What process in living bone is described by Wolff's law?
o   What is the difference between creep and relaxation?
p   Sketch the arrangement of springs and dampers for the Kelvin model of viscoelasticity.
q   What is the equation describing the relationship between stress and applied moment, commonly referred to as the engineer's theory of bending?
r   State one of the fundamental assumptions of the engineer's theory of bending.
s   If a thin longitudinal slit is cut into a hollow tube of circular cross-section the cross-sectional area will be essentially unchanged. Is there any significant change in the torsional stiffness?
t   What are the relationships between body weight, W, and (a) the length, (b) the diameter of a cylindrical body if the scaling process is governed by the law of elastic similarity?

1.9.2. Longer questions (answers are given to some of the questions)

Question 1.9.2.1

Write a brief description of each of the following quantities:

(a)   potential energy;
(b)   kinetic energy;
(c)   strain energy.

Discuss the process of falling to the ground, indicating which of the above energies the body has at different stages in the fall. Draw a sketch illustrating a model that might be used to analyse the process when landing and allowing the knees to flex to cushion the fall. State clearly the assumptions inherent in the model, and discuss whether the assumptions are reasonable.

Derive an expression for the force generated at the feet during the arrest of the fall.

Assuming that the geometrical arrangement of the skeleton and musculature is such that the bones experience a load eight times higher than the load at the foot, and that the fracture load of the tibia is 36 kN, calculate the maximum height from which a 70 kg body might be able to fall before a fracture is likely.

Answer

- *Potential energy* is the potential of a body to do work under the action of a force. It is associated with the position of the body and the force acting on it. It is a relative quantity, because some arbitrary position will be taken as the reference point where the potential energy is zero. Gravitational p.e. $= mgh$.
- *Kinetic energy* is associated with the motion of a body. k.e. $= \frac{1}{2}mv^2$.
- *Strain energy* is associated with the deformation of a body. It is the area under a load/deflection curve, or the integral over volume of the area under a stress/strain curve.

Consider the case of a person dropping from a wall onto level ground. There is a three-stage process.

- There is a period of free-fall during which the potential energy of the body is converted into kinetic energy.

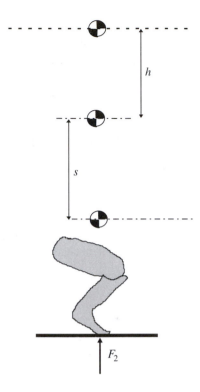

Figure 1.35. *Model of falling, bending knees to arrest.*

- The feet hit the ground and the reaction force from the ground does work to arrest the fall.
- The energy stored as elastic strain energy in the bones and muscles during the arresting process will be released, causing a recoil effect.

The first two stages of the process are indicated in figure 1.35.

We can apply the principle of conservation of energy from the time at which the fall commences to that at which it is arrested and the body is again stationary. At both of these times the kinetic energy is zero. The work done on the body by the reaction to the ground must therefore be equal to the potential energy that the body has lost in the falling process. We have an immediate problem in that we do not know how the reaction force changes with time during the arresting process (F_1 and F_2 are functions of time).

Assume that:

- The body is just a mass concentrated at one point in space, and that the surroundings contain some mechanism for absorbing its energy.
- The instant that the body makes contact with the ground a constant reaction force is applied by the ground, and that this force remains constant until motion is arrested (figure 1.36).

The work done on the body by the reaction force is

$$\text{work} = Fs.$$

The change in potential energy of the body is

$$\Delta U = -mg(h + s).$$

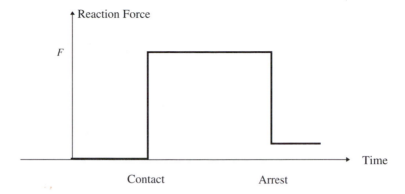

Figure 1.36. *Force versus time on a body during arrest of falling.*

The change in kinetic energy of the body is zero.

For conservation of energy:

$$Fs - mg(h + s) = 0$$

$$F = mg\left(\frac{h}{s} + 1\right)$$

but how big is s?

If we bend our knees, then we know that we can arrest the fall over a distance of perhaps 0.5 m. In this condition we are going to use our muscles, ligaments and tendons to cushion the fall. We are told that the geometrical arrangement of the skeleton and muscles is such that bone forces are eight times higher than external loads, and that the maximum force on the bone is 36 kN. This implies that the maximum force that can be exerted at one foot is 4.5 kN, or 9 kN on the body assuming that force is distributed evenly between the two feet.

Using the assumptions of our current model, this implies that the maximum height from which we can fall and land on bended knees is

$$9000 = 70 \times 10 \times \left(\frac{h}{0.500} + 1\right)$$

$$h \approx 5.9 \text{ m.}$$

Question 1.9.2.2

Determine expressions for the forces in the erector spinae and on the fifth lumbar vertebra in the system illustrated in figure 1.37. Produce a graph of spine elevation against forces for the case $W = 20$ kg and $W_0 = 75$ kg.

Question 1.9.2.3

You are preparing to dive into a pool (see figure 1.38). The mechanical system at the instant of maximum deflection of the diving board is illustrated in the figure. If you are launched 1 m into the air above the static equilibrium position of the board calculate the strain energy that is recovered from the board to achieve this height. Calculate the maximum velocity during the launch assuming that this occurs when the board is level. Calculate the maximum force that is applied to the foot of the diver, assuming that the force is proportional

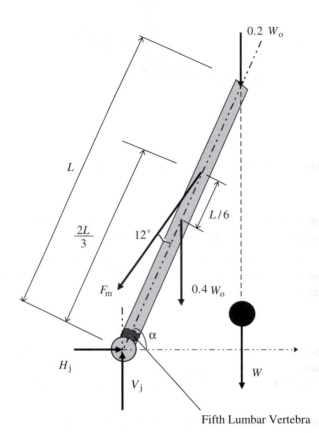

Figure 1.37. *Loads on the spine when lifting.*

Figure 1.38. *Diver on a board.*

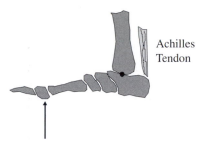

Achilles
Tendon

Figure 1.39. *Diagrammatic illustration of the geometry of the foot and the force on the foot applied by the diving board.*

to the downward deflection of the board and that the maximum deflection of the board is 0.25 m. In any of these calculations you may assume that the mass of the diving board is negligible.

At the instant of maximum load the attitude of the foot is that illustrated in figure 1.39. Estimate the dimensions of your own foot. Calculate in this system the force in your Achilles tendon and the force in your tibia at this instant. Calculate the stress in the Achilles tendon.

Question 1.9.2.4

In a simple model of pole-vaulting we might assume that the kinetic energy of the vaulter that is generated during the run-up is translated into potential energy at the height of the jump. Derive an expression relating the peak velocity in the run-up to the height achieved in the vault. Calculate the energy involved for a person with a mass of 70 kg.

The mechanism for the transfer of energy is assumed to be the pole, and we might assume that at some point in the process the whole of the energy of the vault is contained within the pole as strain energy. Assuming that a 5 m pole (of uniform cross-section) buckles elastically into a shape represented by one half of a sine wave, and that the peak transverse deflection is 1 m, deduce the flexural stiffness (EI) that is required to absorb the energy. Write down an expression for the peak stress in the pole at maximum flexion, and derive an expression for the upper limit on the radius of the pole assuming that it is of circular cross-section.

If the pole is made of a carbon fibre material with a Young's modulus of 100 GPa and a failure stress of 1500 MPa, design a cross-section for the pole to sustain the loads imposed. Calculate the weight of the pole that you have designed if the density is 1600 kg m^{-3}.

Comment on the assumptions, and discuss whether you believe this to be a realistic model. (The strain energy in a beam under flexion can be written as SE $= \frac{1}{2} \int_0^L EI\chi^2 \, dx$).

Answer

Equating potential and kinetic energies,

$$mgh = \tfrac{1}{2}mv^2$$
$$h = \frac{v^2}{2g}.$$

For a height of 6 m, the peak running velocity would have to be 10.8 m s^{-1}; about the peak velocity achievable by man. The energy would be 4.12 kJ.

Assume

$$y = Y \sin \frac{\pi x}{L}.$$

Then

$$\chi \approx \frac{d^2 y}{dx^2} = -\frac{\pi^2 Y}{L^2} \sin \frac{\pi x}{L}.$$

Strain energy

$$SE = \tfrac{1}{2} \int_0^L EI\chi^2 \, dx = \frac{\pi^4 Y^2 EI}{2L^4} \int_0^L \sin^2 \frac{\pi x}{L} dx = \frac{\pi^4 Y^2 EI}{4L^3}$$

$$EI = \frac{4L^3 (SE)}{\pi^4 Y^2}.$$

Assuming that $Y = 1.0$ m,

$$EI = \frac{4 \times 5^3 \times 4120}{\pi^4 1.0^2} = 21\,148 \text{ N m}^2.$$

The bending moment is $EI\chi$,

$$M_{\max} = EI\chi_{\max} = \frac{\pi^2 Y EI}{L^2}.$$

The bending stress in the tube is Mr/I,

$$\sigma_{\max} = \frac{M_{\max} r}{I} = \frac{\pi^2 Y E r}{L^2} = 0.395 Er.$$

If the stress is to be less than two-thirds of its ultimate value for safety, then

$$r < \frac{0.67\sigma_{\text{ult}}}{0.273 Er} = \frac{1000 \times 10^6}{0.395 \times 100 \times 10^9} = 0.253 \text{ m}.$$

If we set the radius to 25 mm the stress is 988 MPa.

Returning to the stiffness requirement:

$$EI = E\pi r^3 t = 21\,148 \quad \rightarrow \quad t = \frac{21\,148}{100 \times 10^9 \times \pi \times 0.025^3} = 0.0043 \text{ m}.$$

The suggested design is a tube of radius 25 mm and wall thickness 4.3 mm.

The weight of this pole would be $2 \times \pi \times 0.025 \times 0.0043 \times 5 \times 1600 = 5.4$ kg.

Question 1.9.2.5

The head of the femur is offset from the axis of the bone, producing a bending moment in the bone when standing in a relaxed position. Calculate the magnitude of this moment for the system shown in figure 1.40 assuming that the offset x is 50 mm and that the body weight is 70 kg.

Write down the general expression for the second moment of area of a cross-section as illustrated in figure 1.41, and show from first principles that the second moment of area of a thin-walled hollow cylinder of radius r and thickness t is $\pi r^3 t$.

For the applied moment calculated, and using the thin-walled approximation, calculate the maximum bending stress in the femur assuming a mean radius of 20 mm and a wall thickness of 12 mm.

Compare the calculated bending stress with the compressive stress that arises due to end load in the femur. Is the bending stress significant?

Figure 1.40. *Moment on femur. Two-leg stance.*

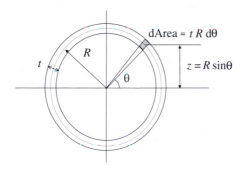

Figure 1.41. *Thin-walled cylindrical section.*

Answer

Force on one hip, two-leg stance $= \dfrac{70 \times 10}{3} = 233$ N: Bending moment on femur $= 233 \times 50 = 11\,667$ N mm.

By definition:

$$I = \int_{\text{area}} z^2 \, \mathrm{d}A$$

$$I = \int_{\text{area}} (R \sin \theta)^2 \, Rt \, \mathrm{d}\theta$$

$$I = \int_{\text{area}} R^3 t \sin^2 \theta \, \mathrm{d}\theta \qquad I = R^3 t \int_0^{2\pi} \sin^2 \theta \, \mathrm{d}\theta$$

$$I = R^3 t \int_0^{2\pi} \left(\frac{1 - \cos 2\theta}{2} \right) \mathrm{d}\theta \qquad I = R^3 t \left[\frac{\theta}{2} - \frac{\sin 2\theta}{2} \right]_0^{2\pi}$$

$$I = \pi R^3 t.$$

For the geometry specified,

$$I = \pi R^3 t = \pi \times 20^3 \times 12 = 302\,000 \text{ mm}^4.$$

The bending stress is calculated from the engineer's theory of bending:

$$\sigma = \frac{Mz}{I}.$$

At the extreme fibre, $z = R + t/2 = 20 + 12/2 = 26$ mm,

$$\sigma = \frac{Mz}{I} = \frac{11\,667 \times 26}{302\,000} = 1.0 \text{ N mm}^{-2}.$$

The failure stress is about 120 N mm^{-2}, and so there is, as expected, a high margin of safety in this condition. For comparison, the compressive stress in the bone is simply the end load divided by the area:

$$\sigma_{comp} = \frac{P}{A} \approx \frac{P}{2\pi Rt} = \frac{233}{2 \times \pi \times 20 \times 12} = 0.15 \text{ N mm}^{-2}.$$

Hence the bending stress is some six or seven times higher than the simple compressive stress, and so bending is important in the analysis of the femur.

Answers to short questions

a   The average pressure under the foot is higher than normal arterial systolic pressure.

b   Bone is weakest when subject to shear stress.

c   The point on a stress/strain curve beyond which permanent strain occurs is the yield point.

d   Tissue is inhomogeneous in terms of mechanical properties.

e   Bone, linear; soft tissue, nonlinear.

f   One where the stress/strain relationship depends upon time.

g   An inductance is equivalent to a dashpot.

h   Maxwell, Voigt and Kelvin models.

i   Approximately one-third of body weight is applied to the head of each femur.

j   No, energy is not conserved during running.

k   Because it can be shown dimensionally that for geometrically similar animals jumping performance should be independent of size.

l   Loading and unloading follow the same path.

m   80 MPa and 1%; see figure 1.5.

n   Stress remodelling is described by Wolff's law.

o   Creep is the process whereby displacement will change with time under a constant stress. Relaxation is the process whereby the stress will fall with time under constant strain; see figure 1.7.

p   See figure 1.10 for the Kelvin model.

q   Stress = applied moment/second moment of area

r   'Plane sections remain plane' and 'stress proportional to strain' are assumed in the engineer's theory of bending.

s   Yes, there will be a large change in the torsional stiffness.

t   (a) Weight proportional to the fourth power of length. (b) Weight proportional to diameter raised to the power 2.67.

BIBLIOGRAPHY

Cochran G V B 1982 *A Primer of Orthopaedic Biomechanics* (Edinburgh: Churchill Livingstone)

Frankel V H and Nordin M 1980 *Basic Biomechanics of the Skeletal System* (Lea and Fabiger)

Fung Y C 1993 *Biomechanics: Mechanical Properties of Living Tissue* 2nd edn (New York: Springer)

Hay J G and Reid J G 1988 *Anatomy, Mechanics and Human Motion* (Englewood Cliffs, NJ: Prentice Hall)

McMahon T A 1984 *Muscles, Reflex and Locomotion* (Princeton, NJ: Princeton University Press)

Miles A W and Tanner K E (eds) 1992 *Strain Measurement in Biomechanics* (London: Chapman and Hall)

Ozkaya N and Nordin M 1991 *Fundamentals of Biomechanics: Equilibrium, Motion and Deformation* (Princeton, NJ: Van Nostrand)

CHAPTER 2

BIOFLUID MECHANICS

2.1. INTRODUCTION AND OBJECTIVES

In Chapter 1 we looked at biomechanical systems, focusing in particular on the fundamental mechanical properties of bone and tissue and on how loads are supported by the skeleton and by the muscles. In this chapter we will investigate some basic fluid-mechanical systems in the human body. Once again the emphasis will be on the use of relatively simple models to describe gross performance characteristics of the systems. The primary difference between a solid and a fluid is that a stationary fluid cannot sustain shear stresses.

There are three scalar parameters that describe the conditions in a small volume of a fluid at rest. Moving fluids have one additional parameter, the velocity vector, which is conveniently expressed in terms of its components in three mutually perpendicular directions.

Table 2.1. *Fluid parameters.*

Variable	Symbol	SI unit
Density	ρ	kg m^{-3}
Temperature	T	K
Velocity	u, v, w	m s^{-1}
Pressure	p	Pa

In our analyses of human fluid-dynamic systems we shall normally consider that densities and temperatures are constant, and so we shall be concerned only with pressure and velocity distributions.

Most of the examples of fluid dynamics discussed in this chapter are drawn from the cardiovascular system. This is simply a reflection of the areas of interest and expertise of the authors, but the basic methodologies described are equally applicable to other areas of biofluid mechanics. Some of the questions that we shall address are listed below.

- What levels of fluid pressure are present in the human body?
- What parameters govern the flow of blood in the arteries and veins?
- Is blood thicker than water?
- How can the fluid-dynamic performance of the heart be measured?
- What are the fluid-dynamic effects of disease processes such as the narrowing of the arteries?

As in Chapter 1 there is material here that should be of interest to all readers.

2.2. PRESSURES IN THE BODY

We can often learn a great deal about the condition and performance of the body by studying the distribution of pressures in and around it. We all know many people who have problems with high blood pressure, and will be aware that blood pressure is taken routinely in the course of a general medical examination. There are many other pressure measurements made by physicians—and sometimes by medical physicists. These include respiratory pressures, bladder pressure, foot pressure, ocular pressure and middle-ear pressure.

We will recognize that pressure has the same units as stress, and indeed that a pressure is just a constant normal stress applied around all surfaces bounding a body. As physicists we will naturally use SI units of N m^{-2}, or Pascals. Our clinical colleagues, however, will demand pressures in millimetres of mercury or sometimes in millimetres (or centimetres) of water, and we need to be familiar with the following conversion factors.

Table 2.2. *Pressure conversion factors.*

	Atmosphere	N m^{-2} (Pa)	mmHg	mmH$_2$O
1 Atmosphere	1	1.01×10^5	760	10 300
1 N m^{-2} (Pa)	9.87×10^{-6}	1	0.0075	0.102
1 mmHg	0.001 32	133	1	13.6
1 mmH$_2$O	9.68×10^{-5}	9.81	0.0735	1

When we measure pressures using devices such as manometers we, in fact, obtain relative pressures. It is customary to quote pressures relative to atmospheric pressure, and we then refer to them as gauge pressures. Hence a pressure that is lower than atmospheric pressure is usually quoted as a negative pressure. Typical levels of pressure in and around the body are presented in table 2.3. All measures are quoted in millimetres of mercury for ease of comparison, but other units are standard for many of the systems.

2.2.1. *Pressure in the cardiovascular system*

There are essentially two circuits in the cardiovascular system. The systemic circuit carries oxygen-rich blood to the major organs, tissue, etc and returns the oxygen-depleted blood to the heart. The pulmonary circuit carries oxygen-depleted blood to the lungs to be recharged and returns oxygen-rich blood to the heart. The pressure in each of the two circuits is not constant. It varies with time and it varies with position—*temporal* and *spatial* variation. The pressure is generated by the muscular action of the heart, and is used to overcome viscous and other resistive forces as it drives the blood around the body. During *systole* the muscle of the heart contracts and blood is ejected into the aorta and pulmonary trunk. During *diastole* the muscle relaxes and the heart fills.

The pressure distribution in the cardiovascular system is governed by the equations of fluid dynamics, and later in this chapter we shall be looking at this variation in more detail. For now a qualitative description of the pressure variations will suffice. The spatial variation of pressure is a decrease along the arteries and veins as blood flows away from the heart. The pressure is used to overcome the viscosity of the blood. All of the branches and orifices in the circulatory system cause further decreases in pressure. We will soon learn how to quantify these losses. At positions close to the heart there is a substantial temporal variation of pressure as the heart beats. The temporal variation of pressure decreases at increasing distance from the heart, so that by the time the blood reaches the smaller vessels there is very little pulsatility. The damping of the pulsatility is attributable to the elasticity of the walls of the vessels in the cardiovascular system.

Table 2.3. *Pressures in the body (some of the values are taken from Cameron and Skofronik (1978)).*

Site		Pressure (mm Hg)
Arterial blood pressure:	systole	100–140
	diastole	60–90
Capillary blood pressure:	arterial end	~30
	venous end	~10
Venous blood pressure:	smaller veins	3–7
	great veins	<1
Cerebrospinal pressure in brain (lying down)		5–12
Gastrointestinal pressure		10–20
Bladder pressure		5–30
Lungs:	during inspiration	minus 2–3
	during expiration	2–3
Intrathoracic cavity (between lung and chest wall)		minus 10
Joints in skeleton		up to 10 000
Foot pressure:	static	up to 1200
	dynamic	up to 7500
Eye		12–23
Middle ear		<1

The blood pressure of a patient is used as a routine clinical measure of health. This pressure is often measured using a device called a sphygmomanometer. A standard procedure for use of this device is outlined below.

- Place cuff around upper arm.
- Rest arm on flat surface just below heart level.
- Place stethoscope on hollow of elbow over brachial artery.
- Inflate cuff to a pressure at which the blood does not flow, usually about 180 mmHg is sufficient.
- Release pressure slowly until sounds of blood squirting through restricted artery are heard. These sounds are called Korotkoff or 'K' sounds. *This is systolic pressure.*
- Continue to release pressure until no further sound is heard. *This is diastolic pressure.*

We will return to the subject of blood pressure and in particular how to measure it, in Chapter 18.

2.2.2. *Hydrostatic pressure*

The pressure in the liquid in a tank varies with the height. At the bottom of the tank the pressure is higher than it is at the top simply because of the weight of fluid supported on the lower layers. The pressure distribution that arises in a stationary fluid due to the action of gravity is called the hydrostatic pressure distribution. The hydrostatic pressure at all points within a stationary continuous fluid varies only with the vertical distance from the surface of the fluid,

$$\frac{\mathrm{d}p}{\mathrm{d}z} = \rho g \qquad \text{hydrostatic pressure distribution}$$

The rate of change of pressure due to the weight of gases is small because the density is small, and we would certainly not be concerned, for example, with the variation of atmospheric pressure outside the body over the height of a person. The hydrostatic pressure distribution is important clinically only with respect to liquids, and to blood in particular. The density of blood is close to that of water, about 1000 kg m^{-3}.

The pressure in a vessel the shape and size of the human body, if filled with blood, would be expected to vary by more than 100 mm of mercury because of the hydrostatic pressure effect when a person is standing. We would expect the pressure in the blood vessels in the skull to be some 35 mmHg lower than that at the heart and that in the blood vessels in the feet perhaps 100 mmHg higher (see figure 2.1). When a physician takes our blood pressure he places the cuff around the upper arm at about the level of the heart. Since typical blood pressure measurements are of the order of 100 mmHg, the hydrostatic effect is clearly an important one. It is not surprising that lying down is beneficial when a person feels faint due to a reduced blood supply to the brain. The pumping pressure of the heart is not excessive compared to the height through which blood must be lifted, even without accounting for any dynamic losses, and we operate on a relatively low factor of safety.

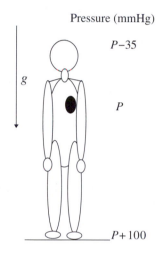

Figure 2.1. *Hydrostatic pressure when standing.*

2.2.3. Bladder pressure

The measurement of pressure in the bladder is a useful diagnostic tool for a range of urological problems. The pressure within the bladder has a passive component due to the viscoelastic nature of the bladder wall, and an active component due to the contraction of the muscular wall (the detrusor muscle). The normal rate of filling of the bladder is about 1 ml s^{-1} from each kidney. With this slow rate of filling, the detrusor muscle stays relaxed and the pressure rises very little, to about 10 cmH$_2$O. The sensation of fullness is triggered by stretch receptors in the wall of the bladder. The emptying of the bladder involves an active contraction of the detrusor muscle which, in a normal subject, will result in a pressure rise of about 25 cmH$_2$O (figure 2.2). A higher pressure in a male subject may be caused by enlargement of the prostate gland, which will increase the resistance to flow through the urethra.

Clinically we might measure these pressures using a catheter inserted through the urethra or a needle through the wall of the bladder (*cystometry*). As physicists we might devise very simple techniques based on

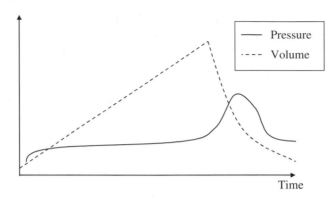

Figure 2.2. *Bladder pressure and volume during filling and voiding.*

principles such as energy conservation. The dynamics of urination present some very interesting problems in the analysis of collapsible elastic tubes, extending into the range of sonic or critically choked flows.

2.2.4. Respiratory pressures

The pressures in the lungs are required only to move air. Typical respiration rates at rest are about 12 breaths per minute, and typical inspiration volume is about 300 ml (see Chapter 17). The mass of the air to be moved is relatively small, and the viscosity of air is small, and so only small pressures are required. Typically the pressure in the lung is less than 5 mmHg below atmospheric for inspiration, and 5 mmHg above for expiration. The variation in pressure is achieved in principle by movement of the diaphragm, but to assist the process the pressure in the intrathoracic cavity, between the wall of the chest and the lung, is sub-atmospheric. This helps to keep the lungs inflated at all times during the breathing cycle.

2.2.5. Foot pressures

At the Royal Hallamshire Hospital in Sheffield one of the services offered by the Department of Medical Physics and Clinical Engineering is a dynamic foot pressure monitoring system. This can be of great benefit in diagnosis and treatment of problems with gait, as well as in the assessment of pathological conditions associated with diseases such as diabetes. In a normal adult the maximum pressure under the sole of the foot is about 100 kPa under static conditions, and perhaps some five times higher under dynamic conditions (see Chapter 20, section 20.2.2, for more detailed information). Higher pressures can indicate a tendency to ulceration, and can be indicative of neuropathological conditions.

2.2.6. Eye and ear pressures

The eyeball contains aqueous humour, a fluid comprised mostly of water. This fluid is produced continuously, and should drain to maintain normal pressure. A blockage in the drain causes an increase in pressure (*glaucoma*) and can lead to a restricted blood supply to the retina. In severe cases permanent damage might ensue. Sticking a finger in your eye to monitor pressure was one of the most common diagnostic techniques used by the local doctor up to the turn of the century. Eye pressure was regarded as a general indicator of physical well-being. Quite sophisticated methods for measuring intraocular pressure have now been developed (see Chapter 18, section 18.2.4).

The pressures in the middle ear are discussed in some detail in the chapters on the senses and on audiology (Chapters 3 and 15, respectively). The middle ear should operate at atmospheric pressure. This pressure is maintained through a vent to the atmosphere (Eustachian tube). The outlet is at the back of the throat, and the tube is not open all of the time. It is opened by actions such as swallowing, yawning and chewing. We can become most uncomfortable if the pressure in the middle ear is too different from atmospheric (positive or negative). This might occur due to a long-term blockage, caused perhaps by a common cold, or due to a sudden change in atmospheric pressure, such as that associated with aircraft take-off or landing.

2.3. PROPERTIES OF FLUIDS IN MOTION: THE CONSTITUTIVE EQUATIONS

Introduction

In Chapter 1, for solids, we sought to describe a material by the relation of a series of dimensionless displacements (strains) to the applied loads per unit area (stresses). The relation could be linear or nonlinear, but we had special terminologies for the linear case. In fluid mechanics we seek similar relationships, but with some important differences at the root of which is our definition that a fluid at rest cannot support a shear stress.

Shear stresses applied to an element of fluid cause a relative movement of the surfaces. Where for the solid we defined a shear strain, for the fluid we are interested in the rate of strain: i.e. how fast one point in the fluid is moving relative to adjacent points. The equation describing the relationship between stress and strain or strain rate is called the *constitutive equation*. A coherent and comprehensive introduction to the constitutive equations, with particular emphasis on the unity of approach in the modelling of solids and fluids, is presented by Fung (1993).

2.3.1. *Newtonian fluid*

The fluid-mechanical equivalent of the Hookean linear elastic solid (strain proportional to stress) is the Newtonian fluid (strain rate proportional to stress). The constant of proportionality is called *viscosity* (μ). In a Newtonian fluid the viscosity is independent of the stress and of the strain rate. The behaviour of a Newtonian fluid is illustrated in figure 2.3.

Algebraically, the shear stress is $\tau = \mu(\mathrm{d}v/\mathrm{d}y)$. The SI units of viscosity are N s m^{-2}, or Pa s, and those of shear stress are Pa. A common unit used in biofluid mechanics for the expression of stress, and particularly for shear stress, is the dyne cm^{-2}, equal to 0.1 Pa.

Many fluids, including water and air, exhibit properties that can reasonably be idealized using the Newtonian viscosity model. The viscosity of water is of the order of 1×10^{-3} Pa s, and that of air is 1.8×10^{-5} Pa s. There is a surprising variation of the viscosity of water with temperature, although we do not notice any difference in flow of hot and cold water in our common experience.

2.3.2. *Other viscosity models*

We should recognize that other viscosity models are possible, just as we can have other models of elasticity. Some alternatives are illustrated in figure 2.4. The simplest model is that of a frictionless or inviscid fluid, which generates no shear stress at any strain rate. This leads to substantial simplifications in the governing flow equations. A Newtonian fluid has a viscosity which is independent of strain rate. A dilatant (shear-thickening) fluid has increasing viscosity with increasing stress. A plastic (shear-thinning) fluid has decreasing viscosity with increasing stress. A Bingham plastic has a finite yield stress, which must be exceeded if fluid is to flow, and thereafter exhibits a linear relationship between stress and strain rate. Toothpaste is often cited as an example of a Bingham plastic.

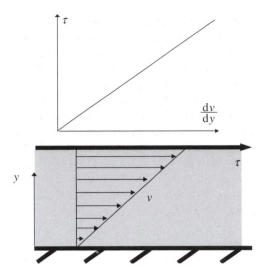

Figure 2.3. *Flow of a Newtonian fluid. Shear stress τ plotted against the shear rate* dv/dy.

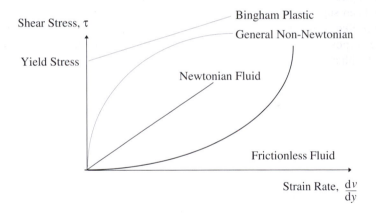

Figure 2.4. *Viscosity models.*

Some materials change viscosity even at a constant strain rate, i.e. they become more or less viscous with time spent in a particular flow regime. A thixotropic fluid is one in which the shear stress decreases with time under a constant strain rate. Blood exhibits this property.

2.3.3. Rheology of blood

Blood is not a homogeneous fluid, but consists of a suspension of cells and platelets in a fluid solution called plasma with about equal proportions of each. Most of the volume of the cellular components is composed of red blood cells, or erythrocytes. The erythrocytes are therefore by far the most important components of the cellular material with respect to the rheological properties of blood. The percentage of the volume occupied by the red cells is called the haematocrit, and a typical value of the haematocrit in a healthy individual would be about 45%. The erythrocytes are biconcave discs, essentially toroidal in shape but with the middle filled in. Marieb (1995) gives the outer diameter of the torus as 7.5 μm and the ring diameter, or cell thickness, as 2 μm. Although these cells are deformable, and are capable of travelling along tubes smaller than their own diameter, the diameter of the red blood cells is about that of the capillaries in the cardiovascular circulation. Adults have about 5 million red blood cells per cubic millimetre of blood. The remainder of the cellular matter is primarily composed of white blood cells, or leukocytes, and platelets. The leukocytes are spherical with diameters of about 7 μm, and their major role is in disease-defence mechanisms. The platelets are fragments of very large parent cells (megakaryocytes), and are much smaller than the red or white cells. They play a key role in clotting processes. Leukocytes make up only about 0.3% of the blood volume and platelets about 0.15%. Neither is present in sufficient volume to have a significant impact on the rheology of blood. The cells (specific gravity 1.1) can be separated from the plasma (specific gravity 1.03) using a centrifuge. The plasma is about 90% water and in isolation behaves as a Newtonian fluid, with a viscosity of 0.0012 Pa s.

Given the biphasic composition of blood, it is not surprising that it does not exhibit the idealized properties of a Newtonian fluid, particularly when it flows in small vessels in which the diameter of a red blood cell (7.5 μm) is a measurable proportion of the tube diameter. Obviously the bulk viscosity of the blood must be a function of the haematocrit, and in principle the viscosity must increase as the haematocrit increases.

An important characteristic of the rheology of blood is that its apparent viscosity reduces as the shear rate increases. A typical plot of viscosity against shear rate for blood with a normal haematocrit of about 45% is illustrated in figure 2.5. At a shear rate of 0.1 s^{-1} the viscosity is more than one hundred times that of water, and it has been suggested that at very low shear rates the blood can sustain a finite shear stress, and thus behave more like an elastic solid than like a fluid. At higher shear rates the viscosity is relatively constant, and for the purposes of analysis of flow in larger vessels at high shear rates blood is often modelled as a Newtonian fluid with a mean viscosity of 0.004 or 0.005 Pa s.

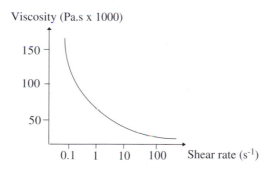

Figure 2.5. *Effect of shear rate on viscosity of blood (Merrill et al (1967) as quoted by Fung (1993)).*

An early model of blood flow, developed to produce a realistic model of behaviour in the lower shear rate regime ($<$10 s^{-1}), was proposed by Casson in the 1950s and is quoted by Fung.

Casson's equation:

$$\sqrt{\tau} = \sqrt{\tau_y} + \sqrt{\eta\dot{\gamma}}$$

where τ_y is the effective shear yield stress and η is a constant.

The magnitude of the yield shear stress is small, about 0.005 Pa for normal levels of haematocrit. Obviously at higher shear rates the yield stress is negligible and Casson's equation has essentially the same form as the Newtonian model. Fung suggests that the shear rate at which the Newtonian model becomes appropriate might be as high as 700 s^{-1} for normal levels of haematocrit.

So is blood thicker than water? The answer must be yes, because even the plasma alone has a viscosity slightly higher than that of water. When the cells are added the apparent viscosity increases to about four times that of water at high shear rates, and possibly by two or more orders of magnitude at low shear rates.

2.3.4.   *Virchow's triad, haemolysis and thrombosis*

No introduction to the rheology of blood would be complete without mention of Virchow's triad. Virchow recognized in the 19th Century that a triad of processes are involved in the phenomena of haemostasis and thrombosis and that to understand them would require knowledge of the interaction of blood at a cellular level with local haemodynamics and with the blood vessel itself. In this chapter we are concerned primarily with the haemodynamics of the flow. We shall concentrate particularly on local pressures and on local shear stresses in the fluid medium. There has been a resurgence of interest in local shear stresses in particular in recent years, as its implication in clotting processes had become increasingly understood. A comprehensive review of the current state of knowledge, covering biochemical as well as rheological and haemodynamic factors, was published by Kroll *et al* in 1996. These authors quote typical average levels of fluid shear stress within the healthy cardiovascular system in a range from 1 to 60 dynes cm^{-2} (0.1–6 Pa), and suggest that in stenotic vessels the levels might be up to an order of magnitude higher. They quote other investigators who have shown that shear stresses greater than 50 dynes cm^{-2} applied to platelet-rich plasma induce morphological changes, secretion and aggregation that are precursors to thrombotic episodes. There is evidence that platelet lysis is minimal at shear stress levels below 250 dynes cm^{-2}, and so it is suggested that platelet activation is responsible for these processes.

2.4.   FUNDAMENTALS OF FLUID DYNAMICS

2.4.1.   *The governing equations*

The equations governing the flow of real fluids in and around bodies of arbitrary geometry are rather complicated, and no general solutions are available. By consideration of the fluid entering and leaving a control volume we can write down two fundamental equations of fluid dynamics. The first equation appeals to the notion of *continuity*: the mass flow into and out of the volume is related to the change of density of the fluid within the volume. The second equation is an expression of Newton's second law, relating the forces (or stresses) on the element to the *momentum* flux into and out of the control volume. We often write the vector momentum equation in terms of the velocity and pressure gradients in three mutually perpendicular directions. In the general case there are more variables than unknowns, and further (thermodynamic) relationships are required to complete the description of the problem. If we assume that the fluid is incompressible, so that the density is constant throughout the fluid domain, and that the fluid is Newtonian, so that the shear strain rate of the fluid is proportional to the shear stress, then the equations are somewhat simplified. The form of the momentum equation based on these assumptions is called the *Navier–Stokes equation*. This, together with the continuity equation, describes the flow of an incompressible fluid. The numerical expressions of the equations are omitted from this introduction, but they are presented in section 2.8.

Unfortunately, even with these assumptions, the equations are nonlinear and cannot readily be solved for any but the simplest geometries. However, we can often gain a strong insight into the performance of fluid systems by making further assumptions to reduce the complexity of the equations to the point at which analytical solutions are available. Different assumptions lead to different simplified equations, and if we are to choose the appropriate equations we will need to be able to recognize the physics of the flow. Fluid mechanics is sometimes regarded as something of a 'black art' because it seems that one needs to know the answer before one can solve the problem!

2.4.2.   Classification of flows

Compressible versus incompressible

An important division of flow problems can be made depending on whether the fluid is compressible or incompressible. If a fluid is incompressible then all terms in the equations that relate to variation in time or space of density are zero, and the Navier–Stokes equations are applicable. It is obvious that liquids such as water and oil are very difficult to compress, and for most problems we assume that liquids are incompressible. It is equally obvious that gases (in biological applications we will be particularly interested in air) are compressible, and the reader is no doubt very familiar with the relationship between the pressure and volume of an ideal gas. However, for problems in fluid dynamics, we can show that the compressibility of a gas has relatively little impact on flow phenomena if the velocities are less than about one-third of the speed of sound in the fluid. For air, then, the maximum velocity that is consistent with the assumption of incompressibility is about 100 m s^{-1}. We encounter very few problems in biomechanics in which velocities exceed a few metres per second, and so for our purposes, in dynamic problems, air is treated as incompressible. We might have to modify this assumption if we want to analyse actions such as sneezing, in which the expelled air can reach remarkably high velocities.

Viscid versus inviscid

It is believed that the Navier–Stokes equation, together with the continuity equation, provides a complete description of the flow of an incompressible Newtonian fluid. Unfortunately the equations are still too difficult to solve for arbitrary geometries with arbitrary boundary conditions. The equations are further simplified if the viscosity of the fluid is ignored (i.e. the flow is assumed to be inviscid). In this case the Navier–Stokes equation reduces to *Euler's equation*. In two and three dimensions Euler's equation can be used to investigate the pressure drag associated with external flow over objects. Analytical solutions are available for some simple geometries, and we can write relatively straightforward computational algorithms to study more general shapes. We might use such a program to look at problems such as the design of a cycle helmet.

Viscous forces within a fluid act to dissipate energy, and thus in a viscous fluid-dynamic system mechanical energy is not conserved. For inviscid flows the balance between potential and kinetic energy yields *Bernoulli's equation*. We can usefully apply Bernoulli's equation, in conjunction with the continuity equation, to get first estimates of the pressure changes associated with area changes in vessels. We can also use it to investigate problems such as the speed of emptying of pressurized vessels. Bernoulli's equation is in fact a form of Euler's equation, integrated along a streamline to make it appear one dimensional. In biofluid mechanics we are concerned mostly with flow in tubes and we might use the one-dimensional Bernoulli equation to obtain initial estimates of the pressure gradients in branching tube systems. We sometimes modify the Bernoulli equation to include a factor to account for viscous losses.

Laminar versus turbulent

In the Navier–Stokes equation the acceleration of the fluid is related to the variations of the pressure and to the viscosity of the fluid. There are thus two contributions to the pressure gradients: one from the accelerations (inertial terms) and one from the viscous stresses (viscous terms). In steady flow the velocity at a point in space is fixed, and the only acceleration terms are associated with convection. The relative magnitudes of the inertial and viscous terms can be related to a dimensionless parameter called the Reynolds number,

$$Re = \frac{vL\rho}{\mu}$$

where v is the velocity, ρ the density, L a length scale and μ the viscosity. The Reynolds number is the major determinant of the flow regime in a fluid-dynamic situation.

In the body we shall be concerned primarily with the flow of fluids in tubes. In this case it is conventional to use the diameter of the tube, d, as the length scale, and the Reynolds number for flow in a tube is

$$Re = \frac{vd\rho}{\mu}.$$

The relationship between the Reynolds number and the characteristics of the flow in a tube are tabulated below.

$0 < Re \leq 1$	Laminar 'creeping' flow, entirely dominated by viscosity.
$1 < Re \leq 2000$	Laminar flow, both viscous and acceleration terms important.
$2000 < Re \leq 10\,000$	Transition to turbulence
$10\,000 < Re$	Turbulent flow, laminar viscosity relatively unimportant.

In many flows even of viscous fluids the effects of viscosity are confined to relatively thin boundary layers attached to solid walls, and gross parameters of the flow are reasonably independent of viscosity. In particular, this is often true of transient and starting flows.

Steady versus unsteady

A major classification of flow will be determined by its condition with respect to time. In steady flow the velocity at all points within a fluid domain is a function of position only, and the velocity and pressure at a point do not, therefore, change with time. This condition is not satisfied for turbulent flow, but we often assume that there is an underlying steady flow field with a local chaotic field superimposed on top of it. The turbulent component of the velocity has no temporal consistency, and appears as noise on a velocity plot. We might perform a 'steady-state' analysis of a turbulent flow, separating the velocity into mean and turbulent components. The analysis of turbulent flows is a numerically and computationally intensive process and is beyond the scope of this chapter. However, the reader should learn to recognize where in biofluid mechanics we might anticipate turbulent flows to arise.

More ordered unsteady flows, such as the regular pulsatile flow of blood in vessels close to the heart or the periodic vortex shedding associated with flow over a cylinder, might yield to analysis with the velocities and pressures solved in terms of harmonic series. These flows might be categorized as temporally disturbed, but they are not turbulent. The flow regime in such systems is characterized by the Reynolds number and by a second parameter, the Strouhal number, given by

$$St = \frac{\omega L}{v}.$$

A composite parameter, combining the Reynolds number with the Strouhal number, has been shown by Womersley to play an important role in the evaluation of oscillatory flows in the cardiovascular system. A comprehensive discussion of Womersley's work in this context is presented by McDonald (3rd edition edited by Nichols and O'Rourke (1990)). The Womersley parameter, α, is

$$\alpha = \sqrt{\pi\,Re\,St}.$$

One special case of unsteady flow is starting flow, covering, for example, a system in which fluid initially at rest in a tube is subjected suddenly to a pressure gradient along its length and left to develop. The fluid distant from the walls simply starts to move as a solid plug, obeying Newton's second law, whilst a region of viscous flow gradually develops from the wall. The viscous region expands to contain the whole of the flow, and the solution is asymptotic to the steady flow solution. In practice, a plug-like flow profile in the body is often an indicator of spatially or temporally undeveloped flow although, as we shall see in section 2.5.2, it might also be an indicator of turbulent flow.

Rigid versus elastic walls

When the walls of a tube are rigid the continuity equation dictates that the flow rate at all cross-sections must be the same. Tubes with elastic walls expand and contract under the action of the internal pressure. This means that they can act as a reservoir, delaying part of the outflow relative to the inflow. Furthermore, the pressure pulse has a finite speed in an elastic tube, and this might be important with respect to the propagation of the pulse from the heart. Simple one-dimensional models of an elastic system can be used to determine the gross characteristics of the flow, and two such models are discussed in section 2.7.

A special and extreme case of flow in elastic tubes occurs when the external pressure is higher than the internal pressure. Under this condition, if the bending modulus of the tube wall is sufficiently low, the vessel will buckle so that the cross-section is no longer circular. There are many examples of flow in collapsible tubes in the body, including some venous flows, airflow in the lungs in some conditions and normal flow in the ureter. Collapsible tubes introduce a whole new set of phenomena including limiting flow, analogous to sonic flow, hydraulic jumps, and self-induced oscillations. A brief review of the literature in this field is included in the comprehensive review of biofluid mechanics presented by Skalak and Ozkaya (1989), who suggest that the Korotkoff sounds monitored in the measurement of blood pressure using a sphygmomanometer are associated with these phenomena. A detailed study of the flow in collapsible tubes is beyond the scope of this chapter.

2.5. FLOW OF VISCOUS FLUIDS IN TUBES

2.5.1. *Steady laminar flow*

Although we can develop an analysis model of the flow of fluid in a tube by starting with the continuity equation and the full Navier–Stokes equations (most conveniently in cylindrical form), it would be a laborious way to proceed for this simple geometry. If we assume that the flow is fully developed, spatially and temporally, then the only component of velocity at any point in the tube is the axial one. In steady flow the fluid in a tube is not accelerating, and an equation relating the pressure gradients to the viscous shear stresses can be derived by consideration of the equilibrium of an annulus of the fluid (see figure 2.6).

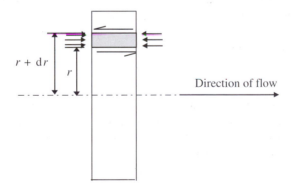

Figure 2.6. *Stresses on an annulus of fluid: developed laminar flow.*

For equilibrium of the annulus of fluid in the axial direction,

$$p\, 2\pi r\, \mathrm{d}r + \tau\, 2\pi r\, \mathrm{d}z = \left(p + \frac{\mathrm{d}p}{\mathrm{d}z}\mathrm{d}z \right) 2\pi r\, \mathrm{d}r + \left(\tau + \frac{\mathrm{d}\tau}{\mathrm{d}r}\mathrm{d}r \right) 2\pi (r + \mathrm{d}r)\, \mathrm{d}z.$$

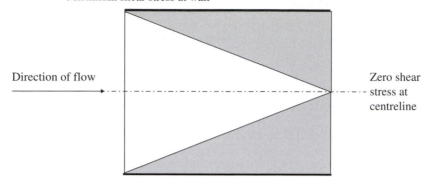

Maximum shear stress at wall

Direction of flow

Zero shear stress at centreline

Figure 2.7. *Distribution of shear stress in laminar flow in a tube.*

Neglecting terms that involve the product of three small quantities:

$$-2\pi r \frac{\mathrm{d}p}{\mathrm{d}z}\mathrm{d}z\,\mathrm{d}r = 2\pi\left(\tau + r\frac{\mathrm{d}\tau}{\mathrm{d}r}\right)\mathrm{d}z\,\mathrm{d}r.$$

Noting that

$$\frac{\mathrm{d}}{\mathrm{d}r}(r\tau) = \tau + r\frac{\mathrm{d}\tau}{\mathrm{d}r}$$

the above equation simplifies to

$$\tau = -\frac{r}{2}\frac{\mathrm{d}p}{\mathrm{d}z}.$$

This equation was developed by Stokes in 1851. The distribution of shear stress is illustrated in figure 2.7.

The shear stress is related to the shear rate through the constitutive equation. For a Newtonian model of viscosity, appropriate for water and perhaps for bulk flow of blood in a larger tube:

$$\tau = \mu\frac{\mathrm{d}v}{\mathrm{d}r}\qquad\text{Newtonian viscosity model.}$$

Substituting into the equation for shear stress:

$$\mu\frac{\mathrm{d}v}{\mathrm{d}r} = \tau = \frac{r}{2}\frac{\mathrm{d}p}{\mathrm{d}z}$$

$$\mathrm{d}v = \frac{1}{2\mu}\frac{\mathrm{d}p}{\mathrm{d}z}r\,\mathrm{d}r$$

$$\int\mathrm{d}v = \frac{1}{2\mu}\frac{\mathrm{d}p}{\mathrm{d}z}\int r\,\mathrm{d}r$$

$$v + C = \frac{1}{2\mu}\frac{\mathrm{d}p}{\mathrm{d}z}\frac{r^2}{2}$$

where C is a constant of integration.

At the wall of the tube the velocity is zero, and so

$$C = \frac{1}{2\mu}\frac{\mathrm{d}p}{\mathrm{d}z}\frac{R^2}{2}.$$

Substituting for C in the equation for v,

$$v + \frac{1}{2\mu}\frac{dp}{dz}\frac{R^2}{2} = \frac{1}{2\mu}\frac{dp}{dz}\frac{r^2}{2}$$

$$v = -\frac{R^2}{4\mu}\frac{dp}{dz}\left(1 - \frac{r^2}{R^2}\right).$$

In terms of pressure *drop* per unit length,

$$v = \frac{R^2}{4\mu}\frac{\Delta p}{L}\left(1 - \left(\frac{r}{R}\right)^2\right).$$

Hence for the Newtonian fluid the velocity profile for laminar flow in a tube is parabolic. The velocity is zero at the wall and reaches a maximum at the centreline (see figure 2.8).

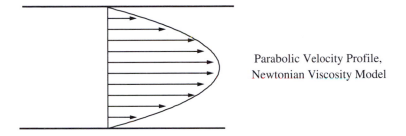

Parabolic Velocity Profile,
Newtonian Viscosity Model

Figure 2.8. *Velocity profile of laminar flow of a Newtonian fluid in a tube.*

By integrating over concentric annuli in the tube we can derive a relationship between volume flow rate, Q, and pressure drop,

$$Q = \frac{\Delta p}{L}\frac{\pi R^4}{8\mu} \qquad \text{(Poiseuille's equation)}.$$

The Poiseuille equation is sometimes re-arranged to express the 'resistance to flow' of a blood vessel. By analogy with Ohm's law, and equating the flow rate to current and the pressure gradient as voltage,

$$\text{resistance to flow,} \qquad K = \frac{\Delta p}{LQ} = \frac{8\mu}{\pi R^4}.$$

This equation gives us an immediate appreciation of the consequences of reduced arterial diameters due to disease processes. A reduction of 30% in the radius of a vessel causes a 400% increase in flow resistance over its length. We shall see later (section 2.6) that there is likely to be an additional irrecoverable pressure loss at a stenosis, further compounding the problem.

We can also express the peak shear stress, at the wall, in terms of volume flow rate,

$$\tau_{max} = -\frac{4\mu Q}{\pi R^3}.$$

There are approximately 100 000 km of blood vessels in the adult human body. The lumen diameter of the aorta, leaving the heart as the first vessel in the systemic circuit, is typically 25 mm. The large arteries have elastic walls and are subjected to substantially pulsatile flow. The smaller (muscular) arteries are more rigid, and range in diameter from about 0.3 to 10 mm. These feed the arterioles, of diameter range 0.3 mm down to

10 μm. The smallest blood vessels are the capillaries, with diameters in the range 8–10 μm. The Poiseuille equation and its derivatives are most applicable to flow in the muscular arteries, but modifications are likely to be required outside this range. The diameters of the capillaries are about those of a red blood cell, and models of capillary flow are normally based on two-phase systems with single-file progression of the erythrocytes along the capillary.

We have demonstrated that the velocity profile of laminar flow in a tube is parabolic for a Newtonian fluid. The maximum shear stress occurs at the walls, and the stress diminishes towards the centre of the tube. This gives us a clue as to what might happen in a fluid represented by the Casson model (section 2.3.4). The shear stress towards the centre of the tube is not sufficient to cause the blood to yield, and a central core simply translates as a rigid body. It can be shown, Fung (1993), that the velocity profile and volume flow rate equations for laminar flow of a Casson fluid can be expressed in a similar form to the Newtonian equations as follows (see figure 2.9).

Axial velocity:

$$v_z = \frac{R^2}{4\eta} \frac{\Delta p}{L} \left(1 - \left(\frac{r}{R} \right)^2 + g(r) \right).$$

Flow rate:

$$Q = \frac{\Delta p}{L} \frac{\pi R^4}{8\mu} f(r).$$

The velocity distribution and flow rate, although of similar form to those for the Newtonian fluid, are modified by functions g and f, respectively. The core flow in the centre of the tube represents the volume in which the shear stress is lower than the yield shear stress, causing the central plug to move as a rigid body. It is, of course, possible for this volume to extend right out to the wall, in which case the fluid bulges under the pressure gradient, but does not flow.

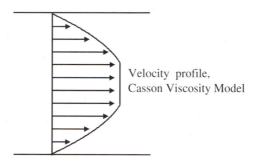

Figure 2.9. *Velocity profile of laminar flow in a tube: Casson viscosity model.*

Although the Casson viscosity model is an improvement over the Newtonian model for blood flow at low shear rates, it is still assumed that the blood can be represented as a homogeneous fluid, albeit with more complex viscous behaviour. In blood vessels of less than about 1 mm in diameter, the inhomogeneous nature of blood starts to have a significant effect on the apparent fluid-dynamic properties. In the 1920s and 1930s Fahraeus and Lindquist investigated the characteristics of blood flow in tubes of diameters in the range 50–500 μm, and they have given their names to two interesting phenomena that they observed in their experiments.

The Fahraeus effect

When blood of a constant haematocrit flows from a large reservoir into a tube of small diameter, the haematocrit in the tube decreases as the tube diameter decreases.

The Fahraeus–Lindquist effect

The apparent viscosity of blood flowing in a tube from a feed reservoir decreases with decreasing tube diameter (in the range 50–500 μm).

Since the viscosity of blood increases with increasing haematocrit, the Fahraeus–Lindquist effect is a consequence of the Fahraeus effect. It is interesting to speculate on why the Fahraeus–Lindquist effect should occur. One possible explanation is that there is a boundary layer close to the wall that is relatively cell-free. Whatever the reason, the effect is important and any study of the flow and physiological characteristics of the microcirculation must take into account the inhomogeneity of blood. Trowbridge (1984) challenged the conventional multi-phase blood models, based essentially on the assumption that the red blood cells are dispersed in the plasma and do not interact with each other. He proposed and developed a continuum mixture model of blood that was able to explain the Fahraeus and Fahraeus–Lindquist effects. Despite this success the model has not achieved widespread recognition.

2.5.2.   Turbulent and pulsatile flows

When the Reynolds number of flow in a tube exceeds about 2000, the ordered laminar flow breaks down into turbulence. It is possible to obtain laminar flow in a tube at higher Reynolds numbers, but only under very carefully controlled conditions. Using pure distilled water flowing in very smooth glass tubes of great length to diameter ratio, laminar flow has been observed at Reynolds numbers up to and exceeding 10 000.

In the chemical engineering industry it is customary to express the pressure loss in tube flow in terms of a friction factor multiplied by the dynamic head. In laminar flow we can readily demonstrate that the friction factor is inversely proportional to the Reynolds number. In turbulent flow there is a sharp rise in the friction factor and the 'resistance to flow' increases significantly. The Reynolds number itself is less important, but the roughness of the wall of the pipe becomes critical. Empirical data are available to determine the friction factor in turbulent flow.

The velocity profile in turbulent flow takes the shape of a truncated parabola, not unlike that associated with laminar flow in the Casson model, although for very different reasons. If the Reynolds number is very high the flat profile might appear to cover the whole of the diameter: it might be practically impossible to measure the velocity close enough to the wall to see the parabolic portion.

It is important to recognize when a flow might become turbulent because it can then no longer be described by the Poiseuille equation and its derivatives. The following guidance is offered on the likely flow regime in physiological systems.

- The flow throughout most of the circulatory system is within the laminar region.
- There is some debate about the nature of flow in the aorta. It is possible that this flow becomes turbulent at or near to peak systole. The flow in the aorta is pulsatile and the onset of turbulence is not determined by the Reynolds number alone. An additional parameter, describing the periodicity of the flow, is required to determine the flow characteristics.
- When heart valves leak in the closed position (regurgitation), the jets that issue from the leak sites can have very high velocities. There is little doubt that such jets are turbulent and that high shear stresses are generated in the ventricular chamber.

Pulsatile flow

It is possible for us to extend the analysis to investigate pulsatile flow, but the mathematics becomes somewhat more difficult and a full discussion is beyond the scope of this chapter. Womersley (see McDonald 1974) did a lot of work in this field and there is a comprehensive discussion of the topic in Chapter 5 of McDonald (Nichols and O'Rourke 1990).

2.5.3.   Branching tubes

When we come to study the anatomy of the circulatory system it will immediately become apparent that there are very many branches as the blood is distributed around the body. The complexity of the circuit makes the wiring harness in a car look relatively simple. Generally, as a vessel branches, each of the distal vessels is smaller in diameter than the parent, or proximal, vessel. The question arises as to how the Reynolds number changes across these branches. Is the flow more or less stable at increasing distance from the heart?

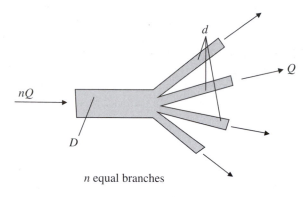

Figure 2.10. *Multiple-branching of an artery from large diameter D to small diameters d.*

We can answer this question very easily by looking at the continuity of flow across the branches. The following analysis is based on that presented in Chapter 2 of Nichols and O'Rourke (1990). Using the terminology of figure 2.10.

Flow in proximal vessel:

$$nQ = \frac{\pi D^2}{4} V.$$

Flow in distal vessels:

$$Q = \frac{\pi d^2}{4} v.$$

Hence for continuity:

$$\frac{v}{V} = \frac{D^2}{nd^2}.$$

The velocity reduction over the branch is inversely proportional to the square of the distal diameter ratio and inversely proportional to the number of branches. Alternatively since nd^2 is proportional to the size of the vascular bed downstream of the branch (the total area of the distal vessels), the reduction in velocity is proportional to the increase in the size of the vascular bed. This proportionality can be used to estimate the change in size of the vascular bed if the mean velocities can be measured.

The ratio of the Reynolds number in a distal vessel to that in the proximal vessel is

$$\frac{Re_d}{Re_p} = \frac{vd}{VD} = \frac{1}{n}\frac{D}{d} = \frac{1}{n}\sqrt{\frac{A_p}{A_d}}.$$

Hence the Reynolds number will remain constant across the branch only if the fractional reduction in diameter (or the square root of the fractional reduction in area) is equal to the number of branches.

The resistance to flow can be determined from the Poiseuille equation. Writing the resistance, K, as the pressure gradient divided by flow,

$$K = \frac{\Delta p}{LQ} = \frac{128\mu}{\pi d^4}.$$

Hence the ratio of distal-to-proximal resistance is

$$\frac{K_d}{K_p} = \frac{D^4}{nd^4} = \frac{1}{n}\left(\frac{A_p}{A_d}\right)^2.$$

At a major arterial bifurcation the diameter reduction is typically about 0.8, corresponding to an area reduction of 0.64 in a distal vessel relative to the proximal vessel. The total area of the two distal vessels taken together is greater than that of the proximal vessel by a factor of 1.28, and this is a measure of the increase in size of the vascular bed. The Reynolds number decreases by a factor of 0.63 over the bifurcation and the resistance increases by a factor of 1.22. In fact, these trends hold true at all observed branches in the vascular system: the Reynolds number decreases and the resistance increases over every branch. When the haemodynamics of flow at branches is studied in detail it is apparent that there are local separations and circulatory flows. These phenomena consume energy, and there are additional irrecoverable pressure losses at each bifurcation.

2.6. FLOW THROUGH AN ORIFICE

The energy required to pump blood around the circulation is provided by the heart. This amazing organ will beat some 30 million times a year, or over a billion times in an average lifetime. The design has to be efficient, and the parts have to be robust. Perhaps the most vulnerable components in the heart are the valves, which control the direction of flow and the timing of the pulse relative to the muscular contractions. There are always pressure losses associated with the expulsion of a fluid through an orifice, and the heart valve is no exception. We can learn something about the state of the heart by investigating the pressure losses through the valves. The detailed physics of the flow through orifices is rather complicated: in general there will be flow separation and reattachment, vortices, recirculation and possibly turbulence too. We might seek to resolve these flows using computational techniques (see section 2.8), but we can gain a strong insight into the gross characteristics of the flow using simple one-dimensional equations.

2.6.1. Steady flow: Bernoulli's equation and the continuity equation

For the one-dimensional analysis we shall assume that there is an effective area of flow at each cross-section, and that the velocity of the flow is constant on any such section,

$$A = A(z).$$

We have an immediate problem in that we do not know the actual cross-sectional area of the jet at any point. There is likely to be a *vena contracta* (marked as 3 in figure 2.11) downstream of the orifice that has a cross-sectional area lower than that of the orifice itself. In practice we shall resolve this difficulty by resorting to empirical data. As a first approximation we might assume that the diameter of the *vena contracta* is that of the orifice.

The first equation to apply is the *continuity equation*: assuming that the flow is incompressible this will yield directly the average velocity at all cross-sections. Note that the continuity equation represents a relationship between flow rate and mean flow velocity at all cross-sections given that we know the area of the

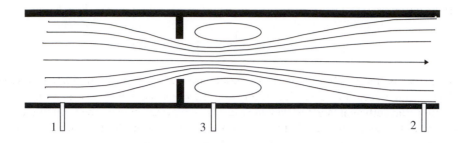

Figure 2.11. *Flow through an orifice.*

cross-section. In fluid dynamics textbooks the upstream velocity is usually taken as the governing parameter, but in biofluid mechanics it is conventional to choose the flow rate,

$$Q = Av = \text{constant}.$$

Hence for a given flow rate, Q,

$$v_1 = \frac{Q}{A_1} \qquad v_2 = \frac{Q}{A_2} \qquad v_3 = \frac{Q}{A_3}$$

or, in terms of the upstream velocity,

$$v_2 = \frac{A_1}{A_2}v_1 \qquad v_3 = \frac{A_1}{A_3}v_1.$$

The second equation is the *momentum equation*. For steady inviscid flow the momentum equation is

$$\rho v \frac{dv}{dz} = -\frac{dp}{dz}.$$

Integrating both sides,

$$\rho \int v\,dv + \int dp = 0,$$

$$\rho\left[\tfrac{1}{2}v^2\right]_{z=0}^{z} + [p]_{z=0}^{z} = 0$$

or

$$p + \tfrac{1}{2}\rho v^2 = \text{constant}.$$

This is an expression of *Bernoulli's equation* for flow in a tube in the absence of gravity. If this equation is integrated over thin slices normal to the axis of the tube, it is apparent that the pressure term is the potential energy of the fluid and the velocity term is the kinetic energy. The equation therefore states that mechanical energy is conserved in the system. From the Bernoulli equation in this form, for a given flow-rate, Q,

$$p_1 - p_3 = \frac{1}{2}\rho\frac{Q^2}{A_1^2}\left(\frac{A_1^2}{A_3^2} - 1\right) \qquad p_2 = p_1$$

or, in terms of the upstream velocity

$$p_1 - p_3 = \frac{1}{2}\rho v_1^2 \left(\frac{A_1^2}{A_3^2} - 1 \right) \qquad p_2 = p_1.$$

In practice we know that the mechanical energy is not conserved. Energy is dissipated by viscous forces and by turbulence. The process of conversion of potential energy to kinetic energy as we approach the *vena contracta* is, in fact, quite efficient, and the pressure there is well represented by the Bernoulli equation. Most of the losses occur as the jet expands again to fill the downstream section. Some, but not all, of the pressure that has been lost is recovered downstream as the flow decelerates (see figure 2.12).

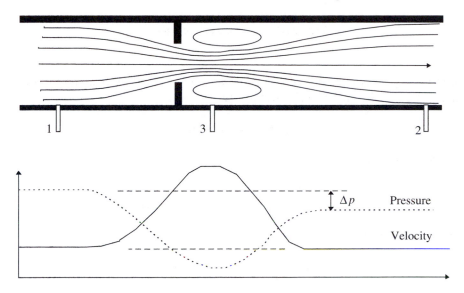

Figure 2.12. *Experimentally observed pressure and velocity distribution across an orifice.*

It is conventional to write the total pressure loss across the orifice in terms of the dynamic head of the fluid:

$$\Delta p = p_1 - p_2 = \frac{1}{2}\rho \frac{Q^2}{A_1^2} \left(\frac{1}{C_d^2} \right)$$

where C_d is a discharge coefficient. If there is no pressure recovery from the condition at the *vena contracta* then $C_d = 1/[(A_1/A_3)^2 - 1]$.

The definition of the discharge coefficient makes more sense when we look at the problem from a different angle. Often we wish to know the flow rate that can be sustained by a particular pressure drop. The pressure drop equation can be written as

$$Q = C_d A_1 \sqrt{\frac{2\Delta p}{\rho}}.$$

The flow through the orifice is directly proportional to the discharge coefficient. An experimentally measured discharge coefficient will be accurate only for the flow rate (or pressure drop) at which it has been measured. In practice, however, the discharge coefficient is remarkably constant over a wide range of flow rates, and the performance of an orifice can be represented in these terms.

These formulae for the pressure drop across an orifice have been adopted with particular reference to the performance of heart valves by Gorlin and Gorlin (1951). The Gorlin equation provides a simple measure of the performance of a valve, expressed as an effective area, and is intended to give an indication of the clinical expediency of corrective surgery,

$$A_c = \frac{Q}{C_d \times 51.6 \times \sqrt{\Delta p}} \qquad \text{(Gorlin equation)}.$$

This equation must be applied with care because it contains a dimensional constant C_d. A critical review of the assumptions inherent in the clinical application of the Gorlin equation, together with a suggested modification for assessment of heart valve substitutes, was published by Tindale and Trowbridge (1986).

One of the criticisms of the Gorlin model has been that it assumes steady flow conditions when, clearly, the flow in the heart is pulsatile. Pulsatile models have been developed (see Yellin and Peskin, 1975) but are beyond the level of this text.

2.7. INFLUENCE OF ELASTIC WALLS

All of the analysis in the preceding sections is based on the premise that the fluid flows in a rigid-walled vessel. Blood vessels in the body, and particularly the larger arteries, are elastic. We shall now explore some simple one-dimensional models in which this elasticity is taken into account.

2.7.1. Windkessel theory

The first theory of flow in the aorta was developed by Frank in 1899 (see Fung). It was assumed that the aorta acts as an elastic reservoir: its instantaneous volume and pressure are related by an equilibrium equation, and the rate of outflow is determined by the peripheral resistance (see figure 2.13).

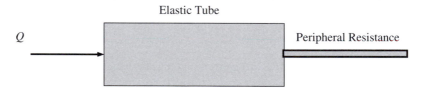

Figure 2.13. *Windkessel model of flow in the aorta.*

The change in volume of an elastic-walled vessel of radius r, thickness t and Young's modulus E under the action of an internal pressure p can be calculated by consideration of the equilibrium of the wall of the tube. If the change in radius is small relative to the original radius, then the change in volume is approximately proportional to the applied pressure,

$$\mathrm{d}Vol = 2\pi r L \frac{pr^2}{Et} = K p.$$

The rate of flow out of the vessel is determined by the peripheral resistance, R,

$$Q_{\text{out}} = \frac{p}{R}.$$

The total flow into the reservoir, Q, is the sum of that required to increase its volume and the outflow:

$$Q = K \frac{\mathrm{d}p}{\mathrm{d}t} + \frac{p}{R}.$$

This is a first-order linear differential equation. We have solved similar equations in Chapter 1 when we looked at models of viscoelasticity. The solution procedures are the same, and solutions for particular forms of the inlet flow (based on the flow from the heart) provide a measure of the anticipated pressure–time waveform in the aorta. An example of the solution of this equation is presented in problem 2.9.2.4.

2.7.2. *Propagation of the pressure pulse: the Moens–Korteweg equation*

A simple model describing the flow of an inviscid fluid in an elastic tube can be developed by consideration of the continuity and momentum equations. For a detailed review of this topic refer to the monograph by Pedley (1980). The basic assumptions that will be made are:

- the fluid is homogeneous and isotropic;
- the fluid is incompressible;
- the fluid is inviscid.

The cross-sectional area at any point in the tube will vary with time as illustrated in figure 2.14. Because the flow is inviscid it is assumed that the velocity, $v_z(z, t)$, does not vary across a cross-section (although of course it varies along the axis and with time). The pressure also is a function of the axial coordinate and of time.

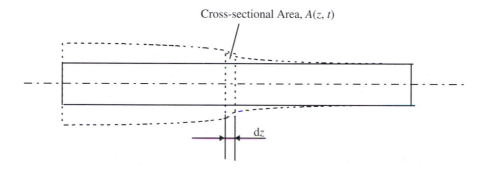

Cross-sectional Area, $A(z, t)$

dz

Figure 2.14. *Wave propagation in an elastic tube.*

The continuity equation is

$$\frac{\partial A}{\partial t} + \frac{\partial}{\partial z}(v_z A) = 0.$$

The momentum equation is

$$\rho\left(\frac{\partial v_z}{\partial t} + v_z \frac{\partial v_z}{\partial z}\right) = -\frac{\partial p}{\partial z}.$$

Expanding the second term in the continuity equation gives

$$\frac{\partial A}{\partial t} + A\frac{\partial v_z}{\partial z} + v_z \frac{\partial A}{\partial z} = 0.$$

This equation can be simplified if it assumed that:

- the pulse is moving slowly (relative to the speed of sound in the fluid) and with a long wavelength.

Under this assumption the last term may be neglected and the continuity equation reduces to

$$\frac{\partial A}{\partial t} + A\frac{\partial v_z}{\partial z} = 0.$$

The fluid is flowing inside an elastic tube, and it is assumed that the pressure at any cross-section can be calculated from the elastic properties and the shape of the tube at that cross-section.

- The pressure $p(z, t)$ is a function of the cross-sectional area $A(z, t)$: $p = P(A)$.

The consequences of this assumption are:

- no account is taken of the stiffness of adjacent cross-sections in determining the distension at a section. The distension at one cross-section has no effect on the distension elsewhere;
- no account is taken of the viscoelastic behaviour of the tube wall.

The differential of area with respect to time in the reduced form of the continuity equation can now be expressed in terms of the differential of pressure with respect to time.

$$\frac{\partial A}{\partial t} + A\frac{\partial v_z}{\partial z} = \frac{\partial p}{\partial t}\frac{\mathrm{d}A}{\mathrm{d}p} + A\frac{\partial v_z}{\partial z} = 0$$

$$\frac{\partial v_z}{\partial z} = -\frac{\partial p}{\partial t}\frac{\mathrm{d}A}{\mathrm{d}p}\frac{1}{A}.$$

A linearized form of the momentum equation can be obtained by discarding the terms associated with the convective acceleration. This is consistent with the assumption that the pulse is propagating relatively slowly. The momentum equation reduces to

$$\frac{\partial v_z}{\partial t} = -\frac{1}{\rho}\frac{\partial p}{\partial z}.$$

Taking the partial differential of the first equation with respect to t and the partial differential of the second equation with respect to z,

$$\frac{\partial^2 p}{\partial t^2}\frac{\mathrm{d}A}{\mathrm{d}p}\frac{1}{A} = -\frac{\partial^2 v_z}{\partial z\,\partial t} = \frac{1}{\rho}\frac{\partial^2 p}{\partial z^2}.$$

This can be expressed in familiar terms as the wave equation (this equation is discussed in more detail in Chapter 3, section 3.4.2):

$$\frac{\partial^2 p}{\partial t^2} = c^2\frac{\partial^2 p}{\partial z^2}$$

where

$$c^2 = \frac{A}{\rho}\frac{\mathrm{d}P(A)}{\mathrm{d}A}.$$

For a thin-walled circular elastic tube with unconstrained ends the hoop stress in the wall under a uniform pressure is

$$\sigma_\theta = \frac{pr_0}{t}.$$

The strain in the wall is σ/E and the change in radius is therefore

$$\delta_r = \frac{pr_0^2}{Eh}.$$

The expanded area is

$$A = \pi r_0^2 \left(1 + \frac{pr_0}{Eh} \right)^2 \quad \text{and} \quad \frac{\mathrm{d}A}{\mathrm{d}p} \approx \frac{2\pi r_0^3}{Eh} \approx \frac{Ad_0}{Eh}.$$

Hence the speed of propagation of the pressure pulse is

$$c = \sqrt{\frac{A}{\rho} \frac{Eh}{Ad_0}} = \sqrt{\frac{Eh}{\rho d_0}}.$$

This is known as the Moens–Korteweg wave speed after two Dutch scientists who developed the linearized theory in 1878. For typical flow in the aorta the speed of propagation of the pulse is about 4 m s^{-1}, rather faster than the peak flow rate of perhaps 1 m s^{-1}. The speed of propagation in a collapsible vein might be as low as 1 m s^{-1}, and this can lead to special phenomena analogous to sonic flow.

2.8.   NUMERICAL METHODS IN BIOFLUID MECHANICS

In the preceding sections of this chapter we have seen how we can gain an insight into the gross fluid-mechanical characteristics of biological systems by introducing assumptions to simplify the physics to the point at which the equations of fluid dynamics become tractable. Unfortunately we are not able to investigate detailed local flow characteristics in any but the very simplest geometries, and we are becoming increasingly aware that local values of parameters such as fluid shear stress hold the key to many biological and clinical phenomena. For example, there appears to be a strong correlation between atherosclerosis in the wall of an artery and the local wall shear stress. Where the shear stresses are high, the endothelial cells stretch and align with the flow, and atherosclerosis does not develop. Where the shear stresses are low, perhaps in regions of separation or recirculatory flow, the endothelial covering is less well structured and there is a tendency to atherosclerosis. There is strong motivation to model more complex geometries such as arterial bifurcations in order to evaluate local fluid-dynamic characteristics.

We cannot hope to achieve general analytical solutions for any complex geometrical shape, and so we seek numerical solutions of the governing equations for these systems. As in structural mechanics, the aeronautical engineers have been there before us. Many of the modern computational fluid-dynamics (CFD) structures, algorithms and software were developed originally in or for the aeronautical industry. Some simplifications, such as that of inviscid flow, can lead to equations that are still worth solving for complex geometries. Special techniques have been developed, particularly for the solution of two-dimensional problems of this type. These are not discussed here because they are not particularly relevant to problems in biofluid mechanics. Turbulence is a phenomenon of primary interest to aerodynamicists, and this is reflected in the number of turbulence models that are available in commercial software and in the richness of the literature in this field.

Moving boundaries (pistons and valves) are of great interest to automobile engineers in the modelling of combustion processes, and again facilities to model events featuring the motion of a boundary have existed for some time in commercial software. Of much more interest to those involved in biofluid mechanics are the problems of the interaction of soft and flexible structures with fluid flow through and around them. Historically structural and fluid-dynamic analyses have been performed by different groups of engineers and physicists, and the two communities had relatively little contact. This situation has changed dramatically in recent years. It is the authors' belief that we stand on the verge of a revolution in our abilities to handle real problems in biofluid mechanics.

2.8.1.   The differential equations

For an incompressible fluid the continuity equation is

$$\frac{\partial u}{\partial x} + \frac{\partial v}{\partial y} + \frac{\partial w}{\partial z} = 0$$

where x, y and z are a set of Cartesian coordinates and u, v and w are the velocities in these three directions. The momentum equation in the x direction for a Newtonian fluid, in the absence of body forces, is

$$\rho\left(\frac{\partial u}{\partial t} + u\frac{\partial u}{\partial x} + v\frac{\partial u}{\partial y} + w\frac{\partial u}{\partial z}\right) = -\frac{\partial p}{\partial x} + \mu\left(\frac{\partial^2 u}{\partial x^2} + \frac{\partial^2 u}{\partial y^2} + \frac{\partial^2 u}{\partial z^2}\right)$$

where t is time, p is the local fluid pressure, ρ is the density and μ is the viscosity of the fluid.

Similar equations can be written for the momentum in the y and z directions, and these are known as the Navier–Stokes equations. Both the continuity and Navier–Stokes equations can be written concisely in vector form. For details of the derivations of these equations the reader is referred to a fluid-dynamics text such as White (1994).

2.8.2.   Discretization of the equations: finite difference versus finite element

The continuity and momentum equations are written in terms of u, v, w and p which, in principle, are continuous (or perhaps even discontinuous) functions of time and of the Cartesian coordinates. If we are to seek numerical solutions we need to choose points in space and time at which solutions are required, and then to relate the spatial and temporal derivatives of the functions to the values of the functions at our chosen points. We will concentrate on a steady-state solution, although in principle the ideas that we shall discuss can be readily extended to the transient case. This statement should not lead the reader to underestimate the formidable difficulties of implementing an appropriate algorithm or of finding a numerical solution.

In the case of a steady-state analysis, we will have a grid of points, or nodes, at which the four variables have (as yet unknown) numerical values. The most obvious approach to the solution of the equations is to implement some sort of *finite-difference* scheme in which the spatial derivatives are related to the values of the function at the nodes. This is most readily accomplished if the nodes form a regular rectangular grid in Cartesian space. We can write four equations for each grid in the fluid domain. We shall need to make some assumptions about conditions either at the boundary or outside the domain because at and close to the boundaries our finite-difference derivatives will require this information. This procedure will generate a series of simultaneous equations that, subject to our imposed boundary conditions, we ought to be able to solve. Unfortunately the momentum equation is nonlinear (it contains products of the velocity with its spatial derivative), and so we would not necessarily expect the solution process to be straightforward. Furthermore the pressure does not appear explicitly in the continuity equation, and this in itself gives rise to numerical difficulties.

The earliest successful CFD codes were based on these finite-difference procedures. Algorithms were quickly developed to allow the nodes to be specified on curvilinear coordinates that followed the local curvatures of the boundaries. These body-fitted coordinates represented a major advance in the analysis capabilities of the software. The body-fitted coordinates represented a mapping of a mathematical rectangular Cartesian space into curvilinear coordinates, but the grid still had to have the fundamental rectangular structure in the mathematical space. This continued to place restrictions on the range of geometries that could be modelled. Complete freedom in geometry modelling in finite-difference codes, enabling the use of unstructured grids, took a little longer.

As both structural and fluid mechanics codes developed apace, some engineers started to look at the application of *finite-element* techniques in fluid mechanics. These techniques handled the problems of curved boundaries almost incidentally, in the specification of the geometrical shape function. More recently spectral element methods have been developed. These combine the advantages of the finite-element method with what is effectively a very high-order shape function, and have been claimed to offer significant benefits in the efficient resolution of local flow fields in biofluid mechanics.

There was a period during the mid to late 1980s when the finite-element-based software was perceived, at least by the authors of this chapter, to have a real advantage for the modelling of complex geometries. This advantage has now all but disappeared with the most recent developments in multi-blocking and multi-gridding in finite-difference and finite-volume codes. The choice of commercial software for the physicist working in the field of biofluid mechanics is increasingly one of personal preference and familiarity. The most important tool is an understanding of the physics of the flow, and the application of a critical and discerning eye to the results from any code. Perhaps a major problem with the current generation of software is that it is often possible to get a numerical result with relatively little understanding, but that the result might be effectively meaningless. It is not without justification that it has been suggested that CFD is an acronym for 'colourful fluid dynamics', or, more cynically, for 'colour for directors'! It is not recommended that any student should start to use computational methods without first developing a real understanding of the material presented in the earlier sections of this chapter.

2.9. PROBLEMS

2.9.1. *Short questions*

a   Will blood pressure measured using a sphygmomanometer on the arm be higher or lower if the arm is raised slightly?

b   What is the approximate ratio of arterial systolic pressure (measured in cm of water) to body height?

c   Is it reasonable to model the bladder as a vessel with pressure proportional to volume?

d   What is the most important distinction between the mechanics of solids and fluids?

e   Is water more or less viscous than air?

f   Does viscosity depend upon strain in a Newtonian fluid?

g   Blood has thixotropic properties. What does this mean?

h   What is a 'normal' value for the haematocrit of blood?

i   Can blood flow be considered as laminar under most circumstances?

j   How does the resistance to blood flow depend upon the effective diameter of a blood vessel?

k   Is blood thicker than water?

l   Is it likely that haematocrit will vary within the cardiovascular system?

m   Does the Reynolds number increase or decrease with distance from the heart?

n   Name one part of the body in which the static pressure is sub-atmospheric.

o   What property of blood does the Casson model of viscosity attempt to describe?

p   State an expression for the Reynolds number governing the flow regime in a pipe.

q   At what Reynolds number does flow in a tube usually cease to be laminar? Is it possible to get laminar flow in a tube at a Reynolds number in excess of 10 000?

r   What property of blood is described by the Fahraeus–Lindquist effect?

s   According to the Poiseuille equation the flow rate, Q, in a tube is proportional to what powers of the radius and length? ($Q \propto R^a L^b$: what are the numerical values of a and b?).

t   Under what circumstances does convective acceleration occur in a steady-state flow situation?
u   Sketch the distribution of pressure along the axis of a tube that contains a restrictive orifice when fluid
       flows through it at a constant rate.

2.9.2.   Longer questions (Answers are given to some of the questions)

Question 2.9.2.1

Write a short paragraph describing the differences between laminar and turbulent flow, and name two physiological conditions under which turbulent flow might arise in the body.

The equation for the velocity, v, at radius r of a Newtonian fluid in a tube of outer radius R under laminar
flow conditions is

$$v = \frac{R^2}{4\mu} \frac{\Delta p}{L} \left(1 - \frac{r^2}{R^2} \right)$$

where μ is the viscosity and $\Delta p/L$ is the pressure drop per unit length.
       Assuming that the shear stress is given by $\tau = \mu(dv/dr)$, derive an expression for the shear stress as a
function of radius. Sketch the distribution of shear stress across the diameter of the tube, and show that it is
zero at the centreline and highest at the wall.

Show by integrating the velocity equation over concentric rings in the tube that the flow rate, Q, is given by
the Poiseuille equation,

$$Q = \frac{\Delta p}{L} \frac{\pi R^4}{8\mu}.$$

Answer

- In laminar flow, elements of the fluid slide smoothly over each other. Small neutral-density particles
  released into the flow follow predictable streamlines through the flow domain. Laminar flow is smooth
  and orderly. Viscosity is important in laminar flow.
- In turbulent flow the fluid becomes disorderly and chaotic. Particle paths become unpredictable, although it might be possible to calculate what will happen on average. The mechanical energy losses
  in turbulent flow are high as cascades of vortices of many length scales intertwine and interact, dis-
  sipating energy as they go. Viscosity in the sense that we have defined it is relatively unimportant in
  turbulent flow—the effective viscosity is determined by the turbulent regime and by length scales of
  vortex cascades within it.
- The flow throughout most of the circulatory system is within the laminar regime. The flow in the aorta
  at peak systole might be turbulent. Turbulent flow might arise when blood vessels are severely narrowed
  by disease, the flow might become turbulent at the constriction, or when heart valves leak, allowing
  high-velocity jets to form. Air can be expelled at very high velocities when sneezing, and the air flow
  is turbulent.

See section 2.5.1.

Question 2.9.2.2

The mean velocity in an artery of internal diameter 5 mm is 0.5 m s^{-1}. It may be assumed that blood has a
viscosity of 0.004 Pa s and a density of 1000 kg m^{-3}. Calculate the Reynolds number and deduce whether
the flow is laminar or turbulent.

Calculate the volume flow rate through the artery in SI units. The blood flow in arteries is normally expressed clinically in terms of litres per minute (1 min^{-1}). Express the flow rate in these units.

Calculate the pressure drop over a 100 mm length of the artery, and calculate the maximum shear stress on the blood in the artery.

The artery described above is narrowed by disease to 70% of its original diameter.

(i) Determine the factors by which the pressure drop and wall shear stress are changed assuming that the volume flow rate is maintained.

(ii) Determine the factor by which the volume flow rate is reduced assuming that the pressure drop is maintained at the level prior to the disease. Also determine the change in wall shear stress under these conditions.

Which of these assumptions do you think is more likely to describe the real physical process? What might be the consequences?

Question 2.9.2.3

Discuss the physical principles underpinning the use of the Gorlin equation as a measure of heart valve performance. How might the cardiologist use this equation to identify a stenotic valve? Assuming that the circulation can be modelled as an electrical circuit featuring two resistances in series, the first representing the aortic valve and the second representing the rest of the system, discuss the consequences of a stenotic valve for the heart and for the circulation given the following conditions:

(i) the heart generates the same pressure (voltage) in the left ventricle as it did before the stenosis;

(ii) the heart generates the same flow rate (current) as it did before stenosis.

Which of these conditions do you think is more likely to describe the real physical process, and why?

Question 2.9.2.4

We wish to investigate the temporal variation of the volume flow into the aorta when the time-averaged flow is 5.4 1 min^{-1}. For the purposes of this exercise we shall define the start of systole as the instant at which positive forward flow first occurs and the end of systole as the instant at which the flow first reverses, and shall assume that the inlet flow in diastole is zero.

Produce two numerical approximations to the waveform:

(i) as an asymmetrical triangular pulse in systole, zero in diastole;

(ii) as a truncated Fourier series.

In the latter case you must judge how many terms are required to represent your flow waveform to sufficient accuracy. Plot a graph showing the two approximations superimposed on your original waveform.

Otto Frank developed a simple model of pulsatile flow in the aorta based on the Windkessel theory. Outline the assumptions in this model and derive the governing differential equation. Assuming that the pressure at the commencement of systole is 120 mmHg, solve for the temporal variation of pressure for either of your approximate waveforms and produce a graph of pressure versus time for the period of one pulse. Discuss the effects of changing the wall stiffness of the aorta and of changing the peripheral resistance.

[Assume a value for the peripheral resistance R of 18.3 mmHg l^{-1} min^{-1} and for the constant K relating the flow to the rate of change of pressure a value of 0.001 92 l mmHg^{-1}.]

Notes on the answer

Section 2.7.1 derives the differential equation relating flow and pressure in the system.

Fourier solution:

$$Q = a_0 + \sum_1^N \left(a_n \cos \frac{2n\pi t}{T} + b_n \sin \frac{2n\pi t}{T} \right).$$

Governing equation, Windkessel model:

$$\frac{dp}{dt} + \frac{p}{KR} = \frac{Q}{K}.$$

Multiplying both sides by an integrating factor, in this case $e^{(1/KR)t}$:

$$e^{(1/KR)t} \left(\frac{dp}{dt} \right) + \frac{1}{KR} e^{(1/KR)t} p = \frac{e^{(1/KR)t}}{K} \left(a_0 + \sum_1^N \left(a_n \cos \frac{2n\pi t}{T} + b_n \sin \frac{2n\pi t}{T} \right) \right)$$

$$\frac{d}{dt} \left(e^{(1/KR)t} p \right) = \frac{e^{(1/KR)t}}{K} \left(a_0 + \sum_1^N \left(a_n \cos \frac{2n\pi t}{T} + b_n \sin \frac{2n\pi t}{T} \right) \right)$$

$$e^{(1/KR)t} p - C = \frac{1}{K} \int e^{(1/KR)t} \left(a_0 + \sum_1^N \left(a_n \cos \frac{2n\pi t}{T} + b_n \sin \frac{2n\pi t}{T} \right) \right) dt.$$

It is not necessary to treat $n = 0$ as a special case, and so for clarity we write

$$e^{(1/KR)t} p - C = \frac{1}{K} \int e^{(1/KR)t} \left(\sum_0^N \left(a_n \cos \frac{2n\pi t}{T} + b_n \sin \frac{2n\pi t}{T} \right) \right) dt$$

$$e^{(1/KR)t} p - C = \frac{e^{(1/KR)t}}{K \left(1/(KR)^2 + 4n^2\pi^2/T^2 \right)} \sum_0^N \left(\left(\frac{a_n}{KR} - \frac{2n\pi b_n}{T} \right) \cos \frac{2n\pi t}{T} \right.$$
$$\left. + \left(\frac{2n\pi a_n}{T} + \frac{b_n}{KR} \right) \sin \frac{2n\pi t}{T} \right)$$

$$p = Ce^{(1/KR)t} + \frac{1}{K \left(1/(KR)^2 + 4n^2\pi^2/T^2 \right)} \sum_0^N \left(\left(\frac{a_n}{KR} - \frac{2n\pi b_n}{T} \right) \cos \frac{2n\pi t}{T} \right.$$
$$\left. + \left(\frac{2n\pi a_n}{T} + \frac{b_n}{KR} \right) \sin \frac{2n\pi t}{T} \right).$$

The constant C can be evaluated from the known pressure at the start of systole.

Answers to short questions

a   The measured blood pressure will be lower if the arm is raised.
b   1. Systolic pressure is approximately the same as body height.
c   No, pressure and volume are not simply related for the bladder.
d   Fluids cannot support a shear stress
e   Water is more viscous than air.
f   No, viscosity does not depend upon strain in a Newtonian fluid.
g   Thixotropic properties means that the shear stress decreases with time under a constant strain.
h   45% is a normal value for the haematocrit of blood.

i   Yes, blood flow can often be considered as laminar.

j   Resistance to the flow of blood is inversely to the fourth power of vessel diameter.

k   Yes, blood is thicker than water.

l   Yes, haematocrit is likely to vary within the cardiovascular system.

m   The Reynolds number decreases with increasing distance from the heart.

n   Static pressure can be sub-atmospheric in the intrathoracic cavity.

o   The Casson model of viscosity includes the concept of a finite yield stress for blood.

p   Reynolds number (Re) = velocity (v) × vessel diameter (d) × density (ρ)/viscosity (μ).

q   Flow usually ceases to be laminar for $Re > 2000$. However, under carefully controlled laboratory conditions flow can be laminar for $Re > 10\,000$.

r   The Fahraeus–Lindquist effect describes the apparent reduction in viscosity of blood as tube diameter increases.

s $a = 4, b = -1$.

t   Convective acceleration can occur for flow through an orifice.

u   Pressure along the axis of a tube with an orifice is as shown. The orifice is marked with an arrow (see figure 2.15).

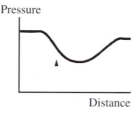

Figure 2.15.

BIBLIOGRAPHY

Cameron J R and Skofronik J G 1978 *Medical Physics* (New York: Wiley)

Fung Y C 1993 *Biomechanics: Mechanical Properties of Living Tissue* 2nd edn (New York: Springer)

Gorlin R and Gorlin S G 1951 Hydraulic formula for calculation of the area of the stenotic mitral valve, other cardiac valves and central circulatory shunts *Am. Heart J.* **41** 1–29

Kroll M H, Hellums J D, McIntire L V, Schafer A I and Moake J L 1996 Platelets and shear stress *Blood J. Am. Soc. Haematology* **88** 1525–41

Marieb E N 1995 *Human Anatomy and Physiology* 3rd edn (Menlo Park, CA: Benjamin-Cummings)

McDonald D A 1974 *Blood Flow in Arteries* (London: Arnold)

Nichols W W and O'Rourke M F 1990 *McDonald's Blood Flow in Arteries* 3rd edn (London: Arnold)

Pedley T S 1980 *The Fluid Mechanics of Large Blood Vessels* (Cambridge: Cambridge University Press)

Skalak R and Ozkaya N 1989 Biofluid mechanics *Ann. Rev. Fluid Mech.* **21** 167–204

Tindale W B and Trowbridge E A 1986 Modification of the Gorlin equation for use with heart valve substitutes *Cardio-vascular Res.* **20** 458–65

Trowbridge E A 1984 The fluid mechanics of blood *J. Inst. Math. Applic.* **20** 89–92

White F M 1994 *Fluid Mechanics* 3rd edn (New York: McGraw-Hill)

Yellin E L and Peskin C S 1975 Large amplitude pulsatile water flow across an orifice *Trans. ASME J. Dynamic Syst. Meas. Control* 92–5

CHAPTER 3

PHYSICS OF THE SENSES

3.1. INTRODUCTION AND OBJECTIVES

We might regard the body as a complex, elegant and sometimes incomprehensible computer. The central processor of this computer is the brain, and the primary connection to this CPU is the spinal cord. These components are called the *central nervous system* (CNS). The CNS receives data through a set of communication ports, processes it and sends instructions through another set of ports. Collectively these ports and their cabling are called the *peripheral nervous system* (PNS). The output devices control muscle activity through the autonomic and somatic nervous systems. The input devices monitor the performance of the body and keep the CNS aware of the external environment. In this chapter we shall focus on these external monitors, represented by the five senses. The importance of these senses to our general well-being cannot be overemphasized. Deprived of sensory input the brain is apt to invent its own: hallucinations of sight, sound and smell are reported by sane and normal volunteers under conditions of sensory deprivation.

Our perception of the outside world depends on our five senses. There are many texts devoted to the study of the senses, and each sense represents a rich field of research in its own right. Damask (1981) devoted a volume of his medical physics text to the external senses, and this chapter draws on his text particularly for the senses of touch, taste and smell. Formally, the five senses are:

Cutaneous sensation	Touch
Gustation	Taste
Olfaction	Smell
Vision	Sight
Audition	Hearing.

The elements of the peripheral nervous system that respond directly to stimuli are called the sensory receptors. For the purposes of the present chapter a receptor is a transducer that produces electrical energy, in this case a nerve impulse, from the form of energy (mechanical, thermal, light) that it is designed to respond to. We shall focus on the physics of the route by which energy is presented to the receptor, and not on the subsequent electrical activity. The process of the initiation of the electrical impulse and the propagation of the signal are covered in more detail in Chapter 16.

The receptors are classified by location, by structure and by function. There are receptors at most locations throughout the body, although the density of receptors varies widely. The internal receptors (interoceptors, viscoreceptors and proprioceptors) monitor position and function of the organs, muscles, joints, ligaments, etc. In this chapter we are concerned primarily with the elements located at or near to the surface of the body, collectively identified as exteroceptors.

The sense of touch is distributed around the surface of the body, although again the distribution of the receptors and the associated sensitivity varies considerably from site to site. The receptors responsible for the sense of touch are relatively simple in structure. The remaining four senses are highly localized, and the whole of the associated organ is often considered to be one complex receptor. The senses (taste, smell, sight and hearing) associated with these complex receptors are referred to collectively as the special senses.

The classification according to function depends on the transduction mechanism of the receptor. Mechanoreceptors transduce mechanical factors such as touch, pressure, vibration and strain into electrical activity. Thermoreceptors respond to temperature, photoreceptors to light and chemoreceptors respond to chemicals in solution. Nociceptors are pain sensors, and indeed most receptors can function as nociceptors if the stimulus level is high enough. Some receptors are designed specifically to respond to the level of a stimulus, whilst others are designed to respond to changes in the level. Thus, for example, some mechanoreceptors respond directly to pressure and some respond to vibration, and there are parallel functions in the other receptors.

The questions which are addressed in this chapter include:

- How close do two pinpricks have to be before we think they are only one?
- What is the displacement of air particles when we speak?
- How does the spatial resolution of our eyes compare with that of a TV camera?
- Why do colours disappear under poor lighting conditions?
- How do we define and compare intensities of sensory stimuli?

You should see this chapter as an introduction to an important part of human physiology. However, we present this information from the point of view of a physical scientist as opposed to a physiologist. In some cases there are associated areas of clinical application, for example in audiology, but for the most part you should see this chapter as simply contributing to your understanding of human function.

3.2. CUTANEOUS SENSATION

3.2.1. *Mechanoreceptors*

The receptors associated with the sense of touch are structurally relatively simple. We shall group nociceptors (pain receptors) and thermoreceptors into this section because of their structural similarity, although strictly the sense of touch might be considered to be a purely mechanical function. The structure of a nerve cell is discussed in detail in Chapter 16, and the categorization of the simple receptors depends on the geometry of the dendritic endings. Free dendritic endings invade and pervade the body tissue much like the roots of a plant do soil. These are amongst the smallest of the nerve fibres, and are distributed almost everywhere in the body. Nociceptors and thermoreceptors are usually of this type, along with some mechanoreceptors. Some free endings twine around the roots of hairs and respond to the bending of the hair, whilst others terminate in flat discs at the base of the epidermis. Both of these types are mechanoreceptors. Encapsulated dendritic endings are contained in a protective capsule of connective tissue, and so have the external appearance of the bulb of a plant. They are almost always mechanoreceptors. There are a number of variations of structure of encapsulated endings, with particular geometries associated with particular types of mechanical stimulus. Some are designed to respond to light pressure, some to higher pressure, some to strain and some to vibration.

Mechanoreceptors respond to the stimulus of a mechanical load. There are several ways in which we might describe the load (directly, or as a pressure or a stress), but fundamentally we are interested in describing the mechanical stimulus to the receptor. In general the load will be time dependent, and the detailed design and structure of the mechanoreceptor will provide an efficient response to some characteristic of the loading function. In particular some mechanoreceptors are designed to respond to the intensity of the load and some are designed to respond to the rate of change of load. The former are described as slowly adapting (SA) receptors and the latter as rapidly adapting (RA) receptors. In the limit an SA receptor simply responds to a static load. In principle we might hope that the rate of generation of impulses might be proportional to the intensity of the stimulus, or at least that it might be constant for a constant stimulus. In practice the response is distinctly non-linear, and furthermore there is a fatigue effect so that the rate is not constant even for constant stimulus. Models of stimulus intensity against response are discussed in more detail with respect to the special senses.

As physicists and engineers we are interested in the detailed structure of the receptors, and in the means by which they might produce a response to different loads. In this brief overview we shall look a little more closely at one of the mechanoreceptors, the Pacinian corpuscle. These receptors are most abundant in the subcutaneous tissue under the skin, particularly in the fingers, the soles of the feet and the external genitalia. They are the largest of the receptors and the approximate shape of a rugby ball. They are typically about 1 mm, and up to 2 mm, long and about half as wide. They resemble an onion in structure, with up to 60 layers of flattened cells surrounding a central core, the whole being enclosed in a sheath of connective tissue. At the centre is a single non-myelinated nerve fibre of up to 10 μm in diameter, which becomes myelinated as it leaves the corpuscle. The Pacinian corpuscle serves most efficiently as a monitor of the rate of change of load rather than to the intensity of the load itself: it is often identified as a vibration transducer. Damask (1981) discusses a simple analytical model constructed by Loewenstein (see Damask 1981) to study the mechanical behaviour of the corpuscle. The model features a series of elastic membranes, ovoid in shape, connected at one end and each filled with a viscous fluid (see figure 3.1).

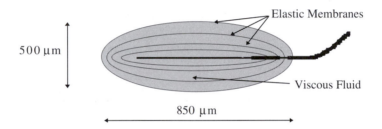

Figure 3.1. *Structure of Pacinian corpuscle (based upon illustrations given in Damask (1981)).*

The model suggests that the mechanical stimulus reaching the core will depend on the rate of loading. This is illustrated by consideration of the application of a pinching load to two concentric ovoids filled with an incompressible viscous fluid. If the load is applied very slowly then the fluid flows so that the pressure within it remains hydrostatic, and the elastic membrane of the outer ovoid stretches along the axis to maintain the enclosed volume of incompressible fluid. The load is supported by this elastic membrane under the fluid pressure. The inner membrane will suffer relatively little change in length under the hydrostatic pressure.

If the load is applied rapidly, the viscous fluid cannot flow sufficiently quickly to maintain the hydrostatic pressure distribution. There will be a local high pressure under the point of application of the load, and therefore some distortion of the circular section of the membrane. This must be accommodated by a stretching along the axis. This process continues through successive layers until finally there is an axial stretching of the innermost membrane, and a corresponding stimulation of the nerve fibre. Hence the mechanical design of the corpuscle protects the central fibre from significant length changes under a constant load, but stretches it under a varying load (see figure 3.2). It therefore represents a mechanism by which a simple strain gauge (a dendritic ending?) might be used to monitor rate of change of strain. This model can be expressed in terms of a combination of the simple spring and damper elements described in the section on viscoelasticity in Chapter 1. If the relative values of the spring stiffnesses and the dashpot viscosity coefficients are properly chosen, then the model can represent quite accurately the measured performance of the Pacinian corpuscle.

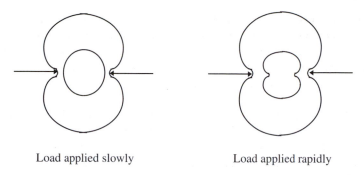

Load applied slowly   Load applied rapidly

Figure 3.2. *Effect of rate of application of load on Pacinian corpuscle.*

For a distributed sense such as the sense of touch we are interested not only in the response of a single receptor, but also in the density of receptors and in the resulting spatial discrimination of the sense. We might investigate the spatial discrimination by stimulating adjacent sites, and establishing the minimum distance between the sites for which we can distinguish between the stimuli. Assuming that the brain can somehow identify signals from individual receptors, this simple experiment will give us a simple approximation of the density of the receptors. The results are surprising: the density is rather lower than we might expect. Marieb (1991) illustrates the spatial sensitivity of the skin at various sites using this simple two-point test, and the results, expressed as a spatial density of receptors, are tabulated below.

Table 3.1. *Density of mechanoreceptors in human skin (from Marieb (1991)).*

Site	Spatial discrimination (cm)	Spatial density (receptors per cm^2)
Tip of tongue	0.2	25
Tip of index finger	0.5	4
Lips	0.75	2
Edge of tongue	1	1
Palm	1.2	0.7
Forehead	2.5	0.16
Back of hand	3.2	0.1
Upper surface of foot	4.0	0.06
Neck	5.5	0.03
Back	6.8	0.02

3.2.2. Thermoreceptors

Thermoreceptors respond to the stimulus of heat energy. As for the mechanoreceptors, there are surprisingly few thermoreceptors in the human skin. A typical density is of the order of 5 to 10 receptor sites per cm^2 over most of the surface. Some receptors are designed to monitor steady-state temperature, whilst others respond efficiently to rapid changes of temperature. Subdividing further, some of the steady-state receptors, the 'hot detectors', respond to high temperatures, and some, the 'cold detectors', respond to low temperatures. The density of cold detectors in the skin is generally an order of magnitude higher than that of the hot detectors. Experiments on the steady-state detectors in animals have indicated that the detectors are relatively inactive (although not completely so) at normal skin temperature and that they start to fire nerve impulses as the temperature is maintained at other levels.

3.2.3. *Nociceptors*

The primary function of nociceptors is to signal pain when the intensity of stimulation exceeds a particular threshold. Although most receptors initiate a pain response if the intensity is too great, some receptors are specifically designed to start to signal at the pain threshold. For example, some thermal nociceptors transmit only at steady temperatures exceeding 42 °C and some transmit only at temperatures below 10 °C. There are rapid pain receptors, transmitting through myelinated fibres, that let us know very quickly that something is wrong. The conduction velocity of these nerve fibres is up to 30 m s^{-1}. There are also persistent pain receptors, mostly non-myelinated, that react more slowly but maintain the sensation of pain. These receptors and the associated fibres are amongst the smallest and slowest nerve fibres in the body, with conduction velocities down to 2 m s^{-1} and below.

3.3. THE CHEMICAL SENSES

The senses of gustation and olfaction, or taste and smell, have much in common, and they are often referred to together as the chemical senses. They are dependent on the chemical stimulation of special nerve cells called chemoreceptors, which respond to chemicals in an aqueous solution. They are regarded as the most primitive of the special senses. In some primitive life-forms, such as protozoa, there is no distinction between taste and smell. Our perception of taste in particular is often a composite sense, in which we supplement true taste information with cues from our sense of smell. For this reason great care must be taken in the design of experiments to separate taste from smell. One simple expedient is the use of a well-designed nose clamp. A subject who is blindfolded and nose-clamped cannot tell the difference between grated apple, onion and turnip!

3.3.1. *Gustation (taste)*

Taste and anatomy

The chemoreceptors responsible for the sense of taste are called *taste buds*. We have about 10 000 of them, most of which are located on the tongue. Different areas of the tongue react most effectively to particular tastes, suggesting that the taste buds have some degree of specificity (see figure 3.3). Every taste bud appears to be able to react to every taste, but the level of response varies. There is no obvious structural difference between the taste buds that respond best to different tastes. The task of transmission of the taste sensation to the brain is shared between two of the cranial nerves—the seventh (facial) and the ninth (glossopharyngeal).

The taste buds are amongst the most dynamic cells in the body. They are shed and replaced about every 7–10 days.

Classification of tastes

We normally classify tastes into four categories (see table 3.2). These categories are convenient but they are neither complete nor are they foolproof. The classification of tastes in this way is rather arbitrary. In taste tests some compounds appear to fall into different categories depending on concentration. Even common salt, sodium chloride, tastes sweet at very low concentrations, close to the threshold of taste. Potassium chloride also tastes sweet at very low concentrations, becomes bitter as concentration increases and then salty or sour

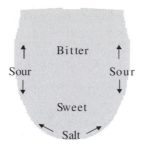

Figure 3.3. *Location of taste sites on the human tongue. Salt sensors are at the tip of the tongue.*

at high concentration. Some models feature many more categories in attempts to classify tastes: the simplest is extended to add alkaline and metallic to the four basic tastes.

Table 3.2. *Classification of tastes.*

Category	General	Specific
Sweet	Many organic compounds	Sugars, saccharin
Salt	Many inorganic salts, metal ions	Sodium chloride, potassium iodide
Sour	Acids, hydrogen ion (H$^+$)	Acetic acid (vinegar), citric acid (lemon)
Bitter	Many alkaloids	Caffeine, nicotine, quinine, strychnine

Models of gustation

We have already established that taste is a chemical sense, and it should not therefore be surprising that analytical models based on ideas from chemistry should be applicable. A very simple *chemical equilibrium model of taste intensity*, developed and explored by Damask following Beidler (see Damask 1981), is outlined below.

In order for a sensation of taste to occur, it is necessary for there to be a stimulus and a receptor. It is reasonable to assume that the stimulus must interact with the receptor and that during this period the stimulus and the receptor must form a complex of some kind. In Damask's model, it is assumed that the intensity of a taste is determined by the number of stimulus/receptor complexes that are present at any one time. The system is described by two rate constants, K_1 and K_2, describing the rates at which the complexes are formed and broken. Assuming that the concentration, C, of the stimulus remains constant, and that there are a finite number, N, of available receptor sites, Z of which are bonded into complexes at any one time, the chemical equilibrium model can be described as follows:

$$\text{Stimulus} \quad + \quad \text{Receptor sites} \quad \overset{K_1}{\underset{K_2}{\rightleftharpoons}} \quad \text{Stimulus/receptor complexes}$$
$$C \quad + \quad (N - Z) \qquad\qquad Z$$

The rate of change of Z is determined by the probability of stimulus molecules approaching receptor sites at any particular time, the rate at which the combination occurs, and the rate at which the complexes are dissociated. Damask states the equation in the following form:

$$\frac{\mathrm{d}Z}{\mathrm{d}t} = K_1 C(N - Z) - K_2 Z.$$

For a steady-state equilibrium condition, the rate of change of Z is zero, and the equilibrium constant, K, is defined as

$$K = \frac{K_1}{K_2} = \frac{Z}{C(N - Z)}.$$

Assuming that the magnitude of the neural response, R, is proportional to the number of stimulus/receptor complexes that are present, then

$$R = aZ$$

where a is a constant of proportionality.

The maximum response will occur when all of the receptors are used, and the saturation response is

$$R_{\text{sat}} = aN.$$

The relationship between the equilibrium constant and the neural response is therefore

$$K = \frac{R}{C(R_{\text{sat}} - R)}.$$

Rearranging this equation,

$$\frac{C}{R} = \frac{C}{R_{\text{sat}}} + \frac{1}{K R_{\text{sat}}}.$$

This equation is called *the taste equation*, and it gives us an idea of the likely form of the relationship between concentration of stimulant and taste intensity. Damask quotes data from Beidler demonstrating that the equation does appear to be valid for a range of sodium salts. He develops the theory further, studying the energies of the system, to show that the attachment between the stimulus and the receptor is likely to be a physical one rather than a chemical bond: the energies of formation and dissolution of the complex are too low for a chemical reaction. Beidler quotes values of 2.17 for the saturation response and 9.8 for the equilibrium constant for the taste of sodium chloride. The results of the taste equation using these values are illustrated in figure 3.4.

The model describes and predicts the observed saturation effect of the taste response, but it does not offer any clues as to why our sense of taste can fatigue following repeated exposure to a single type of taste.

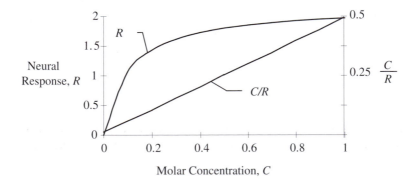

Figure 3.4. *Neural response and concentration/response ratio according to the taste equation.*

3.3.2.   *Olfaction (smell)*

Smell and anatomy

The chemoreceptors responsible for the sense of smell, the olfactory receptors, are located in the epithelial lining of the roof of the nasal cavity. There are millions of these receptors, orders of magnitude more than the number associated with taste. In other land animals the sense of smell is often very highly developed: it is associated with survival. Man has a relatively poor sense of smell, and part of the reason for this is the geometrical design of the olfactory system. The olfactory receptors are situated in a side passage, off the main pathway of air into the lungs. This appears to be an inefficient design with respect to collection of odorous molecules, and limits the acuity of our sense of smell. We can, when we wish, collect more molecules by sniffing, drawing more air upwards onto the sensitive area.

Unlike the taste buds, which consist of epithelial cells, the olfactory receptors are actually neurones. They are unique in that they are the only neurones in the body that are replaced continually in adult life. They typically last for about 60 days. The discovery that the olfactory receptors regenerate is potentially of great importance. Before this discovery it was generally held that all neurones in the adult animal were irreplaceable: the fact that they are not gives hope that we might be able to develop techniques for neural regeneration in all sorts of situations, including trauma rehabilitation. The task of transmission of the sensation of smell to the brain is performed by the first cranial nerve. Since this is its only function, it is often called the olfactory nerve.

Classification of smells

The sense of smell is rather more developed than the sense of taste, and an adequate classification of smells has proved relatively difficult to develop. There are several theories of olfaction, and no doubt each contains an element of the true, complex picture. Many of the theories are underpinned by chemical models that are beyond the scope of this text. One appealing and simple model suggests that there are seven basic odours: camphoric, musky, floral, pepperminty, ethereal, pungent and putrid. More complex models suggest 30 or more primary odours, and more recent work suggests that there are a thousand or more separately identifiable odours.

Thresholds of smell

Our sense of smell is very sensitive. Damask quotes references showing the threshold concentration for ethyl mercaptan as 4×10^8 molecules per cm^3. Given that there are about 5×10^{19} molecules of nitrogen per cm^3, this suggests that we can detect ethyl mercaptan in air at a concentration of about one molecule in 10^{11}. We might consider this to be quite impressive, but the threshold of smell for a typical dog is about one thousand times lower, and for some substances, such as acetic acid, their threshold is some seven or eight orders of magnitude lower than our own.

Models of olfaction

Our sense of smell is much more sophisticated than our understanding as to how it works. There is no well-accepted theory of the process of olfaction, although complex chemical theories have been proposed.

3.4.   AUDITION

In this section we will investigate the sense of hearing. This is a particularly important sense for the medical physicist: in many hospitals the clinical service of audiology is provided or supported by medical physics personnel. Chapter 15 is devoted to audiology, with particular emphasis on clinical aspects of the topic, including clinical tests, equipment and aids. This complements the description of the basic science of audition presented in the current chapter. Chapter 7 also contains some fundamental information on wave properties. Between these chapters, we will attempt to answer the following questions.

- What do we hear?
- How can we measure it?
- How do we hear it?
- How good is our hearing?
- What can go wrong?
- What remedies are available?

We can profitably start with a review of some of the terms associated with the science of audition. Audition itself is the faculty or sense of hearing. Audiology is the study and measure of hearing, including the detection and definition of hearing defects. Acoustics is the science of sound. A pure-tone sound wave is one in which the oscillations of pressure and displacement can be described in terms of a simple sine wave. The pure-tone sound wave is described *quantitatively* by its frequency and intensity. The waveform of all other sounds can be built up from linear combinations of pure tones. A sound wave is described *subjectively* in terms of its pitch, loudness and quality.

3.4.1. Physics of sound

A sound wave is a mechanical wave that moves through a medium as particles in the medium are displaced relative to each other. It should therefore be anticipated that the mathematical description of sound will have similarities with the descriptions of other mechanical dynamic systems. The simplest possible dynamic system is the *harmonic oscillator*, consisting of a point mass attached to earth by an ideal massless spring (see figure 3.5).

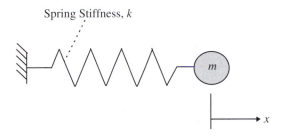

Spring Stiffness, k

Figure 3.5. *Simple harmonic oscillator.*

If the mass is displaced from the equilibrium position by an amount x, then a force will be developed in the spring that tries to restore the system to the equilibrium condition. This restoring force, F, acting on the mass is proportional to the extension of the spring, and acts in the positive x direction when the spring is compressed,

$$F = -kx.$$

Newton's second law applied to the mass yields the equation

$$-kx = F = ma = m\ddot{x}$$

where ˙ denotes differentiation with respect to time.

Rearranging this equation and introducing a constant (for the system), $\omega = \sqrt{k/m}$,

$$\ddot{x} + \omega^2 x = 0.$$

The solution of this equation is sinusoidal in form. The general solution of the equation is:

$$x = D \cos \omega t + E \sin \omega t.$$

The two constants, D and E, will be determined by the initial conditions, i.e. by how much the mass is displaced from the equilibrium condition at the time at which measurements commence, and by the velocity that is imparted to the mass at that time. There are several alternative ways in which this solution can be expressed, and very often in medical physics and biomedical engineering other forms of the solution are most valuable. In Chapter 13 we investigate alternative ways of writing the same solution. One method that we find particularly useful is to express it in terms of a single sinusoid: in this expression the cosine and sine amplitude constants, D and E, are replaced by a composite amplitude, C, and a phase shift, ϕ,

$$x = C \sin(\omega t + \phi).$$

We might also express the equation and its solution in terms of complex exponentials, and this form of expression will also have its place. For the description of sound waves we conventionally use the phase form described above.

Simple harmonic motion is often initiated by displacing the mass to a given position and then letting go. Under these conditions the initial velocity is zero. Substituting these initial conditions into the governing equation, we find that the phase shift is 90°, and $x = C \cos \omega t$.

In a sound wave the particles in the medium act much like the mass in the simple harmonic oscillator. It is convenient to consider the motion of small elements of the medium. The principle of the motion of sound waves can be demonstrated clearly in one dimension by consideration of a tube of air excited by an oscillating piston at one end.

The wave equation

The wave equation for longitudinal mechanical waves is derived in a number of texts. We follow a similar derivation to that given by Keller *et al* (1993). We can consider the position of a particle, or in this case an element of fluid, as it travels through time and space (see figure 3.6). Assume that the displacement in the axial direction of an element of fluid initially at position x is $\psi(x, t)$.

At any particular point in time, the element that was initially situated between x and $(x + \delta x)$ has moved as illustrated in figure 3.7. Assuming that the cross-sectional area of the tube is constant and equal to A, the element of fluid changes in volume as follows:

initial volume,

$$V_0 = A\,\delta x$$

volume at time t,

$$V = A(\delta x + \psi(x + \delta x, t) - \psi(x, t)).$$

Hence the change in volume of this element of fluid is

$$\Delta V = A(\psi(x + \delta x, t) - \psi(x, t))$$

and the fractional change in volume is

$$\frac{\Delta V}{V} = \frac{A(\psi(x + \delta x, t) - \psi(x, t))}{A\,\delta x}.$$

In the limit, as $\delta x \to 0$, the term on the right approaches $\partial \psi / \partial x$.

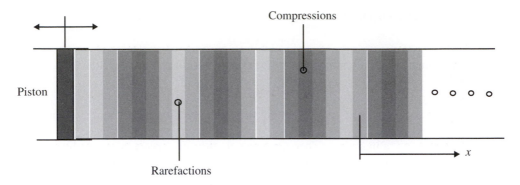

Figure 3.6. *Progression of a sound wave by compression and rarefaction of a medium.*

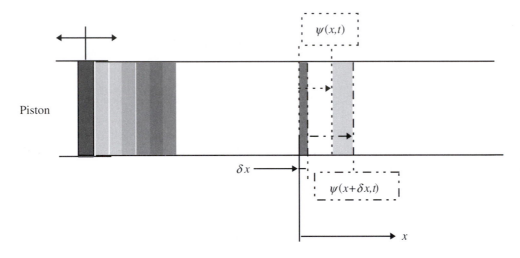

Figure 3.7. *Motion of an element of fluid within a sound wave.*

The change in pressure associated with the change in volume is dependent on the adiabatic bulk modulus of the fluid, B. The bulk modulus is defined as

$$B = -V\frac{dp}{dV} = -\frac{\Delta p}{(\Delta V/V)}.$$

Substituting for $\Delta V = V$ and rearranging (assuming that the amplitude of vibration is small),

$$\Delta p(x, t) = -B\frac{\partial \psi}{\partial x}.$$

Now consider the forces that are acting on the element of fluid at position x to $x + \delta x$ (see figure 3.8). The forces on the areas at each end are the pressures there multiplied by the cross-sectional areas.

The net force acting on the element, positive to the right, is

$$F = A\,\Delta p(x, t) - A\,\Delta p(x + \delta x, t).$$

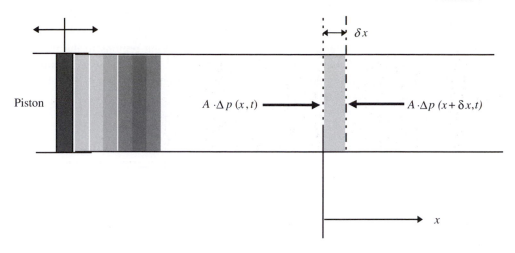

Figure 3.8. *Forces on an element of fluid.*

Substituting for Δp,

$$F = AB\left(\frac{\partial \psi}{\partial x}\bigg|_{x+\delta x} - \frac{\partial \psi}{\partial x}\bigg|_{x}\right)$$

where the derivatives are evaluated at the positions x and $(x + \delta x)$ as indicated.

The mass of the element of fluid is the density multiplied by its volume: $m = \rho A \delta x$. Newton's second equation applied to the element of fluid gives

$$AB\left(\frac{\partial \psi}{\partial x}\bigg|_{x+\delta x} - \frac{\partial \psi}{\partial x}\bigg|_{x}\right) = F = ma = m\frac{\partial^2 \psi}{\partial t^2} = \rho A \,\delta x \frac{\partial^2 \psi}{\partial t^2}$$

$$B\frac{1}{\delta x}\left(\frac{\partial \psi}{\partial x}\bigg|_{x+\delta x} - \frac{\partial \psi}{\partial x}\bigg|_{x}\right) = \rho \frac{\partial^2 \psi}{\partial t^2}.$$

In the limit, as $\delta x \to 0$, the second term on the left-hand side approaches $\partial^2 \psi / \partial x^2$. Hence,

$$\frac{\partial^2 \psi}{\partial t^2} = c^2 \frac{\partial^2 \psi}{\partial x^2}, \qquad \text{where} \quad c = \sqrt{\frac{B}{\rho}}.$$

This equation is the classical *wave equation*, and in this context it describes the position of the element of fluid relative to its equilibrium position.

A solution of the wave equation is

$$\psi(x, t) = C \cos(kx - \omega t)$$

where C is the amplitude of the displacement, representing the maximum excursion of the element of fluid from its starting position, k is the wavenumber, equal to $(2\pi/\lambda)$ where λ is the wavelength and ω is the angular frequency of the wave (in rad s^{-1}; be careful!).

Hence it is apparent that the motion of the element of fluid is described by a similar equation to that for the mass in the harmonic oscillator.

The speed at which the wave propagates is determined by the constant c in the wave equation. By substituting the above solution back into the wave equation, it can be shown that the constants k and ω are related to each other by the equation $c = \omega/k$.

Pressure and velocity distributions

The wave moves along the tube by alternately compressing and rarefying regions in the medium. In the compressed regions the medium is denser and in the rarefied regions it is less dense. This implies that the pressure is higher in the compressed regions than in the rarefied regions. A series of pressure pulses will move along the tube as the wave travels (see figure 3.9).

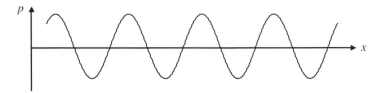

Figure 3.9. *Pressure distribution along the axis at an instant in time.*

The Eulerian approach to mechanics is to describe a system in terms of the values of field variables at each point in space. An obvious field variable to choose to represent this system is the pressure at a point along the axis. The pressure will vary along the axis and it will vary with time as the wave moves through the medium. It seems reasonable to anticipate that the pressure waveform might be represented in terms of a sine function, and in fact an expression for the pressure can be obtained from its relationship with ψ, developed in the preceding analysis,

$$\Delta p(x, t) = -B \frac{\partial \psi}{\partial x} = P \sin(kx - wt)$$

where P, the amplitude of the pressure change, is given by $P = CBk$.

The velocity, u, is the derivative of the position with respect to time,

$$u = \frac{\partial \psi}{\partial t} = C\omega \sin(kx - \omega t).$$

The ratio of pressure to velocity is therefore

$$\frac{p}{u} = \frac{CBk \sin(kx - \omega t)}{C\omega \sin(kx - \omega t)} = \frac{Bk}{\omega} = \frac{B}{c} = \rho c.$$

This ratio is constant for a given medium at a given temperature and so the pressure and the velocity are in phase in a sound wave. This is different from the simple harmonic oscillator, in which force and velocity are 90° out of phase.

Speed of sound in air

The speed of sound is determined by the bulk modulus and the density of the medium through which it travels,

$$c = \sqrt{\frac{B}{\rho}}, \qquad \text{where} \quad B = -V \frac{\mathrm{d}p}{\mathrm{d}V} = -\frac{\Delta p}{(\Delta V / V)}.$$

In the study of audition we are interested primarily in the speed of sound in air, and so we shall develop an equation for the speed of sound in a gaseous medium.

The equation of state for an ideal gas is $pV = nRT$. Assuming that the pressure and temperature changes in the gas occur so rapidly that heat cannot be transferred between adjacent elements (i.e. the process in each element is adiabatic), the pressure and volume are related as follows:

$$pV^\gamma = \text{constant}$$

where $\gamma = (c_p/c_v)$ is the ratio of the specific heat at constant pressure to that at constant volume.

Now, since $pV\gamma$ is constant,

$$0 = \frac{d(pV^\gamma)}{dV} = V^\gamma \frac{dp}{dV} + \gamma p V^{\gamma - 1}.$$

Then,

$$\frac{dp}{dV} = -\frac{\gamma p}{V}$$

and the adiabatic bulk modulus, B, is

$$B = -V \frac{dp}{dV} = \gamma p.$$

The speed of sound in an ideal gas is therefore

$$c = \sqrt{\frac{B}{\rho}} = \sqrt{\frac{\gamma p}{\rho}}.$$

The density of the gas can be expressed in terms of the molecular weight $\rho = nM/V$. Then,

$$c = \sqrt{\frac{\gamma p V}{nM}} = \sqrt{\frac{\gamma RT}{M}}.$$

At room temperature (293 K), the speed of sound in air is approximately 340 m s^{-1}.

Intensity of sound

It has been shown that a pure-tone sound wave can be described in terms of a single sinusoidal function, representing either the displacements of elements in the medium or the local pressure at points along the axis, depending on the chosen viewpoint. In either case the wave is characterized by its amplitude and its frequency. It is useful to generate a measure of a sound wave which quantifies the amount of energy that is associated with the transmission of the wave. The intensity of the sound wave is defined as the energy which passes through a unit area perpendicular to the direction of propagation per unit time (see figure 3.10).

It can be shown (e.g. Keller *et al* 1993, pp 827–8) that the intensity of the plane sound wave can be expressed as

$$I(t) = p \frac{\partial \psi}{\partial t} = -B \frac{\partial \psi}{\partial x} \frac{\partial \psi}{\partial t}$$
$$= -B(-Ck \sin(kx - \omega t))(C\omega \sin(kx - \omega t)),$$
$$I(t) = BC^2 \omega k \sin^2(kx - \omega t).$$

In terms of the pressure,

$$I(t) = \frac{\omega}{Bk}(P \sin(kx - \omega t))^2 = \frac{\omega p^2}{Bk} = \frac{p^2}{\rho c}.$$

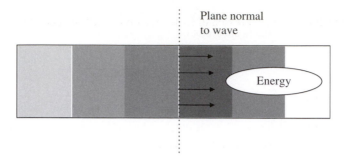

Figure 3.10. *Intensity is energy passing through a plane in unit time.*

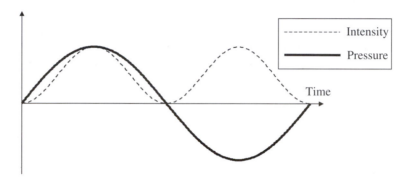

Figure 3.11. *Pressure amplitude and intensity of a pure-tone sound wave.*

The intensity of the wave is a function of time, and fluctuates as the wave passes through the plane (see figure 3.11). The medical physicist will be interested primarily in the average intensity of the sound over the whole of the period. The time average over one complete cycle of the square of a sine wave is 0.5. The average intensity of the sound wave is therefore

$$\bar{I} = \frac{\omega P^2}{2Bk} = \frac{P^2}{2\rho c}.$$

Hence the average intensity of the sound wave can be expressed in terms of the amplitude of the pressure wave and the density and speed of sound in the fluid medium. It is independent of the frequency of the wave.

For the plane wave considered here the intensity remains constant in the direction of propagation. In the case of a spherical wave radiating from a point source, the intensity of the wave diminishes with distance, and in fact is inversely proportional to the square of the distance from the source.

Units

The intensity of the sound wave has been defined as the energy which passes through a unit area perpendicular to the direction of propagation per unit time. Its derived SI units are therefore joules per square metre per second (J m^{-2} s^{-1}), or watts per square metre (W m^{-2}). When we come to study the human ear we shall find that this is a rather cumbersome unit of measurement, because we can hear sound intensities over an enormous range, from something of the order of 1×10^{-12} W m^{-2} to something in excess of 1 W m^{-2}. For convenience, sound intensity is often expressed in terms of a unit called the decibel (dB). The sound intensity level in dB is given by:

$$\beta = 10 \log \frac{\bar{I}}{\bar{I}_0}.$$

- The decibel is actually a measure of the relative intensity of the sound compared with a reference intensity.
- The reference intensity is taken as $\bar{I}_0 = 1 \times 10^{-12}$ W m^{-2}, which is the lowest intensity sound (of frequency 1 kHz) that the average human can hear.

Perception of sound intensity

From the definition of the decibel, the ear should register any sound of greater than 0 dB intensity. Also from the definition, an increase of one order of magnitude in sound intensity (W m^{-2}) will cause an increase in sound intensity level (dB) of 10. In principle, the intensity level of a sound (which we have seen is independent of frequency) should provide a good measure of its perceived loudness. In practice, the human ear is more sensitive to some frequencies than to others, and so the subjective loudness is not determined entirely by the intensity level. Indeed, our perception of sound is limited to a particular band of frequencies, between about 20 Hz and 20 kHz. We shall return to this topic in section 15.2.4 of Chapter 15.

Characteristic impedance, acoustic impedance and an electrical analogy

It has been shown that the product ρc is one expression of the ratio of the pressure on an element of fluid to the speed of that element. In these terms, an alternative expression of the average intensity is

$$\bar{I} = \frac{P^2}{2(p/u)}.$$

The characteristic impedance, Z (sometimes called the acoustic impedance), is defined as $Z = p/u = \rho c$.

  The characteristic impedance is a property of the medium through which the sound wave is travelling, and not of the wave itself. It is a measure of the ratio between the driving force (the pressure) on the elements in the wave and the velocity of the elements. It has units of kg s^{-1} m^{-2}, or N s m^{-3}.

  The intensity of the wave can be expressed in terms of the characteristic impedance,

$$\bar{I} = \frac{P^2}{2Z}.$$

In many texts the characteristic impedance is called the acoustic impedance of the medium. We shall reserve acoustic impedance to describe the resistance to flow of a particular system. The volume flow is the velocity of the element multiplied by its cross-sectional area. The acoustic impedance, defined as pressure per unit flow, has dimensions of kg s^{-1} m^{-4}, or N s m^{-5}.

  An electrical analogy for the acoustic impedance is the relationship between driving force (voltage) and flow of electrons (current). By analogy with the electrical system, the unit of acoustic impedance (kg s^{-1} m^{-4}, see above) is sometimes referred to as the acoustic ohm.

3.4.2. *Normal sound levels*

We have investigated the constitution of a sound wave, and have looked at how the wave might be described in terms of units of measurement. It is time now to introduce some numbers. This will serve to put the theory into context, and to give us some feel for the quantities that we are dealing with. We might be surprised at how small the pressures and displacements associated with sound waves are. The sound intensity can be expressed in terms of the amplitude of the pressure pulse as described previously. In summary,

$$\beta = 10 \log \frac{\bar{I}}{I_0}, \qquad \text{where} \quad \bar{I} = \frac{P^2}{2\rho c} \qquad \text{and} \quad \bar{I}_0 = 1 \times 10^{-12} \text{ W m}^{-2}.$$

The relationship between sound intensities and pressure amplitudes over the range normally encountered in a human environment is illustrated in table 3.3. The types of sound and possible physical consequences of the various intensities are illustrated in table 15.1.

Table 3.3. *Relationship between intensity level, intensity and pressure in a sound wave.*

Average intensity level (dB)	Average intensity (W m^{-2})	Average pressure (Pa)	Peak pressure (Pa)
160	1×10^4	2000	3000
140	1×10^2	200	300
120	1×10^0	20	30
100	1×10^{-2}	2	3
80	1×10^{-4}	0.2	0.3
60	1×10^{-6}	0.02	0.03
40	1×10^{-8}	0.002	0.003
20	1×10^{-10}	0.0002	0.0003
0	1×10^{-12}	0.00002	0.00003

Thus a sound at a level of 120 dB, which will cause physical discomfort, has a mean pressure amplitude is about 20 Pa, or approximately 0.02% of 1 atm. The mean pressure amplitude generated by someone whispering in your ear (20 dB) is not much more than one billionth of an atmosphere.

The particles that transmit the sound waves oscillate about an equilibrium position. The magnitude of the oscillation for a particular sound intensity is dependent on the frequency of the sound. The peak amplitude of displacement (C) of the particles in the sound wave is related to the peak pressure (P) as follows:

$$C = \frac{P}{Bk} = \frac{P}{\omega \rho c} = \frac{P}{2\pi f \rho c}.$$

For a sound wave of frequency 1 kHz (about where the ear hears best), and air of density ρ equal to 1.3 kg m^{-3}, the magnitude of displacement causing discomfort is approximately 10 μm. The peak displacement of the particles excited by a whisper is of the order of 1×10^{-10} m, or about the diameter of one atom.

3.4.3.   *Anatomy and physiology of the ear*

The ear serves as a transducer, converting the mechanical energy of the incoming sound wave into electrical energy for onward transmission to the brain. In this context it is similar to a microphone, which also converts the mechanical oscillation of a sound wave into electrical signals. In fact the ear is a remarkably good microphone, able to accommodate sound intensities that differ by a factor of 10^{12}. A cross-section of the ear is illustrated in figure 3.12.

The primary function of the ear, the conversion of mechanical to electrical energy, is accomplished by the inner ear. In principle, the mechanical sound energy is presented to the entrance to the inner ear, and an electrical signal is transmitted onwards by the cochlear nerve. We shall find, however, that the pressure

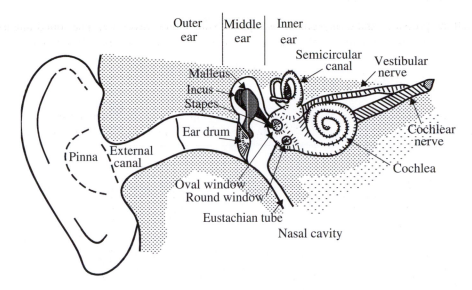

Figure 3.12. *Anatomy of the ear. This diagram is not to scale; the middle- and inner-ear structures have been enlarged for clarity. (Redrawn from J L Flanagan (1972), Speech Analysis: Synthesis and Perception (Berlin: Springer).)*

fluctuations associated with the incoming sound waves are rather small, and that an amplification device is needed if sounds at our lowest threshold of intensity are to be heard. Furthermore, the inner ear is filled with fluid, and the fluid has a rather different impedance to that of the air through which the incoming sound wave arrives. This would mean that a large proportion of the energy of the incoming sound wave would simply be reflected back from the interface. The middle ear serves both to amplify the incoming sound pressures and to match the impedance of the inner and outer ear structures. The outer ear is primarily a collecting and funnelling device, although its physical characteristics do give rise to a particularly efficient response to sound waves with frequencies of around 3–4 kHz.

Outer ear

The outer ear consists of the pinna and the ear canal, the inner boundary of which is the eardrum. The pinna (external auricle) of the human serves little useful purpose—apart from being somewhere to hang the spectacles and miscellaneous decorative devices. In some animals the shape is such that it funnels sound into the ear; in some, such as the elephant, a very large pinna provides a high surface area to serve as a heat exchanger. The ear canal (auditory canal, external canal) is roughly the shape of a tube, approximately 30 mm in length and 6 mm in diameter. The air in a tube will resonate at particular frequencies determined by the length of the tube and by the boundary conditions at the ends of the tube (see figure 3.13). The ear canal is closed at one end, by the eardrum, and open at the other.

Standing waves in a tube open at one end occur at frequencies given by

$$v_n = (2n - 1)\frac{v}{4l} \qquad n = 1, 2, 3, \ldots .$$

Assuming the velocity of sound in ambient air to be 340 m s^{-1}, the fundamental frequency of the ear canal is approximately 2.8 kHz. The human ear is most sensitive to sounds at about this frequency.

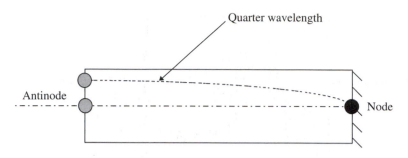

Figure 3.13. *Standing waves in the ear canal.*

The eardrum (tympanic membrane, tympanum) forms the boundary between the outer ear and the middle ear. It vibrates in response to incoming sound waves, and transmits the vibrations into the small bones in the middle ear. It is a thin membrane of tissue, approximately 0.1 mm thick, with a cross-sectional area of about 60 mm^2. It can be damaged (ruptured or perforated) by sounds of high intensity, above about 160 dB, but in common with other tissues, it does have the capacity to heal over a period of time.

Middle ear

The middle ear is an air-filled cavity in the temporal bone of the skull. Its outer boundary is the eardrum and its inner boundaries are the oval window and the round window at the entrance to the cochlea. The cavity of the middle ear has an air vent to the pharynx through the Eustachian tube. The purpose of this venting is to equalize the air pressures on either side of the eardrum. The vent is not open at all times, but is opened by actions such as chewing, swallowing and yawning. The system works well most of the time, but tends to break down when we subject ourselves to conditions for which we were not designed. During rapid descent in an aircraft or in a lift the external air pressure builds up quite quickly, causing the eardrum to deflect inwards. If the external pressure exceeds the pressure in the middle ear by 8000 Pa (8% of 1 atm) then you will probably feel pain. Furthermore, when the pressures are not equalized the sensitivity of the ear is reduced. Pressure inequalities are most likely to occur when the Eustachian tube is compressed and closed by inflamed tissue from conditions such as the common cold.

The primary mechanical components of the middle ear are the ossicles (small bones). They each have a distinctive shape from which they derive their common names. They are the hammer (malleus), the anvil (incus) and the stirrup (stapes).

The function of the middle ear is to amplify the pressures from the incoming sound waves and to match the impedances of the air-filled ear canal and the fluid-filled cochlea; figure 3.14.

The lever system provided by the ossicles increases the force at the oval window relative to that at the eardrum, and reduces the amplitude of displacement. This adjustment of the relative pressures and displacements works as an impedance-matching system, the purpose of which is to maximize the transfer of power from the air in the ear canal, which is light and compliant, to the fluid in the inner ear, which is heavy and stiff. If it were not for the impedance-matching system, much of the energy of the sound wave would simply be reflected back out of the ear. The middle ear is capable of producing a reasonable match of the impedances of the inner and outer ear over a range of frequencies from about 400 Hz to about 4000 Hz. At frequencies lower than 400 Hz the middle ear is too stiff, and at frequencies above 4000 Hz its mass is too great. It will be shown that the acuity of hearing drops off outside this frequency range as the sound energy is no longer transmitted effectively into the inner ear.

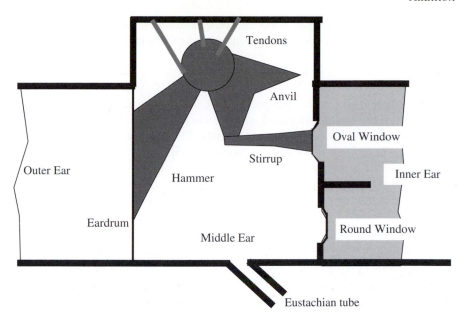

Figure 3.14. *Components of the middle ear.*

To illustrate the need for an impedance-matching device within the ear, we might consider the transmission and reflection of the energy of a wave at a typical fluid/liquid interface. The pressure amplitudes of transmitted and reflected waves at a discontinuity normal to the direction of the wave can be shown to be

$$P_{\text{trn}} = \frac{2\alpha}{\alpha + 1} P_{\text{inc}} \qquad P_{\text{ref}} = \frac{\alpha - 1}{\alpha + 1} P_{\text{inc}}$$

where

$$\alpha = \frac{Z_2}{Z_1} = \frac{\rho_2 c_2}{\rho_1 c_1}$$

is a parameter describing the relative impedances of the media at either side of the discontinuity. The characteristic impedance of water is approximately 1.48×10^6 kg m^{-2} s and that of air is approximately 430 kg m^{-2} s. Hence for a plane sound wave travelling through air and meeting water, the parameter α takes a value of 3442. The intensity of the transmitted wave is therefore only about one-tenth of 1% of that of the incident wave, and most of the energy of the wave is simply reflected back from the interface.

The primary mechanism that the middle ear uses for impedance matching is a mechanical pressure amplification system. The lever arm from the fulcrum to the centre of the eardrum is greater than that to the oval window by about 30%, and so the force is increased by 30%. Furthermore, the oval window has a cross-sectional area of only about 3 mm^2, compared with the effective cross-sectional area of the eardrum of approximately 60 mm^2. The anticipated pressure amplification that might be achieved by the system is therefore

$$\frac{p_{\text{oval\_w}}}{p_{\text{eardrum}}} = 13 \times \frac{60}{3} \approx 25.$$

Typical pressure amplifications measured experimentally on human subjects are around 17. The domestic cat has a mechanism that achieves a pressure amplification of around 60.

It has been shown that the efficiency of the ear in transmitting the incident sound energy is about 60% but would be only about 3% without the matching provided by the middle ear.

Inner ear

The inner ear (see figure 3.12) is a multi-chambered cavity in the temporal bone of the skull. It is said that it is the best protected of the organs associated with the sense. It consists of several parts, including the cochlea, the saccule, the utricle and the semi-circular canals. The function of the saccules is not known, and functions of the utricle and semi-circular canals are associated with our senses of position and balance. Only the *cochlea* has a role in the hearing process. The role of the cochlea is to convert the mechanical energy presented as a mechanical vibration at the oval window into electrical energy, transmitted onwards through the auditory nerve. We can study the anatomy of the cochlea and can thus gain some clues as to how this energy conversion might occur.

The cochlea is essentially a tapered tube approximately 30 mm in length. The tube is not straight but is wound into a spiral shape, so that its external appearance is much like the shell of a snail. The diameter of the tube tapers downwards from the outside of the spiral, at the interface to the middle ear, to the apex. The tube is divided along its length into three chambers. The two outer chambers, the scala vestibuli and the scala tympani, are connected at the apex of the cochlea through an opening called the helicotrema. The fluid in the scalae is called perilymph. The vibration of the oval window, induced by the stirrup, gives rise to a wave travelling inwards away from the oval window. The outer boundaries of the cochlea are rigid, and, since the fluid is basically incompressible, some flexibility is required to allow the system to move under the applied loads. This flexibility is provided by the round window at the outer boundary of the scala tympani. The cochlear duct, sandwiched between the two scalae, contains a fluid called endolymph.

The membrane separating the scala tympani from the cochlear duct is called the basilar membrane (see figure 3.15), and the motion of this membrane plays a key role in the generation of the nerve impulses associated with the sense of hearing. The basilar membrane runs the full length of the spiral, and spans across from a bony shelf to a ligament. The membrane contains many fibres, each approximately 1–2 μm in diameter, that run spanwise across it. The length of these fibres varies from approximately 75 μm near to the oval window up to about 475 μm at the helicotrema. Inside the cochlear duct, and supported on the basilar membrane, is the *organ of Corti*. It is this organ that accomplishes the transformation of mechanical to electrical energy.

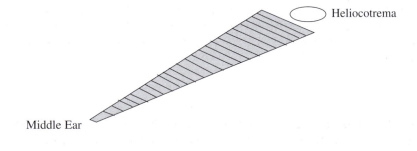

Figure 3.15. *Developed view of the basilar membrane.*

The organ of Corti contains hair cells, rooted on the basilar membrane, which are connected to nerve fibres. These cells have projecting hairs or fibres that bridge across to the tectorial membrane. The hairs are approximately 8–12 μm in diameter. As the basilar membrane moves in response to the pressure waves, it bends or shears the hairs and these stimulate the nerve cells to produce electrical impulses. We can therefore understand the process by which the stimulation of the cells occurs, but it is rather more difficult to understand the detail of the correspondence between the mechanical input and the electrical output.

Neural response

We can measure the individual response of a single neurone to the stimulation of a sound wave. One measurement of interest is the threshold intensity which will cause a just measurable response in the nerve. The threshold intensity of the neurone varies with frequency, and the results are often presented graphically in the shape of tuning curves such as that illustrated in figure 15.2. This curve is typical of the response obtained experimentally. There is a characteristic frequency to which the neurone is particularly sensitive, in the sense that it responds to sounds of a low intensity at this frequency, and this is represented by the minimum on the tuning curve. For the majority of the nerves, the threshold intensity at the characteristic frequency is about 10–15 dB, although a significant minority have higher thresholds. The slope of the curve away from the minimum tends to be quite steep, implying that, although a sound of a frequency other than the characteristic frequency can also excite a response from the nerve, its intensity needs to be much greater.

The response of the nerve fibre, in terms of its firing rate, to a sound of a given frequency but variable intensity can also be investigated. Pickles (1988) describes the shape of the curve as 'sigmoidal'. The saturation response typically occurs at about 50 dB above the threshold response. There is a substantial portion of the range over which the response might reasonably be approximated as linear.

3.4.4. Theories of hearing

In this section we shall explore the transformer mechanism of the ear in a little more detail. We know that the ear transforms mechanical sound energy into electrical activity. What is the correlation between the frequency and intensity of the incident sound at the cochlea and the resultant nerve impulses that are transmitted to the brain? Many eminent scientists have addressed this problem, and there is still no wholly satisfactory model. Much depends on how we try to divide the processing of the data between the ear itself and the brain. A comprehensive review of the current state of knowledge is presented by Pickles (1988). There is little doubt that the encoding of intensity information is in terms of the frequency of nerve impulses. As discussed in the previous section, there is clear evidence that a single neurone fires at higher frequency as the intensity of the stimulus increases. The nature of the encoding of frequency information is rather more hotly debated. Essentially two alternative mechanisms have been proposed. In the first, collectively known as place theories, the frequency is identified by the position within the cochlea of the responding nerve. In the second, known as the temporal theories, the frequency of the sound is identified from the timing of the nerve impulses. Many audiologists consider that both mechanisms are required to describe the range of human hearing adequately.

Place theories

It is instructive to consider some of the early theoretical models, and to explore their limitations, as we try to understand this complex phenomenon. The first attempts to describe the function of the cochlea were made by Helmholtz in the mid-19th Century. He observed that the fibres in the basilar membrane vary in length, being shorter near the oval window and longer near the apex. A comparison can be made between this array of fibres and the wires in a harp or piano. Assuming that the fibres are all in tension, then each fibre has a resonant frequency that can be calculated from the laws of mechanics. Each fibre will thus have a characteristic and sharply tuned response to an input vibration frequency. Helmholtz's resonator theory suggested that those fibres tuned to the particular frequency of the incoming sound wave would resonate, and that the pitch of the wave is known from the position within the cochlea at which resonance occurs. A primary objection to this theory is the fact that the response of the ear to sounds does not correspond to the anticipated behaviour of a resonating system. In particular, once the fibre had been excited at its resonant frequency it would continue to resonate for some time after the exciting source has ceased, and this is not consistent with observation. When the theory was first proposed it was impossible to prove or disprove, because accurate measurements of the cochlear dimensions and of the tension in the basilar membrane were unavailable. It has more recently been shown that the resonator theory cannot adequately explain the range of human hearing.

A major contributor to our current understanding of the hearing process was Georg von Bekesy (1900–1970), a very active researcher in the mid-decades of the 20th Century. His work and studies earned a Nobel prize, awarded in 1961. He showed that the cochlea is a critically damped mechanical system, and that the motion of the basilar membrane is in the form of a wave moving away from the oval window. The amplitude of vibration of the basilar membrane varies along its length, and the position of maximum amplitude is a function of frequency. At low frequencies (25 Hz), the whole of the membrane is vibrating, and the peak amplitude of vibration is at the far end, near the helicotrema. At high frequencies (3 kHz), the peak amplitude of vibration occurs near to the oval window. This suggests that Helmholtz's notion of the importance of the position within the cochlea of the peak response is correct, but that the response is not one of resonance. The place theory of von Bekesy has received widespread acceptance. Place theories require that the brain can separately identify signals from clusters of nerve fibres in order to discriminate pitch, but they are attractive because the encoding of pitch and frequency are distinctly separate: the position of the responding nerve within the cochlea determines the pitch of the sound and its firing frequency determines the intensity.

Temporal theories

The *telephone theory*, proposed by Rayleigh and others, is based on the premise that the cochlea serves simply as a microphone, and transmits into the auditory nerves signals of the same form as that of the incident pressure wave. The basic idea is that each nerve cell generates a signal at intervals of time that depend on the amplitude of pressure, generating more impulses at higher pressures. This theorem runs into difficulties because of the limits of response of the nerve cells. Each cell can transmit no more than 1000 pulses per second, and so the ear would not be able to recognize and resolve sounds of greater than 1 kHz frequency. In fact the actual resolution would be very much lower still because the frequency of the pulses would have to be used to describe the amplitude of the wave, and several pulses would be necessary to define one sine wave. The basic telephone theory is clearly untenable, but the primary tenet that the brain should relate the frequency of incoming pulses to the frequency of the incident waves is the basis of more modern theories.

The *volley theory* is essentially an extension of the telephone theory. The assumption that a single nerve cell generates sufficiently many impulses to describe the incident wave is replaced by the assumption that the wave is assembled from an integration of the responses of a group of cells. The volley theory requires the brain to do some intelligent combination (integration) of incoming pulses as a post-processing operation. There is evidence that the brain does conduct such post-processing operations. If both ears are excited simultaneously, but with slightly different frequencies, the difference frequency is heard (this phenomenon is called binaural beats). Since the difference frequency does not arise mechanically in either ear, the combination must occur in the brain.

3.4.5.   *Measurement of hearing*

Loudness versus intensity

We have seen that the intensity of sound waves, defined simply as a rate of flow of energy through an area, can be measured objectively. The intensity can be expressed in terms of the amplitude of the wave and of a characteristic of the medium through which the fluid travels, independently of the frequency of the wave. If any device were able directly to measure sound intensity levels, then there would be no frequency dependence. All practical measurement devices, however, contain mechanical and electrical or electronic components that limit the range and accuracy of their function. The ear is no exception to this rule, although it is a particularly fine piece of equipment—able to resolve and recognize sound intensities over a remarkably wide range. It has been suggested earlier that the frequency range of the ear might be limited by the mechanical properties of the middle ear: its ability to match the impedance of the air and fluid on either side is limited by its mass and stiffness.

In practice the ear responds best to a range of frequencies from about 1–6 kHz. Outside this range the response deteriorates progressively until, outside the range 20 Hz to 20 kHz, even sounds of very high intensity are inaudible. The apparent loudness of a sound is a subjective quantity that varies from person to person. A sound at the optimum frequency (for hearing) of about 3 kHz will seem much louder than one of the same intensity at, say, 100 Hz. The measurement of the subjective loudness of a sound is part of the science of psychophysics, discussed in more detail in section 3.6. The unit of subjective loudness is called the *phon*. It is instructive to study the relationship between sounds measured in phons and the objective measure of intensity level (dB). It would be too arbitrary to ask a subject to judge when, for example, one sound was twice as loud as another, but it is reasonable to ask when two sounds are of equal loudness. A reference level is therefore required for each loudness level. It is taken that a loudness of n phons corresponds to a sound intensity level of n dB at a frequency of 1 kHz. It is then possible to find the sound intensity level at other frequencies that seems equally loud, and to plot contours of equal loudness on a graph of intensity level against frequency. The result for the average human ear is illustrated in figure 15.7.

The most important range of hearing, at least with respect to quality of life, is that from about 20–60 dB, encompassing the range of normal speech. The 40 phon loudness contour therefore has special significance. When comparing sounds in a human environment it is probably more appropriate to measure in phons rather than in dB. The form of the 40 phon contour as a function of frequency can be expressed algebraically, and a factor (frequency dependent) can be derived by which subjective and objective sound intensities differ. It is possible to scale recorded sounds by the inverse of this factor, giving a loudness rather than an intensity readout. This scaling process (sometimes called a filter) is obviously directly applicable only to sounds of 40 phons, but it does give a realistic appreciation of loudness over a reasonably wide range. The unit of loudness measured in this way is called the dBA. It is possible to filter based on other contours, giving units of dBB or dBC, but only dBA is in common usage.

Hearing tests

A measure of speech comprehension is the most desirable feature of a hearing test. Tests are used in which speech is presented to the subject at a range of intensities and their ability to understand is recorded. Speech audiometry is a valuable test of hearing, although the results depend not only on the hearing ability of the subject but also upon their ability to comprehend the language which is used. Sounds other than speech are also used: a tuning fork can be used by a trained person to assess hearing quite accurately. Sources of sound such as rattles are often used to test a child's hearing: the sound level required to distract the child can be used as evidence of their having heard a sound. Two commonly used hearing tests are pure-tone audiometry and middle-ear impedance audiometry (tympanometry). Both of these tests use clinical electronic instruments, and they are discussed in detail in Chapter 15. Pure-tone audiometry relies on the co-operation of the subject to report on the audibility of sound over a range of frequencies and intensities to establish hearing threshold. Middle-ear impedance audiometry produces an objective measurement of the function of the middle ear.

Hearing defects and correction

The problems of detection and classification of hearing defects and the specification and design of hearing aids are addressed in Chapter 15 under the heading of audiology.

3.5.   VISION

In this section we will investigate the sense of vision. Vision is arguably our most important sense: certainly a complete loss of vision would have the most profound effect on the lifestyle of the average person. The study of vision does not normally fall within the remit of the medical physicist: all large hospitals have ophthalmology departments devoted to this particular science. He or she might nevertheless be involved in research in this sphere, often together with an ophthalmic colleague. The study of light, and particularly of lasers and their applications, is part of the role of the medical physicist. The science of lenses, or geometrical optics, is an integral part of any undergraduate physics course. We shall adopt a similar structure for this section to that used for the study of audition. Our fundamental questions will be:

- What do we see?
- How can we measure it?
- How do we see it?
- How good is our sight?
- What can go wrong?
- What remedies are available?

3.5.1.   Physics of light

The nature of light

The nature of light itself is much more difficult to understand than that of sound. Newton, in the 17th Century, was convinced that light was comprised of tiny mass-less particles, whereas Huygens, working at the same time, argued that it must be a wave of some sort. We now know that light exhibits properties of both particulate and wave motion, but it was not until the beginning of the 19th Century that the wave properties of light were finally established. The major problem then was that we could only understand vibration when a mechanical medium was available to vibrate. It was known that light propagated through space and that therefore, unlike sound, no current mechanical theory could describe its motion. Maxwell derived the mathematical theories of electromagnetic waves, and demonstrated that light exhibited the properties expected of electromagnetic radiation. Old ideas die hard, and for a long time it was postulated that the whole of space was permeated by an 'ether' that was capable of transmitting the vibration of light without having any mechanical effect on bodies such as planets travelling through it. It was only when scientists proved unable to detect the ether that it was finally accepted that there could be no mechanical underpinning of the theories of electromagnetic waves.

The mathematical form of the 'one-dimensional' electromagnetic wave equations, derived from Maxwell's equations, is identical to that of the one-dimensional wave equation that we developed for sound waves in the previous section, with the displacement ψ replaced by the electrical and magnetic field vectors, E and B. The solution of this classical wave equation is essentially the same as it was for sound, and the properties of the solution are the same. All of the concepts that we are familiar with from the study of sound, such as frequency and wavelength, interference and superposition, apply equally well to light.

The wavelength of electromagnetic radiation has no theoretical bound, but we are most familiar with the range from about 10^{-14} m to about 10^8 m, as illustrated in figure 3.16.

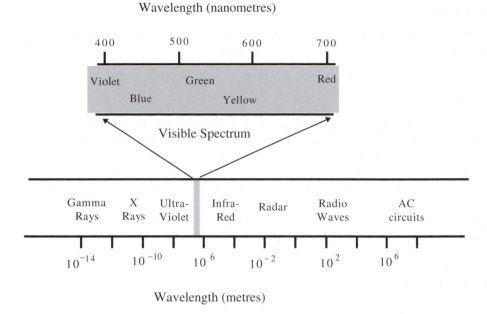

Figure 3.16. *The electromagnetic spectrum.*

We use electromagnetic radiation over this whole spectrum in various applications of medical physics, and many of them are covered in other chapters of this book. Visible light occupies only a narrow band from about 400–700 nm. For comparison, the wavelength of sound in the audible range (20 Hz to 20 kHz) in air is 17 m to 17 mm.

Speed of light

The speed at which light travels in a vacuum is approximately 3×10^8 m s^{-1}, or 186 000 miles per second. Its speed in a transparent medium is always less than this, and the ratio of the speed in vacuum to that in the medium is known as its *index of refraction, n,*

$$n = \text{index of refraction} = \frac{\text{speed of light in vacuum}}{\text{speed of light in medium}} = \frac{c}{v}.$$

The indices of refraction of some common media are listed in table 3.4.

Table 3.4. *Typical index of refraction of some common media.*

Medium	Index of refraction
Air	1.0003
Water	1.3
Glass	1.5

For most practical purposes the speed of sound in air is taken as the same as that in vacuum.

Refraction and reflection at a boundary

When light travels from one medium into another its frequency cannot change but its velocity must. The wavelength changes to accommodate the change in velocity. At a fixed frequency, wavelength is proportional to speed and therefore the ratio of the wavelength in vacuum to the wavelength in a medium is also equal to the index of refraction. If the light meets a boundary at an oblique angle then its direction must change if the wavelength is to change by the appropriate factor. The angle through which the light is refracted can be calculated from simple geometry, and the relationship between the angle of incidence and the angle of refraction is determined by Snell's law.

Snell's law:

$$n_1 \sin \theta_1 = n_2 \sin \theta_2.$$

Figure 3.17 is a little misleading because it implies that all of the light that reaches the boundary is refracted and continues into the second medium. This is not true, and some percentage of the incident light is reflected away from the surface. The reflected light, which has been omitted from the figure for clarity, leaves at an angle which is simply the reflection of the incident angle in the normal to the boundary plane. The relative intensities of the incident, reflected and refracted light depend on the properties of the media and on the angle of incidence. An investigation of Snell's law by substituting some values of the angle of incidence and the relative indices will quickly reveal that it is easy to get a value for $\sin \theta_2$ that has a magnitude of greater than unity. There is no angle that satisfies this requirement, and the physical consequence is that all incident light is reflected back into the first medium (total reflection). It is possible to use the principle of total reflection to keep light trapped within a medium, if the angle at which the light strikes its surfaces is always greater than the critical angle for total reflection. This is called total internal reflection, and is the basis of the design of optical fibres.

A consequence of Snell's law is that a spherical boundary will focus parallel light rays onto a single point. This property is the basis of the geometrical optics of the eye, and of our design of corrective optical devices for defective vision. Two portions of spherical boundaries, as represented by a simple biconvex lens, will also focus parallel incident light rays to a point (see figure 3.18).

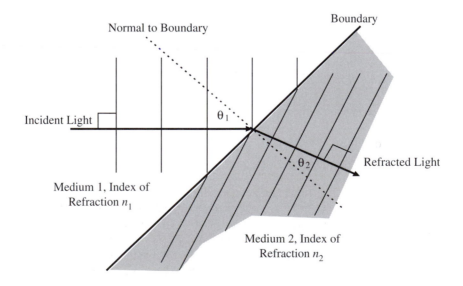

Figure 3.17. *Refraction of light at a boundary between media, $n_2 > n_1$.*

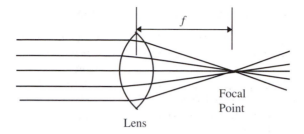

Figure 3.18. *Focusing of light by a biconvex lens.*

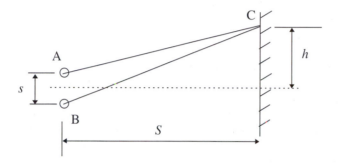

Figure 3.19. *Conditions for constructive and destructive interference.*

If the index of refraction of the material of a thin lens is n_2 and that of the surrounding medium is n_1, the focal length, f, of the lens is given by

$$\frac{1}{f} = \left(\frac{n_2 - n_1}{n_1}\right)\left(\frac{1}{R_a} - \frac{1}{R_b}\right)$$

where R_a is the radius of curvature of the first surface of the lens and R_b is that of the second surface, measured from the same side of the lens. This equation is known as the lens-makers' equation.

Interference of light waves

Light waves interact and interfere with each other in just the same way as do sound waves. The relative phase of the waves determines whether the interference is constructive, increasing the intensity, or destructive, reducing the intensity. Constructive and destructive interference can be observed experimentally by looking at the patterns produced by two phase-locked point sources separated by a known distance as illustrated in figure 3.19. The point sources are at A and B, and they are separated by a distance s. A screen is placed parallel to and at a distance S from the line connecting the two sources. Light falling at a point C on the screen has to travel a different distance from each source. From simple geometry, assuming that the distance of the screen from the sources is large compared to all other distances, the difference in lengths of the paths BC and AC is

$$\delta = \mathrm{BC} - \mathrm{AC} \approx \frac{hs}{S}.$$

There will be constructive interference when this distance represents a whole number of wavelengths of the light. Hence the conditions for constructive interference are

$$\frac{hs}{S} = n\lambda.$$

This implies that the distance between bright peaks will be $S\lambda/s$. Dark troughs corresponding to destructive interference lie half-way between the bright peaks.

This experiment gives us a way to measure the wavelength of light by examining the spacing between regions of constructive and destructive interference. The problem of ensuring that the sources are phase-locked is readily overcome: use the same source! This is achieved by replacing the sources A and B with very thin slits that are a long way from the primary source. Each of the slits acts as a line source of light. The resulting light and dark bands on the screen are referred to as fringes. This experiment, performed by Young and often called 'Young's slits', was the first to demonstrate clearly the wave properties of light.

Diffraction

In the preceding section we have looked at how light rays from two slits interfere. The width of each slit is taken to be infinitesimal, so that each represents a line source. By considering a real slit of finite width as an aggregation of very much thinner slits we can develop an expression for the intensity of the light falling on a screen as a function of position along the screen. Once again we find that there are fringes of light. The central fringe is the brightest, and the brightness of each successive fringe moving away from the centre is less than that of the previous one. The spreading of the light from the slit is called diffraction, and the pattern of fringes on the screen is the diffraction pattern. The width of the bright central fringe can be taken as a measure of the diffraction. The half-angle, θ, at which the beam appears to diverge can be approximated by the relationship

$$\theta = \sin^{-1}\left(\frac{\lambda}{w}\right)$$

where λ is the incident wavelength and w is the width of the slit.

Complete diffraction occurs when the width of the slit approaches the wavelength of the incident wave. For light waves, with a wavelength down at sub-microns, the slits need to be down at this width. In our everyday experience we usually deal with objects on much larger length scales: light is not diffracted and appears to travel in straight lines. In contrast, sound waves in the range of hearing have a wavelength of at least 17 mm and quite large sources (such as one's mouth) behave as point sources. The physical process of diffraction does have consequences for our acuity of vision. The pupil is an aperture of finite size, and diffraction occurs as light passes through it. Consequently, there will always be some diffusion of the intensity of 'a light ray' about the nominal point at which it strikes the retina. Diffraction at a circular aperture produces a similar effect, but with rings rather than parallel fringes.

3.5.2.   Anatomy and physiology of the eye

The human eye is approximately spherical in shape, and its main features with respect to the sense of vision are the cornea, the iris, the lens and the retina. The first three are all focusing and filtering devices that serve to present a sharp image to the retina. In common with the components of the outer and middle ears, their function might be regarded as essentially mechanical, if such a description can be applied to a process involving light. The transduction of the light energy into electrical energy is the function of the photoreceptors in the retina. A cross-section through the human eye is illustrated in figure 3.20.

The following anatomical description of the eye is based on that of Marieb (1991), although errors are our own! The basic structure of the eye is much like that of a football. The wall of the eyeball consists of

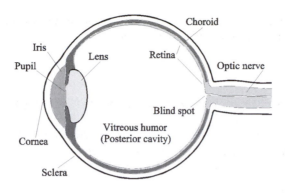

Figure 3.20. *Anatomy of the human eye.*

three layers, called tunics. The only interruption of these tunics is in the posterior portion where the optic nerve enters the eye. The outermost layer is a fibrous tunic composed of dense connective tissue. Its two components are the cornea and the sclera. The cornea, at the front of the eye, has its collagen fibres regularly arranged, and it is transparent. It is curved, and represents the first of the focusing components of the eye. It does not contain blood vessels (and can therefore be transplanted with relatively little risk of rejection), but does contain nerve fibres and is sensitive to irritation. The sclera, or white of the eye, is a tough, opaque sheath to which the muscles of the eye are connected. The next layer is a vascular tunic called the uvea. Its three components are the choroid, the ciliary body, and the iris. The choroid contains a large number of small blood vessels, supplying the other tunics as well as itself. It also contains pigments that help to prevent reflection of light within the eye, which would tend to blur the images registered. The ciliary body contains the muscles that support and focus the lens, and also the capillaries that secrete fluid into the anterior segment of the eyeball. The iris is the characteristically coloured part of the eye. Its musculature is constructed to allow a change in the diameter, from about 1.5 to 8 mm, of the central aperture, the pupil, thus controlling the amount of light that can enter the posterior segment. Circumferentially orientated muscles contract the pupil and radially orientated ones expand it.

The innermost layer is the retina, which itself consists of two layers: an outer single-cell layer of pigmented cells and an inner neural layer containing the photoreceptors. The former reduces reflection and the latter is the primary transducer of the visual system. We shall explore the nature of the photoreceptors in more detail later. Much of the processing of the image is performed locally by the neurones within the retina. This is not generally true of the other complex receptors, which simply transmit information to the brain for analysis. For this reason the retina is sometimes considered to be a remote outpost of the brain. There is an area where the optical nerve enters the eyeball that has no photoreceptors, creating a blind spot in our field of vision. Adjacent to this is an area of about 3 mm^2 called the macula lutae, over which the retina is relatively thin. Within the macula lutae is the fovea centralis, a pit-like depression the size of a pin-head (about 0.4 mm diameter). The fovea is on the optical axis of the lens, and is the site of our best visual acuity.

The two chambers of the eye, the anterior and posterior segments, are filled with fluid. The anterior segment is filled with an aqueous humour, a clear plasma-like fluid that is continually drained and replenished. The pressure within the anterior segment is maintained at about 16 mmHg. The posterior segment is filled with vitreous humour, a clear viscous gel that is not replenished.

The lens is the component that does the final fine focusing of the incident light onto the retina. It is biconvex, but not symmetrically so. Its general shape, and in particular the curvatures of the surfaces, can be changed by the action of the ciliary muscles thus providing a variable-focus capability. The media within which the lens is contained have similar indices of refraction to the lens itself, and its effective index of refraction is only about 1.07. For this reason most of the focusing is done at the cornea, at the external air interface. As we get older the lens loses its flexibility, and is not able to adopt the higher curvatures. The result is that we have difficulty in focusing on objects that are close to the eye, and we might need reading glasses. This loss of accommodation of the lens is called presbyopia. A second deterioration mechanism that affects the lens, again often associated with age, is a loss of clarity. Clouded or opaque spots within the lens are called cataracts. In extreme cases the lens might be surgically excised and replaced, possibly with a prosthetic component.

Photoreceptors

There are two types of photoreceptors in the retina, called rods and cones. The rods are distributed over the whole of the retina, with the exception of the immediate area of entrance of the optical nerve. There are about 125 million rods in the retina, or on average something like $100\,000$ mm^{-2} of its surface. The density is highest in an annulus about 30° from the fovea, with a gradual reduction away from this location. There are no rods in the fovea. They have a low (but variable) intensity threshold, and thus respond to low levels of incident light. Although the photoreceptors themselves are non-renewable, active elements within them are renewed and replenished throughout our lifetime.

Our night vision is almost entirely attributable to the rods. The light-sensitive pigment within the rods is called rhodopsin, or visual purple: the wavelength/absorption spectrum of rhodopsin matches almost exactly the wavelength/sensitivity spectrum of the human eye. The peak sensitivity of the rods is to light of a wavelength of about 510 nm, within the blue/green range. Upon exposure to light, a chain of chemical reactions known as the rhodopsin cycle is initiated. This cycle begins with the chemical dissociation of rhodopsin on the absorption of a photon of light, and finishes with the recombination of the resulting chemical components back into rhodopsin. The action potential occurs very rapidly, at an early point in the cycle, and thereafter the recovery process to rhodopsin takes 20–30 min. The highest sensitivity of vision is associated with the maximum supply of rhodopsin in the rods, and thus adaptation of the eye to night vision takes up to 30 min. When the eye is in this condition it is called scotopic, or dark-adapted. An essential element of the rhodopsin cycle is the availability of vitamin A, and deprivation of this vitamin leads to nyctalopia, or night blindness. The opposite of scotopic is photopic (light-adapted) vision. More than 100 rods are connected to a single ganglion cell, and the brain has no way of identifying which of them has generated an action potential. The visual acuity associated with the rods is therefore relatively low, although the sensitivity is high.

There are about 7 million cones. Their density is highest at the fovea centralis, and diminishes rapidly across the macula lutae: outside this region the density of cones is insufficient for useful function. Within the fovea there are no rods and the cones are closely packed together, at a density of up to $140\,000$ cones per mm^2. The diameter of a single cone is about 2 μm. The peak sensitivity of the cones is to light with a wavelength of about 550 nm, in the yellow–green region. The sensitivity of the cones is low relative to that of the rods, perhaps by four orders of magnitude. However, within the fovea there is almost a one-to-one correspondence between cones and ganglions, and resolution is very high. The cones are responsible for our colour vision, although the actual mechanism by which colour is perceived is still unknown. It is seductive to assume that the cones function in a similar manner to the rods, and that there are chemical equivalents to the rhodopsin cycle occurring within the cones. Many of the phenomena associated with our colour vision can be explained by the existence of three types of cone each containing a different photopigment, responding optimally to light of one particular wavelength. Unfortunately we have yet to isolate any of them. Nevertheless the so-called trichromatic theories are most valuable in furthering the understanding of our perception of colour.

3.5.3.   Intensity of light

The intensity of sound is defined in terms of an energy flux, or power, per unit area, and has dimensions of $W \, m^{-2}$. In pure physics texts the properties of light, like sound, are often described in terms of the plane wave, and the intensity of light is similarly defined. In texts devoted to light and to optics, the conventional system of units is defined in terms of the behaviour of a point source. The energy flux from a point source is denoted by Φ and its SI unit is the watt. For a plane wave it was convenient to think in terms of energy flux through a unit area normal to the plane, but for a point source, from which the energy radiates in all directions, an alternative definition of intensity is more appropriate. The intensity of a point source, I, is defined in terms of energy flux per unit solid angle, with SI units of watts per steradian ($W \, sr^{-1}$). There are 4π steradians about a point, and the intensity of a uniform point source, radiating energy equally in all directions, of power 1 W is therefore $1/4\pi \, W \, sr^{-1}$. The energy per unit area emitted from a surface, denoted M, is called the exitance. If the surface is very small, or if we measure at distances far removed from it, it behaves like a point source. We define the radiance of a surface to be the intensity per unit projected area of the surface along any given line, or the intensity divided by the effective area that we can see when we look back towards the radiant surface.

The properties of electromagnetic radiation depend on its wavelength: for example, we can only see radiation in the wavelength spectrum from about 400–700 nm. If we are to understand and predict the effects of electromagnetic radiation, we need to know not only the spatial and temporal distribution of its energy but also its spectral distribution. A quantity evaluated over an increment in a given spectral range is conventionally assigned a subscript appropriate to the spectrum, so that, for example, the spectral radiant exitance over a small increment $d\lambda$ around a wavelength λ is denoted by M_λ.

The radiation of electromagnetic energy from the surface of a black body is governed by Planck's radiation law, which can be expressed as follows:

$$M_\lambda = \frac{2\pi \, hc^2}{\lambda^5 (\exp(hc/kT\lambda) - 1)}$$

where M_λ is the exitance at wavelength λ over an interval $d\lambda$ ($W \, m^{-2} \, m^{-1}$); h is the Planck constant, $6.63 \times 10^{-34} \, J \, s$; k is the Boltzmann constant, $1.38 \times 10^{-23} \, J \, K^{-1}$; c is the speed of light in vacuum, $3.00 \times 10^8 \, m \, s^{-1}$; T is the temperature of the body (K); λ is the wavelength of the emission (m).

This formula defines the energy emission spectrum. Energy is emitted over all wavelengths, and the spectral distribution depends on the temperature of the body. Figure 3.21 illustrates the emission spectrum at 2045 K. This temperature has a special significance which we shall discuss later. We might, however, note that most of the energy falls at wavelengths above the visible range, in the infrared region. Inspection of the radiation law reveals that the peak in the spectrum moves towards the lower wavelengths as temperature is increased, so that at, say, 5000 K a larger percentage of the energy will be within the visible range.

The total radiant exitance can be calculated by integration over all wavelengths (from zero to infinity). The integration of Planck's radiation law yields the Stefan–Boltzmann law for a black body:

$$\text{total radiant exitance} = \int_0^\infty M_\lambda \, d\lambda = \sigma T^4$$

where the Stefan–Boltzmann constant, $\sigma \approx 5.67 \times 10^{-8} \, W \, m^{-2} \, K^{-4}$. (We might be more familiar with the version of this law that includes an emissivity constant, ε, which describes the proportion of energy emitted by a real body relative to that from the ideal black body.)

For sound waves we had two measures of intensity, the first the actual physical quantity in $W \, m^{-2}$ and the second (loudness) the perceived intensity. For light we distinguish between radiometric quantities, describing the physical energy levels in the wave, and photometric quantities, describing their visually evaluated equivalents. Conventionally the subscript 'e' is used to denote a radiometric quantity and 'v' to denote

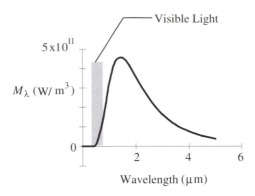

Figure 3.21. *Spectral radiant exitance of a black body at 2045 K.*

a photometric quantity. The unit of photometric power is the lumen (lm), and that of photometric intensity is the candela (cd). One candela is equal to one lumen per steradian, and therefore a uniform point source of power 1 lm has a luminous intensity of $1/4\pi$ cd. The photometric equivalent of radiance is luminance.

Scientists have been experimenting on vision for a very long time, and the first unit of measurement used for the intensity of light was the candle, equal to the flux of light energy per steradian from a standard candle. There are a number of obvious problems with this unit, from the problem of definition of a standard candle through to a proper appreciation of the wavelength spectrum of the energy coming from it. When the study of vision started to come into line with the science of physics, the need for a more appropriate and scientific unit was recognized. The photometric units of light intensity were formalized when the first satisfactory black-body emitters were developed in the middle of the 20th Century. The first such devices operated at the solidification temperature of platinum, 2045 K, and the units adopted were based on this standard. The emission spectrum for a black body at this temperature is illustrated in figure 3.21. The candela, chosen to represent an intensity approximately equal to the older unit of the candle, was formally defined as the luminous intensity of a black body at 2045 K radiating into a semi-infinite space from an area of $1/60$ cm^2. It is interesting to note that most of the energy emitted by this body is at wavelengths outside the visible spectrum: the proportion of the energy emission that lies within the visible spectrum is only about 1% of the total.

The relationship between radiometric and photometric units is not straightforward. We know that the eye can only detect electromagnetic radiation within the visible spectrum, but the issue is further complicated because it is not uniformly sensitive to all wavelengths within this range. There is a peak of sensitivity at around 555 nm for photopic vision and at around 510 nm for scotopic vision, and a rapid deterioration away from these wavelengths. The ratio, $V(\lambda)$, of the sensitivity at a particular wavelength to the peak sensitivity, C, is called the spectral luminous efficiency. Standardized curves are available to describe $V(\lambda)$, and the form of the curve is illustrated in figure 3.22.

Figure 3.22 illustrates the distribution for photopic vision, defined as vision adjusted to a luminance of 10 cd m^{-2}: that for scotopic vision, adjusted to a luminance of 0.001 cd m^{-2}, is shifted towards the shorter wavelengths. The difference between the luminous efficiency spectra for photopic and scotopic vision leads to a phenomenon called the Purkinje effect, which describes an increased sensitivity to the green/blue end of the spectrum and a reduced sensitivity to the red/orange end as brightness diminishes. Hence as night falls, red flowers appear to lose their colour, and the colour of the foliage becomes dominant.

The spectral luminous flux is the spectral radiant flux multiplied by the spectral luminous efficiency, and the total luminous flux, or luminance, is again the integral over all wavelengths. The luminous efficacy

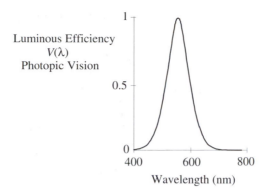

Figure 3.22. *Spectral luminous efficiency versus wavelength: photopic vision.*

of electromagnetic radiation is the ratio of its total luminance (lm) to its total radiance (W). The value of the peak sensitivity (at 555 nm for photopic vision) can be found by calculation if the form of the efficiency curve is known: the radiant exitance of the black body at 2045 K is known from Planck's radiation law and its luminance is known by definition—the two together imply a peak sensitivity, measured in W lm^{-1}. At the peaks of luminous efficiency, the efficacy of monochromatic radiation at 555 nm in photopic vision is 673 lm W^{-1} and that at 510 nm in scotopic vision is 1725 lm W^{-1}. For black-body radiators the spectrum is known from Planck's radiation law and therefore the efficacy can be determined. The efficacy of a black body at 2045 K is about 2 lm W^{-1}, and it rises to a peak of about 100 lm W^{-1} at about 7000 K.

Photons

The energy associated with the light is not delivered in a continuous stream, but in discrete 'packets', or *photons*. The amount of energy in a photon depends upon the wavelength of the light, but it is very small and for most practical purposes associated with vision the discrete nature of the energy delivery is not important.

3.5.4.   Limits of vision

Visual acuity

If the angle between two light rays passing through the optical centre is too small, we will not be able to distinguish between them with respect to location. The minimum angle at which resolution is just possible is called the visual angle, and the inverse of the visual angle, measured in minutes of arc, is our visual acuity. The most commonly applied test of visual acuity has been the Snellen chart. One version of this chart consists of a series of 11 rows of letters, progressively smaller from the top. When it is viewed from a distance of 20 ft the letters on the eighth line are just distinguishable by a person of good vision: the distinguishing characteristics of the letters on this line form a visual angle of 1 min of arc at 20 ft. A person with 20/20 vision can read the letters of the eighth line at 20 ft. The lines above the eighth are marked with greater distances, again at which they are just discernible by the person with good vision. A person with 20/40 vision can read at 20 ft the line (in fact the fifth) that with good vision is readable at 40 ft. Note that the visual acuity expressed as a ratio is dimensionless, and the distances could equally well be expressed in metres or any other unit. Many Snellen charts are now marked in metres and 6/6 vision (recorded at 6 m) is the equivalent of 20/20 when measured in feet. There are alternative tests, based for example on patterns of lines or grids of squares. The Snellen chart, in particular, is quite crude because some letters are distinguishable just by their general shape, and are therefore easier to read than others. All tests of this type are very dependent on the lighting conditions, and more sophisticated tests using machines in which the luminance is carefully controlled are available for clinical diagnosis. Nevertheless, the simple tests do give an immediate and inexpensive measure of visual acuity. If the colours on the Snellen chart are reversed, so that the letters are white and the background is black, our measured acuity is very much lower.

Under ideal conditions a person with excellent vision might achieve a visual acuity of two, implying that their visual angle of resolution is 0.5 min. It is interesting to investigate the absolute limits of visual acuity based on the geometry and spatial distribution of cones on the retina. Our most acute vision occurs when light is focused on the fovea. Each of the cones in this region has its own communication line down the optic nerve, and we can distinguish between the activities of two adjacent cones. This tells us the spatial resolution available at the retina, and gives an upper bound on our visual acuity. For the normal eye the optical centre is about 17 mm in front of the retina, and the cones are 2 μm apart. This implies that the maximum visual angle is about 0.4 min, and the upper bound on visual acuity is 2.5. It would seem, then, that those of us with the best visual acuity are limited by the density of the cones.

Given the above calculation, we might ask the question as to whether the human eye is optimally developed with respect to the size and density of cones. In fact, we can show by a simple calculation that the acuity is also at the limit determined by diffraction of light from the aperture in the iris. We would need more modification than just a higher cone density to achieve higher visual acuity.

Our peripheral vision is much less acute than that at the fovea for the reasons already discussed. There is an order of magnitude reduction in visual acuity at $10°$ of arc from the fovea. The brain appears to be able to compensate for this by scanning the scene in front of us and building up a high-resolution image within the brain.

Visual sensitivity

What is the lower threshold of light energy required to stimulate the visual process? This problem was investigated very successfully by Hecht in the middle part of this century. He devised several ingenious and well-controlled experiments to measure the sensitivity of the eye. As discussed earlier, the rods are much more sensitive than the cones. In terms of luminance, the cones do not function below about 0.001 cd m^{-2}, and our vision is entirely dependent on the rods. The optimal sensitivity of the rods is to light at a wavelength of about 510 nm. Hecht directed light at this wavelength at an area of high rod concentration (away from the fovea). He demonstrated that, for a number of observers, the average threshold was about 100 photons arriving at the cornea. He further calculated that only about 48% of these would arrive at the retina: 4% would be reflected at the cornea, 50% of those remaining would be absorbed in the media within the eye. Of the 48% getting through, 20% would be absorbed by the rhodopsin to create a visual stimulus (the remainder would either have been absorbed by the neural components before reaching the photoreceptors or would miss the rods entirely and be absorbed by the black pigment behind). In total then, only about 10% of the light arriving at the retina, or about ten photons, actually generates the visual stimulus. This is a very small amount of energy, and once again it can be demonstrated that little could be done to improve our visual sensitivity without a very major re-design. The background level of stimulus, arising from the statistical distribution of thermal energy within the molecules in the rods, is such that no further reduction in threshold is achievable. The signal-to-noise ratio would be unacceptably high.

A great incentive for the exploration of the limits of vision, with respect to both acuity and sensitivity, was the development of the television. Much of the literature in the middle part of the 20th Century has immediate application in this field.

3.5.5.   Colour vision

It has already been stated that our colour vision is associated with the cones. Colour is a psychophysical property of light, in that it is associated with visual perception. There are two attributes of a light wave that we would expect to govern our perception of it. The first is the wavelength, and the second is the intensity. When we mix light together, we would further expect the spectral composition of the resulting combination to be important. This effectively gives three parameters that we might use to describe a colour. In addition, we might anticipate that the duration of exposure to the light might also be important, and indeed we have already discussed the processes of light- and dark-adaptation, which obviously depend on time of exposure to the stimulus. Putting aside the time element, the remaining three parameters have been represented by Munsell as a double cone, as illustrated in figure 3.23.

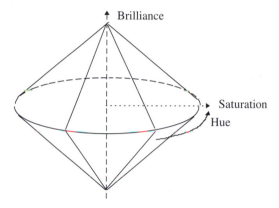

Figure 3.23. *Munsell's colour description in terms of brilliance, saturation and hue.*

The vertical axis represents the intensity of the colour. The circumferential coordinate represents the hue: it is dependent on the wavelength, and is what we would normally think of as the determinant of colour. The horizontal axes define the saturation of the colour, and reflect the spectral composition. At the outer extreme the light is of only one pure colour, and at the inner extreme all wavelengths are present. The vertical axis represents brilliance, which is a property of intensity. At the bottom there is no light, and the colour is black. At the top all wavelengths are present at maximum intensity, and the resulting colour is white. All other points on the vertical axis represent shades of grey. Although this particular representation might be useful to aid our understanding of the qualities of colour, it is less useful in predicting the outcome of combining different colours. The rules of combination are not simple.

Pre-dating the Munsell description is the Young–Helmholtz trichromatic theory. Young originally observed that all colours could be made up from three primary ones: his chosen primaries were red, green and blue. He postulated that the eye contained three different types of nerve receptor, and that the brain made up composite colours by combination of signals. Helmholtz did not initially accept Young's theory, because he was aware that some colours could not apparently be produced from pure monochromatic (single-wavelength) primaries. He later realized that the receptors might not be 'pure' in that they might have overlapping spectral response curves, and that this could explain the discrepancies in experimental results. Essentially the trichromatic theory uses the 'concentration' of three colours as a mathematical basis rather than three parameters such as brilliance, saturation and hue. Simple rules for the combination of colours using addition and subtraction can readily be developed on this basis.

The choice of primaries for a trichromatic combination is arbitrary, and any three 'independent' colours will serve the purpose. Suppose that we take red, green and blue as the primaries. The Young–Helmholtz theory suggests that we can write any colour, C, of intensity c as a linear sum of the three primaries,

$$cC \equiv rR + gG + bB.$$

The intensities of the colours (c, r, g and b) can be measured in any standard photometric units, such as lumens. The total light flux must be the sum of the components, and so $c = r + g + b$.

A standard colour-matching experiment is to project a spot of a colour, cC, at a particular intensity onto a screen, and to focus on one spot next to it three filtered beams producing each of the three primary colours. By adjusting the intensities of the three primaries, we expect to be able to produce a match of the original colour. For many colours this is true, but it turns out that there are some colours that cannot be matched in this way, the basis of Helmholtz's early rejection of Young's trichromatic theory. A saturated blue–green is one example of a colour that cannot be produced by a combination of red, green and blue light. What is possible, however, is to refocus the red beam so that it falls onto the original blue–green spot, and then to match the resulting colour with a combination of blue and green. Although we have been unable to demonstrate that we can satisfy the trichromatic equation as written, we have been able to satisfy the following:

$$cC + rR \equiv gG + bB.$$

In principle the two equations are identical, except that we have to accommodate the notion of a negative coefficient of a colour in the trichromatic equation.

Chromaticity diagrams

Chromaticity diagrams provide a two-dimensional representation of a colour. The parameter that is sacrificed is brightness. It is assumed that the basic determinants of colour are the *relative* intensities of the three chosen primaries. We can write the trichromatic equation in terms of relative intensities if we assume that one unit of colour is produced by a particular relative mixture of the primaries, irrespective of their absolute magnitudes. In this case a relative form of the trichromatic equation can be written;

$$C = \frac{r}{r+g+b}R + \frac{g}{r+g+b}G + \frac{b}{r+g+b}B.$$

Changing each of the intensities by the same factor will produce the same colour. The three coefficients are not independent: any one of them can be obtained by subtracting the sum of the other two from unity. This means that we can choose any pair of the coefficients as the independent variables and represent the colour as a point in a single plane. The resulting graph is called a chromaticity diagram. If we choose red and green primaries as the independents, we obtain the diagram illustrated in figure 3.24.

All colours can be represented on this diagram, each occupying a point in the plane. We should recognize that some colours will not lie within the triangle shown because negative coefficients would be required to produce them.

Chromaticity diagrams have many uses beyond the simple description of a colour space. We have already established that real colours lie within a defined envelope. Straight lines on the chromaticity diagram have special properties: all colours on a straight line can be represented as a linear combination of the two monochromatic colours at its extremes. The monochromatic colours either side of a line through the white spot are complementary (they add together to give white), and we can devise simple experiments to 'see' complementary colours when they are not really there. If we stare at a red object for some time, and then look at a white piece of paper, we see a blue–green image of the object. This fits in with the trichromatic theory, because we would anticipate that the photochemicals associated with red vision have been 'used up', and therefore we will get a higher response to the white light from the photosensors associated with the blue and green reception.

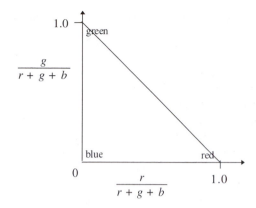

Figure 3.24. *Chromaticity diagram based on red, green and blue primaries.*

The triangular envelope formed by the connection with straight lines of points on a chromaticity diagram defines all of the colours that can be produced using an additive combination of the colours represented by the points. This has immediate application in the fields of television, cinema and colour printing. The choice of primaries determines the range of colours available.

Deficiencies in colour vision can be identified by the pattern of lines connecting different colours that an individual cannot distinguish from one another. Particular deficiencies have characteristic signatures. One example is dichromatic colour blindness, which results from the absence or malfunction of one of the conjectured three types of cone: the 'missing element' is identified from the pattern.

A comprehensive discussion of 'the perception of light and colour' can be found in Padgham and Saunders' book of that title, published by Bell and Sons, London, in 1975.

3.6. PSYCHOPHYSICS

Throughout our discussions of the senses we have had to distinguish between objective, measurable quantities (frequency, concentration) and subjective quantities (loudness, taste, intensity). The investigation of the relationships between objective and subjective measures is the science of psychophysics. An introduction to this topic is presented by Damask (1981). We shall begin by introducing some of the terminology of psychophysics.

- Absolute stimulus threshold

Lower limit: point at which a stimulus can just be detected. Since not every occurrence will be detected, an arbitrary success rate must be established for this threshold. Often the point at which 75% of applied stimuli are detected is taken as the absolute stimulus threshold.

- Terminal stimulus threshold

Upper limit: point at which no further increase can be detected. This might be a point of saturation, or, on a different scale, it might be a point at which a measure of a stimulus exceeds the human range (e.g. 20 kHz frequency of sound).

- Just noticeable differences (JND)

Differential limit: differences that can just be detected.

> *limen*:   unit in psychophysics, equal to the JND at each point on a scale.
> *supraliminal*: detectable change, greater than 1 limen.
> *subliminal*:   undetectable change, less than 1 limen.

The subliminal stimulus has received a good deal of publicity in recent years, particularly with respect to subliminal visual cues in advertising.

3.6.1.   Weber and Fechner laws

In comparing the intensities of two stimuli, the first thing that we might do is try to determine whether they are subjectively the same. We can easily tell the difference between the light from a single candle and that from two, but we would not be able to tell from our perception of light intensity the difference between illumination from 1000 candles or 1001. Similarly, we might readily distinguish between weights of 1 g and 2 g, but not between weights of 1000 g and 1001 g. Clearly the difference in intensity of stimulus that we are just able to perceive, the JND, is a function of the intensity of the stimulus itself. Weber, working in the middle of the last century, observed that the JND is not a function of the arithmetical difference but of the ratio of stimulus magnitudes. He showed experimentally that the JND is, in fact, directly proportional to the intensity of the stimulus. Damask (1981) expresses this observation in terms of our formal definitions: the difference limen is a constant fraction of the stimulus. The basic experimental method required to test Weber's law is to investigate the range of actual stimulus intensity that results in the same perceived intensity. The fixed quantity is therefore the perception, and the actual stimulus is the variable.

Fechner argued that if this proportionality applied to the JND, it should also apply to any other increment of sensation and extended Weber's law to the supraliminal range. He introduced a scaling factor to the JND to represent any increment of sensation. The incremental form of Fechner's law is

$$dS = k \frac{dI}{I}$$

where k is a scaling factor and dS is an increment of perception, or a limen.

Integrating this equation

$$S = k \log I + C.$$

This logarithmic relationship between stimulus and intensity is often called the Weber–Fechner relationship, although strictly speaking it represents an extension of Weber's law beyond its intended range of application.

Without loss of generality, the constant could be written as $(-k \log I_0)$, and the equation is then

$$S = k \log \frac{I}{I_0}.$$

The form of this equation should be familiar: the intensity level of a sound, in decibels, was expressed as the logarithm of the ratio of intensities. The decibel scale was conceived to compress the range of numbers for convenient shorthand reference. There was no suggestion that there was any immediate psychophysical correspondence between the intensity level and the perceived intensity (loudness) of the sound. The Weber–Fechner relationship provides another motivation for representing sounds on the decibel scale.

When we design an experiment to test the validity of these laws we have to be very careful to ask the right questions of the subject. The basis of Fechner's law is that an increment in perception is equal to an increment in magnitude of the sensation. We should therefore construct a scale of perceived intensity and ask the subject to judge where on the scale the presented stimulus lies. We should not make the scale too subtle, or use descriptions that might be confusing. Wright (1964) (see Damask) reports on tests on the sense of smell performed by Fieldner. In these tests the subject was asked to put a smell into one of just six categories: no smell, barely perceptible, distinct, moderate, strong or overpowering. The experimental results, averaged over five or six subjects, showed good agreement with the Weber–Fechner relationship. The concentration of odorant in the air ranged from 0.000 01 to 100 000 parts per million, indicating that Fechner's law was valid over a wide range of stimulus intensity in this experiment.

3.6.2. Power law

The validity of the Weber and Fechner laws was investigated in some detail by Stevens in the late 1950s and early 1960s. He found that although the Weber–Fechner relationship did describe accurately the results of some psychophysical experiments, such as apparent length, it was less accurate for others, such as perceived brightness. He suggested that, rather than fix the perception intensity and vary the applied stimulus intensity until the subject notices a difference, it might be more appropriate to present two stimuli of known intensity and ask the subject to compare the two by assigning a ratio to the perceived intensities. For example, the subject might state that one light was 'twice as bright' as another. Stevens then proposed that the relationship between stimulus intensity and perceived intensity might take the form

$$\frac{dS}{S} = n\frac{dI}{I}.$$

Hence the increment in sensation is assumed to be proportional to the intensity of the sensation. The integrated form of this equation is

$$\log S = n \log I + K.$$

As for the Weber–Fechner relationship, the constant K can be written as a logarithm with no loss of generality, and an alternative expression of Stevens' equation, often described as the power law is:

$$S = \left(\frac{I}{I_0}\right)^n$$

where I_0 is a reference intensity.

Stevens (1961) investigated experimentally a variety of stimuli, and deduced the exponent of the power law for each. A subset of his results is summarized in table 3.5.

Table 3.5. *Exponents in power law of psychophysical response: Stevens (1961).*

Perceived quantity	Exponent	Stimulus
Loudness	0.6	Binaural
Brightness	0.5	Point source
Smell	0.55	Coffee
Taste	1.3	Salt
Temperature	1.0	Cold on arm
Vibration	0.95	60 Hz on finger
Duration	1.1	White noise stimulus
Pressure	1.1	Static force on palm
Heaviness	1.45	Lifted weight
Electric shock	3.5	60 Hz through fingers

3.7. PROBLEMS

3.7.1. *Short questions*

a   What is a Pacinian corpuscle?
b   At what elevated temperature do many thermal pain sensors react?
c   About how many taste buds are there on 1 cm^2 of the tongue?
d   Is sound mainly a longitudinal or transverse vibration?
e   About how long will it take for sound to be transmitted between two people having a conversation?
f   Does the speed of sound increase or decrease with increasing pressure?
g   About what sound intensity in dB will give rise to negative pressures?
h   What is the approximate value of the resonant frequency of the ear canal?
i   Is sound in the air mainly transmitted or reflected at a water surface?
j   At which end of the basilar membrane in the ear do low frequencies dominate?
k   About what wavelength has green light?
l   What component of a Pacinian corpuscle is essential in making it sensitive to the rate at which a pressure is applied?
m   Is the size of a cone in the fovea greater or less than the wavelength of green light?
n   Does the retina have a fairly uniform spatial resolution?
o   For what colour has the eye the best sensitivity?
p   Why do objects appear more green as night falls?
q   What is a Snellen chart and what is it used to test?
r   Will yellow text on a blue background offer good or bad visibility?
s   What type of relationship between stimulus and response does the Weber–Fechner equation predict?
t   Do rods or cones have the higher sensitivity to light and which gives us colour vision?

3.7.2. *Longer questions (answers are given to some of the questions)*

Question 3.7.2.1

When you meet somebody you know in the street you say hello and give his or her name in about 0.5 s. Consider the problems your sensory/central/motor system has in making this response and draw some conclusions about how your brain and nervous system must be organized.

Answer

This is an essay question. Time delays in the eye, optic nerve, air, the ear, the auditory nerve, the brain, the nerves supplying the vocal chords, neuro-muscular delays, sound generation and air transmission should be considered. Approximate values should be given. The bandwidth of the audio and visual signals is also relevant and the conclusion should be drawn that the nervous pathways must be parallel systems if they are to carry the signals in the required time.
      The pattern recognition involved in the brain recognizing a face or a spoken word will take a significant time. Again a parallel solution is required, although just how the brain does the job is not clear.

Question 3.7.2.2

Describe the process by which sound is transmitted from the air to the fluid in the cochlea. Explain the function and effect on the sound of the outer and middle ear. Finally, describe the process by which sound is converted to nerve impulses in the eighth cranial nerve and say what function the inner ear appears to perform in the analysis of sound.

Question 3.7.2.3

We are designing an implanted hearing aid in which a small magnetic mass is to be attached to the oval window of the inner ear. This mass is then made to vibrate by applying an external alternating magnetic field.

Estimate the amplitude of vibration required from the mass when a sound level of 60 dB at 1 kHz is experienced. Assume the ossicles have an effective mass of 100 mg and that only inertial forces need to be considered. Assume also that the eardrum has an area of 60 mm² and that all the incident energy is used in vibrating the ossicles. (SPL in dB is referred to an intensity of 1×10^{-12} W m^{-2}.)

Answer

If we consider a mass m made to vibrate with an amplitude a then we can determine the power required to force this vibration:

$$\text{displacement} = x \sin \omega t$$

$$\text{force } m\ddot{x} = -ma\omega^2 \sin \omega t$$

$$\text{average work done over one cycle} = \frac{1}{2\pi/\omega} \int_0^{2\pi/\omega} ma^2\omega^2 \sin^2 \omega t \, dt$$

$$= \frac{\omega}{4\pi} \int_0^{2\pi/\omega} ma^2\omega^2 (1 - \cos 2\omega t) \, dt = \tfrac{1}{2}ma^2\omega^2$$

$$\text{average power} = \frac{ma^2\omega^3}{4\pi}.$$

Now the incident power for a SPL of 60 dB is given by

$$10 \log \frac{\bar{I}}{I_0} \qquad \text{where} \quad I_0 = 10^{-12} \text{ W m}^2$$

over the 60 mm² area of the eardrum:

$$\text{incident power} = \frac{60}{10^6} 10^{-6} = 60 \text{ pW}.$$

If the whole of this is absorbed in vibrating the ossicles, of mass 100 mg, then

$$60 \times 10^{-12} = \frac{ma^2\omega^3}{4\pi}$$

$$a \cong 5.5 \text{ nm}.$$

Question 3.7.2.4

Estimate the size of the optic nerve. Assume that it is circular and packed with circular fibres that conduct action potentials at 50 m s^{-1}.

Assume that there are 125 million rods and that 100 are connected to a single nerve fibre. There are 7 million cones and a one-to-one correspondence of these to nerve fibres. Use the 'rule of thumb' that nerve conduction velocity in metres per second is given by 5 × diameter (μm).

What factors may have put your estimate in error?

Answer

125 million rods will be connected to 1.25 million nerves and 7 million cones connected to the same number of nerves giving a total of 8.25 million nerves. Each nerve axon will be of diameter of 10 μm. The total area of nerve axons will be $\pi(5 \times 10^{-6})^2 \times 8.25 \times 10^6$ m^2.

Neglecting the packing effect of the circular fibres within the circular cross-section nerve bundle of diameter D,

$$D \cong 30 \text{ mm.}$$

We have overestimated the diameter because we have neglected the packing effect. In addition there will probably be some pre-processing of the signals within the retina so there will be fewer nerve fibres than we have estimated. The assumed velocity and nerve fibre diameter may also have been overestimated.

Question 3.7.2.5

We appear to be able to see a high spatial resolution image across our whole field of vision. How is this possible given the fact that only the macula of the retina has a very high density of receptors?

Question 3.7.2.6

How does the eye adapt to falling light levels and what compromises in performance result?

Question 3.7.2.7

A sound wave incident on a boundary between media is partly transmitted and partly reflected. By consideration of pressure equilibrium at the boundary and conservation of energy, write two equations relating the three pressures in the incident, transmitted and reflected waves. Hence prove the expressions stated in section 3.3. What would be the effect of adding a mass at the boundary and what is the implication concerning the mass of the eardrum?

Answers to short questions

a   A Pacinian corpuscle is an ovoid-shaped mechanoreceptor. They sense touch and are placed in the subcutaneous tissues.

b   Many pain sensors begin to react at temperatures greater than 42 °C.

c   There are about 1000 taste buds in 1 cm^2 of tongue.

d   Sound is mainly a longitudinal vibration.

e   Sound will take about 1/1000th of a second to travel between two people if they are about 30 cm apart.

f   The speed of sound decreases with increasing pressure.

g   186 dB is where the peak pressure is equal to 1 atm.

h   The ear canal is resonant at about 3 kHz.

i   Most of the incident sound in air is reflected at a water boundary because of the mismatch in impedances.

j   Low frequencies dominate at the far end of the basilar membrane, away from the oval window. The high frequencies are absorbed close to the window.

k   Green light has a wavelength of about 500 nm.

l   A viscous fluid held within membranes in a Pacinian corpuscle makes the sensor sensitive to the rate of change of applied pressure.

m   Greater. The cones have a diameter of about 2 μm.

n   No the retina does not have uniform sensitivity. The spatial resolution of the eye is best in the central region—the macula.

o   The eye is most sensitive to green light.

p   The Purkinje effect causes objects to appear greener in falling light levels. Peak retinal sensitivity moves towards shorter wavelengths at low light levels.

q   A Snellen chart contains different sized letters and is used to test visual acuity.

r   Yellow text on a blue background will offer good visibility.

s   The Weber–Fechner equation describes a logarithmic relationship between stimulus and response.

t   Rods have the higher sensitivity; cones are responsible for colour vision.

BIBLIOGRAPHY

Damask 1981 *External Senses* (*Medical Physics* vol 2) (New York: Academic)

Flanagan J L 1972 *Speech Analysis: Synthesis and Perception* (Berlin: Springer)

Keller F J K, Gettys W E G and Skove M J 1993 *Physics, Classical and Modern* 2nd edn (New York: McGraw-Hill)

Marieb E N 1991 *Human Anatomy and Physiology* 3rd edn (Menlo Park, CA: Benjamin Cummings)

Padgham and Saunders 1975 *The Perception of Light and Colour* (London: Bell)

Pickles J O 1988 *An Introduction to the Physiology of Hearing* 2nd edn (New York: Academic)

Stevens S S 1961 The psychophysics of sensory function *Sensory Communication* ed W A Rosenblith (Cambridge, MA: MIT Press)

CHAPTER 4

BIOCOMPATIBILITY AND TISSUE DAMAGE

4.1. INTRODUCTION AND OBJECTIVES

It is impossible to deny the major impact made by the application of physics and engineering to medicine. There can be few of us who will not have experienced this impact directly having had a dental filling or an x-ray for a broken bone. We are likely to know someone who has had a hip joint replacement or an ultrasound scan during the course of pregnancy. Most of these technologies are all well established and the benefits judged to outweigh any potential risks.

However, no treatment is completely without hazard and the balance of risks and benefits can be difficult to assess. It is not just the potential impact of new treatments that must be considered. Our increasing understanding of the interactions between these technologies and the cells, tissues and systems of the human body can lead us to regularly reassess each situation and treatments must be continuously re-evaluated as clinical information accumulates. The impact of some potential hazards are obvious especially where safety-critical systems are concerned (see Chapter 22). Others may be more subtle and longer reaching and may affect tissues at some distance from the treatment site. If a problem occurs it may not be possible to simply remove a device or withdraw a treatment. The concern arising from the potential hazards of mercury poisoning from dental amalgams is a good example of this, as removal of the fillings by drilling will increase the patient's exposure to any toxic effects.

In this chapter we will identify some of the potential biological consequences of these technologies. We will address the following questions:

- How do tissues interact with biomaterials?
- How can the interaction be assessed?
- How can the effects of adverse interactions be minimized?

When you have finished this chapter, you should be aware of:

- The meaning of the term 'biocompatibility'.
- The range of applications of biomaterials within the human body.
- The effects of the biological environment on an implant.
- The potential tissue response to an implant.
- Some of the methods employed in the testing of implant materials

Biocompatibility is not an easy subject to address. It is not even easy to define, but there is no doubt that the subject is important. It is impossible to discuss the subject of biocompatibility without using the language of biology and a reasonable background in biology is assumed for those reading this chapter. In considering tissue damage we can look at the cellular level and the effect of one cell upon another. We can also look at the effects of external energy sources such as electromagnetic radiation and ultrasound. Chapters 7 and 8 deal with ultrasound and non-ionizing radiation, respectively. Ionizing radiation is the subject of Chapter 5. However, we make no attempt to cover the subject of damage at the cellular level from ionizing radiation. Radiobiology is an important subject which we cannot hope to cover in just a few pages. The bibliography at the end of this chapter gives references to some texts on this subject.

4.1.1. Basic cell structure

Cells are the building blocks from which we are made. They are usually transparent when viewed through a light microscope and have dimensions usually in the range 1–100 μm. The study of cells is called cytology and has been made possible by improvements in the light microscope and, more recently, developments such as those of the phase-contrast microscope, the electron microscope and surface-scanning microscopes.

Almost all cells have a surrounding membrane of lipids and proteins that is able to control the passage of materials into and out of the cell. The membrane is about 10 nm (10^{-8} m) thick and its ability to sort out different ions is the basis of all electrophysiological changes. The most obvious structure within the cell is the nucleus. The nucleus contains materials vital to cell division, such as deoxyribonucleic acid (DNA), the genetic material of the cell. The DNA carries and transmits the hereditary information of a species in the form of genes. The process of cell division is called mitosis. The nucleus is surrounded by fluid called cytoplasm within which are other structures such as mitochondria endoplasmic reticulum and the Golgi apparatus. All these structures contribute to the complicated process of cell metabolism.

Atoms can be joined together by a process of electronic interactions. The forces which hold them together are called chemical bonds. A covalent bond is formed when two atoms share a pair of electrons, but two atoms can also share just one electron and they are then held together by electrostatic attraction as an ionic bond. The ionic bond is much weaker than a covalent bond. There are also other weak bonds such as the hydrogen bond and van der Waal bonds, which are common in biological materials.

Chemical bonding is a complicated although well developed subject, but it is easy to appreciate the importance of electrons to biological structures in that they form the 'glue' which holds everything together. Obviously ionizing radiation is going to interfere with these chemical bonds. The energy present in ionizing radiation eventually appears as heat if it is absorbed in tissue. The amount of heat is very small (see section 5.6) and it is not a likely method of tissue damage. It is just possible that local heating, when an ionizing particle is absorbed, may cause damage but the more likely method of damage is to sensitive targets within the structure of biological molecules. It is the process of ionization which causes most damage and it is for this reason that the units of dose measurement (see Chapter 5) are based on the number of electrons released by the radiation.

4.2. BIOMATERIALS AND BIOCOMPATIBILITY

Before considering the use of biomaterials in medicine we first need to agree on what it is we are discussing. When browsing through the wealth of literature written on implant devices and artificial organs you will find a variety of definitions to describe a biomaterial. There is no single 'correct' interpretation and many reflect the author's interest in a particular biomaterial application. Jonathan Black (see Black 1992) has coined an appropriate generic definition which fits within the broad overview covered in this current chapter. He states that:

> Biomaterials are materials of natural or man-made origin that are used to directly supplement or replace the functions of living tissue.

A biomaterial which is able to exist within the physiological environment without any significant adverse effect to the body, or significant adverse effect on the material is said to be 'biocompatible'. As you will see later, there is a wide spectrum of potential degrees of interaction and this definition can be interpreted in many ways.

4.2.1.   Uses of biomaterials

The use of man-made materials to replace diseased or damaged tissues and organs has become an increasingly important area in current medical practice (see table 22.1). Whilst the success rates of procedures involving the use of biomaterials have increased, we cannot congratulate ourselves on the introduction of a novel idea as the history of biomaterials goes back hundreds of years. The use of metal for the fixation of broken bones was first reported in 1775. Two French physicians used metal wire to fix a fractured humerus. Unfortunately, the wire was not sterile and the patient died due to infection (Guthrie 1903).

The majority of developments have taken place during the last 30 years and the study of biomaterials is now a discipline in its own right. Implant design is not simply the domain of the engineer but requires input from chemists, biologists and physicists. We have now reached the stage when there are few parts of the human body which cannot be at least partially replaced by man-made devices (see figure 4.1).

Figure 4.1. *Pacemaker, bioprosthetic heart valve, hip and knee joint replacements that were not available in the time of Abraham Lincoln.*

In addition to implantable devices there are a number of extracorporeal systems which come into contact with the patient's blood, these include the artificial kidney (haemodialyser) and the blood oxygenator (see sections 22.3.3 and 22.3.2, respectively).

Some devices such as fracture fixation plates are intended for short-term use and are usually removed once the broken bone has healed. Other implants, for example heart valves (see section 22.3.1), vascular grafts and prosthetic joints, have been developed for permanent use.

Our expectations are increasing in the case of permanent implants. At one time, it was sufficient that the implant should alleviate pain and restore some function for a few years. We now expect 20 years or more of functional reliability. To avoid premature failure of a device we need to manufacture them from durable materials which elicit an appropriate host response.

4.2.2. Selection of materials

The range of different materials needed, with the correct chemical and physical properties to produce suitable artificial replacements, is enormous. It ranges from the polymers required to make tubes for use as blood vessels, tear ducts or an oesophagus, to the flexible materials with high tensile strength required for tendon replacements and to the porous resorbable materials for use as bone substitutes.

The selection of a suitable material for a particular application is of paramount importance. In making this choice we must consider the physical and chemical properties of the material and have an understanding of the way in which these may change once the implant is introduced into the biological environment. At first, it was believed that an ideal biomaterial was one which remained chemically inert. We now know that this is often inappropriate. Modern biomaterials are often developed with the intention of generating an 'appropriate' host response. In this context, control of the interface between the biomaterial and the patient's natural tissue is extremely important.

Many factors must be considered when choosing a biomaterial. Implants which are subject to repeated loading must be manufactured from materials which have adequate mechanical strength in terms of static and fatigue loads (see section 1.2.1). In order to establish this we must have an understanding of the loads, cycle rates and number of loading cycles to be encountered. For a cardiac prosthesis the number of cycles may be relatively easy to estimate but the loads acting on the individual components of the valve are difficult to determine. For a hip joint it can be very difficult to estimate the number of cycles but the loads are easier to estimate.

The complexity of this situation can be illustrated with reference to the anterior cruciate ligament in the knee. This ligament is situated within the knee joint where it is attached to both the femur (thigh bone) and the tibia (shin bone). It is one of the important ligaments which maintain the stability of the knee. Replacement of this ligament is often carried out as a result of a sports injury. The patient is likely to be relatively young with an ideal requirement that the implant should function satisfactorily for the rest of their life. The consequences of this are shown in table 4.1, which assumes a reasonable level of activity. If the patient was an enthusiastic sportsperson the demands may be considerably higher.

In some cases properties may be chosen which, on first consideration, may seem less than ideal. It seems logical to produce a hip joint prosthesis which is capable of withstanding the highest possible loads without fracturing. In practice, this may pose a problem for the clinician as, if the joint is subject to an unusually high load, as might be the case during a fall for example, the bone may fracture around the implant and complex surgery may be required to repair the damage. In this context there is an added complication as many hip joint prostheses are used to repair fractured hips in elderly ladies who have poor quality bone due to osteoporosis. In this case the bone may be too weak to support the prosthesis. Other physical properties to consider include stiffness, hardness, corrosion resistance and wear resistance.

Any material which is to be implanted must be sterilized before use. Common methods of sterilization include: chemical sterilization, dry heat, steam sterilization, gas (ethylene oxide) and gamma irradiation. The

Table 4.1. *Requirement for an anterior cruciate ligament replacement. (Adapted from Chen and Black 1980).*

Anterior cruciate ligament replacement
Permanent replacement
Post-traumatic replacement: patient age 35–48 yrs
Estimated life expectancy: 40 yrs
Mechanical conditions: strain (range of maximum): 5–10%
loads: moderate activity level, including recreational jogging

Activity		Peak load (N)	Cycles/year	Total cycles
Stairs:	ascending	67	4.2×10^4	1.7×10^6
	descending	133	3.5×10^4	1.4×10^6
Ramp walking:	ascending	107	3.7×10^3	1.5×10^5
	descending	485	3.7×10^3	1.5×10^5
Sitting and arising		173	7.6×10^4	3.0×10^6
Undefined		<210	9.1×10^5	3.6×10^7
Level walking		210	2.5×10^6	1.0×10^8
Jogging		630	6.4×10^5	2.6×10^7
Jolting		700	1.8×10^3	7.3×10^5
Totals			4.2×10^6	2.9×10^8

process specified for a particular material must be chosen with care as the method may affect the properties of the material.

4.2.3. Types of biomaterials and their properties

Initially, relatively few materials were used and the artificial parts were of relatively simple designs. For example, stainless steel was widely used. In the intervening years there has been an exponential increase in both the number of clinical applications and the range of materials used.

A material must be chosen for a specific use on the basis of its properties. These properties, whether mechanical, physical or chemical in nature, are related to the inherent structure of the material. Materials are held together by interatomic bonds. These may be, in order of strength, ionic, metallic, covalent or van der Waals bonds.

There are three major categories of biomaterial; metals, polymers and ceramics/carbons. We will consider these in turn.

Metals

These have a regular arrangement of atoms forming a crystalline structure. Interatomic bonds are formed due to the electrostatic attraction of free electron clouds. These bonds are strong but lack directionality.

Molten metals solidify around several foci resulting in the formation of grains. Grains are highly ordered regions with almost exclusively metallic bonding. Grain boundaries contain impurities and are sites of inherent weakness. Strength is inversely related to grain size; the smaller the grain sizes, the stronger the material. The method of preparation of the material affects the properties of the finished metal. Cast metals have larger grains and therefore tend to be weaker. In wrought metals the grain structure is deformed. This gives increased strength.

Alloys are often used in preference to pure metals for the manufacture of medical devices. Formation of an alloy involves the addition of one or more elements to the parent metal. This leads to improved mechanical properties.

Complex geometry and anatomical restrictions which limit the maximum size of an implant may create regions of high stress in the finished device. In orthopaedic implants, the elastic moduli of bone and the implant may vary by an order of magnitude, with the result that most of the mechanical load is carried by the implant. For this reason, high strength materials are needed.

The strength of the material must be maintained in an extremely hostile environment. Exposure of a metallic implant to the biological environment can be likened to exposure of a car body to seawater. Resistance to corrosion in an aqueous chloride-containing environment is an essential requirement for metallic implants.

Types and physical properties of metals employed

The primary use of metals is in the manufacture of orthopaedic implants. Stainless steels were among the first types of metals to be used clinically. Corrosion resistance can be improved by the addition of molybdenum. Currently, stainless steel or titanium alloys are commonly used for orthopaedic purposes. Cobalt–chromium, cobalt–chromium–molybdenum and cobalt–nickel–chromium–molybdenum alloys are also used.

All metallic implant materials are denser, stronger and have higher elastic moduli than bone (see section 1.2.2). The elastic modulus of compact human cortical bone is 14 000–21 000 MPa which is 10 times less than that of stainless steel and five times less than that of the Ti alloys. Titanium and titanium alloys also have the advantage of lower densities when compared with stainless steels. This is coupled with high strength and a very low rate of corrosion. The lower density reduces the weight of the implant and hence a reduced awareness of the implant by the patient. The lower modulus of elasticity gives a 'springier' implant and reduces stresses around the implant, as it is more able to flex with the bone. This is very important to the well-being of the surrounding bone. Bone needs to be exposed to mechanical stress for the maintenance of bone density. In situations where most of the load is taken by an implant, the surrounding bone is resorbed and loosening of the implant may result.

We have yet to develop a metal implant which has properties which are anything like the bone with which it interacts. Under physiological loading conditions bone behaves as a composite material, whilst metals behave as simple elastic materials. In common with all biological tissues, bone is a viscoelastic material, thus we must model its mechanical behaviour in terms of combinations of elastic and viscous elements (see Chapter 1, section 1.2.4).

Polymers

Polymers are long-chain high-molecular-weight materials formed from repeating units or monomers. These are joined by covalent bonds which, unlike metallic bonds, are highly directional but not very strong. Strength can be enhanced by covalent bonds formed between chains, ionic bonds between charged side groups and by van der Waals bonds (the diffuse attraction between hydroxide groups (OH^-) and hydrogen, oxygen and nitrogen atoms).

The properties of polymers can be tailored to suit a specific purpose by a number of different means:

- by combining different monomers to form copolymers;
- by controlling the extent of polymerization;
- by controlling the extent of cross-linking (ionic bonds) between adjacent polymeric chains;
- by incorporation of chemical additives (these lubricate movement between polymer chains and improve flexibility).

Conduits manufactured from woven or knitted polymers such as Dacron and Teflon have been used as replacement blood vessels since the 1950s. Other types of polymer are used in many different procedures including use as bulk implants to fill spaces in plastic surgery and use as tubes and fibres in tendon replacement. Porous polymers have been used to facilitate tissue ingrowth. Polymers are made in the form of porous solids by the addition of foaming agents during polymerization or as porous fabrics made by felting or weaving. Porous materials of these types are used for: vascular grafts, septal defect correction and sutures. Typical materials include: Teflon, Orlon, silk and polypropylene.

In terms of physical properties polymers are viscoelastic (see section 1.2.4), where deformation is both time and temperature dependent.

Biomedical polymers can be classified as either elastomers or plastics. *Elastomers* are long-chain molecules and, as such, can withstand large deformations. When unloaded they return to their original dimensions, e.g. Silastic (silicone rubber), Esthane (polyurethane). *Plastics* are more rigid. They are sub-divided into two types: thermoplastic and thermosetting plastics. Thermoplastics can be reheated and remodelled like wax. Teflon (PTFE), PVC, cellophane, cellulose acetate and nylon are all examples of this group. *Thermosetting plastics* cannot be re-used, as the chemical reactions occurring as they form are irreversible. Epoxy resins are examples of thermosetting plastics. These are used as encapsulants for implantable electronic devices such as pacemakers.

Specific polymers used in medical devices

Some examples of polymers commonly used as biomaterials are given below.

Polyethylene. Polyethylene is available with a range of densities. The tensile strength, hardness, and chemical resistance of the material increases with density. Ultra-high molecular weight (UHMW) and ultra-high molecular weight high-density (UHMWHD) polyethylenes (molecular weights in the range of $1.5–2 \times 10^6$) and are used for acetabular cups in artificial hips or for the bearing surfaces of artificial knee prostheses. The material can be machined or moulded.

Polymethylmethacrylate (PMMA). This is used for a diverse range of applications from bone cement used to fix joint prostheses to bone, to dialyser membranes and implantable lenses.

Polyethyleneterephthalate or Dacron. This is used, as a machinable solid, to manufacture supporting frames for heart valve prostheses and as a fibre. It can be woven or knitted for use as blood vessels and frame covers.

Silicone rubber. This is found in mammary prostheses, finger joint prostheses and catheters.

Hydrogels. These are hydrophilic polymers (have an affinity for water). Water exists in at least three structural forms dependent on the concentration of water in the hydrogel. Biocompatibility of the material is influenced by the form predominating. Hydrogels are now widely used as electrodes for ECG/EKG and other electrophysiological measurements (see section 9.2.4 on electrodes).

Polyhydroxyethyl methacrylate (PHEMA). When dry this material is rigid but if immersed in an aqueous solution it absorbs water to form an elastic gel. Up to 90% of the weight of the hydrated polymer may be water, depending on the fabrication techniques used. This material is transparent and very pliable when wet, but easily machined when dry. These properties make it an ideal material for the manufacture of contact lenses.

Many hydrogels are weak mechanically. For this reason they are often grafted onto tougher materials such as silicone rubber, polyurethane and PMMA. A number of different surface grafting techniques may be used, e.g. chemical initiation or gamma radiation. Other examples of hydrogels include polyvinyl alcohol (PVA). PVA is an excellent candidate for synthetic articular cartilage because of its hydrophilic properties, high tensile strength, wear resistance and permeability.

Ceramics

Most ceramics are ionically bound structures comprising one or more metallic elements with one or more non-metallic elements. The exceptions are the covalently bound materials such as pyrolytic carbon.

Ceramics have a high-melting point and are hard, rigid and have a high compressive strength. A major drawback is their brittle nature. This can be reduced by increasing the purity and the density of the material or by reducing the grain size. They are used in joint prostheses, as bone substitutes and in dentistry. Resorbable ceramics are designed to be slowly replaced by bone. They are usually porous and are able to act as a temporary scaffold for bony ingrowth.

The use of carbon in the manufacture of medical devices is becoming increasingly common. A number of different types are available:

- Vitreous carbon: this can be moulded, enabling complex shapes to be made.
- Pyrolytic carbon: this is normally coated onto a graphite substrate but the technology has been developed enabling blocks of pyrolytic carbon to be produced. These can then be machined.
- Diamond-like carbons: these are relatively new materials, developed for use as coatings on metallic or polymeric substrates.
- Carbon fibres: these have a high tensile strength and can be incorporated into polymers to form carbon-fibre-reinforced materials.

Carbon-based materials are strong and are isotropic. They also have the advantage of low thrombogenicity. Specific types can be manufactured which have values of elastic modulus and density similar to those of bone. They have become quite widely used. For example, carbon-fibre-reinforced PTFE is used as artificial articular cartilage. Polysulphone carbon fibre is used for the manufacture of bone plates. Unlike metal bone plates, these have a similar stiffness to the natural bone. Pyrolytic carbon is used in the manufacture of heart valve prostheses (see Chapter 22).

4.3. MATERIAL RESPONSE TO THE BIOLOGICAL ENVIRONMENT

The properties of a material generally relate to the behaviour of that material when subjected to various types of energy, such as mechanical, thermal, electromagnetic and chemical. If one part of a structure displays a weakness with respect to the applied energy, then the material will be susceptible to deterioration or failure. Changes in the material are likely to occur after implantation. These may be beneficial or they may lead to failure of the device. The response of a material may also vary with the site of implantation. For example, pH will vary with anatomical site as illustrated in table 4.3. Other variables that may affect an implant include pO_2, pCO_2 and temperature.

Table 4.2. *pH at some sites in the body.*

Tissue component	pH
Gastric contents	1.0
Urine	4.6–6.0
Intracellular fluid	6.8
Interstitial fluid	7.0
Venous blood	7.1
Arterial blood	7.4

Toxicity is an important factor. The possible toxicity of a biomaterial is a major consideration. For example, it is essential that manufacturing procedures ensure even and complete polymerization of polymeric materials. The presence of residual monomers due to incomplete polymerization may act as an irritant. Polymerization may be complete on the surface of the material, whilst monomers remain more deeply within the bulk of the material; subsequent machining may expose unpolymerized monomers. In the case of metallic implants release of corrosion products in the form of metal ions may affect cell metabolism

4.3.1. Metals

From the engineering point of view metallic implants may fail in two general ways:

- they may corrode, or
- fracture.

As we will see these two factors are not mutually exclusive and corrosion frequently leads to fracture of a device. In addition, metallic implants or appliances may fail due to wear, yielding, loosening, biological incompatibility or infection.

Metallic corrosion

Early implants were associated with chronic inflammatory reactions due to grossly corroded ferrous alloys. Even mild corrosion may necessitate removal of the implant. Symptoms range from local tenderness, to acute pain, reddening and swelling, indicating that the tissue is reacting to the implant.

Metals are inherently susceptible to corrosion since corrosion products represent lower-energy states. Reactions between metals and aqueous environments are electrochemical in nature, involving the movement of metal ions and electrons. The oxidation of a metal (acting as an anode), requires an equivalent cathodic reaction. For implants the main cathodic reaction is the reduction of dissolved oxygen to form hydroxide ions,

$$O_2 + 2H_2O + 4e^-_{metal} \gg 4OH^-.$$

Wounds and crevices around implants have very low oxygen concentrations. In this environment, reduction of water may occur,

$$2H_2O + 2e^-_{metal} \gg H_2 + 2OH^-.$$

For metals there is a potential difference between the metal and a solution which contains only metallic ions in equilibrium with hydrogen. The electrochemical series lists the normal electrode potentials of the elemental metals referenced to a standard hydrogen electrode (see section 9.2.2). The metals with the most positive potential are the noble metals and these are the least reactive.

The potentials obtained in sea water provide valuable data (see table 4.3). Sea water contains many of the ions present in tissue fluids, therefore the order is the same as would be expected *in vivo*. Metals high in the electrochemical series tend to ionize readily, causing an electrical double layer with a negative charge. The nobler metals do not ionize as easily and may carry a positive surface charge. The galvanic series in sea water can be used to predict qualitatively how two metals will affect each other when in contact in the physiological environment.

This is important when designing multi-component devices where the use of more than one metal may be a problem if electromotive potential differences exist. The worst condition occurs when the cathodic metal is large (in the case of a bone plate, for example) and the anodic metal is small (in this case, the screw).

A common type of corrosion is crevice corrosion. The gaps which are found between orthopaedic plates and screw heads are ideal areas for crevice corrosion to occur, particularly in anaerobic conditions. The rate of corrosion is enhanced when the oxygen concentration is low.

Table 4.3. *Galvanic series of metals and alloys in sea water.*

WORST

    Magnesium
    Zinc
    Low-alloy steel
    Stainless steel
    Copper
    Nickel
    Silver
    Stainless steel, passivated[†]
    Titanium

BEST

[†] Passivation involves immersion of the metal in an oxidizing agent, e.g. nitric acid. This thickens the oxide layer on the surface of the metal.

Implant conditions remain the dominant factors. In the hostile physiological environment there are many ions present and the metal surface is free to form passivating oxide films. Present-day materials owe their corrosion resistance to the formation of oxides or hydroxides on the surface, which prevent further chemical reactions. For example, stainless steel has a surface film of chromium oxide. The corrosion resistance of titanium is due to a tightly adhering layer of titanium oxide. For the film to be effective a continuous layer must be maintained.

A special type of equilibrium diagram has been developed to demonstrate how metals behave under various conditions of pH and potential. These are called Pourbaix diagrams (see Pourbaix 1966) and one for chromium is shown in figure 4.2. Pourbaix diagrams show which compounds are stable at particular values of pH and potential. In practice, metals are selected from those which naturally have a minimal reactivity or those in which the tendency is suppressed by the formation of an oxide film. The upper line marked O_2 in figure 4.2 is that for oxygen and the lower line marked H_2 is for hydrogen. The region between the lines is that where water is stable. Above the line oxygen is released and below the line hydrogen is released. However, if water is replaced by 1N hydrochloric acid then the passive region in the middle shrinks radically as shown. The chloride anions can react with the metal ions to raise the effective solubility of chromium. The diagram shows that pure chromium would be stable in neutral conditions such as found in the bile duct or urinary tract, but would be unsatisfactory in the stomach where the pH could be low. Pourbaix diagrams are a useful way of predicting corrosion in a general way, but they have many limitations.

Fracture

Fracture of a metallic implant is almost always due to fatigue. This is caused by repeated loading at a level for which failure would not normally occur under static loading. It is a particular problem for devices which are subjected to repeated heavy loads. All metals suffer from fatigue.

The fatigue strength of a metallic component depends on a number of factors which include:

- the type of loading;
- the presence of surface defects or stress concentrations;
- the material structure;
- the presence of defects such as inclusions in the material;

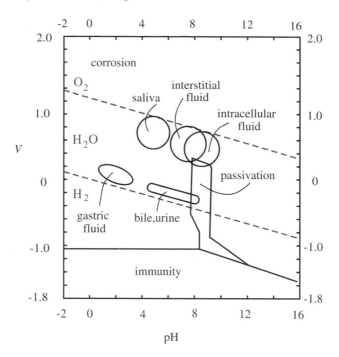

Figure 4.2. *A Pourbaix diagram for chromium in water (1N Cl⁻). (Redrawn from Black 1992 Fundamentals of Biocompatibility (New York: Dekker).)*

- coarse grain size;
- environmental factors.

Fatigue may originate at sites of stress concentration such as a notch or hole. This may be due to a poor design including sharp re-entrant corners, for example. Unlike metals, bone is not very susceptible to fatigue failure, probably because of the constant turnover of bone components.

Fatigue life is reduced by the presence of a corrosive environment—'environmental stress corrosion'—a small crack propagates more quickly due to a form of crevice corrosion at the tip of the crack. Thus, for a long life the implant should have no stress concentrations, it should be manufactured with a good surface finish and it should have good corrosion resistance.

Yield wear and loosening

A material implant may undergo permanent deformation if the applied load exceeds the yield point. This may or may not be important. In the case of orthopaedic implants deformation may be preferable to fracture of the device. Wear is not usually important in terms of failure but wear particles may cause local tissue reactions. There is also experimental evidence to show that cobalt–chromium particles may induce malignant tumours.

Loosening can be very important in the case of orthopaedic joint prostheses. Areas of high stress lead to bone resorption and areas of low stress lead to bone atrophy. The correct stress levels must be maintained for proper bone development and remodelling to occur. Loosening may occur due to improper stress distribution and may lead to pain and the need for removal of the implant. This is not really a metal problem—in the case of joint prostheses loosening is more often due to failure of the supporting cement. Loosening was more common in older designs of prostheses which had metal-on-metal bearing surfaces. These devices gave rise to higher frictional torque than those with metal/plastic contact areas.

4.3.2. Polymers and ceramics

In contrast to metals, polymeric materials do not have a natural tendency to degrade under physiological conditions. Degradation can occur but usually requires UV light, heat and ionizing radiation. As these are not normally encountered in the body, the only factor of any significance is hydrolysis. This can be very important. Polymers can be categorized in order of inertness ranging from hydrophobic, where the material is essentially non-hydrolysable, to hydrophilic where the material is hydrolysable.

Hydrolysis causes fragmentation and crazing in nylon, polyesters and some polyurethanes. Nylon is easily degraded losing 40% of its original tensile strength after 17 months implantation and 80% after 3 years. Degradation can be random, where large fragments are produced, or systematic where individual monomers are detached.

Polymers 'age' if they are not in thermodynamic equilibrium and they can change their molecular order with time. The chemical and physical properties of polymeric materials are derived mainly from the nature of the monomer and the extent of cross-linking between chains. Chemical stability depends on the strength of the chemical bonds in the molecule and their availability to the surface. Steric effects may give some protection. Physical factors such as the degree of crystallization may also have an effect; the more crystalline the polymer, the less likely it is to swell and the less susceptible it is to degradation.

The physical properties of polymers may change dramatically when the material is exposed to the biological environment. The nature of these changes will depend on the molecular make up of the material in question. For example, in some situations, ingress of biological fluid may have a plasticizing effect, reducing the elastic modulus of the material. In other cases leaching of inbuilt plasticizers may increase the material stiffness. Like metals, polymers also suffer from fatigue failure. Absorption of blood lipids proved a problem with early models of heart valve with silastic poppets. With time, this caused swelling of the poppet and changes in material properties, leading to crazing of the surface and finally fragmentation.

Some polymers are designed to be degradable, or to allow for leaching of chemicals, but this must progress at a controlled rate. *Biodegradation* is the gradual breakdown of a material mediated by specific biological activity. Resorbable polymers are soluble implant materials which dissolve in the body. They have four main applications and are used as adhesives, sutures, as drug delivery devices and as resorbable implants which serve as temporary scaffolding. Degradation occurs by a number of processes including dissolution, hydrolysis and enzymatic degradation. Obviously degradation products must be non-toxic. Ideally the breakdown products should be naturally occurring compounds.

Some examples of biodegradable materials are:

Polyethylene oxide/polyethylene tetraphalate (PEO/PET). These are used for temporary mechanical support. Once implanted they undergo hydrolysis with a predictable rate of degradation. Whilst breakdown products are not naturally occurring compounds they do not appear to be toxic.

Polyglycolic acid (PGA) and polylactic acid (PLA). These degrade to give breakdown products which are found naturally in tissues. PLA has been used for resorbable bone plates.

Sutures. Are an important application of polymeric materials. Natural materials such as silk and cat-gut are rapidly resorbed. PGA/PLA copolymer has also been used for resorbable sutures. It is important that the rate of resorption be predictable as the continued presence of a suture is no indication of its mechanical strength. Studies indicate that the effective mechanical properties are lost long before resorption occurs as indicated by weight loss or radioactive labelling.

Tissue adhesives. May be preferable to sutures for some applications. In the case of plastic surgery, scarring may be reduced. Cyanoacrylates are used in this context.

Drug delivery applications. Polymers are ideal for localized drug delivery. The carrier material is engineered in such a way that the drug is released at a controlled dosage over long time periods.

Ceramics

There are three types of bioceramics, classified with respect to their degree of interaction with the body:

- Nearly inert (for example, alumina and pyrolytic carbon).
- Totally resorbable (for example, calcium phosphate).
- Controlled surface activity, bind to tissues (for example, Bioglass, which comprises 45% silicon oxide, 25.5% sodium oxide, 6% phosphorus pentoxide—the latter component encourages new bone formation at the implant/bone interface).

Carbons

Here the effect of exposure to the biological environment depends on the form of carbon used. Pyrolytic carbons are not believed to suffer from fatigue failure and have excellent resistance to chemical attack. The results for carbon fibre tendon replacements, however, are less encouraging with fatigue failure reported to be a significant problem.

4.4.   TISSUE RESPONSE TO THE BIOMATERIAL

Most of our current understanding of the interactions between a biomaterial and the tissue has been obtained by a trial and error approach. The biological response induced in tissues, blood or body fluids is obviously of primary concern. Biological responses include changes in structure, cell metabolism and development at the cellular level. Tissues at some distance from the implant may also be affected.

The adjective 'biocompatible' should be used with caution. No material can be considered universally biocompatible for what is potentially an acceptable performance in one situation may be unacceptable in another. In a simplistic way it is logical to assume that the most desirable state of biomaterial interaction with tissue is a minimal response from the tissues. Materials which elicit such a minimal response have been described as 'biocompatible'. However, this term is inappropriate when used without reference to a specific application.

There are numerous factors which influence biocompatibility. These are not always related to the type of biomaterial, as a material may display 'poor' biocompatibility in one situation but not in another. Therefore a material cannot be described as biocompatible without precise reference to the conditions in which it is employed.

Biocompatibility is a two-way interaction and we need to consider both the effect of the biological environment on the material (4.3) together with the effect of the presence of the material and any break-down products on the surrounding tissues. In the case of the effects of materials on the tissue, we need to consider both local effects and possible systemic effects. The features of the local response can be determined by techniques of microscopy and analytical biochemistry, however the possibility of remote or systemic effects is more difficult to investigate. Systemic effects can either be acute, for example a dramatic loss of blood pressure may occur as bone cement is inserted when implanting a hip joint, or chronic, for example from the slow accumulation of implant-derived products in some distant organ. This latter phenomenon may be very important. Distribution and storage of metal ions or low-molecular-weight organic compounds can give rise to immunological effects and tumour formation. Often it is difficult to positively link such effects with the implant, which can be located at a different site from that where the symptoms occur. One example of this is the controversy surrounding breast implants. The use of silicone implants have been linked with the onset, some years later, of autoimmune disease. As yet there is no conclusive evidence to support a link. However, a number of research programmes are investigating this potential problem.

4.4.1. The local tissue response

The local response to an implanted material can be thought of in terms of the normal processes of wound healing. Changes to this process are dependent on the nature and toxicity of the material and the extent to which it is affected by the physiological environment.

Immediately after insertion in the tissue a cellular response will be elicited by the trauma. This involves an acute inflammatory response. If there is relatively little chemical interaction between the implant and the tissue, the inflammatory response will rapidly subside. Macrophages will clear up dead cells and other debris and the normal wound healing process will follow.

If there is an extensive interaction, a chronic inflammatory state will develop. Cells are constantly recruited from the bloodstream to the site of the implant. Release of degradative enzymes from the cells leads to the creation of an extremely hostile local environment. This often precipitates further release of degradation products which in turn stimulates further enzyme release. This reaction may result in oedema, pain and, in severe cases involving death of the surrounding tissues, a lack of a supporting framework leading to loosening of the implant. A wide spectrum of activity may be observed between these two extremes.

4.4.2. Immunological effects

Tissues may react to 'foreign' materials by an immune reaction. This is an extremely complex process and is covered only briefly. The foreign material or *antigen* can stimulate a number of systems to give a variety of responses in an attempt to eliminate the antigen. Substances which are not naturally antigenic can become so if linked to a suitable protein.

Soluble products diffuse from the implant into the blood or tissue fluids to produce systemic effects. These products can be recognized by small lymphocytes (B-cells). This results in the production of plasma cells which secrete immunoglobins (antibodies). If the antigen is complexed with a specific antibody it is cleared from the circulating system by macrophages.

A second mechanism acts against particulate antigens allowing macrophages to ingest and destroy the particulate debris.

4.4.3. Carcinogenicity

In 1941 Turner discovered that a disc of Bakelite (phenol formaldehyde) implanted in a rat for 2 years induced, or appeared to induce, the formation of a tumour. Repeated experiments showed that almost 50% of similar animals also developed tumours around such discs within 2 years. During the last 50 years, numerous studies have been performed to investigate the incidence of this so-called solid-state carcinogenesis and the mechanisms by which it occurs.

The possibility that implants can give rise to tumours is clearly very significant. The low incidence of tumours reported clinically indicates that the phenomenon may be species specific.

4.4.4.   Biomechanical compatibility

Although the chemical (or biochemical, or electrochemical) interaction between a material and tissues is the dominant factor controlling biocompatibility, physical and mechanical factors are also known to be important. These factors include:

- The effect of implant shape. The shape of an implant can influence the tissue response. For example, the reaction to triangular cross-sections is found to be greater than that to circular cross-sections. This is not surprising, as sharp corners will produce high stresses.
- The effect of surface finish. Surface morphology has been shown to have a significant influence on the tissue response.
- The role of surface porosity in tissue ingrowth. This has been investigated widely and the conditions under which different types of tissue can be encouraged to grow into porous surfaces have been established.

4.5.   ASSESSMENT OF BIOCOMPATIBILITY

Many different approaches are possible to the assessment of biocompatibility. Each has advantages and disadvantages. Brown (1988), when reviewing the corrosion and wear of biomaterials, concludes by saying that 'If anything can be concluded from this review of corrosion and wear of implant materials, it is that there are many different test methods and different results. Each investigator has a rationale for the method used and the interpretation of the results.'

4.5.1.   In vitro models

In vitro models are essential for the initial screening of biocompatibility when large numbers of materials need to be investigated. A number of *in vitro* methods exist, designed to suit the needs of particular types of materials and applications. The advantages of a controlled system of study are numerous, including rapidity of assessment, a high level of sensitivity to toxic substances and the availability of human blood for study.

Haemocompatibility

A wide range of different tests has been proposed for the evaluation of blood compatibility. It is not possible to consider all of these here. Further details are available in more specialized texts (see Silver and Doillon 1989).

A number of simple tests can be carried out to obtain an initial prediction of material suitability. These involve placing a known volume of whole blood on the surface of the specimen. Clotting usually occurs within a short period of time (1 s to 1 min). Clotting time, weight of clot, number of adherent platelets and a range of clotting factors are measured. Reduction or extension of the normal clotting time indicates activation or deactivation of the clotting system by the material. A prolonged clotting time with exposure to the biomaterial in itself is not necessarily a problem as long as the clotting process is not irreversibly inactivated or the clotting factors removed by contact with the biomaterial under test. This can be tested by exposing blood to the biomaterial followed by re-exposure to a material known to activate the clotting process, for example glass. If a clot forms on the glass within 3–5 min the biomaterial is deemed blood compatible.

Protein absorption can also be used to give an indication of blood compatibility. Analysis of the deposition of plasma proteins can be carried out using ^{125}I labelled proteins (albumin, fibrinogen or gamma globulins). Preferential deposition of albumin is thought to decrease platelet adhesion and activation.

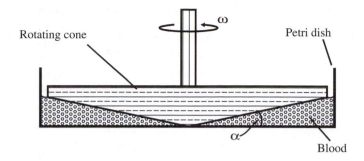

Figure 4.3. *Evaluation of blood under controlled shear forces. Diagram of a cone–plate viscometer employing a standard Petri dish. A cone of shallow angle (α between 0.5° and 5°) in contact at its tip with the centre of the dish rotates at ω rad s^{-1}.*

Rheological test chambers use a spinning disc (figure 4.3) to evaluate the compatibility of the material surface with blood under a range of controlled shear rate conditions.

It is essential that all tests are carried out in parallel with suitable control materials with known haemocompatibility properties.

Toxicity tests

These can be carried out using cell cultures. Cells such as mouse connective tissue cells (fibroblasts) can be grown under closely maintained conditions in a Petri dish containing the test material. Toxicity is estimated by the degree of inhibition of cell growth or by measuring cell death. In the case of polymers toxicity may be due to low molecular weight additives (for example, plasticizers) or residual monomers leaching from the material. To test this the bulk material is incubated in a polar (water or physiological saline) or a non-polar solvent (polyethylene glycol 400) and the extract tested for cytotoxity.

4.5.2.   In vivo models and clinical trials

Materials which are shown to behave favourably in *in vitro* tests for haemocompatibility are fully evaluated in an *in vivo* model. The use of animal models is complicated by differences in clotting properties between the human and animal systems. A typical test is the vena cava ring test. A ring lined with the test material is introduced into the vena cava of a dog for a fixed time period (usually up to 2 weeks). The ring is then removed and the deposition of clot determined. This involves weighing the clot or estimating the area of the vessel lumen which remains open to flow. Clot deposition on the surface of the material is not the only consideration. If the clot is poorly adherent to the material surface or if the deposit is particularly friable, fragments may become detached. The degree of fragmentation (embolization) is then measured by determining the amount of clot which is filtered out by the smaller blood vessels downstream of the test material, in this case the fragments will be found within the lungs.

In order to increase the reproducibility of tests, and the range of materials which can be examined, it is often necessary to utilize an animal model system. Non-invasive techniques include the hamster cheek pouch, particularly useful in the evaluation of dental biomaterials, in which specimen retention is achieved via use of a collar placed around the neck of the hamster. Histological assessment may be performed following sacrifice.

The most common procedures adopted are those involving implantation of the test material directly into the tissues of a series of animals, sacrifice and examination being performed at suitable time intervals.

Toxicity may be assessed by implanting materials under the skin of rats or mice and the tissue response evaluated by histology.

Whilst all of the above studies provide invaluable information, the final evaluation can only be carried out after human implantation. Obviously these trials must be carried out under carefully controlled conditions if meaningful information is to be obtained.

4.6. PROBLEMS

4.6.1. Short questions

a   Which is the stronger out of covalent and ionic bonds?
b   Estimate the range of human cell sizes.
c   Is the 'ideal' biomaterial always one which is chemically inert?
d   Can polymeric materials degrade with time after implantation?
e   Is 10 kN a reasonable estimate of the peak load on a ligament in the lower limb?
f   Are cell membranes typically 1 μm, 10 nm or 1 nm in thickness?
g   Does the heart beat about 10^7, 10^8 or 10^9 times in a lifetime?
h   What is the most common type of bone cement?
i   Would the 'ideal' hip joint implant be as strong as possible?
j   What constituents of chromosomes carry hereditary information?
k   What is the main characteristic of a hydrogel polymer?
l   Explain what passivation of stainless steel involves.
m   Where is the most likely site for interaction between ionizing radiation and matter?
n   List three of the factors that influence the fatigue strength of a metal.
o   Will a metal with a large grain size be weaker or stronger than one with small grain size?
p   Give a definition of a biomaterial.
q   What is a Pourbaix diagram and for what is it used?
r   Why may it be undesirable to have two different metal components in an implant?
s   What is a systemic effect?
t   Does the shape of an implant matter and if so why?
u   Are there tests for and an agreed definition of biocompatibility?
v   Does glass activate the blood clotting process?

4.6.2. Longer questions

Question 4.6.2.1

Using the natural artery as a model, define the properties required for the development of an artificial blood vessel which is intended to replace segments of the femoral artery. Justify your conclusions.

Question 4.6.2.2

A polymer manufacturer has produced a new material which he believes is suitable for the manufacture of this type of artificial vessel. If necessary, he is willing to modify the polymer to suit your particular requirements.

Construct a flow chart indicating the protocol you would follow to test the suitability of the new material for clinical use and to provide information to the manufacturer if the initial material proves unsuitable.

Question 4.6.2.3

List the properties which are (i) *essential* and (ii) *preferable*, for a material to be used in the manufacture of plates for fracture fixation.

Answers to short questions

a   Covalent bonds are stronger than ionic bonds.
b   1–100 μm is the range of human cell sizes.
c   No, it is considered best if a biomaterial produces an appropriate response from the host.
d   Yes, polymers can degrade. Indeed, some are designed to be biodegradable.
e   No, most peak loadings on ligaments are less than 1 kN.
f   Cell membranes are typically 10 nm in thickness.
g   The heart will beat about 1.5×10^9 times in a lifetime.
h   Polymethylmethacrylate is used as a bone cement.
i   No, it may be desirable that the implant is no stronger than the surrounding tissues.
j   Genes are the genetic components in chromosomes.
k   Hydrogels are hydrophilic, i.e. they have an affinity for water.
l   Passivation of stainless steel involves immersion in nitric acid to thicken the surface oxide layer.
m   The chemical bonds that depend upon electrons are the most likely site of interaction between ionizing radiation and matter.
n   Surface finish, inclusions, grain size, stress concentrations and environment will all affect the fatigue strength of a metal. This list is not exhaustive.
o   A metal with a large grain size will be weaker than one with a small grain size.
p   Biomaterials are materials of natural or man-made origin that are used to directly supplement or replace the functions of living tissue.
q   A Pourbaix diagram plots pH against potential and helps understand how metals will behave under different conditions.
r   Two different metals can give rise to differences in electrochemical potential and hence corrosive currents.
s   In a systemic effect the whole of the body can be affected by the reaction to an implant.
t   The shape of an implant is important. Corners can give rise to high mechanical stresses and hence tissue responses.
u   No, there are no agreed definitions and tests for biocompatibility.
v   Yes, glass will activate the blood clotting process.

BIBLIOGRAPHY

Alpen E L 1997 *Radiation Biophysics* (New York: Academic)
Black J 1992 *Biological Performance of Materials: Fundamentals of Biocompatibility* (New York: Dekker)
Brown S A 1988 Biomaterials, corrosion and wear of *Encyclopaedia of Medical Devices and Instrumentation* ed J G Webster (New York: Wiley)
Chen H and Black S A 1980 *J. Biomed. Mater. Res.* **14** 567
Guthrie G W 1903 Direct fixation in fracture *Am. Med.* **5** 5
Hall E J 1978 *Radiobiology for the Radiologist* 2nd edn (New York: Harper and Row)
Johns H E and Cunningham J R 1983 *The Physics of Radiology* 4th edn (Springfield, IL: Thomas)

Krasin F and Wagner H 1988 Biological effects of ionising radiation *Encyclopaedia of Medical Devices and Instrumentation* ed J G Webster (New York: Wiley)

Meyn R E and Withers (eds) 1980 *Radiation Biology in Cancer Research* (New York: Raven)

Pourbaix M 1966 *Atlas of Electrochemical Equilibrium in Solution* (Oxford: Pergamon)

Silver F and Doillon C 1989 *Biocompatibility—Interactions of Biological and Implantable Materials* vol 1 *Polymers* (New York: VCH)

CHAPTER 5

IONIZING RADIATION: DOSE AND EXPOSURE—MEASUREMENTS, STANDARDS AND PROTECTION

5.1. INTRODUCTION AND OBJECTIVES

Ionizing radiation has been studied very extensively and as a result very well defined standards exist for how measurements should be made and what quantities of radiation are likely to be a hazard. The hazards from non-ionizing electromagnetic radiation are much less well understood and as a result there is, in many cases, no agreement on 'safe' exposures.

In this chapter we consider how ionizing radiation interacts with the human body, how radiation can be quantified and what levels are likely to be encountered both in the environment and in health care. Radiation dosimetry is a mature subject and many excellent textbooks are available to readers who want a more detailed coverage; see, for example Greening (1992). This chapter is an introduction to the subject.

Some examples of the questions we hope to answer in this chapter are:

- How is ionizing radiation absorbed in the human body?
- How can we measure ionizing radiation?
- What are the biological effects of exposure to ionizing radiation?
- What are the risks associated with exposure to ionizing radiation?
- How does an ionization chamber work?
- What are the approximate radiation doses associated with cosmic radiation?
- What are the allowable radiation doses to radiation workers?
- Does radon cause significant radiation exposure?

This chapter is not heavily mathematical and should be easily accessible to all our readers. It forms an introduction to ionizing radiation and as such should be read before Chapter 6 on Nuclear Medicine and Chapter 21 on Radiotherapy.

5.2. ABSORPTION, SCATTERING AND ATTENUATION OF GAMMA-RAYS

It is very important to understand how ionizing radiation is absorbed as it affects all the uses of radiation in medicine. People often refer to γ-rays as photons. A photon may be described as a 'bundle' or 'particle' of radiation. Use of the term arose from Einstein's explanation of the photoelectric effect where he considered that light could only travel in small packets and could only be emitted and absorbed in these small packets, or photons. The only difference between light photons and γ-ray photons is that the γ-ray has much higher energy (E), and therefore a much higher frequency (ν). $E = h\nu$ where h is Planck's constant.

5.2.1. *Photoelectric absorption*

A γ-ray can be absorbed by transferring all of its energy to an inner orbital electron in an atom of the absorber (see figure 5.1). The electron is ejected from the atom and the γ-ray disappears as it has lost all of its energy, and it never had any mass. This is not the end of the story as the atom is now left with a vacant inner electron orbit, which it will fill with one of the outer electrons. When it does this it releases a small amount of energy in the form of a characteristic x-ray photon. The x-ray is called a characteristic photon because its energy is characteristic of the absorbing material. The x-ray photon has a fixed energy because orbital electrons have fixed energies which correspond to the orbit which they occupy.

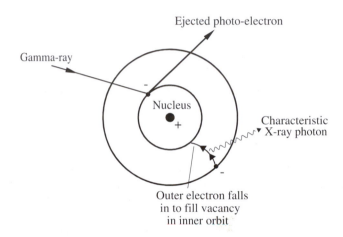

Figure 5.1. *Absorption of a γ-ray by the photoelectric process.*

Photoelectric absorption is the most likely form of absorption when the incident γ-ray has a fairly low energy. The lower the energy of the photon, the more likely it is to be absorbed by the photoelectric process. The gamma photons produced by ^{133}Xe used in lung scanning have an energy of 0.081 MeV and almost all will be absorbed by the photoelectric process in the sodium iodide detector, whereas the 1.53 MeV photons produced by ^{40}K will be absorbed by other processes.

5.2.2. *Compton effect*

The Compton effect is named after an American physicist who, in 1922, showed how photons can be scattered by outer or free electrons in an absorber. The photoelectric effect is an interaction of photons with the inner electrons, whereas the Compton effect is an interaction with the outer electrons which are not tightly bound to an atom.

What happens is that the photon collides with an electron and so gives some of its energy to it (see figure 5.2). If the collision is 'head on' the photon has its direction of travel reversed and it loses the maximum amount of energy, but if the collision is only a glancing one the energy given to the recoil electron will be much less. A single γ-ray may undergo several collisions, losing some energy on each occasion, and eventually be absorbed by the photoelectric effect.

The actual loss of energy as a result of Compton scattering depends upon the angle through which the γ-ray is scattered and may be calculated from the laws of conservation of momentum and energy. The γ-ray and the scattered electron are considered in the same way as two balls which collide and exchange energy.

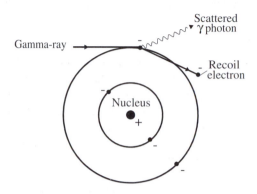

Figure 5.2. *The process of Compton scattering of a γ-ray photon.*

If the incident γ-ray photon has an energy of E_0 and the scattered γ-ray has an energy E, then because energy is conserved

$$E_0 = E + m_0 c^2 \left[\frac{1}{(1 - \beta^2)^{1/2}} - 1 \right] \tag{5.1}$$

where the second term on the right is the energy of the recoil electron. $m_0 c^2$ is the energy associated with the rest mass (m_0) of the electron and the term in brackets is the relativistic term which takes into account the change of mass with velocity. $\beta = v/c$ where v is the velocity of the recoil electron and c is the velocity of light.

The requirement for momentum to be conserved gives two equations, one for the component in the direction of the γ-ray and the other for the direction at right angles to this,

$$\frac{E_0}{c} = \frac{E}{c} \cos \phi + \frac{m_0 v}{(1 - \beta^2)^{1/2}} \cos \theta \tag{5.2}$$

$$0 = \frac{E}{c} \sin \phi - \frac{m_0 v}{(1 - \beta^2)^{1/2}} \sin \theta \tag{5.3}$$

where θ is the angle between the scattered γ-ray and the direction of the incident γ-ray, and ϕ is the angle between the direction taken by the recoil electron and the direction of the incident γ-ray. This gives us four unknowns E, β, ϕ and θ, but only three equations. However, we can eliminate β and ϕ so that we can see how E will change with the scattering angle θ.

Rearranging equations (5.1)–(5.3) we can obtain:

$$\frac{E_0 - E + m_0 c^2}{c} = \frac{m_0 c}{(1 - \beta^2)^{1/2}} \tag{5.4}$$

$$\frac{E_0}{c} - \frac{E}{c} \cos \phi = \frac{m_0 \beta c}{(1 - \beta^2)^{1/2}} \cos \theta \tag{5.5}$$

$$\frac{E}{c} \sin \phi = \frac{m_0 \beta c}{(1 - \beta^2)^{1/2}} \sin \theta. \tag{5.6}$$

Squaring and adding equations (5.5) and (5.6),

$$\left[\frac{E_0}{c}\right]^2 + \left[\frac{E}{c}\right]^2 - \frac{2E_0 E}{c^2}\cos\phi = \frac{m_0^2\beta^2 c^2}{(1-\beta^2)} = \frac{m_0^2 c^2}{(1-\beta^2)} - m_0^2 c^2. \tag{5.7}$$

Squaring equation (5.4) we obtain

$$\left[\frac{E_0}{c}\right]^2 + \left[\frac{E}{c}\right]^2 + m_0^2 c^2 - \frac{2E_0 E}{c^2} + 2m_0(E_0 - E) = \frac{m_0^2}{(1-\beta^2)}. \tag{5.8}$$

Subtracting equation (5.7) from equation (5.8):

$$2m_0(E_0 - E) - \frac{2E_0 E}{c^2}(1 - \cos\phi) = 0$$

$$\frac{1}{E} = \frac{1}{E_0} + \frac{1}{m_0 c^2}(1 - \cos\phi). \tag{5.9}$$

Now if $\phi = 0$, i.e. the γ-ray is not scattered at all, then $E = E_0$.

However, if $\phi = 180°$ then maximum scattering occurs and

$$\frac{1}{E} = \frac{1}{E_0} + \frac{2}{m_0 c^2}. \tag{5.10}$$

We can put in values for $m_0 = 0.9107 \times 10^{-31}$ kg, $c = 2.998 \times 10^8$ m s^{-1} to find the energy in joules. To convert to electron volts we can use the fact that 1 eV $= 1.6 \times 10^{-19}$ J. However, it is simpler to recognize that $m_0 c^2$ is the rest mass of an electron, which has an equivalent energy of 0.51 MeV. Therefore if the incident γ-ray has an energy of 1 MeV, then the maximum energy loss will be given by

$$\frac{1}{E} = \frac{1}{1} + \frac{2}{0.51}$$

thus giving E as 0.203 MeV. The γ-ray loses about 800 keV of its energy by scattering.

However, if the incident γ-ray only has an energy of 200 keV the loss by scattering will only be about 90 keV.

5.2.3.   Pair production

This method of absorption is less important than the Compton and photoelectric effects because it only happens for high-energy gamma photons, which are not often encountered in medicine. If the gamma photon has sufficient energy then it can be absorbed by an atomic nucleus in the absorber and results in the production of an electron and a positron (see figure 5.3). This is a case of energy being converted into mass. The mass of the electron and positron is such that 1.02 MeV is needed to produce the pair of particles (from $E = mc^2$).

If the incident gamma photon has more than 1.02 MeV of energy then the excess simply increases the velocity of the electron and the positron. The positron will not live very long because if it meets an electron it can combine with it to produce two photons of 0.51 MeV. These gamma photons are called annihilation radiation. Positron emitters are the basis of the technique of positron-emission tomography (PET), which uses the synchronous detection of the two 0.51 MeV gamma photons to localize the emitter (see section 12.2.6).

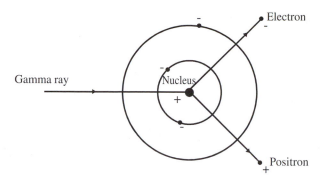

Figure 5.3. *Interaction of a γ-ray with a nucleus to produce an electron/positron pair.*

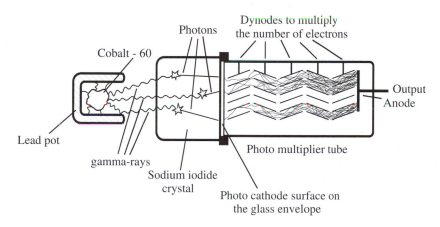

Figure 5.4. *A ^{60}Co source produces γ-rays which are detected by the scintillation counter. Pulses at the output have an amplitude proportional to the intensity of the scintillations.*

5.2.4. Energy spectra

The effect of all three absorption processes can be seen if we look at the size of the flashes of light produced in a sodium iodide crystal. Remember that the size of the flashes of light is determined by the energy of the electrons produced in the crystal by the three absorption processes. γ-rays from the ^{60}Co shown in figure 5.4 will produce thousands of flashes in the sodium iodide crystal. The graph of figure 5.5 shows how many flashes occur as a function of the size of the flashes, which corresponds to the energy of the absorbed electrons.

All three absorption processes are illustrated in figure 5.5, which is called an energy spectrum. Starting from the right of the spectrum: the peaks at 1.15 and 1.33 MeV are caused by photoelectric absorption of the γ-rays of this energy emitted by cobalt-60; the very small peak at 0.51 MeV is caused by the annihilation radiation which results from the pair-production process; the broad peak around about 0.2 MeV is caused by Compton scattering and corresponds to the energy of the recoil electrons. The very sharp peak at 88 keV is caused by the lead pot in which the cobalt was placed. The γ-rays from the cobalt are absorbed by a photoelectric process in the lead which then emits characteristic x-ray photons. These photons then travel to

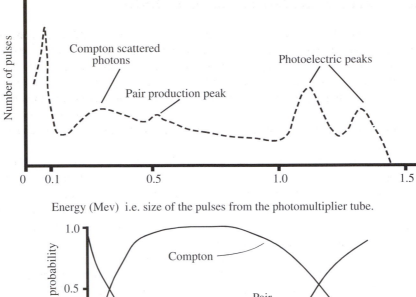

Figure 5.5. *(a) The spectrum of pulse amplitudes produced by the scintillation counter of figure 5.4; (b) the relative probability of different absorption mechanisms for γ radiation in carbon.*

the sodium iodide crystal and are absorbed by a photoelectric process. In figure 5.5(*b*) the relative probability of photoelectric, Compton scattering and pair-production absorption occurring in carbon is given. The Compton and photoelectric absorption as a function of energy would be very different in a material such as lead, so that calculation of the total attenuation by an absorber is not a trivial task.

5.2.5.   *Inverse square law attenuation*

It is misleading to think of the absorption processes we have talked about as the only processes which will reduce the intensity of a beam of gamma photons. Just as for light, the intensity of radiation falls off as you move away from the source. If the radiation from a source can spread in all directions, then its intensity will fall off in inverse proportion to the distance squared.

  This result is expected as the total number of photons emitted by a source will be spread out over the surface area of the sphere, of radius r, surrounding the source. The surface area of a sphere is $4\pi r^2$. As an example, should you be unlucky enough to have the 5000 curie (1.85×10^{14} Bq or 185 TBq) source from a ^{60}Co therapy machine fall on the floor in front of you, then, if you were 1 m away, you would receive a lethal dose of radiation in about 2 min. If you moved 4 m away then it would take about 32 min to receive a fatal dose.

5.3. BIOLOGICAL EFFECTS AND PROTECTION FROM THEM

Within three months of the discovery of x-rays it was noticed that they could cause conjunctivitis and by 1905 it was known that exposure could result in sterility. By 1920, half the employees of the London Radium Institute had low blood cell counts. Ionizing radiation is a possible hazard both in radiotherapy and nuclear medicine departments. The hazards we need to consider are those that result from exposure to the radiation from a source of ionizing radiation, and those that might arise from the radioactive contamination either of a person or the surroundings in which they might be working. Contamination is unlikely to be a problem in a radiotherapy department but it is obviously a possibility in nuclear medicine where unsealed sources of radioactive material are being handled.

This section deals with protection from ionizing radiation which forms the part of the electromagnetic spectrum beyond the far ultraviolet. The energy of this radiation is above about 10 eV. The radiation encountered in a medical physics department is unlikely to have an energy exceeding about 10 MeV, although higher-energy radiation can be produced and cosmic radiation includes particles with much greater energies.

Ionizing radiation is more hazardous than non-ionizing radiation simply because the ionization can interfere directly with the structure of atoms and molecules (see section 4.1.1).

In general the higher the energy of the radiation, the greater the hazard. One reason for this is the penetration of the radiation into tissue. Table 5.1 shows how the $D_{1/2}$ in water changes with the energy of the incident γ-rays. $D_{1/2}$ is the 'half-value thickness' (HVT), the thickness needed to reduce the intensity of the radiation to 50%. The literature on radiation protection also includes the term 'one-tenth value' layer, which is the thickness required to reduce the radiation intensity to 10%. The energies of x-rays which might be encountered from various sources are shown in table 5.2.

The most basic rules of radiation protection are that you should either move away from the source of radiation or put an absorber between yourself and this source. We noted in the previous section that, should you be unlucky enough to have a 5000 curie (185 TBq) source from a cobalt therapy machine fall on the floor

Table 5.1. *This shows how the 'half-value thickness in water changes with the energy of incident γ-rays.*

Energy	$D_{1/2}$ in water
10 keV	0.14 cm
100 keV	4.1 cm
1 MeV	9.9 cm
10 MeV	32.2 cm

Table 5.2. *Approximate energies of x-rays which are produced by five different types of equipment.*

Source	X-ray energy
Colour TV set	20 kV
Dental x-ray equipment	50 kV
Diagnostic medical x-ray and superficial radiotherapy	50–150 kV
Radiotherapy: orthovoltage	200–500 kV
cobalt-60	
Linear accelerator	1–20 MV

only 1 m away from you, you might receive a lethal dose in about 2 min. However, if you moved 4 m away then it would take 32 min to receive the same dose. About 30 mm of lead would be needed to afford you the same protection as moving away to 4 m.

Moving away from a source is not always the best way of reducing your exposure to radiation. At low energies, lead screening is very effective and it is for this reason that lead aprons are used for radiation protection in diagnostic x-ray departments. If the radiation has an energy of 100 keV then about 1 mm of lead or 10 cm of brick will reduce the intensity by a factor of about 1000. You would need to increase the distance between yourself and the radiation source thirty-fold to achieve the same reduction in radiation intensity.

The design of areas such as x-ray rooms to afford the best protection to the staff is not a simple procedure, but it is one of the responsibilities of radiation physics staff. There are internationally agreed standards, or codes of practice, for people working with ionizing radiation and it is the responsibility of the medical physics department to apply these.

5.4.   DOSE AND EXPOSURE MEASUREMENT

In order to be able to protect people from ionizing radiation it is obviously necessary to measure the radiation to which they may be exposed, and thus quantify exposure. Within the radiation field close to a radiation source there will be a *fluence* of particles which has been defined as $\mathrm{d}N/\mathrm{d}a$ where N is the number of particles incident on a sphere of cross-sectional area a. The energy carried by the particles is also important so that an *energy fluence* has also been defined in terms of the radiant energy incident on the cross-section of the sphere. A sphere is used so that the direction of incidence of the radiation is not relevant. Fluence and energy fluence can be used to define units of *exposure* to radiation. The unit of *exposure* used to be the roentgen (R) which was defined as 'That quantity of radiation which will release an electrical charge of 2.58×10^{-4} coulombs in one kilogram of dry air'. This represents the charge on about 1.3×10^{15} electrons. Note that exposure refers to the amount of ionization produced in air, and is not directly related to the energy absorbed in other materials, such as tissue. However, many of the instruments which are used to detect ionizing radiation rely upon the measurement of ionization in air and are calibrated in R or mR. The Roentgen is an old unit. The new unit has no name but is simply the value in Coulombs per kilogram ($C\,kg^{-1}$),

$$1 \text{ roentgen (R)} = 2.58 \times 10^{-4} \text{ C kg}^{-1} \text{dry air.}$$

A more useful concept when assessing the effect of radiation on tissue is to consider the energy deposited in the tissue: the *radiation dose*. Radiation dose is measured in two ways: the energy absorbed by tissue exposed to radiation can be measured, i.e. the 'absorbed dose'; alternatively, account can be taken of the fact that some types of radiation are more damaging than others by defining a 'dose equivalent'—the same dose will then produce the same biological damage whatever the type of radiation. These two definitions are explained in the next sections.

5.4.1.   Absorbed dose

The absorbed dose is measured in terms of the energy absorbed per unit mass of tissue. Energy is measured in joules and mass in kilograms. The unit of dose is the 'gray' (Gy) where

$$1 \text{ Gy} = 1 \text{ J kg}^{-1} \text{ of tissue.}$$

There is an old unit of dose which is still sometimes used, the rad,

$$1 \text{ rad} = 0.01 \text{ Gy} = 0.01 \text{ J kg}^{-1} \text{ of tissue.}$$

You should be clear in your mind just what absorbed dose means: if 1000 particles are completely absorbed in 1 kg of tissue, then the energy absorbed will be 1000 times the energy of each particle. Radiation energy is usually measured in keV or MeV, but you can convert these energies to joules:

$$1 \text{ J} = 6.2 \times 10^{18} \text{ eV}.$$

A dose of 1 Gy means that 6.2×10^{18} eV of energy have been absorbed in 1 kg of tissue. This could arise from 6.2×10^{12} x-ray photons of energy 1 MeV or any other combination of numbers of particles and energies.

Absorbed dose is difficult to measure. It is usually calculated by first measuring the exposure and then calibrating from a knowledge of the mass absorption coefficients for air and tissue.

5.4.2. *Dose equivalent*

The unit of dose equivalent is that dose which gives the same risk of damage or detriment to health whatever the type of radiation. This unit is called the sievert (Sv):

$$1 \text{ Sv} = 1 \text{ J kg}^{-1} \text{ tissue} \times \text{constant}.$$

There is an old unit of dose equivalent which is still used, the rem,

$$1 \text{ rem} = 0.01 \text{ Sv} = 0.01 \text{ J kg}^{-1} \text{ tissue} \times \text{constant}.$$

You should note that both the gray and the sievert are expressed as a number of joules per kilogram because they both involve measuring the energy absorbed in unit mass of tissue. The dose equivalent in sieverts is obtained by multiplying the dose in grays by a constant:

$$\text{dose equivalent (Sv)} = \text{absorbed dose (Gy)} \times \text{constant}.$$

The constant, called the 'radiation weighting factor', depends upon the type of radiation. For x- and γ-rays the constant is 1, for neutrons it is 10 and for α-particles it is 20. *Likely exposure to radiation in a medical physics department is almost always to β, x- or γ-rays. For these radiations, doses measured in grays and sieverts are numerically the same.* Table 5.3 gives some idea of the size of the units of dose.

Table 5.3. *Typical figures for x- and γ-ray doses for five different conditions.*

Dose due to background radiation in 1 year (this can vary greatly from place to place and arises from cosmic radiation, radioactive material in the surroundings and man-made radiation).	1 mSv	(0.1 rem)
Level set as the maximum dose to the general population in 1 year (a higher dose is sometimes allowed in 1 year provided the 5 year average does not exceed 1 mSv).	1 mSv	(0.1 rem)
Level set as the maximum dose to people who work with radiation. (50 mSv is the maximum in any one year.)	20 mSv (5 year average)	(2.0 rem)
Dose exposure which will cause nausea sickness and diarrhoea in some people	0.5 Gy	(50 rad)
Dose exposure which will kill many people in the few months following exposure	5 Gy	(500 rad)

5.5. MAXIMUM PERMISSIBLE LEVELS

Maximum permitted doses set in the various codes of practice are expressed in units of dose equivalent. The International Commission on Radiological Protection (ICRP) recommends the maximum annual dose equivalent for radiation workers as 50 mSv (5 rem) with a 5 year average of less than 20 mSv. Larger doses are permitted to specified parts of the body. For members of the public, the recommended maximum whole-body dose is 1 mSv (0.1 rem) averaged over 5 years. The maximum permitted dose levels have been reduced over the last 70 years—in 1931, the maximum permitted level was 15 mSv (1.5 rem) per week—and it is possible that further reductions will be made. The maximum dose levels apply to occupational exposure only and do not include radiation exposure of the worker for medical purposes.

It should be appreciated that even small doses do have long-term effects and it is these effects which are the cause of continuing controversy in setting 'safe' levels. These biological effects can only be expressed in statistical terms as the chance that a genetic change, a leukaemia or some other cancer might develop over a given period of time. The assessment of risks is complicated because there are also natural causes of these changes. The existence of long-term effects is the reason why young people, and in particular the unborn foetus, are subject to the greatest risk from ionizing radiation and are therefore the subject of specific radiation protection measures. For example, it is recommended that, under the '10 day rule', women are only exposed to diagnostic x-ray procedures during the 10 days following menstruation when pregnancy is unlikely.

5.5.1. *Environmental dose*

We are exposed to radiation from many sources during life. Sources include cosmic radiation, natural radioactivity in the ground and man-made radioactivity. Table 5.4 quantifies the body dose to which these sources of radiation can give rise. You should compare these values with the maximum permitted levels given in the previous section.

Table 5.4. *The doses given in this table correspond to six different situations, but are only approximate values as doses can vary widely.*

Cosmic radiation	200 μSv (20 mrem) over 1 year
Natural radioactive materials such as uranium in the ground	300 μSv (30 mrem) over 1 year
Naturally occurring radioactive materials within the body, e.g. ^{40}K	300 μSv (30 mrem) over 1 year
Chest radiograph	500 μSv (50 mrem) skin dose from one x-ray procedure
Coronary angiogram	20 mSv (2 rem) skin dose from one x-ray procedure
Nuclear power station	<1 mSv (100 mrem) over 1 year 1 km from the station

Exposure to cosmic radiation will increase with altitude because of absorption in the atmosphere. It is increased by about 50% in going from sea level to a height of 1000 m. γ-rays from buildings and from the ground can vary by a factor of 10 depending mainly upon the presence of uranium in the local rock. The largest contribution to overall radiation dose is exposure to naturally occurring radon. It is estimated to represent about 40% of the total dose, but the levels of exposure to radon can vary by about a factor of 300, depending upon local geology and atmospheric conditions. The radioactive content of food is estimated to contribute 10% to average dose and the medical uses of radiation contribute a further 10% to the average population dose.

5.5.2. Whole-body dose

The maximum permitted doses of 20 mSv (2 rem) for radiation workers and 1 mSv (0.1 rem) for the general public have already been explained. The basis for these levels is the risk of biological damage. Because it is possible to measure very low levels of radiation and to quantify the hazard, it is easy to exaggerate radiation hazards when making comparisons with other hazards of life. Table 5.5 is given to help you understand the relative risks. The figures are given in terms of an equal risk of causing death in 1 year.

Table 5.5. *All these activities carry the same risk. They give a 1 in 20 000 chance of causing death in 1 year. (Data from E E Pochin (1974), Community Health, b.2.)*

Exposure to 5 mSv (0.5 rem) whole-body radiation
Smoking 75 cigarettes
Travelling 2500 miles by motor car
Travelling 12 500 miles by air
Rock climbing for 75 min
Canoeing for 5 h
Working in a typical factory for a year
Being a man aged 60 for 16 h
Being a man aged 30 for 20 days

5.5.3. Organ dose

If you swallow a radioactive isotope then there may be a hazard to a particular part of your body, and therefore maximum permitted doses are specified for particular organs. We know from the use of radiopharmaceuticals that certain isotopes are preferentially absorbed by specific organs: it is the basis of imaging techniques in nuclear medicine. The organ within which an isotope is absorbed, and also the rate at which it is excreted by the body, depend upon the chemical form of the isotope.

It is possible to calculate the dose equivalent absorbed by a particular organ when a particular radioactive compound is ingested. On the basis of this result a maximum permitted body burden can be defined which will give rise to an equivalent dose below the annual maximum permitted dose. This maximum permitted quantity will be different for every radioactive compound.

Calculations of organ dose are very important considerations when new radiopharmaceuticals are introduced, but the calculations are outside the scope of this book.

5.6. MEASUREMENT METHODS

If we are to use ionizing radiation it is obvious that we must have a method of detecting and then recording its presence. As we cannot detect ionizing radiation directly, we rely on the radiation interacting with another material and producing an effect which we can detect. The story of how the radioactivity of uranium was first detected illustrates the point.

Shortly after Roentgen's announcement of the discovery of x-rays in 1895, a French physicist called Henri Becquerel took a piece of uranium salt (actually potassium uranyl sulphate) and inadvertently placed it close to an unexposed photographic film. The film itself was wrapped in opaque paper but Becquerel found that, after leaving the uranium salt close to the film for one or two days, the film was blackened at the points where it was closest to the uranium. It was the interaction between the radiation from the uranium and the photographic emulsion which allowed Becquerel, accidentally, to detect the radiation.

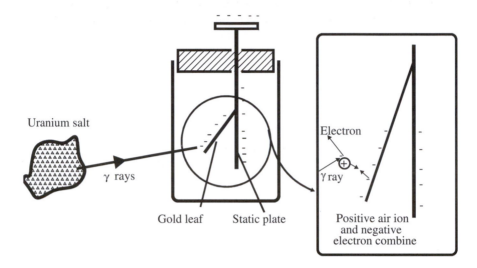

Figure 5.6. *A gold leaf electroscope irradiated by γ-rays from a uranium salt. The expanded diagram on the right shows how ionization of the air can remove negative charge from the electroscope.*

Becquerel used another form of interaction between ionizing radiation and matter to detect the radiation emitted by the uranium salt. He used a gold leaf electroscope (see figure 5.6), which measures the electric charge placed upon it. If there is an excess of electrons on the electroscope then the negatively charged electrons will repel each other and so the gold leaf will be repelled by the static metal plate. If the electrons can escape from the electroscope then they will do so, but they cannot pass through the surrounding air. However, Becquerel found that, if the uranium salt was left close to the electroscope, the gold leaf fell and the electrons were apparently escaping. He explained this by saying that the ionizing radiation from the uranium was able to remove electrons from the atoms in the air surrounding the gold leaf; so ionizing the air. Electrons from the gold leaf could then be attracted by, and combine with, the positive air ions, thereby escaping from the electroscope. The negative air ions would be repelled by the gold leaf.

Within a year of the discovery of x-rays, Becquerel had already used two of the methods which are still commonly used to detect ionizing radiation. In both cases it is an interaction of the radiation with something else which allows it to be detected.

If a person is being exposed to radiation, we do not want to have to wait a year before knowing whether they have exceeded the maximum permitted dose equivalent. We need to know the *dose rate*, so that we can calculate the accumulated dose and give a warning if the dose rate is very high. Monitoring equipment will often be calibrated in terms of mrad h^{-1} or μGy h^{-1}. It is easy to calculate that the maximum dose of 20 mSv in a year corresponds to an x-ray dose rate of about 10 μSv h^{-1} over a 40 h week. Spread over the whole year it represents a dose rate of about 2 μSv h^{-1}, which we can compare with normal background of about 0.1 μSv h^{-1}. These levels of dose rate can easily be measured using Geiger–Müller (G-M) tubes or scintillation counters. Dose rate can also be measured using an ionization chamber, which is more accurate and less affected by the energy of the radiation than either the G-M tube or scintillation counter monitors. Ionization chamber systems also have the advantage that they can measure high dose rates which would saturate the other monitors.

Standard instruments for measuring dose rate are almost invariably ionization chamber systems. You might think that it would be possible to measure the heating effect of radiation. As the units of dose are $J\,kg^{-1}$ it should be easy to calibrate a calorimeter directly in terms of dose. However, the temperature rises involved are so small that this type of dose measurement is not feasible, except for very high dose rates. For example, $1\,Gy\,s^{-1}$ (i.e. $100\,rads\,s^{-1}$) corresponds to $1\,J\,s^{-1}\,kg^{-1}$. The temperature rise in a given mass is

$$\frac{\text{energy}}{\text{mass} \times \text{specific heat}}.$$

For our energy of $1\,J\,s^{-1}\,kg^{-1}$ we obtain

$$\text{temperature rise} = 0.000\,24\,°C\,s^{-1}.$$

Even after $100\,s$ we only have a temperature rise of $0.024\,°C$.

It is desirable to be able to monitor both dose and dose rate with an instrument which can be worn on the body. Pocket dosimeters which use an ionization chamber are available, and also G-M tube instruments with a dose rate alarm. Solid-state detector and scintillation counter systems are also manufactured but all these instruments are relatively expensive and not applicable to routine dose measurements on large numbers of people.

Currently the cheapest, and therefore the most commonly used, personal monitors use either film or thermoluminescent dosimetry. The accuracy required from dose monitors is not high. However, the dosimeters used in radiotherapy treatment planning must be accurately calibrated. National standard primary dosimeters exist and local secondary standards are calibrated against these.

5.6.1.　*Ionization chambers*

Electricity cannot flow through air because there are no free electrons or ions to carry the current. An electric current is simply a flow of electrons or ions. However, if some of the atoms in the air are ionized, then free electrons are produced and an electric current can flow. In a flash of lightning, the very high potential gradient between the cloud and the ground is sufficient to ionize the air and so allow current to flow. In an ionization chamber, it is the ionizing radiation which frees electrons in the air filling the chamber and thus allows a current to flow.

In figure 5.7, the potential, V, which would typically be $100\,V$, is applied across the metal plates contained within the ionization chamber which is usually filled with air at atmospheric pressure. The chamber may be sealed or open to the atmosphere, where in the latter case a correction is applied to measurements to account for air pressure variations. When the chamber is exposed to x-rays, positive and negative ions are produced. The positively charged ions are attracted to the negative plate, and the negative ions are attracted to the positive plate, thus allowing a current to flow through the chamber. The current is measured by the electrometer which is nothing more than a sensitive ammeter; this is needed because the currents to be measured are often of the order of $10^{-9}\,A$. This corresponds to 6×10^9 electrons per second, but it is still quite difficult to measure.

If the $10^{-9}\,A$ is made to flow through a $1\,M\Omega$ resistance then we can use a voltmeter to measure the resulting $1\,mV$ signal. It is interesting to ask whether Johnson noise in the resistance will be important when making the measurement. In Chapter 9, section 9.3.2, we show that the thermal Johnson noise is given by

$$V^2 = 4kTR\,\mathrm{d}f$$

where V is the thermal noise, R the resistance, T the absolute temperature, k Boltzmann's constant and $\mathrm{d}f$ the bandwidth. For a resistance of $1\,M\Omega$, at $23\,°C$ and a bandwidth of $5\,kHz$, we can determine V as

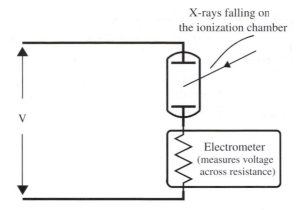

Figure 5.7. *Circuit diagram of an ionization chamber and an electrometer which is used to measure the ionization current.*

9.1 μV rms. This will only contribute about 1% noise to the measurement of the 1 mV developed across the 1 MΩ resistance. More important sources of error are likely to be thermal EMFs and drifts in the amplifier off-set potential.

Ionization chambers are used to measure the x-ray output both of therapy and diagnostic x-ray generators and also in making accurate measurements of patient x-ray dose. The chamber does not have to be cylindrical. It can be made in the form of a well into which the source of activity is placed. This type of well chamber is often used for measuring the activity of radiopharmaceuticals prior to injection.

5.6.2. *G-M counters*

The G-M tube, which was invented in 1929 by the Germans Hans Geiger and Walther Müller, is a very sensitive form of ionization chamber, indeed it is so sensitive that it can detect single ionizing particles which enter the tube. The construction (figure 5.8) is similar to many ionization chambers, with a central wire electrode inside a hollow metal tube. It differs from an ionization chamber in being filled with a gas such as argon or neon rather than air. The gas is at a pressure about one-fifth of atmospheric pressure.

Incident ionizing radiation will produce free electrons within the tube and these will be attracted towards the central electrode which is held at a positive potential. The potential which is applied is larger than that used in an ionization chamber and is usually several hundred volts.

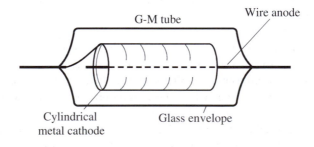

Figure 5.8. *Construction of a Geiger–Müller tube.*

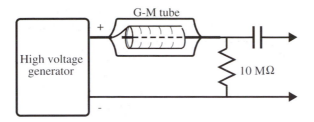

Figure 5.9. *Circuit diagram to show how a G-M tube is used to record ionizing radiation. Positive electrical pulses are produced across the 10 MΩ resistor.*

The electrons attracted towards the central anode are accelerated by the potential and gain sufficient energy to cause further ionization, thus causing a chain reaction. When all the electrons produced hit the central anode they can cause photons (of visible light or ultraviolet radiation) to be emitted, which can cause yet more ionization in the gas of the chamber. The net result is that the original incident γ-ray can produce about 10^5 electrons in the chamber and this is quite easily measured as a pulse of current lasting about 1 μs. Figure 5.9 shows how a G-M tube counter can be connected so that pulses of about 10 V can be obtained. The capacitor shown in the diagram is used to isolate the recording circuit from the high voltage applied to the tube.

This description of a G-M tube operation is only a simple one and does not deal with the movement of the positive ions in the tube. These travel much more slowly than the electrons and, as a result, the tube takes quite a long time to recover from the recorded pulse. This gives rise to what is called a 'dead time' for the tube. This is quite important in the practical uses of G-M tubes, because it limits the number of events which can be recorded each second (see section 12.2.2 for a discussion of dead time).

5.6.3.   Scintillation counters

Scintillation counters are the type used in gamma cameras and isotope scanners. The basic principle is that the ionizing radiation is made to produce a flash of light for each event and the flashes are converted into electrical pulses in a photomultiplier tube. This was illustrated in figure 5.4.

The very early workers with radioactivity knew that some substances, such as zinc sulphide and diamond, were luminescent when exposed to x-rays. The substance which is now most commonly used is sodium iodide (activated by the introduction of a small percentage of thallium), which has two major advantages: it can be made into large crystals, and it is transparent so that the flashes of light can escape to be counted. Before the flash of light is produced the incident x- or γ-ray must be absorbed by one of three processes (see section 5.2), and its energy transferred to an electron. This moving electron will then lose its energy by a series of interactions with the sodium iodide molecules. These interactions will cause the electrons in the sodium iodide molecules to be excited, which means that they are raised to a higher-energy level. When the excited electrons fall back to their lower-energy levels the surplus energy is released as light photons. The intensity of each flash of light is in proportion to the energy which the electron produced by the γ-ray acquired, and so a scintillation counter is able to measure not only the number of γ-rays absorbed but also their energy. To understand how this is done you need to know how the γ-ray is absorbed and how the intensity of the flashes of light is measured.

The flashes of light produced by α-particles hitting a surface coated in zinc sulphide can actually be seen by the human eye in a very dark room. However, the flashes produced by γ-rays in a sodium iodide crystal are very weak and, in any case, it is obviously impossible to try counting thousands of flashes by eye. The photomultiplier shown in figure 5.4 is able to amplify the flashes of light and give an electronic pulse in proportion to the brightness of each flash. The light from the sodium iodide crystal hits the photocathode which is coated in a material which emits electrons when it absorbs light photons (photoelectric effect). These electrons are accelerated towards the nearby electrode, which is held at a positive potential with respect to the photocathode. When the electrons strike the first electrode, called a dynode, they can eject many more electrons from the metal and these electrons are then attracted towards the next dynode which is held at a higher positive potential. This process goes on for all the other dynodes and at each one there is an amplification of the number of electrons so that only one electron from the photocathode can produce perhaps 10^7 at the final anode. The 'venetian blind' construction of the dynodes ensures that the electrons will always be attracted to the next dynode, and do not have a direct path to the final dynode in the chain.

This whole process takes place in a few millionths of a second (μs) so that the output from the anode resulting from one electron at the photocathode will be 10^7 electrons in, say, 10 μs, i.e. 10^{12} electrons per second. This corresponds to a current of about 0.2 μA, which is quite easy to measure.

Pulse height analysis

The size of the signal from a scintillation detector is, on average, proportional to the energy of the radiation absorbed by the detector. However, if the signals are measured from a set of γ-rays of the same energy, a spread of values will be observed because of limitations in the processes by which the γ-ray energy is converted to the electrical pulse. Nevertheless, the signal sizes will cluster around an average value. If the distribution of values is plotted it will take on a Gaussian shape and for the γ-ray energies of the radioisotopes used in clinical work, the spread of the signal values will typically be a few per cent of the average values. The exact percentage will depend on the geometric structure of the detector (for example, it will be larger in a gamma camera detector than a well counter, see section 6.4) and the energy of the γ-ray being detected.

This proportionality allows us to selectively count γ-rays which have a particular energy. There are several situations where we might want to do this. In sample counting (Chapter 6) we may have a situation in which the sample contains more than one radioisotope. These will have been chosen to emit γ-rays of significantly different energies. We would like to be in a position to count the γ-rays from these isotopes separately. In imaging, the radiation reaching the detector is contaminated by radiation scattered (Compton scattering) within the patient and we would like to eliminate this radiation.

We can do this because the electrical signal produced by the detector is proportional to the energy of the γ-ray. The size of the signal (represented as a voltage level) can be compared electronically with a chosen reference voltage level (representing some γ-ray energy). If the signal is larger than this level a pulse is generated and this pulse passed onto a counter (or further on in the imaging process in the case of a gamma camera). If the signal is less than this level, no pulse is generated and the signal is rejected. We can go further than this. We can simultaneously compare the signal to a second reference voltage representing a different γ-ray energy. If the signal strength is greater than this value we can reject the pulse. This process is illustrated in figure 5.10.

Assuming that we know the proportionality between the signal size and γ-ray energy, which we can find by calibrating the detector with γ-rays of known energy, we can choose the energy levels associated with these two levels. The gap between the two levels is known as the energy window. It is fairly common to set the energy levels in terms of the lower level (or possibly the mean level) and a window size, rather than upper and lower levels separately. If a small window size is used the number of γ-ray events which pass through the window represent the number of γ-rays from the γ-ray source between energies E and $E + \Delta E$, where ΔE can be as small as we choose. If we keep ΔE constant but change E and plot the number of γ-rays detected in a fixed time interval against the value of E we end up with a spectrum. This represents the relative number of γ-rays emitted from a γ-ray source as a function of energy. From a simple γ-ray emitter such as ^{99m}Tc (see Chapter 6) which only emits a single energy (140 keV) γ-ray, the spectrum would look like figure 5.11.

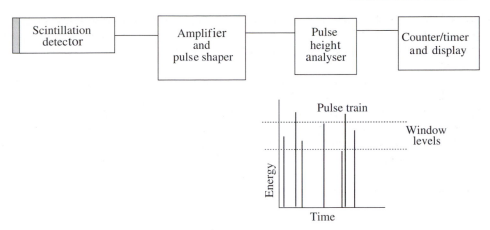

Figure 5.10. *Selection of pulses falling within a window set by a pulse height analyser.*

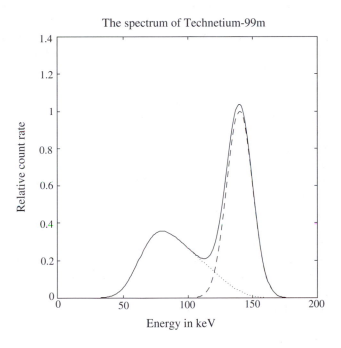

Figure 5.11. *A typical spectrum from* ^{99m}Tc. *Full curve, total spectrum; broken curve, photopeak; dotted curve, scatter spectrum. Note that the scatter spectrum and the photopeak spectrum overlap and so cannot be completely separated by pulse height analysis.*

The peak represents the distribution of signals produced by 140 keV γ-rays completely absorbed by a photoelectric interaction in the detector. The tail comes from γ-rays which have either been scattered in the source by Compton scattering (as is the case with radioisotope imaging) or have undergone scattering in the detector. In the former case a lower-energy γ-ray is detected. In the latter case the scattered ray may escape from the detector but the energy left behind produces a signal. For a 140 keV γ-ray this last type of event is of fairly low probability but for higher-energy γ-rays can be quite significant. In any case, the tail can largely (but not completely) be rejected by choosing an appropriate window.

If more than one gamma energy is present then provided the photopeak energies are far enough apart the photopeaks can be separately counted using pulse height analysis. However, the scatter spectrum from the higher-energy isotope will cross the photopeak spectrum of the lower-energy isotope and this contribution to this photopeak cannot be removed simply by pulse height analysis.

5.6.4. Film dosimeters

Films are easy to use for detecting radiation, and they have the advantage that they tell you where the radiation interacted with the film; however, the way they work is actually quite complicated.

The photographic films used are the same as those used in normal light photography except that the emulsion, containing silver bromide attached to a gelatin base, is about ten times thicker. The reason for the thicker emulsion is that the x-rays have a greater chance of interacting with the silver bromide than they would have in a thin layer.

The way in which the incident radiation interacts with the crystals of silver bromide is as follows: the incident x- or γ-ray will be absorbed, probably by the photoelectric process, and its energy imparted to an electron. This electron will produce other free electrons by ionization and these free electrons can be trapped in what are termed 'sensitivity specks'. These specks are actually faults in the crystal lattice formed from the silver and bromine ions which are suspended in the emulsion. The negatively charged electrons at the sensitivity specks are able to attract the positive silver ions and separate them from the bromine ions. The bromine atoms escape into the gelatin and the atoms of silver are left behind.

The method by which the distribution of the silver atoms is made visible is just the same as in conventional photography. The film is first put in a developer which will increase the number of silver particles by a reaction which uses the silver produced by the radiation exposure as a catalyst. If the film is left like this then subsequent exposure to light will reduce more of the silver bromide to silver. To prevent this happening the film is immersed in a fixer which removes the remaining silver bromide before the film is washed and then dried.

The two major medical uses of films for the detection of ionizing radiation are:

- For recording images in radiography: the x-ray image.
- For measuring exposure of people to ionizing radiation: the film badge.

Film badges

The blackening of a film is measured as the optical density (see figure 5.12):

$$\text{density} = \log_{10} \frac{I_0}{I}$$

where I_0 is the incident light intensity and I is the intensity after passing through the film. If the intensity is reduced from 100% to 10%, then the density is $\log_{10} 10 = 1$. If the intensity is reduced from 100% to 1%, then the density is $\log_{10} 100 = 2$. The range of densities normally encountered is 0.2 to about 2.5. A density of 2 is very black; a density of 1 is such that the writing on the page of a book can just be read in normal room lighting through a film of this density.

Personnel monitoring film badges consist of a small film in a plastic holder. After exposure the film is developed and fixed, following which the density of any blackening can be measured. The optical density

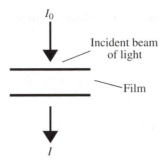

Figure 5.12. *A photographic film reduces the light intensity from I_0 to I. The ratio of these determines the optical density (see text).*

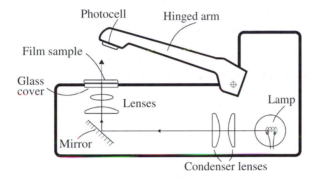

Figure 5.13. *Principle of operation of a film densitometer. The hinged arm allows the photocell to be placed in contact with the film to measure the light transmitted from the collimated beam produced by the lens system.*

measurement is made using a densitometer consisting of a light source and a light-sensitive detector (see figure 5.13).

The case of the typical film badge shown in figure 5.14 contains six special filter areas which can be used to give some information on the type of radiation which has been recorded. This can be important if we wish to know the source of the radiation to which a person has been exposed. The six special areas on the film badge illustrated in figure 5.14 are:

1. A filter made of an alloy of aluminium and copper (Dural), 1 mm thick.
2. A filter made of 0.7 mm cadmium plus 0.3 mm lead.
3. A filter of 0.7 mm tin plus 0.3 mm lead.
4. An open window.
5. A thin plastic window of 50 mg cm^{-2} (500 g m^{-2}).
6. A thick plastic window of 300 mg cm^{-2} (3000 g m^{-2}).

High-energy radiation will not be significantly attenuated by the windows and so the whole of the film will be uniformly exposed. However, low-energy radiation will suffer attenuation with the result that there will be greater exposure of the film under the thin windows than under the Dural and tin windows. It is not possible to measure the energy of the radiation to any great accuracy by this method, but it is possible to tell the

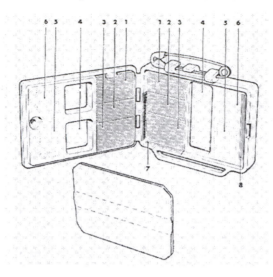

Figure 5.14. *The AERE/RPS film badge for personal monitoring. (Courtesy of the Harwell Laboratory.) Filter types: 1, 1 mm Dural; 2, 0.7 mm Cd + 0.3 mm Pb; 3, 0.7 mm Sn + 0.3 mm Pb; 4, open window; 5, 500 g m⁻² plastics; 6, 3000 g m⁻² plastics; 7, 0.3 mm Pb; 8, 0.4 g indium. (Filter 7 is a strip of lead which prevents leakage of radiation around the main filters. The indium, when included, would be used to record neutron dose following a reactor incident.)*

difference between a film exposed to 100 kV x-rays in diagnostic radiography and one exposed to the higher energies used for radiotherapy. The filter containing cadmium is used to estimate doses arising from exposure to neutrons since neutrons produce gamma radiation when they interact with cadmium. Filter 7 is a small strip of lead to reduce errors due to leakage of radiation around the main filters. The thin plastic filter (5) attenuates beta rays and a comparison of the blackening under this filter with that under the open window enables beta ray doses to be estimated.

Films are usually monitored at intervals of four weeks and records are kept of the accumulated dose for all radiation workers. Certainly radiographers and the staff working in the nuclear medicine section of a medical physics department will be issued with film badges. Some departments operate their own radiation monitoring service, issuing and developing their own films, but it is much more common for films to be issued and processed by a larger central monitoring service. Automated equipment is used in these centres so that large numbers of films can be handled economically.

Radiation dose is measured by relating the optical density to the exposure. A film cannot be used to record very small doses as the film blackening is insignificant. There is also a limit to the maximum dose which can be recorded before no further blackening is produced. Most films in use can record a minimum dose of 0.2 mGy (20 mrad) and a maximum dose of 0.1 Gy (10 rad). The smallest dose which can be recorded accurately, if 12 films are issued each year, is 2.4 mSv. This should be compared with the 1 mSv set by ICRP for exposure to the general public.

5.6.5. *Thermoluminescent dosimetry (TLD)*

Many crystalline materials can absorb ionizing radiation and store a fraction of the energy by trapping electrons at impurity atoms and at crystal lattice flaws. The incident ionizing radiation frees electrons, which are then trapped in the crystal structure. If the crystal is heated then this stored energy can be released, as the trapped electrons move to a lower-energy level, with the emission of quanta of radiation. This radiation can be in the visible spectrum.

One application of this phenomenon is in the dating of art objects. Crystals such as quartz and feldspar are found in pottery which will therefore store energy arising from environmental radiation. A 1000 year old vase might well have absorbed a dose of 2 Gy (200 rad) from background radiation and the natural radioactive materials such as uranium, thorium and potassium within the vase. If a very small sample (typically 30 mg) of the pottery is heated up to 500 °C, then a quantity of light will be emitted in proportion to the absorbed dose of radiation.

Obviously if an estimate can be made of the background dose each year, then an estimate can be made of the age of the vase. The assumption is made that the measured radiation dose is that acquired since the pottery was fired.

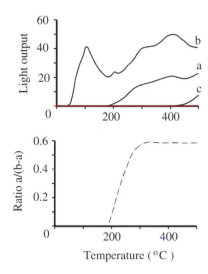

Figure 5.15. *Thermoluminescent glow curves for the 'Affecter' amphora (Ashmolean Museum). a, natural thermoluminescence from the vase; b, natural thermoluminescence plus thermoluminescence induced by 22.5 Gy (2250 rad) of radiation; c, background incandescence. The ratio $a/(b-a)$ is given in the lower part of the figure. (After S J Fleming 1971 Naturwissenschaften* **58** *333.)*

Figure 5.15 shows the thermoluminescent glow curves for samples taken from a Greek vase. The three upper traces show the light output as a function of temperature for a sample taken straight from the vase, a sample irradiated with a dose of 22.5 Gy (2250 rad) and a sample which has previously been heated to 500 °C. The temperature at which an electron breaks away from the crystal lattice depends upon the nature of the imperfection. The higher the temperature at which light is emitted by the sample, the stronger is the binding of the electron to the imperfection. Even ambient temperatures can release some of the electrons over a period of time; this is the reason for the absence of any light output from the sample of pottery below about 200 °C.

The curve in the lower part of figure 5.15 shows the ratio of the natural thermoluminescence to the laboratory-induced thermoluminescence; the ratio is stable for temperatures above 300 °C and this ratio enables the absorbed dose in the sample to be calculated. In the example given, the dose is found to be 12.8 Gy (1280 rad), which corresponds fairly well with the 2500 year age of the vase. A fake vase would be younger and have accumulated a much smaller dose.

Equipment

The equipment needed to make thermoluminescent dosimetry (TLD) measurements consists of a chamber, in which the sample can be heated, and a photomultiplier detector to record the light output from the sample. The chamber is heated in a reproducible manner and the temperature of the sample can be recorded. In order to stop oxidation of the sample when it is heated to 500 °C, the inside of the chamber is filled with nitrogen. If reproducible measurements are to be made from TLD equipment, then great care has to be taken in the preparation and positioning of the sample. Even finger marks can contaminate a sample, so these must be prepared under clean conditions.

It should be possible to make measurements of medical x-ray doses using samples of old vases: in practice a higher light output and more consistent results are obtained by using lithium fluoride powder. Lithium fluoride has an atomic number close to that of soft tissue, so the variation of absorbed dose with radiation energy will be similar to that of tissue. 10 mg samples of lithium fluoride powder can be used; this is contained in a sachet which is first exposed in the radiation field to be measured, and then is removed from the sachet and placed in the oven where the light emitted can be viewed by the photomultiplier tube. An alternative to powder in sachets is to use Teflon discs which contain lithium fluoride in suspension. In both cases glow curves have to be obtained for a set of samples exposed to known doses of radiation so that calibration curves can be obtained.

TLD is becoming the method of choice for monitoring both therapeutic and diagnostic x-ray doses. It is convenient to use and can be used to measure quite small doses.

5.7. PRACTICAL EXPERIMENT

5.7.1. *Dose measurement during radiography*

Objective

To use an ionization chamber, electrometer and integrator to measure the radiation dose during radiography.

Theory

The x-ray dose received by a patient during radiography depends greatly upon the particular procedure. A typical chest x-ray might only give a dose of 0.5 mSv (50 mrem), whereas an intravenous pyelogram can result in a total dose of 10 mSv (1 rem). Systems such as image intensifiers are used primarily to enable x-ray images to be produced at the lowest possible dose rates.

You can measure the dose rate during patient exposure by using a small ionization chamber. The current from the chamber is used to charge an integrating capacitor and the output is recorded using an electrometer. The final voltage across the capacitor is proportional to the integral of the chamber current and can be calibrated directly in mSv or mrem.

Method

You should first familiarize yourself with the use of the ionization chamber and electrometer.

Try to obtain measurements of exposure dose during several radiographic procedures. Your ionization chamber should be placed between the patient and the x-ray head. You should, of course, seek the advice of a radiographer in placing the chamber such that it will not interfere with the usefulness of the x-ray image obtained.

Note. If TLD equipment is available, then this technique could be used in parallel with the ionization chamber, and the dose measurements compared. TLD has the considerable advantage that only the small sachet of thermoluminescent material is placed in the x-ray beam. This can be placed in positions which an ionization chamber cannot reach and it causes minimal interference with the radiographic procedure.

5.8. PROBLEMS

5.8.1. *Short questions*

a   Do x- or γ-rays have the higher energy?
b   Is the photoelectric process of γ-ray absorption dominant at low or high energies?
c   What is Compton scattering?
d   What might cause a peak at 0.51 MeV in an energy spectrum?
e   What is the most likely form of γ-ray absorption at an energy of 500 keV?
f   What energy is usually considered as the boundary between ionizing and non-ionizing radiation?
g   What is the unit of 'absorbed dose' and how is it defined?
h   What is a 'radiation weighting factor'?
i   Why is the unborn foetus considered to have a higher risk from ionizing radiation than is an adult?
j   Which is more likely to kill you in any one year, a radiation dose of 20 mSv or smoking 500 cigarettes?
k   What type of instrument is usually considered as setting the standard for radiation dose rate measurements?
l   Is there gas inside a G-M tube?
m   What does a photomultiplier tube actually multiply?
n   How does thermoluminescent dosimetry work?
o   What type of radiation forms the hazard from radon?
p   Can medical x-rays such as a chest radiograph form a significant fraction of the maximum annual dose level set for exposure of the general public.
q   Would a G-M counter be useful in assessing radon levels?
r   What is the approximate gamma detection efficiency of a G-M tube counter?
s   Photographic films are used to produce the conventional x-ray. What other uses can films be put to in the applications of ionizing radiation?

5.8.2. *Longer questions (answers are given to some of the questions)*

Question 5.8.2.1

A well collimated beam of gamma radiation from a ^{60}Co source is directed onto a glass-walled rectangular tank filled with water. The tank has the dimensions $400 \times 400 \times 400$ mm^3. Describe the energy spectrum of γ-ray energies which will be seen by a scintillation counter as it is rotated in a horizontal plane around the centre of the water tank. Estimate the ratio of the incident and exit radiation intensities along the path of the γ-ray beam.

Question 5.8.2.2

Describe how Geiger–Müller and sodium-iodide scintillation counters work. In each case say what the advantages and disadvantages are, and explain where each technique is most useful.

Discuss the means by which personal radiation dose might be minimized during the process of manipulation of a unsealed γ-ray emitting substance.

Answer

The G-M tube description should explain the role of the anode and cathode and the multiplication effect that enables single ionizing events to be detected. The fact that there is a threshold voltage that must be applied across the tube should be explained and also the placement of the working voltage on the plateau. Advantages include the ability to detect single events and simplicity. Disadvantages include the dead time that limits count rate, the poor efficiency in detecting γ-rays and the inability to distinguish the energy of different events.

The scintillation counter description should include the light-tight crystal and the photomultiplier tube. The advantages should include the high sensitivity to γ-rays and the ability to provide an energy spectrum. Disadvantages include high cost and the inability to be used for the measurement of β radiation.

The G-M tube is useful for simple surveys and for the measurement of β and γ contamination. Scintillation counters are widely used in γ-ray counting and the measurement of energy spectra.

The discussion of the handling of unsealed sources should include possible external and internal radiation hazards and the statutory limits to dose. The use of shielding and maximizing distance should be discussed. Trial runs of procedures can minimize handling times. Sharing work between people and the techniques for dose monitoring should be discussed. Waste handling is also important.

Question 5.8.2.3

Describe the mechanisms by which gamma radiation loses energy when it interacts with tissue. How does the relative importance of the mechanisms vary with the energy of the gamma radiation?

Answers to short questions

a γ-rays have a higher energy than x-rays.
b   The photoelectric process is dominant at low energies
c   Compton scattering is the process by which a γ-ray interacts with the outer orbital electrons of an absorber. Some of the energy of the γ-ray is passed to the outer electron which recoils from the collision.
d   A peak at 0.51 MeV in an energy spectrum may be caused by annihilation radiation when an electron combines with a positron.
e   Compton scattering is the most likely process of γ-ray absorption at 500 keV.
f   10 eV is usually taken as the boundary between ionizing and non-ionizing radiation.
g   The unit of absorbed dose is the gray. The dose that causes an energy of 1 J to be absorbed in a mass of 1 kg.
h   Radiation weighting factor is a factor that takes into account the fact that some radiation, such as neutrons and α-particles, are more damaging than others, even though they impart the same energy to the tissue.
i   The existence of long-term effects from radiation and damage to tissues during the process of development *in utero* makes the unborn foetus particularly vulnerable to the effects of ionizing radiation..
j   Smoking 500 cigarettes in a year carries a greater risk to health than a radiation dose of 20 mSv.
k   An ionization chamber system is usually considered as a standard for radiation dose rate measurement.
l   Yes, usually argon or neon at about 20% of atmospheric pressure is used inside a G-M tube.
m   The number of electrons is multiplied in a photomultiplier tube.
n   TLD works by measuring the energy trapped in some crystalline materials when they absorb ionizing radiation. The energy is released by heating the crystalline material and measuring the light emitted.

o α-particles are emitted by radon.

p   Yes, a chest radiograph might involve a dose of 0.5 mSv. The annual dose level set for the general population is 1 mSv.

q   No, a G-M counter could not be used to assess radon levels. It is not sensitive to α-particles.

r   The counting efficiency of a G-M tube is a few per cent.

s   Films are used for measuring exposure to ionizing radiation. They can also be used for contact radiography.

BIBLIOGRAPHY

Bomford C K, Kunkler I H and Sherriff S B 1993 *Walter and Miller's Textbook of Radiotherapy* 5th edn (Edinburgh: Churchill Livingstone)

Greening J R 1992 *Fundamentals of Radiation Dosimetry* (Bristol: IOP Publishing)

Martin A and Harbison S A 1986 *An Introduction to Radiation Protection* 3rd edn (London: Chapman and Hall)

Meredith W J and Massey J B 1977 *Fundamental Physics of Radiology* (Bristol: Wright)

Sumner D 1987 *Radiation Risks: an Evaluation* (Wigtown: Tarragon)

CHAPTER 6

RADIOISOTOPES AND NUCLEAR MEDICINE

6.1. INTRODUCTION AND OBJECTIVES

Nuclear medicine consists of a range of diagnostic and therapeutic procedures that use radioisotopes. For diagnostic work it is usually necessary for the radioisotope to emit γ-rays because these can penetrate tissue and be detected outside the body, and therefore the distribution in the body can be determined. In some cases, if the diagnostic test involves measuring samples taken from a patient, it may be possible to use a β-particle emitter. For therapeutic work it is usually more desirable for the radioisotope to emit β-particles, since these have a short range in tissue and can deliver a high radiation dose to the location of the radioisotope. For example, radioiodine localizes in the thyroid gland and the dose to the local thyroid tissue can be used to treat a variety of thyroid diseases. This chapter will concentrate on the diagnostic uses of radioisotopes.

The questions we intend to address include:

- Can we attach radioactive labels to specific body components?
- Can we localize a radioactive isotope within the body?
- How do tracers disappear from the body?

At the end of this chapter you should:

- Understand the principles of the use of radioisotopes in diagnosis.
- Be able to describe some examples of the use of radioisotope imaging
- Be able to define the difference between static and dynamic gamma camera studies.
- Appreciate the very wide range of clinical applications of radioactive isotopes.

This chapter sets out to give a general description of nuclear medicine. It is not mathematical and is suitable to most levels of reader.

6.1.1. Diagnosis with radioisotopes

There are five stages in the use of radioisotopes which define their use in clinical diagnosis. These are:

- Identify a clinical problem.
- Find a biochemical substance whose distribution in the patient as a function of space or time will depend on whether the particular disease being investigated is present or absent.
- Label this substance with an appropriate radioisotope to produce a radiotracer.

- Measure the distribution of the tracer with an external detector.
- Interpret the results.

A simple example is the use of radioiodine in the diagnosis of thyroid disease. The thyroid gland normally extracts iodine from the blood as this is used in the production of thyroid hormones. Radioactive iodine is chemically identical to normal iodine and so if some radioactive iodine is given orally a fraction of it will appear in the thyroid gland after a period of time. After a certain period of time, a normally functioning gland will have taken up on average a certain percentage of the administered iodine, whereas an overactive gland will have taken up more than a normal amount. The amount taken up can be measured using external detectors and by comparing this amount with the range of uptakes measured on normal subjects a diagnosis can be made on whether the gland is overactive or not.

In this case the radioisotope used is a simple chemical. In most other cases the biochemical has a more complex molecule. One of the atoms in this molecule can be replaced by a radioactive atom to form a radioactive molecule. Hopefully this does not change the biological behaviour of the molecule too much. On administration to the patient the molecule is processed by the appropriate biochemical routes. If these are upset by disease the distribution of the biochemical in either space or time or both will be upset and this fact can be used to diagnose the presence of disease.

6.2. ATOMIC STRUCTURE

In order to understand what a radioisotope is we need to remind ourselves briefly about the structure of the atom. All atoms consist of a nucleus of protons and neutrons, with electrons in orbit around the nucleus. The difference between elements is the numbers of the fundamental particles, i.e. the protons, neutrons and electrons, which the atoms contain. Each atom consists of a nucleus surrounded by a cloud of electrons. Each nucleus is made up of a mixture of two types of nuclear particle, protons and neutrons. The conventional symbols for atoms use a superscript for the atomic mass (neutrons + protons), and a subscript for the number of protons. An example is $^{14}_{7}N$, which describes an atom of nitrogen, atomic mass number 14, with seven protons. The subscript is often omitted.

Protons have a positive electric charge and electrons have a negative charge but, as the number of protons is usually equal to the number of electrons, the net electric charge is zero. If an atom gains or loses an electron, so that the numbers of electric charges no longer balance, then the resulting atom is called an ion. The chemical properties of an atom are determined by the way the electrons surround the atom and their number. As the number of electrons is equal to the number of protons, so the number of protons uniquely defines the chemical element. The number of neutrons can vary.

The lightest atom is hydrogen which has only one proton and one electron. Uranium (^{238}U) has 238 protons and neutrons and there are other atoms which are even heavier.

6.2.1. Isotopes

All the nuclei of the atoms of one particular element have the same number of protons, but the number of neutrons can vary. Isotopes are atoms of the same element which have different numbers of neutrons. They are referred to by their atomic mass number. The neutrons and protons are often referred to generically as nucleons.

Stable iodine ^{127}I has a mass number of 127. In addition, there are small quantities of ^{128}I found in nature. Other isotopes of iodine exist. ^{131}I is also an isotope of iodine, but unlike ^{127}I and ^{128}I it is not stable. Associated with each type of nucleus is a binding energy. If it is possible for a lower energy to be achieved by a rearrangement of the nucleons, possibly by ejecting some of them or by turning a neutron into a proton or vice versa or simply by an internal rearrangement of the nucleons, then this will eventually happen. This process is accompanied by the ejection of the energy difference in some form or another, either as a particle or a photon. An isotope that undergoes a transformation of this form, called a decay, is known as a radioisotope. Almost all elements have radioisotopes. From uranium upwards there are no stable isotopes. Within the bulk of the periodic table there is one element, technetium, which also has no stable isotopes. Coincidentally a radioisotope of this element is widely used in nuclear medicine.

^{131}I is the isotope of iodine that is often used for treating an overactive thyroid gland, but it is not the only radioisotope of iodine. ^{123}I and ^{125}I are other radioisotopes of this element. All the isotopes of iodine have 53 protons in the nucleus, but the number of neutrons can be 70, 72 or 78 to give the three isotopes, i.e. ^{123}I, ^{125}I and ^{131}I.

As a general rule there are about equal numbers of neutrons and protons in a nucleus but, in heavier atoms, a greater proportion of neutrons have to be added to maintain the stability of the atom.

6.2.2. Half-life

Some radioisotopes can be found in nature. Many others can be made artificially. We mentioned that ^{131}I was unstable. This does not mean that the atom is so unstable that it will fall apart immediately, but merely that it has a tendency to be unstable. There is no way of predicting when a given atom will decay. However, in a given time, on average, a certain proportion of a group of atoms will become unstable and disintegrate. From the observation that each atom will decay completely independently of the other atoms we can show that the number of atoms which disintegrate in unit time is directly related to the number of atoms remaining. We can express this mathematically:

$$\frac{\mathrm{d}N}{\mathrm{d}t} = -\lambda N$$

where λ is a constant and N is the number of atoms. We can rearrange the equation to give

$$\frac{\mathrm{d}N}{N} = -\lambda \, \mathrm{d}t$$

which on integration gives $\log_e N = -\lambda t + k$, where k is a constant.

If $t = 0$ corresponds to $N = N_0$, then $k = \log_e N_0$ and so

$$N = N_0 \mathrm{e}^{-\lambda t}.$$

This is the basic equation that governs the decay of unstable or radioactive isotopes. The larger the decay constant, λ, the more quickly the isotope will decay. The rate of decay is usually expressed as the half-life of the isotope, which is the time it takes for half of the original number of atoms to decay. It is quite easily shown that $T_{1/2}$ is given by $\log_e 2/\lambda = 0.693/\lambda$. In this case the decay equation becomes

$$N = N_0 \mathrm{e}^{-(0.693t/T_{1/2})}.$$

The half-lives of different isotopes can range from millions of years to small fractions of a second. Half-life is a very important factor in the choice and use of isotopes in medicine. Naturally occurring radioisotopes such as ^{238}U, with a half-life of 4.5×10^{10} years, have been decaying since the Earth was formed. Radioisotopes such as ^{14}C, with a half-life of 5760 years, are formed naturally (in the atmosphere by cosmic radiation). Most of the radioisotopes used in nuclear medicine have much shorter half-lives and are made artificially.

6.2.3. Nuclear radiations

There are many types of nuclear radiation which can accompany the decay of an atom. The following list covers the main types:

- x-rays
- γ-rays (gamma-rays)
- β-particles (electrons)
- positrons
- neutrons
- α-particles.

x- and γ-rays are photons, similar to visible radiation but of much higher energy. They carry no electric charge and simply carry away the energy difference between the original atom and what is left after the decay. There is no fundamental difference between x-rays and γ-rays. x-rays are photons produced when an electron changes its energy state in the cloud surrounding the atom. γ-rays are produced by the decay of a radioactive atom when a nucleon changes its energy state within the nucleus. Usually γ-rays have a much higher energy than x-rays, but there is some overlap in energy ranges. As these photons carry no charge they do not easily interact with the atoms of the material through which they may be travelling, and so are not easily absorbed within matter.

Beta-particles are fast moving electrons. An electron is produced when a neutron in the nucleus of an atom converts to a proton. An electron is produced to balance the electric charge and the energy released by this transformation is carried from the nucleus as the kinetic energy of the emitted electron. Even though they travel very fast, because the electron has a negative charge they will interact with the electric charges associated with atoms when they come close to them and the result is that they soon transfer their energy to the material through which they are travelling. A β-particle might be able to travel about 1 m in air before it loses its kinetic energy, but in tissue it will only travel a millimetre or so.

Positrons. These particles have the same mass as an electron but have a positive electric charge. They are produced within a nucleus when a proton is converted to a neutron. Their ability to travel through air and tissue is the same as a β-particle of the same energy. When a positron has been slowed down to about the average energy of the electrons in the material through which it is travelling it will interact with an electron. The pair of particles will be converted into two γ-rays which move off in opposite directions. These γ-rays have a characteristic energy of 0.511 MeV (the energy associated with the rest mass of the electron) and are essential to the technique of positron-emission tomography (see section 12.2.6).

   In general, if an unstable nucleus is rich in neutrons it will attempt to convert a neutron to a proton with the emission of an electron. If the converse is true and the nucleus is rich in protons then it will convert a proton to a neutron with the emission of a positron. In the former case the atom will move up the periodic table. In the latter case it will move down. In some cases a cascade of electrons will be emitted. Sometimes after an electron has been emitted there is still some excess energy in the nucleus, but this is then ejected as a γ-ray. Invariably γ-ray emission is preceded by electron emission, but these may be sufficiently separated in time for this not to be immediately obvious.

Alpha-particles are actually doubly ionized helium atoms with a mass of 4, consisting of two neutrons and two protons. They have a positive charge corresponding to the two protons. Their relatively high mass and charge cause α-particles to be stopped easily by collisions so that the range of an α-particle in air is only a few centimetres. Even a piece of paper is sufficient to stop most α-particles.

Neutrons are as heavy as protons but, because they have no electric charge, they can travel large distances before they are exhausted. Neutrons can be produced by a reactor and they have been used both for treatment of disease and also in neutron activation analysis. However, they are not used routinely in most medical physics departments and will not be discussed here.

6.2.4.   Energy of nuclear radiations

The energy of any nuclear radiation can be related to the mass which was lost when the atom decayed and emitted the radiation. The equation relating the energy, E, to the mass loss, m, is given by Einstein's formula

$$E = mc^2$$

where c is the velocity of light.

The units in which energy is expressed are electron volts (eV), thousands of electron volts (keV), or millions of electron volts (MeV). The electron volt is defined as the energy which a unit charge, such as an electron, will receive when it falls through a potential of 1 V.

The energy of the x-rays used in diagnostic radiology goes up to about 100 keV and the linear accelerators used in the treatment of malignant tumours produce x-rays with energies of the order 10 MeV or more. Actually, x-ray energies in diagnostic imaging and therapy are usually expressed in kV and not keV. The beam of x-rays do not all have the same energy and the kV refers to the accelerating potential which is applied to the x-ray generator.

Each nuclear decay has associated with it a very specific energy. For example, the main γ-ray energy associated with the decay of ^{131}I is precisely 364 keV. The principal β-particle energy associated with this isotope is 610 keV, but somewhat confusingly if the energies of these particles are measured they are found to cover a range of energies because a second particle, a neutrino, is also emitted with the electron and this takes a variable amount of the energy away. As the neutrino cannot be detected this energy is invisible but it is there. In decays which involve the emission of a γ-ray, the energy of the ray is the same for every disintegration and indeed by accurately measuring the energy of the γ-ray it is possible to identify which radioisotopes are present.

Gamma-rays and β-particles can both deposit their energy in tissue through which they are travelling. The γ-ray tends to lose its energy either all at once or in large lumps (see Chapter 5), the energy being transferred to an electron that then behaves rather like a β-particle. A β-particle loses its energy continuously as it passes through the tissue (or any other material). The energy is transferred to electrons which are ejected from their atoms or molecules and this tends to result in broken chemical bonds and the formation of ions. For the latter reason this process is called ionization. The destruction of critical chemical substances within the tissue (such as DNA) is thought to be the major cause of the biological effects which radiation can produce. A certain minimum energy is required to remove an electron from an atom. If the incident radiation does not have sufficient energy it will not be capable of causing ionization.

There is not a well defined, minimum energy which is required to cause ionization as this depends upon the particular atom to be ionized. However, we do know that ionization in air requires about 30 eV and it is normally assumed that at least 10 eV is needed to cause ionization in tissue. This means that radio waves, microwaves and most of the visible spectrum do not cause ionization, but the far ultraviolet, x- and γ-rays are capable of causing ionization.

6.3. PRODUCTION OF ISOTOPES

6.3.1. *Naturally occurring radioactivity*

There are several radioactive isotopes which are present in the ground and in the atmosphere which contribute to what is called 'background radiation'. This background radiation is important as it is a source of interference when measurements of radioactivity are being made.

Uranium is one of the radioactive elements present in the ground and, as traces of uranium are found in most rocks, there is background radiation everywhere. There are variations between rocks; for example, granite contains a relatively high concentration of uranium so that cities such as Aberdeen, which are built upon granite, have a relatively high background radiation.

^{238}U has a very long half-life, about 4.5×10^{10} years, but when it decays it produces atoms of much shorter half-life until, after about 20 stages, it becomes the stable lead isotope ^{206}Pb. One of these stages is the element radium which was the first radioactive isotope to be used for the treatment of disease. Another of the stages is radon, which appears in the atmosphere as a radioactive gas. In the morning when the air is very still, radon gas coming from the ground can accumulate in quite high concentrations in the air. There are usually about 10^6 radon atoms in each cubic metre of air.

Cosmic radiation

Another contribution to the background radiation is radiation which comes from the rest of the universe. Much of this radiation is absorbed by the atmosphere or deflected by the Earth's magnetic field and so never reaches the Earth, but quite a significant amount does reach ground level.

The energy of cosmic radiation is very high and it can therefore penetrate large amounts of screening. Lead, which is often used to protect people from radioactive isotopes and x-rays, is totally ineffective at stopping cosmic rays which can penetrate through the Earth to the bottom of mines. The average energy of cosmic rays is about 6000 MeV. Fortunately the total number of cosmic rays is relatively small, but they do add a significant amount to the background radiation which affects radiation detection equipment.

6.3.2. *Man-made background radiation*

This is the radiation which is emitted from isotopes which have escaped into the atmosphere either from atomic bombs or from the industrial uses of atomic energy. When many atomic bomb tests were being carried out, the 'fall-out' was considerable and the radioactive isotopes which were produced in the nuclear explosions could be detected over the entire surface of the Earth. However, atomic bomb fall out reached a peak in the early 1960s and it is now only a very small contribution to the normal background radioactivity. The contribution from the industrial uses of nuclear energy is also very small, except in very close proximity to installations such as nuclear power stations.

6.3.3. *Induced background radiation*

Because it is so energetic, cosmic radiation can produce interesting atomic changes in the atmosphere. One of these changes results in the production of carbon-14, which is a radioactive isotope of carbon. An interesting use of this isotope is in the technique of ^{14}C dating, which is not directly relevant to medical physics but forms an educational example. ^{14}C is produced in the upper atmosphere when neutrons in cosmic radiation interact with atmospheric nitrogen. The reaction that takes place is called a neutron/proton interaction:

$$^{14}_{7}\text{N} + ^{1}_{0}\text{n} \rightarrow ^{14}_{6}\text{C} + ^{1}_{1}\text{H}.$$

Note that the total atomic mass and the total number of protons are the same on both sides of the equation. The ^{14}N and the neutron come together and produce ^{14}C and the proton ^{1}H. ^{14}C is radioactive and, in fact, decays into ^{14}N with the production of a β-particle, so that the cosmic rays continuously make radioactive carbon from the non-radioactive nitrogen in the atmosphere.

Now, the small amounts of ^{14}C in the atmosphere are rapidly oxidized to produce radioactive carbon dioxide which circulates in the atmosphere. When plants absorb this carbon dioxide for photosynthesis the ^{14}C becomes incorporated in them. All living things therefore have a certain amount of radioactive carbon mixed in with their stable isotopes of carbon which are ^{12}C and ^{13}C.

When the plant dies, fresh ^{14}C is no longer added, and so the radioactive carbon slowly decays with the half-life of ^{14}C, which is about 5700 years. The amount of radioactive carbon, expressed as a fraction of the total carbon in the plant, will therefore be a maximum at the time the plant dies and will fall steadily as time passes. This is the basis of ^{14}C dating where the β-particles, which the ^{14}C emits, are counted and so related to the total amount of ^{14}C present.

Measuring the β-particles which the ^{14}C emits is not easy but it is possible to date articles by this method to an accuracy of about 50 years in 2000 years.

6.3.4.   Neutron reactions and man-made radioisotopes

The isotopes used in medicine can only be used in small quantities and yet they must produce an easily measured amount of radiation. Their specific activity must be high. The unit of activity is the becquerel and this will be defined and explained in the next section.

The easiest way to produce isotopes of high specific activity is to bombard a substance with neutrons and so produce nuclear reactions; the example given in the previous section, where ^{14}C was produced by cosmic neutrons interacting with nitrogen, was a nuclear reaction. Molybdenum-99 (^{99}Mo) is produced by the interaction of neutrons with the stable isotope molybdenum-98:

$$^{98}\text{Mo} + \text{n} = {}^{99}\text{Mo} + \gamma$$

(this is called an nγ reaction)

The source of the neutrons could be a nuclear reactor and many radioactive isotopes are produced by irradiating a stable isotope with neutrons in a reactor. Neutrons are produced in the reactor by a chain reaction in which uranium emits neutrons which cause the fission of other uranium atoms and so release further neutrons. The neutrons produced by the reactor are called slow neutrons because they have been slowed down by the graphite moderator inside the core of the reactor. They are slowed down deliberately because this increases the chance that they will interact with an atom of uranium rather than shooting straight past. The reaction which takes place is called 'neutron capture'.

Isotopes can also be produced using a cyclotron. Protons (hydrogen nuclei), deuterons (nuclei of the hydrogen isotope ^{2}H) or α-particles (helium nuclei) can be accelerated in a cyclotron and used to bombard certain elements. Neutrons can be knocked out of the nuclei of the bombarded element and radioactive products produced. These are generally positron-emitting radioisotopes. Self-contained cyclotrons for the production of the 'biological' isotopes ^{11}C, ^{13}N and ^{15}O and also the isotope ^{18}F, which can be usefully labelled onto a glucose analogue, can be purchased. All these isotopes have relatively short half-lives and with the exception of ^{18}F must be generated fairly close to the point at which they are to be used. ^{18}F can be transported over limited distances.

6.3.5.   Units of activity

In section 6.2.2 the idea of a radioactive half-life was explained. The half-life indicates how quickly a radioactive isotope decays, but it does not tell you how much of the isotope is present. This is measured by the number of atoms which decay each second and is called the amount of activity. The unit of activity is the becquerel. One becquerel (Bq) equals one disintegration per second. An older unit is the curie (Ci). $1 \, \mu\text{Ci} = 37$ kBq. For the quantities of radioisotopes used in diagnostic tests the curie is a rather large unit and the becquerel is a rather small unit. For example, for a typical diagnostic dose of

$$10 \text{ mCi} = 370 \text{ MBq (mega-becquerel)}.$$

As an isotope decays the activity decreases, so that, for example, 100 MBq of ^{123}I will have decayed to 50 MBq after one half-life, i.e. 13 h, to 25 MBq after another 13 h, and so on. It should be appreciated that activity is a measure of the rate of disintegration and not a measure of the mass of the isotope.

6.3.6. *Isotope generators*

Most radioactive isotopes can be purchased from organizations such as Amersham International and delivered either by road or rail. However, there are a number of isotopes which have such short half-lives that they have decayed away by the time they arrive. An example of this is ^{99m}Tc which is used for a range of diagnostic imaging procedures and has a half-life of 6 h.

Fortunately there is a solution to this problem. Many radioactive isotopes decay to produce a second radioactive isotope which is called a daughter product. If the mother isotope has a long half-life then, even though the daughter might have a short half-life, the daughter product will be produced continuously and can be separated from the mother when required.

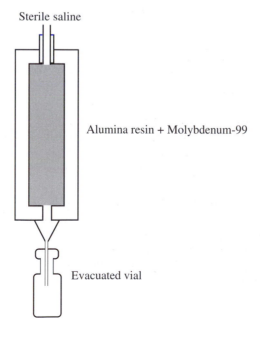

Figure 6.1. *A ^{99m}Tc generator.*

^{99m}Tc is produced from the mother isotope ^{99}Mo as follows:

$$^{99}_{42}\text{Mo} = {}^{99m}_{43}\text{Tc} + \beta\text{-particle}$$

$$\downarrow$$

$$^{99}_{43}\text{Tc} + \gamma.$$

The process by which the technetium is obtained from the cow or isotope column which contains the mother isotope, is shown schematically in figure 6.1.

Molybdenum, in the form of ammonium molybdate, is adsorbed onto an alumina column which is held in a glass tube. To obtain the daughter product, sterile saline is passed through the column and the technetium, in the form of sodium pertechnetate, is eluted (flushed out) into the lead pot. It takes some time for the concentration of daughter product to build up once the column has been milked so that the process cannot be repeated for several hours.

6.4.   PRINCIPLES OF MEASUREMENT

All radionuclide tests involve the use of small quantities of a radioactive biochemical, a radiotracer. For example, the volume of the blood can be measured by injecting a known amount of coloured dye and measuring the dilution, and kidney function can be assessed by measuring the chemical concentration of administered substances in the urine. Radioactive labels are now used instead of a dye because they have the advantage that very small quantities of tracer can be measured and are potentially much more sensitive. Most importantly using radiotracers allow us to measure many things that cannot be measured using chemical tests.

The diagnostic use of radioisotopes in medicine involves imaging, non-imaging surface-counting methods and measurement of samples. The principles of the imaging system currently in use, the gamma camera, are described in section 12.2.1 of Chapter 12. Some imaging applications are given later in this chapter (section 6.7). Non-imaging techniques will be discussed first.

6.4.1.   Counting statistics

All diagnostic radioisotope measurements, whether imaging or non-imaging, involve counting the number of γ-rays (or β-particles) detected in a given interval of time and all use the scintillation principle for detection, either a sodium iodide detector, described in section 5.6.3, or liquid scintillators (see section 6.4.2).

If we count the number of γ-rays detected in a fixed time emitted from a radioactive sample and then repeat this count over and over again we will find that the number detected varies. This variation arises from the random way that radioactive atoms disintegrate. It can be shown that if the expected number of counts over time t is c (which could be determined by averaging over many measurements) then the probability of actually measuring m counts is

$$p(m) = \frac{e^{-c}c^m}{m!}.$$

This probability distribution is a lattice function in that it is only defined at points corresponding to integer values of m. We can only get an integer (whole) number of counts. We cannot observe fractional γ-rays. This distribution is called the *Poisson distribution*. The integral of this function over all m (including zero) is unity.

$p(m)$ is the probability of getting m counts. It can be shown that the average number of counts is c and the variance in the number of counts is also c. This is an important property of the Poisson distribution.

For small values of c the distribution is not symmetric about c. For large c the distribution becomes effectively symmetric about c and can be reasonably represented by a normal distribution of mean value c and variance c. We can generally assume that this is the case for most counting experiments. In this case, if we observe a count of m it is usual to suppose that the error associated with this measurement, or the standard deviation, is given by $\sqrt{m}$. The fractional error is clearly given by $(1/\sqrt{m})$, which gets smaller as m increases. From this we can deduce that the relative accuracy of a measurement increases as the number of counts increases. Collecting as many counts as possible in any counting experiment, whether it is measuring a sample in a well counter or collecting a radionuclide image, increases accuracy.

Sometimes we may wish to express a measurement as counts per second. If we collect m counts in t seconds the count rate is $R = m/t$. Note that the standard deviation of this is not $\sqrt{R}$. If we multiply a random variable by a multiplier k, the variance of the variable is multiplied by k^2 so that the standard deviation of R is $(\sqrt{m})/t$. The fractional error is that associated with m, as it should be. The total number of counts determines the fractional error.

Suppose we perform an experiment in which we collect m counts and then a separate independent experiment in which we collect b counts. What is the variance on the total $m + b$? We could treat these two experiments as a single one in which the number of counts is $m + b$. In this case we know that $m + b$ is Poisson distributed and that the variance is $m + b$ so that the variance of $m + b$ is given by summing the variance of m and the variance of b. This is actually a general result of adding two independent random variables; the variance of the sum is the sum of the variances whatever their distributions, but it is particularly obvious in this case.

Less obvious from a simple argument like the one above, but equally true, is that the variance of $m - b$ is also the sum of the variances, i.e. $m + b$. In this case $m - b$ is not Poisson distributed (it is possible for $m - b$ to be negative). This is an important result since we often have a situation where a measurement m is contaminated by a 'background' component b (for example, cosmic radiation or in an image radiation from surrounding tissues). If we can measure b independently, then we can subtract it from m to obtain an estimate of $m - b$, the value we actually want to measure. However, in this case the fractional error on $m - b$ is given by $\sqrt{(m+b)}/(m-b)$ which is greater than $1/\sqrt{(m-b)}$ (the two become equal when $b = 0$).

If the expected value of b is of the same size as the expected value of $m - b$, then the fractional error of $m - b$ is $\sqrt{3/(m-b)}$. To match the fractional error obtainable in the absence of b, the sample would need to be counted three times longer, with a corresponding background count. The presence of a large background signal has severely degraded the accuracy of the measurement of $m - b$. Things get worse as b becomes even larger relative to $m - b$. To accurately measure low-activity samples it is necessary to make the background as low as possible.

We do not need to count the sample and background for the same length of time. For example, suppose we count the sample for a time T_m and the background for a time T_b. Then, expressing the measurements as count rates we have a fractional error in the corrected count rate $(m/T_m - b/T_b)$ of

$$\frac{\sqrt{m/T_m^2 + b/T_b^2}}{m/T_m - b/T_b}.$$

For a given total counting time $T_m + T_b$ this fractional error can be minimized provided m and b are known. In analysing images the size of the regions of interest being used plays the same role as collecting time in computing the fractional error on the count density of a background-subtracted measurement.

6.4.2. Sample counting

The quantity of radioactivity in a sample, for example a blood sample, is measured using a scintillation detector. If the energy of the emitted radiation is high enough, and is gamma radiation, then the activity in the sample is measured using a sodium iodide crystal coupled to a photomultiplier (PM) tube. The principles behind the sodium iodide detector are given in section 5.6.3 of Chapter 5. In order to maximize sensitivity it is usual to surround the sample as far as possible by crystal. There are two ways in which this is currently done,

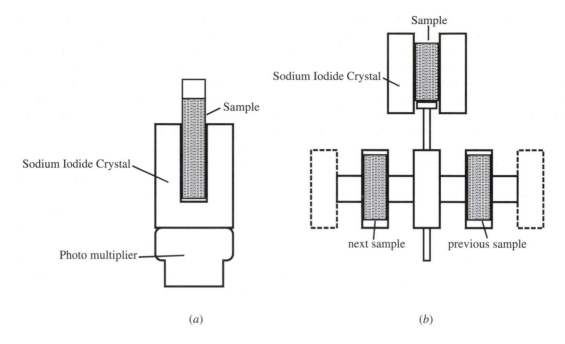

Figure 6.2. *(a) A well counter; (b) a through counter. In both cases the sample is surrounded by crystal.*

either by use of a hole in the crystal closed at one end (figure 6.2(*a*)), called a 'well' counter, or a complete hole through the crystal (figure 6.2(*b*)) called a 'through' counter. Both configurations have advantages and disadvantages.

The crystal in either configuration is typically cylindrical with a diameter of 75 mm or more and with a similar thickness. The hole might have a diameter of 20 mm and a depth of 50 mm in the case of a well counter. The sodium iodide crystal has to be contained in a light-tight container in order to exclude ambient light. Sodium iodide crystals are also hygroscopic (they absorb water vapour from the air) so that, in addition, the container has to be hermetically sealed.

Although it is still possible to find sample counters in which the samples are loaded into the detector manually, most sample counting is performed automatically. This is because it is usual in a busy laboratory for many samples to be counted each day (or overnight). An automatic sample counter can be loaded up with many (tens to hundreds) of samples. These are then automatically loaded one-by-one into the detector and counted for preset times or counts. If necessary the sample can be removed and a background count made. The count values are logged automatically by a computer and any operations such as background subtraction, decay correction and cross-channel correction (see below) can be done automatically if required. The results can be printed out or passed on in digital form for further analysis. Pulse height analysis is used to exclude as much background radiation as possible and the detector is shielded to keep the background levels as low as possible. This is especially important in automatic counting systems since the counter will contain many samples waiting to be counted so the local background radiation levels (as far as the current sample being counted is concerned) are relatively high. These machines take much of the tedium out of sample counting and allow samples to be counted for a long time, thus maximizing sensitivity.

In both cases the radioactive sample to be counted is placed in a plastic vial of standard size. When the time comes for this sample to be counted it is picked up from the input stream of samples by a robot mechanism and placed in the well counter. When counting is complete it is removed by the same method and deposited in the output stream. In the case of the through counter the sample is positioned below the detector and lifted into position by a small piston. This mechanism is less complex than a lifting arm, which leads to a mechanically more reliable system, and sample changing may be faster. However, this configuration is potentially more sensitive to increased background radiation from other adjacent samples in the sequence (see figure 6.2(b)) unless great care is taken with shielding. Adequate shielding is easier to produce in the case of the well configuration. The detector is surrounded as far as possible by thick lead.

Pulse height analysis is used to exclude as much background radiation as possible by appropriate selection of energy windows. If a single isotope is being counted the use of pulse height analysis to minimize background counts is fairly straightforward. It is unlikely that the spectrum of background radiation will have any clear photopeaks and so the smaller the window around the photopeak can be made, consistent with counting as many photopeak counts as possible, the smaller the background contamination. There will be an optimum width of the photopeak window for any particular counting experiment, although it is usually not worth determining in detail what it is. Calibration experiments should allow sensible windows to be determined. Some isotopes emit several γ-rays of different energies and most sample counters will have facilities for setting more than one window simultaneously, one for each photopeak.

Cross-channel corrections

In some cases, for example haematological measurements using iron (^{59}Fe) and chromium (^{51}Cr), two different isotopes are present simultaneously and there is a need to measure the amount of each of them independently in the same sample. In this case the γ-ray energy associated with ^{51}Cr (323 keV) is lower than those associated with ^{59}Fe (1.1 and 1.3 MeV). Although the contribution from chromium γ-rays can be completely eliminated from the determination of the iron activity by the use of an appropriate lower pulse height analyser setting on the ^{59}Fe spectrum there will always be a component of scatter from the ^{59}Fe spectrum within the ^{51}Cr photopeak window. In order to correct for this it is necessary to count a standard sample of ^{59}Fe separately in both the window over the ^{59}Fe photopeaks and also in the ^{51}Cr photopeak. Suppose we obtain F counts/Bq/s in the ^{59}Fe photopeak window and S counts/Bq/s in the ^{51}Cr photopeak window. When we are counting ^{51}Cr we will obtain C counts/Bq/s in the ^{51}Cr photopeak window and (hopefully) no counts in the ^{59}Fe window. If significant numbers of counts are obtained in this case then something is probably wrong.

Suppose we now measure a sample containing A Bq of ^{51}Cr and B Bq of ^{59}Fe. The counts/s in the ^{59}Fe window (M_{Fe}) and in the ^{51}Cr window (M_{Cr}) are

$$M_{\text{Fe}} = BF$$
$$M_{\text{Cr}} = AC + BS$$

from which we can obtain

$$B = \frac{M_{\text{Fe}}}{F}$$

and

$$A = \frac{M_{\text{Cr}} - (M_{\text{Fe}}/F)S}{C}.$$

Note that the correction to the chromium counts only needs the ratio of F to S which we can obtain by counting an iron sample of arbitrary activity.

If there is a contribution from the lower-energy isotope to the higher-energy channel (perhaps because of a small additional higher-energy photopeak) then the above correction can be modified to take this into account. Indeed, cross-channel corrections for any number of isotopes can be dealt with in this way. The general formulation of this problem can be most compactly expressed in matrix form. Note that the estimate A is obtained by subtraction of two random variables. The variance of AC is given by

$$\text{Variance}\,(AC) = M_{Cr} + \frac{S^2}{F^2} M_{Fe}$$

and clearly if the iron counts are large compared to the chromium counts then the error on the estimate of the chromium activity could be significantly increased. If possible things should be arranged so that the counts from the lower-energy isotope are as large as possible compared to those from the higher-energy isotope.

Quality issues

Since automatic sample counters are allowed to operate on their own for extended periods of time, the most crucial issue is one of stability. By this we mean that, if a sample of known activity and γ-ray energy is counted at a time t_1 and then again at a later time t_2 where the difference between these times is significantly larger than the time taken to count a whole batch of samples, the counts should not be significantly different, once any physical decay has been allowed for. If these measurements are different then the validity of any measurements made in the intervening period is compromised. There are many reasons why changes can occur. Most of them are electronic. The voltage applied to the photomultiplier (PM) tube(s) needs to be very stable. The voltages defining the energy windows need to be stable. There should be no significant drift in PM tube gain for any reason. Other potential problems are contamination of the scintillation detector due to the presence of a leaky vial, or one that has become externally contaminated and not properly cleaned, and changes in background activity due to external factors. Problems can arise, especially with through crystals, if radiation from an active sample next in line for counting can 'leak' through the shielding to the detector. This problem should have been detected when the counter underwent acceptance testing if it is a design flaw, but it could conceivably develop later if shielding is dislodged in some way, perhaps by movement.

Whatever the reason, procedures must be in place to detect any deviations from stable sensitivity. Periodic counting of a standard sample during the process of counting a large batch of samples and measuring the background activity periodically (by inserting empty vials in the batch of samples to be counted) are the best ways of establishing stability during the measurement procedure and validating the reliability of the results obtained. If this cannot be done, then measurement of standard sources before and after counting a batch of samples will indicate any significant changes in performance during counting. However, this approach will not detect transient changes in sensitivity or background. If a problem is detected which can be rectified then provided the physical half-life of the samples is long enough, the batch of samples can be recounted. In some cases if a relatively short half-life tracer is being counted this may not be possible.

On a longer time scale the photopeak windows should be regularly calibrated. The equipment will invariably show some drift over a very long time scale and although the counter may have some facilities to automatically correct for this, maximum sensitivity can only be achieved with optimal photopeak windows. Since cross-channel correction factors are at least partially dependent on the sample, these should ideally be recomputed every time they are required. Fortunately this is simply done by including in the batch of samples being counted a separate pure sample of each isotope being counted. The counts in each of the channels for these samples can be used to compute the cross-channel corrections. Remember that since these corrections are computed from sample counts they will also have errors associated with them and so sufficient counts need to be collected relative to the counts found in the patient samples being counted for these errors to be negligible.

For either type of detector the sensitivity of the measurement depends on the volume of the sample. While both the well counter and the through crystal attempt to surround the sample as much as possible there will always be some residual sensitivity to the exact position of the sample within the counting volume. This must be the same for all samples that are going to be compared in some way. Although in principle a calibration curve can be produced for this effect, for example by progressively increasing the volume of a sample of known activity and plotting the counts/s against volume, this is generally not particularly desirable. A better approach is to have all samples of the same volume. In many counting situations a crucial assumption is that all the volumes are the same (for example, counting blood samples taken at different times) and any errors in pipetting sample volumes will contribute to the overall accuracy of measurement. If volume cannot be measured to 1% there is little point in counting to this accuracy.

An additional assumption is that the sample is homogeneous. If this is not the case then sensitivity errors can arise. For example, if two samples are being compared, one with all the activity precipitated to the bottom of the vial and the other with the same activity distributed homogeneously through the sample volume, different counts will generally be obtained. Counting solid samples is particularly fraught with difficulties and wherever possible such samples should be powdered, homogenized or dissolved in liquid.

6.4.3.   *Liquid scintillation counting*

The fact that the sodium iodide crystal has to be contained in a sealed can means that very low-energy radiation cannot penetrate the container. The normal sodium iodide scintillation counter cannot be used to count γ-rays of energy below about 20 keV or β-particles of energy less than about 500 keV.

In cases where we want to measure radioisotopes which emit poorly penetrating radiations, we can bring the radioactivity and the scintillator into intimate contact by using a liquid scintillator. Sample and scintillator are mixed together in a glass or plastic vial and the scintillations are viewed with a photomultiplier tube. By this technique important biological isotopes such as tritium ^{3}H and carbon ^{14}C, which emit β-rays of energy 15 and 155 keV, respectively, can be counted.

Scintillators and solvents

The radioactive sample and the scintillator must be dissolved in the same solvent. The most common solvent is toluene. It is transparent to the light flashes emitted by the scintillator, and does not interfere with the scintillation process.

The scintillators are usually rather complex organic molecules whose names are abbreviated to acronyms such as PPO and POPOP. They are excited by transfer of energy from the β-particles emitted by the sample, and then emit photons of visible or ultraviolet light. Sometimes two scintillators are used together. The primary scintillator is the one already mentioned. The secondary scintillator absorbs the photons emitted by the primary scintillator and emits photons of its own at a different wavelength. This is done in order to match the photon wavelength to the optimum wavelength for the photomultipliers.

Counting equipment

The basic counting equipment is a photomultiplier tube which views the scintillations within the sample vial. Obviously the PM tube and sample must be housed in a light-tight box.

The flashes of light produced in the sample are of very low intensity and so methods of reducing background counts have to be employed. The major cause of the background in liquid scintillation counting is not cosmic rays or other external radiation but is the random thermal emission of electrons by the photocathodes of the PM tubes. This is reduced firstly by cooling the equipment to about 5 °C, so that thermal emission is reduced, and secondly by using a coincidence counting system.

Two PM tubes are used to simultaneously view the sample and only pulses which appear simultaneously from both tubes are counted. The genuine scintillations will cause a pulse from both tubes, whereas thermal emission from the individual PM tube cathodes will not be synchronized.

Most liquid scintillation counting systems will automatically handle several hundred samples which are counted in turn. The samples and the PM tube assembly are contained in a chest freezer and the power supplies and counting electronics are mounted above the chest. Most systems can be programmed to count each sample in turn and to print out the results. The outline of a typical liquid scintillation counter is shown in figure 6.3.

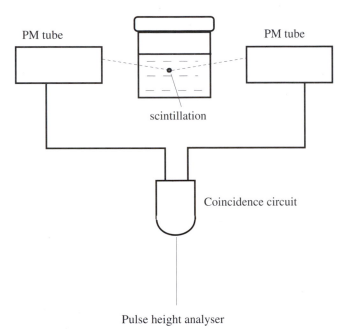

Figure 6.3. *Construction of a liquid scintillation counter. Only pulses appearing simultaneously at both PM tubes are counted. This reduces the number of background counts.*

Calibration and quenching

It is obviously necessary to be able to relate the number of counts obtained from a sample to the amount of activity present in the sample. This is the sensitivity or counting efficiency. Unfortunately the counting efficiency varies from sample to sample because of a process called quenching. This arises because of the presence of the sample material, which can affect chemically the production and detection of the scintillations. For example, if the material is coloured this may well absorb some of the scintillation radiation.

Quenching reduces the amplitude and number of the scintillations, either by interfering with the transfer of energy from the β-particles to the scintillator or by interfering with the transfer of light through the sample to the PM tubes. Both these processes depend upon the chemical composition of the sample being counted. The effect of quenching can be illustrated by plotting a spectrum of the pulses coming from the PM tubes. Figure 6.4 shows the spectra obtained as more and more of a non-radioactive sample is added to the counting

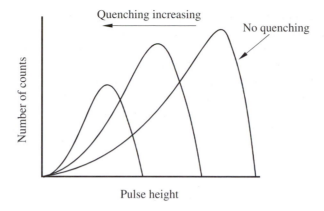

Figure 6.4. *The three pulse amplitude spectra show how both the number and the amplitude of the recorded scintillations are reduced by quenching within the sample.*

vial and the quenching increases. There are three common methods of taking quenching into account and so calculating the counting efficiency; these will be described briefly.

Internal standard

In this method the sample is first counted, then a known amount of radioactivity (the same isotope) is added and the sample is counted again. If C_s is the first count and C_{s+i} the count obtained when activity A is added, then

$$X = \frac{C_s - B}{C_{s+i} - C_s} A$$

where X is the activity of the sample and B is the background count. This method works well, but has the disadvantage that every sample must first be counted, then a known amount of activity added by pipette and then the sample counted again. This is very time consuming and the pipetting must be carried out very carefully if accurate results are to be obtained. A further disadvantage is that the sample has been contaminated by the internal standard and so cannot be recounted.

Channels' ratio method

This method involves measuring the shape of the quench curve for a particular sample and then correcting the value for the measured activity of the sample. The shape of the quench curve is described by using two pulse height analysers to give the ratio of the counts in two channels (figure 6.5(*a*)). The ratio of the counts in channel A to that in channel B will decrease as the quenching increases. This ratio depends upon the shape of the curve, but is not affected by the amplitude.

The channels' ratio, value R, and the total figure for the counts in both channels can be used to determine the activity of the sample from the quench correction curve shown in figure 6.5(*b*). The curve is produced by using the internal standard method to make a number of samples covering a wide range of quench factors. The same amount of radioactivity is pipetted into each sample but a range of volumes of a quenching agent, such as acetone or carbon tetrachloride, are added. For an unknown sample, the efficiency is determined from the quench correction curve and the activity calculated as counts divided by efficiency.

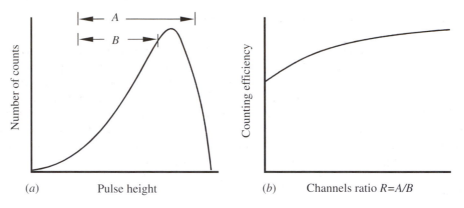

(a)   Pulse height   (b)   Channels ratio $R=A/B$

Figure 6.5. *(a) The ratio R of the counts in channels A and B decreases as quenching increases, but is not affected by the overall amplitude of the spectrum; (b) a quench correction curve. The channel ratio R is determined as shown in (a) and plotted against the counting efficiency.*

External standards method

If a source of γ-rays is brought close to the liquid scintillation sample, then scintillations will be produced by electron scattering and absorption of the γ-rays. The number of scintillations recorded will depend upon the quenching within the sample. By this means a correction factor can be obtained and used in just the same way as the channels' ratio is used as a correction factor.

^{137}Cs or ^{241}Am are often used as the source. The sample is first counted without the source present. The source is then brought close to the sample and another count taken. The ratio of this count to the count obtained in a sample with no quenching is the correction factor.

Correction curves have to be constructed using samples of known activity and a range of quenching in the same way as is done for the channels' ratio method. The correction curve gives the counting efficiency and thus enables the activity of the unknown sample to be calculated.

Errors

There is more scope for errors with liquid scintillation counting than with conventional external scintillator counting. Similar considerations apply with respect to control of sample volumes and homogeneity. In addition, the effects of quenching must be considered. The following short list covers some of the major quality control problems:

- Samples must be prepared very carefully and the quantities of solvent, sample and scintillator accurately dispensed.
- The counting vials must be chemically clean and also be uncontaminated by radioactive materials.
- The vials should not be exposed to bright lights prior to counting. The light can cause chemiluminescence in the vial and its contents and so give spurious counting results.
- Remember that radioactive decay is a random process and so the fewer the counts obtained the less accurate are the results. If only 1000 counts are obtained then the statistical accuracy will be approximately +30 counts, i.e. +3%. If 1% accuracy is required then 10 000 counts are needed.

6.5. NON-IMAGING INVESTIGATION: PRINCIPLES

Non-imaging investigations, as the name implies, do not use a gamma camera (see section 12.2.1). There are two technical consequences of this. The first is that, in general, much lower doses of activity can be used than for imaging investigations. Gamma cameras are not very sensitive devices, whereas well and surface counters can have high counting efficiencies. Whereas imaging doses are of the order of hundreds of MBq, non-imaging doses are typically of the order of hundreds of kBq, i.e. lower by two to three orders of magnitude. This reduces the radiation dose to the patient considerably. The second technical consequence is that it is possible to use isotopes, such as ^{59}Fe, which could not be used with a gamma camera. Set against this is the fact that many of the isotopes are less 'friendly' than ^{99m}Tc in that they emit β-particles or have long half-lives. Nevertheless, the radiation dose is generally substantially lower than those delivered during imaging procedures.

The fact that an image is not available means that information has to be obtained by other means. Some of these are outlined below.

6.5.1. Volume measurements: the dilution principle

Radioisotope tracers can be used to measure the volume of an inaccessible space. Here the term space does not indicate that there is a hole filled with something, but that there is a region of the body which is accessible to the particular tracer being used. An example of such a space is the plasma within the blood vessels and examples of volume measurements are measuring the volume of plasma and of the red cells in the body. Another space is the extracellular space. A general principle which can be used to measure these volumes, provided some general assumptions are valid, is the dilution principle.

The principle is simple. If activity A is injected, and a sample volume v contains activity a, then the dilution volume V is given by

$$V = A\frac{v}{a}$$

where v/a is the volume per unit activity in the space being investigated and since the total activity is A the volume of this space follows directly.

There are at least two assumptions which have to be met before this result can be relied on. These are:

- The distribution of the tracer has to be uniform within the space before the sample is withdrawn. Errors may arise because not enough time was left for the tracer to mix uniformly. However, more subtle effects can also be present. For example, in the case of red cell distribution within the blood the haematocrit (the fraction of red cells in the blood) is different in the large vessels (usually including the vessel from which the blood sample is taken) from that in the peripheral circulation.
- The space must not be leaky. In the case of measurement of plasma volume the plasma does diffuse through the vessel walls into the extracellular space. If there is too long a delay between administration of the tracer, which is labelled to plasma, and the taking of a sample, the volume reached by the tracer will be larger than the plasma volume within the blood. There is a conflict here between sampling early to avoid leakage and sampling late to ensure mixing.

The volume withdrawn should always be small compared with the volume being measured, otherwise the volume remaining will be significantly smaller than that being measured. This condition is usually met.

6.5.2. Clearance measurements

When making the measurements of a dilution volume we assumed that none of the radioactive material escaped from the volume before our sample was taken. In many cases a tracer is lost from a space within the body, or from the body itself through excretion and the rate of this loss may be required. Examples are the removal of tracer by the kidneys and the removal of ageing red blood cells by the spleen. In the simplest model of this process, such as water in a tank with a leak in the bottom, the rate of removal of the tracer from the space being sampled is proportional to the amount of activity remaining in the space

$$\frac{\mathrm{d}A(t)}{\mathrm{d}t} = -kA(t).$$

Integrating we obtain

$$A(t) = A_0\mathrm{e}^{-kt}.$$

The constant k is called the rate constant; it is often used to quantify the clearance of a radioactive substance from the body. If a substance is cleared rapidly from the body then k is large, whereas a substance retained in the body will give a small value of k.

The clearance of a substance from the body is often described in terms of a biological half-life. This is given by $(\log_e 2)/k$ and is the time taken for half the radioactivity to be cleared from the body. The term biological half-life is used since it is determined by biological processes rather than physical ones. It is additional to the physical decay of the tracer which must be corrected for before the biological half-life can be measured (see previous comments on p 166).

This simple model is useful for deriving a clinical indicator of whether a biological process is proceeding at a normal rate. However, it is a simplification. In most cases the model of the biological processes is more complex than this. For example, if kidney function is being measured by measuring the loss of a tracer from the blood, as well as being removed by the kidney the tracer may also initially diffuse into the extracellular space. Later on when the concentration in the blood has fallen the tracer will diffuse back, so raising the concentration in the blood. These processes can usually be modelled using compartmental analysis, which is a generalization of the simple model above, to deal with the situation where there are many interconnecting compartments, with different rate constants between the compartments. An example of a compartmental model is given in figure 6.9 in section 6.7.2. However, unlike the simple case where it is easy to measure k, if the model is at all complex it becomes difficult to determine the multitude of rate constants and compartment sizes with any accuracy. Compartmental analysis is outside the scope of this book.

You should also be aware that the above model is not always applicable. A particular example is that of measuring the rate of removal of red blood cells from the blood. Red blood cells are created in the bone marrow, live approximately 120 days and then are removed by the spleen and liver. Red blood cells can be labelled with a tracer. Normally there are roughly equal numbers of cells of any age (up to 120 days) in the blood and these get equally labelled. If blood cells were removed from the blood in a random fashion, independent of age, then the process would be mathematically equivalent to radioactive decay and the amount of tracer in the blood would decay exponentially. However, in normal blood only the old cells are being removed so the tracer theoretically disappears in a linear fashion rather than an exponential one. This is slightly modified by the fact that the labelling process may make the labelled cells more fragile so that they are prone to early and random removal, which leads towards an exponential disappearance, but for a normal subject we would not expect a pure exponential loss. In some diseases where cells are damaged and can be removed from the blood at any age the loss of tracer is exponential.

It is always important to be aware of the validity of the model being used.

6.5.3.   *Surface counting*

In some non-imaging tests it is appropriate to measure the uptake of tracer within specific regions of the body, for example the liver or spleen, especially as a function of time. This could be done with a gamma camera but this is not usually sensitive enough to record enough counts from the low amounts of activity given to the patient in such tests and in any case detailed information about the distribution within the organ is not really required. What is needed is a measure of the global uptake in the organ. Measures of absolute uptake are rarely required but rather the way the uptake changes with time.

Such measurements are made with a surface counter. This consists of a sodium iodide detector of reasonable size (7.5 cm diameter and the same thickness) behind a single-hole collimator. Typical dimensions of the hole are 7.5 cm diameter and 5 cm in length, although different laboratories may use different designs. The increased thickness of the crystal relative to a gamma camera means that these detectors have a higher sensitivity to higher-energy gamma radiation, such as that from ^{59}Fe, than does a gamma camera.

Measurements are made by placing the front face of the collimator against the skin surface of the patient over the organ of interest and counting the emitted γ-rays for a fixed period of time (see figure 6.6). As the organ cannot be seen its position has to be estimated from surface features, often the ribs. The counter will often only view a portion of the organ if it is large. Conversely if the organ is small then appreciable amounts of surrounding tissue will be included. This need not be a great disadvantage but if measurements over time are being made it is important to make sure that the counter is replaced in exactly the same position each time. This can be most simply achieved by drawing a circle on the patient's skin around the collimator on the first day of counting with an indelible marker, and repositioning the collimator over this circle on subsequent days. Some sort of surface marking is desirable as reproducibility is important.

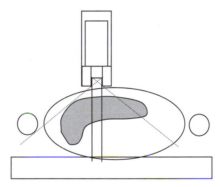

Figure 6.6. *A surface counter. The region of the patient directly below the collimator is seen with maximum sensitivity. The region between the broken and full lines (the penumbra) is seen with less sensitivity.*

The sensitivity of the counter should be regularly checked by counting a standard source placed in a standard position in front of the collimator. It is usual to fix the geometry by using a custom-made Perspex jig to position the counter and the source.

6.5.4.   *Whole-body counting*

If estimates of total-body activity are required then a simple surface counter is inadequate. To measure the absorption of, for example, bile acids it is necessary to administer a radiolabelled acid orally and measure the uptake of this acid into the body. To do this, the activity throughout the whole body needs to be measured. This needs to be done just after the tracer has been taken and then after a suitable period of time (typically 7 days). The distribution of tracer within the body will be very different on the two occasions. On the first day the entire tracer will be in the gut. On the seventh day that tracer which has been absorbed will be distributed elsewhere within the body. Any tracer not absorbed will have been excreted.

The total-body activity can be measured with a whole-body counter. The principle feature of such a system is that the sensitivity should be fairly independent of the distribution of tracer within the body of the patient. One system which can be used is a 'scanning shadow shield' system. In this system the patient is scanned between a set of, say, ten scintillation detectors, five above and five below the patient. This arrangement produces a result which is fairly insensitive to the detailed distribution of tracer within the patient, producing a total count related to the total activity within the body. The overall sensitivity will vary from patient to patient, depending on their size, but for tests which compare measurements on the same subject this is not a critical problem. Other geometries have also been used.

Whole-body counters are less common now than they used to be. An alternative approach which has been explored with some success is to use a gamma camera without a collimator. If a scanning couch is available the patient can be scanned in a similar manner to the shadow shield system, with similar results. The relatively thin crystal of gamma cameras means that they are significantly less sensitive to high-energy radiation than special purpose whole-body counters, and the background activity with the gamma camera environment may not be as low as that in a special purpose whole-body counting area. This makes them inappropriate for counting very low levels of activity.

6.6.   NON-IMAGING EXAMPLES

A variety of tests can be performed in a non-imaging laboratory. The majority are related to measuring some aspect of haematological function, sometimes several at once. The remainder are associated with gastro-intestinal function and kidney function. Most of these tests will be provided by most nuclear medicine laboratories. The detailed protocols for performing these tests will be held locally and these should be consulted for details. An extensive list of nuclear medicine tests, both imaging and non-imaging, is given in table 6.1 at the end of this chapter. This section will outline briefly a few of the more common procedures to illustrate some of the principles of non-imaging investigations.

Many of the haematological and related procedures require the labelling of blood components. For example, red cells can be labelled with at least two different isotopes, ^{51}Cr and ^{99m}Tc. The plasma protein human serum albumin (HSA) can be labelled with any of the iodine isotopes (usually ^{125}I) and also with ^{99m}Tc. White cells can be labelled with an indium tracer (^{111}In) and a ^{99m}Tc compound. Transferrin can be labelled with iron isotopes. Procedures are available for all of these. They require skill and an attention to sterile procedures as the labelled products, initially taken from the patient (HSA is usually an exception) will be re-injected back into the subject.

6.6.1.   Haematological measurements

The body contains approximately 5 l of blood, which consists mainly of red blood cells (erythrocytes) white cells (leucocytes), platelets (thrombocytes), and extracellular fluid or plasma. Blood serves a number of functions. It conveys oxygen from the lungs to the cells and carbon dioxide in the reverse direction. It carries glucose, amino acids, fats and vitamins from the gastrointestinal tract to the cells. It transports waste products to the excretory organs. It defends cells against foreign bodies. It also transports hormones for body control and maintains body temperature.

There are about 5×10^9 red cells, 7×10^5 white cells and 3×10^8 platelets in each ml of blood. The percentage by volume occupied by the red cells is the haematocrit. The red cells are about 8 μm in diameter, the white cells are larger and the platelets smaller. The density of blood is about 5% greater than water. The *red blood cells* are manufactured continuously in the red bone marrow (mainly in the ribs, vertebrae and the ends of the limb bones) and are removed by the liver and spleen when they die. The mean lifespan of red blood cells is about 120 days, whereas the white cells and platelets have a much shorter life. The *plasma* occupies about 60% of the blood volume and can be separated by centrifuging the blood. If the blood is allowed to clot then the clear fluid left is called serum; this consists of plasma minus the fibrin used in the coagulation process.

This information is very brief and basic, but even the basic facts about blood can be clinically useful. The number of red blood cells, their production rate, their lifespan and the total blood volume are all potentially useful. Isotope tests can supply this information.

Haemoglobin contains iron and constitutes approximately 35% of red cell volume. For this reason, the study of iron metabolism is important to the investigation of haematological diseases such as anaemia. The normal body contains about 4 g of iron and, of this, 75% is contained in the blood. Haemoglobin is the substance which carries oxygen around the body and so is obviously important.

Iron is absorbed from food in the gut. The amount of iron in the plasma is small, but the transferrin in the plasma is used to carry the iron to the bone marrow. The iron is used to produce haemoglobin in the bones and this is released in the red blood cells which circulate around the body. When the red cells die the iron is released and carried back by the plasma to the bone marrow. This internal circulation of iron is carried on at a rate of about 40 mg/day which is a much greater rate than that by which iron is either obtained from food or excreted by the body. Only about 1 mg/day is absorbed from food (see figure 6.7).

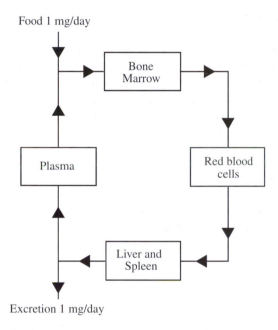

Figure 6.7. *A simple model of iron metabolism.*

Various parameters of the haematological system, such as red cell and plasma volumes, red cell survival and iron utilization can be measured using radioisotope techniques. Some examples are given in the following sections.

Red cell volume and plasma volume

Measurement of red cell volume and plasma volume can each be carried out independently. However, it is often useful to carry out these measurements together. These volumes are measured using the dilution principle.

The procedure is outlined here in some detail to illustrate general principles but you should consult your own departmental procedure for full details.

The procedure begins by taking a sample of blood ($\sim$20 ml) from the patient and separating out the red cells from the plasma using a centrifuge. The plasma is discarded. The fraction of blood by volume which consists of red cells (the haematocrit) is typically of the order of 45% in a normal subject. The red cells are labelled by the simple addition of radioactive sodium chromate, followed by a room temperature incubation period. A dose of 400 kBq is used, but not all of this will end up on red cells. After the incubation period the cells are washed and rewashed and then resuspended in saline to bring the sample volume back to the original value. 90 kBq of ^{125}I-HSA (purchased labelled) is then added.

A known volume of the mixture is then removed ($\sim$20 ml) and reinjected into the patient through an arm vein. Sufficient should be left in the vial to allow a standard to be prepared. Most non-imaging procedures require the preparation of a standard to be produced so that a measure of the administered dose is available. After 20 min, which should be sufficient to allow the radiolabelled cells and plasma to be distributed through the vascular system (the homogeneity requirement), 20 ml of blood are drawn from the opposite arm and distributed equally between two containers. The opposite arm is used because the site of injection on the other arm is likely to be contaminated with radiotracer. As the blood injected is very active compared to the blood removed, even small amounts of contamination at the injection site will result in significant contamination of the blood sample if blood is drawn from this site or close to it.

Blood from one of these samples is used to determine the haematocrit (the technique is simple but will not be described here). A substance called saponin is then added to this sample which ruptures (lyses) the red cell membranes. The cell contents disperse into the plasma and the sample effectively becomes homogeneous. This is important to minimize the volume effects associated with sample counters described above. A sample of 2 ml is withdrawn from this sample tube and placed in a counting tube.

The other sample is centrifuged at fairly high speed and 2 ml of plasma drawn off and placed in a counting tube. In the meantime, returning to the preparation of the standard, the material left behind to prepare a standard is lysed with saponin and 1 ml withdrawn. This is diluted with saline to 100 ml and then 1 ml withdrawn and placed in a counting vial. The dilution is necessary because the volume injected is much more active than the volume removed (because of dilution in the body!) and there is a risk that the count rate from this sample could be too high for the sample counter to deal with.

We now have three samples, two from the blood and a standard. Let the ^{51}Cr count rate from the standard be S. This is the counts in 2 ml of standard. But this was from a 100:1 dilution of 1 ml of the injected blood. So 1 ml of injected blood would have 50 times this count rate. In fact, we injected 20 ml of blood so that the count rate of the injected blood would be 1000 times that of the 2 ml standard. We now know the activity injected. It is (in terms of count rate) $S \times 1000$. The same argument applies for the ^{125}I measurement.

In the case of the plasma things are fairly straightforward. If the activity of the plasma sample is P counts/s the activity per unit volume is $P/2$. It follows from the dilution principle that the plasma volume (PV) is

$$\text{PV} = \frac{S \times 1000 \times 2}{P}.$$

The situation for the red cells is a little more tricky. To obtain the red cell volume (RCV) we would need to count pure red cells in the sample drawn from the blood. In fact, we counted lysed blood. The 2 ml sample corresponded to ($2 \times$ haematocrit) ml of red cells, so for the RCV we compute

$$\text{RCV} = \frac{S \times 1000 \times 2 \times \text{haematocrit}}{R}$$

where R is the count rate for this sample.

This is not quite correct because the haematocrit measured from a venous sample, as it is here, is too large. The haematocrit in the small peripheral vessels is lower than in the large vessels and so the venous haematocrit is an overestimate. To compute a more accurate estimate of RCV it is necessary to multiply the RCV value by 0.91.

There are many practical details left out here, such as the need to prevent coagulation of blood samples. When counting, cross-channel correction must be used to compensate for the ^{51}Cr scatter counts in the ^{125}I channel. These volumes in normal subjects are dependent on age, body height and weight and thus computed values can be assessed against normal ranges tailored to the individual patient.

The accuracy of these measurements depends as much on the accuracy of the volume measurements as on the counting statistics. A good laboratory technique is important.

Ferrokinetics

The behaviour of iron in the body is outlined in figure 6.7. The absorption of iron can be measured using a whole-body counter. A dose of ^{59}Fe (ferric chloride) with a loading dose of non-radioactive ferrous sulphate and ascorbic acid is administered to the patient. The body activity is then measured using the whole-body counter. One week later the body activity is measured again. Any iron not absorbed will have been excreted in the faeces. The ratio of these two measurements is a measure of the absorption of iron from the gut.

The fate of iron absorbed into the body can be investigated using ^{59}Fe. It should be obvious that the test to be described is not attempted immediately after a measurement of iron absorption.

On absorption into the body iron is bound to transferrin. Transferrin in plasma can be labelled with ^{59}Fe. A labelled plasma sample is injected into the patient. Iron is removed from the plasma by the bone marrow for incorporation into red cells so the first thing that can be measured is the rate at which this occurs. This is done by taking blood samples at frequent intervals over the 90 min following the injection of tracer, measuring the activity in the plasma and plotting the activity as a function of time. The $T_{1/2}$ derived from this curve is a measure of plasma iron clearance (PIC). If the plasma iron concentration P_{Fe} is measured in one of the plasma samples (this is a biochemical measurement of the total iron concentration per ml of plasma) then the plasma iron turnover (PIT) can be computed from

$$PIT = \frac{P_{Fe} \times 0.693}{T_{1/2}}$$

in whatever units P_{Fe} and $T_{1/2}$ were measured. The radioactive iron should eventually appear in the red cells. Blood samples taken over the two weeks following the administration of the iron should show increasing amounts of ^{59}Fe in the red cells. As there will be negligible radioactive iron in the plasma by this time the blood samples can be lysed and counted. Extrapolating the PIC curve back to zero time gives a measure of the iron per unit volume of plasma at time zero and by correction with the haematocrit the activity per unit volume of blood. The activity in the blood at day N compared to the activity injected on day zero, as calculated above, gives a measure of the fraction of iron that has been incorporated into the blood by day N. If red cell volume and plasma volume are available then these should be used instead of the haematocrit.

It is also useful to have information about the sites where ^{59}Fe is distributed. This can be obtained using a surface counter. Counts are recorded over the sacrum (the part of the spine joining onto the pelvis)—for bone marrow, the heart—for the blood pool, and the liver and spleen as potential sites of red cell destruction. Measurements are taken frequently in the first few hours and then less frequently up to 8 h or so and then on a daily basis for another week. In a normal subject the counts over the sacrum rise during the first few days as iron is removed for red cell production and then fall during the following days as red cells leave the marrow. Conversely the counts over the heart fall during the first day as iron is removed from the plasma and then rise again later as iron re-enters the blood. In normal subjects the liver and spleen are fairly neutral over this time scale. The curves will deviate from these expected patterns in explicable ways in the presence of different diseases.

This test can be further enhanced by administering ^{51}Cr labelled red cells. These are administered after measuring the plasma iron clearance. The RCV can be computed as outlined above and red cell survival can be measured by observing the rate at which labelled cells disappear from the circulation. Regular blood samples are taken over a few days and a plot of the fraction of labelled cells remaining as a function of time is produced, from which a survival half-life can be calculated. It is important to use cross-channel correction as red cells labelled with ^{59}Fe will be re-entering the blood during this period.

If it is believed that there may be significant blood loss into the gut (this may be a cause of anaemia) then faecal samples can be collected and counted to determine fractional losses.

Significant amounts of information about ferrokinetics and blood cell viability can be obtained from the sorts of measurements described above. They require a careful laboratory technique with attention being paid to obtaining reliable measurements. This is true both for the sample counting components of these tests and the surface-counting components. Careful preparation of the labelled plasma and red cells is particularly important.

6.6.2. *Glomerular filtration rate*

The kidneys clear waste products from our bodies and, if we are not to be poisoned by our own waste, need to function properly. Many substances are filtered from the blood through the glomerulus in the kidney and an important indicator of kidney function is the rate at which such substances are removed, the glomerular filtration rate (GFR). This can be measured by injecting a radiolabelled substance which is filtered by the kidney and not reabsorbed or metabolized in any way, and measuring the clearance of this substance. A suitable substance is ethylenediaminetetraacetate (EDTA) labelled with ^{51}Cr.

The procedure is fairly straightforward. A dose of the tracer is injected into the patient. An identical dose drawn from the same vial is set aside to form the standard. As the volumes injected are fairly small, the volume injected is determined by weight. The syringe with the dose is weighed and then weighed again after the dose is given, the difference being the dose given. This is done for both the patient syringe and the standard syringe so that the dose administered can be calculated (or rather the count rate associated with the dose). Blood samples are taken at 3, 4, 5 and 6 h after injection. The times at which these are taken, relative to the injection time, are accurately recorded. The plasma is separated off and 1 ml volumes taken for counting. The standard dose is diluted 100:1 and a 1 ml sample taken for counting.

From the counts derived from the 1 ml standard dose and the relative amounts of tracer administered to the patient and used in the standard (determined from the weight data for the two syringes), the activity injected into the patient can be calculated as a count rate A. The other samples are counted to obtain the count rates at the various times and these are plotted on log–linear paper. The plot is normally reasonably exponential over this time period, i.e. the graph is a straight line. The $T_{1/2}$ is determined from this graph and the graph is then extrapolated back, assuming an exponential model, to time $t = 0$. This gives a measure of plasma activity per unit volume at time zero. From this and the injected activity we can immediately determine the distribution volume (DV).

Given the half-time $T_{1/2}$ the rate of removal of the tracer can be computed in appropriate units from

$$\text{GFR} = \frac{\text{DV} \times 0.693}{T_{1/2}}.$$

The exponential curve would be acceptable if the tracer remained solely within the blood system. However, there is substantial movement into the extracellular space (figure 6.9 of section 6.7.2) and this complicates things. The curve of tracer removal from the blood is more closely described by a double-exponential model. In some cases it may be necessary to use a double-exponential model. Fitting such a model to the data is theoretically straightforward, but the computed values of the parameters of the components will become unreliable if the two components are fairly similar.

A small experimental correction to the calculated GFR to allow for the fact that there is more than one compartment is

$$\text{GFR}_c = a\text{GFR} - b(\text{GFR})^2$$

where $a = 0.99$ and $b = 0.0012$. This can produce a correction of the order of 10%.

6.7. RADIONUCLIDE IMAGING

Many uses of radioisotopes involve the use of imaging techniques. The distribution of a radioisotope within the human body can be determined using a gamma camera, which is described in detail in Chapter 12. In the above examples of the use of radioisotopes in simple *in vivo* tests the isotope has been administered in effectively elemental form. An exception was the case of red-cell survival where the isotope was labelled on to red cells before injection into the patient. In the case of imaging the use of a radioisotope in its elemental form is much rarer. In fact, many common radioisotope imaging tests use the same radioisotope, ^{99m}Tc, but labelled to a variety of different pharmaceuticals. The choice of radiopharmaceutical depends on the disease being investigated. ^{99m}Tc has some useful physical properties which make it very suitable for imaging. In particular, it has a relatively short half-life (6 h) which means that it disappears fairly rapidly from the patient after the imaging investigation is complete, and it can be generated locally in large quantities from a generator. In addition, the γ-ray emitted is not accompanied by additional β particles which limits the radiation dose to the patient, and the energy of the γ-ray at 140 keV is suitable for imaging. Other radioisotopes can be used for imaging, but few are as satisfactory as ^{99m}Tc. Technetium has been labelled to a variety of substances and new radiopharmaceuticals are constantly being produced. However, sometimes it has not proved possible to label an appropriate pharmaceutical with this isotope and an alternative has to be found. ^{111}In is an alternative which has been used for a variety of applications. In fact, the 6 h half-life of ^{99m}Tc can be a disadvantage in cases where the distribution of tracer has to be followed over a longer time period. In this case a longer-lived radioisotope must be used.

The use of radioisotopes in imaging is an enormous subject and is covered by many specialist texts which you should consult for further information. In this chapter only a few examples of its use will be given. A table of available radiopharmaceuticals is given in table 6.1 in section 6.8.

6.7.1. Bone imaging

Bone consists largely of a matrix of calcium phosphates. A variety of phosphate compounds have been developed which are rapidly incorporated into bone when introduced via the bloodstream. It has proved possible to label these compounds with ^{99m}Tc and thus visualize the bones in the body with a gamma camera. A compound widely used for bone imaging is ^{99m}Tc-methylene diphosphonate (^{99m}Tc-MDP), although others have been developed and new bone-seeking radiopharmaceuticals are still under investigation. MDP and other similar substances deposit in the normal bone, allowing the skeleton to be imaged. They also localize in bone tumours, around fractures and metabolic disorders such as Paget's disease. They do not localize in dead or necrosed bone. The mechanism of increased uptake appears to depend on a mixture of increased vascular flow and osteoblastic activity, i.e. deposition of new bone, and both of these are increased around bone tumours and fractures. In the case of fractures increased bone deposition around the site of fracture is to be expected as the bone heals and rejoins. Bone tumours are usually associated with breakdown of the bone, oseteolytic activity, and this will eventually result in the reduction of the density of the bone to the point where it breaks. However, pathological osteolytic activity seems to be associated with osteoblastic activity, and the latter at the site of a tumour results in increased deposition. There is also increased blood flow to the tumour site. Deposition is irreversible over the duration of the investigation.

The uptake of MDP is not immediate, but generally within an hour or so of injection most of the tracer that is going to be taken up is already in the bone. The remainder is in the blood and tissues. This tracer is then excreted by the kidneys. As it is removed the contrast between the bone and surrounding tissue increases, and so image quality can be improved by extending the period between administration of the radiopharmaceutical and imaging the patient. This is usually a few hours. If the normal function of the kidneys is reduced for any reason they may be seen on the images. The bladder is often seen and should be emptied if the pelvic region is being imaged.

Typically 400 MBq of ^{99m}Tc-MDP are administered to the patient intravenously. This quantity of tracer allows a complete scan of the whole skeleton to be completed in 15–20 min using a gamma camera with a whole-body imaging facility. Figure 6.8 shows an image taken with a dual-headed camera showing anterior and posterior views of a patient with normal uptake of tracer.

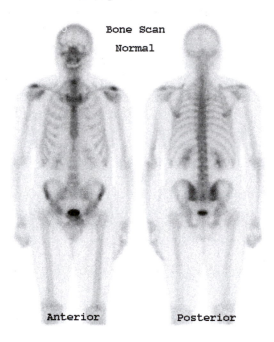

Figure 6.8. *The normal distribution of ^{99m}Tc-MDP in the skeleton. These images were taken with a whole-body scanning facility. In this case the lower limbs were not scanned.*

6.7.2. Dynamic renal function

The ability of the kidneys to clear waste substances from the body through urine is essential to good health. Kidney function can be impeded for a variety of reasons, ranging from infection to the presence of a kidney stone. Some substances are extracted from the blood via the glomerulus along with water. As they pass though the tubules most water is reabsorbed back into the body, thus leaving the extracted substances in a relatively concentrated form in the urine. Some other substances are excreted through the tubules. It is therefore possible, in principle, to distinguish between glomerular and tubular excretion and radiopharmaceuticals exist which are excreted by different routes. In practice, interest is generally in overall function and the choice of which tracer is used to image the excretory function of the kidney is rarely critical. Two radiopharmaceuticals, ^{99m}Tc-DPTA, which is extracted from the blood through tubular excretion and ^{99m}Tc-MAG3, which is extracted through glomerular excretion, are in current use, the latter becoming more common. The rest of this section assumes MAG3 is the tracer being used.

The function of the kidney is determined by imaging the way the tracer moves through the kidney as a function of time. The gamma camera is placed to view the kidneys from a posterior direction and 100 MBq of MAG3 is injected. A sequence of images are collected and stored on computer in digital form. The initial distribution of the tracer is relatively rapid and basically consists of the tracer being distributed around the blood pool. Images are usually collected rapidly in this phase of the investigation, either every 1 or every 2 s. During this phase all the tracer is in the blood, and vascular organs such as the liver and the kidneys can be visualized against the less vascular background. Failure to visualize the kidneys in this phase may indicate reduced flow to the organs. After 30–40 s from injection, image collection slows down, with each image being collected over a period of typically 20 s. During the first few minutes of this phase the kidneys, if they are functioning normally, steadily remove tracer from the bloodstream via glomerular filtration. The passage of tracer from the glomerulus to the pelvis of the kidney takes about 3 minutes and during this period following the first arrival of tracer along the renal artery, no tracer leaves the kidney. The kidney therefore accumulates tracer and the total activity in the kidney increases. The rate at which the tracer is removed is proportional to the concentration of tracer in the blood. As tracer is removed this concentration decreases and thus the rate of removal decreases. A plot of the amount of tracer in the kidney as a function of time would therefore show an approximately exponential uptake of tracer with a characteristic half-time. It is only approximate because tracer also leaves the blood for the tissue, and returns to the blood later from the tissue and this complicates the mathematics. A model describing the behaviour of the tracer is shown in figure 6.9.

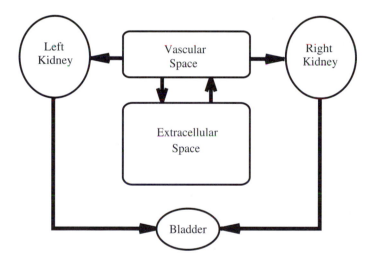

Figure 6.9. *The behaviour of ^{99m}Tc-MAG3 in the body. The tracer is injected into the vascular space. The tracer is filtered from the blood as it passes through the kidneys and is then excreted to the bladder. Tracer also moves from the vascular space into the extracellular space when the concentration in the former exceeds the concentration in the latter. When the concentration of tracer in the vascular system falls below that in the extracellular space tracer flows back again.*

After approximately 3 min the first tracer to arrive at the kidney reaches the pelvis and leaves for the bladder. By this time the rate of input has slowed considerably and thus the rate of outflow, related to the initial concentration in the blood, will exceed the rate of inflow. The total activity now falls. Images are usually collected over a period up to 25 min following the injection of tracer. A sequence of such images is shown in figure 6.10.

Figure 6.10. *A sequence of images taken every 20 s of the flow of ^{99m}Tc-MAG3 through the body. The sequence runs from the top left-hand corner along the rows to the bottom right-hand corner. The tracer is initially in the vascular system, but is then extracted by the kidneys which become visible. The tracer then passes out to the bladder.*

The shape of the curve representing the amount of tracer in the kidney as a function of time is given in figure 6.11. If the transit time through the kidney is τ seconds and the activity concentration in the artery supplying the blood to the kidney is $a(t)$ then the total activity in the kidney at time t is

$$K(t) = \int_{t-\tau}^{t} a(t)\, dt.$$

Given that $a(t)$ decays approximately in an exponential manner $K(t)$ rises in an exponential manner. This curve would have a sharp peak at time $t = \tau$, but in practice there are many parallel routes through the kidney and a spectrum of transit times. This tends to blur the peak as shown in figure 6.11.

As the image data are collected and stored in numerical form it is easy, at least in principle, to extract a curve of the form of figure 6.11. A region is defined on the image, usually manually, and the total image intensity within that region computed for each image in the sequence. For a gamma camera the total intensity in a region is equal to the total number of γ-rays detected in the region and this in turn is reasonably proportional to the activity in the kidney while that image was being collected. Unfortunately the kidney is contained within surrounding tissue which, while not actively accumulating tracer, does contain tracer within the blood and diffused from the blood. This activity will contribute γ-rays to the image in the defined region and a correction needs to be made. Usually a second region is defined close to but not overlapping the kidney which can be used to investigate how the activity in the surrounding tissue varies with time. An appropriately scaled version of the curve from this region is subtracted from the curve derived from the kidney region, to reveal a curve representing the behaviour of tracer in the kidney. Accurate determination of the scaling factor is difficult. Usually a ratio of the areas of the two defined regions is used.

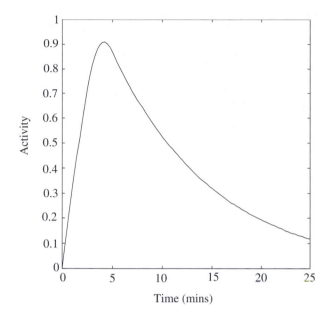

Figure 6.11. *An idealized kidney curve showing initial uptake of tracer followed by washout. The time to the peak of the curve is typically a few minutes.*

If kidney function is impaired in some way then the kidney curves will be distorted from their normal shape. In some cases the outflow of the kidney is impaired and this means that tracer cannot escape. If obstruction is substantial the activity level in the kidney will continue to rise with time to the end of the study. If there is partial obstruction a net excess of outflow may eventually be observed, but the point at which the curve reaches a maximum value may be significantly delayed relative to normal. Figure 6.12 shows curves and regions and images from a study in which there is significant obstruction in the right kidney. As well as inspecting these curves visually it is possible to make quantitative measurements on them, for example measuring the time to the peak or the relative rate of uptake of the two kidneys, which can aid in the diagnosis of the clinical problem being investigated.

6.7.3. *Myocardial perfusion*

The heart is a powerful structure of muscle which pumps blood, containing oxygen and nutrients, around the body. There are four chambers to the heart. Although all chambers must function well, the most important and the most critical is the left ventricle. In order to continue to function the left ventricle must receive a continuous supply of blood, and this is supplied by three major routes. If the blood supply to the heart down any of these routes is interrupted for even a relatively short period of time then the heart wall supplied by that route will die and form scar tissue. Such an interruption can be caused by a blood clot forming and then lodging in the supplying artery. The dead scar tissue represents an infarction. In other cases the artery wall becomes coated by fatty deposits and the lumen of the vessel is reduced in area. Under resting conditions there may still be sufficient blood flowing through the vessel to adequately perfuse the heart wall. However, if there is a need for increased blood supply to the body, for example during exercise, then there will also be a need for increased blood to the heart wall and it may not be possible to deliver oxygenated blood at a

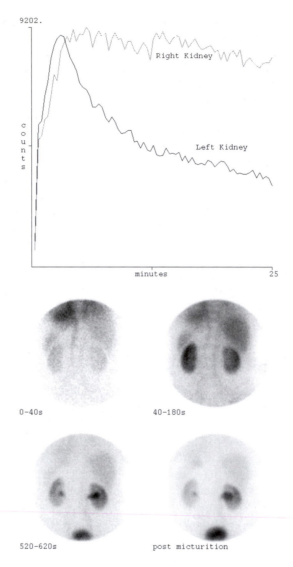

Figure 6.12. *A processed kidney study showing the background subtracted curves from the two kidneys. The outflow from the right kidney is partially obstructed and this is reflected in the shape of the kidney curve and the visual appearance of the kidney. The left kidney is normal.*

sufficient rate through the narrowed vessel to meet the needs of the heart wall. The tissue supplied by the diseased vessel becomes deoxygenated and painful. Such a condition is known as reversible ischaemia.

Infarction is permanent but ischaemia may be treatable, either by bypassing the diseased section of blood vessel with an arterial graft or opening the lumen by other techniques. Several radiopharmaceuticals exist which, when injected into the body will accumulate in the heart walls in proportion to the perfusion of the muscle. By imaging the wall after such an injection, if there are any regions which are under-perfused then these will show reduced or absent uptake of tracer relative to the perfused regions of the wall. Injecting the tracer immediately after exercise or stress induced by other means, such as an appropriate drug, will maximize the differences between normal and abnormal regions. If these regions also show on a second investigation with the tracer injected after the subject has rested an infarct is suspected. If the regions of reduced uptake seen on the first (stress) image are not seen on the second (rest) image then ischaemia is indicated.

For many years the tracer of choice was thallium, which deposits in the cardiac muscles according to perfusion. The tracer was administered during or immediately after stress and the patient imaged immediately afterwards. The patient was allowed to rest for several hours during which the tracer would redistribute to the rest condition. A diagnosis was made by comparing the two images. Usually three images were taken, an anterior, a left anterior oblique image and a lateral or nearly lateral image. From these three views the presence and anatomical location of any areas of reduced perfusion were determined. The sensitivity of the method was reduced by the presence of activity in adjacent structures and although substantial perfusion defects could clearly be seen, small defects were more problematic. With the development of reliable SPET systems (see section 12.2.5) the emphasis changed to the use of ^{99m}Tc substances, of which there are several available. These have the advantage of providing higher count rates, which are important for SPET, and lower tissue attenuation because of the higher energy. Although they can accumulate in adjacent structures such as the liver, in three dimensions the overlap on the image is substantially reduced. In general, they do not redistribute in the way that thallium does. In fact, it is desirable to wait for up to 1 h after injection before beginning imaging. Although the patient needs to be injected during or immediately after stress, as with thallium imaging, imaging does not need to be carried out immediately, which has many operational advantages. In order to obtain a rest study a second injection is needed. If this is to be done on the same day then the activity from the stress study will not have disappeared and will obviously be present during the rest study. This means that ischaemic areas may still be visualized on the rest study, even if the rest image on its own would look normal. The usual approach to this problem is to arrange for twice as much activity to be given in the rest study as for the stress study. The contribution from the stress study will be swamped by the rest study distribution, although at the expense of the stress study having to be performed with reduced activity. A second approach is to do the two investigations on different days, allowing sufficient time between investigations for the activity from the first study to have decayed to negligible levels. With ^{99m}Tc 24 h is sufficient. The principal disadvantage with this method is administrative inconvenience.

After administration of 400 MBq of, for example ^{99m}Tc-tetrofosmin and a delay of 1 h the patient is scanned with a SPET gamma camera. There are several protocols used by different centres. Ideally the detector takes a sequence of views around the patient, with the detector remaining as close to the patient as possible at all times, over a full 360°. From these data transverse tomographic images are constructed representing consecutive slices through the patient in a plane normal to the long axis of the patient. Ideally some form of attenuation correction should be applied, but this is difficult and is generally not done in routine practice. If the sequence of images are inspected it is clear that, because the heart is closer to the front of the chest (actually the left anterior oblique wall) than the back of the chest, views taken from around the right posterior oblique position do not show the heart well, whereas views around the left anterior oblique position do show the heart well. Given that the scan must be completed in a reasonable time (if only because of discomfort to the patient) it seems reasonable to devote the scan time to collecting images over only 180°, centred on the left anterior oblique position, rather than the full 360°. Although technically this incomplete data set could produce additional image artefacts, it seems to perform reasonably well and is widely used. If a dual-headed SPET system is used, with detectors fixed 180° apart, this problem does not arise.

The heart is in an inconvenient orientation for viewing as a three-dimensional data set. The long axis of the left ventricle is oriented approximately 45° to the left of centre and 30° downwards. Most computer systems associated with SPET gamma cameras allow manual reorientation of the heart so that the heart can be displayed as a sequence of slices normal to the long axis. Once this orientation is achieved (the short-axis view) the image volume is resliced to produce two data sets orthogonal to this. It is usual to display the rest image data and the stress image data together in these three orthogonal views (figure 6.13) so that they can

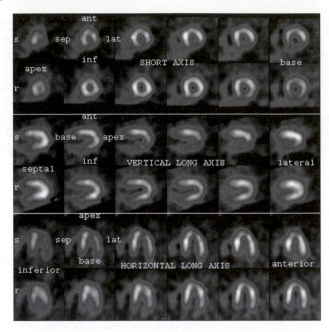

Figure 6.13. *Slices normal to the long axis of the left ventricle. A SPET myocardial examination. The three-dimensional image of the left ventricle has been displayed as slices normal to the long axis of the ventricle and slices in two orthogonal planes to this. The top row in each of the three sections is a stress image and the second row a rest image of the same subject. There is a region of decreased perfusion in the inferior wall of the heart in the stress image which is perfused in the rest image, indicating ischaemic heart disease in this territory. This figure appears in colour in the plate section.*

be directly compared. Figure 6.13 shows a stress study with a clear inferior wall defect which reperfuses on the rest study, indicating that it is an ischaemic area.

6.7.4.   *Quality assurance for gamma cameras*

In the early days of radionuclide imaging with gamma cameras a key function of quality control measures was to allow the optimization of the performance of the gamma camera. These days a modern gamma camera should be electronically and mechanically stable and it will generally be beyond the competence of medical physics staff to retune a gamma camera if performance falls outside acceptable limits. The function of quality control for a modern gamma camera is to ensure that the performance of the camera is within accepted limits and that no specific performance problems have arisen as a result of electronic or mechanical failure. As with all nuclear medicine equipment reliability and repeatability of performance are key issues.

The aim of quality control measures is to detect any significant loss of performance. There are a few basic measurements that should be made on a routine basis.

Uniformity

This is the simplest quality check that can be made. The gamma camera detector is exposed to a uniform flux of γ-ray photons. The image produced should be uniform. Early gamma cameras were unable to produce uniform images under these conditions largely because of the fact that an array of photomultiplier tubes are used to detect γ-ray events in the crystal and this inevitably introduced non-unformities of response. However, this problem, along with the problem of varying energy sensitivity across the field of view, has now been largely eliminated using front-end digital techniques (see Chapter 12).

The image can be inspected visually and can be analysed numerically. Visual inspection is quite sensitive and will usually identify if a problem is present. Various standard numerical analyses are used.

The source of radiation for the illumination of the crystal without a collimator (for measuring intrinsic uniformity) is usually a point source. This cannot produce a completely flat flux of photons but provided it is placed far enough away, is usually adequate. To produce a flux uniform to 1% over a gamma camera of 50 cm diameter the source would need to be placed 2.5 m away which may be more than the gamma camera room can accommodate.

Ideally sufficient photons must be collected to produce a random fluctuation of less than the expected intrinsic deviations from uniformity. At the 1% level this requires at least 10 000 photons per pixel. For an image of the order of 50 000 pixels a total of 500 million counts would be needed which at 50 000 counts/s would require nearly 3 h! In fact, we do not expect significant non-uniformities to occur over the distance of the few mm represented by such a small pixel and thus less counts and coarser pixels can be used. However, it is important to appreciate the need to collect adequate numbers of counts for a uniformity measurement.

Illuminating the crystal with a uniform flux of activity will detect any intrinsic detector uniformity problems. Of equal importance, though less often explicitly investigated, is the uniformity of the collimator. This needs to be at least as uniform as the detector. While it is reasonable to assume that if any non-uniformity problems are going to appear they will be associated with the detector, it is also true that collimator damage can occur which results in the regular hole pattern, especially with foil collimator designs, being distorted. Measurement of uniformity with the collimator in position requires the use of a uniform flood source of some sort. Use of a tank filled with a radioactive fluid is one possibility but achievement of a uniform mixture in a thin tank is difficult and it is virtually impossible to confirm that uniformity has been achieved. If a uniform source is required it is best to use a commercially supplied solid source. This is unlikely to have the same energy as the isotopes being imaged, but this is probably not important for quality control purposes. A common source, a 'flood' source, is based on ^{57}Co which has a relatively long half-life (270 days) and a γ-ray energy of 120 keV, which is close to that of ^{99m}Tc. This can be used to detect any collimator non-uniformities, given that the intrinsic uniformity is satisfactory.

A third approach to producing uniform flood data, which has gained some ground, is to scan a line source across the detector. A line source consists of a straight thin hollow tube filled with isotope. The concentration of tracer in such a source can be made uniform. However, the requirements on the tube for uniformity of cross-sectional area are quite stringent, as is the requirement for a constant scanning speed.

At one time it was fashionable to use uniformity measurements to correct for the non-uniformities present in images. It is now well understood that this is not an appropriate thing to do since it implies that the source of non-uniformity is due to sensitivity variations across the detector. In fact, non-uniformities can arise from other sources, such as spatial nonlinearity, which cannot be corrected in this way. Uniformity correction using flood data is now rightly frowned upon.

Spatial nonlinearity

The gamma camera should image a straight line as a straight line. Early gamma cameras failed to do this and this was, in fact, one source of non-uniformity (see Chapter 11). Spatial nonlinearity can be measured by imaging a set of parallel line sources. These can be slits cut into a thin lead sheet (if ^{99m}Tc is being imaged) which is laid on the face of the crystal (no collimator) to produce a slit phantom. The phantom is then illuminated using a point or flood source. The nonlinearity can be measured in any direction by rotating the phantom. Normally two directions would be measured. Visual inspection is normally quite good at detecting non-straight images. The deviations from linearity can be quantified by appropriate analysis of the images produced.

Theoretically spatial nonlinearity should produce non-uniformity although measuring non-uniformity is not a particularly sensitive way of detecting spatial nonlinearity.

Resolution

Intrinsic resolution can be inspected using a slit phantom. Provided the slit is smaller than the intrinsic resolution of the detector, a measurement of the full width half maximum (see section 11.4) of the image of the slit can be used to measure resolution along the line of the slit. Total system resolution, including the collimator, can be measured using a line source of activity. Note that it is quite possible for intrinsic resolution to be quite non-uniform in the presence of uniform sensitivity. It is therefore possible, in principle, for the resolution to vary from point to point without any obvious changes in sensitivity uniformity.

A more informal check of intrinsic resolution is to use a phantom consisting of arrays of equally spaced holes. The phantom is divided into six sections with each section containing a pattern of holes of a given size and spacing. The working resolution can be checked by noting the size of the holes that can just be resolved. This test will detect any gross reduction in performance.

Pixel size

Pixel size can be measured by computing the number of pixels between the centres of the images of two point sources placed a known distance apart. It should be done along both axes of the image. Pixel size is important for some quantitative measurements. An error of 3% along each axis can lead to a 10% error in pixel area.

SPET quality control

In addition to the quality control issues outlined above, SPET imaging has several problems of its own. An important requirement for all tomographic imaging is that the boundary data, the profiles, are consistent. This means that they must all have come from the same object through the Radon transform (see section 11.8.1). In SPET this is never the case. However, it is important to distinguish between factors which can be easily corrected, such as centre-of-rotation errors, and factors which cannot be simply corrected, such as the effects of attenuation and scatter. In addition to the issues of detector uniformity and resolution, the principal sources of error arise from the mechanical relationship of the detector to the space in which the image is being reconstructed.

In SPET the detector rotates around the patient. An axis of rotation is defined. It is important that the normal to the point on the detector representing the centre of the digitized image passes through the axis of rotation (the centre of rotation). It is also important that the tangential axis of the detector be normal to the axis of rotation at all positions around the axis of rotation, and that the axial axis of the detector be parallel to the axis of rotation of the system. In short, it is important that any axis of importance is either at a right angle to the axis of rotation, parallel to it or passes through it. The angular position of the detector should also agree with the position claimed by the angular position sensors.

Most of these are mechanical issues and most of them can be detected by analysing images of point sources scanned by the system. The most important error which can arise is the centre-of-rotation error since this can be quite large but can be corrected during image reconstruction. If a line is drawn from the axis of rotation to the detector such that it is normal to the detector face, the point at which it touches the detector face should be the centre of the image formed by that detector. This may not be the case for a variety of reasons. The detector system (crystal and PM tubes) may be offset in the detector head, or the event positioning system may have some electronic offsets. Offset in the axial direction can be tolerated; the net effect will be to shift the reconstructed image along the axis of rotation, but offsets in the tangential direction will cause image artefacts. The cure is simple. Each collected image is shifted along the tangential direction by the appropriate amount before image reconstruction.

To determine this amount a point source in air is scanned and the sinogram extracted (see figure 11.12(*b*) for an example of a sinogram). Let the intensity along the *i*th row of the sinogram be p_{ij}. Then the average of this function is formed as

$$I_i = \frac{\sum_{\text{all } j} i \, p_{ij}}{\sum_{\text{all } j} p_{ij}}.$$

If there is no centre-of-rotation offset the average of I_i over all rows of the sinogram is zero. If there is an offset, the average will be the value of this offset. Once this has been calculated it should be used to correct all SPET data before reconstruction.

Hopefully it should be adjusted by the manufacturer to be zero. If it drifts from this value this should be taken of evidence that something is wrong.

Computer systems

All modern gamma systems include a computer system for data acquisition and analysis. Software for the routine analysis of image data is usually included with the system and many systems allow the development of new analysis routines by the user. The degree to which this is possible varies from system to system. A potentially serious problem is the quality control of such software. A program will not 'drift' in the way that the gamma camera can do, but there is still an important issue of whether the software is supplying correct or at least reliable answers. The software may contain 'bugs' or logical errors which do not stop it executing but will result in erroneous answers. These may not be obvious (we can usually assume that obvious errors will have been detected by the developer and other users) and may only occur for unusual data inputs. In principle, such errors could be detected by analysing standard image data, and proposals along this line have been made, but unless it is possible to exhaustively test the software with a wide range of inputs this may not be a very sensitive way of detecting more subtle errors. No simple answer to this problem has been devised.

Output from the analysis of image data is rarely accompanied by an estimate of the error associated with the analysis. This should not be confused with ranges deriving from between-subject variations in physiology. If a method of analysis requires that regions of interest are drawn then different users will draw different regions and will obtain different answers when the analysis is completed. It is important to appreciate that the analysis being used may amplify small differences in procedure and produce results which differ significantly from user to user, even if the underlying methodology is correct. Any procedure which involves subjective input from a human should be used by a range of people (on the same data) to assess the errors associated with this input. It may be appropriate to tolerate an increase of the objective error of the analysis by simplifying part of it (for example, by simplifying background subtraction) if this can be shown to result in a larger reduction in the error arising from the subjective input.

Software quality control needs considerable development.

6.8.   TABLE OF APPLICATIONS

This chapter has only attempted to introduce the subject of nuclear medicine. Table 6.1 lists a wide range of clinical problems where specific nuclear medicine techniques can offer help.

Table 6.1. *Radiotracers used in diagnosis.*

Study	Clinical problem
Endocrine glands	
Thyroid	
^{99m}Tc pertechnetate	Thyroid nodules, goitre, hyperthyroidism
^{123}I Iodide	As above plus ectopic thyroid tissue
^{131}I Iodide	Primary and secondary differentiated thyroid cancer
^{99m}Tc (V) DMSA	Medullary thyroid cancer
Parathyroid	
^{201}Tl and ^{99m}Tc or	Localization of parathyroid adenoma or hyperplastic glands
^{99m}Tc isonitrile and ^{123}I	
Adrenal	
^{123}I MIBG	Location of phaeochromocytoma and secondary deposits, neuroblastoma, carcinoid tumour, medullary thyroid cancer
^{75}Se cholesterol	Adrenal cortical function, adrenal tumour detection
Cardiovascular system	
Myocardial perfusion	Myocardial ischaemia or infarction assessment of
^{201}Tl or ^{99m}Tc compounds, e.g. tetrofosmin	perfusion following myocardial revascularization
Gated blood pool	Ventricular function, e.g. in ischaemic heart
^{99m}Tc RBC or albumin	disease or in response to exercise or drugs
first pass ^{99m}Tc perfusion compounds	Abnormal wall motion
	Ventricular aneurysm
Central nervous system	
Cerebral blood flow	Cerebral ischaemic disease, dementias, epilepsy
^{99m}Tc HMPAO	
Brain scintigraphy	Blood–brain barrier abnormalities
^{99m}Tc pertechnetate	Subdural haematoma
	Primary and secondary tumours
	Response to therapy
Isotope cisternography	CSF leaks
^{111}In-DTPA	Shunt patency
Gastrointestinal tract	
Oesophageal reflux and transit	Reflux oesophagitis
Labelled non-absorbable	Motility disorders of the oesophagus
compounds (^{99m}Tc, ^{111}In, ^{113m}In)	
Gastric emptying	Assessment of disease or emptying abnormality
Labelled non-absorbable compounds	after gastric surgery
(^{99m}Tc, ^{111}In, ^{113m}In)	Drug and anaesthetic effects
Meckel's diverticulum ^{99m}Tc pertechnetate	Detection of Meckel's diverticulum
G I bleeding ^{99m}Tc RBC or ^{99m}Tc colloid	To demonstrate the site of acute gastrointestinal bleeding
Bile acid absorption ^{75}Se HCAT	Bile acid malabsorption

Table 6.1. *Continued.*

Study	Clinical problem
Haematology	
Red cell mass ^{51}Cr RBC	Diagnosis of anaemia and polycythaemia
Splenic sequestrian ^{51}Cr RBC	Investigation of anaemia and splenomegaly
Faecal blood loss ^{51}Cr RBC	Contribution of gastrointestinal bleeding to anaemia
Vitamin B$_{12}$ absorption ^{57}Co and ^{58}Co labelled B$_{12}$	Investigation of macrocytic anaemia
Ferrokinetic studies ^{59}Fe	Investigation of anaemia and quantitation of haemolysis
Platelet survival studies ^{111}In-labelled platelets	Investigation of thrombocytopenia
Red cell turnover ^{51}Cr RBC	Detection of gross haemolysis
Plasma volume ^{125}I-labelled HSA	Investigation of anaemia and polycythaemia
Whole-body iron loss ^{59}Fe	Quantitation of blood loss
Inflammation and Infection	
Detection of infection ^{111}In white blood cells ^{99m}Tc HMPAO white blood cells Labelled anti-white cell antibodies	Localization of infection and inflammation including assessment of inflammatory bowel disease
^{67}Ga citrate	Localization of infection or inflammation
Liver	
Liver imaging ^{99m}Tc colloid	Morphological studies
Hepatobiliary imaging ^{99m}Tc IDA	Acute cholecystitis, bile leaks, biliary atresia, reflux of bile into stomach and oesophagus
Renal tract	
Renal imaging ^{99m}Tc DMSA	Localization of morphology of kidneys, scarring, bilateral function
Renal function ^{99m}Tc DTPA, MAG3, orthoiodohippurate	Renal perfusion, function, evaluation of obstruction, ureteric reflux, kidney transplant progress
Respiratory	
Ventilation studies (^{81m}Kr, ^{133}Xe, ^{127}Xe) gases	Diagnosis of pulmonary embolus (PE) (in conjunction with perfusion studies below) Assessment of regional lung disease
Ventilation studies ^{99m}Tc aerosols	As ventilation studies above, lung permeability
Perfusion studies ^{99m}Tc MAA	Diagnosis of PE (in conjunction with above two)
Skeleton	
Bone imaging ^{99m}Tc phosphates and phosphonates	Detection of metastatic deposits, stress fractures, sports injuries, infection in bones, painful prostheses, Paget's disease
Tumour imaging and function	
Non-specific ^{67}Ga	Detection of tumours Evaluation of residual disease
Specific labelled antibodies ^{131}I, ^{111}In, ^{123}I, ^{99m}Tc	Detection and evaluation of tumours

Table 6.1. *Continued.*

Study	Clinical problem
Tumour imaging and function	
Somatostatin receptors	Detection of tumours
^{111}In or ^{123}I labelled	expressing somatostatin receptors
Metabolism	Differentiating tumours
^{18}FDG, labelled amino acids	Evaluation of disease activity

6.9.   PROBLEMS

6.9.1.   *Short problems*

a   What is a nucleon?
b   To what does the quantity 'MBq' refer?
c   How many stable isotopes has technetium?
d   In a scintillation counter what is the role of the photomultiplier tube?
e   Which component of a single-channel analyser selects whether a detected gamma pulse is to be counted or not?
f   How many half-lives are required for activity to fall below 1% of the initial activity?
g   What is the approximate half-life of ^{14}C?
h   The chemical symbol ^{99}Tc represents what element?
i   On what common principle do α, β and γ detectors depend?
j   Can α radiation be stopped by a piece of paper?
k   What is the mass of an α-particle?
l   Are 10 MeV x-rays used in diagnostic radiology?
m   Can you shield yourself from cosmic rays?
n   What is the principle of carbon dating?
o   What does quenching mean in liquid scintillation counting?
p   What is the minimum number of counts you would require if you want an accuracy of 1% in assessing the activity of a sample?
q   How is the rate constant related to the biological half-life when measuring the clearance of a radioactive material from the body?
r   To what will 10 MBq of ^{99m}Tc have decayed after 12 h?
s   What is the radionuclide ^{99m}Tc-MDP used for?
t   Would 10 GBq of ^{99m}Tc-MDP be a typical patient dose?
u   How could you estimate the total volume of water in the body?

6.9.2.   *Longer problems (answers are given to some of the questions)*

Question 6.9.2.1

The activity in a radioactive sample is often determined by counting the number of γ-rays or β-particles emitted from the sample in a predetermined period of time. The accuracy of the measurement depends on the number of decay events detected. If the expected number of events is N then the actual number follows a Poisson distribution of mean value N and variance N. The fractional error is therefore $1/\sqrt{N}$ and thus accuracy increases with the number of events detected.

Measurements are often contaminated by 'background' events such as cosmic rays and if the number of background events in the counting time is B, the actual number of detected events is $M = N + B$. B can be estimated by removing the sample from the counter and counting the number of background events detected in the counting time. Subtracting this from the number of events M detected while the sample is in place gives an estimate of N.

M has an error associated with it and B has an error associated with it. Calculate the error associated with $M - B$.

Suppose M is collected for unit time and B is the number of background events collected over a time T. Then N would be given by $N = M - B/T$. For given expected values of M and B what value of T minimizes the fractional error of N?

Question 6.9.2.2

In order to estimate the uptake of radioisotope in an organ, for example a kidney, a region is drawn round the organ as seen on a radionuclide image and the total events contained in this region are summed. This number is M and this value contains a contribution from tissues outside the organ ('background counts'). For the purposes of this question the region has unit area. To correct for background (uptake in tissue outside the organ) a second region is drawn over an adjacent area. The fractional area of this region, relative to the unit area of the organ region, is A. The number of events in this region is B.

Give an estimate of the total number of events contributed by the organ alone and an estimate of the error on this number.

Often the background counts are scaled by a factor designed to account for the fact that the organ may displace some background tissue and therefore unscaled background subtraction is inappropriate. This scaling factor is K. Assume that K is of the order of 1. Derive values of K which give systematic errors of the same size as the random errors arising from the random nature of radioactive emission. If $M = 300$ and $B = 50$ and $A = 1$, what range of K values does this represent?

Question 6.9.2.3

With the aid of diagrams, describe the physical principles behind the scintillation detector. Describe how such a detector can be coupled to appropriate components to create a single-channel analyser. How can this arrangement be used to obtain a gamma energy spectrum?

A scintillation crystal 6mm thick absorbs 40% of incident mono-energetic gamma photons. Calculate the crystal thickness required to absorb 95% of the incident flux.

Answer

A scintillation detector system is shown in figure 6.14 and the block diagram of a single-channel analyser in figure 6.15. This type of system is described in section 5.6.3.

A gamma spectrum can be obtained by incrementing the threshold (with fixed window width) across the entire energy range.

We can determine the crystal thickness by considering the absorption process given by

$$I = I_0 e^{-\mu x}$$

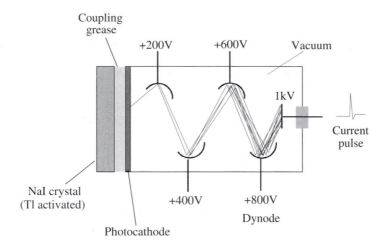

Figure 6.14. *Scintillation detector.*

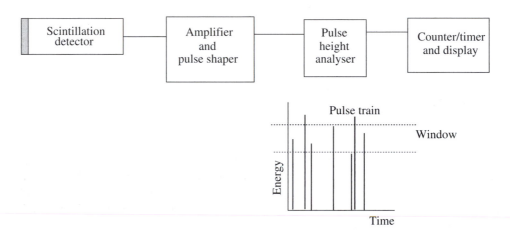

Figure 6.15. *Single-channel analyser.*

where μ is given by

$$\mu = \frac{\log_e(I_0/I)}{x} = 0.085 \qquad \text{when} \quad x = 6 \text{ mm} \quad \text{and} \quad I/I_0 = 0.6.$$

For $I/I_0 = 0.05$, x can be calculated as 35 mm.

Question 6.9.2.4

If the probability of obtaining m counts in a γ-ray counting experiment is $P(m)$, where $P(m)$ is from a Poisson distribution, show that the sum of the probabilities for all values of m is equal to 1.

Answer

The probability of getting m counts when the expected number is c is given by (the Poisson distribution)

$$P(m) = e^{-c}\frac{c^m}{m!}.$$

The sum over all m is

$$\sum_0^\infty P(m) = e^{-c}\left(1 + c + \frac{c^2}{2!} + \frac{c^3}{3!} + \cdots\right) = e^{-c}e^c = 1$$

since the term in parentheses is the series expansion of e^c.

Question 6.9.2.5

In a measurement of GFR 1 ml of ^{51}Cr-EDTA is injected into the patient. An identical volume is diluted 1000:1 and the count rate from 1 ml is 7000 counts/s. The counts/s from 1 ml of plasma at 3, 4, 5 and 6 h are

Time	3	4	5	6
Counts	99	57	41	26

Calculate the volume of distribution and the GFR from these data.

Answer

The data plot back to an activity at time zero of about 360 counts/s. The half-life is 1.6 h. The GFR is therefore 140 ml min^{-1} and the volume of distribution 19.4 l.

Question 6.9.2.6

A radioactive sample contains N atoms of a radioisotope. In 1 s M atoms undergo decay. What is the half-life of this radioisotope? ^{98}Mo decays to ^{99m}Tc with a half-life of 67 h. ^{99m}Tc decays with a half-life of 6 h. Equilibrium is reached when the rate of production of ^{99m}Tc equals the rate of decay of ^{99m}Tc. Calculate the relative activities of the two radioisotopes when equilibrium is reached. Calculate the relative numbers of Mo and Tc atoms present at equilibrium.

Answer

The probability of an atom decaying in unit time is

$$\frac{dN}{N} = -\lambda = -\frac{M}{N}.$$

The half-life is therefore

$$T_{1/2} = 0.693\frac{N}{M}.$$

Equilibrium is the point at which the rate of decay of Mo is the same as the rate of decay of Tc, thus the activities at equilibrium are identical. Then

$$N_{Tc}\lambda_{Tc} = N_{Mo}\lambda_{Mo}$$

so that

$$\frac{N_{Mo}}{N_{Tc}} = \frac{\lambda_{Tc}}{\lambda_{Mo}} = 11.17.$$

Question 6.9.2.7

A new kidney function tracer is found which does not diffuse from the blood into the extracellular space. In a particular patient both kidneys remove tracer from the blood at equal rates. This causes the activity in the blood to disappear with a half-time of 15 min. Assuming for simplicity that there is a single transit time for each kidney (rather than a spectrum of times), if the right kidney has a transit time of 3 min and the left kidney a transit time of 4 min, what will be the relative peak uptake of the two kidneys?

The uptake of the left kidney now drops to one-half of its former value. What is the relative peak uptake of the two kidneys?

Answer

As there is no diffusion into the ECS the loss of tracer from the blood is mono-exponential. The λ for this is 0.693/15 min^{-1}. The uptake of tracer in each kidney is given by

$$U(t) = A \int_0^t e^{-\lambda \tau} \, d\tau$$

and uptake continues for the period of the transit time, when washout begins.

   In the first case A is the same for both kidneys. The relative peak uptakes are

$$\frac{U_L}{U_R} = \frac{\int_0^4 e^{-\lambda \tau} \, d\tau}{\int_0^3 e^{-\lambda \tau} \, d\tau} = 1.304.$$

   If the left kidney has its uptake reduced by half then the rate of removal of tracer from the blood is reduced. The value of λ is decreased to 75% of the previous case (and the half-time extended by 33%). The ratio of peak uptakes is now

$$\frac{U_L}{U_R} = 0.5 \frac{\int_0^4 e^{-\lambda \tau} \, d\tau}{\int_0^3 e^{-\lambda \tau} \, d\tau} = 0.655.$$

Answers to short questions

a   A nucleon is a generic term for a neutron or proton.
b   'MBq' refers to mega-becquerels—millions of disintegrations per second.
c   Technetium has no stable isotopes.
d   A photomultiplier tube is used for conversion of light to an electrical current.
e   A pulse height analyser is used to select whether a detected gamma pulse is counted or not.
f   Seven half-lives are required for activity to fall below 1% of the original activity.
g   The approximate half-life of ^{14}C is 2500 years.
h ^{99}Tc is the symbol for the element technetium
i   Alpha, beta and gamma detectors usually depend upon ionization.
j   Yes, alpha radiation can be stopped by a single piece of paper.
k   An α-particle has a mass of four units.
l   No, 10 MeV x-rays are not used in diagnostic radiology.
m   No, cosmic rays are so energetic they will penetrate even down a deep mine shaft.
n   Carbon dating depends upon measurement of the decay of carbon-14 which has been 'locked into' old objects.
o   Quenching is the reduction of counting efficiency by the sample or contaminants. For example, if the sample is coloured it may reduce the light detected.

p At least 10 000 counts are required if an accuracy of 1% is required in counting the activity of a sample.
q The biological half-life is equal to 0.693/rate constant.
r 10 MBq of ^{99m}Tc will have decayed to 2.5 MBq after 12 h if we take the half life as 6 h.
s ^{99m}Tc-MDP is used in bone scanning.
t No, 10 GBq is too large a patient dose.
u Total-body water volume can be estimated by measuring the dilution of a dose of tritium (^{3}H).

BIBLIOGRAPHY

Behrens C F, King E R and Carpenter J W J 1969 *Atomic Medicine* (Baltimore, MD: Williams and Wilkins)
Belcher E H and Vetter H 1971 *Radioisotopes in Medical Diagnosis* (London: Butterworths)
Harbert J C, Neumann R D and Eckelmans W C (eds) 1996 *Nuclear Medicine: Diagnosis and Theory* (Stuttgart: Thiéme Med.)
Henkin R E 1996 *Nuclear Medicine: Principles and Practice* (St Louis, MO: Mosby)
Horton P W 1982 *Radionuclide Techniques in Clinical Investigation* (Bristol: Hilger)
McAlister J M 1979 *Radionuclide Techniques in Medicine* (Cambridge: Cambridge University Press)
Maisey M 1997 *Nuclear Medicine: a Clinical Introduction* (London: Update Books)
Parker R P, Smith P H S and Taylor D M 1978 *Basic Science of Nuclear Medicine* (Edinburgh: Churchill Livingstone)
Sharp P F, Dendy P P and Sharp W I 1985 *Radionuclide Imaging Techniques* (New York: Academic)
Sorenson J A and Phelps M E 1987 *Physics in Nuclear Medicine* (Philadelphia, PA: Saunders)

CHAPTER 7

ULTRASOUND

7.1. INTRODUCTION AND OBJECTIVES

This chapter is concerned mainly with the basic physics of ultrasound, but it also provides background material for several different areas. The propagation of pressure waves through different media underlies audiology, ultrasound techniques and, to a lesser extent, blood pressure measurement.

We will consider some of the physics of the generation and propagation of ultrasound in tissue. A description of the practical implementations of ultrasound imaging is given in Chapter 12 (section 12.3). Use of Doppler-shifted ultrasound for blood flow measurement is described in Chapter 19 (section 19.7).

Some of the questions which we will address are:

- Is sound a transverse or longitudinal vibration?
- Is energy transmitted within a sound wave?
- Why does inhaling helium make the pitch of a voice rise?
- Under what circumstances is sound refracted and reflected in the same way as light?
- Can we actually produce simple sound field patterns?
- How big should an ultrasound transducer be?

Ultrasound is sound which is at too high a pitch for the human ear to detect. It is widely used in medical diagnosis because it is non-invasive, and there are no known harmful effects at the power levels that are currently used. It is most commonly used in the same way that bats use ultrasound for locating obstacles when flying at night. The bat transmits a short burst of ultrasound and listens for the echo from surrounding objects. The time taken for the echo to return is a measure of the distance to the obstacle. If the ultrasound is transmitted into the body, the interfaces between the different structures in the body will produce echoes at different times. A display of echo size against time (an *A scan*) gives information about the position of the structures. This is used, for instance, to locate the mid-line of the brain. If the position and orientation of the transmitter and receiver are known, and the echoes are used to intensity modulate a display, a two-dimensional map of the structures within the body can be obtained. This is a *B scan* which can be used, for instance, to determine the orientation and well being of the foetus within the uterus.

The ultrasound frequency will be altered if it is reflected from moving structures. This is the *Doppler effect*. The most common use of this effect is the measurement of blood velocity using a continuously transmitted beam of ultrasound. This is dealt with in Chapter 19.

There is some mathematics in this chapter. Skip it, if you must, but make sure you read sufficient at least to answer the short questions at the end.

7.2. WAVE FUNDAMENTALS

At audible frequencies, sound waves are capable of moving the eardrum, and thus giving rise to the sensation of hearing. At higher (ultrasonic) frequencies, the sound is not audible, but can be used to image materials and measure the velocity of moving objects. Sound waves are usually defined as a compressional disturbance travelling through a material. This definition is actually too limiting, as any material which will support some shear (e.g. a solid or a liquid with non-zero viscosity) will support transverse waves as well as longitudinal (compressional) waves. However, the transverse waves are usually of much less importance than longitudinal waves in biological materials.

The physics of ultrasound waves is no different from those for audible sound and the equations relating to these were derived in section 3.4.1 of Chapter 3. The classical wave equation was derived as

$$\frac{\partial^2 \psi}{\partial t^2} = c^2 \frac{\partial^2 \psi}{\partial x^2}$$

where $\psi(x, t)$ gives the displacement of an element of fluid initially at point x and where

$$c = \sqrt{\frac{B}{\rho}}.$$

c is the speed of the wave, B the bulk modulus of the medium and ρ the density.

In an ideal gas, where $(PV)^\gamma$ is constant, c is given by

$$c = \sqrt{\frac{B}{\rho}} = \sqrt{\frac{\gamma P}{\rho}} \qquad (\gamma \cong 1.4 \text{ for air}).$$

We can answer one of the questions posed in the introduction straight away: 'Why does inhaling helium make the pitch of the voice rise?' ρ will be less for helium than for air and so, from the above equation, c will be higher than in air. Now for any transmitted wave $c = f\lambda$, where f is the frequency and λ the wavelength. The wavelength is fixed by the size of the vocal cavity and therefore frequency will increase in proportion to velocity.

Radiation pressure

A sound wave travelling through a medium exerts a static pressure on any interface across which there is a decrease in intensity along the direction of propagation. The mechanism is disputed, but the result can be used to measure the power output of ultrasound transducers. Power = force × velocity, so the transducer power can be found by using a force balance. A force of 6.7×10^{-4} N is generated by the complete absorption of the beam from a 1 W transducer in water, where the ultrasound velocity is 1500 m s^{-1}.

7.3. GENERATION OF ULTRASOUND

7.3.1. *Radiation from a plane circular piston*

We will take a plane circular piston as an example of a transducer. The method used for deriving the pressure distribution generated by the transducer can be applied to any shape of transducer. We make use of Huygen's principle, divide the surface of the transducer into infinitesimal elements, calculate the pressure due to one element, and then integrate over the surface of the transducer. The full derivation is not given here, because it is far too long, and the interested reader is referred to a standard acoustics text (e.g. Kinsler *et al* 1982).

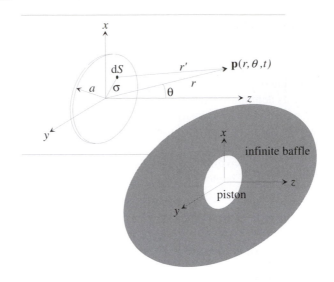

Figure 7.1. *A circular piston that can move back and forth to generate a pressure wave.*

In this discussion, we often refer to ultrasound transducers, but you should remember that the theory is equally applicable to sound that we can hear.

We start by assuming that a piston (i.e. a rigid circular plate) of radius a is mounted on a rigid infinite baffle (figure 7.1), and use the notation of Kinsler *et al.* The piston is moving normal to the baffle (i.e. in the Z direction) at a speed $U_0 \exp(j\omega t)$. Each infinitesimal element, of area dS, acts as a point source on the piston. It can be shown that the pressure generated by a point source at a distance r' with respect to time t is given by

$$p(r, t) = j\rho_0 c \frac{k U_0 \, dS}{2\pi r'} e^{j(\omega t - kr')} \tag{7.1}$$

where $U_0 \, dS$ is the strength of the source, k is the wavenumber $2\pi/\lambda$, and ρ_0 is the density of the undisturbed medium. The total pressure generated by the piston is therefore given by

$$p(r, \theta, t) = j\frac{\rho_0 c U_0 k}{2\pi} \int_S \frac{e^{j(\omega t - kr')}}{r'} \, dS \tag{7.2}$$

where the integral is over the surface $\sigma \le a$.

We will consider first the axial response of the transducer, and then examine the beam profile. For points on the axis, $r' = \sqrt{r^2 + \sigma^2}$. The axial response can be found by evaluating the integral in equation (7.2) with $r' = \sqrt{r^2 + \sigma^2}$, and the elemental area dS taken as an annulus of radius σ and width $d\sigma$:

$$p(r, 0, t) = j\frac{\rho_0 c U_0 k}{2\pi} e^{j\omega t} \int_0^a \frac{\exp(-jk\sqrt{r^2 + \sigma^2})}{\sqrt{r^2 + \sigma^2}} 2\pi \sigma \, d\sigma.$$

The integral can be evaluated, as the integrand is a perfect differential:

$$\frac{\sigma \exp(-jk\sqrt{r^2 + \sigma^2})}{\sqrt{r^2 + \sigma^2}} = -\frac{d}{d\sigma}\left(\frac{\exp(-jk\sqrt{r^2 + \sigma^2})}{jk}\right)$$

so the complex acoustic pressure is given by

$$p(r, 0, t) = \rho_0 c \, U_0 e^{j\omega t}\left[e^{-jkr} - \exp(-jk\sqrt{r^2 + a^2})\right].$$

The pressure amplitude on the axis is given by the magnitude of $p(p, 0, t)$, which after some tedious manipulation can be shown to be:

$$P(r, 0) = 2\rho_0 c\, U_0 \left| \sin \left\{ \frac{kr}{2} \left[\sqrt{1 + \left(\frac{a}{r}\right)^2} - 1 \right] \right\} \right|. \tag{7.3}$$

This is plotted in figure 7.2 for a piston radius of five wavelengths.

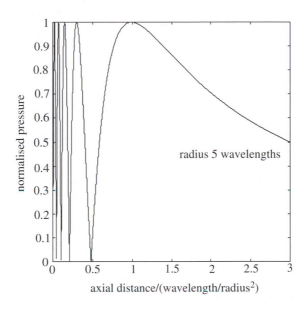

Figure 7.2. *Pressure amplitude on the axis of a circular transducer.*

If $r/a \gg 1$, i.e. the axial distance is large compared to the piston radius, this expression reduces to $P(r, 0) = 2\rho_0 c U_0 \sin(ka^2/4r)$. The pressure amplitude on the axis oscillates between 0 and $2\rho_0 c U_0$, with the final maximum at $r = a^2/\lambda$. If, in addition, $r/a \gg ka$, i.e. the axial distance is large compared to both piston size and radius, the axial pressure amplitude is asymptotic: $P(r, 0) = \frac{1}{2}\rho_0 c U_0(a/r)ka$. The pressure distribution has two different regions—the near field (also called the *Fresnel zone*) at $r < a^2/\lambda$, which displays strong interference effects, and the far field (also called the *Fraunhofer zone*) at $r > a^2/\lambda$, where the pressure displays a $1/r$ dependence. Calculation of the off-axis pressure distribution in the near field is difficult, and is discussed in Wells (1977).

The pressure distribution in the far field is interesting, and is clearly related to the light intensity distribution produced by a circular aperture. We can impose the restriction that $r \gg a$, and treat the piston as a series of line sources of different lengths (figure 7.3). We could reach the same result by integrating the elemental area dS over the piston. The strength of each source is $dQ = U_0\, dS = 2U_0 a \sin \phi\, dx$.

The pressure for this line source on an infinite baffle is

$$dp = j\rho_0 c\, \frac{U_0}{\pi r'} ka \sin \phi\, e^{j(\omega t - kr')}\, dx.$$

In the far field, with $r \gg a$, the distance r' is approximately

$$r' \approx r - a \sin \theta \cos \phi = r + \Delta r.$$

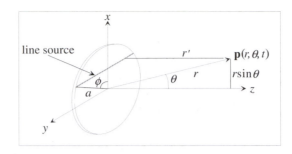

Figure 7.3. *Piston treated as a series of line sources.*

Inserting this into equation (7.3), and letting $r' \to r$ in the denominator, gives

$$p(r, \theta, t) = \mathrm{j}\rho_0 c \frac{U_0}{\pi r} ka\, \mathrm{e}^{\mathrm{j}(\omega t - kr)} \int_{-a}^{a} \mathrm{e}^{\mathrm{j}ka \sin \theta \cos \phi} \sin \phi \, \mathrm{d}x.$$

We can replace $\mathrm{d}x$ by making use of $x = a \cos \phi$ to give

$$p(r, \theta, t) = \mathrm{j}\rho_0 c \frac{U_0}{\pi} \frac{a}{r} ka\, \mathrm{e}^{\mathrm{j}(\omega t - kr)} \int_{0}^{\pi} \mathrm{e}^{\mathrm{j}ka \sin \theta \cos \phi} \sin^2 \phi \, \mathrm{d}\phi.$$

The integral has a real and imaginary part. The imaginary part vanishes, by symmetry, and the real part is a first-order Bessel function:

$$\int_{0}^{\pi} \cos([ka \sin \theta] \cos \phi) \sin^2 \phi \, \mathrm{d}\phi = \pi \frac{J_1(ka \sin \theta)}{ka \sin \theta}$$

thus giving:

$$p(r, \theta, t) = \mathrm{j}\rho_0 c U_0 \frac{a}{r} ka\, \mathrm{e}^{\mathrm{j}(\omega t - kr)} \left[\frac{J_1(ka \sin \theta)}{ka \sin \theta} \right]. \tag{7.4}$$

We can relate this to the previous result for the asymptotic case. If $\theta = 0$, the bracketed term is unity. The exponential term gives the time variation of the pressure, and the remaining term is the same as the asymptotic expression for the axial case. All of the angular dependence is in the bracketed term. If we examine a table of first-order Bessel functions, we find that $J_1(x) = 0$ for $x = 3.83, 7.02, 10.17$, etc. In other words, the beam energy is confined to lobes, with zeros at angles defined by the values of x for which $J_1(x) = 0$. Conventionally, for an ultrasound transducer, the side lobes are ignored, and the axial spread of the beam is taken as the angular size of the main lobe. The centre lobe reduces to zero for angles given by $\theta = \sin^{-1}(3.83/ka) = \sin^{-1}(0.61\lambda/a)$. For a transducer that is 10 wavelengths in diameter, the beam diverges in the far field with a half-angle of $7.0°$ (in the near field, the beam can be considered to be parallel).

If $a/\lambda < 0.61$ (i.e. the transducer is less than 1.22 wavelengths in diameter), $J_1(x) \neq 0$ for any real values of θ, so there are no side lobes. In the limit as $ka \to 0$ the transducer radiates spherical waves. Figure 7.4 shows the beam shape for a piston that is 10 wavelengths in diameter.

Although circular pistons are a good representation of loudspeakers and the classical ultrasound transducer, they have been replaced for many ultrasound applications by arrays of rectangular transducers (see section 12.3.4). It can be shown that the normalized power emitted in a direction (θ, ϕ) relative to the axis for a rectangular piston is given by

$$D(\theta, \phi) = \frac{\sin[(kl_x/2) \sin \theta]}{(kl_x/2) \sin \theta} \frac{\sin[(kl_y/2) \sin \phi]}{(kl_y/2) \sin \phi}$$

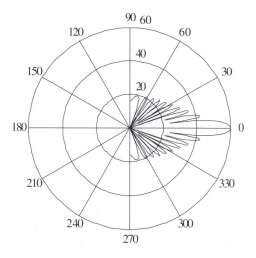

Figure 7.4. *The beam shape around a circular transducer of radius equal to 5 wavelengths. The amplitude is shown radially in dB.*

where l_x and l_y are the lengths of the sides of the plate, and θ and φ are the angular deviations from the axis in the planes parallel to l_x and l_y, respectively.

It is clear from an examination of the directivity function that it is the product of the distribution from two line sources at right angles. There is also an obvious resemblance between the circular and rectangular plates—the rectangular case is described by a zero-order Bessel function, and the circular case by a first-order Bessel function, both with an argument involving the plate size in terms of the wavelength, and the angle.

The shape of the pressure distribution in the far field is determined by the relative phase of the wave emitted by each infinitesimal element. Focused ultrasound transducers have been made by utilizing curved plates, or by placing an acoustic lens in front of the transducer. In principle, given a transducer made of many elements, and a desired beam pattern, it should be possible to solve the inverse problem and calculate the phase relationship of each element that would generate the desired field. On a relatively coarse scale, this is how a phased array transducer works. By suitably controlling the phase relationship between the (usually rectangular) elements, the ultrasound beam can be both steered and focused.

The elements in ultrasound transducers are usually used to both transmit and detect ultrasound. It can be shown (see Kinsler *et al* 1982) that a reversible acoustic transducer has the same directional properties for both transmission and reception.

This discussion has considered only continuous sine wave excitation, which is actually less common than the transmission of pulses. If discrete pulses of ultrasound are emitted, information on the range of a reflector can be found from the time taken for the pulse to travel to the reflector and return to the transducer. In principle, the effect of pulsing the continuous system could be found by convolving the continuous sine wave with a square pulse of unit amplitude. This attractive simplicity is not met with in the real world, where the physical characteristics of the transducer have to be taken into account. The principal determinants of the pulse shape are the resonant frequency and Q (the resonant frequency divided by the half-power bandwidth) of the transducer.

7.3.2. *Ultrasound transducers*

We are all familiar with the use of loudspeakers to generate audible sound. The most common type uses a stiff conical diaphragm as an approximation to a piston—the relationship between wavelength and loudspeaker size is left for the reader to explore. The driving force is the interaction between the magnetic field of a coil mounted on the cone and a fixed permanent magnet.

Ultrasound transducers are less familiar. We need a device which will act as a piston at frequencies of a few megahertz, with dimensions of the order of a few wavelengths. Table 7.1 gives the velocity of sound in some biological media, from which it can be calculated that the wavelength at 1 MHz is of the order of 1.5 mm in soft tissue. Acoustic impedance is the product of the density and velocity of the sound for the medium and is therefore very low for air.

Table 7.1. *The acoustic impedance and velocity of sound for different media.*

	Acoustic impedance (kg m^{-2} s^{-1})	Velocity of sound (m s^{-1})
Air	0.0004×10^6	330
Water at 20 °C	1.48×10^6	1480
Soft tissue	1.63×10^6	1540
Muscle	1.70×10^6	1580
Bone	7.80×10^6	4080

Ultrasound transducers are commonly made of piezoelectric materials. Quartz and tourmaline are naturally occurring piezoelectric materials, which change their shape when subjected to an electric field. The converse effect is also obtained: mechanical deformation of the crystal produces an electric field across it. Piezoelectric crystals are therefore suitable both for transmitting and receiving transducers. In practice, at frequencies which are used for medical diagnosis, an artificial material (lead titanium zirconate, PZT) is used. PZTs are ceramic ferroelectrics, which are hard, inert and impervious to water. The precise chemical composition will alter the physical and piezoelectric properties of the material, so different materials are chosen for different applications. The ceramic can be manufactured in any desired shape, so that focused transducers can be made by making the surface a section of a sphere. The frequency at which the transducer can be used is determined by the thickness of the crystal. A typical lead titanium zirconate (PZT) transducer operating at its fundamental resonant frequency would be half a wavelength thick (about 2 mm thick for a 1 MHz transducer).

Figure 7.5 shows the construction of a typical transducer. The electrodes are evaporated or sputtered onto each face of the transducer, and the backing material is chosen to give the required frequency characteristics. The backing material is usually designed to have the same acoustic impedance as the transducer, so that there is no reflection at the boundary, and to absorb ultrasound energy. A typical material is an epoxy resin loaded with tungsten powder and rubber powder, which might have an absorption coefficient of about 8 dB cm^{-1}. If the ultrasound is to be transmitted as pulses, it is usual to use the same transducer as both transmitter and receiver. If the ultrasound is to be transmitted continuously, as is the case for continuous wave ultrasonic Doppler blood flow measurement, separate transmitting and receiving crystals are mounted side by side on backing blocks separated by an acoustic insulator.

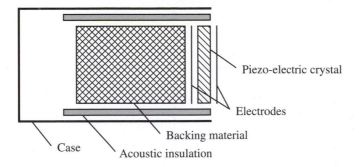

Figure 7.5. *Simplified cross-section of an ultrasound transducer.*

7.4.   INTERACTION OF ULTRASOUND WITH MATERIALS

In the current context, for audible sound, we are usually only interested in the transmission through air. Environmental and architectural acoustics, which deal with noise propagation in the world in general and in buildings in particular, are important and interesting fields in their own right, but we refer interested readers to Kinsler *et al* (1982).

Of more importance to us is the interaction of ultrasound with biological tissue, because this interaction is the basis of both ultrasound imaging techniques and Doppler blood velocity measurement. The interaction of ultrasound with tissue is complex, and what follows is a summary.

7.4.1.   Reflection and refraction

An acoustic wave incident on a boundary between two media of different acoustic impedance will give rise to reflected and transmitted waves. For most of this discussion we will assume that the boundary is plane and large compared to the wavelength, and that we are dealing with a boundary between two fluids. The latter assumption implies that the media cannot support shear waves, which is largely true for soft tissue. The pressure amplitudes and intensities of the reflected and transmitted waves depend on the characteristic acoustic impedances and the angle of incidence.

A plane wave (see figure 7.6) in medium 1 (acoustic impedance $Z_1 = \rho_1 c_1$) is incident on the boundary (at $x = 0$) with medium 2 (acoustic impedance $Z_2 = \rho_2 c_2$). The complex pressure amplitudes of the incident, transmitted and reflected waves are P_i, P_t and P_r, respectively. We define the *pressure* transmission and reflection coefficients T_p and R_p as $T_p = P_t/P_i$ and $R_p = P_r/P_i$. We have shown in section 3.4.1 that the intensity of a plane progressive wave is given by $I = p^2/2\rho c$, so the *intensity* transmission and reflection coefficients are real and given by

$$T_I = \frac{I_t}{I_i} = \frac{\rho_1 c_1}{\rho_2 c_2}|T_p|^2 \qquad \text{and} \qquad R_I = \frac{I_r}{I_i} = |R_p|^2.$$

The power transmitted by the beam is the product of the intensity and the cross-sectional area. The reflected beam always has the same cross-sectional area as the incident beam, whereas the cross-sectional area of the transmitted beam is, in general, different. The *power* transmission and reflection coefficients are therefore given by

$$T_\pi = \frac{A_t}{A_i}T_I = \frac{A_t}{A_i}\frac{\rho_1 c_1}{\rho_2 c_2}|T_p|^2 \qquad \text{and} \qquad R_\pi = R_I = |R_p|^2.$$

Energy in the beam is conserved, therefore $R_\pi + T_\pi = 1$.

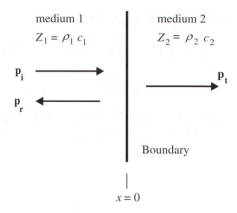

Figure 7.6. *Reflection and transmission of normally incident plane waves.*

We will first deal with the case of normal incidence, shown in figure 7.6. The incident, reflected and transmitted waves are defined by

$$p_i = P_i e^{j(\omega t - k_1 x)}$$
$$p_r = P_r e^{j(\omega t + k_1 x)}$$
$$p_t = P_t e^{j(\omega t - k_2 x)}.$$

Note the reversal of the sign of x for the reflected wave, and the different wavenumbers k_1 and k_2 in the two media, as a result of the different phase speeds c_1 and c_2. We can now state the boundary conditions which have to be satisfied. The acoustic pressures on both sides of the boundary are equal (i.e. there is no net force acting to displace the boundary), and the particle velocities normal to the boundary are equal (i.e. the two fluids do not separate). The boundary conditions are

$$p_i + p_r = p_t$$
$$u_i + u_r = u_t \qquad \text{at } x = 0.$$

Dividing pressure by velocity gives

$$\frac{p_i + p_r}{u_i + u_r} = \frac{p_t}{u_t} \qquad \text{at } x = 0. \tag{7.5}$$

A plane wave has $p/u = \pm \rho c$, so this is a statement of the continuity of the normal specific acoustic impedance across the boundary. The incident, reflected and transmitted waves must all satisfy the relationship *pressure/velocity = acoustic impedance*, therefore

$$Z_1 = p_i/u_i = -p_r/u_r \qquad \text{and} \qquad Z_2 = p_t/u_t$$

Substituting into equation (7.5) gives

$$Z_1 \frac{p_i + p_r}{p_i - p_r} = Z_2.$$

Re-arranging, and making use of $1 + R_p = T_p$, gives the reflection and transmission coefficients:

$$R_p = \frac{Z_2 - Z_1}{Z_2 + Z_1} \qquad \text{and} \qquad T_p = \frac{2Z_2}{Z_2 + Z_1}.$$

The *intensity* reflection and transmission coefficients are thus given by

$$R_I = \left(\frac{Z_2 - Z_1}{Z_2 + Z_1}\right)^2 \quad \text{and} \quad T_I = \frac{4Z_1 Z_2}{(Z_1 + Z_2)^2}. \tag{7.6}$$

For normal incidence, all the beams have the same cross-sectional area, so the power coefficients are the same as the intensity coefficients.

In practice, of course, we are much more likely to see oblique incidence on the boundary than normal incidence. We will therefore now consider oblique incidence, again for the boundary of two fluids that do not support shear waves (figure 7.7).

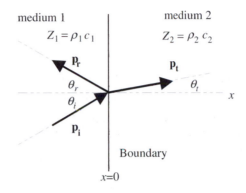

Figure 7.7. *Reflection and transmission of an oblique incident beam.*

The incident, reflected and transmitted waves now subtend angles of θ_i, θ_r and θ_t, respectively, at the x-axis. The equations for the incident, reflected and transmitted waves now include the angles θ,

$$p_i = P_i e^{j(wt - k_1 x \cos\theta_i - k_1 y \sin\theta_i)}$$
$$p_r = P_r e^{j(wt + k_1 x \cos\theta_r - k_1 y \sin\theta_r)} \tag{7.7}$$
$$p_t = P_t e^{j(wt - k_2 x \cos\theta_t - k_2 y \sin\theta_t)}$$

(note that θ_t is complex, for reasons that will become clear).

Once again, the pressure at the boundary is continuous

$$P_i e^{-jk_1 y \sin\theta_i} + P_r e^{-jk_1 y \sin\theta_r} = P_t e^{-jk_2 y \sin\theta_t}.$$

This condition applies for all values of y, so the exponents must be the same. We can therefore find the relationship between the angles of incidence, reflection and refraction by equating the exponents,

$$\sin\theta_i = \sin\theta_r \quad \text{and} \quad \frac{\sin\theta_i}{c_1} = \frac{\sin\theta_t}{c_2}.$$

These will be recognized from optics, the first states that the angles of incidence and reflection are equal, and the second is Snell's law (see section 3.5.1). As the exponents are equal, continuity of pressure at the boundary gives the previous expression,

$$1 + R_p = T_p. \tag{7.8}$$

The second boundary condition is continuity of the normal component of boundary velocity,

$$u_i \cos\theta_i + u_r \cos\theta_r = u_t \cos\theta_t.$$

Substituting $u = p/Z$ and making use of $\theta_i = \theta_r$, and remembering the definitions of R_p and T_p gives

$$1 - R_p = \frac{Z_1 \cos \theta_t}{Z_2 \cos \theta_i} T_p.$$

T_p can be eliminated by making use of equation (7.8) to give

$$R = \frac{Z_2/Z_1 - (\cos \theta_t / \cos \theta_i)}{Z_2/Z_1 + (\cos \theta_t / \cos \theta_i)}. \tag{7.9}$$

This is referred to as the Rayleigh reflection coefficient. From Snell's law we can find the angle of transmission in terms of the velocity of sound:

$$\cos \theta_t = \sqrt{1 - (c_2/c_1)^2 \sin^2 \theta_i}.$$

Three cases can be distinguished, according to the relative velocities of sound in the two media and the associated angle of incidence.

- $c_1 > c_2$. The angle θ_t is real and less than the angle of incidence—the beam is bent towards the normal for all angles of incidence.
- $c_1 < c_2$. A critical angle of incidence exists for which $\theta_t = 90°$, given by $\sin \theta_c = c_1/c_2$. The angle of transmission is real but the beam is bent away from the normal for $\theta_i < \theta_c$.
- $c_1 < c_2$ and $\theta_i > \theta_c$. We now find that $\sin \theta_t > 1$, i.e. $\cos \theta_t$ is pure imaginary (hence the requirement to have θ_t complex). We can show, using equation (7.7), that

$$p_t = P_t e^{\gamma x} e^{j(\omega t - k_1 y \sin \theta_i)}$$

where

$$\gamma = k_2 \sqrt{(c_2/c_1)^2 \sin^2 \theta_i - 1}.$$

The transmitted wave propagates parallel to the boundary with an amplitude that has an exponential decay perpendicular to the boundary. The incident wave is totally reflected. Once again, the cross-sectional areas of the incident and reflected beams are equal, so the power reflection coefficient is $R_\pi = |R_p|^2$ as before. The power transmission coefficient can be determined from $R_\pi + T_\pi = 1$ and equation (7.9):

$$T_\pi = \begin{cases} \dfrac{4(Z_2/Z_1) \cos \theta_t / \cos \theta_i}{(Z_2/Z_1 + \cos \theta_t / \cos \theta_i)^2} & \theta_t \text{ real} \\ 0 & \theta_t \text{ imaginary.} \end{cases} \tag{7.10}$$

If the media can support shear waves, the boundary conditions must be satisfied parallel to the boundary, as well as perpendicular to the boundary. The resulting analysis is complex, particularly for anisotropic media such as tissue, and will not be pursued here. The shear wave velocity is always less than the longitudinal wave velocity, so that a shear wave can still be propagated into the second medium when the angle of incidence is equal to the critical angle. As the angle of incidence is further increased, the shear wave will approach the boundary. It is thus possible to propagate a shear wave along the boundary by a suitable choice of angle of incidence. This has been used as a means of measuring the mechanical properties of the skin.

7.4.2. Absorption and scattering

A pressure wave travelling through a medium will decrease in intensity with distance. The loss of intensity is due to the divergence of the beam, scattering which is not specular, mode conversion at interfaces and absorption. Beam divergence and scattering reduce the intensity but not the energy of the beam, whereas mode conversion and absorption both result in the beam energy being converted to another form. The most important contribution to absorption comes from relaxation processes, in which the ultrasound energy is converted into, for instance, vibrational energy of the molecules of the tissue. This is not a one-way process—there is an exchange of energy. If an increase in pressure compresses the tissue, work is done on the tissue which is recovered as the pressure drops and the tissue expands. In general, the returned energy is out of phase with the travelling wave, so the beam intensity is decreased. The lost energy is converted to heat. The absorption is frequency dependent and varies for different materials. As the absorption usually increases with frequency, it is normally quoted in terms of dB cm^{-1} MHz^{-1}, but this should not be taken to mean that the absorption is strictly linear with frequency. Typical values are given in table 7.2. The very high absorption in bone and lung tissue means that they are effectively opaque to ultrasound, and structures which are behind them will be hidden. As the absorption rises with frequency, there will be a maximum depth for detecting echoes with ultrasound of a particular frequency. Frequencies of 5–10 MHz can be used for scanning the eye, but the upper limit for the abdomen is 2–3 MHz.

Table 7.2. *The absorption of ultrasound by different body tissues.*

Tissue absorption (dB cm^{-1} MHz^{-1})	
Blood	0.18
Fat	0.63
Liver	1.0
Muscle	1.3–3.3 (greater across fibres)
Bone	20
Lung	41

The scattering of the beam is dependent on the relative size of the scattering objects and the wavelength of the ultrasound. The earlier treatment assumed that a plane wave was incident on a plane boundary that was large compared to the wavelength. For 1 MHz ultrasound in soft tissue, the wavelength is 1.54 mm, so the large object scenario would correspond to scattering from the surface of major organs within the body. If the irregularities in the scattering surface are about the same size as the wavelength, diffuse reflection will occur. If the scatterers are very small compared with the wavelength, then Rayleigh scattering will take place, in which the incident energy is scattered uniformly in all directions, with a scattering cross-section proportional to $k^4 a^6$, where a is the radius of the scatterer. Obviously, with this form of scattering, very little energy will be reflected back to the transducer. Red blood cells are about 8–9 μm in diameter and act as Rayleigh scatterers. In practice, the scattering from small objects is complex, because there are many scatterers, and the scattering cannot be treated as a single-body problem. Even in blood, which is a fluid, 36–54% by volume of the blood is red cells, and the mean distance between cells is only 10% of their diameter.

The signal which is received by the transducer will be greatly attenuated due to all these mechanisms. Energy will be absorbed in propagating the mechanical vibrations through the tissue, and energy will be lost by scattering at every interface. Refraction at interfaces will divert the ultrasound away from the transducer, and the divergence of the ultrasound beam will also reduce the received energy. The received echoes used to form a typical abdominal scan may be 70 dB below the level of the transmitted signal, and the signals from moving red blood cells may be 100–120 dB below the transmitted signal. The attenuation will be roughly proportional to the frequency of the ultrasound and to the distance that the ultrasound has travelled through the tissue. It is usually necessary to compensate for the increasing attenuation of the signals with distance by increasing the amplification of the system with time following the transmitted pulse. For more detailed information on the diagnostic uses of ultrasound refer to Wells (1977) and McDicken (1991).

7.5. PROBLEMS

7.5.1.   Short questions

a   What is the difference between an *A scan* and a *B scan*?
b   Does a sound wave exert a force on the surface of an absorber?
c   What name is attached to the far field of an ultrasound transducer?
d   What is the approximate velocity of sound in tissue?
e   Could a transducer of diameter 20 mm give a narrow beam of ultrasound in tissue at a frequency of 1 MHz?
f   What property have tourmaline and quartz that makes them suitable for ultrasound transducers?
g   Does ultrasound use mainly transverse or longitudinal vibrations?
h   Is the velocity of transverse shear waves greater or less than longitudinal waves?
i   Does the absorption of ultrasound increase or decrease with increasing frequency?
j   Is the absorption of ultrasound greater in muscle than in blood?
k   What is Rayleigh scattering?
l   Typically how much smaller are the ultrasound echoes from blood than from liver and muscle?
m   If the velocity of sound in tissue is 1500 m s^{-1} what is the wavelength of 10 MHz ultrasound?
n   What is the approximate attenuation of a 2 MHz ultrasound signal in traversing 10 cm of tissue?
o   What range of ultrasound frequencies are used in medicine?
p   Does the velocity of ultrasound depend upon the density and viscosity of the medium?
q   Does Snell's law apply to acoustic waves?
r   Define the Q of an ultrasound transducer.
s   What happens to the energy that is lost as a beam of ultrasound is absorbed?
t   What is the purpose of using phased arrays of ultrasound transducers?

7.5.2.   Longer questions (answers are given to some of the questions)

Question 7.5.2.1

We will assume that red blood cells can be considered as spheres of diameter 10 μm. At what frequency would the wavelength of sound be equal to the diameter of the cells? Would it be feasible to make an ultrasound system which operated at this frequency?

Question 7.5.2.2

What gives rise to specular reflection and to refraction of a beam of ultrasound?

If 4% of the intensity of an ultrasound beam is reflected at the boundary between two types of tissue, then what is the ratio of the acoustic impedances of the two tissues? Assume the beam is at normal incidence to the boundary.

Now assume that the angle of the beam is increased from the normal and that the transmitted intensity falls to zero at an angle of 60°. Calculate the ratio of the velocity of sound in the two tissues. If the ratio of the densities is $\sqrt{3}$:1 then which of the two tissues has the higher impedance?

Answer

Specular reflection occurs at the boundary between two tissues that have different acoustic impedances. Refraction will occur as a result of a difference in sound velocities in the two tissues.

Equation (7.6) gave the reflection coefficient R_I as

$$R_I = \left(\frac{Z_2 - Z_1}{Z_2 + Z_1}\right)^2$$

therefore

$$\left(\frac{Z_2 - Z_1}{Z_2 + Z_1}\right) = (0.04)^{1/2} = \pm 0.2$$

and so either $4Z_2 = 6Z_1$ or $4Z_1 = 6Z_2$. *The ratio of the acoustic impedances is thus either 2:3 or 3:2*

Snell's law was used following equation (7.9) to give

$$\cos\theta_t = \sqrt{1 - (c_2/c_1)^2 \sin^2\theta_i}.$$

The transmitted intensity will fall to zero when the term inside the square root sign becomes negative. The term is zero when

$$\frac{c_1}{c_2} = \sin\theta_i = \frac{\sqrt{3}}{2}.$$

Therefore, the ratio of the velocity of sound in the two tissues $c_1:c_2$ is $\sqrt{3}:2$

The ratio of the densities is either $1:\sqrt{3}$, i.e. $\rho_2 > \rho_1$ or $\sqrt{3}:1$, i.e. $\rho_2 < \rho_1$.

Now the acoustic impedance is the product of the density and the velocity. Therefore the ratio $Z_1:Z_2 = \rho_1 c_1 : \rho_2 c_2$ and $Z_1:Z_2$ is either 1:2 or 3:2.

In the first part of the question we showed that the ratio of the acoustic impedances was either 2:3 or 3:2, so the answer must be 3:2. i.e. $\rho_2 < \rho_1$.

The tissue onto which the beam falls first has the higher impedance.

Question 7.5.2.3

We design an ultrasound transducer to operate at 1.5 MHz and it has a radius of 5 mm. Using the derivations given in section 7.3.1 estimate the fall of intensity of the ultrasound beam between distances of 10 and 15 cm along the axis of the beam. Express your answer in dB.

List any assumptions that you make.

Answer

If we assume that the radius of the transducer (a) is much less than the distances along the axis (r), then we can simplify equation (7.3) to give

$$P(r, 0) \propto \sin\frac{ka^2}{4r}$$

where k is the wavenumber equal to $2\pi/\lambda$.

If the velocity of sound in tissue is assumed to be 1500 m s^{-1} then λ is 1 mm.

At $r = 10$ cm: $P_{10} = \sin\dfrac{2\pi\,10^3 \times 25 \times 10^{-6}}{4 \times 10^{-1}} = 0.383.$

At $r = 15$ cm: $P_{15} = \sin\dfrac{2\pi\,10^3 \times 25 \times 10^{-6}}{6 \times 10^{-1}} = 0.259.$

The reduction in intensity is thus

$$20\log\frac{P_{15}}{P_{10}} = -3.4 \text{ dB}.$$

Question 7.5.2.4

We are to use an ultrasound transducer of diameter 15 mm, operated at 1.5 MHz to obtain a Doppler signal from the aortic arch. The intention is to use this to measure cardiac output. If the distance between the probe and the blood vessel is 10 cm then calculate the width of the ultrasound beam at the blood vessel. Would the beam width be sufficient to completely insonate the cross-section of the blood vessel if the diameter of the vessel is 16 mm?

Answer

Following equation (7.4) it was shown that an ultrasound beam will diverge in the far field with a half angle given by $\sin^{-1}(0.61\lambda/a)$ where λ is the wavelength and a is the radius of the transducer.

In our case the wavelength is 1 mm if the velocity of sound is assumed to be 1500 m s^{-1}.

Therefore, the half angle of the beam is given by

$$\sin^{-1}\left(\frac{0.61 \times 10^{-3}}{7.5 \times 10^{-3}}\right) = 4.67°.$$

The far field starts at $r = a^2/\lambda = 56$ mm. The beam width at a distance of 100 mm is given by $15 + 2(100 - 56)\tan(4.67°) = 22.2$ mm.

This is greater than the diameter of the blood vessel so that the vessel would be fully insonated and we would get Doppler information from the whole of the cross-section.

Answers to short questions

a   An *A scan* displays echo size against time. In a *B scan* the echo size is used to modulate display intensity on a two-dimensional map of the tissue.

b   Yes, sound does exert a force on an absorbing surface.

c   The Fraunhofer zone is the name attached to the far field of an ultrasound transducer.

d   The velocity of sound in tissue is approximately 1500 m s^{-1}.

e   Yes, the wavelength is 1.5 mm. The transducer diameter is more than 10 times the wavelength so a narrow beam of ultrasound should be produced.

f   Tourmaline and quartz have piezoelectric properties.

g   Ultrasound is mainly a longitudinal vibration.

h   The velocity of transverse shear waves is less than that of longitudinal waves.

i   Absorption of ultrasound increases with frequency.

j   Yes, ultrasound has a higher absorption coefficient in muscle than in blood.

k   Rayleigh scattering is the scattering of ultrasound by objects much smaller than a wavelength, e.g. red blood cells.

l   The ultrasound echoes from blood are about 100 times (40 dB) smaller than from muscle.

m $c = f\lambda$, therefore $\lambda = 1500/(10 \times 10^6) = 150\ \mu$m.

n   Attenuation is approximately 1 dB cm^{-1} MHz^{-1}, therefore there will be 20 dB attenuation in traversing 10 cm of tissue.

o   The range of ultrasound frequencies used in medicine is usually 1–10 MHz.

p   Velocity depends upon the density and the elasticity of the medium.

q   Yes, Snell's law can be used to understand acoustic waves.

r   The *Q* of a transducer is the resonant frequency divided by the bandwidth at half the maximum output.

s   The energy lost as a beam of ultrasound is absorbed will appear as heat.

t   Phased arrays are used to both steer and focus an ultrasound beam.

BIBLIOGRAPHY

Bushong S and Archer B 1991 *Diagnostic Ultrasound: Physics, Biology and Instrumentation* (St Louis, MO: Mosby)

Docker M and Dick F (eds) 1991 *The Safe Use of Diagnostic Ultrasound* (London: BIR)

Duck F, Baker A C and Starritt H C 1999 *Ultrasound in Medicine* (Bristol: IOP Publishing)

Evans J A (ed) 1988 *Physics in Medical Ultrasound* (London: IPSM) reports 47 and 57

Kinsler L E, Frey A J, Coppens A B and Sanders J V 1982 *Fundamentals of Acoustics* (New York: Wiley)

Lerski R A 1988 *Practical Ultrasound* (New York: IRL)

McDicken W N 1991 *Diagnostic Ultrasonics: Principles and Use of Instruments* 3rd edn (Edinburgh: Churchill Livingstone)

Wells P N T 1977a *Biomedical Ultrasonics* (New York: Academic)

Wells P N T 1977b *Ultrasonics in Clinical Diagnosis* (Edinburgh: Churchill Livingstone)

Zagzebski J 1996 *Essentials of Ultrasound Physics* (St Louis, MO: Mosby)

CHAPTER 8

NON-IONIZING ELECTROMAGNETIC RADIATION: TISSUE ABSORPTION AND SAFETY ISSUES

8.1. INTRODUCTION AND OBJECTIVES

An understanding of the interaction of electromagnetic radiation with tissue is important for many reasons, apart from its intrinsic interest. It underpins many imaging techniques, and it is essential to an understanding of the detection of electrical events within the body, and the effect of externally applied electric currents. In this chapter we assume that you have a basic understanding of electrostatics and electrodynamics, and we deal with the applications in other chapters. Our concern here is to provide the linking material between the underlying theory and the application, by concentrating on the relationship between electromagnetic fields and tissue. This is a complex subject, and our present state of knowledge is not sufficient for us to be able to provide a detailed model of the interaction with any specific tissue, even in the form of a statistical model. We have also limited the frequency range to $<10^{16}$ Hz, i.e. radio waves to ultraviolet. Higher frequencies (ionizing radiation) were covered in Chapter 5.

Some of the questions we will consider are

- Does tissue conduct electricity in the same way as a metal?
- Does tissue have both resistive and capacitive components?
- Do the electrical properties of tissue depend upon the frequency of electromagnetic radiation?
- Can any relatively simple models be used to describe the electrical properties of tissue?
- What biological effects might we predict will occur?

When you have finished this chapter, you should be aware of

- The main biological effects of low-frequency electric fields.
- How tissue electrical properties change with frequency.
- The main biological effects of high-frequency fields such as IR and UV.
- How surgical diathermy/electrosurgery works
- Some of the safety issues involved in the use of electromedical equipment.

There is some mathematics in this chapter, mainly in sections 8.2 and 8.3. These two sections assume some knowledge of the theory of dielectrics. However, the rest of the chapter should be understandable to all of our readers. The purpose of including the first two sections is to give a theoretical basis for understanding the rest of the chapter which covers the practical problems of tissue interactions with electromagnetic fields. If our understanding of these interactions is to be quantitative then we need to have a theoretical baseline.

8.2.   TISSUE AS A LEAKY DIELECTRIC

If two electrodes are placed over the abdomen and the electrical impedance is measured between them over a wide range of frequencies then the results obtained might be as shown in figure 8.1. The results will depend somewhat upon the type and size of electrodes, particularly at the lowest frequencies, and exactly where the electrodes are placed. However, the result is mainly a function of the tissue properties. The impedance always drops with increasing frequency. This chapter is concerned first with trying to explain why tissue impedance changes with frequency in this way. It is an important question because unless we understand why tissue has characteristic electrical properties then we will not be able to understand how electromagnetic fields might affect us. We will start in this section by considering tissue as a lossy dielectric and then look at possible biological interactions with electromagnetic fields in later sections.

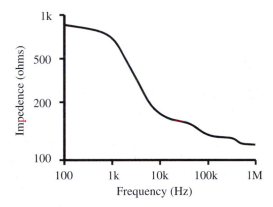

Figure 8.1. *Electrical impedance as a function of frequency for two electrodes placed on the abdomen.*

We are familiar with the concept of conductors, which have free charge carriers, and of insulators, which have dielectric properties as a result of the movement of bound charges under the influence of an applied electric field. Common sense tells us that an insulator cannot also be a conductor. Tissues though contain both free and bound charges, and thus exhibit simultaneously the properties of a conductor and a dielectric. If we consider tissue as a conductor, we have to include a term in the conductivity to account for the redistribution of bound charges in the dielectric. Conversely, if we consider the tissue as a dielectric, we have to include a term in the permittivity to account for the movement of free charges. The two approaches must, of course, lead to identical results.

We will begin our exploration of the interaction between electromagnetic waves and tissue by exploring the properties of dielectrics. We are familiar with the use of dielectrics which are insulators in cables and electronic components. A primary requirement of these dielectrics is that their conductivity is very low ($<10^{-10}$ S m^{-1}). Metals and alloys, in which the conduction is by free electrons, have high conductivities ($>10^4$ S m^{-1}). Intermediate between metals and insulators are semiconductors (conduction by excitation of holes and electrons) with conductivities in the range 10^0–10^{-4} S m^{-1}, and electrolytes (conduction by ions in solution) with conductivities of the order of 10^0–10^2 S m^{-1}. Tissue can be considered as a collection of electrolytes contained within membranes of assorted dimensions. None of the constituents of tissue can be considered to have 'pure' resistance or capacitance—the two properties are inseparable.

We start by considering slabs of an ideal conductor and an ideal insulator, each with surface area A and thickness x (see figure 8.2). If the dielectric has relative permittivity ε_r then the slab has a capacitance $C = \varepsilon_0 \varepsilon_r A / x$. The conductance of the slab is $G = \sigma A / x$, where the conductivity is σ. It should be borne

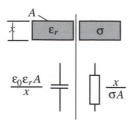

Figure 8.2. *Slabs of an insulator, on the left, and a conductor, on the right. The capacitance of the insulator and the resistance of the conductor are given.*

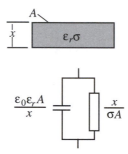

Figure 8.3. *Tissue with both capacitive and resistive properties in parallel. The capacitance and resistance of the two arms are marked.*

in mind that the conductivity σ is the current density due to unit applied electric field (from $J = \sigma E$), and the permittivity of free space ε_0 is the charge density due to unit electric field, from Gauss' law. The relative permittivity $\varepsilon_r = C_m/C_0$, where C_0 is the capacitance of a capacitor *in vacuo*, and C_m is the capacitance with a dielectric completely occupying the region containing the electric field. This background material can be found in any book on electricity and magnetism. In tissue, both of these properties are present, so we take as a model a capacitor with a parallel conductance, as shown in figure 8.3.

The equations $C = \varepsilon_0 \varepsilon_r A/x$ and $G = \sigma A/x$ define the static capacitance and conductance of the dielectric, i.e. the capacitance and conductance at zero frequency. If we apply an alternating voltage to our real dielectric, the current will lead the voltage.

Clearly, if $G = 0$, the phase angle $\theta = \pi/2$, i.e. the current leads the voltage by $\pi/2$, as we would expect for a pure capacitance. If $C = 0$, current and voltage are in phase, as expected for a pure resistance. For our real dielectric, the admittance is given by $Y^* = G + j\omega C$, where the * convention has been used to denote a complex variable (this usage is conventional in dielectric theory).

We can, as a matter of convenience, define a generalized permittivity $\varepsilon^* = \varepsilon' - j\varepsilon''$ which includes the effect of both the resistive and capacitive elements in our real dielectric.

ε' is the real part and ε'' is the imaginary part.

We can relate the generalized permittivity to the model of the real dielectric by considering the admittance,

$$Y^* = G + j\omega C = \frac{A}{x}(\sigma + j\omega\, \varepsilon_0\varepsilon_r).$$

By analogy with an ideal capacitance C which has admittance $j\omega C$, we can define the complex capacitance C^* of the real dielectric,

$$C^* = \frac{Y^*}{j\omega} = \frac{A}{x}\left(-\frac{j\sigma}{\omega} + \varepsilon_0\varepsilon_r\right) = \frac{A}{x}\varepsilon_0\varepsilon^* = \varepsilon^* C$$

i.e.

$$\varepsilon^* = \varepsilon_r - \frac{j\sigma}{\omega\varepsilon_0}$$

thus

$$\varepsilon' = \varepsilon_r \quad \text{and} \quad \varepsilon'' = \frac{\sigma}{\omega\varepsilon_0}.$$

From this it can be seen that we can consider the properties of our non-ideal capacitor as being the result of inserting a dielectric with a relative permittivity ε^* in an ideal capacitor C. The real part ε' is the relative permittivity ε_r of the ideal capacitor, and the imaginary part $j\varepsilon''$ is associated with the resistive properties. We now have a means of handling real dielectrics which is analogous to that for ideal dielectrics.

We can also consider the admittance in terms of a complex conductivity,

$$Y^* = G + j\omega C = \frac{A}{x}(\sigma + j\omega\varepsilon_0\varepsilon_r) = \frac{A}{x}\sigma^*$$

i.e.

$$\sigma^* = \sigma + j\omega\, \varepsilon_0\varepsilon_r.$$

The complex permittivity and complex conductivity are related by

$$\sigma^* = j\omega\,\varepsilon^*\varepsilon_0.$$

We are thus able to relate the behaviour of the conductivity and permittivity. Note that as the frequency tends to zero, the complex conductivity becomes purely real, and in the high-frequency limit, the complex permittivity becomes purely real. We would thus expect the conductivity to be dominant at low frequencies, and the permittivity to be dominant at high frequencies.

8.3. RELAXATION PROCESSES

If there is a redistribution of charges within the dielectric as a result of applying the electric field, the response to a step function of the electric field will be a function of time, and the equations for the capacitance C and conductance G will only be true in the limit as time goes to infinity. In other words, ε and σ are a function of frequency, and have a characteristic time constant which is called a *relaxation time*. There are a number of models of relaxation processes, all of which give qualitatively similar results. We will outline the Debye model.

8.3.1. *Debye model*

When a capacitor is charged, the applied electric field E either creates or orientates electrical dipoles within the material. This process is called polarization. By making the assumption that the process of polarization

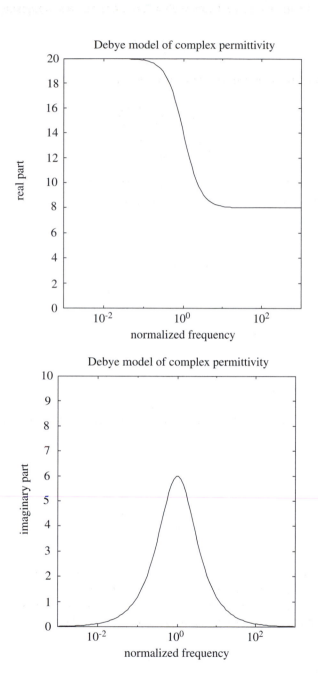

Figure 8.4. *The real (top) and imaginary (bottom) parts of the complex permittivity given by the Debye equations.*

has an exponential approach to its final value it can be shown that the real and imaginary parts of the complex permittivity will be given by

$$\varepsilon^* = \varepsilon_\infty + \frac{(\varepsilon_s - \varepsilon_\infty)}{1 + j\omega\tau}.$$

Now $\varepsilon^* = \varepsilon' - j\varepsilon''$, so by equating real and imaginary parts we find

$$\varepsilon' = \varepsilon_\infty + \frac{(\varepsilon_s - \varepsilon_\infty)}{1 + \omega^2\tau^2}$$

$$\varepsilon'' = \frac{(\varepsilon_s - \varepsilon_\infty)\omega\tau}{1 + \omega^2\tau^2}$$

where ε_∞ and ε_s are the permittivities at infinite and zero (static) frequencies, respectively. These two equations are the Debye equations. The real and imaginary parts of the complex permittivity are plotted in figure 8.4.

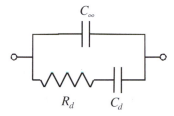

Figure 8.5. *Electrical circuit equivalent of the Debye model.*

The same results would be obtained for an equivalent circuit as shown in figure 8.5,

$$C_\infty = \frac{\varepsilon_0\varepsilon_\infty}{k}$$

$$R_d = \frac{k\tau}{\varepsilon_0(\varepsilon_s - \varepsilon_\infty)}$$

$$C_d = C_s - C_\infty = \frac{\varepsilon_0(\varepsilon_s - \varepsilon_\infty)}{k}$$

(k is a geometrical factor).

However, when measurements of the relative permittivity are made over a wide range of frequencies, it is found that the results do not agree with the predictions from the Debye model. In particular, the measured values spread over a wider range of frequencies, and the value of ε'' is too low. Cole and Cole (1941) introduced an empirical modification to match the models to the data. Their model is widely used, and will be described next.

8.3.2. Cole–Cole model

The two figures given in figure 8.6 show the real and imaginary parts of a relaxation process. The full curves with the steeper gradients are the prediction from the Debye model. The less steep curves show a typical measurement from actual tissue. The agreement between theoretical and measured values is not very good. Cole and Cole (1941) proposed an alternative formula, which was similar to the Debye equations but had an addition term called alpha.

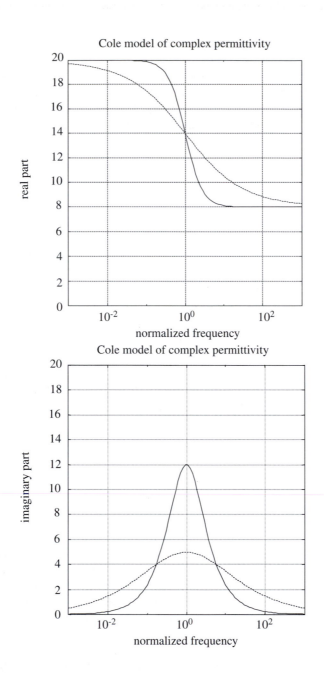

Figure 8.6. *The real (top) and imaginary (bottom) parts of the Cole and Debye models for a relaxation process. The less steep curves are for the Cole model and agree well with many measurements made on biological tissues.*

Cole proposed that the real distribution could be modelled by

$$\varepsilon^* - \varepsilon_\infty = \frac{(\varepsilon_s - \varepsilon_\infty)}{1 + (j\omega\tau)^{(1-\alpha)}}. \tag{8.1}$$

Alpha can be chosen to give a good fit to measured data. It can be seen that, if $\alpha = 0$, the equation reduces to the Debye model. (A word of warning, both $(1 - \alpha)$ and α are used as the exponent by different workers!)

It can be shown that the real and imaginary components of the complex permittivity are given by the following equations, which reduce to the values derived for the Debye model when $\alpha = 0$:

$$\varepsilon' = \varepsilon_\infty + \frac{(\varepsilon_s - \varepsilon_\infty)\left[1 + (\omega\tau)^{1-\alpha}\sin(\alpha\pi/2)\right]}{1 + 2(\omega\tau)^{1-\alpha}\sin(\alpha\pi/2) + (\omega\tau)^{2(1-\alpha)}}$$

$$\varepsilon'' = \frac{(\varepsilon_s - \varepsilon_\infty)(\omega\tau)^{1-\alpha}\cos(\alpha\pi/2)}{1 + 2(\omega\tau)^{1-\alpha}\sin(\alpha\pi/2) + (\omega\tau)^{2(1-\alpha)}}.$$

It is worth emphasizing at this point that the Cole–Cole model is a model of the measured data. It has been applied very successfully to a wide variety of materials and interactions over the past 60 years, but it does not give any information about the underlying causes of the phenomena being measured.

A number of workers use a form of the Cole equation written in terms of impedance instead of a complex permittivity. The impedance Z is given by

$$Z = R_\infty + \frac{R_0 - R_\infty}{1 + (jf/f_c)^{1-\alpha}}$$

where R_0 and R_∞ are the resistances at zero frequency (i.e. DC) and infinity, respectively. f_c is often referred to as the characteristic frequency. It should be emphasized that the characteristic frequency is not the same when the analysis is carried out in terms of the complex permittivity.

A simple interpretation of the above equation is in terms of a circuit where a resistance S is in series with a capacitor C and this combination is placed in parallel with a resistance R. In this case $R_0 = R$ and $R_\infty = RS/(R + S)$. It can be shown that f_c is given by $1/(2\pi C(R + S))$. The circuit is the same as that shown for the Debye model in figure 8.5, but with the resistors and capacitors interchanged.

8.4.   OVERVIEW OF NON-IONIZING RADIATION EFFECTS

If electromagnetic radiation is incident on the human body then it will be absorbed in discrete amounts or quanta. This amount may be sufficient for ionization of atoms to occur. The equation $E = h\nu$, where h is Planck's constant (6.624×10^{-34} J s), gives the relationship between the energy E of the radiation and the frequency ν. Quantized radiation absorption is an important concept where we are dealing with ionizing radiation, but it becomes less important where lower-energy radiations are concerned. Table 8.1 lists some of the major components of the electromagnetic spectrum and the associated energies and frequencies. An energy of about 10 eV is required for ionization of an atom to occur.

Because the human body is at a temperature of 37 °C (310 K), i.e. normal body temperature, all the atoms will be moving around and the associated thermal energy is given by kT, where k is Boltzmann's constant ($k = 1.38 \times 10^{-23}$ J deg^{-1}) and T is the temperature. This is about 0.03 eV. It corresponds to a frequency of 7×10^{12} Hz, which is in the infrared part of the spectrum. At lower frequencies than the infrared the quantized energy states are so close together that absorption is normally considered as a continuum. In this region we do not normally consider absorption by single atomic events but by a continuous energy loss. There is a grey area between the 0.03 eV thermal energy and the 10 eV required for ionization of atoms. Single-photon events may occur in this region. Table 8.2 lists some of the possible types of absorption that

Table 8.1. *Principal components of the electromagnetic spectrum and associated wave and energy parameters.*

Components	Wavelength in air (m)	Frequency (Hz)	eV
γ-radiation	1.3×10^{-13}	2.3×10^{21}	10^7
Ultraviolet	10^{-7}	3×10^{15}	13
Visible light	5×10^{-7}	6×10^{14}	2.6
Infrared	10^{-6}	3×10^{14}	1.3
Microwaves	10^{-2}	3×10^{10}	1.3×10^{-4}
Radio waves	10	3×10^7	1.3×10^{-7}

Table 8.2. *Approximate activation energies for some possible biological interactions with radiation in the infrared to ultraviolet part of the spectrum. (Selected values taken from a table given by S F Cleary in J G Webster (ed) Encyclopedia of Medical Devices and Instrumentation (New York: Wiley). Reprinted by permission of John Wiley & Sons, Inc.)*

Effect	eV	Frequency (GHz)
Ionization	10	2.4×10^6
Covalent-bond disruption	5	1.2×10^6
Photoconduction	1–4	$(2.4–9.6) \times 10^5$
Dielectric relaxation of proteins	0.4	9.6×10^4
Dielectric relaxation of water (25 °C)	0.2	4.8×10^4
Hydrogen bond disruption	0.1–0.2	$(2–5) \times 10^4$
Thermal energy (38 °C)	0.03	7.2×10^3

may occur in this region. All of these interactions might occur and be of biological significance in the infrared (IR) to ultraviolet (UV) part of the spectrum.

Most of the electromagnetic radiations to which we are exposed are of lower energy than the 0.03 eV thermal energy level. Some examples are given in table 8.3. Most of these radiations are most likely to be absorbed in tissue as heat. There is a very large difference between the total amount of heat involved in the absorption of ionizing and non-ionizing radiations. For example, the sun exposes us to an energy field of approximately 10^3 J s^{-1} m^{-2} and can raise our body temperature by several degrees, whereas the temperature rise resulting from a lethal exposure to gamma radiation will be only about one-hundredth of a degree centigrade.

It might seem that the electromagnetic radiations given in table 8.3 are most unlikely to cause significant biological interactions. However, it is possible that absorption by many photons might raise energy levels sufficiently to disrupt molecular structures. The whole area of hazards from electromagnetic fields is a controversial one, and one that is very poorly understood. The three most important interactions between electromagnetic fields and tissue are electrolysis, neural stimulation and heating. There are other interactions such as the proton and molecular resonances which occur in a DC magnetic field but these are not thought to cause any significant biological effects.

In the sections that follow we will consider the biological interactions that are reasonably well understood. We will consider them in terms of the frequency of the electromagnetic radiation, starting with very low frequencies.

Table 8.3. *Some sources of electromagnetic fields.*

Source	Frequency range	Intensity range
Lightning	1 Hz–1 kHz	10 kV m^{-1}
Short-wave and microwave diathermy	27 MHz 2.450 GHz	>2 kV m^{-1} (in air)
Surgical diathermy/ electrosurgery	0.4–2.4 MHz	>1 kV m^{-1} (in air)
Home appliances	50–60 Hz	250 V m^{-1} max. 10 μT max.
Microwave ovens	2.45 GHz	50 W m^{-2} max.
RF transmissions	<300 MHz	1 W m^{-2} max.
Radar	0.3–100 GHz	100 W m^{-2} max.
High-voltage cables	50–60 Hz	>10 kV m^{-1}
Portable phones	500 MHz typical	>1 W m^{-2}

8.5.   LOW-FREQUENCY EFFECTS: 0.1 Hz–100 kHz

8.5.1.   *Properties of tissue*

We now return to the properties of tissue that we considered more theoretically in section 8.2. Biological tissue contains *free charge carriers* so that it is meaningful to consider it as an electrical conductor and to describe it in terms of a conductivity. *Bound charges* are also present in tissue so that dielectric properties also exist and can be expected to give rise to displacement currents when an electric field is applied. These properties might arise as electronic or nuclear polarization in a non-polar material, as a relative displacement of negative and positive ions when these are present or as a result of a molecular electric dipole moment where there is a distribution of positive and negative charge within a molecule. These effects may be described in terms of a relative permittivity (dielectric constant).

In addition to the above two passive electrical properties, biological tissue contains mechanisms for the active transport of ions. This is an important mechanism in neural function and also in membrane absorption processes, such as those which occur in the gastro-intestinal tract.

Conductivity is the dominant factor when relatively low-frequency (less than 100 kHz) electric fields are applied to tissue.

Frequency-dependent effects

The electrical properties of a material can be characterized by an electrical conductivity σ and permittivity ε. If a potential V is applied between the opposite faces of a unit cube of the material (see figure 8.7) then a conduction current I_c and displacement current I_d will flow, where

$$I_c = V\sigma \qquad I_d = \frac{dV}{dt}\varepsilon\varepsilon_0$$

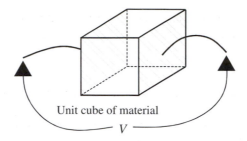

Figure 8.7. *Potential V applied to a unit cube of material with conductivity σ and permittivity ε.*

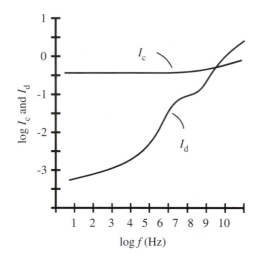

Figure 8.8. *The change in conduction current I_c and displacement current I_d for a typical tissue. Consider what the shape of the graphs might be for a pure resistance or pure capacitance.*

and where ε_0 is the dielectric permittivity of free space with the value $8.854 \times 10^{-12}\,\mathrm{F\,m^{-1}}$. If V is sinusoidally varying then I_d is given by

$$I_d = V\,2\pi f\,\varepsilon\varepsilon_0$$

where f is the frequency of the sinusoidal potential.

   Both conductivity and permittivity vary widely between different biological tissues but figure 8.8 shows a typical frequency variation of I_c and I_d for soft biological tissue. I_c increases only slowly with increasing frequency and indeed at frequencies up to 100 kHz conductivity is almost constant. I_d increases much more rapidly with increasing frequency and above about 10^7 Hz the displacement current exceeds the conduction current. Permittivity decreases with increasing frequency and there are, in general, three regions where rapid changes take place. The region around 10 Hz is generally considered to arise from dielectric dispersion associated with tissue interfaces such as membranes; the region around 1 MHz is associated with the capacitance of cell membranes; the region around 10^{10} Hz represents the dielectric dispersion associated with polarizability of water molecules in tissue (see Pethig (1979) and Foster and Schwan (1989) for more information). Inspection of figure 8.8 shows that for relatively low frequencies (less than 100 kHz) the displacement current is likely to be very much less than the conduction current and it is therefore reasonable to neglect dielectric effects in treating tissue impedance.

Resistivity of various biological tissues

There is a large literature on the electrical resistivity of biological material (see Duck (1990) for a compendium of data). Units can often be confusing. In most cases the resistivity or conductivity is given. However, the properties of membranes may be quoted as a resistance or capacitance per cm^2 or m^2. There are many discrepancies in the reported values for tissue resistivity, but this is not surprising in view of the great difficulties in making measurements *in vivo* and the problems of preserving tissue for measurement *in vitro*. Table 8.4 gives typical values for a range of biological materials at body temperature (resistivity can be expected to fall by 1–2% $°C^{-1}$).

Many tissues contain well-defined long fibres, skeletal muscle being the best example, so that it might also be expected that conductivity would be different in the longitudinal and transverse directions. This is indeed the case, and it has been shown that the transverse resistivity may be 10 times greater than the longitudinal resistivity. We will investigate this in the practical experiment described in section 8.9.2.

Table 8.4. *The electrical resistivity of a range of tissues.*

Tissue		Resistivity (Ω m)	Frequency (kHz)
CSF		0.650	1–30
Blood		1.46–1.76	1–100
Skeletal muscle:	longitudinal	1.25–3.45	0.1–1
	transverse	6.75–18.0	"
Lung:	inspired	17.0	100
	expired	8.0	"
Neural tissue:	grey matter	2.8	"
	white matter	6.8	"
Fat		20	1–100
Bone		>40	"

8.5.2. Neural effects

If low-frequency currents are passed between a pair of electrodes placed on the skin then a current can be found at which sensation occurs. In general, this *threshold of sensation* rises with increasing frequency of applied current, as shown in figure 8.9. *Three fairly distinct types of sensation* occur as frequency increases.

- At very low frequencies (below 0.1 Hz) individual cycles can be discerned and a 'stinging sensation' occurs underneath the electrodes. The major effect is thought to be *electrolysis* at the electrode/tissue interface where small ulcers can form with currents as low as 100 μA. The application of low-frequency currents can certainly cause ion migration and this is the mechanism of iontophoresis. Current densities within the range 0–10 A m^{-2} have been used to administer local anaesthetics through the skin, and also therapeutic drugs for some skin disorders. The applied potential acts as a forcing function that can cause lipid soluble drugs to penetrate the stratum corneum. Sweat ducts are the principal paths for ion movement.
- At frequencies above 10 Hz, electrolysis effects appear to be reversible and the dominant biological effect is that of *neural stimulation*. If the electrodes are placed over a large nerve trunk such as the

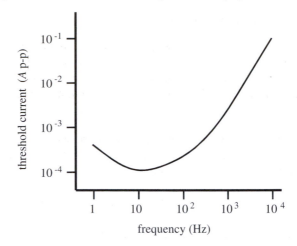

Figure 8.9. *Threshold of sensation as a function of frequency for an electric current applied between 5 mm wide band electrodes encircling the base of two adjacent fingers. (Result from one normal subject.)*

ulnar or median, then the first sensation arises from the most rapidly conducting sensory fibres. If the amplitude of the current is increased, then more slowly conducting fibres are stimulated and motor contractions occur. Stimulation over a nerve trunk arises as a result of depolarization at a node of Ranvier. The capacitance of a single node is of the order 10 pF such that a charge of 10^{-12} C is required to remove the normally occurring polarization potential of about 0.1 V. 10^{-12} C can be delivered as a current of 10^{-9} A for 1 ms. However, when the current is delivered through relatively distant surface electrodes only a very small fraction of the current will pass into a particular node of Ranvier. It is therefore to be expected that the threshold shown in figure 8.9 will fall as the electrodes are moved closer to a nerve trunk.

Propagation of a nerve action potential is controlled by the external currents which flow around the area of depolarization and so depolarize adjacent areas. This depolarization of adjacent areas is not instantaneous, as time is required to remove the charge associated with the capacitance of the nerve membrane. If the frequency of an applied external field is such that an action potential cannot be propagated within one cycle then neural stimulation will not occur. It is for this reason that the threshold of sensation rises as the frequency of the applied current is increased.

- At frequencies above about 10 kHz the current necessary to cause neural stimulation is such that *heating of the tissue* is the more important biological effect. Displacement currents are usually negligible within the range 10–100 kHz and therefore the $I^2 R$ losses are dominant.

The major biological effects within our frequency range of interest are therefore electrolysis, neural stimulation and heating. In figure 8.9 the threshold sensation is given in terms of the total current which is passed between a pair of surface electrodes. The threshold will depend upon the electrode area as there is ample evidence to show that current density rather than current is the important parameter. However, the relative magnitude of the three effects we have considered is not changed when current density rather than current is used. A typical value of current density at threshold and 50 Hz is 2 A m^{-2}.

8.5.3.   Cardiac stimulation: fibrillation

Electromedical equipment is a possible source of hazard to the patient. In many cases the patient is directly connected to the equipment so that in cases of a fault electrical current may flow through the patient. The response of the body to low-frequency alternating current depends on the frequency and the current density. Low-frequency current (up to 1 kHz) which includes the main commercial supply frequencies (50 Hz and 60 Hz) can cause:

- prolonged tetanic contraction of skeletal and respiratory muscles;
- arrest of respiration by interference with the muscles that control breathing;
- heart failure due to ventricular fibrillation (VF).

In calculating current through the body, it is useful to model the body as a resistor network. The skin can have a resistance as high as 1 MΩ (dry skin) falling to 1 kΩ (damp skin). Internally, the body resistance is about 50 Ω. Internal conduction occurs mainly through muscular pathways. Ohm's law can be used to calculate the current. For example, for a person with damp skin touching both terminals of a constant voltage 240 V source (or one terminal and ground in the case of mains supply), the current would be given by $I = V/R = 240/2050 = 117$ mA, which is enough to cause ventricular fibrillation (VF).

Indirect cardiac stimulation

Most accidental contact with electrical circuits occurs via the skin surface. The threshold of current perception is about 1 mA, when a tingling sensation is felt. At 5 mA, sensory nerves are stimulated. Above 10 mA, it becomes increasingly difficult to let go of the conductor due to muscle contraction. At high levels the sustained muscle contraction prevents the victim from releasing their grip. When the surface current reaches about 70–100 mA the co-ordinated electrical control of the heart may be affected, causing ventricular fibrillation (VF). The fibrillation may continue after the current is removed and will result in death after a few minutes if it persists.

Larger currents of several amperes may cause respiratory paralysis and burns due to heating effects. The whole of the myocardium contracts at once producing cardiac arrest. However, when the current stops the heart will not fibrillate, but will return to normal co-ordinated pumping. This is due to the cells in the heart all being in an identical state of contraction. This is the principle behind the defibrillator where the application of a large current for a very short time will stop ventricular fibrillation.

Figure 8.10 shows how the let-go level varies with frequency. The VF threshold varies in a similar way; currents well above 1 kHz, as used in diathermy, do not stimulate muscles and the heating effect becomes dominant. IEC 601-1 limits the AC leakage current from equipment in normal use to 0.1 mA.

Direct cardiac stimulation

Currents of less than 1 mA, although below the level of perception for surface currents, are very dangerous if they pass internally in the body in the region of the heart. They can result in ventricular fibrillation and loss of pumping action of the heart.

Currents can enter the heart via pacemaker leads or via fluid-filled catheters used for pressure monitoring. The smallest current that can produce VF, when applied directly to the ventricles, is about 50 μA. British Standard BS5724 limits the normal leakage current from equipment in the vicinity of an electrically susceptible patient (i.e. one with a direct connection to the heart) to 10 μA, rising to 50 μA for a single-fault condition. Note that the 0.5 mA limit for leakage currents from normal equipment is below the threshold of perception, but above the VF threshold for currents applied to the heart.

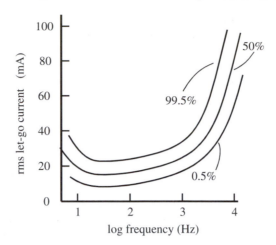

Figure 8.10. *Percentage of adult males who can 'let go' as a function of frequency and current.*

Ventricular fibrillation

VF occurs when heart muscle cells coming out of their refractory period are electrically stimulated by the fibrillating current and depolarize, while at the same instant other cells, still being in the refractory period, are unaffected. The cells depolarizing at the wrong time propagate an impulse causing other cells to depolarize at the wrong time. Thus, the timing is upset and the heart muscles contract in an unco-ordinated fashion. The heart is unable to pump blood and the blood pressure drops. Death will occur in a few minutes due to lack of oxygen supply to the brain. To stop fibrillation, the heart cells must be electrically co-ordinated by use of a defibrillator.

The threshold at which VF occurs is dependent on the current density through the heart, regardless of the actual current. As the cross-sectional area of a catheter decreases, a given current will produce increasing current densities, and so the VF threshold will decrease.

8.6.   HIGHER FREQUENCIES: >100 kHz

8.6.1.   *Surgical diathermy/electrosurgery*

Surgical diathermy/electrosurgery is a technique that is widely used by surgeons. The technique uses an electric arc struck between a needle and tissue in order to *cut* the tissue. The arc, which has a temperature in excess of 1000 °C, disrupts the cells in front of the needle so that the tissue parts as if cut by a knife; with suitable conditions of electric power the cut surfaces do not bleed at all. If blood vessels are cut these may continue to bleed and current has to be applied specifically to the cut ends of the vessel by applying a blunt electrode and passing the diathermy current for a second, or two or by gripping the end of the bleeding vessel with artery forceps and passing diathermy current from the forceps into the tissue until the blood has coagulated sufficiently to stop any further bleeding. *Diathermy can therefore be used both for cutting and coagulation.*

The current from the 'live' or 'active' electrode spreads out in the patient's body to travel to the 'indifferent', 'plate' or 'patient' electrode which is a large electrode in intimate contact with the patient's body. Only at points of high current density, i.e. in the immediate vicinity of the active electrode, will coagulation take place; further away the current density is too small to have any effect.

Although electricity from the mains supply would be capable of stopping bleeding, the amount of current needed (a few hundred milliamperes) would cause such intense muscle activation that it would be impossible for the surgeon to work and would be likely to cause the patient's heart to stop. The current used must therefore be at a sufficiently high frequency that it can pass through tissue without activating the muscles. A curve showing the relationship between the minimum perceptible current in the finger and the frequency of the current was given in figure 8.9.

Diathermy equipment

Diathermy machines operate in the radio-frequency (RF) range of the spectrum, typically 0.4–3 MHz. Diathermy works by heating body tissues to very high temperatures. The current densities at the active electrode can be 10 A cm^{-2}. The total power input can be about 200 W. The power density in the vicinity of the cutting edge can be thousands of W cm^{-3}, falling to a small fraction of a W cm^{-3} a few centimetres from the cutting edge. The massive temperature rises at the edge (theoretically thousands of °C) cause the tissue fluids to boil in a fraction of a second. The cutting is a result of rupture of the cells.

An RF current follows the path of least resistance to ground. This would normally be via the plate (also called dispersive) electrode. However, if the patient is connected to the ground via the table or any attached leads from monitoring equipment, the current will flow out through these. The current density will be high at these points of contact, and will result in surface burns (50 mA cm^{-2} will cause reddening of the skin; 150 mA cm^{-2} will cause burns). Even if the operating table is insulated from earth, it can form a capacitor with the surrounding metal of the operating theatre due to its size, allowing current to flow. Inductive or capacitive coupling can also be formed between electrical leads, providing other routes to ground.

8.6.2. Heating effects

If the whole body or even a major part of the body is exposed to an intense electromagnetic field then the heating produced might be significant. The body normally maintains a stable deep-body temperature within relatively narrow limits (37.4 ± 1 °C) even though the environmental temperature may fluctuate widely. The normal minimal metabolic rate for a resting human is about 45 W m^{-2} (4.5 mW cm^{-2}), which for an average surface area of 1.8 m^2 gives a rate of 81 W for a human body. Blood perfusion has an important role in maintaining deep-body temperature. The rate of blood flow in the skin is an important factor influencing the internal thermal conductance of the body: the higher the blood flow and hence, the thermal conductance, the greater is the rate of transfer of metabolic heat from the tissues to the skin for a given temperature difference. Blood flowing through veins just below the skin plays an important part in controlling heat transfer. Studies have shown that the thermal gradient from within the patient to the skin surface covers a large range and gradients of 0.05–0.5 °C mm^{-1} have been measured. It has been shown that the effect of radiation emanating from beneath the skin surface is very small. However, surface temperatures will be affected by vessels carrying blood at a temperature higher or lower than the surrounding tissue provided the vessels are within a few millimetres of the skin surface.

Exposure to electromagnetic (EM) fields can cause significant changes in total body temperature. Some of the fields quoted in table 8.3 are given in volts per metre. We can calculate what power dissipation this might cause if we make simplifying assumptions. Consider the cylindrical geometry shown in figure 8.11 which represents a body which is 30 cm in diameter and 1 m long (L). We will assume a resistivity (ρ) of 5 Ω m for the tissue. The resistance (R) between the top and bottom will be given by $\rho L / A$ where A is the cross-sectional area. $R = 70.7 \, \Omega$.

For a field of 1 V m^{-1} (in the tissue) the current will be 14.1 mA. The power dissipated is 14.1 mW which is negligible compared to the basal metabolic rate.

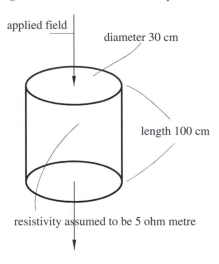

Figure 8.11. *The body modelled as a cylinder of tissue.*

For a field of 1 kV m^{-1}, the current will be 14.1 A and the power 14.1 kW, which is very significant. The power density is 20 W cm^{-2} over the input surface or 200 mW cm^{-3} over the whole volume.

In the above case we assumed that the quoted field density was the volts per metre produced in tissue. However, in many cases the field is quoted as volts per metre *in air*. There is a large difference between these two cases. A field of 100 V m^{-1} in air may only give rise to a field of 10^{-5} V m^{-1} in tissue.

8.7.   ULTRAVIOLET

We now come to the border between ionizing and non-ionizing radiation. Ultraviolet radiation is part of the electromagnetic spectrum and lies between the visible and the x-ray regions. It is normally divided into three wavelength ranges. These define UV-A, UV-B and UV-C by wavelength.

A	315–400 nm
B	280–315 nm
C	100–280 nm

The sun provides ultraviolet (mainly UV-A and UV-B) as well as visible radiation. Total solar irradiance is about 900 W m^{-2} at sea level, but only a small part of this is at ultraviolet wavelengths. Nonetheless there is sufficient UV to cause sunburn. The early effects of sunburn are pain, erythema, swelling and tanning. Chronic effects include skin hyperplasia, photoaging and pseudoporphyria. It has also been linked to the development of squamous cell carcinoma of the skin. Histologically there is subdermal oedema and other changes.

Of the early effects UV-A produces a peak biological effect after about 72 h, whereas UV-B peaks at 12–24 h. The effects depend upon skin type.

The measurement of ultraviolet radiation

Exposure to UV can be assessed by measuring the erythemal response of skin or by noting the effect on micro-organisms. There are various chemical techniques for measuring UV but it is most common to use physics-based techniques. These include the use of photodiodes, photovoltaic cells, fluorescence detectors and thermoluminescent detectors such as lithium fluoride.

Therapy with ultraviolet radiation

Ultraviolet radiation is used in medicine to treat skin diseases and to relieve certain forms of itching. The UV radiation may be administered on its own or in conjunction with photoactive drugs, either applied directly to the skin or taken systemically.

The most common application of UV in treatment is psolaren ultraviolet A (PUVA). This has been used extensively since the 1970s for the treatment of psoriasis and some other skin disorders. It involves the combination of the photoactive drug psoralen, with long-wave ultraviolet radiation (UV-A) to produce a beneficial effect. Psoralen photochemotherapy has been used to treat many skin diseases, although its principal success has been in the management of psoriasis. The mechanism of the treatment is thought to be that psoralens bind to DNA in the presence of UV-A, resulting in a transient inhibition of DNA synthesis and cell division. 8-methoxypsolaren and UV-A are used to stop epithelial cell proliferation. There can be side effects and so the dose of UV-A has to be controlled. Patch testing is often carried out in order to establish what dose will cause erythema. This minimum erythema dose (MED) can be used to determine the dose used during PUVA therapy.

In PUVA the psoralens may be applied to the skin directly or taken as tablets. If the psoriasis is generalized, whole-body exposure is given in an irradiation cabinet. Typical intensities used are $10\,\mathrm{mW\,cm^{-2}}$, i.e. $100\,\mathrm{W\,m^{-2}}$. The UV-A dose per treatment session is generally in the range 1–$10\,\mathrm{J\,cm^{-2}}$.

Treatment is given several times weekly until the psoriasis clears. The total time taken for this to occur will obviously vary considerably from one patient to another, and in some cases complete clearing of the lesions is never achieved. PUVA therapy is not a cure for psoriasis and repeated therapy is often needed to prevent relapse.

8.8. ELECTROMEDICAL EQUIPMENT SAFETY STANDARDS

An increasing number of pieces of electrically operated equipment are being connected to patients in order to monitor and measure physiological variables. The number of patients where direct electrical connection is made to the heart through a cardiac catheter has also increased. As a result, the risk of electrocuting the patient has increased. To cope with this danger, there are now internationally agreed standards of construction for patient-connected electrical equipment.

The recommendations for the safety and constructional standards for patient-connected equipment are contained in an international standard drawn up by the International Electrotechnical Commission (IEC). The standard has the reference IEC601 and has been the basis for other standards drawn up by individual countries. For example, it is consistent with BS 5724 which is published by the British Standards Institute. IEC 601 is a general standard and other standards are devoted to specialized pieces of equipment, such as cardiographs. These formal documents are excessively detailed. In this section we will consider some of the background to IEC 601.

8.8.1. *Physiological effects of electricity*

Electricity has at least three major effects that may be undesirable—electrolysis, heating and neuromuscular stimulation. Nerve stimulation is potentially the most dangerous effect, as the nervous system controls the two systems that are essential to life—the circulation of the blood and respiration. We considered the electrical stimulation of cardiac muscle in section 8.5.3 and the process of neural stimulation by electricity is described in some detail in section 10.2 of Chapter 10.

Electrolysis

Electrolysis will take place when a *direct current* is passed through any medium which contains free ions. The positively charged ions will migrate to the negative electrode, and the negatively charged ions to the positive electrode. If two electrodes are placed on the skin, and a direct current of 100 μA is passed beneath them for a few minutes, small ulcers will be formed beneath the electrodes. These ulcers may take a very long time to heal. IEC 601 defines *'direct current'* as a current with a frequency of less than 0.1 Hz. Above this frequency, the movement of ions when the current is flowing in one direction appears to be balanced by the opposite movement of the ions when the current flow is reversed, and the net effect is that there is no electrolysis. IEC 601 limits the direct current that can flow between electrodes to 10 μA.

Neural stimulation

There is normally a potential difference of about 80 mV across a nerve membrane. If this potential is reversed for more than about 20 μs, the neurone will be stimulated and an action potential will be propagated along the nerve fibre. If a sensory nerve has been stimulated, then a pain will be felt, and if a motor nerve has been stimulated, then a muscle will be caused to contract. The major hazards are the stimulation of skeletal and heart muscle, either directly or by the stimulation of motor nerves. Stimulation becomes increasingly difficult at frequencies above 1 kHz. The co-ordinated pumping activity of the heart can be disrupted by electric currents which pass through the heart. This is called fibrillation (see section 8.5.3) and can continue after the current is removed.

Stimulation through the skin

Nerves are stimulated by a current flow across the nerve membrane, so that the voltage needed to cause stimulation will depend on the contact impedance to the body. If alternating current at 50/60 Hz is applied through the body from two sites on the skin, the effect will depend on the size of the current. At about 1 mA, it will just be possible to feel the stimulus. At about 15 mA, the skeletal muscles will be stimulated to contract continuously, and it will not be possible to release an object held in the hands. As the current is further raised, it becomes increasingly painful, and difficult to breathe, and at about 100 mA ventricular fibrillation will begin. Currents up to 500 mA will cause ventricular fibrillation which will continue after the current stops flowing, and burns will be caused by the heating of the tissue. At currents above 500 mA the heart will restart spontaneously after the current is removed—this is the principle of the defibrillator.

  To put these figures into perspective, the impedance of dry skin is about 10–100 kΩ. Mains supply voltage of 240 V applied directly to the skin would therefore give a current of between 2.5 and 25 mA; i.e. above the threshold of sensation, and possibly sufficiently high to cause objects to be gripped (see figure 8.10). If a live electrical conductor had been gripped, the physiological shock would cause sweating, and the contact impedance could drop to 1 kΩ. This would give a current of 250 mA, causing ventricular fibrillation. Good contact to wet skin could give a contact impedance of 100 Ω, causing a current of 2.5 A to pass.

  It is unlikely that electromedical equipment would pass sufficient current to cause ventricular fibrillation, even when the equipment had a fault. The main source of currents of this magnitude is unearthed metalwork which could become live at mains supply potential. This, of course, may not be part of the patient-connected equipment, but could be a motorized bed or a light fitting. IEC601 limits the current flow through contact to the skin to 0.5 mA with a single fault in the equipment.

Direct stimulation of the heart

If a current is passed through two electrodes which are attached to, say, the arms, the current will be distributed throughout the body. Only a very small fraction of the current will actually flow through the heart. Obviously, ventricular fibrillation will be caused by a much lower current if it is applied directly to the heart. Experiments have shown that currents of about 100 μA can cause ventricular fibrillation if applied directly to the ventricular wall. It should be noted that this is well below the threshold of sensation for currents applied through the skin, so that sufficient current to cause fibrillation could be passed from a faulty piece of equipment through an operator's body to a cardiac catheter, without any sensation being felt by the operator.

IEC601 limits the current from equipment which can be connected to the heart to 10 μA under normal operating conditions, and 50 μA with a single fault. This requires a high impedance (>5 MΩ) between the patient connections and the power source within the equipment. This is difficult to achieve without electrical isolation of the patient connections. The high impedance is then due to the small capacitance between isolated and non-isolated sections.

Tissue heating

Neural tissue is not stimulated by high-frequency electrical currents, whose major effect is that of heating (see section 8.6). Frequencies between 400 kHz and 30 MHz are used in surgical diathermy/electrosurgery to give either coagulation, or cutting. Induced currents at 27 MHz, or at microwave frequencies, are used by physiotherapists for therapy.

The local effect of heating depends on the tissue, the time for which it is heated, the contact area, and the blood flow. Current densities of less than 1 mA mm^{-2} are unlikely to cause damage. Burns have been produced by a current density of 5 mA mm^{-2} for 10 s. Greater current densities than this can be achieved if the earth plate on an electrosurgery machine is not correctly applied.

The time of exposure, the depth of the tissue, and the blood flow will all affect the tissue damage from bulk heating. There are codes of practice for exposure to microwave and radio-frequency radiation. These usually limit the power levels for continuous exposure to 0.1 mW mm^{-2}, which is well below the thermal damage level. Physiotherapy machines use considerably higher power levels for tissue heating without obvious damage to the tissue.

8.8.2.   Leakage current

A possible cause of a hazardous current is that of *leakage current*. Figure 8.12 illustrates how a leakage current can arise. A patient is shown connected to two different pieces of electromedical equipment, both of which make a connection between mains supply ground/earth and the patient. Consider what would result if the earth/ground connection is broken in the piece of equipment A. A current can now flow from the live mains supply, through any capacitance between this connection and the equipment case, to the patient and return to earth/ground via equipment B. The capacitance between the live supply mains and the equipment case arises mainly from the proximity of the live and earth/ground wires in the supply cable and from any mains transformer within the equipment. These capacitances are marked as C_1 and C_2 in figure 8.12. If any interference suppressors are included in the mains supply to the equipment these can also contribute a capacitance between mains/supply and the equipment case.

If the mains supply is given as $a \sin \omega t$ then the leakage current L_c will be given by

$$L_c = C \frac{dV}{dt} = a\omega \cos \omega t$$

$$C = C_1 + C_2.$$

If $C = 1$ nF then:

- if the mains supply is at 230 V rms, then $L_c = 10^{-9} \times 2 \times \pi \times 50 \times 230 = 72$ μA rms.
- if the mains supply is at 110 V rms, then $L_c = 10^{-9} \times 2 \times \pi \times 50 \times 110 = 41$ μA rms.

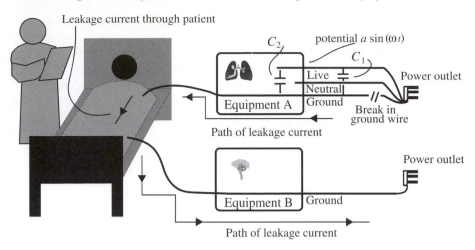

Figure 8.12. *Two monitors connected to a patient illustrating how a leakage current can arise.*

Table 8.5. *Permissible leakage currents (in mA rms) in different classes of equipment. IEC 601 also limits to 1 mA the current measured in the earth wire for permanently installed equipment; 5 mA for portable equipment which has two earth wires, and 0.5 mA for portable equipment with a single earth wire.*

Type of equipment →	B & BF		CF	
Condition →	Normal	Fault	Normal	Fault
Case to earth	0.1	0.5	0.01	0.5
Patient to earth	0.1	0.5	0.01	0.05
Mains to patient	—	5	—	0.05
Electrode AC	0.1	0.5	—	0.05
Electrode DC	0.01	0.5	0.01	0.05

Leakage current is the current that can be drawn from the equipment to earth, either under normal operating conditions or under single-fault conditions. The single-fault conditions are specified, and include reversal of the line and neutral connections, breakage of the earth wire or breakage of the live or neutral wires with the earth connected. The intention is to limit the maximum current that can be passed through the patient. It is not possible to make a machine that is perfectly insulated, because there will always be stray capacitances between different parts of the machine, which will act as conductors for alternating currents. Table 8.5 gives the permitted leakage currents for different types of equipment.

Electromedical equipment is subdivided into three types. Type B equipment is intended for connection to the skin of the patient only. Type BF equipment is also intended only for connection to the patient's skin, but has floating input circuitry, i.e. there is no electrical connection between the patient and earth. Type CF equipment also has floating input circuitry, but is intended for use when a direct connection has been made to the patient's heart.

Measurement of leakage current

Leakage current can be measured using the standard test circuit shown in figure 8.13. The *test lead* is connected to the case of the piece of equipment to be tested and the voltage generated across the 1 kΩ load is measured using an AC voltmeter. Note that a capacitance of 0.15 μF is added across the 1 kΩ load resistance. The effect of this is to increase the current required for a given reading of the AC voltmeter as frequency is increased. The impedance presented by an R and C in parallel is simply given by $R/(1 + jR\omega C)$, and the magnitude of this impedance by $R/(1 + R^2\omega^2C^2)^{1/2}$. The leakage current required to produce a given reading on the AC voltmeter can thus be determined as the inverse of this magnitude. This current is shown as a function of frequency in figure 8.14.

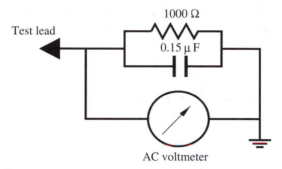

Figure 8.13. *The standard test circuit for measuring leakage current. The test load has a constant impedance of 1000 Ω at frequencies of less than 1 kHz, and a decreasing impedance at higher frequencies. The 50/60 Hz leakage current is given by the AC voltage divided by 1000 Ω.*

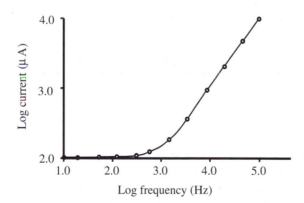

Figure 8.14. *The current required, as a function of frequency, to give a reading of 100 mV across the test load shown in figure 8.13.*

Figure 8.14 is interesting as it shows that as frequency increases then a higher and higher current can be applied before a hazard occurs. This is because it becomes increasingly difficult to stimulate a nerve as frequency increases. It might be thought that high currents at high frequencies are unlikely to arise. However, it was shown in section 8.6 that such currents can arise as a result of electrosurgery/surgical diathermy.

8.8.3. *Classification of equipment*

It is obviously wasteful, and sometimes impractical, to design all equipment to the most stringent specification. Therefore, IEC 601 classifies the equipment according to the intended use. The majority of electromedical equipment is class I equipment. This equipment is contained within a metal box which is connected to earth (figure 8.15). All the exposed metal parts of the equipment must be earthed. The connection to earth, and the provision of a fuse in the live wire from the mains electricity supply, are the two essential safety features for class I equipment. Figure 8.16 shows the complete mains supply circuit, including the sub-station transformer which reduces the high voltage used for electricity transmission to the lower voltage used to power equipment. In the UK, one side of the secondary of the transformer is connected to earth literally, by means of a conductor buried in the ground. This end of the transformer winding is called 'neutral' and the other end is called 'live' or 'line' and will have a potential of 240/110 V with respect to earth. Consider what would happen if you were touching the metal case of the equipment and your feet were electrically connected, through the structure of the building, to earth.

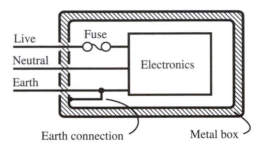

Figure 8.15. *General layout of equipment in an earthed metal case with a fuse in the live lead.*

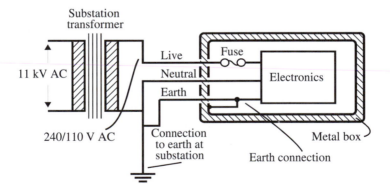

Figure 8.16. *Block diagram showing the neutral wire connected to earth at the substation transformer.*

If the live wire broke inside the instrument and touched the unearthed case, the case would then be at 240/110 V with respect to earth, and the current path back to the sub-station would be completed through you and the earth, resulting in electrocution. If the case were earthed, then no potential difference could exist between the case and earth, and you would be safe. Because of the low resistance of the live and earth wires, a heavy current would flow, and this would melt the fuse and disconnect the faulty equipment from the mains supply.

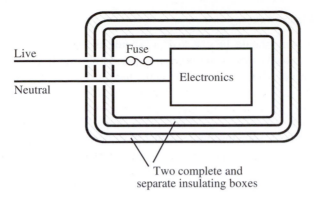

Two complete and
separate insulating boxes

Figure 8.17. *General layout of double-insulated equipment showing the two complete insulating layers and no earth wire.*

Class IIA equipment does not have exposed metal work, and class IIB equipment is double-insulated (figure 8.17). All the electrical parts of double insulated equipment are completely surrounded by two separate layers of insulation. Because there is no possibility of touching any metal parts which could become live under fault conditions, double-insulated equipment does not have an earth lead. Many pieces of domestic electrical equipment are double insulated, e.g. hair dryers, electric drills and lawn mowers.

8.8.4. *Acceptance and routine testing of equipment*

All electromedical equipment which is purchased from a reputable manufacturer should have been designed to comply with IEC 601, and should have undergone stringent tests during the design stage to ensure that it will be safe to use. However, mistakes can be made during the manufacture of the equipment, and the equipment might have been damaged during delivery. All new equipment should therefore have to pass an acceptance test before it is used clinically. The acceptance test should be designed to check that the equipment is not obviously damaged and is safe to use—it is not intended as a detailed check that the equipment complies with the standards.

It is desirable that all equipment be checked regularly to ensure that it still functions correctly. Unfortunately, this is usually not possible. Defibrillators are used infrequently, but must work immediately when they are needed. They must be checked regularly to see that the batteries are fully charged, that they will charge up and deliver the correct charge to the paddles, and that the leads, electrodes and electrode jelly are all with the machine. Category CF equipment, which is intended for direct connection to the heart, should also be checked at regular intervals.

Visual inspection

A rapid visual inspection of the inside and the outside of the equipment will reveal any obvious damage, and will show whether any components or circuit boards have come loose in transit. Any mains supply voltage adjustment should be checked, as should the rating of the fuses and the wiring of the mains plug. It is obviously sensible to avoid using equipment that is damaged. Mechanical damage can destroy insulation and reduce the clearances between live parts of the equipment. The most common mechanical damage is caused by people falling over the power supply lead or moving the equipment without unplugging it. Examination of the condition of the power lead and the patient connections should be automatic. If the wires have been pulled out of the plug, the earth wire might be broken, so avoid using the equipment until it has been checked.

Earth continuity and leakage current

The integrity of the earth wire should be checked, as should the impedance to earth of all the exposed metalwork. All the earth leakage current measurements specified by the IEC standard should be made with the equipment in its normal operating condition and under the specified single-fault conditions.

Records

A formal record should be kept, showing the tests that have been made and the results of measurements such as leakage current. Each time the equipment is routinely tested or serviced, the record can be updated. This could give early warning of any deterioration in the performance of the equipment and possibly allow preventative maintenance to be undertaken before the equipment breaks down.

No equipment, or operator, is infallible, but care and common sense in the use of patient-connected equipment can prevent many problems arising. The more pieces of equipment that are connected to a patient, the greater the risk that is involved. It is sensible to connect the minimum number of pieces of equipment at the same time. If the equipment has an earth connection to the patient (i.e. it is not category BF or CF equipment), then the power leads should be plugged into adjacent sockets, and, if possible, all the earth connections to the patient should be made to the same electrode. In theory, all the power supply earth connections should be at the same potential. In practice, a fault either in a piece of equipment connected to the mains or in the mains wiring can cause a current to flow along the earth wire. As the earth wire has a finite (though small) resistance, there will be a potential difference along the earth wire. This could amount to tens of volts between opposite sides of a ward, and could give the patient an electric shock if two earth electrodes were attached to different parts of the mains earth wire.

Cardiac catheters

The patient with a cardiac catheter is extremely vulnerable to electric currents. The current which will cause fibrillation if applied directly to the heart is lower than the threshold of sensation for currents applied to the skin, so that an operator touching a catheter could inadvertently pass a lethal current from a faulty piece of equipment. Only category CF equipment should be connected to patients with cardiac catheters. Great care should be taken to see that there is no accidental connection between earth and the catheter or any tubing and pressure transducers connected to the catheter. The catheter and the connections to it should only be handled using dry rubber gloves, to preserve the insulation.

Ulcers and skin reactions

Ulcers caused by electrolysis at the electrodes should not be seen unless the equipment is faulty. If the patient complains of discomfort or inflammation beneath the electrodes, check the DC current between the electrode leads—it should be less than 10 μA. Skin reactions are very occasionally caused by the electrode jelly and should be referred to the medical staff. Skin reactions are uncommon if disposable electrodes are used, but may be seen with re-usable metal plate electrodes that have not been properly cleaned or that have corroded.

Constructional standards and the evaluation of equipment

The various standards set out in great detail the construction standards to which electromedical equipment should be built. It is impractical to check that all equipment meets every detail of these standards. In practice, the equipment management services which are operated by many medical engineering departments attempt to check that the technical specification of the equipment is adequate for its function, and that the more important requirements of the standards are fulfilled. This is not as simple as it might appear to be. It is essential for the people who run the evaluation to be familiar with the task that the equipment has to perform, and to know what measurement standard it is possible to achieve. For this reason, equipment evaluation services are associated with fairly large departments which are committed to the design and development of clinical instrumentation which is not commercially available.

The technical evaluation may be followed by a clinical evaluation, in which several clinicians use the equipment routinely for a period of time. The clinical evaluation will reveal how easy the equipment is to use, whether it will stand up to the rather rough handling it is likely to receive in a hospital, and whether any modifications to the design are desirable. Discussion between the evaluation team and the manufacturer will then, hopefully, result in an improved instrument.

It should be emphasized that equipment produced for a special purpose within a hospital must conform to international safety standards. The performance and construction of specially made equipment should be checked by someone who has not been involved with either the design or the construction of the equipment, and the results of the tests should be formally recorded in the same way as for commercially produced equipment.

8.9. PRACTICAL EXPERIMENTS

8.9.1. *The measurement of earth leakage current*

This experiment involves the deliberate introduction of faults into the equipment. *Be careful.* Mains power electricity is lethal. Do not alter any connections unless the mains is switched off at the mains socket, as well as the instrument ON/OFF switch. Do not touch the case or controls of the instrument when the power is switched on. Do not do this experiment on your own.

Objective

To check that the earth leakage currents from an isolated ECG monitor meet the standards laid down for IEC 601.

Equipment

An isolated ECG/EKG monitor with a standard set of input leads.
A digital voltmeter with a sensitivity of at least 1 mV rms, preferably battery powered.
A means of altering the connections to the mains lead. It is not good practice to rewire the mains power plug incorrectly. Test sets are available which have a switch to alter the connections to a mains socket. Alternatively remove the mains plug, and use a 'safebloc' to connect the mains leads.
A test load as shown in figure 8.13.

Method

- Get someone who is involved with acceptance testing or evaluation of equipment to explain the layout of the standard, and make sure you understand the section on leakage currents.
- Connect all the ECG/EKG input leads together, and connect them to earth via the DVM and the test load.
- Measure the earth leakage current. This is the 'no-fault' current from patient to earth.
- Repeat the measurement for the specified single-fault conditions (i.e. no earth connection, live and neutral interchanged, etc).

- Do all the measurements shown in table 8.5 for normal and single-fault conditions. For isolated equipment one of these tests involves connecting the 'patient' to the mains power supply. *Be careful!* Use a 1 MΩ resistor between the mains supply and the patient connections so that the maximum current is limited to 240 μA (for a 240 V mains power supply).
- When you have finished, check that the mains power plug is correctly wired.
- If you were acceptance testing this ECG/EKG monitor, would you pass it on this test?

8.9.2.   Measurement of tissue anisotropy

Objectives

To show how a tetrapolar measurement of tissue resistivity can be made *in vivo*.
To measure the longitudinal and transverse resistivity of the arm.

Equipment

A battery powered current generator delivering 1 mA p–p at 20 kHz. (This can consist of a square wave generator, low-pass filter and simple voltage/current generator.)
Six Ag/AgCl electrodes, four lead strip electrodes and a tape measure.
An oscilloscope with a differential sensitivity down to at least 1 mV per division up to 20 kHz.

Methods and calculations

Longitudinal measurement

- Use the current generator and two of the Ag/AgCl electrodes to pass a current of 1 mA down the right arm of a volunteer. These are the drive electrodes and should be connected with one on the back of the hand and the other near the top of the arm.
- Place the other two Ag/AgCl electrodes 10 cm apart along the forearm. These are the receive electrodes. Place a third ground electrode on the back of the arm. Connect these electrodes to the differential input of the oscilloscope.
- Measure the distance between the receive electrodes and the average circumference of the section of forearm between the electrodes.
- Use the oscilloscope to measure the potential drop down the forearm and hence calculate the longitudinal resistivity of the tissue in Ω m.

Transverse measurement

- Remove the electrodes used above. Now place the lead strip electrodes as the drive and receive electrodes as shown in the figure 8.18. They should be placed around a single transverse section through the forearm.
- Use the oscilloscope to measure the voltage between the receive electrodes and hence calculate the transverse resistivity of the arm.

Conclusions

Did you get similar values for the longitudinal and transverse resistivities? If not then can you explain the difference? Are the measured values reasonable on the basis of the known values for the resistivity of muscle fat and bone? Would you have obtained different values on a subject with a fatter or thinner arm?

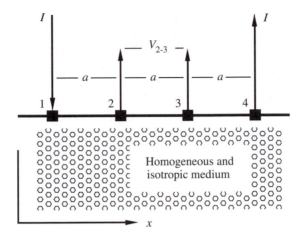

Figure 8.18. *Four electrodes placed on a conducting medium.*

Background notes

Tetrapolar measurement: how to measure resistivity

Consider four electrodes on the surface of a homogeneous, isotropic, semi-infinite medium as shown in figure 8.18. If positive and negative current sources of strength I are applied to electrodes 1 and 4 then we may determine the resulting potential V between electrodes 2 and 3. The four electrodes are equally spaced and in a straight line. The current I will spread out radially from 1 and 4 such that the current density is $I/2\pi r^2$ where r is the radial distance. The current density along the line of the electrodes x, for the two current sources is given by

$$\text{current density} = \frac{I}{2\pi x^2} \quad \text{and} \quad \frac{I}{2\pi(x-3a)^2}.$$

If we consider an element of length Δx and cross-sectional area Δa of the tissue with resistivity ρ then

$$\Delta R = \rho \frac{\Delta x}{\Delta a}$$

and the potential drop ΔV along this element is

$$\Delta V = \left[\frac{I}{2\pi x^2}\right]\Delta a\, \rho \frac{\Delta x}{\Delta a}.$$

Integrating between electrodes 2 and 3 we obtain

$$V_{2\text{-}3} = \int_a^{2a} \frac{\rho I}{2\pi}\left(\frac{1}{x^2}\right)\mathrm{d}x \qquad V_{2\text{-}3} = \frac{\rho I}{4\pi a}.$$

This is the potential from the first current source. An identical potential will be obtained from the second source so we can apply the superposition principle to obtain

$$V_{2\text{-}3} = \frac{\rho I}{2\pi a} \qquad \rho = \frac{2\pi a}{I}V_{2\text{-}3}.$$

The resisitivity of the medium can therefore be simply obtained by measuring $V_{2\text{-}3}$.

Longitudinal

If we can approximate the conductor to a cylinder then it is very easy to measure the longitudinal resistivity. If a current I is passed along the cylinder as shown in figure 8.19 and the potential gradient between points 1 and 2 is V then the resistivity ρ is given quite simply as

$$\rho = \frac{Va}{IL}$$

where a is the cross-sectional area of the cylinder and L is the distance between points 1 and 2.

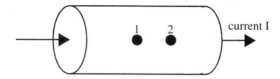

Figure 8.19. *Electrode geometry to make a measurement of longitudinal resistivity.*

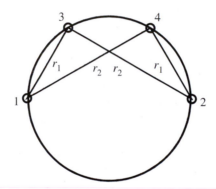

Figure 8.20. *Electrode geometry to make a measurement of transverse resistivity.*

Transverse

Consider a limb with electrodes placed diametrically opposite and parallel to the axis of the limb as shown in figure 8.20. If the electrodes are very long then the current paths are constrained to flow in the plane shown and so we can consider a homogeneous slab of thickness h into which the current enters at points 1 and 2 and we can measure the current which results between points 3 and 4.

It can be shown that the potential on the circumference is given by

$$V = \frac{\rho I}{\pi h} \log_e \frac{r_1}{r_2}$$

where ρ is the resistivity, I the current flowing, h the length of the electrodes and r_1 and r_2 the distances as shown in figure 8.20.

The potential between points 3 and 4, when these are symmetrically placed with respect to 1 and 2, is

$$V_{3-4} = \frac{2\rho I}{\pi h} \log_e \frac{r_1}{r_2}$$

and if the electrodes are equally spaced,

$$V_{3-4} = \frac{2\rho I}{\pi h} \log_e \sqrt{3}.$$

The transverse resistivity ρ may thus be calculated. A correction can be applied for the presence of the relatively insulating bone at the centre of the arm and also for the current which flows beyond the edge of the electrodes, but the correction factor is relatively small.

8.10. PROBLEMS

8.10.1. *Short questions*

a   Does tissue conduct electricity in a similar way to electrons in a metal?
b   Does tissue have both resistive and capacitive properties?
c   What is the main biological effect of low-frequency (100 Hz to 1 kHz) electric fields?
d   How would you expect the impedance of tissue to change as the measurement frequency is increased?
e   Is tissue permittivity dominant at very high or very low frequencies?
f   What is a relaxation process in relation to tissue impedance?
g   Is the Cole equation based upon a physical model of tissue?
h   What does the parameter α determine in a Cole equation?
i   What is the approximate energy (in eV) of UV radiation?
j   What is the approximate energy (in eV) due to thermal motion at room temperature?
k   Are microwaves more or less energetic than UV?
l   Would you feel a current of 10 mA at 100 Hz applied through electrodes to your arm?
m   Is lung tissue more or less conductive than brain tissue?
n   Could a current of 100 μA at 60 Hz cause cardiac fibrillation?
o   What is a current above the 'let-go' threshold?
p   Has UV-B a wavelength of about 30 nm?
q   What is a leakage current?
r   What two effects limit the current that can be safely injected into the body?
s   Why is a four-electrode measurement to be preferred to a two-electrode measurement?
t   The current required to stimulate nerves varies with frequency. Does the curve have a minimum and, if so, at what frequency?
u   What is the effect of passing a DC current through tissue?

8.10.2. *Longer questions (Answers are given to some of the questions)*

Question 8.10.2.1

Calculate the leakage current which might flow through a person who touches the roof of a car placed beneath an overhead power cable carrying a potential of 600 kV. Assume that the cable is 10 m above the car and that the roof is 2 m by 2 m. Is this current likely to pose a hazard to the person? List any assumptions you need to make ($\varepsilon_0 = 8.85 \times 10^{-12}$ F m^{-1}).

Question 8.10.2.2

Consider the situation of 60 Hz electrical current flowing through an adult human body for 1 s. Assume the electrical contacts are made by each hand grasping a wire. Describe the possible physiological effects of the current on the body as a function of current amplitude. Give the relative importance of each of the effects. What would be the effect of increasing or decreasing the duration of the current?

Question 8.10.2.3

It is claimed that monkeys can hear microwaves. Suggest a possible mechanism by which this might take place and how you could test the possibility.

Estimate the power deposited in a body which is placed between the plates of a 27 MHz short-wave diathermy unit which generates 2 kV. Assume that the capacitance between each plate and the body is 10 pF, that the part of the body between the plates can be represented as a cylinder of diameter 20 cm and length 30 cm and that the resistivity of tissue is 5 Ω m. Is this power likely to cause significant tissue heating?

Answer

It should be explained that the most likely effect is one of heating. If tissue were nonlinear then it might be possible that transmitted microwave signals could be rectified and so produce a low-frequency signal within the cochlea of the monkey.

The resistance of the body segment is $(5 \times 0.3)/(\pi 0.1^2) = 47.7$ Ω.

The capacitive path of 2×10 pF is equivalent to 5 pF which at 27 MHz presents an impedance of

$$\frac{1}{2\pi \times 27 \times 10^6 \times 5 \times 10^{-12}} = 1178 \ \Omega.$$

The current that flows will therefore be dominated by the capacitance and the current calculated as

$$I = C\frac{dV}{dt} = 5 \times 10^{-12} \times 2\pi \times 27 \times 10^6 \times 2 \times 10^3 = 1.7 \text{ A}.$$

If we assume the 2 kV was rms then the power deposited in the tissue will be $1.7 \times 1.7 \times 47.7 = 138$ W.

This is likely to cause significant heating. Bear in mind that the body normally produces about 50 W from metabolism. An additional 138 W is very significant.

Question 8.10.2.4

Discuss the statement that 'Ohm's law applies to tissue'.

In an experiment a current of 100 mA at a frequency of 500 kHz is passed along a human arm. It is found that the resistance of the arm appears to fall with time. Can you explain this observation?

Question 8.10.2.5

A collaborating clinician wishes to investigate the likelihood of burns being produced when a defibrillator is used on a patient. A defibrillator produces an initial current of 100 A which then decays exponentially with a time constant of 1 ms. If the resistance presented by the body is 100 Ω, then calculate the energy deposited in the body by a single discharge of the defibrillator and also calculate the average temperature rise of the body. Assume a value of 4000 J kg^{-1} $^\circ$C^{-1} for the specific heat capacity of tissue. List any further values which you assume in deriving your answer.

Answer

If the time constant is 1 ms and the resistance 100 Ω then the storage capacitor must be 10 μF. The energy stored will be 0.5 C V^2 which we can determine as follows:

$$0.5 \times 10^{-5} \times (100 \times 100)^2 = 500 \text{ J.}$$

It would be possible to find the same answer by integrating $V \times I$ over the exponential decay curve.

If the 500 J was spread throughout the body (actually an unreasonable assumption), then the average temperature rise would be $500/4000 \times 50 = 0.0025\,°$C. for a 50 kg person. The temperature rise underneath one electrode might, of course, be much larger because the effective tissue mass where the heat is deposited might be only 10 g in which case the temperature rise would be 12.5°C.

Question 8.10.2.6

Some people claim to feel depressed in thundery weather. Discuss the claim and describe how you might make an investigation to test the claim.

If the phenomenon is true then speculate on some possible physical as well as psychological explanations for the effect.

Answer

Thundery weather is associated with high potential fields underneath thunderclouds so that it is just possible that these have an effect on the brain. The field is a DC one so that perhaps some charge movement might occur in tissue and if the field changes in amplitude then AC effects such as relaxation phenomena might occur. However, it should be pointed out that the energies involved in low-frequency electric fields are very small and are very unlikely to produce a biological effect as opposed to a small biological interaction. It should also be pointed out that the other factors associated with thundery weather, heat, humidity, fear and darkness might be even more relevant.

To investigate the claim it would be necessary to carry out a large epidemiological survey with information on the incidence of thundery weather built in. If it were thought that electric fields were relevant then it would be necessary to devise some means of dosimetry for this.

Question 8.10.2.7

A clinical colleague who is interested in hydration comes to talk to you about a method of measuring intracellular and extracellular spaces in her patients. She suggests that it might be done by making body impedance measurements, first at a low frequency and then at a high frequency.

Can you describe a possible method of measuring the ratio of extracellular and intracellular spaces using electrical impedance? Outline a possible method and point out any practical or theoretical problems which you could foresee.

Answer

The Cole–Cole model given in section 8.3.2 has been used to justify a model based upon the assumption that current flows through the extracellular space alone at low frequencies but through both the extracellular and intracellular spaces at high frequencies. The Cole equation (equation (8.1)) can be put in terms of resistances as

$$Z = R_\infty + \frac{(R_0 - R_\infty)}{1 + (\mathrm{j}f/f_c)^{(1-\alpha)}} \tag{8.2}$$

where R_∞ and R_0 are the resistances at very high and very low frequencies, respectively; f is frequency, f_c is the relaxation frequency and α is a distribution constant.

By measuring the impedance Z over a range of frequencies it is possible to determine R_0 and R_∞ and hence estimate the extracellular and intracellular volume ratios. However, there are significant errors in this technique which assumes that tissue can be represented by a single equation. In practice, tissue consists of many types of tissues, structures and molecules, all of which will give rise to different relaxation processes.

Question 8.10.2.8

Draw three graphs, in each case showing how displacement and conduction currents change as a function of frequency for a resistance, a capacitance and a typical piece of tissue. Explain the shape of the graphs for the piece of tissue and, in particular, the reasons for the form of the frequency range around 1 MHz.

Question 8.10.2.9

Surgical diathermy/electrosurgery equipment is perhaps the oldest piece of electromedical equipment in general use. Discover what you can of its origin and describe the principles of the technique.

Answers to short questions

a   No, the conduction in tissue utilizes ions as the charge carriers.

b   Yes, tissue has both resistive and capacitive properties.

c   Neural stimulation is the main biological effect at low frequencies.

d   The impedance of tissue always falls with increasing frequency.

e   Tissue permittivity is dominant at very high frequencies.

f   Relaxation is a process whereby the application of an electric field causes a redistribution of charges so that the electrical properties of the tissue will change with time.

g   No, the Cole equation is empirical and not based upon a physical model.

h α determines the width of the distribution of values for the resistors and capacitors in a Cole model.

i   UV has an energy range of approximately 5–100 eV.

j   0.03 eV is the approximate energy of thermal motion at room temperature.

k   Microwaves are less energetic than UV.

l   Yes, you would feel a current of 10 mA at 100 Hz.

m   Lung tissue is less conductive than brain tissue.

n   Yes, 100 μA at 60 Hz would cause fibrillation if it was applied directly to the heart.

o   'Let-go' current is a current that will cause muscular contraction such that the hands cannot be voluntarily released from the source of current.

p   No, the wavelength of UV-B is about 300 nm.

q   Leakage current is a current that can flow accidentally through the body to ground as a result of poor design or malfunction of equipment which the person is touching.

r   Stimulation of neural tissue (principally the heart) and $I^2 R$ heating are the main factors that limit the current that may be safely applied to the body.

s   A four-electrode measurement measures only material impedance and does not include the electrode impedance. A two-electrode measurement cannot distinguish between tissue and electrode impedances.

t   Yes, there is a minimum in the graph of neural stimulation threshold versus frequency. This is at about 50–60 Hz.

u   DC passed through tissue will cause electrolysis.

BIBLIOGRAPHY

IEC 601 1988 *Medical Electrical Equipment; 601-1* 1991, 1995 *General Requirements for Safety & Amendments* (Geneva: International Electrotechnical Commission)

Cole K S and Cole R H 1941 Dispersion and absorption in dielectrics *J. Chem. Phys.* **9** 341–51

Diffey B L 1982 *UV Radiation in Medicine* (Bristol: Hilger)

Duck F A 1990 *Physical Properties of Tissue* (London: Academic)

Foster K R and Schwan H P 1989 Dielectric properties of tissue and biological materials: a critical review *Crit. Rev. Biomed. Eng.* **17** 25–104

Hawk J L M 1992 Cutaneous photobiology *Textbook of Dermatology* ed Rock, Wilkinson, Ebling, Champion (Oxford: Blackwell)

Macdonald J R (ed) 1987 *Impedance Spectroscopy* (New York: Wiley)

McKinlay A F, Harnden F and Willcock M J 1988 *Hazards of Optical Radiation* (Bristol: Hilger)

Moseley H 1988 *Non-Ionising Radiation: Microwaves, Ultraviolet Radiation and Lasers* (*Medical Physics Handbook* vol 18) (Bristol: Hilger)

Parrish J A, Fitzpatrick T B, Tanenbaum L and Pathak M A 1974 Photochemotherapy of psoriasis with oral methoxsalen and longwave ultraviolet light *New England J. Med.* **291** 1207–11

Pethig R 1979 *Dielectric and Electronic Properties of Biological Materials* (New York: Wiley)

Schwan H P 1957 Electrical properties of tissue and cell suspensions *Advances in Biological and Medical Physics* vol V ed L J H Tobias (New York: Academic) pp 147–224

Webster J G (ed) 1988 *Encyclopedia of Medical Devices and Instrumentation* (New York: Wiley)

CHAPTER 9

GAINING ACCESS TO PHYSIOLOGICAL SIGNALS

9.1. INTRODUCTION AND OBJECTIVES

This chapter addresses the problem of gaining access to physiological parameters. How can we make measurements of the electrical activity of a nerve or the movement of the chest wall during cardiac contraction?
Other questions we hope to answer are:

- Can we eavesdrop on the electrical signals produced by the body?
- What limits the smallest electrical signals which we can record?
- Does the body produce measurable magnetic fields?
- Can we gain access to the pressures and flows associated with fluid movement within the body?

When you have finished this chapter, you should be aware of:

- the need for conversion of physiological signals from within the body into a form which can be recorded.
- how electrodes can be used to monitor the electrical signals produced by the body.
- the significance of ionic charge carriers in the body.
- some invasive and non-invasive methods of measuring physiological signals.

We start by considering how we can convert a physiological parameter into an electrical signal which we can record and display. A transducer is a device which can change one form of energy into another. A loudspeaker may be thought of as a transducer because it converts electrical energy into sound energy; an electric light bulb may be considered as transducing electrical energy into light energy. Transducers are the first essential component of almost every system of measurement. The simplest way to display a measurement is to convert it into an electrical signal which can then be used to drive a recorder or a computer. However, this requires a transducer to change the variable to be measured into an electric variable.
A complete measurement system consists of a transducer, followed by some type of signal processing and then a data recorder.

9.2. ELECTRODES

Before any electrophysiological signal can be recorded, it is necessary to make electrical contact with the body through an electrode. Electrodes are usually made of metal but this is not always the case, and indeed there can be considerable advantages in terms of reduced skin reaction and better recordings if non-metals are used. If we are to be accurate then we should regard an electrode as a transducer as it has to convert the ionic flow of current in the body into an electronic flow along a wire. However, this is still electrical energy so it is conventional to consider separately the measurement of electrical signals from the measurement of non-electrical signals using a transducer.

A very common type of electrode is the Ag/AgCl type, which consists of silver whose surface has been converted to silver chloride by electrolysis. When this type of electrode is placed in contact with an ionic conductor, such as tissue, then the electrochemical changes which will occur underneath the electrode may be either reversible or irreversible. The following equation describes the reversible changes which may occur:

$$Ag + Cl^- \Leftrightarrow AgCl + e^-. \tag{9.1}$$

The chloride ions are the charge carriers within the tissue and the electrons are the charge carriers within the silver. If irreversible changes occur underneath the electrode then an electrode potential will be generated across the electrode. This type of potential is sometimes referred to as a polarization potential. It is worth considering a little further the generation of potentials across electrodes because they are important sources of noise in electrical measurements.

9.2.1. Contact and polarization potentials

If a metal electrode is placed in contact with skin via an electrolyte such as a salt solution, then ions will diffuse into and out of the metal. Depending upon the relative diffusion rates, an equilibrium will be established which will give rise to an electrode potential. The electrode potential can only be measured through a second electrode which will, of course, also have a contact potential. By international agreement electrode potentials are measured with reference to a standard hydrogen electrode. This electrode consists of a piece of inert metal which is partially immersed in a solution containing hydrogen ions and through which hydrogen gas is passed. The electrode potentials for some commonly used metals are shown in table 9.1.

Table 9.1. *Electrode potentials for some metals.*

Iron	−440 mV
Lead	−126 mV
Copper	+337 mV
Platinum	+1190 mV

These potentials are very much larger than electrophysiological signals. It might be thought that, as two electrodes are used, the electrode potentials should cancel, but in practice the cancellation is not perfect. The reasons for this are, firstly, that any two electrodes and the underlying skin are not identical and, secondly, that the electrode potentials change with time.

The changes of electrode potential with time arise because chemical reactions take place underneath an electrode between the electrode and the electrolyte. These fluctuations in electrode potential appear as noise when recording a bioelectric signal. It has been found that the silver/silver chloride electrode is electrochemically stable and a pair of these electrodes will usually have a stable combined electrode potential less than 5 mV. This type of electrode is prepared by electrolytically coating a piece of pure silver with silver chloride. A cleaned piece of silver is placed in a solution of sodium chloride, a second piece is also placed in the solution and the two connected to a voltage source such that the electrode to be chlorided is the anode. The silver ions combine with the chloride ions from the salt to produce neutral silver chloride molecules that coat the silver surface. This process must be carried out slowly because a rapidly applied coating is brittle. Typical current densities used are 5 mA cm^{-2} and 0.2 C of charge is passed.

If two steel electrodes are placed in contact with the skin then a total contact potential as high as 100 mV may be obtained. Any recording amplifier to which the electrodes are connected must be able to remove or amplify this potential, without distortion of the bioelectric signal which is also present.

Polarization is the result of direct current passing through the electrodes and it results in an effect like that of charging a battery, i.e. the electrode potential changes. The electrode contact potential will give rise to a current flow though the input impedance of any amplifier to which the electrode is connected and this will cause polarization. Electrodes can be designed to reduce the effect of polarization but the simplest cure is to reduce the polarization current by using an amplifier with a very high input impedance.

9.2.2. *Electrode equivalent circuits*

We have already mentioned that two electrodes placed on the skin will generate an electrode potential. The electrical resistance between the pair of electrodes can also be measured. Now an electrode will present a certain resistance to current flow simply by virtue of the area of contact between the electrode and the tissue. Before we can calculate this resistance we need to consider some of the fundamentals of how electric current flows in a conducting volume.

The electric field vector E is related to the scalar potential, ϕ, by

$$E = -\nabla\phi \tag{9.2}$$

where ∇ is the gradient operator.

From Ohm's law

$$J = \sigma E \tag{9.3}$$

where J is the current density vector field and σ is the conductivity which we will assume to be a scalar. Now if we have current sources present, which of course we have if we are injecting current, then the current source density I_v is given by the divergence of J,

$$\nabla J = I_v \tag{9.4}$$

substituting from (9.2) into (9.3) and then into (9.4) we obtain

$$\nabla J = I_v = -\sigma\nabla^2\phi.$$

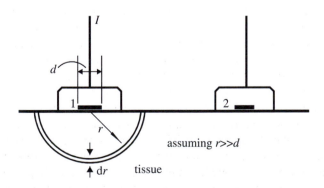

Figure 9.1. *Two electrodes placed in contact with a volume of tissue which is assumed to be both homogeneous and semi-infinite in extent.*

For a region where the conductivity is homogeneous, but which contains a source of density I_v, the following equation results:

$$\nabla^2 \phi = -\frac{I_v}{\sigma}.$$ (9.5)

This is Poisson's equation and allows us to determine the current density given by the potential field ϕ. For a single current source I_v it can be written in integral form as follows:

$$\phi = \frac{1}{4\pi\sigma} \int \frac{I_v}{r} \, dV.$$ (9.6)

We can use this form of Poisson's equation to determine the potential field set up by the current I injected by an electrode. If we take electrode 1 in figure 9.1 and apply a current I then, if we consider the tissue to be homogeneous and semi-infinite in extent, the current will spread out radially and the current density will be $I/2\pi r^2$. If we consider the hemisphere of tissue of thickness dr at radius r from the electrode then the potential drop dV across this element is given by

$$dV = \frac{I\rho}{2\pi r^2} \, dr$$

where ρ is the resistivity of the medium.

If we fix the potential at infinity as zero then the potential at any radius r is given by

$$V(r) = I \int_r^\infty \frac{\rho}{2\pi r^2} \, dr = \frac{\rho I}{2\pi r}$$

We can immediately see that the potential rises rapidly close to the electrode so that a point electrode will have an impedance which approaches infinity.

An electrode of finite area a can be modelled as a hemisphere of radius $\sqrt{a/2\pi}$ and the potential on the electrode will be given by

$$V = \frac{\rho I}{\sqrt{2\pi a}}$$

and the electrode impedance by

$$\text{electrode impedance} = \frac{\rho}{\sqrt{2\pi a}}.$$ (9.7)

This is sometimes referred to as the spreading resistance and is a useful way of calculating the approximate impedance presented by a small electrode. It is left to the student to extend the above method of working to calculate the resistance which will be measured between two electrodes placed on the skin. The potential gradients produced by the current from the two electrodes can be superimposed.

The spreading resistance should not be confused with the impedance of the electrode/tissue interface. If we use the above method to calculate the resistance of two electrodes, of 1 cm diameter, placed on the forearm we obtain an answer of about 200 Ω for an arm of resistivity 2 Ω m. However, even when the skin has been well cleaned and abraded, the actual magnitude of the impedance measured is likely to be at least 1 kΩ. The difference between the 200 Ω and the 1 kΩ is the impedance of the electrode/skin interfaces.

It is not a simple matter to calculate the total impedance to be expected between two electrodes placed on the skin. Both skin and tissue have a complex impedance. However, the impedance measured between a pair of electrodes placed on the skin will be similar to the impedance of the equivalent electrical circuit shown in figure 9.2. The values of the components will depend upon the type of electrodes, whereabouts on the body they have been applied, and how the skin was prepared. For a high-frequency sine wave voltage applied to the electrodes, the impedance of the capacitance, C, will be very small, and the total resistance is that of

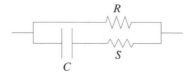

Figure 9.2. *A simple equivalent circuit for a pair of electrodes applied to skin.*

R and *S* in parallel. At very low frequencies, the impedance of the capacitance is very high and the total resistance will be equal to *R*. It is a relatively simple experiment to determine the values of the components in an equivalent circuit by making measurements over a range of sine wave frequencies. This can be done after different types of skin preparation to show the importance of this procedure.

The circuit given in figure 9.2 is not the only circuit which can be used; the circuit has approximately the same electrical impedance as a pair of electrodes on the skin, but the components of the circuit do not necessarily correspond to particular parts of the electrodes and tissue, i.e. this is an example of a model which fits the data, and not a model of the actual electrodes and skin. An even better fit can be obtained using the Cole equation given in section 8.3 and question 8.10.2.7 of Chapter 8.

We have spent some time considering the electrical contact presented by an electrode placed on the body. The reason for this is that the quality of the contact determines the accuracy of any electrical measurement which is made from the body. Any measurement system has to be designed to minimize the errors introduced by electrode contact.

9.2.3. Types of electrode

There is no clear classification of electrodes, but the following three groups include most of the commonly used types:

- *microelectrodes*: electrodes which are used to measure the potential either inside or very close to a single cell;
- *needle electrodes*: electrodes used to pass through the skin and record potentials from a small area, such as a motor unit within a muscle; and
- *surface electrodes*: electrodes applied to the surface of the body and used to record signals such as the ECG and EEG.

Microelectrodes are not used routinely in departments of medical physics and biomedical engineering. They are electrodes with a tip small enough to penetrate a single cell and can only be applied to samples of tissue. A very fine wire can be used, but the smallest electrodes consist of a tube of glass which has been drawn to give a tip size as small as 0.5 μm diameter; the tube is filled with an electrolyte such as KCl to which a silver wire makes contact. Microelectrodes must be handled with great care and special recording amplifiers are used in order to allow for the very high impedance of tiny electrodes.

Needle electrodes come in many forms but one type is shown in figure 9.3. This needle electrode is of a concentric type used for electromyography. A fine platinum wire is passed down the centre of the hypodermic needle with a coating of epoxy resin used to insulate the wire from the needle. The way in which the needle is connected to a differential amplifier, to record the potential between the tip of the platinum wire and the shaft of the needle, is shown. The platinum wire tip may be as small as 200 μm in diameter. This electrode is used for needle electromyography as it allows the potentials from only a small group of motor units to be recorded.

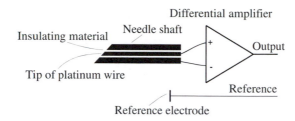

Figure 9.3. *A concentric needle electrode showing the connections to the recording amplifier.*

Needle electrodes must be sterilized before use and they must also be kept clean if they are to work satisfactorily. Some electrodes are suitable for sterilization by autoclaving, but others must be sterilized in ethylene oxide gas. This form of sterilization requires the needles to be placed in the ethylene oxide gas at 20 psi (140 kPa) for 1.5 h at a temperature of 55–66 °C. The articles must be left for 48 h following sterilization before use; this allows for spore tests to be completed and any absorbed gas to be cleared from the article. Cleaning of the electrodes applies particularly to the metal tip where a film of dirt can change the electrical performance of the electrode; it is possible for dirt on the tip to give rise to rectification of radio-frequency interference, with the result that radio broadcasts can be recorded through the electromyograph.

The earliest types of *surface electrode* were simply buckets of saline into which the subject placed their arms or legs. A wire was placed in the bucket to make electrical contact with the recording system. There are now hundreds of different types of surface electrode, most of which can give good recordings if correctly used. The most important factor in the use of any type of electrode is the prior preparation of the skin. There are electrodes in experimental use where an amplifier is integrated within the body of the electrode and no skin preparation is required if the capacitance between the electrode and the skin is sufficiently large. However, these types of electrode are expensive and have not yet been adopted in routine use.

9.2.4.   Artefacts and floating electrodes

One of the problems with nearly all surface electrodes is that they are subject to movement artefacts; movement of the electrode disturbs the electrochemical equilibrium at the electrode/tissue interface and thus causes a change in electrode potential. Many electrodes reduce this effect by moving the contact between metal and electrolyte away from the skin. Figure 9.4 shows how this can achieved by having a pool of electrolyte between the silver/silver chloride disc and the skin. The electrolyte is usually in the form of a gel or jelly. Movement of the electrode does not disturb the junction between metal and electrolyte and so does not change the electrode potential.

9.2.5.   Reference electrodes

There are some situations where we wish to make a recording of a steady or DC voltage from a person. For instance, steady potentials are generated across the walls of the intestines and, as these potentials are affected by intestinal absorption, their measurement can be useful diagnostically. Measurement of acidity, i.e. pH or hydrogen ion concentration, requires that a special glass electrode and also a reference electrode are connected to the test solution (see section 9.5.4). Figure 9.5 shows the construction of a silver/silver chloride reference electrode which has a stable contact potential of 343 mV. This electrode is stable because the interchange of ions and electrons is a reversible process as was described in section 9.2. The chlorided silver wire makes contact with 0.01 molar solution of KCl which also permeates the porous plug at the base of the electrode.

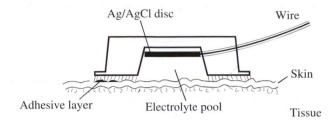

Figure 9.4. *This floating electrode minimizes movement artefacts by removing the silver/silver chloride disc from the skin and using a pool of electrode jelly to make contact with the skin.*

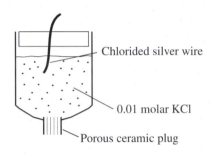

Figure 9.5. *A silver/silver chloride stable reference electrode.*

The plug is placed in contact with the potential source. In an alternative electrode mercurous chloride replaces the AgCl; this is often called a calomel electrode. Both types of electrode give a reference which is stable to about 1 mV over periods of several hours.

9.3. THERMAL NOISE AND AMPLIFIERS

The body produces many electrical signals such as the ECG/EKG and the EEG. The amplitudes of these signals are given in table 9.2. In order to record these we need to know both their sizes and how these relate to the noise present in any measurement system. In the limit, the minimum noise is that produced by the

Table 9.2. *The typical amplitude of some bioelectric signals.*

Type of bioelectric signal	Typical amplitude of signal
ECG/EKG	1 mV
EEG	100 μV
Electromyogram EMG	300 μV
Nerve action potential NAP	20 μV
Transmembrane potential	100 mV
Electro-oculogram EOG	500 μV

thermal motion of the electrons and ions in the measurement system. In this section we will consider thermal noise and signal amplification.

9.3.1. *Electric potentials present within the body*

Some electric fish generate pulsed electric potentials which they use to stun their prey. These potentials are generated by adding thousands of transmembrane potentials placed in series which can give potentials as high as 600 V and currents up to 1 A. It has also been shown that electric fish use electric current to navigate, which requires that they are able to sense quite small currents flowing into their skin; certainly some fish will align themselves with a current of only a few microamperes. Humans do not have such a system.

If small reptiles are wounded, then an electric current can be shown in the water surrounding the animal; similar currents have been shown to flow from the stump of an amputated finger in human children. These currents may be associated with the healing process and be caused by the potential which normally exists between the outside and inside of our skin, which is a semi-permeable membrane. The size of these potentials is up to 100 mV and they are the largest potentials which are likely to be encountered from the human body.

Whilst most bioelectric signals range in amplitude from a few microvolts up to a few millivolts, it should be born in mind that much smaller signals can be expected in some situations. For example, the typical amplitude of a nerve action potential measured from the surface of the body is given as 20 μV in table 9.2. This is what might be expected for a recording from the ulnar nerve at the elbow when the whole nerve trunk is stimulated electrically at the wrist. However, the ulnar nerve is relatively superficial at the elbow. If a recording is made from a much deeper nerve then very much smaller signals will be obtained. Also, the whole ulnar nerve trunk might contain 20 000 fibres. If only 1% of these are stimulated then the recorded action potential will be reduced to only 200 nV.

9.3.2. *Johnson noise*

Noise is important in any measurement system. If the noise is larger than the signal then the measurement will be of no value. Certain types of noise can be minimized if not eliminated. Noise caused by electrode movement or interference caused by nearby equipment falls into this category. However, noise caused by thermal movement of ions and electrons cannot be eliminated. Movement of the charge carriers represents an electric current which will produce a 'noise voltage' when it flows in a resistance. We must be able to calculate the magnitude of this noise so that we can predict what signals we might be able to measure.

An electrical current along a wire is simply a flow of electrons. Most currents which we might wish to measure represent a very large number of electrons flowing each second. For example,

$$1 \text{ A} \cong 6 \times 10^{18} \text{ electrons per second}$$
$$1 \text{ } \mu\text{A} \cong 6 \times 10^{12} \text{ electrons per second}$$
$$1 \text{ pA} \cong 6 \times 10^{6} \text{ electrons per second.}$$

Now the electrons will not move smoothly but will have a random movement, similar to the Brownian motion of small particles in a fluid, which increases with the temperature of the conductor. In a classic paper in 1928 J B Johnson showed that the mean square noise voltage in a resistor is proportional to the temperature and the value of the resistance. In a paper published at the same time H Nyquist started from Johnson's results and derived the formula for the noise as follows.

Consider two conductors I and II each of resistance R and connected together as shown in figure 9.6. The EMF due to thermal motion in conductor I will cause a current to flow in conductor II. In other words power is transferred from conductor I to conductor II. In a similar manner power is also transferred from conductor II to conductor I. Since at equilibrium the two conductors are at the same temperature it follows from the second law of thermodynamics that the same power must flow in both directions.

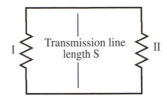

Figure 9.6. *Two conductors I and II assumed to be in thermal equilibrium.*

It can also be shown that the power transferred in each direction must be the same at all frequencies as otherwise an imbalance would occur if a tuned circuit was added between the two conductors. Nyquist's conclusion was, therefore, that the EMF caused by thermal motion in the conductors was a function of resistance, frequency range (bandwidth) and temperature.

Nyquist arrived at an equation for the EMF as follows. Assume that the pair of wires connecting the two conductors in figure 9.6 form a transmission line with inductance L and capacitance C such that $(L/C)^{1/2} = R$. The length of the line is s and the velocity of propagation v. At equilibrium there will be two trains of energy flowing along the transmission line.

If, at an instant, the transmission line is isolated from the conductors then energy is trapped in the line. Now the fundamental mode of vibration has frequency $v/2s$ and hence the number of modes of vibration, or degrees of freedom, within a frequency range f to $f+\mathrm{d}f$ will be $[(f+\mathrm{d}f)2s/v - f2s/v] = 2s\,\mathrm{d}f/v$ provided s is large. To each degree of freedom an energy of kT can be allocated on the basis of the equipartition law where k is Boltzmann's constant. The total energy within the frequency range $\mathrm{d}f$ is thus $2skT\,\mathrm{d}f/v$ and this will be the energy which was transferred along the line in time s/v. The average power, transferred from each conductor to the transmission line within the frequency interval $\mathrm{d}f$, during the time interval s/v, is thus $2kT\,\mathrm{d}f$.

The power transferred to the conductors of total resistance $2R$ by the EMF (V) is given by $V^2/2R$ and thus

$$V^2 = 4kTR\,\mathrm{d}f. \qquad (9.8)$$

This is the formula for the *Johnson noise* where V is the rms value of the thermal noise voltage, R is the value of the resistance in ohms, T is the absolute temperature, k is the Boltzmann constant $(= 1.37 \times 10^{-23}\ \mathrm{W\ s\ deg^{-1}})$ and $\mathrm{d}f$ is the bandwidth of the system in hertz.

We can see how this might be applied to calculate the thermal noise to be expected from two electrodes applied to the body. If the total resistance measured between two electrodes is 1 kΩ and the recording system has a bandwidth of 5 kHz in order to record nerve action potentials, then the thermal noise voltage at a temperature of 23 °C will be

$$(4 \times 10^3 \times 1.37 \times 10^{-23} \times 300 \times 5 \times 10^3)^{1/2} = 0.286\ \mu\mathrm{V\ rms}$$

This is the rms noise voltage. The actual voltage will fluctuate with time about zero, because it is caused by the random thermal movement of the electrons, and the probability of any particular potential V occurring could be calculated from the probability density function for the normal distribution. This is given by

$$p(V) = \frac{1}{\sigma\sqrt{2\pi}} \exp\left[-\frac{V^2}{2\sigma^2}\right] \qquad (9.9)$$

where σ is the rms value of V. However, it is sufficient to say that the peak-to-peak (p–p) value of the noise will be about 3 times the rms value, so that for our example the thermal noise will be about 0.85 μV p–p. This value of noise is not significant if we are recording an ECG/EKG of 1 mV amplitude, but it becomes very significant if we are recording a nerve action potential of only a few microvolts in amplitude.

9.3.3.   *Bioelectric amplifiers*

In order to record the bioelectric potentials listed in table 9.2 amplification is required. The simplest form of amplifier is as shown in figure 9.7 and uses a single operational amplifier. It is a *single-ended amplifier* in that it amplifies an input signal which is applied between the input and 'ground' or 'earth'.

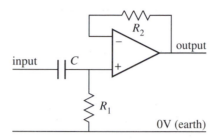

Figure 9.7. *A single-ended amplifier.*

The resistor R_1 is required to allow the 'bias current' to flow into the non-inverting (+) input of the operational amplifier and R_2 is required to balance R_1 so that the bias currents do not produce a voltage difference between the two inputs of the amplifier. Unfortunately, R_1 then defines the maximum input impedance of the amplifier. The *input impedance* is an important consideration in bioelectric amplifiers because it can cause attenuation of a signal which is derived from electrodes with high impedances. For example, if the two electrode impedances were 10 kΩ and the input impedance of the amplifier was 1 MΩ then 1% of the signal would be 'lost' by the attenuation of the two impedances. The impedance presented by the electrodes is termed the *source impedance* which must be very much less than the *input impedance* of the amplifier. Source impedance is seen to be even more important when we consider differential amplifiers shortly.

There is also capacitor C introduced in figure 9.7 in series with the input signal. This is introduced because of another property of electrodes. We explained in section 9.2.2 that contact and polarization potentials arise within electrodes and that these DC potentials can be very much larger than the bioelectric signals which we wish to record. Capacitor C blocks any DC signal by acting as a high-pass filter with the resistor R_1. This function is usually referred to as *AC coupling*. AC coupling will also cause some attenuation of the signal which may be important. We can determine this attenuation produced by R_1 and C by considering the transfer function between V_{in} applied to C and V_{out} the voltage across R_1,

$$\frac{V_{out}}{V_{in}} = \frac{R_1}{R_1 + 1/j\omega C} = \frac{R_1(R_1 + j/\omega C)}{(R_1^2 + 1/\omega^2 C^2)}$$

$$\left|\frac{V_{out}}{V_{in}}\right| = \frac{1}{\sqrt{1 + 1/R_1^2\omega^2 C^2}} \tag{9.10}$$

$$= \frac{1}{\sqrt{1 + (\omega_0^2/\omega^2)}} \qquad \text{where} \quad \omega_0 = 1/R_1 C. \tag{9.11}$$

Inserting values into equation (9.10) shows that if C is 1 μF and R_1 is 1 MΩ, then the attenuation of a 1 Hz signal will be 1.25%. This might be a significant attenuation for an ECG/EKG which will have considerable energy at 1 Hz.

Unfortunately, even with C added, this type of amplifier is not suitable for recording small bioelectric signals, because of interference from external electric fields. An electrode has to be connected to the amplifier

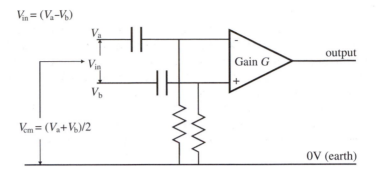

Figure 9.8. *A differential amplifier.*

via a wire and this wire is exposed to interfering signals. However, the interference will only appear on the input wire to the amplifier and not on the 'ground' wire which is held at zero potential. An elegant solution to this problem is to use a *differential amplifier* as shown in figure 9.8. The input to this type of amplifier has three connections marked '+', '−' and 'ground'. The signal which we wish to record is connected between the '+' and '−' points. Now both inputs are exposed to any external interfering electric field so that the difference in potential between '+' and '−' will be zero. This will not be quite true because the electric fields experienced by the two input wires may not be exactly the same but if the wires are run close together then the difference will be small. Differential amplifiers are not perfect in that even with the same signal applied to both inputs, with respect to ground, a small output signal can appear. This imperfection is specified by the *common mode rejection ratio* or CMRR. An ideal differential amplifier has zero output when identical signals are applied to the two inputs, i.e. it has infinite CMRR. The CMRR is defined as

$$\text{CMRR} = 20 \log \left[\frac{\text{signal gain}}{\text{common-mode gain}} \right] \tag{9.12}$$

where, using the terminology of figure 9.8, signal and common-mode gains are given by

$$\text{signal gain} = \frac{V_{\text{out}}}{V_{\text{in}}} = \frac{V_{\text{out}}}{(V_a - V_b)}$$

$$\text{common-mode gain} = \frac{V_{\text{out}}}{V_{\text{cm}}} = \frac{V_{\text{out}}}{(V_a + V_b)/2}$$

In practice V_{cm} can be as large as 100 mV or even more. In order to reject this signal and record a signal V_{in} as small as 100 μV a high CMRR is required. If we wish the interfering signal to be reduced to only 1% of V_{out} then

$$\text{required signal gain} = \frac{V_{\text{out}}}{100 \, \mu\text{V}}$$

$$\text{required CM gain} = \frac{V_{\text{out}}/100}{100 \, \text{mV}}$$

$$\text{CMRR} = 20 \log \frac{V_{\text{out}} 10^2 \times 10^{-1}}{10^{-4} V_{\text{out}}} = 100 \, \text{dB}.$$

It is not always easy to achieve a CMRR of 100 dB. It can also be shown that electrode source impedances have a very significant effect on CMRR and hence electrode impedance affects noise rejection. The subject of differential amplifiers and their performance is considered in more detail in section 10.3.3 of the next chapter.

AC or DC coupling

The AC coupling shown in both figures 9.7 and 9.8 degrades the performance of the amplifiers. If the input impedance and bias current of the operational amplifiers is sufficiently high then they can be connected directly to the input electrodes, without producing significant electrode polarization. DC offsets will of course occur from the electrode contact potentials, but if the amplifier gain is low (typically <10) this is not a significant problem. The offset can be removed by AC coupling at a later stage.

However, there are safety arguments against the use of DC coupling. If a fault arises in the operational amplifier then it is possible for the power supply voltage to be connected directly to the patient and so give rise to a hazard. DC currents will cause electrolysis and result in tissue necrosis. AC coupling avoids this problem and is often used. Nonetheless DC coupling is also often used.

Other sources of error

Amplifiers can give rise to many types of error in addition to those already considered. For example, there is a limit to the maximum voltage that an amplifier can produce. This is usually set by the power supply voltages. When the output required from the amplifier approaches the power supply voltages then saturation or clipping will occur. This may occur gradually or be a sudden limit to the voltage output.

Another source of error arises because of the construction of the output stages of some amplifiers. Often different transistors are used to produce the positive and negative output components of a signal. Problems can arise when the output signal is close to 0 V when neither of the two output transistors is conducting. A dead zone can arise and cross-over distortion occurs.

Finally, hysteresis can occur whereby, when a sudden rise in voltage is followed by a sudden fall, the output of the amplifier may lag behind the input differently for positive and negative inputs.

Noise

Finally in this section we will consider briefly the effect of an amplifier on the thermal noise which was described in section 9.3.2. We calculated the thermal noise generated by electrodes with a source resistance of 1 kΩ as 0.286 μV rms over a bandwidth of 5 kHz. If these electrodes are connected to an amplifier of gain G then the noise will simply be amplified and we will obtain a noise of 0.286G μV rms at the output. However, this is the situation for a perfect amplifier. Amplifiers are not perfect and will actually add some noise to the output signal. The noise performance of amplifiers is specified in many ways, but these will not be described here. The most simple *noise figure* gives the increase in the noise output which is contributed by the amplifier. This is specified in dB. If the noise figure is 4 dB then the actual noise output in our example will be given by

$$\text{noise at output} = 10^{4/20}G \times 0.286 \ \mu\text{V} = G \times 0.453 \ \mu\text{V rms}.$$

9.4.   BIOMAGNETISM

Neural tissue produces electrical potentials within the body and these potentials give rise to electrical currents in tissue. These currents give rise to electrical signals such as the ECG/EKG, but they will also give rise to magnetic fields. Such biomagnetic fields certainly exist but they are not measured in routine health care, simply because they are difficult to record and understand. Biomagnetic fields from the heart were first recorded as recently as 1963 by Baule and McFee. In this section we will use some basic physics to assess what biomagnetic signals are likely to exist and then consider some techniques for measuring these signals.

9.4.1. *Magnetic fields produced by current flow*

Ampére working near Lyon between 1820 and 1825 was the first person to investigate the interaction between coils carrying electric currents. He found that two parallel wires carrying a current in the same direction experienced an attractive force, whereas if the current was flowing in opposite directions in the two wires the force was one of repulsion. The interaction was described as being the result of a magnetic induction B produced by the current flow.

A magnetic induction or magnetic flux density of B will exert a force on either a current-carrying wire or a magnet. The unit of magnetic flux density is the *tesla* (T) which will exert a force of 1 N on a wire of length 1 m carrying a current of 1 A. Tesla was a Serb who was born in Croatia, but lived much of his life in the USA. He was offered, but refused, the Nobel prize for physics! An older unit of magnetic flux is the gauss (G); 10^4 G = 1 T. The experiments of Ampére and others showed that the magnetic flux density dB at one position produced by a current i in an element of wire ds at a second position is given by the formula

$$dB = \left(\frac{\mu_0}{4\pi r^3}\right) i \,(ds \wedge r) \tag{9.13}$$

where r is the vector joining the two positions, r is the distance between the two positions and μ_0 is the permeability of free space with a value of $4\pi \times 10^{-7}$. r, s and B are mutually orthogonal. The magnetic field H is also used and this is given by

$$B = \mu_0 H.$$

We can put the tesla into context by considering some simple geometries. If we have a wire of length ds metres carrying a current of 1 A, then the magnitude of B at a point 1 m away in a direction at right angles to ds will be $\mu_0 ds/4\pi$ T. A more practical example is in terms of B at the centre of a plane circular coil of one turn carrying a current of 1 A. Here, r is the radius of the coil and the element ds is always perpendicular to r. Hence, we can integrate equation (9.13) to give the magnitude of B at the centre of the coil as

$$B = \int_0^{2\pi r} \frac{\mu_0}{4\pi r^2} \, ds = \frac{\mu_0}{2r} \tag{9.14}$$

So the magnetic flux density at the centre of a single-turn coil of diameter 1 m carrying a current of 1 A will be μ_0, i.e. $4\pi \times 10^{-7}$ T. The magnetic flux density produced by the Earth is about 5×10^{-5} T, so to produce a field at the centre of the coil equal to that of the Earth would require $500/4\pi \cong 40$ A or 40 turns carrying 1 A, i.e. 40 A turns.

9.4.2. *Magnetocardiogram (MCG) signals*

Calculation of the magnetic field produced by the currents which circulate around the heart is very difficult. The whole of the myocardium generates electrical potentials and this source is, of course, moving during the cardiac cycle. In addition, the currents which flow and hence the magnetic fields produced will depend upon the conductivity of the tissue which surrounds the heart. Geselowitz (1980) gives a full derivation of the magnetocardiogram (MCG). We can attempt to estimate the magnetic fields which might be produced during the cardiac cycle by making some very sweeping assumptions.

Firstly, we will assume that the tissue which surrounds the heart is homogeneous and infinite in extent. Secondly, we will assume a value for the potential generated across the heart during cardiac systole and hence derive the current required to produce this potential. The geometry which we will assume is given in figure 9.9, where four points a, b, c, d are considered. We will determine what current passed between points a and d will give rise to potential u between points b and c. The separation of the four points is s.

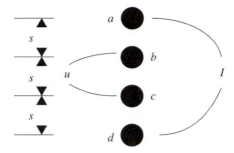

Figure 9.9. *Geometry assumed in finding the currents flowing during cardiac contraction.*

Consider a spherical shell of radius r and thickness dr centred on point a. The potential drop dv across this shell when current I is injected at point a will be given by

$$dV = \frac{I\rho \, dr}{4\pi r^2}$$

where ρ is the resistivity of the medium. The potential developed between points b and c, V_{bc}, will be given by

$$V_{bc} = \int_s^{2s} \frac{I\rho}{4\pi r^2} \, dr = \frac{I\rho}{8\pi s}. \tag{9.15}$$

An identical potential will be produced between the same two points by the current $-I$ injected at point d. Potentials are additive and hence, the total potential u generated between points b and c will be $2V_{bc}$. Now we can put in some typical values for the heart. If we assume that all four points lie on the heart then s might be at most 50 mm. The potential generated by the heart can be estimated as about 4 mV, from measurements made of the ECG/EKG close to the heart. If we assume the surrounding tissue to have a resistivity of 2 Ω m then we can determine the current I required to generate 4 mV,

$$I = \frac{8\pi s V_{bc}}{\rho} = \frac{8\pi \times 5 \times 10^{-2} \times 2 \times 10^{-3}}{2} = 1.26 \text{ mA}. \tag{9.16}$$

So the current surrounding the heart during cardiac systole is likely to be about 1 mA. We can now ask the question as to what magnetic flux density will be produced by this current. Using equation (9.15) we can determine the magnitude of B at a distance of 20 cm from the heart on the assumption that the current flowing along 5 cm, at the heart, is 1 mA,

$$B = \frac{Is\mu_0}{4\pi r^2} = \frac{10^{-3} \times 5 \times 10^{-2} \times 4\pi \times 10^{-7}}{4\pi \times 4 \times 10^{-2}} = 125 \text{ pT}. \tag{9.17}$$

Therefore, the magnetic flux which we might expect to be generated by the heart during cardiac systole is about 125 pT. The fields actually recorded from the heart are typically 50 pT so our simple model gives a plausible answer (cf table 9.3). 125 pT is a very small field in comparison to that produced by the Earth, which we quoted in the previous section as about 50 μT. Nonetheless, the magnetic field produced by the heart can be detected and a typical recording looks very much like an ECG/EKG. It does contain different information to that contained in the ECG/EKG, but it has yet to establish itself as a diagnostic method.

Table 9.3. *Typical amplitudes of several biomagnetic signals.*

Biomagnetic signal	(pT)
Magnetocardiogram	50
Fetal MCG	1–10
Magnetoencephalogram	1
Evoked fields	0.1
Magnetomyogram	10
Magneto-oculogram	10
Earth's field	50×10^6

In the next sections we consider some ways of measuring small magnetic flux densities.

9.4.3. Coil detectors

The simplest method of measuring a magnetic field is to record the voltage induced in a coil, but this method can only be used to measure changing magnetic fields. Faraday's law of magnetic induction can be expressed mathematically as

$$V = -\frac{\mathrm{d}}{\mathrm{d}t} \int \boldsymbol{B} \cdot \mathrm{d}\boldsymbol{s} \tag{9.18}$$

where V is the voltage induced by the magnetic flux density $\boldsymbol{B}$, and the integral is taken over any area bounded by the coil. We can insert typical values for the magnetocardiograph to see whether we might be able to detect the signal levels given in section 9.4.2. If the magnitude of $\boldsymbol{B}$ is 50 pT and this change occurs during cardiac systole which lasts about 100 ms, then the voltage induced in a coil of radius r, equal to 5 cm, and with N turns, equal to 10^3, will be

$$\text{induced voltage} = N\pi r^2 \frac{\mathrm{d}B}{\mathrm{d}t} = \frac{10^3 \times \pi \times 25 \times 10^{-4} \times 50 \times 10^{-12}}{10^{-1}} \cong 4 \times 10^{-9} \text{ V}.$$

This is a very small voltage and is much smaller than the thermal noise voltages which we considered in section 9.3.2. The coil we have considered is 'air cored'. Now we can use materials such as iron, ferrites or special alloys with a high magnetic permeability to increase the induced voltage. Magnetic permeabilities of 10^3 can be obtained quite easily and even higher values are found for some materials. We can also increase the number of turns by perhaps a factor of 10 so that a total increase of 10^4 might be achieved, giving us a voltage of 4×10^{-5}, i.e. 40 μV. Our conclusion is that it will be difficult to detect the MCG using a coil as the detector, but not impossible. Indeed, the first recording of the MCG was made, by Baule and McFee, using a coil detector.

9.4.4. Interference and gradiometers

There is a further problem to be overcome in the detection of the magnetocardiograph and that is interference. The magnetocardiograph is typically 50 pT in amplitude, but typical values for the magnetic fields produced by power wiring and equipment in a modern building are 5×10^5 pT at 50/60 Hz. It is possible to screen the room in which the magnetocardiograph is being recorded, but this is expensive and not very effective. Rooms with copper sheet in the walls are very effective at attenuating electric fields but not magnetic fields. Large thicknesses of ferromagnetic material or sheets of aluminium or expensive mu-metal are required to provide significant attenuation of magnetic fields.

One method to reduce interference uses a gradiometer to measure the spatial gradient of the magnetic field. Equation (9.17) showed that the magnitude of B 20 cm from the heart would be 125 pT, but this field falls off as an inverse square law. Now the spatial differential of this field will fall off as an inverse cube law. The ratio of B to the spatial gradient of B is equal to $r/2$,

$$\frac{B}{\mathrm{d}B/\mathrm{d}r} = \frac{r}{2}. \tag{9.19}$$

By measuring the spatial gradient of B it is possible to discriminate against large but distant interfering sources. A source of interference 2000 cm away might give rise to a field of 1250 pT in contrast to our MCG of only 125 pT, but the spatial differential of the interference will be 1.25 pT cm^{-1}, whereas that of the MCG at 20 cm from the heart will be 12.5 pT cm^{-1}. We have made an improvement of a factor of 100 in our signal-to-noise ratio. However, we have also lost a little because any random noise generated in the coils will be increased by a factor of $\sqrt{2}$ as two coils are now contributing to the noise.

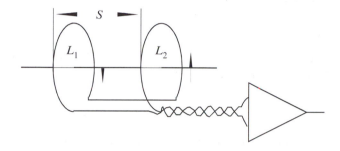

Figure 9.10. *A magnetic field gradiometer using two detectors to measure the field gradient.*

A gradiometer can be implemented as shown in figure 9.10 using two coils placed axially, at a distance S apart and in opposite phases. The same principle can be used with any type of magnetic field detector. There is a similarity between the use of a gradiometer to reduce magnetic field interference and the use of differential amplifiers to reduce electric field interference. A gradiometer is very useful in reducing interference, but it may still not be sufficient to allow a signal from the body to be detected. In some cases second-order gradiometers are used. These use three detectors to determine the second differential of B. This will fall off as r^2 and so offer even higher discrimination against distant objects. Third-order gradiometers have also been constructed.

9.4.5. Other magnetometers

Accurate magnetic field measurements can be used in many measurement situations. For measurements to an accuracy of about 1% *Hall-effect* probes are adequate. The Hall effect is the production of a transverse voltage in a current-carrying conductor when placed in a magnetic field. The conductor is usually a semiconductor and commercial probes can be obtained for measurements in the range 10 μT to 1 T.

Flux-gate magnetometers offer higher sensitivity than coils or Hall-effect probes and have the advantage of DC sensitivity. They depend upon the sensitivity of a metal or ferrite core to a magnetic field. Two coils are wound onto the core and the first coil is driven with an AC current which is sufficient to magnetically saturate the core. The second (sensor) coil detects the changes in magnetic induction. With no external magnetic field the sensor coil detects a waveform which consists of a fundamental and odd harmonics. However, an external magnetic field causes asymmetric magnetization and hence distortion of the sensor waveform which then exhibits even harmonics. This is illustrated in figure 9.11, which shows two waveforms and the associated

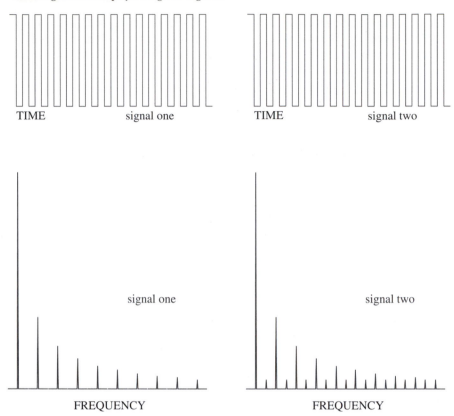

Figure 9.11. *Two waveforms as a function of time. Signal one is a square wave; signal two looks identical but has a mark-to-space ratio changed by 2.5%. The lower plots show the frequency transforms of the two waveforms. Signal one shows only odd harmonics whereas, signal two also shows even harmonics because of the asymmetry.*

Fourier transforms. The small asymmetry in signal one is not visible in the figure but gives rise to the very obvious appearance of even harmonics in the Fourier transform of the signals. In the flux-gate magnetometer a current feedback technique is used to compensate for the effects of the external field and thus minimize the even harmonics. The signal which has to be fed back is thus proportional to the external magnetic field. This type of device can offer a sensitivity of about 4 pT over a 100 Hz bandwidth.

Superconducting quantum interference devices (SQUIDs) allow very low-noise amplifiers to be made, but they need to be operated at liquid helium temperature (4.2 K). They are expensive devices, but they do allow voltages as small as 10^{-15} V to be recorded.

A SQUID device is illustrated in figure 9.12 and uses a superconducting Josephson junction. A SQUID magnetometer consists of a flux transformer and a SQUID amplifier immersed in liquid helium in order to maintain superconductivity.

Flux transformers consist of one or more coils. A single-loop transformer can be used for DC magnetic field measurement. Two or more coils are used in gradiometers. The beauty of the SQUID is that its output depends periodically on the external magnetic flux present inside the SQUID ring so that feedback can be used to give a digital output proportional to the magnetic field. SQUID magnetometers have noise levels

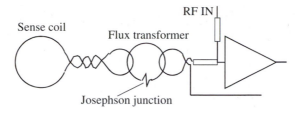

Figure 9.12. *Diagram of a SQUID magnetometer.*

below 5 fT/$\sqrt{\text{Hz}}$. One femtotesla is 10^{-15} T so that the noise over a bandwidth of 100 Hz will be less than 5×10^{-14} T, i.e. 0.05 pT. This is very much lower than the amplitude of the MCG which we determined as about 125 pT. SQUIDS have sufficient sensitivity to record signals from the brain as well as from the heart.

9.5. TRANSDUCERS

We argued earlier that electrodes should be considered as transducers because they allow an ionic current in the body to be converted into an electronic current that we can record using wires. We now come to transducers that really do convert a parameter such as pressure or temperature into an electronic signal.

Pressure transducers

Pressure measurement was introduced in section 2.2 of Chapter 2. Pressure transducers are an important tool in both medical physics and biomedical engineering and hence they are dealt with separately in Chapter 18.

9.5.1. *Temperature transducers*

Measurement of body temperature is relatively easy if the measurement site is accessible. The control of temperature is made from the base of the brain where the medulla senses any change in temperature and then controls the blood flow in order to maintain thermal equilibrium. The control of body core temperature is tight, with a normal range of about 37.0–37.5 °C. However, temperature elsewhere in the body can vary considerably and the arms may only be at 32 °C.

Measurement of core temperature is difficult unless a needle containing a transducer can be placed well inside the body. If a single measurement is needed then a mercury thermometer inserted into the mouth is certainly the simplest measurement; because of the hazards attached to mercury an electronic probe is now more likely to be used. Techniques have been developed to measure temperature close to the eardrum which has been shown to give the best approach to core temperature. The technique is now commonly used, but can give inaccurate results if not used very carefully. Most patient monitoring systems use either a thermistor or a thermocouple to measure temperature. The transducer probe can be used to give oral or rectal temperature, or that under the armpit (the axilla).

The *thermistor* is a semiconductor device which exhibits a large change in resistance with temperature. The resistance usually falls with temperature and the change may be as great as 5% per °C. The thermistor can be made as small as 0.5 mm diameter so that, if necessary, one can be mounted at the end of a hypodermic needle.

The equation which relates resistance and temperature is of the form

$$S = ae^{-bt}$$

where S is resistance, t is temperature and both a and b are constants.

The measuring circuit for a thermistor transducer is simply a bridge and a differential amplifier. The circuit shown will not give a linear output change with temperature because of the thermistor characteristics. We can calculate how the output voltage of a simple resistance bridge changes with thermistor resistance. Using the notation of figure 9.13:

$$\text{output voltage} = V_a - V_b = V\frac{(R - S)}{2(R + S)}.$$

This equation is plotted in figure 9.14, which also shows the thermistor characteristic. The curves shown in figures 9.14(a) and (b) have similar slopes. If the two curves are combined (figure 9.14(c)), an approximately linear relationship between the output voltage and thermistor temperature results.

To obtain a better correction for the thermistor nonlinearity, another resistance is often added between V_a and V_b. Analysis of this circuit is not trivial, but it is possible to obtain a response which is linear to within $0.1\,^{\circ}\text{C}$ over the temperature range 25–$45\,^{\circ}\text{C}$. This is adequate for most patient monitoring systems.

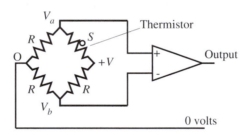

Figure 9.13. *The output from this resistive bridge and differential amplifier will be proportional (but not linearly) to the thermistor resistance.*

An alternative temperature transducer is a *thermocouple*. If two different metals are in contact then electrons will diffuse across the junction, but not in equal numbers. This results in a potential between the metals. Because diffusion increases with temperature, the potential increases with temperature and indeed the potential is linearly related to temperature over a wide temperature range. If accurate temperature measurements are required, then a thermocouple is better than a thermistor. For a particular pair of metals, the junction potential is always the same at a particular temperature, and therefore thermocouple probes can be interchanged without causing errors. The disadvantage of a thermocouple is that a reference junction is required at an accurately maintained temperature. The reference may be melting ice, an accurate oven or a semiconductor reference element.

Thermocouple systems can be made to have an accuracy of better than $0.01\,^{\circ}\text{C}$, but careful design is needed. A copper/constantan thermocouple will give an output of about $40\ \mu\text{V}$ for a $1\,^{\circ}\text{C}$ change in temperature and therefore only $0.4\ \mu\text{V}$ for $0.01\,^{\circ}\text{C}$. The amplifier which is used to record this voltage must have an input offset voltage drift and a noise level which is much less than $0.4\ \mu\text{V}$. This can be achieved, but an operational amplifier with this performance may be quite expensive.

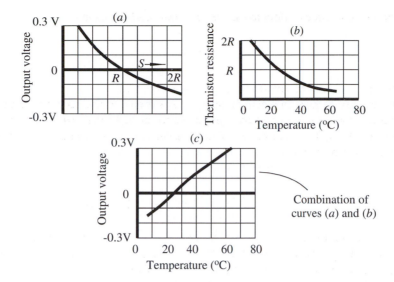

Figure 9.14. *An analysis of the circuit shown in figure 9.13. In (a) the output voltage is shown as a function of S; in (b) the thermistor resistance S is shown as a function of temperature; in (c) curves (a) and (b) have been used to show the approximately linear relation between output voltage and temperature.*

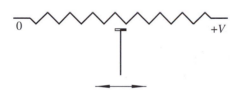

Figure 9.15. *A linear displacement transducer. The potential on the wiper is proportional to its position along the resistance.*

9.5.2. *Displacement transducers*

A displacement transducer will give an electrical output in proportion to a change in position (see figure 9.15). A linear displacement transducer may be connected to a motorized syringe so that the movement of the syringe can be recorded. There are very many other applications of displacement transducers in biomedical equipment.

There are also a very wide variety of displacement transducers. The simplest transducer is an electrical resistance with a sliding contact; the volume control on a car radio is a displacement transducer which converts rotary motion into an electrical voltage. *Linear variable differential transformer* (LVDF) transducers consist of a transformer with a core which can move linearly. By exciting the primary with a sine wave and measuring the induced voltage in the secondary, a signal proportional to the position of the core is obtained. *Capacitive transducers* simply measure displacement by measuring the capacitance between two closely spaced plates— the capacitance is inversely proportional to the plate separation. *Optical methods* based upon counting the interference fringes produced by mixing the transmitted light with the light reflected from the target used to be very expensive, but can now use semiconductor laser diodes at relatively low cost. *Digital transducers* which give a pulse output corresponding to each increment of displacement can be more accurate and durable than analogue transducers.

9.5.3. Gas-sensitive probes

Transducers can be made to give an electrical output proportional to the partial pressure of a specific gas. The design of these transducers is a very large subject which includes the design of probes to measure gases dissolved in blood as well as respiratory gases. Blood gas analysis is used particularly in neonatal care units; the level of CO_2 and O_2 in the blood can change rapidly in infants and the supply of gases to an incubator must take these values into account. Too little oxygen can cause death and too much oxygen can give rise to retinal and brain damage.

The volume of blood available for repeated measurements of blood gases is small and this causes many difficulties in measurement. However, instruments such as the mass spectrometer and the flame photometer are being developed to the point where they can be used for the continuous estimation of blood gases. Only one technique will be described here; this is the use of an oxygen membrane electrode for the measurement of P_{O_2} from a blood sample. A sample of about 100 μl is required.

P_{O_2} is the *partial pressure of oxygen*. Consider a sample of air at atmospheric pressure, i.e. 760 mmHg or 101.3 kPa. Dry air contains 21% of oxygen by volume and so the partial pressure will be

$$101.3 \times 0.21 = 21.3 \text{ kPa (159.6 mmHg).}$$

Whole blood has a mean water content of 850 g l^{-1}, so if air is equilibrated with blood in a test tube at 37 °C it then becomes saturated with water vapour, which has a partial pressure of 6.25 kPa (47 mmHg) at 37 °C. The P_{O_2} will now be

$$0.21 \times (101.3 - 6.25) = 20.0 \text{ kPa (149.7 mmHg).}$$

This pressure is the same in the blood as in the air because they are in equilibrium. P_{O_2} is sometimes called oxygen tension and normal values are: 10.7–12.0 kPa (80–90 mmHg) for arterial blood and 5.3–6.7 kPa (40–50 mmHg) for venous blood.

A blood oxygen transducer is shown in figure 9.16. In the centre of the transducer is a platinum wire set in glass which will react with oxygen to release charge carriers and so give rise to an electrical current. The oxygen in the blood sample diffuses through the Teflon membrane and then reacts with the platinum wire. The reduction reaction is

$$O_2 + 2H_2O + 4e^- = 4OH^-.$$

An electrical current will flow in proportion to the amount of O_2 diffusing through the membrane. For the current to flow, a potential must be applied in order to attract the electrons from the platinum electrode.

Before a current can flow a circuit has to be complete and so a silver/silver chloride electrode is immersed in the KCl as an indifferent electrode. Typically a potential of 0.8 V is applied, with the platinum negative, and a current of the order 10 nA is obtained. A larger electrode will give a larger current and a membrane with a high permeability will give a high current, but in both cases the oxygen removed from the sample is increased and so a larger sample will be required. The aim is to make the measurement without significantly altering the P_{O_2} in the sample. To measure a current of 10 nA a very high input impedance amplifier is required with a bias current which is much less than 10 nA.

Oxygen electrodes of this type are widely used for blood gas analysis. In a number of automated systems, the blood sample in a capillary is sucked past a P_{O_2} electrode and also a P_{CO_2} and then a pH electrode. All these transducers are fragile and, if accurate results are to be obtained, they must be kept clean and calibrated.

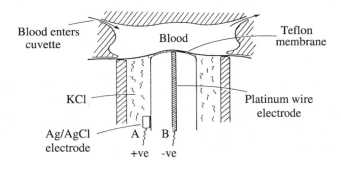

Figure 9.16. *A blood oxygen transducer. The current flow between A and B is proportional to the oxygen diffusion through the Teflon membrane.*

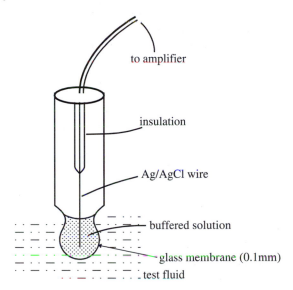

Figure 9.17. *A glass pH electrode.*

9.5.4. pH electrodes

An important indicator of chemical balance in the body is the pH of the blood and other fluids. pH is a measure of H^+ concentration which was devised by the Danish biochemist Peter Sorensen, and is given by

$$pH = -\log_{10}[H^+].$$

It is a measure of the acid–base balance of a fluid. A neutral fluid has a pH of 7. Lower pH values indicate increasing acidity. Higher pH values indicate increasing alkalinity. The pH of arterial blood is usually about 7.4. Gastric juices have a typical pH of 1.5, but values vary greatly with food content and acid secretion.

The most common electrode for the measurement of pH is the glass electrode shown in figure 9.17. It consists simply of a glass membrane which is permeable to H^+ ions and across which a Nernst potential (see section 16.1.3) is generated in proportion to the difference in H^+ concentration. We can calculate the potential to be expected from equation (16.1).

In order to measure the potential generated across the glass membrane electrical contact must be made to both sides of the membrane. Now the Ag/AgCl wire inserted into the buffer solution, as shown in figure 9.17, gives access to the potential on the inside of the glass bulb. In order to measure the potential on the outside a second reference electrode has to be inserted into the fluid. The reference electrode can take the form of the one illustrated in figure 9.5. However, in many commercial systems the glass electrode and the reference electrode are combined into a single probe.

The glass membrane is permeable to H^+ ions, but nonetheless it will have a high electrical resistance. Typical values are 100–500 MΩ. This resistance will form a source resistance to any amplifier used to record the potential generated across the glass membrane and hence a very high input impedance is required for the amplifier. However, input impedances of 10^{12} Ω can be obtained using MOSFET devices for amplification.

9.6.   PROBLEMS

9.6.1.   *Short questions*

a   Is a typical magnetocardiogram signal larger or smaller than the magnetic field surrounding the Earth?

b   What is the most important disadvantage of a SQUID detector for measuring biomagnetic signals?

c   What is Johnson noise?

d   Are most bioelectric amplifiers single ended or differential types?

e   What interference does an amplifier with a high common-mode rejection ratio discriminate against?

f   Is a typical magnetocardiogram 50 pT, 50 nT or 50 μT in amplitude?

g   What limits the smallest signals that can be recorded?

h   What is meant by an electrode or contact potential?

i   Do electrode potentials usually change with time?

j   What is electrode chloriding?

k   Will a large electrode have a lower or higher impedance than a small one?

l   How small can the tip of a microelectrode be made?

m   What is meant by the term 'floating electrode'?

n   About how large is the action potential recorded from just one fibre of the ulnar nerve at the elbow?

o   What is meant by the term 'AC coupling'?

p   Where in the body is the temperature most stable?

q   What is a thermistor?

r   How does a glass pH electrode work?

s   What is pH?

t   What reference electrode is used when measuring standard electrode potentials?

9.6.2.   *Longer questions and assignments (answers are given to some of the questions)*

Question 9.6.1.1

We wish to track a probe as it is moved in a single plane over the surface of the body. The probe is mounted on an arm, but is free to move over the surface of the body. Sketch a design using two transducers which will enable the coordinates of the probe to be obtained. Explain how you might extend the design to allow movement of the probe in three dimensions.

Answer

One linear displacement transducer and one rotary transducer can be used to give the information which is needed. The probe can be mounted on an arm and a displacement transducer used to give a voltage in proportion to the distance, r, between the probe and the centre of rotation of the arm, marked as O in figure 9.18. A rotary displacement transducer is connected to measure the rotation of the arm, i.e. angle a.

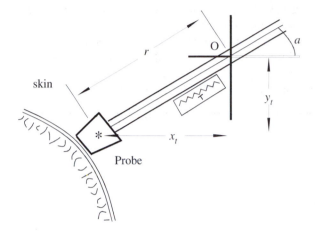

Figure 9.18. *The probe can be moved in a single plane over the surface of the skin in this diagram. Angle a is given by a rotary transducer; distance r is given by a linear displacement transducer. From these, the position and angle of the probe are found.*

The coordinates of the probe will be x_t and y_t where

$$x_t = r \cos a \qquad \text{and} \qquad y_t = r \sin a.$$

$\sin a$ and $\cos a$ can be obtained from a, either by using a digital rotary transducer which gives out a number of digital pulses in proportion to the angle of rotation, or by passing the analogue output of the transducer via an A–D converter into the computer which will calculate $\sin a$ and $\cos a$.

Question 9.6.2.2

Describe a three-component electronic model for the impedance of an electrode in contact with skin. Indicate the significance of each component.

An electrode is characterized by placing it in contact with saline in a bath, along with an Ag/AgCl electrode with a much greater surface area and a known half-cell potential of 0.252 V. The DC voltage between the electrodes is measured as 0.555 V with the test electrode negative. The magnitude of the impedance between the electrodes is measured as a function of frequency at low currents. The results are as shown in table 9.4 and figure 9.19. From this data determine the approximate values of the equivalent circuit components for the electrode under test and the half-cell potential of the electrode. State any assumptions which you make.

Answer

An R, S and C model as given in figure 9.2 should be given. It should be pointed out that the low-frequency impedance (R) will depend very much upon skin preparation that affects the local electrochemistry where

Table 9.4. *Electrode impedance measured over the range 10 Hz to 100 kHz (see figure 9.19).*

Frequency (Hz)	Impedance magnitude (Ω)	Impedance angle (deg)
10	29 860	−5.4
27.8	28 990	−14.7
77.4	24 090	−35.9
215.4	13 080	−62.2
599	5 170	−74.7
1 669	1 940	−71.5
4 642	840	−52.2
12 915	555	−25.4
35 900	507	−9.7
100 000	500	−3.5

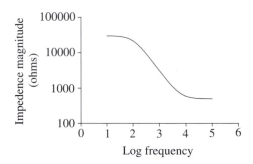

Figure 9.19. *Electrode impedance measured over the range 10 Hz to 100 kHz.*

the electrode connects to the tissue. S will be a function of the area of the electrode and the resistivity of the tissue. A large electrode will give a lower value for S. If the tissue has a relatively high resistivity then S will be high. C depends upon the interfacial capacitance and will again be proportional to the electrode area. It should be pointed out that this is a very approximate model. In practice, the three components are found to be functions of frequency.

The half-cell capacitance is simply given by 0.252 − 0.555 = −0.303 V.

The low-frequency impedance is approximately 30 kΩ and hence this is the value for R. The very high-frequency impedance is about 500 Ω and will be given by R in parallel with S, i.e. $RS/(R + S)$.

Therefore,

$$S = \frac{500R}{R - 500} = 508 \ \Omega.$$

The impedance (Z) of R, S and C is given by

$$\frac{1}{Z} = \frac{1}{R} + \frac{1}{S - j/\omega C}$$

at the higher frequencies we can neglect $1/R$ and hence

$$Z = S - \frac{j}{\omega C} \quad \text{and} \quad |Z| = (S^2 + 1/\omega^2 C^2)^{1/2}.$$

Taking the tabulated value for the impedance at 1669 Hz we obtain

$$1940 = \left(508^2 + \frac{1}{(2\pi \times 1669C)^2}\right)^{1/2}.$$

This gives C as 51 nF.

Therefore the approximate values of the model parameters are: $R = 30$ kΩ, $S = 508$ Ω, $C = 51$ nF.

Question 9.6.2.3

Calculate an approximate expression for the spreading impedance of a small circular electrode placed in contact with skin.

Two electrodes, each of 20 mm diameter, are placed on two hands and the resistance between them is measured as 1.5 kΩ at a frequency of 50 kHz. Does this allow us to determine the resistance of the arms? (Assume the resistivity of the body to be 4 Ω m.)

Question 9.6.2.4

What is meant by the term 'linear response' in describing the performance of a transducer or electronic circuit? Sketch the input/output relationship for systems which display the following characteristics:

(i)   Cross-over distortion or dead zone.
(ii)   Saturation or clipping.
(iii) Hysteresis.

Question 9.6.2.5

A circular coil of 1000 turns, resistance 3 kΩ and diameter 20 mm is placed in a 60 Hz magnetic field of flux density 15 nT p–p. The coil is connected to a perfect amplifier with a gain of 1000 and a bandwidth 200 Hz. Describe the signal which will be observed at the output of the amplifier. If this signal has a Fourier transform applied then describe the appearance of the transform.

Question 9.6.2.6

What are the three parameters that determine the thermal noise as predicted by the Johnson noise formula?

Discuss how these parameters can be changed in particular bioelectric and biomagnetic measurement situations in order to improve measurement accuracy.

A second-order gradiometer is used to discriminate against a source of interference that is 15 m away from our magnetic field detector. The signal from the heart that we wish to record is only 10 cm from the detector. If the interfering magnetic flux is 2000 times larger than our cardiac signal then what signal-to-noise ratio might we hope to achieve in our recording?

Answer

Bandwidth, resistance and absolute temperature determine noise power in the Johnson noise formula.

Bandwidth can be reduced to minimize noise provided that the signal frequency components are not attenuated.

Resistance is usually the resistance of electrodes and this can be minimized by using as large electrodes as possible. Temperature is difficult to change but in SQUID magnetometers the detector is cooled to liquid helium temperature.

The field will fall off as r^{-2}. The second-order gradiometer will give a gradient that falls off as d^{-4}. The ratio of the distances of the interference and the heart is 15/0.1, i.e. 150. The ratio of the two signals will therefore be reduced by $150 \times 150 = 22\,500$.

The SNR will therefore be $22\,500/2000 = 11.25$ to 1. *This is 21 dB.*

Question 9.6.2.7

We want to attempt to record the spontaneous action potentials travelling normally along the ulnar nerve. We estimate the size of these to be 10 nV (10^{-8} V) rms. If we use two electrodes to record the signal and these present a combined resistance of 500 Ω, then calculate the signal-to-noise ratio to be expected. Assume that we use a perfect amplifier and that the only significant source of noise is thermal noise, given by the Johnson noise formula $V^2 = 4kTR\,\mathrm{d}f$ where k is Boltzmann's constant (1.37×10^{-23} W s deg^{-1}), T is the absolute temperature, R the source resistance and $\mathrm{d}f$ the bandwidth in hertz. Assume that the temperature is 27 °C and the recording bandwidth is 10 kHz.

Are we likely to succeed in measuring the action potentials?

Answer

The noise can be calculated as 286.7 nV. SNR is 0.0349 or -29.1 dB.

The signal-to-noise ratio (SNR) could be improved by reducing the bandwidth to match the expected signal, but the SNR is so poor that we are unlikely to succeed in measuring the action potentials.

Question 9.6.2.8

Outline the need for differential amplifiers when recording bioelectric signals.

If we wish to record an ECG of amplitude 1 mV in the presence of interference that causes a common-mode voltage of 100 mV to appear on the inputs of the amplifier, then what common-mode rejection ratio (CMRR) do we require if we need a signal-to-noise ratio of better than 20/1 on our recording?

Answer

The need arises because most bioelectric signals are smaller than the interference that arises from 50/60 Hz and radio-frequency signals. By having two non-ground-referenced inputs that will see the same interference a differential amplifier can be used to discriminate against interference. The definition of common-mode rejection ratio (CMRR) was explained in section 9.3.3.

We want to reduce the 100 mV interference to 1/20 of the ECG, i.e. 0.05 mV. This is a reduction of 2000/1, i.e. 66 dB CMRR.

Answers to short questions

a   The magnetocardiogram signal is smaller than the Earth's magnetic field.
b   A major disadvantage of a SQUID detector is the need to supercool the junction.
c   Johnson noise is the thermal noise caused by movement of electrons and ions.
d   Nearly all bioelectric amplifiers are differential types.
e   Common-mode interference is the most common type of interference.

f   50 pT is a typical amplitude for the magnetocardiogram.

g   Thermal noise limits the smallest signal that can be recorded.

h   An electrode or contact potential is the potential generated when a metal electrode is placed in contact with an electrolyte such as tissue.

i   Electrodes usually change with time because active changes in the skin cause changes in the electro-chemical equilibrium.

j   By coating a silver electrode with silver chloride electrode polarization can be reduced. The AgCl is deposited by passing a current through an Ag electrode placed in a salt solution.

k   A larger electrode will have a lower impedance than a smaller one.

l   Microelectrodes of <0.5 μm in tip diameter can be produced.

m   In floating electrodes the interface between the electronic and ionic conductors is removed from the skin. This produces a much more stable electrode contact as the skin is an unstable environment.

n   If there are 20 000 fibres in the ulnar nerve and the total recorded action potential is typically 10 μV, then each fibre contributes about 0.5 nV.

o   AC coupling uses a capacitor at the input of an amplifier in order to block DC potentials. It forms a high-pass filter with the input impedance of the amplifier.

p   The most stable temperature is at the base of the brain.

q   A thermistor is a semiconductor device whose resistance depends upon temperature.

r   A glass pH electrode measures the potential produced across a glass membrane by the difference in hydrogen ion concentrations across the membrane.

s   pH is the logarithm of the hydrogen ion concentration.

t   A standard hydrogen electrode is used as the reference when measuring electrode potentials.

BIBLIOGRAPHY

Baule G M and McFee R 1963 Detection of the magnetic field of the heart *Am. Heart J.* **66** 95–6

Geddes L A 1972 *Electrodes and the Measurement of Bioelectric Events* (New York: Wiley)

Geselowitz D B 1971 An application of electrocardiographic lead theory to impedance tomography *IEEE Trans. Biomed. Eng.* **18** 38–41

Geselowitz D B 1980 Computer simulation of the human magnetocardiogram *IEEE Trans. Magn.* **16** 812–917

Johnson J B 1928 Thermal agitation of electricity in conductors *Phys. Rev.* **32** 97–109

Nyquist H 1928 Thermal agitation of electric charge in conductors *Phys. Rev.* **32** 110–13

Plonsey R and Fleming D G 1969 *Bioelectric Phenomena* (New York: McGraw-Hill)

CHAPTER 10

EVOKED RESPONSES

10.1. TESTING SYSTEMS BY EVOKING A RESPONSE

Diagnosis of disease can be very difficult as a single disease might well affect several parts of the body. For example, renal failure will affect the composition of blood and this in turn can cause a range of symptoms elsewhere in the body. In this particular example the analysis of a blood sample will be an important part of the diagnostic process. However, if we wish to test specifically the function of the kidneys then a more invasive test is required. A common test is to inject a pharmaceutical which is specifically absorbed by the kidneys and to follow the process of absorption. A radioactive label can be attached to the pharmaceutical and then a gamma camera used to image the uptake and excretion by the kidneys (see section 6.9.2). Thus renal function can be tested by applying a stimulus and then observing the response which the stimulus evokes.

In other diseases the symptoms may be specific to a particular organ which can again be tested by applying a stimulus and observing the response. In testing a system by seeing how it responds to a stimulus, we need to know how the organ would respond when functioning normally. In some cases the response may be obvious; if we apply a sound to the ear then the response may be a statement from the subject that they have heard the sound. However, in other cases such as the response to an injected drug the response may be very difficult to define.

The questions we hope to answer in this chapter are:

- Can we assess the performance of a biological system by measuring the response to a stimulus?
- Can we distinguish signal from noise?
- Can we be sure that a measured response is real?
- Can interpretation by computer be more reliable than a human observer?

When you have finished this chapter, you should be aware of:

- What responses can be evoked by sensory input to the human body.
- How evoked responses can be used to characterize a physiological process.
- Some methods which can be used to enhance the 'signal-to-noise' ratio.
- The sources of interference which may be encountered when recording a response.

The chapter follows logically from Chapter 9, as evoked potentials are simply a way of gaining access to physiological signals. Testing a system by applying a stimulus and then measuring a response is an extremely powerful way of testing function. Later in this chapter we will consider specific examples of evoked electrical responses from the body, but we will start by considering a simple physical system and how it responds to a stimulus.

10.1.1.   *Testing a linear system*

A linear system is one where the components of the system behave in a proportional manner. For example, in an electrical system the current will double if we double the applied voltage and in a mechanical system the acceleration will double if we double the applied force. Mathematicians like linear systems because they can be described relatively simply in mathematical terms (see section 13.6). Unfortunately, most biological systems are nonlinear. However, we will start by considering a linear system in part because, like the mathematicians, we can handle them and in part because we can learn something about how to test systems by applying a stimulus.

Consider a linear system contained in a box as shown in figure 10.1. We will apply an input $I(t)$ to the system and this will result in an output $O(t)$. If the system was the lower limb then $I(t)$ might be the stimulus applied by a patella hammer and $O(t)$ might be the resulting vertical movement of the foot. What we want to do is to find a function which will allow us to determine $O(t)$ if we know $I(t)$. We want to know the function F where

$$O(t) = F(I(t)). \tag{10.1}$$

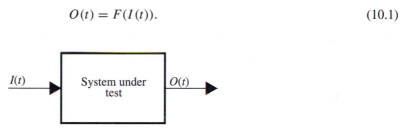

Figure 10.1. *A general system with input $I(t)$ and output $O(t)$.*

Systems are very commonly described in terms of a frequency response. The concept and mathematics relating to frequency analysis are described in Chapter 13. One way of measuring the frequency response of a system is to apply sinusoidal inputs and measure the resulting outputs over a wide frequency range. This can be a very time consuming procedure and carries the risk that the system under test will change whilst the measurements are being made. Fortunately there is an alternative stimulus which can be applied.

Perhaps the simplest stimulus, which we can apply to test a system, is a single pulse. One such pulse can be described mathematically as having zero duration, infinite amplitude but an integral over time of unity. This particular pulse is called a *delta function* ($\delta(t)$). This function was first used by Dirac and has an interesting property, which we can derive as follows. The Fourier integral (see Chapter 13, section 13.4.2) of $\delta(t)$ is given by

$$g(f) = \int_{-\infty}^{+\infty} \delta(t)\,e^{-2\pi j f t}\,dt.$$

Now as the delta function is zero everywhere except at time $t = 0$, the above integral becomes

$$g(f) = \int_{-\infty}^{+\infty} \delta(t)\,1\,dt$$

which from the definition of the delta function is unity. The conclusion which we reach is thus that the Fourier transform of the delta function is the same at all frequencies. In other words, the spectrum of a very sharp pulse is of equal amplitude at all frequencies. For this reason a very short pulse is an easy way of interrogating an unknown system because we can apply all frequencies simultaneously. In practice of course we cannot produce an infinitely short pulse, but we can often get quite close to this ideal.

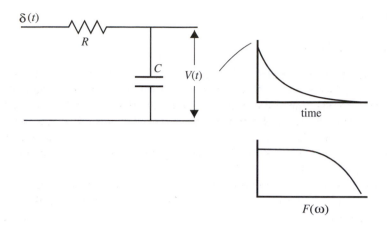

Figure 10.2. *A delta function $\delta(t)$ applied to a low-pass filter, the resulting output $V(t)$ and the associated Fourier spectrum $F(\omega)$.*

We can test the above idea by applying a delta function to the low-pass filter circuit shown in figure 10.2,

$$\frac{q}{C} + R\frac{dq}{dt} = 0$$

where q is the charge on capacitor C. The solution to this differential equation is

$$q = q_0\,e^{-t/RC}$$

where q_0 is the charge imparted by the input delta function. The integral of the input voltage is unity and hence

$$R\int i\,dt = Rq_0 = 1$$

therefore

$$V(t) = \frac{q(t)}{C} = \frac{1}{CR}e^{-t/RC}.$$

Now the Fourier transform of this is

$$F(\omega) = \int_0^\infty \frac{1}{CR}e^{-t/CR}e^{-2\pi jft}\,dt$$

$$= \frac{1}{CR}\frac{1}{(1/CR + 2\pi jft)} = \frac{1}{(1 + 2\pi jfCR)}.$$

We can recognize this as the transfer function $(1/j\omega C)/(R + 1/j\omega C)$ of the circuit given in figure 10.2.

By measuring the response of the system to the delta function $\delta(t)$ and taking the Fourier transform, we obtain the frequency response of the system. This is a very powerful method of characterizing a system. Unfortunately, most biological systems are not linear and there may be a time delay imposed between stimulus and response. The time delay can be taken into account as it can be shown that the Fourier transform of a function of time $g(t)$ shifted in time by τ is simply the Fourier transform of $g(t)$ multiplied by $e^{2\pi jf\tau}$. The nonlinearity cannot be taken into account and so its significance needs to be considered in each particular case.

10.2. STIMULI

10.2.1. Nerve stimulation

Because the body is a good conductor of electricity, and because our nerves and muscles function electrically, we would expect to see physiological effects when current is applied to the body. These effects can be a source of hazard, but they can also be utilized both for the diagnosis and treatment of disease. The three most important physiological effects of electricity are electrolysis, neural stimulation and heating. Only neural stimulation will be considered in this section. A more detailed consideration of the biological effects of electricity was given in Chapter 8 and the electrophysiology of neural stimulation will be further described in Chapter 16 (section 16.5).

If a current of sufficient amplitude is passed between a pair of surface electrodes then muscles will contract. They contract because a stimulus is being introduced to the nerve fibres, which supply the muscles. If the current used is an alternating current then a graph can be drawn which shows the frequency of the current against the amplitude which is necessary to cause muscular contraction. Figure 10.3 shows such a graph and illustrates that the lowest threshold for stimulation is within the 10–50 Hz range. Above about 200 kHz (each cycle lasting 5 μs) stimulation is almost impossible. A current lasting at least 50 μs and preferably as long as 20 ms is needed to stimulate nerve fibres.

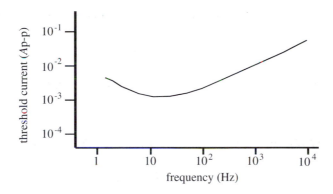

Figure 10.3. *The current needed to cause muscle stimulation beneath a pair of surface electrodes on the arm is shown as a function of the frequency of the alternating current. (Result from one normal subject.)*

If, instead of an alternating current, we now use a short pulse of current, another graph can be drawn in which the current needed to cause stimulation is plotted against the duration of the current pulse. This graph (figure 10.4) shows that, unless an excessively high current is to be applied, a pulse lasting at least 50 μs, and preferably as long as 2 ms, is needed to stimulate the nerve fibres. This result is consistent with the conclusion drawn from the application of an alternating current. However, there is one difference: that stimulation only occurs underneath one of the surface electrodes. The stimulation occurs underneath the electrode at which the pulse is seen as a negative pulse. We can explain this by considering what is actually happening to a nerve fibre when a current is applied.

A nerve will be stimulated when the transmembrane potential is reversed by an externally applied current. If, where the current enters the nerve, the externally applied current flows from positive to negative, then the transmembrane potential will be increased, and where the current leaves the nerve, the transmembrane potential will be reduced. Stimulation will occur where the current leaves the nerve. If the positive electrode is called the anode and the negative the cathode, then stimulation starts underneath the cathode.

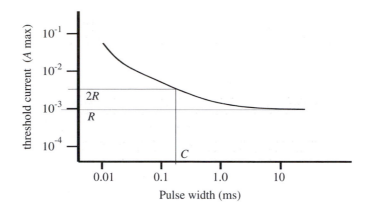

Figure 10.4. *A similar graph to that of figure 10.3 but using pulse stimulation. R is called the rheobase and C the chronaxie. (Result is for one normal subject.)*

10.2.2.   *Currents and voltages*

For a nerve to be stimulated, all that is necessary is that sufficient current flows out of the nerve to initiate reversal of the transmembrane potential. Indeed, the potential need only be reduced by about 30% for an action potential to be generated. Because the membrane has capacitance it will require a finite charge to change the transmembrane potential; this is why the stimulation current must flow for a certain minimum time before stimulation occurs. Experimentally we can measure the charge which is necessary to cause muscular stimulation when using different positions of electrode; the results are shown in table 10.1.

Table 10.1. *The charge required to stimulate a motor nerve depends on the distance between the electrode and the nerve.*

Position of the electrodes	Approximate charge required for stimulation
Inside a nerve axon	10^{-12} C
On the surface of a nerve trunk	10^{-7} C
On the skin several mm from the nerve	10^{-6} C

When the electrode is inside the nerve, the applied current has to remove the charge from the membrane capacitance associated with one node of Ranvier as illustrated in figure 10.5. A node of Ranvier is the gap between the bands of myelin which surround a nerve axon. This capacitance is approximately 10 pF and the polarization potential will be about 0.1 V, so that the charge to be removed to stimulate the nerve will be given by

$$\text{charge to be removed} = 10^{-11} \times 0.1 = 10^{-12} \text{ C.}$$

This figure is in good agreement with experimental results for stimulation via microelectrodes (see section 9.2.3). Much more charge is required to stimulate a nerve from surface electrodes simply because most of the current goes through the tissue surrounding the nerve and only a small fraction passes into the nerve fibres. The energy required to stimulate a nerve will depend upon the resistance through which the current has to pass; for a pair of electrodes directly over a nerve the energy required is about 10^{-4} J. This corresponds

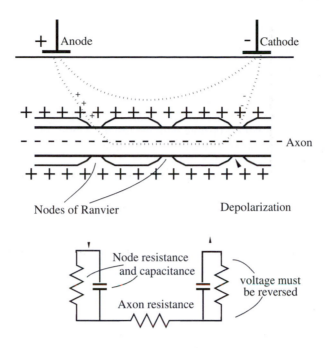

Figure 10.5. *A nerve axon can be stimulated by the current passing between the two surface electrodes. Stimulation occurs underneath the cathode, where the transmembrane potential is reduced.*

to 10 mA flowing for 100 μs through an electrode resistance of 10 kΩ. This energy may be compared with the energy stored in a small battery, which has a capacity of 500 mA hour and an output of 9 V,

Energy in small battery = power × time = 9 V × 0.5 A × 3600 s = 16 200 J.

Energy required to stimulate a nerve via surface electrodes = 10^{-4} J.

Even a small battery has sufficient stored energy to stimulate a nerve more than a 100 million times. If the nerve is not superficial (close to the skin) then much larger currents may be required, and if we wish to give a strong stimulus to a large mass such as the heart then an energy of several joules may be required.

The energy required to stimulate all types of nerve fibre is not the same. A large nerve fibre has a lower threshold to stimulation than a small fibre, i.e. less energy is required, and it also has a faster conduction velocity. If the nerve supply to a muscle is cut then it will slowly die. An electrical stimulus can still be used to stimulate the muscle directly, rather than by stimulating the nerve, but the threshold is increased. This is the basis of a diagnostic test in which the threshold for stimulation is measured. The measurement can be explained by reference to figure 10.4, where the threshold current was recorded as the pulse width of the stimulus was changed. If the muscle is denervated then the curve is moved to the right, which simply means that a longer pulse width and therefore greater energy is required to stimulate the muscle. To quantify this change the rheobase and the chronaxie are measured. *Rheobase* is the minimum current which will stimulate the muscle whatever the width of the stimulating pulse. It is marked as R on figure 10.4. *Chronaxie*, marked as C, is the pulse width such that the threshold current is twice the rheobase.

These two measurements are not now widely used but they form the basis of a technique which is sometimes referred to as 'strength duration curves'.

10.2.3.   *Auditory and visual stimuli*

We have only considered the input of electrical stimuli so far. However, any of the senses can be stimulated and responses recorded. A brief mechanical stimulus can be used to elicit a reflex response in testing the integrity of the nervous system. A brief auditory stimulus can be used and a response recorded either from the ear or the brain. If the response does not require the subject to say whether or not they have heard the stimulus then the measurement is said to be an objective test of hearing. Such tests are important in testing young children or adults who will not co-operate with the tester. A brief visual stimulus can be used in a similar way and responses recorded either from the eye or the brain. In principle, the senses of taste and smell could be assessed by applying a short stimulus, although this might be quite difficult in practice.

   The audio equivalent of a *delta function* stimulus is a very short click which will contain a wide range of frequencies. This type of stimulus is used in some objective hearing tests where the audiologist simply wishes to know if the auditory system is working. However, if the audiologist wishes to know which sound frequencies have been heard then the single-click stimulus is not effective. In principle, as we have shown, the click contains all frequencies and we should be able to carry out a Fourier transform of the response and so obtain the transfer function of the auditory system. However, the nonlinearity of the cochlea and brain make this method inaccurate. The type of stimulus which is often used in practice is shown in figure 10.6(a) and consists of a short burst of a single audio frequency. By testing with different audio frequencies the response of the ear can be obtained. It is interesting to consider the frequency content of the type of stimulus shown in figure 10.6(a). A continuous sinusoid will have a Fourier transform which is a single spike at the frequency of the sinusoid, but a short burst will have a wider bandwidth. We can show this as follows.

   Consider the signal shown in figure 10.6(a) as given by $y = y_0 \cos 2\pi f_0 t$, which starts at time $-T$ and stops at time T. Now the Fourier transform of this is given by

$$y(f) = \int_{-\infty}^{\infty} y_0 \cos 2\pi f_0\, e^{-2\pi \mathrm{j} f t}\, dt = \int_{-\infty}^{\infty} \frac{y_0(e^{2\pi \mathrm{j} f_0 t} + e^{-2\pi \mathrm{j} f_0 t})\, e^{-2\pi \mathrm{j} f t}}{2}\, dt$$

$$= \frac{y_0}{2}\left[\frac{e^{2\pi \mathrm{j}(f_0 - f)t}}{2\pi \mathrm{j}(f_0 - f)} + \frac{e^{-2\pi \mathrm{j}(f_0 + f)t}}{-2\pi \mathrm{j}(f_0 + f)}\right]_{-\infty}^{+\infty}.$$

Since $y(t)$ is zero for t less than $-T$ and greater than $+T$, the limits can be taken between these times and we obtain

$$y(f) = \frac{y_0}{2}\left[\frac{e^{2\pi \mathrm{j}(f_0 - f)T} - e^{-2\pi \mathrm{j}(f_0 - f)T}}{2\mathrm{j}\pi(f_0 - f)} - \frac{e^{2\pi \mathrm{j}(f_0 + f)T} - e^{2\pi \mathrm{j}(f_0 + f)T}}{2\mathrm{j}\pi(f_0 + f)}\right]$$

since $(e^{\mathrm{j}\alpha} - e^{-\mathrm{j}\alpha})/2\mathrm{j} = \sin \alpha$,

$$y(f) = y_0 T\left[\frac{\sin 2\pi(f_0 - f)T}{2\pi(f_0 - f)T} + \frac{\sin 2\pi(f_0 + f)T}{2\pi(f_0 + f)T}\right]. \tag{10.2}$$

This is seen to be the sum of two functions, each of the form $(\sin x)/x$, one centred on frequency $(f_0 + f)$ and the other on frequency $(f_0 - f)$. Now it can be shown that for any real function the negative frequency range of the Fourier transform can be ignored since it can be derived from the positive frequency range.

   The form of $y(f)$ is sketched in figure 10.6(b) for the situation where six cycles of the sinusoid are contained within the period $2T$. It is not difficult to show that as T becomes longer the Fourier transform gets sharper and indeed becomes infinitely sharp for a continuous waveform. However, as T becomes shorter the Fourier transform becomes wider and in the limit when T tends to zero the spectrum becomes flat, i.e. we have the transform of a delta function. Using equation (10.2) it is possible to determine the range of frequency components contained within any duration of stimulus used.

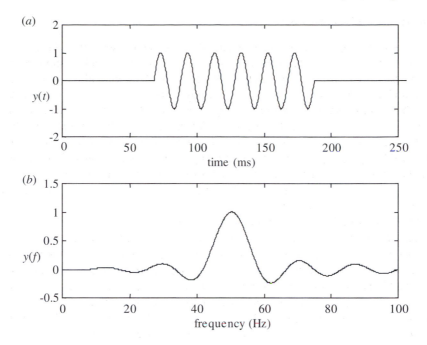

Figure 10.6. *(a) An audio stimulus consisting of six cycles of a cosine wave of frequency 50 Hz. (b) Fourier transform of the signal shown in (a).*

The visual equivalent of a delta function stimulus might be a flash of light and such a stimulus is often used in testing the visual system. It is also possible to modulate the intensity of the light to obtain a stimulus analogous to that shown in figure 10.6(*a*), but of course the light intensity cannot go negative so the actual waveform will be raised upon a pedestal and the Fourier transform will not be the same as we derived in equation (10.2). The eye actually has a very limited response in time so it is not possible to sense modulation of intensity above frequencies of about 20 Hz. The eye and brain do, of course, have the ability to detect patterns and it is found that large responses can be evoked from the brain by presenting pattern stimuli. Fourier transforms are applicable in this situation by considering the spatial instead of temporal frequencies present in a stimulus (see Chapter 11). However, the transforms have to be carried out in two dimensions. A spatial delta function will be a single line in one dimension but a point in two dimensions.

10.3. DETECTION OF SMALL SIGNALS

Bioelectric signals and amplifiers

Almost every part of the body produces electrical signals. They are not just by-products but are essential control signals to make us function and, for that reason, electrophysiological measurements contain useful diagnostic information. The amplitudes of some of these signals were given in table 9.2. The signals are small and even one of the largest, the ECG/EKG, only offers a peak power of about 10^{-9} W to drive a recording system. All of these signals must be amplified before they can be recorded.

In this section we consider what limits the accuracy with which we can record small signals and how we can optimize the recording accuracy.

10.3.1.   Bandwidth and signal-to-noise ratios

The bandwidth of an amplifier is the frequency range over which the gain remains constant. In practice, it is normally quoted as the -3 dB bandwidth which is the frequency range over which the gain is not less than 3 dB below the maximum gain. The bandwidth of an amplifier must be sufficient to handle all the frequency components in the signal of interest. Some common physiological signals and their approximate frequency contents are given in table 10.2.

Table 10.2. *This shows the approximate frequency content of some common physiological signals. Where the low-frequency limit is listed as DC, this means that there are steady components in the signal.*

ECG	0.5 Hz–100 Hz
EEG	0.5 Hz–75 Hz
Arterial pressure wave	DC–40 Hz
Body temperature	DC–1 Hz
Respiration	DC–10 Hz
Electromyograph	10 Hz–5 kHz
Nerve action potentials	10 Hz–10 kHz
Smooth muscle potentials	0.05 Hz–10 Hz
(e.g. signals from the gut)	

It is desirable to make the bandwidth of an amplifier as narrow as possible without losing any information from the signal. There are two reasons for this: the first is to exclude unwanted signals that do not fall within the pass band of the amplifier and the second is to reduce noise. Purely random noise has equal power in equal frequency intervals, i.e. the heating effect of noise between frequencies 1 and 2 Hz is the same as that between frequencies 5 and 6 Hz or indeed 991 and 992 Hz. It follows that the noise power between 1 and 5 Hz is five times that between 1 and 2 Hz; therefore, reducing the bandwidth by a factor of five will reduce the noise power by a factor of five, and the noise voltage by $\sqrt{5}$. To optimize the quality of a recording the bandwidth should be reduced so that it is as narrow as possible, consistent with not causing distortion to the signal.

10.3.2.   Choice of amplifiers

Amplification can be either of voltage or current. In general, it is taken for granted that voltage amplification is required and that the amplifier will supply sufficient current to drive a recording device.

The amplifier is unable to distinguish between signal and *noise*, and will therefore amplify them equally. The amplifier will also contribute some noise which can be measured by short circuiting the input connections: any noise at the output must then be produced only by the amplifier. The amount of noise at the output will depend upon the gain of the amplifier, and is normally quoted in terms of the equivalent noise at the input of the amplifier (noise referred to input, RTI).

$$\text{noise (RTI)} = (\text{noise at output})/(\text{amplifier gain}).$$

The noise referred to the input is a useful measurement because it can be compared directly with the size of an input signal. If we are measuring an ECG of 1 mV, and need a signal-to-noise ratio of 40 dB (i.e. 100:1), the noise referred to the input of the amplifier must be less than 1 mV/100 = 10 μV.

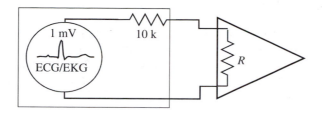

Figure 10.7. *The amplifier of input resistance R will affect the signal that is recorded via electrodes whose resistance is shown as 10 kΩ.*

Obviously the amplifier *gain* required will be large if the signal to be amplified is small. An EEG amplifier will have a higher gain than an ECG/EKG amplifier, because the EEG is only about 100 μV in amplitude. The gain of an amplifier is the ratio of the output and input voltages. In an ideal amplifier the gain is independent of frequency, but in a real amplifier this is not the case; thus, the frequency response of the amplifier has to be matched to the frequency content of the signal.

The *input resistance* of the amplifier is the load that the amplifier presents to the signal. In figure 10.7 the ECG can be considered to be a voltage source of 1 mV in series with a source resistance of 10 kΩ. The resistance represents the tissue and electrode impedances. Clearly, the source resistance and the input resistance (R) of the amplifier form a potential divider. If $R = 10\,$kΩ, then the signal measured across R will be 0.5 mV, i.e. half of its true value. To reduce the error in the measured voltage to 1%, the input resistance of the amplifier must be $100\times$ the source resistance, i.e. $R = 1\,$MΩ. When making any electrophysiological measurement the source impedance must be known and the amplifier input resistance must be very much greater than this. Some very fine microelectrodes have source resistances as high as 10 MΩ, so that an amplifier with an input resistance greater than 1000 MΩ is required.

Input resistance can be measured very simply if a signal generator is available. A signal is applied to the amplifier directly and the output noted. A resistance is now inserted in series with the input signal; if this resistance is equal to the input resistance then the output signal will be exactly halved in amplitude. A resistance is selected by taking gradually increasing values and when the output is halved then the resistance connected is equal to the input resistance of the amplifier.

10.3.3. Differential amplifiers

The concept of the differential amplifier was introduced in section 9.3.3 of the last chapter. Nearly all bioelectric amplifiers are differential because of their ability to reject interference. A simple differential amplifier has three inputs: the signal is applied between the '+' input and the '−' input, and the common connection is placed anywhere on the body. If we were to use it to record an ECG from the lead II position then the '+' connection would go to the right arm, the '−' connection to the left leg and the common connection to the right leg. The common connection is connected to 0 V in the amplifier. In an electrically isolated amplifier this would only be an internal reference, but in a non-isolated amplifier it would be connected to mains earth. Both input connections will see the same common-mode voltage, but will see different signal voltages. The common-mode voltage is the signal which is common to both inputs and it will usually be an interfering signal such as that caused by the mains power supply.

It is convenient to construct a bioelectric differential amplifier from operational amplifiers and one of the best configurations is that shown in figure 10.8. This circuit is often referred to as an 'instrumentation amplifier configuration'. Using the same terminology for the inputs as was used in section 9.3.3 and the

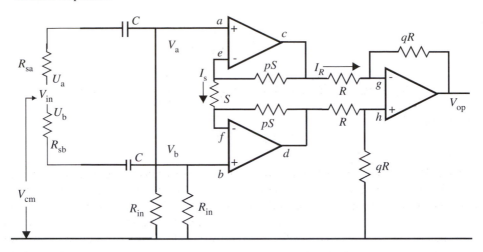

Figure 10.8. *Circuit for a differential bioelectric amplifier. We wish to amplify the input voltage V_{in}, but reject the interference V_{cm} that is common to both inputs.*

additional terminology given in figure 10.8 we can determine the gain of the system. Remember that the two 'rules' which we can use to analyse an operational amplifier circuit are:

A　there will be no voltage between the inputs of an operational amplifier;
B　no current will flow into the inputs of the operational amplifier.

We may require AC coupling, as was explained in section 9.3.3, but if C is sufficiently large and we neglect the source resistances R_{sa} and R_{sb}, then the input voltages U_a and U_b will appear at the two inputs a and b. Using rule A we can thus obtain

$$V_e = V_a \qquad \text{and similarly} \qquad V_f = V_b$$

therefore,

$$I_s = \frac{(V_a - V_b)}{S}$$

and

$$V_c = V_e + I_s pS = V_a + \frac{(V_a - V_b)pS}{S} = V_a(1 + p) - V_b p$$

and similarly

$$V_d = V_f - I_s pS = V_b - \frac{(V_a - V_b)pS}{S} = V_b(1 + p) - V_a p.$$

Therefore,

$$V_c - V_d = (V_a - V_b)(1 + 2p). \tag{10.3}$$

The gain of the first stage is $(1 + 2p)$.
　　Using rules B and A we can determine

$$V_g = V_h = V_d \frac{q}{(q + 1)}$$

therefore

$$I_R = \frac{V_c}{R} - \frac{V_d q}{R(q+1)}$$

and

$$V_{op} = V_g - I_R q R = \frac{V_d q}{(q+1)} - V_c q + V_d \frac{q^2}{(q+1)} = (V_d - V_c) q$$

substituting for $(V_d - V_c)$

$$V_{op} = -q(1+2p)(V_a - V_b) \qquad (10.4)$$

the overall gain is thus $-q(1+2p)$.

Note that if $V_a = V_b$ then the output is zero, i.e. the amplifier rejects common-mode signals. This is, in fact, a very elegant circuit. The clever part is the first stage which provides high differential gain but only unit gain to common-mode signals. It achieves this without the requirement to carefully match any resistors.

Unfortunately, nothing is perfect and in the case of the circuit of figure 10.8 problems can arise because of the resistors marked R_{in}, which determine the input resistance. These resistors cannot be too large because they have to carry the bias current for the operational amplifiers. We will consider their significance in two situations.

Case 1. Source impedances high but equal

Ideally R_{sa} and R_{sa} will be very much greater than R_{in}. There will be no voltage drop across R_{sa} and R_{sb} so that $(V_a - V_b)$ will equal V_{in}. However, consider the case where R_{sa} is equal to R_{sb}, but where they are not very much greater than R_{in},

$$V_a = U_a \frac{R_{in}}{(R_{in} + R_{sa})} \quad \text{and} \quad V_b = U_b \frac{R_{in}}{(R_{in} + R_{sb})}. \qquad (10.5)$$

We can use equation (10.4) to obtain the overall gain as

$$\text{gain to input } (U_a - U_b) = -q(1+2p)\frac{R_{in}}{(R_{in} + R_{sa})}.$$

The gain has therefore been reduced by the factor $R_{in}/(R_{in}+R_{sa})$. However, when $U_a = U_b$ then V_a is still equal to V_b and the output will be zero, so the circuit will still reject common-mode signals. The common-mode gain is zero.

Case 2. Source impedances unmatched

Equation (10.5) still applies, but the overall gain will be given by

$$\text{gain to input } (U_a - U_b) = -q(1+2p) R_{in} \left(\frac{U_a}{(R_{in} + R_{sa})} - \frac{U_b}{(R_{in} + R_{sb})} \right).$$

So the gain is now a function of R_{sa}, R_{sb} and R_{in}. Even more importantly, when $U_a = U_b$ the output will not be zero. When $U_a = U_b$ the gain to this input will be

$$\text{gain to common-mode input} = -q(1+2p) R_{in} \left(\frac{1}{(R_{in} + R_{sa})} - \frac{1}{(R_{in} + R_{sb})} \right).$$

We can then use equation (9.12) to determine the common-mode rejection ratio (CMRR) as

$$\text{CMRR} = 20 \log \left(\frac{(R_{in} + R_{sa})(R_{in} + R_{sb})}{R_{in}(R_{sb} - R_{sa})} \right). \qquad (10.6)$$

Typical values for R_{sa} and R_{sb} when recording from simple surface electrodes might be $R_{sa} = 5$ kΩ and $R_{sb} = 7$ kΩ. If the input resistance R_{in} is 10 MΩ then the CMRR will be 74 dB. If we have a common-mode signal at 50/60 Hz of 100 mV and a 1 mV ECG/EKG signal, then if $q = 10$ and $p = 10$ we will have an output ECG/EKG of 210 mV and noise of 4.2 mV at 50/60 Hz. The unwanted interference will be about 2% of the signal size. This is acceptable. However, if we were recording an evoked potential of only 10 μV amplitude, the interference would be two times larger than the signal and this is not acceptable.

There are many techniques by which the CMRR of bioelectric amplifiers can be increased. However, in most cases these techniques aim simply to increase the input impedance of the amplifier. One powerful technique is called *boot-strapping* and reduces any current flow through R_{in} by driving both ends of the resistance with the common-mode voltage. It is left to the student to look into this technique.

10.3.4. *Principle of averaging*

Many physiological signals are very small. A small signal can be amplified electronically, but in so doing noise will also be amplified and in many cases the 'noise' may be larger than the signal. By 'noise' we mean any signal which interferes with the signal which is required. It may be electronic noise generated in the equipment or it may be another physiological signal. The EMG often produces noise on an ECG record.

As was shown in the previous section, differential amplifiers can be used to reduce interference. Filters can also be used to reduce noise and interference which has frequency components outside the signal bandwidth. Yet another common technique is that of *averaging*. The principle of an averager is simply that by repeating a measurement many times all the results can be summed together. The signal will sum in proportion to the number of repetitions, whereas if the noise is random it will sum more slowly.

Consider the problem of determining the effect of a new drug which is claimed to reduce blood pressure. The drug can be given and then the systolic blood pressure measured at regular intervals. Unfortunately blood pressure is not constant; any form of mental, physical and certainly sexual stimulus will cause the blood pressure to fluctuate. How is it possible to separate any change in blood pressure caused by the drug from the other factors which cause changes? One way to separate the effects is to utilize the fact that any response to the drug will follow in sequence after the administration of the drug, whereas the other changes in blood pressure may not be related to the time at which the drug was given. If the drug is given to many patients then an average can be taken of all the graphed responses. In figure 10.9 four responses have been summed to illustrate the effect. The fall in systolic pressure in response to the drug is clearer when the four responses are summed and might be even clearer if 20 records were used. The reason for this is that the response always follows the drug at the same time interval, whereas the other variations occur at any time and will therefore cancel to some extent. This all sounds very reasonable, but we must be very cautious in our interpretation of the result. The different patients may not respond in the same way to the drug and there may also be a placebo effect. However, we will assume for now that all the patients respond in the same way and that the 'noise' is not related to the administration of the drug. We will ask the question as to how much noise cancellation we might expect.

Noise cancellation

If a measurement m is repeated n times, the mean result can be expressed as $\bar{m} \pm s$ where s is the standard deviation of the measurement. If we now repeat this set of measurements many times then we can obtain the standard deviation on the mean $\bar{m}$, which is usually quoted as $s/\sqrt{n}$. Therefore, if a measurement is repeated n times then the variability of the mean answer will fall in proportion to $\sqrt{n}$. This simple conclusion is actually based upon many assumptions. It requires that the noise is random and that the randomness is distributed in a particular way. It can be shown to be true for a Poisson distribution (see Chapter 13). The reduction in the noise in proportion to $\sqrt{n}$ should be seen as the best result which can be obtained.

systolic blood pressure (mmHg)

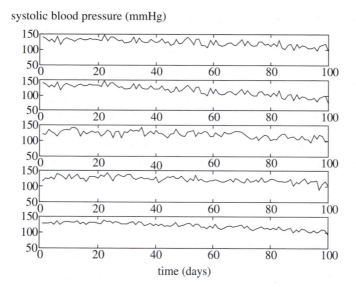

time (days)

Figure 10.9. *The top four recordings show systolic blood pressure, corresponding to four different patients to whom a drug has been administered on day 20. At the bottom the average of the four traces allows the small fall in pressure to be seen.*

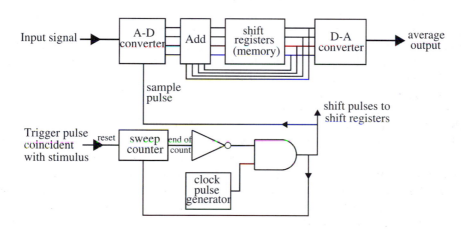

Figure 10.10. *Block diagram of a digital signal averager. The input signal is sampled sequentially following receipt of a trigger pulse. The stored sample values are added to the values already held by the bank of shift registers to give the sum of many responses.*

The technique of averaging is widely used in the measurement of EEG evoked responses. It enables small responses to a sound stimulus or a visual stimulus to be observed and used in clinical practice. In these cases the EEG is assumed to be random and not temporally related to the stimulus. Improvements in noise reduction of 20 dB are easily obtained when summing several hundred responses, but improvements beyond 40 dB are not usually obtained.

Figure 10.10 shows, in the form of a block diagram, how a signal averager can be constructed. The input signal which follows the stimulus is converted to digital form and stored in a memory. When repeated stimuli are given, the responses are added to those which are already in the digital memory. When the required number of signals has been added then the total response may be displayed.

Signal averaging is now often carried out on a computer with the averaging sequence determined in software.

10.4.   ELECTRICAL INTERFERENCE

The subject of electromagnetic compatibility (EMC) has received considerable attention with the increase in use of computers and mobile communications devices, which can both generate and be affected by electrical interference. This section does not attempt to deal with the whole subject of EMC but covers the different ways in which electromedical measurements might be subject to interference. The subject of EMC is considered further in section 22.2.2 on safety-critical systems.

Electrical interference can be a problem when making almost any physiological measurement. Electrophysiological signals such as the ECG, EEG and EMG are particularly susceptible to interference because they are such small electrical signals. Laboratory instruments, nuclear medicine equipment and analogue inputs to computing equipment can also be subject to interference from nearby electrical machinery such as lifts, air conditioning plant and cleaning equipment. Electronic equipment such as computers, monitors and mobile telephones can also generate interference. This section will explain the different types of interference, how it can arise and what you can do to reduce the effect upon your measurement.

10.4.1.   Electric fields

Any electric charge produces an associated electric field, whose magnitude is inversely proportional to the distance from the charge. A wire connected to the mains supply will produce an alternating electric field which can cause interference to any electrophysiological measurement. We can explain how this arises by considering a human body placed fairly close to the mains supply wiring. Figure 10.11 shows the electrical capacitance between a body and the mains supply wiring, and also the earth. Electrical capacitance is a convenient way of describing how an electric field can allow two objects to interact. If the electric field is alternating, then a displacement current, I, can flow through the capacitance, C, where

$$I = C \, dV/dt.$$

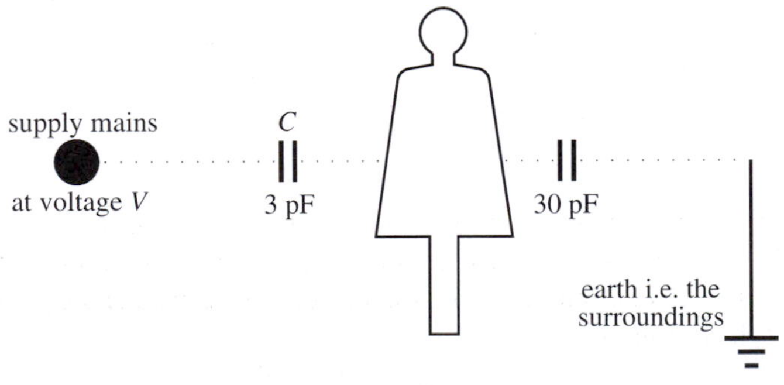

Figure 10.11. *Capacitance between a body and the supply mains and between the body and earth allows a small alternating current to flow.*

V is the voltage of the mains supply, which in most of Europe is 240 V rms, at a frequency of 50 Hz and in North America 110 V rms, at a frequency of 60 Hz. Therefore, in Europe the voltage is

$$V = \sqrt{2} \times 240 \times \sin(\omega t) \qquad \text{and} \qquad dV/dt = \sqrt{2} \times 240\omega \times \cos(\omega t).$$

The rms current which will flow through a capacitor of 3 pF is

$$C \, dV/dt = 0.23 \; \mu\text{A rms}.$$

This current will flow through the body and then return to earth through the 30 pF capacitance. Because the 3 and 30 pF capacitors form a potential divider, the voltage across the 30 pF will be 22 V rms. Twenty-two volts is a very large signal when compared with most electrophysiological signals and so this type of electric field interference is very important. The actual values of capacitance depend upon the size of the human body, how close it is to the surroundings, and how close the mains supply wiring is, but the values given are approximately correct for a person standing on the ground about a metre away from an unscreened mains supply cable.

You can observe the effect of electric field interference by holding the input connection to an oscilloscope amplifier. A 50 or 60 Hz trace will be seen as current flows from mains wiring, through your body and returns to earth through the input resistance of the oscilloscope amplifier. If the current is 0.23 μA rms and the input resistance 10 MΩ then a voltage of 2.3 V rms, i.e. 6.5 V p–p, will be seen.

The electricity mains supply is the most common source of electric field interference. However, some plastics materials and man-made fibres can carry an electrostatic charge which gives rise to a large electric field. Because the field is constant no interference is caused to measurements such as the ECG and EEG. However, if you walk around within a room then you may pass through different static electric fields and so a voltage can be produced on the body. This type of interference can be a problem when slowly changing potentials, such as pH measurement, are being recorded or when the patient may be moving during an exercise test. Computer VDUs are also a common source of electric field interference. Voltages of several kV are used for scan deflection on CRT monitors and these can cause very significant interference if the patient is close to the screen. Laptop computers use relatively low-voltage displays and produce less interference. Unfortunately the very fast processors in modern laptop computers can often give rise to considerable radio-frequency interference.

10.4.2. *Magnetic fields*

A magnetic field is produced around a conductor carrying an electric current and if there is another conductor nearby then the magnetic field will induce a current in this conductor. You will remember from Faraday's law (see section 9.4.3) that the induced voltage is proportional to the rate of change of the magnetic field. The wire carrying the current may be mains supply cable and the wiring in which a voltage is induced may be the input connections to a physiological recorder. The interaction between these two is described as the mutual inductance.

In order to appreciate the sizes of the signals involved, consider two single-turn coils placed 5 m apart (see figure 10.12). We will assume that 1 A p–p of AC current flows in the left-hand coil and then calculate the voltage which will be induced in the right-hand coil. We can extend equation (9.14) to give the field on the axis of the left-hand coil as

$$B = \frac{\mu_0}{2} \frac{r^2}{(r^2 + x^2)^{3/2}}$$

where *r* is the radius of the coil and *x* is the distance along the axis. Derivation of this equation involved calculating the solid angle subtended by the coil as a function of *x*. As *x* becomes greater than *r*, *B* soon falls

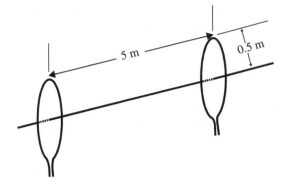

Figure 10.12. *One ampere of sinusoidal 50 Hz current flowing in the coil on the left will induce 0.3 μV in the coil on the right.*

off as x^3. For the values of x and r given in figure 10.12

$$B = \frac{\mu_0}{2} \frac{0.25}{(25.25)^{3/2}} = 1.238 \times 10^{-9} \text{ T.}$$

If we assume that this field is uniform over the area of the right-hand coil then from equation (9.18) we obtain the voltage induced as

$$\text{voltage induced} = \pi r^2 \frac{dB}{dt} = \pi r^2 \omega \sin(\omega t)(1.238 \times 10^{-9})$$

where, ω is the angular frequency of the AC current flowing in the left-hand coil. If this is assumed to be at 50 Hz then the voltage induced can be calculated as 0.305 μV.

A 0.3 μV interference signal is unlikely to be important and, indeed, magnetic field interference is not usually a significant problem. However, we noted that the magnetic flux falls off as the cube of the distance from the coil. If the patient is very close to a source of magnetic interference then the induced voltages can be large. Very significant interference can be found close to large mains supply transformers, where the large currents involved and the number of turns of wire inside the transformers can give rise to large interfering fields.

The oldest unit of magnetic field is the gauss, the current unit is the tesla; when measuring very small fields the γ (gamma) is sometimes used.

$$1 \gamma = 10^{-9} \text{ T} = 10^{-5} \text{ G.}$$

10.4.3.   Radio-frequency fields

Radio-frequency fields are electromagnetic fields and, in principle, need not be considered separately from electric fields and magnetic fields. The reason for considering them separately is that high-frequency fields are propagated over long distances. Fortunately they do not often give rise to problems of interference in electromedical equipment although it is a growing problem with the use of mobile telephones and many other local communications systems.

Any rapidly alternating current in a wire will give rise to the radiation of electromagnetic waves. The source of the rapid oscillations may be an electronic oscillator, as in a radio transmitter, or it may be the rapid surge in current, which occurs when a steady current is switched. Interference may arise from a radio transmitter, from the rapid current switching inside computers or from the switch contacts on a piece of electrical equipment. This last type of interference can be troublesome as switching transients can cause radio-frequency currents to be transmitted along the mains supply and so into an item of measurement equipment where interference is caused.

Radio-frequency fields are measured in $V \, m^{-1}$. This figure is simply the voltage gradient in the air from a particular radio-frequency source. The radio-frequency fields which arise from radio and television transmitters are usually only a few $mV \, m^{-1}$ unless measurements are made very close to a transmitter. Interference at this level is relatively easy to eliminate because it is at a high frequency, which can be removed with a filter, but problems can arise if there is a transmitter very close to where measurements are being made, e.g. a mobile phone being used nearby. A common source of interference in hospitals is surgical diathermy equipment and physiotherapy diathermy equipment. Both of these pieces of equipment generate radio-frequency signals at quite a high power level which can give considerable interference on electrophysiological recordings.

10.4.4. Acceptable levels of interference

It would be nice if all interference could be eliminated but, in practice, decisions have to be made as to what interference is acceptable and what is not. For example, a hospital may be built and include facilities for electromyography and for electrocardiography. Will these rooms need to be screened and will special precautions have to be taken in their design? Will the surgical diathermy/electro surgery equipment in the operating theatres below the EMG room cause interference? Another situation may arise when monitoring equipment is to be installed in an old hospital where interference levels are high. How high a level can be tolerated before interference begins to affect the usefulness of the patient monitoring equipment? In table 9.2 of Chapter 9 the typical amplitudes of the most common bioelectric signals were listed. The EEG and ECG/EKG ranged from 100 μV to 1 mV. Interference of 5 μV will be negligible on an ECG/EKG but certainly visible on an EEG. We will take an interference level of 5 μV as our 'acceptable' level.

We will consider, in turn, *electric fields*, *magnetic fields* and *radio-frequency fields*.

Electric fields

Figure 10.13 shows a differential amplifier connected to a person through electrodes with a resistance of 1 kΩ. These are well-applied electrodes; poorly applied electrodes might give a resistance of 100 kΩ or more. The 100 Ω resistance shown is the resistance of the tissues within the body. If a 50 Hz interfering potential appears on both signal leads then the differential amplifier should reject the interference. However, if current flows through the body then the voltage drop across the 100 Ω cannot be rejected as it appears between the two signal leads. If we are to limit this interference to 5 μV then we can only allow an interfering current of 0.05 μA to flow. In section 10.4.1 we showed that a current of 0.23 μA could arise from a nearby unscreened mains supply cable. If we are to limit our interference to be less than 5 μV then we should aim to have interference which gives rise to less than 0.05 μA of current between a body and supply mains earth. This means that when we hold the input connection of an oscilloscope with a 1 MΩ input resistance, the 50 Hz waveform should be less than 50 mV amplitude. It is very difficult to meet this standard if supply mains cabling is not screened or enclosed in a metal conduit. It is most unlikely to be achievable in a domestic situation.

The limit of 5 μV interference may be rather too stringent if only an ECG/EKG is to be recorded. However, an interference of 5 μV will be quite noticeable on an EEG. The interference obtained will depend upon the path which the current takes through the body. Some electrode positions may give greater interference than others. Another source of noise may be the leads to the amplifier which, if they are not perfectly screened, will pick up electrical interference. If the two signal wires are identical then they will both carry the same

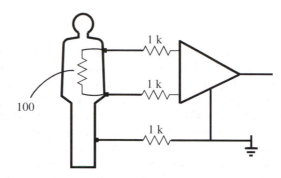

Figure 10.13. *The 1 kΩ resistors represent the electrodes connecting the body to the amplifier. The 100 Ω resistor represents tissue resistance.*

interference and the amplifier should reject the voltages. But if we assume that the cables attract interference of 0.25 μA m^{-1} and that there is a difference in unscreened length of 20 mm then there will be a difference in the interference currents of 5×10^{-9} A. If the electrodes have a difference in resistance of 1 kΩ, then there will be 5 μV of interference.

To put these levels in context, the 50 Hz interference currents which could be drawn from a body to earth found in three typical situations were:

- Laboratory with unscreened mains supply, 1.2 μA.
- New laboratory with screened mains, 0.03 μA.
- An electrically screened room, 0.006 μA.

A suggested limit for EEG measurement would be 0.05 μA, which implies that a laboratory with screened mains cables is needed.

Magnetic fields

A standard can be set for the maximum tolerable *magnetic field* by seeing what field will give 5 μV of interference. If a 500 mm diameter circle of wire is taken as equivalent to the leads to an ECG machine, then the magnetic field which will induce 5 μV is found to be about 10^{-7} T. Again to put this into context, the measured field in several situations was as follows:

- Laboratory with unscreened mains supply, 2×10^{-8} T.
- An electrically screened room, 2×10^{-8} T.
- 2 m from a sub-station transformer, 2×10^{-6} T.
- 6 m from a sub-station transformer, 2×10^{-7} T.

Two important points follow from these figures. Firstly it can be seen that a screened room is no better than a normal laboratory. This is because most screened rooms are effective for electric fields but not for magnetic fields. Secondly, the laboratory with unscreened mains cabling is quite good enough. The reason for this is that, because nearly all mains supply cables are twin cables with the same current flowing in opposite directions in the two conductors, the magnetic fields produced cancel each other out.

Another important practical point can be made if interference from magnetic fields is to be minimized. A voltage can only be induced in a loop of wire. Therefore, if the wires connecting the patient to the EEG machine are run as close together as possible and as close to the skin as possible then the size of the loop will be minimized. This can be a very important practical point in many recording situations.

Radio-frequency fields

It is very difficult to set a standard for acceptable radio-frequency interference because some physiological measurement procedures are much more likely to be subject to interference than others. If EEG and ECG/EKG equipment is well designed then very high levels of radio-frequency fields can be tolerated; it is even possible to record an ECG from a patient in the operating theatre where the surgical diathermy/electrosurgery equipment is connected to the same patient. The interfering field in this situation may be several tens of V m^{-1}. However, one technique which is particularly susceptible to radio-frequency interference is electromyography. Fields of approximately 10 mV m^{-1} can be rectified by surface contamination of an EMG needle electrode and cause interference. Radio transmissions are a particular source of interference and speech or music can appear on the EMG record. Careful cleaning of the electrodes with good screening and earthing can often reduce the interference, but there are some situations where a screened room is necessary.

Where there is a choice of rooms which could be used for sensitive electrophysiological measurements then a useful standard to adopt for radio-frequency interference is 10 mV m^{-1}. Levels of interference below this level are very unlikely to be a source of trouble and much higher levels can be tolerated with well-designed equipment. The following list gives some idea of the fields which may be found in practice:

- 50 m from a 100 W VHF transmitter, 10 V m^{-1}.
- 1500 m from a 100 W VHF transmitter, 20 mV m^{-1}.
- Inside a screened room, 10 μV m^{-1}.
- 6 m from surgical diathermy operating at 400 W and at 400 kHz, 10 mV m^{-1}.

10.4.5.   Screening and interference reduction

A room can be screened to reduce both electric fields and radio-frequency fields but screening is expensive. The door to the room has to be made of copper and electrical contacts made all round the edge of the door. Windows cannot be effectively screened so that screened rooms must be artificially lit at all times. Fortunately, in nearly all cases, good equipment design and careful use make a screened room unnecessary.

A number of pieces of advice for the reduction of interference can be given:

- Whenever possible use earthed equipment and use screened cables both for patient connections and the mains supply. A screened cable will remove nearly all electric field interference.
- Do not use fluorescent lights close to a patient if you are making an ECG/EKG or an EEG recording. Fluorescent lights produce a lot of electric field interference from the tube and this interference can extend for about 2 m in front of the light. Tungsten lighting is much better.
- Do not trail mains supply leads close to the patient or the patient leads. The magnetic field around the cables will induce interference.
- Avoid making measurements close to an electricity supply transformer. Even the measurement equipment may contain a transformer so that the patient should not be placed too close to the equipment.
- Computers and monitors produce interference and should not be placed very close to the patient.
- Faulty leads and badly applied patient electrodes are the most common source of interference.
- If you are involved with the specification of electricity supply wiring then ask that all the cables be run in earthed metal conduit.
- Finally, remember that all interference falls off in intensity with distance so that to move the place of measurement may well be the most expedient solution to an interference problem.

10.5.   APPLICATIONS AND SIGNAL INTERPRETATION

10.5.1.   Nerve action potentials

An evoked response can be recorded from a nerve following electrical stimulation. The response has the form shown in figure 10.14 and, by measuring the time between the stimulus and the response, a conduction time and velocity can be determined. The average nerve conduction velocity is about 50 m s^{-1} or 112.5 mph. This velocity is actually very slow and certainly far slower than the speed at which electrical signals travel down a length of wire, which approaches the speed of light. However, the slow conduction velocity of nerves does not appear to handicap the human body and it does make measurement of the velocity very easy. 50 m s^{-1} is the conduction velocity of myelinated nerve fibres, but there are also non-myelinated fibres which transmit signals even more slowly.

   The subject of the measurement of nerve action potentials is covered in some detail in Chapter 16. The purpose of this section is to consider some of the difficulties in making and analysing these measurements.

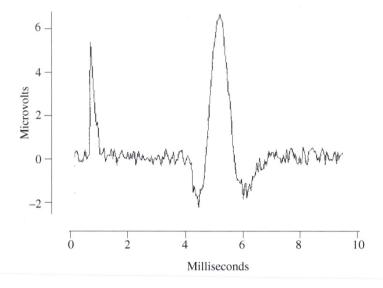

Figure 10.14. *A nerve action potential recorded from electrodes placed over the median nerve at the antecubital fossa, following a supra-maximal stimulus to the wrist.*

   One of the difficulties in measuring nerve action potentials is their small size. The signal shown in figure 10.14 has a maximum amplitude of 5 μV and this is typical of such signals when recorded from surface electrodes. In section 9.3.2 we calculated the thermal or Johnson noise which will appear across a 1 kΩ resistance within a bandwidth of 5 kHz as 0.286 μV rms, i.e. approximately 1 μV p–p. 1 kΩ is typical of the source resistance presented by surface electrodes placed on the skin. Our 'signal-to-noise' ratio (SNR) is thus only about 5:1 and so our recording will be of poor quality. Fortunately, signal averaging is a very appropriate technique to apply as a nerve can be stimulated repeatedly and an average response obtained. As the likely time delay between stimulus and response will be at most about 10 ms (50 cm conduction at 50 m s^{-1}), the stimulus can be repeated at more than 10 pps without any overlap of responses. If we average 1000 responses then the SNR might improve by a factor of about 30 ($\sqrt{1000}$) so that the SNR of the recording will be 150:1.

   Bear in mind that this is the best result which will be obtained. We have assumed that the noise is simply thermal noise and not interference either from another signal such as the electromyograph or from an external electromagnetic field. Our assumption may be valid if great care has been taken with the recording and the patient is well relaxed. Another assumption which we have made is that the noise is not correlated with the stimulus. This is a valid assumption about the thermal noise but it is not valid for any direct interference between the stimulus and the recording amplifier. Yet another assumption which we have made is that the signal is stationary, i.e. that the response is the same to every stimulus. This is usually a good assumption for repeated neural stimulation but not always true for muscle stimulation. It may not be a reasonable assumption if the stimulus to the nerve is not consistent. If the position of the electrodes with respect to the nerve changes during the recording then, as slower nerve fibres have a higher threshold to stimulation, the nerve fibre population which is stimulated may change.

10.5.2. EEG evoked responses

If any stimulus is presented to the body then it is likely that the EEG will be affected; if there is a stimulus then there will be a response. However, the response is usually small and lost in the background EEG, so that the technique of signal averaging must be used to extract the response from the noise. The technique is widely used for recording responses to visual and auditory stimuli. Chapter 15 will consider audio-evoked responses in some detail.

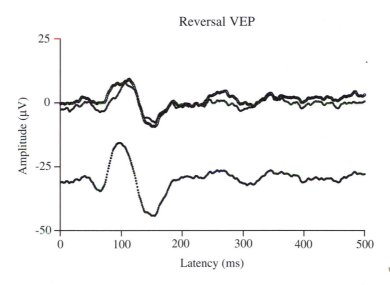

Figure 10.15. *An EEG response evoked by an alternating chequerboard pattern visual stimulus. The two upper traces are successive averages to show that the response is consistent. The lower trace is the sum of the two upper traces. (Figure provided by Professor Colin Barber, Queen's Medical Centre, Nottingham.)*

Figure 10.15 shows a visual EEG evoked potential (VEP) recorded in response to an alternating chequer-board stimulus. A VEP can be recorded in response to a simple repeated flash of light, but a larger response is usually obtained if a visual pattern is recorded. By alternating the dark and light squares in a chequerboard pattern a repeated response can be obtained. The response shown in figure 10.15 is approximately 50 μV in amplitude. This is well above the thermal noise levels but the 'noise' in this case is the EEG.

The electroencephalograph (EEG) is typically 100 μV in amplitude and so our SNR is less than unity. Averaging can be applied but the constraints are more severe than in the case of nerve action potential recording. The response of the brain to a visual stimulus continues for about a second so that it is not possible to repeat a stimulus more frequently than about 0.5 pps without a risk of interference between successive responses. If we want to collect 1000 responses from which to produce an average record then we would need to record for 2000 s, i.e. more than 0.5 h. The subject is quite likely to fall asleep in this time and this will of course change the EEG and the evoked response. As a result we are usually limited to a recording of only a few hundred responses and an associated improvement in SNR of about 20. If the SNR before averaging was 0.5/1 then the SNR of the averaged response might be 10/1. This is not a very high SNR and places a constraint on the usefulness of such signals.

The assumptions which we have had to make in using averaging to obtain an EEG evoked response are firstly, that the background EEG and noise is a random signal and not correlated with the stimulus and secondly, that the evoked response is stationary. Obviously the EEG is not a completely random signal because if it were it would contain no information. However, the EEG is a complex signal and the amplitude distribution of the signal has been shown to be approximately Gaussian. The assumption that the background is not correlated with the stimulus may not be reasonable if the stimulus can be directly received by the EEG recording amplifier. The last assumption concerning the stationarity of the evoked response is a very difficult one to decide because the brain will almost certainly adapt in some way to the stimuli which are being presented.

10.5.3. Measurement of signal-to-noise ratio

Because evoked potential recordings often have a poor SNR, the question often arises as to the significance of an apparent response. Noise and response are often difficult to distinguish. One way to tackle this problem is to make several recordings at different levels of stimulus. In this way it can be seen whether the response appears to be related to the stimulus. The rms value of the averaged waveform can be obtained at different levels of stimulus. However, this does not always give a clear answer. The noise may be related to the stimulus in some way; for example, a subject may be much less relaxed and hence produce EMG interference when a stimulus is given. Another problem is that of 'stimulus artefact' where part of the stimulus interferes directly with the recorded signal. One solution which can be adopted is referred to as 'add–subtract averaging'.

If we consider the recorded response to the nth stimulus as $R(t)$ then we obtain the averaged signal A_{signal} as

$$A_{\text{signal}} = \frac{1}{N} \sum_{n=1}^{n=N} R_n(t) \tag{10.7}$$

however, we can also produce an average by alternate subtraction and addition of responses. This average A_{noise} should contain no signal as the alternate responses should cancel, but the noise will be unchanged as the reversal of alternate segments should not affect the amplitude distribution,

$$A_{\text{noise}} = \frac{2}{N} \sum_{n=1}^{n=N} R_n(t)_{n \text{ even}} - \frac{2}{N} \sum_{n=1}^{n=N} R_n(t)_{n \text{ odd}}. \tag{10.8}$$

In practice A_{signal} and A_{noise} will be vectors of say m values recorded in time T so we can obtain the rms values (variances) as

$$\text{rms signal} = \left[\frac{1}{m} \sum_{1}^{m} (A_{\text{s}} - \bar{A}_{\text{s}})^2 \right]^{1/2}$$

$$\text{rms noise} = \left[\frac{1}{m} \sum_{1}^{m} (A_{\text{n}} - \bar{A}_{\text{n}})^2 \right]^{1/2}.$$

$\bar{A}_{\text{n}}$ should of course be zero if the noise is really Gaussian.

We can now obtain the SNR as (rms signal)/(rms noise).

10.5.4. Objective interpretation

We can simply use the SNR measured as in the last section to decide whether a response is present or absent. We have the mean values ($\bar{A}_s$ and $\bar{A}_n$) and the associated variances on the means so we can use normal statistical tests to attach a probability value to any difference between the variances. However, this takes no account of the shape of any possible response. There may be a very clear response localized in time and yet the SNR may be poor.

One way in which we can take the shape of the response into account is to test for the consistency of the average response. We can split the data into two parts and hence obtain two averaged responses $A_{\text{signal}}(p)$ and $A_{\text{signal}}(q)$,

$$A_{\text{signal}}(p) = \sum_{n=1}^{n=N/2} R_n(t) \quad \text{and} \quad A_{\text{signal}}(q) = \sum_{n=N/2+1}^{n=N} R_n(t).$$

Very often it will be obvious by eye that there is a similar response in the two averaged traces. However, if we wish to obtain objective evidence then we can calculate the correlation coefficient r between the two averages and test for the significance of the correlation.

Splitting the data into two in this way is a very common technique, but the interpretation of a significant value for r needs to be carried out with caution. Figure 10.14 showed that a stimulus artefact can occur because of the direct pick-up of the stimulus by the recording amplifier. This artefact will be present in both signal averages and will give a significant correlation even in the absence of a true evoked potential.

If the shape of the expected response is known then a technique called 'template matching' can be used. A vector describing the expected response is correlated with the averaged signal obtained from equation (10.6) over a range of time delays τ to give $r_s(\tau)$. The same correlation is performed for the noise signal obtained from equation (10.8) to give $r_n(\tau)$. $r_s(\tau)$ can then be searched to see whether there are values which are outside the range of values contained in $r_n(\tau)$.

10.6. PROBLEMS

10.6.1. Short questions

a   What is meant by the phrase 'an evoked electric response'?

b   What is the maximum reduction in noise we might expect by reducing the bandwidth of a recording system from 9 kHz to 1 kHz?

c   Name two techniques by which evoked potentials can be separated from background activity.

d   Why does stimulation of a nerve occur underneath the cathode?

e   What is a Dirac delta function and why can it be useful?

f   Could a pulse width of 1 μs be used to stimulate a nerve?

g   Is 10^{-4} J typical of the energy required to stimulate a nerve through the skin?

h   Give a definition of rheobase.

i   Why is it not possible to use a delta function (a very large pulse of very short duration) of sound to test hearing?

j   What happens to the frequency transform of a short packet of sine waves as the duration is reduced?

k   How can bandwidth be controlled to optimize the quality of a recording?

l   Does the source resistance need to be large or small if a high-quality bioelectric recording is to be made?

m   What are the two rules that enable the function of most operational amplifier circuits to be deduced?

n   Does a differential amplifier work best with identical electrodes placed on the body or one large electrode and one much smaller probe electrode?
o   What is the principle of averaging?
p   Will the noise reduce in proportion to n^2 in averaging? n is the number of summed responses.
q   Could you use your computer as a signal averager and if so what extra equipment might you need?
r   What causes the 50/60 Hz trace on an oscilloscope when you touch the input probe?
s   Could interference of 10 μV at 50/60 Hz be seen on an EEG recording?
t   Why is stationarity important in the context of signal averaging?
u   What improvement in SNR might result from averaging 100 responses?
v   What is the purpose of an add–subtract average?

10.6.2. Longer questions (answers are given to some of the questions)

Question 10.6.2.1

We wish to make an audio-evoked response measurement which will test the frequency specificity of the receptors within the ear. Suggest how this might be done by measuring the response to repeated pure-tone bursts of varying durations.

Question 10.6.2.2

An instrumentation amplifier as shown in figure 10.8 is used to record a nerve action potential with an amplitude of 5 μV. If there is supply mains interference which causes a common-mode voltage of 300 mV,

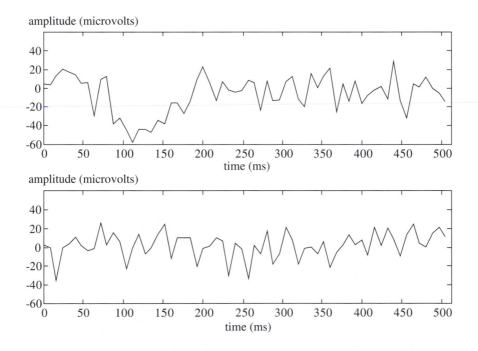

Figure 10.16. *The average response (upper trace) and the 'add–subtract' response (lower trace).*

$R_{in} = 10$ MΩ, and the source impedances R_{sa} and R_{sb} are 2 and 2.2 kΩ, respectively, then calculate the signal-to-noise ratio obtained at the output of the amplifier.

Answer

-1.6 dB.

Question 10.6.2.3

The traces in figure 10.16 (see table 10.3) show the average EEG evoked response and 'add–subtract' response obtained when testing the hearing of a young child. Determine the probability that there is a significant signal in the average response. The digitized values of the traces are given in the table.

Table 10.3. *The data used to produce the traces in figure 10.16.*

Time (ms)	Average	+/− average	Time (ms)	Average	+/− average
0	4.4	3.0	256	8.1	−33.1
8	3.9	−0.6	264	6.3	2.3
16	13.6	−35.2	272	−24.0	−7.1
24	20.5	−0.1	280	7.9	17.1
32	17.4	3.8	288	−13.6	−18.0
40	13.9	10.6	296	−12.3	−6.6
48	5.0	1.9	304	7.0	21.6
56	6.2	−3.3	312	12.1	7.4
64	−28.9	−0.9	320	−11.4	−18.4
72	9.5	26.1	328	−19.8	−0.8
80	12.1	3.0	336	15.7	0.2
88	−38.0	16.0	344	0.5	−7.0
96	−32.1	6.1	352	12.2	6.2
104	−44.4	−22.7	360	20.9	−21.2
112	−57.9	−0.1	368	−25.3	−5.0
120	−44.2	13.7	376	4.7	3.1
128	−43.4	−6.9	384	−14.0	12.9
136	−47.1	−0.7	392	7.9	2.6
144	−34.3	14.3	400	−16.8	7.5
152	−38.3	24.8	408	−7.3	−8.2
160	−15.5	−11.9	416	−2.2	21.4
168	−15.4	10.2	424	2.3	2.2
176	−27.2	9.9	432	−11.4	20.2
184	−14.1	10.1	440	28.2	8.4
192	9.0	−20.1	448	−13.0	−9.1
200	22.7	−1.1	456	−32.0	13.5
208	5.7	1.6	464	4.0	24.4
216	−13.2	9.6	472	1.3	4.5
224	6.7	6.6	480	11.9	0.2
232	−1.6	−30.4	488	−0.3	15.0
240	−4.1	4.4	496	−5.4	21.5
248	−2.4	−2.3	504	−14.2	12.0

Answer

On a parametric F test the two variances are significantly different ($p = 0.002$). See section 13.5 for more information on applied statistics.

Question 10.6.2.4

The normal signal traffic along the median nerve is estimated to involve 0.1% of the fibres at any one time and location. It is suggested that by correlating the signals recorded over a whole nerve trunk at two sites that these signals can be detected. Discuss the problems involved in making such a measurements and what might be the chances of success. Assume that if all the fibres are stimulated at a single time that an action potential of 10 μV is obtained.

Question 10.6.2.5

Given that the stray capacitance from the body to ground is 150 pF and from the body to the 240 V power supply is 1.5 pF, estimate the displacement current in the body. Explain how this current can lead to contamination at an amplifier output. Indicate how a differential amplifier with a good CMRR can improve the situation.

A non-isolated differential amplifier has a balanced CM input impedance of 15 MΩ, a CMRR of 100 dB and a differential gain of 1000. When used to record an ECG between the two arms the electrode contact impedances at 50 Hz are 20 kΩ (right arm), 50 kΩ (left arm) and 100 kΩ (right leg). Calculate the rms amplitude of the interference at the output of the amplifier. State any required assumptions.

Answer

0.113 μA; 2.3 V.

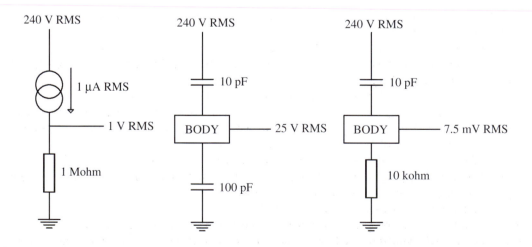

Figure 10.17. *Sketch of the three configurations described in question 10.6.2.6.*

Question 10.6.2.6

Estimate the 50/60 Hz potential on the body, under the following three conditions:

(a)   Holding an oscilloscope probe with an impedance to earth of 1 MΩ.
(b)   No electrical connection to the body.
(c)   Connected to earth through an electrode with an impedance of 10 kΩ.

In each case draw a sketch of the configuration.

Answer

The three configurations are sketched in figure 10.17. Estimated values for the potential on the body are given. The results are for a 240 V supply at 50 Hz. The results should be reduced by about a factor of 2 for a supply at 110 V at 60 Hz.

Answers to short questions

a   An electric evoked potential is the term used to describe an electrical event, usually recorded from the scalp, in response to stimulation of one of the senses.
b   A bandwidth reduction by a factor of 9 might give a factor of 3 reduction in noise, i.e. 6 dB. This would be achieved if the noise was Gaussian.
c   Differential amplification and averaging can be used to reduce noise.
d   Because the current flow underneath the cathode will reduce the transmembrane potential. The current underneath the anode will increase the transmembrane potential.
e   The delta function proposed by Dirac is a pulse of zero duration but infinite amplitude. However, the integral with time is unity. It has the property of equal energy at all frequencies and so is useful as a means of testing the frequency response of a recording system.
f   No, pulses of duration greater than about 20 μs are required to stimulate a nerve.
g   Yes, 10^{-4} J is typical of the energy required to stimulate a nerve through the skin.
h   The rheobase is the minimum current that will stimulate a muscle, whatever the duration of the stimulating pulse.
i   A delta function cannot be used to test hearing because the cochlea and brain are nonlinear and will not respond to a very short duration stimulus.
j   The frequency transform will become broader as the duration of the stimulus is reduced.
k   The bandwidth should be reduced to be as narrow as possible, consistent with causing no distortion of the signal to be recorded.
l   The source resistance needs to be as small as possible to optimize a recording.
m   Most operational amplifier circuits can be understood by following the consequences of assuming: firstly, that there will be no voltage between the inputs of the operational amplifier and secondly, that no current will flow into the inputs of the operational amplifier.
n   Differential amplifiers work best with balanced inputs, i.e. identical electrodes.
o   The principle of averaging is to repeat a measurement and add the responses. The signals should add together linearly but the noise should sum more slowly.
p   No, the noise will reduce as $n^{1/2}$ in signal averaging.
q   Yes you could use a computer as an averager but you would need an input A–D card.
r   Electrical current flowing through the capacitance between your body and the supply mains wiring gives rise to the 50/60 Hz trace on an oscilloscope when you touch the input probe.
s   Yes, the EEG is typically 50 μV in amplitude so that 10 μV is significant.

t   Unless the signals are stationary the response to each stimulus will not be the same.
u   100 averages might improve SNR by 40 dB.
v   An add–subtract average should result in noise alone as the signals should sum to zero.

BIBLIOGRAPHY

Bendat J S and Piersol A G 1971 *Random Data: Analysis and Measurement Procedures* (New York: Wiley-Interscience)
Horowitz P and Hill W 1989 *The Art of Electronics* (Cambridge: Cambridge University Press)
Lynn P A 1989 *An Introduction to the Analysis and Processing of Signals* (London: MacMillan)
Lynn P A and Fuerst W 1990 *Introductory Digital Signal Processing with Computer Applications* (New York: Wiley)

CHAPTER 11

IMAGE FORMATION

11.1. INTRODUCTION AND OBJECTIVES

Medical imaging is an important diagnostic tool. There are a variety of medical imaging techniques, each imaging a different physical or physiological property of the body. For example, x-ray images, both planar and tomographic (x-ray CT), are images of the distribution of the linear attenuation coefficient at an average x-ray energy within the human body. This is a property which is largely a function of tissue density, at least for soft tissues, but is also a function of tissue composition. Thus bone has a higher attenuation coefficient than can be explained simply by increased density because it contains significant amounts of relatively high atomic number elements. X-ray images are largely images of *anatomy*. Radioisotope imaging (RI) produces images of the distribution of a chemical labelled with a γ-ray-emitting isotope. The chemical is distributed according to physiological *function* so the image is primarily an image of function, although as function is distributed anatomically it also produces recognizably anatomical images. Ultrasonic imaging (UI) produces images related to changes in the acoustic impedance of tissues, again largely anatomical in nature, although Doppler flow imaging (DFI) produces functional images of flow. Magnetic resonance imaging (MRI) produces images of proton density, largely a function of the water content of tissue, and images of relaxation times which depend on the environment of the protons. These are largely anatomical. Some functional images, such as flow images, can also be produced and there are prospects of images of other elements apart from hydrogen. Other less familiar imaging techniques produce images of other properties. For example, in electrical impedance tomography (EIT) the electrical properties of tissue are imaged. Each of the imaging techniques is unique, in the sense that a unique physical property is being imaged, although in some cases different techniques can be used to make the same diagnosis. Figure 11.1 illustrates the distinction between functional and anatomical imaging by comparing RI and MRI images of the head.

One big development which has taken place over the past few years is the development of digital imaging. Some of the newer imaging techniques, such as x-ray CT and MRI, are intrinsically digital in that conversion of the captured data to digital form is essential for the imaging to work. In other cases, such as planar x-ray and planar radioisotope imaging, direct exposure of film (analogue imaging) is being replaced by digital imaging where the captured image is in digital form, even if it is ultimately reconverted back to a film image. The demise of film as an imaging medium in medical imaging has been predicted now for some time and although it has proved remarkably resilient there can be no real doubt that its future is limited, at least as far as large imaging departments are concerned. Film is expensive and is bulky to store. Digital techniques of film storage are generally reliable and space efficient. Improvements in computer technology, image capture and image display have brought us close to the point where imaging departments are likely to become completely digital. Conversion of all images to a common digital form highlights the fact that although images can be created by many different mechanisms they share a common theoretical description.

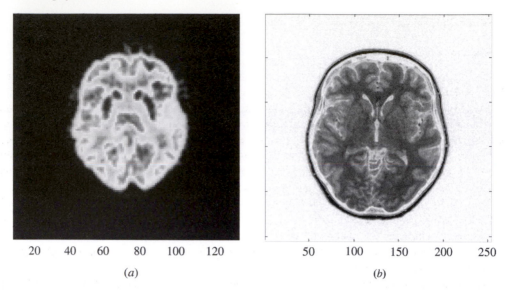

20   40   60   80   100   120   50   100   150   200   250

(*a*)   (*b*)

Figure 11.1. *(a) A PET image of the brain of a normal subject. This is an image of function, since the colour is related to the uptake of tracer (^{18}FDG) in the brain tissues and this is a measure of physiological activity (glucose metabolism). (b) An MR image of approximately the same slice. This shows the anatomy of the brain. Both these images are images of the same object, but imaging a different aspect of the brain. This figure appears in colour in the plate section.*

This chapter will describe image formation in terms of a general theory, which can in most cases be applied to all imaging systems. The specific details of the individual imaging techniques will be left to Chapter 12. The questions we hope to answer include:

- Can we write down an equation that describes an image?
- How can we define image quality?
- What is meant by describing an image as linear?
- Are there factors that are common between imaging systems?
- How can an image filter be used?
- What does back-projection mean?
- What is a digital image?

This chapter covers the basic theory of medical imaging and is quite mathematical. Readers who wish to leave the mathematics should move on to Chapter 12. However, this chapter is in many ways an introduction to Chapter 14 on image processing, so readers who want to come to terms with that chapter should at least get to the point where they can answer the short questions in section 11.10.1.

11.2. BASIC IMAGING THEORY

11.2.1. Three-dimensional imaging

Most of the images we meet in everyday life are essentially two dimensional. This may be a surprising thing to say given that the world is three dimensional (3D), but what we see are collections of surfaces, and these are two dimensional, although they may be curved in the third dimension. The retina is of course only able to produce 2D images, but our brains are able to assemble these into 3D visualizations. To cope with the world visually we only need to form and understand two-dimensional images. However, an increasing number of medical images are essentially three dimensional. Even if the individual images produced by the imaging device are two-dimensional slices through the object, stacking images of successive slices will produce three-dimensional images. We need three-dimensional images since the object, the patient being imaged, is genuinely three dimensional, not simply a collection of surfaces. The technology to form images of this (visually) unusual object has been available in various forms for 30 years, although such images have only been routinely available for 20 years or so. In line with these developments this chapter will generally take a three-dimensional view of imaging, although where appropriate or convenient we will switch to two- or even one-dimensional representations.

Although the physical and biochemical processes being imaged, and the mechanism by which images are formed, differ between the different types of imaging, it is possible to develop a mathematical description of the imaging process. From this a methodology for understanding and analysing the performance of an imaging system can be derived, which is largely independent of the physical details of the imaging process.

The distribution of the property being measured within an object can be represented as a three-dimensional function, where the value of the function represents the value of intensity of the property at a point in space. An example would be the function $\mu(x, y, z)$ representing the distribution of the linear attenuation coefficient or the function $a(x, y, z)$ representing the distribution of radioactivity inside a patient. Similarly, an image of the object can be represented as a function. These two functions are linked by the imaging system. In its most abstract form the imaging process can be represented by a transformation

$$g = T(f) \tag{11.1}$$

where f is the object distribution, T is the imaging process and g is the resulting image. In most texts on imaging f and g are usually represented as two-dimensional functions, but in medical imaging the objects being imaged, people, are three dimensional and so it is more appropriate to treat f and g as three-dimensional functions.

The object f as represented here is rather abstract. A real object, such as a liver, can be imaged in many ways. f may consist of a distribution of attenuation coefficients but, if an appropriate radiotracer has been administered, it may also be seen as a distribution of the radioisotope. f may also be a distribution of ultrasonic scatterers. None of these 'objects' completely represents the real liver, but each represents a different aspect of it. An imaging system, such as a CT scanner or a SPET camera, can only image one aspect of the complete object. The image g is an image of the partial object represented by f. The real object is much more complex than any of the possible f objects, which is why there is increasing interest in trying to bring the various representations together using techniques of multi-modality imaging.

We will find it useful to talk of two distinct spaces. One space is object space. This is clearly a real space, since the object exists, but is also a mathematical space. We will also want to speak of an image space. This can also be a real space, such as a piece of film, but it is also a mathematical space.

11.2.2. Linear systems

The detailed form of equation (11.1) can be derived from the physics of the imaging process. This then forms a complete description of this process and allows the output of the imaging system (the image) to be predicted for any input (the object). However, unless this description can be simplified it may not be very useful. One powerful simplification, which is not in fact true for many imaging systems but is often used, is to assume that the imaging system is linear.

An object to be imaged is represented as a three-dimensional function $f(x, y, z)$, where the value of this function can be thought of as a brightness or intensity value. Physically this may be the concentration of radioactivity at each point in space, or the density of protons and so on. If the function is scaled or multiplied by a scalar value A we can think of the overall intensity as increasing if A is greater than 1 and decreasing if A is less than 1.

Now, suppose an object f_1 is presented to an imaging system and an image g_1 produced. If the object brightness is scaled by A then the first condition for linearity is that the image brightness will also be scaled by A,

$$Ag_1 = A(T(f_1)) = T(Af_1).$$

If a second object, scaled in this case by B, is presented to the imaging system then

$$Bg_2 = T(Bf_2).$$

If both objects are now presented simultaneously to the imaging system then if

$$g = (Ag_1 + Bg_2) = T(Af_1) + T(Bf_2) = AT(f_1) + BT(f_2) \tag{11.2}$$

i.e. the image of both objects together is the same as adding the images of the individual objects, the system is said to be *linear*.

In fact, few medical imaging systems are actually linear. Perhaps unexpectedly the only imaging modality which can be considered linear, provided care is taken over definitions, is (digital) radioisotope imaging. This is because if the concentration of radioactivity in one part of the object is changed this change does not affect the way other parts of the object are imaged. In the case of x-ray CT, if the linear attenuation coefficient of part of the image is changed, for example by replacing soft tissue with bone, this will change the spectrum of the x-rays passing through the object at this point and this will in turn affect the apparent attenuation coefficients of other parts of the object. In the case of MRI, the homogeneity of the magnetic field can be affected by changing the material being imaged and this can result in a nonlinear response. In the case of ultrasonic imaging bone will create ultrasonic shadows.

One simple nonlinearity is produced if images are recorded on film, since the film can saturate and thus increasing the object intensity can result in the image intensity not increasing by the same amount. Most medical imaging processes have some degree of nonlinearity, although this is often small. An example of an imaging technique which is very nonlinear is electrical impedance tomography. However, linearity has such useful properties that it is usual to assume it holds, at least for part of the imaging process, for example the imaging process prior to exposure of the image onto film.

11.3.   THE IMAGING EQUATION

The most important feature of linearity is that it allows the image of a complex object to be decomposed into the summation of the images of a set of simple objects. These objects are *point objects* and are represented mathematically by the delta (δ) function. This is a function which has the property of being zero everywhere except at the point x_0, y_0, z_0, where it becomes infinite, subject to the constraint that

$$\iiint \delta(x - x_0, y - y_0, z - z_0) \, \mathrm{d}x \, \mathrm{d}y \, \mathrm{d}z = 1.$$

The δ function was introduced in section 10.1.1 and appears again in section 13.4.2 of Chapter 13. For all the requirements of this chapter, the delta function can be thought of as the limit of a well-behaved function such as a Gaussian

$$\delta(x, y, z) = \lim_{d \to 0} \frac{1}{\sqrt[3/2]{2\pi d^2}} e^{-(x^2+y^2+z^2)/2d^2}.$$

In two dimensions these equations would be

$$\iint \delta(x - x_0, y - y_0) \, dx \, dy = 1$$

and

$$\delta(x, y) = \lim_{d \to 0} \frac{1}{2\pi d^2} e^{-(x^2+y^2)/2d^2}$$

so three dimensions is not much more complex than two.

The delta function has an interesting property which is represented in the *sifting integral*,

$$f(x, y, z) = \iiint f(\xi, \eta, \zeta) \, \delta(x - \xi, y - \eta, z - \zeta) \, d\xi \, d\eta \, d\zeta.$$

In this integral the δ function picks (sifts) out the value of the function f at the point $\xi = x, \eta = y, \zeta = z$. Now the image g is given by

$$g = T(f)$$
$$= T\left(\iiint f(\xi, \eta, \zeta) \, \delta(x - \xi, y - \eta, z - \zeta) \, d\xi \, d\eta \, d\zeta \right).$$

The right-hand side of this equation is similar to the right-hand side of equation (11.2), where f is interpreted as a scaling factor at the point (ξ, η, ζ) and δ as an object function situated at the point (ξ, η, ζ). It then follows from the linearity definition of equation (11.2) that

$$g(x, y, z) = \iiint f(\xi, \eta, \zeta) \, T(\delta(x - \xi, y - \eta, z - \zeta)) \, d\xi \, d\eta \, d\zeta.$$

The expression $T(\delta(x - \xi, y - \eta, z - \zeta))$ represents the result of imaging a point object. This can be written as

$$h(x - \xi, y - \eta, z - \zeta; \xi, \eta, \zeta) = T(\delta(x - \xi, y - \eta, z - \zeta)).$$

The notation used here is a little complex but can be interpreted as follows. The object is at the point ξ, η, ζ in object space. Assuming that there is no overall scaling of the spatial dimensions between object and image, i.e. one unit of distance in object space is equal to one unit of distance in image space, the image of the object is *centred* at the point $x = \xi, y = \eta, z = \zeta$ in image space but may spread around this point, i.e. may no longer be a delta function. This spread about a central point is represented by the fact that the coordinates of h are expressed in terms of distances from a central point and by the fact that h has non-zero values (unlike the δ function) at points other than $x = \xi, y = \eta, z = \zeta$. In the general case the *shape* of h depends on the position of h in the imaging space. The notation for h reflects this with the parameters after the semicolon representing this position dependence. Inserting h into the above integral equation gives

$$g(x, y, z) = \iiint f(\xi, \eta, \zeta) \, h(x - \xi, y - \eta, z - \zeta; \xi, \eta, \zeta) \, d\xi \, d\eta \, d\zeta \qquad (11.3)$$

and this equation is the *superposition integral* or in the current context the *imaging equation*. In two dimensions this equation is

$$g(x, y) = \iint f(\xi, \eta) \, h(x - \xi, y - \eta; \xi, \eta) \, d\xi \, d\eta$$

and in one dimension

$$g(x) = \int f(\xi) \, h(x - \xi; \xi) \, d\xi.$$

Again, we can see that the equations have a similar form in all dimensions.

11.3.1. The point spread function

h is the image of a point object. In general, h will not be a delta function since all imaging systems produce a degraded representation of the object, i.e. they blur the object. Unlike the delta function h will be spread to a smaller or greater extent around a central point (ξ, η, ζ) and this leads to one of the common names for h, the *point spread function* (PSF). An equivalent but slightly more formal term is the point response function (PRF). These are equivalent terms and are freely interchangeable. Completely describing h for a linear imaging system represents a complete description of the imaging process, apart from effects due to noise which will be considered later.

Figure 11.2(*a*) represents three point-like objects in a two-dimensional object plane and figure 11.2(*b*) represents the images produced by an imaging system. The images of the objects are blurred relative to the original objects. The amount of blurring is a function of position.

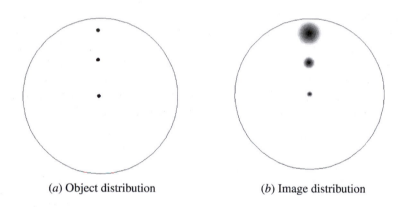

(*a*) Object distribution   (*b*) Image distribution

Figure 11.2. *(a) Object distribution; (b) image distribution. Three point objects in the object space are shown on the left. The images shown on the right illustrate that the point spread function (PSF) is a function of position within the image.*

11.3.2. Properties of the PSF

Before the advent of various forms of digital imaging it would have been correct to assert that h must always be non-zero. The energy transmitted through the imaging system at a point could not be negative. However, digital imaging techniques, especially three-dimensional techniques, can include computational operations which produce an h which can have negative values so this is no longer strictly true. h is usually unimodal and decreases from a maximum value to zero in what is hopefully a fairly small distance. The exact shape of h will depend on the imaging system. h is often modelled as a Gaussian function,

$$h(x - \xi, y - \eta, z - \zeta) = S \frac{1}{\sqrt[3/2]{2\pi d(\xi, \eta, \zeta)^2}} \, e^{-((x-\xi)^2 + (y-\eta)^2 + (z-\zeta)^2)/2d(\xi,\eta,\zeta)^2}$$

although this is only a model, and rarely represents the true state of affairs.

11.3.3. Point sensitivity

The integral over h is the *point sensitivity* function,

$$S(\xi, \eta, \zeta) = \int\int\int h(x - \xi, y - \eta, z - \zeta; \xi, \eta, \zeta) \, \mathrm{d}x \, \mathrm{d}y \, \mathrm{d}z.$$

This represents the total signal collected from a unit point object and may well be different at each point in space. Although it is often convenient to assume that S is independent of position this is rarely the case. Objects are usually more visible at some parts of the image than at others.

11.3.4. Spatial linearity

The Gaussian as a model of h is very seductive, but it can obscure some subtleties in the imaging process. The imaging process can be visualized as one in which the point object is projected into image space and then transformed with the PSF. This seems to suggests that the peak point of h is at the point of projection of the object, i.e. (ξ, η, ζ), but this need not be the case. The PSF can be shifted from this point so that h is asymmetrically positioned about the point (ξ, η, ζ). We can incorporate this into h by writing

$$h = h(x - \xi - s_\xi, y - \eta - s_\eta, z - \zeta - s_\zeta; \xi, \eta, \zeta)$$

where (s_ξ, s_η, s_ζ) are shift values representing this effect. These, in general, will be functions of (ξ, η, ζ). The effect of having an h of this form is that the image of a straight line may no longer be straight. In this case the imaging system demonstrates spatial nonlinearity, but this use of the term nonlinear must be distinguished from the original use of the term nonlinear. A system which demonstrates spatial nonlinearity as described above can still be linear in the original sense of equation (11.2). One way of looking at the effect of spatial nonlinearity is to image a regular grid of line objects. After imaging with a system which exhibits spatial nonlinearity the grid will no longer be regular. This is illustrated in figure 11.3

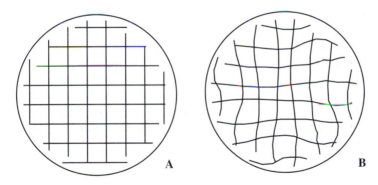

Figure 11.3. *Spatial nonlinearity. B is the image of A.*

Finally, although the Gaussian is a symmetric function, h does not have to be symmetric. If h is not symmetric then the Gaussian model breaks down.

A complete description of h provides a complete description of the imaging process. However, if h is at all complex this may not be very useful. In principle, h can be measured at every point in space using a point object. These measurements will then completely characterize the imaging system and from them the image of any object can, in principle, be derived by calculation. In practice this is rarely done. What we often need from h are some general measures of imaging performance and these are usually derived by making some additional simplifications to the imaging equation.

11.4. POSITION INDEPENDENCE

We can only handle the mathematics of images if we make simplifying assumptions. As long as these are reasonable then we may obtain useful conclusions. One assumption we can make is that the shape of h is independent of position. This means in practice we can take (s_ξ, s_η, s_ζ) to be zero, since any shift will be the same for all points and so can be ignored. We may also assume that the integral over h (i.e. S) is also independent of position. Then the imaging equation reduces to

$$g(x, y, z) = \iiint f(\xi, \eta, \zeta) \, h(x - \xi, y - \eta, z - \zeta) \, d\xi \, d\eta \, d\zeta. \tag{11.4}$$

In two dimensions this equation is

$$g(x, y) = \iint f(\xi, \eta) \, h(x - \xi, y - \eta) \, d\xi \, d\eta$$

and in one dimension

$$g(x) = \int f(\xi) \, h(x - \xi) \, d\xi.$$

This equation is known as the *convolution equation* and is such an important integral that it has a notation of its own. There is actually no completely universal notation. The one we prefer is

$$g = f \otimes h$$

but others can be found. The convolution integral has some useful properties as will be shown below. One immediately useful property is that only one h needs to be specified to completely characterize the imaging system and the convenience of this usually outweighs any doubts concerning the accuracy of the assumption of position independence. The imaging system is characterized by extracting one or two parameters from h. These include the integral over h, a measure of system sensitivity, and a measure of the width of h. This is usually the width of the function at a level one half of the maximum amplitude of the function. This measure is called the full-width at half-maximum height (FWHM).

Figure 11.4 gives a representation of the convolution equation in one dimension. The object is shown as a series of point objects (figure 11.4(*a*)), each of which is replaced by an appropriately positioned and scaled h function. All these functions are added together to produce the image function (figure 11.4(*b*)).

Consider the image of one point object centred at ξ (figure 11.5). It has an amplitude proportional to $f(\xi)$. The contribution of this signal at the point x is given by

$$f(\xi) \, h(x - \xi)$$

and the total signal at the point x is given by summing over the contributions from all the N point objects

$$g(x) = \sum_{i=1}^{N} f(\xi_i) \, h(x - \xi_i)$$

which in the limit becomes the convolution integral, in this case in one dimension.

The important conclusion of this section is thus that, if we know the point spread function (PSF) and the imaging system is position independent, then we can totally characterize the image system by using the convolution integral.

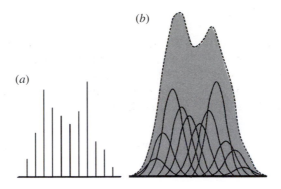

Figure 11.4. *(a) An object decomposed into δ functions which are then transformed (b) into point spread functions and added together to give the shaded profile, the image.*

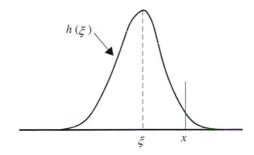

Figure 11.5. *The contribution of the object centred at the point ξ to the image at x.*

11.4.1. Resolution

We can attempt to describe a PSF by measuring the FWHM as shown in figure 11.6. The FWHM is often taken as a measure of *resolution*. For a symmetric Gaussian function with width parameter d it can easily be shown that

$$\text{FWHM} = 2.36d.$$

The concept of resolution comes from the idea that if two point objects are close together images will run together in a way which makes it impossible, at least visually, to determine whether there are two discrete objects present or simply one extended object. As the objects are moved apart a point will come at which it is clear that there are two objects because two peaks of intensity are seen. This is shown informally in two dimensions in figure 11.7, where the images of two objects run together so that the image of two objects cannot be easily distinguished from the image of one object.

The exact distance apart that two objects need to be before they can be distinguished cannot be specified uniquely. In practice, it will depend on the shape of the PSF.

For the Gaussian function it turns out that if two Gaussians of equal intensity spaced one FWHM apart are combined, the intensity at the midpoint between them dips just below (by about 6%) the value at the peaks of the Gaussians (figure 11.8). The Gaussians are said to be just resolved under these circumstances and therefore the FWHM is often taken as a measure of the resolution of the imaging system. The 6% dip is only true for Gaussians and if h is not Gaussian the FWHM will not have the same significance. However, it

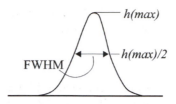

Figure 11.6. *The FWHM height of the PSF.*

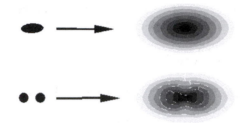

Figure 11.7. *Imaging the objects on the left produces very similar images on the right.*

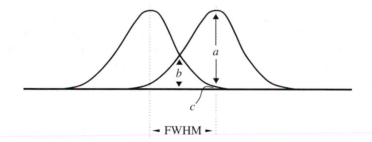

Figure 11.8. *When the Gaussians are one FWHM apart (a + c) is 6% larger than 2b.*

is a useful indicator of the limits of resolution and is easily determined, but it should be recognized that it is not a precise method of comparing the resolution of two imaging systems.

There are other ways of assessing resolution, each of which leads to a definition not always completely compatible with the FWHM definition. One other common definition is to image an object consisting of a set of parallel lines and determine how close the lines have to be before they can no longer be seen as lines. A phantom based on this approach is a popular way of visualizing the resolution of an imaging system. The resolution is then expressed as the number of *line pairs per unit length* which can just be seen. This is not a very objective test and really dates from the time when all images were photographic. The number of lines per cm or inch that can be resolved by a camera is still often used as a measure of resolution. With the advent of digital imaging, where images can be manipulated freely, it is no longer as useful a measure as the more objective FWHM. It does provide a quick visual assessment of the way resolution can vary with position for an imaging system.

In practice h may (will) not be Gaussian, and other additional measures of shape, such as a full-width at one-tenth maximum height (FWTM) may also be quoted.

11.4.2.   *Sensitivity*

The integral under the PSF is the *point sensitivity* (see section 11.3.2) and is generally a function of ξ, η and ζ. However, it is often convenient to assume that it is independent of position in which case it can be taken outside the integral,

$$g(x, y, z) = S \iiint f(\xi, \eta, \zeta)\, p(x - \xi, y - \eta, z - \zeta; \xi, \eta, \zeta)\, \mathrm{d}\xi\, \mathrm{d}\eta\, \mathrm{d}\zeta$$

where p is the normalized version of h, i.e. the integral over p is unity. In practice, S is rarely independent of position and so the above equation must be written as

$$g(x, y, z) = \iiint f(\xi, \eta, \zeta)\, S(\xi, \eta, \zeta)\, p(x - \xi, y - \eta, z - \zeta; \xi, \eta, \zeta)\, \mathrm{d}\xi\, \mathrm{d}\eta\, \mathrm{d}\zeta.$$

Looking at this equation we can see that a non-uniform point sensitivity effectively generates a new object fS which becomes the object being imaged.

In practice point sensitivity is rarely measured. What is measured is some sort of area sensitivity. If the object f is of uniform intensity then the image of this object is given by

$$U(x, y, z) = \iiint S(\xi, \eta, \zeta)\, p(x - \xi, y - \eta, z - \zeta; \xi, \eta, \zeta)\, \mathrm{d}\xi\, \mathrm{d}\eta\, \mathrm{d}\zeta$$

and the intensity measured at a point in this image is not S but the value of the integral, and this represents imaging the function S with p. If S is not constant then U will not be constant, but they will not in general be the same. U will be smoother than S. However, a more subtle result is that, if p is not position independent, then U can be non-uniform even if S is uniform. This arises whenever

$$\iiint p(x - \xi, y - \eta, z - \zeta; \xi, \eta, \zeta)\, \mathrm{d}\xi\, \mathrm{d}\eta\, \mathrm{d}\zeta \neq \iiint p(x - \xi, y - \eta, z - \zeta; \xi, \eta, \zeta)\, \mathrm{d}x\, \mathrm{d}y\, \mathrm{d}z.$$

This result can arise, for example, if the imaging system exhibits spatial nonlinearity since one effect of this is to cause the images of equally spaced objects to cluster towards certain points. The classic example of this is the way images of point objects can shift towards the positions of photomultiplier tubes in a gamma camera. A non-uniform U may therefore be attributed to variations in point sensitivity across the imaging device rather than spatial nonlinearity.

Why is this important? A pointwise division of the image g by U is often used to try and correct for these sensitivity variations. The above analysis shows that this is not the correct thing to do. Having said that, in the case where there is no spatial nonlinearity or S varies slowly with position it may be a fairly reasonable thing to do. Slowly in this context means that the value of S does not change appreciably over distances comparable with the FWHM. However, when a non-uniform U is due to spatial nonlinearity it is a completely inappropriate thing to do. Such operations should only be performed once the imaging process is well understood and they are seen to be reasonable. Even if division by U is theoretically appropriate, U must be determined under conditions similar to those used for collecting g; for example, for radioisotope imaging, similar scattering conditions.

If the imaging system obeys the convolution equation (equation (11.4)) then most of these restrictions disappear. Although most imaging systems do not strictly obey the convolution equation over the whole of the image space, over small enough regions of the image it is a reasonable approximation. The important thing is to be aware that it is an approximation and to use the convolution model with caution.

Measurements of sensitivity need to be taken to calibrate the imaging system, allowing the image intensity to be interpreted in terms of the units of the physical properties being measured. Sensitivity may also be measured to enable the performance of imaging systems to be compared, although in this case the measurements are meaningless without some measure of image noise.

11.4.3. Multi-stage imaging

An image may itself form the input to an imaging device. We may take an image and wish to manipulate it or the basic imaging process may be in two or more stages. If our initial imaging of the object distribution f is given by

$$g = T(f)$$

and the subsequent process carried out on g is given by

$$k = W(g)$$

then there exists an overall imaging operation given by

$$k = Z(f) = W(T(f)).$$

If the system is linear then

$$W(T(f)) = T(W(f)).$$

Some imaging processes have two stages. For example, radionuclide imaging can be easily divided into discrete steps. Sometimes it may be conceptually useful to divide the imaging process into steps, even if this does not correspond to any clear physical division. An example of this might be the following. Many imaging devices are rarely position independent, although variations in the shape of the PSF with position are small. The imaging process can conceptually be divided into an initial stage representing the position-dependent processes and a second stage representing a position-independent stage. This division can provide a justification for subsequent image manipulations using spatial filters, for example, which strictly depend on the imaging process being position independent.

11.4.4. Image magnification

It is sometimes possible to convert a position-dependent imaging situation into a position-independent situation, or at least to improve things. A classic example is the case of image magnification. In the strict sense of the above results this imaging situation is position dependent, since it can be represented as an imaging situation in which the vector (s_ξ, s_η, s_ζ) is very position dependent, even when the shape of h is independent of position.

However, in this case a simple transformation converts this imaging situation to position independence. In this case h can be written as $h(x - M\xi, y - M\eta, z - M\zeta)$, where we will assume that the magnification M is the same in all directions. The imaging equation becomes

$$g(x, y, z) = \iiint f(\xi, \eta, \zeta) h(x - M\xi, y - M\eta, z - M\zeta) \, d\xi \, d\eta \, d\zeta.$$

We replace $M\xi$ by ξ' and similarly for the other coordinates and, remembering that $d\xi = d\xi'/M$, etc, we find

$$g(x, y, z) = \frac{1}{M^3} \iiint f\left(\frac{\xi'}{M}, \frac{\eta'}{M}, \frac{\zeta'}{M}\right) h(x - \xi', y - \eta', z - \zeta') \, d\xi' \, d\eta' \, d\zeta'$$

which is now in true convolution form. This equation represents expanding the linear dimensions of the object by M and then imaging with a position-independent PSF. In two dimensions this equation is

$$g(x, y) = \frac{1}{M^2} \iint f\left(\frac{\xi'}{M}, \frac{\eta'}{M}\right) h(x - \xi', y - \eta') \, d\xi' \, d\eta'.$$

11.5. REDUCTION FROM THREE TO TWO DIMENSIONS

So far imaging has been represented as a three-dimensional process, largely because the object being imaged is three dimensional. Historically most medical images, for example planar x-rays, have been two dimensional. We can derive the relationships between the object and these images by integrating along one axis. We shall take the z-axis to be the axis along which we integrate. Then

$$g_{2D}(x, y) = \int g(x, y, z) \, dz.$$

Expanding g gives

$$g_{2D}(x, y) = \int \left(\iiint f(\xi, \eta, \zeta) h(x - \xi, y - \eta, z - \zeta; \xi, \eta, \zeta) \, d\xi \, d\eta \, d\zeta \right) dz.$$

Since this equation is linear, the integration over z can be taken as the inner integral and performed first

$$g_{2D}(x, y) = \iiint f(\xi, \eta, \zeta) w(x - \xi, y - \eta; \xi, \eta, \zeta) \, d\xi \, d\eta \, d\zeta$$

where

$$w(x - \xi, y - \eta; \xi, \eta, \zeta) = \int h(x - \xi, y - \eta, z - \zeta; \xi, \eta, \zeta) \, dz$$

We can do this since h is the only function on the right-hand side which contains z. w is still a three-dimensional function since it is a function of ζ. Clearly the relationship between f and g_{2D} is complex. If the variables in w can be separated so that

$$w = w_{xy}(x - \xi, y - \eta; \xi, \eta) \, w_z(\zeta)$$

then

$$g_{2D}(x, y) = \iiint f(\xi, \eta, \zeta) w_z(\zeta) w_{xy}(x - \xi, y - \eta; \xi, \eta) \, d\xi \, d\eta \, d\zeta$$

$$= \iint \left(\int f(\xi, \eta, \zeta) w_z(\zeta) \, d\zeta \right) w_{xy}(x - \xi, y - \eta; \xi, \eta) \, d\xi \, d\eta$$

$$= \iint k(\xi, \eta) w_{xy}(x - \xi, y - \eta; \xi, \eta) \, d\xi \, d\eta.$$

We have managed to reduce the imaging equation to a true two-dimensional form. In this case the equivalent two-dimensional function $k(\xi, \eta)$ is a weighted integration over the z or depth direction. The variable separation condition we needed for the above implies that the shape of the depth-integrated PSF, i.e. w_{xy}, is independent of depth. This is generally not the case, although in some circumstances it may be a reasonable approximation.

Although complete separation of the variables is in most cases not possible, we may be able to partially separate the variables by taking out a sensitivity term. We can write

$$w = v(x - \xi, y - \eta; \xi, \eta, \zeta) \, a_z(\zeta)$$

where the integral over v is unity for all ζ. An example of this is where a_z represents γ-ray attenuation in radionuclide imaging. Then,

$$
\begin{aligned}
g_{2D}(x, y) &= \iint f(\xi, \eta, \zeta) \, a_z(\zeta) \, v(x - \xi, y - \eta; \xi, \eta, \zeta) \, d\xi \, d\eta \, d\zeta \\
&= \iint k(\xi, \eta, \zeta) \, v(x - \xi, y - \eta; \xi, \eta, \zeta) \, d\xi \, d\eta \, d\zeta.
\end{aligned}
$$

In this case the function k is taken to include the effects of attenuation and it is this function which is transformed with v.

The above results show that the relationship between the three-dimensional object and the two-dimensional planar image can be rather more complex than the straightforward three-dimensional case. Two dimensions is not a simple version of three dimensions in medical imaging and this is why we need to keep in mind the fact that medical imaging is really a three-dimensional imaging process.

11.6. NOISE

All imaging systems are noisy. If repeat images are taken of the same object they will vary from image to image. This image variation is essentially unpredictable and is called noise. The sources are varied and depend on the imaging system. For example, the principal source of noise in a radioisotope image is the Poisson noise associated with the radioactive decay process. In magnetic resonance imaging the principal noise is electronic noise.

Many of the modern imaging modalities are made up of a sequence of imaging steps. For example, x-ray CT consists of a data collection stage followed by an image reconstruction stage. There is usually some stage in the imaging sequence at which the noise is introduced and is then modified by subsequent stages. Although the effects of these subsequent stages on the noise can be reversed there is always a point at which the noise becomes an irreversible addition to the image.

At a point in the image the noise will have a mean value and a variance. Representing the added noise function by $n(x, y)$ then the expected mean value $\mu(x, y)$ is given by

$$\mu(x, y) = \langle n(x, y) \rangle.$$

The variance of the noise will be given by

$$\sigma_n^2(x, y) = \langle (n(x, y) - \mu(x, y))^2 \rangle.$$

There will, in general, be correlation between the noise at adjacent points in the image,

$$\langle (n(x_1, y_1) - \mu(x_1, y_1))(n(x_2, y_2) - \mu(x_2, y_2)) \rangle = \tau_n(x_1, y_1, x_2, y_2).$$

This is the autocorrelation function (see section 13.3.6) of the noise.

Simple 'white' noise has $\mu_n = 0$, with σ_n independent of position and $\tau_n = 0$, i.e. the mean value of the noise is zero everywhere, the amplitude of the random noise is uniformly distributed over the image but is uncorrelated between points. If μ_n, σ_n and τ_n are independent of position the noise is said to be stationary.

In the presence of noise the imaging equation becomes

$$g(x, y, z) = \iiint f(\xi, \eta, \zeta) h(x - \xi, y - \eta, z - \zeta; \xi, \eta, \zeta) \, d\xi \, d\eta \, d\zeta + n(x, y, z).$$

Much of image processing concerns the use of methods which seek to eliminate the effects of noise while retaining the detail in the image (see Chapter 14).

11.7. THE FOURIER TRANSFORM AND THE CONVOLUTION INTEGRAL

The relationship between the object and the image is a complex one. As we have seen, a significant simplification arises if we can assume that the PSF has a shape which is independent of position in all directions. Under this approximation the imaging equation simplifies to the convolution equation. As we have seen this assumption has all sorts of benefits. One particular benefit which we are about to explore is that the convolution equation takes on an even simpler form if we express it in terms of the Fourier transforms of its component functions. This is very important as it makes a range of image manipulations and analyses possible.

11.7.1. The Fourier transform

A discussion of the imaging process is incomplete without visiting the Fourier transform. The Fourier transform (FT) of a function $f(x, y, z)$ is defined as

$$F(u, v, w) = \iiint f(x, y, z) \, e^{-2\pi j(ux + vy + wz)} \, dx \, dy \, dz$$

where $e^{-2\pi j(ux + vy + wz)}$ is the complex exponential function, $j = \sqrt{-1}$ and the integrations are over all space ($-\infty$ to ∞).

The *inverse* Fourier transform relates $F(u, v, w)$ to the function $f(x, y, z)$ by

$$f(x, y, z) = \iiint F(u, v, w) \, e^{+2\pi j(ux + vy + wz)} \, du \, dv \, dw.$$

The Fourier transform is also defined in one dimension

$$F(u) = \int f(x) \, e^{-2\pi j ux} \, dx$$

$$f(x) = \int F(u) \, e^{+2\pi j ux} \, dx$$

and in two dimensions

$$F(u, v) = \iint f(x, y) \, e^{-2\pi j(ux + vy)} \, du \, dv$$

$$f(x, y) = \iint F(u, v) \, e^{+2\pi j(ux + vy)} \, dx \, dy \tag{11.5}$$

and indeed in any number of dimensions. The properties of the FT are similar in all dimensions, but since most applications of the FT in medical imaging to date involve the use of the two-dimensional FT, the rest of this section will concentrate on this. It must be stressed that there are applications of the three-dimensional FT in image reconstruction and these will become more common as new image reconstruction techniques are developed, but we will not deal with these here.

Real and imaginary components

In general, $f(x, y)$ can be a complex function but in most cases where it represents an image or an object it is real. $F(u, v)$, on the other hand, is usually complex and can be written as

$$F(u, v) = F_R(u, v) - jF_I(u, v)$$

where

$$F_R(u, v) = \iint f(x, y) \cos 2\pi(ux + vy) \, dx \, dy$$

$$F_I(u, v) = \iint f(x, y) \sin 2\pi(ux + vy) \, dx \, dy.$$

This result comes from expanding the equation for the transform given above.

Substituting F_R and F_I into the definition of $f(x, y)$ gives, when $f(x, y)$ is real,

$$f(x, y) = \iint F_R(u, v) \cos 2\pi(ux + vy) + F_I(u, v) \sin 2\pi(ux + vy) \, du \, dv. \tag{11.6}$$

This is an important result since it shows that $f(x, y)$ can be decomposed into a summation (an integral is the limit of a summation) of weighted *sine* and *cosine* functions.

In the one-dimensional case $\cos 2\pi ux$ (and $\sin 2\pi ux$) is easily represented. If the wavelength (the distance from peak to peak) is L then $u = 1/L$, where u is the spatial frequency of the cosine wave (figure 11.9).

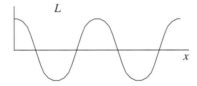

Figure 11.9. *A cosine wave.*

The two-dimensional form of the cosine wave is also easily represented. It is a wave travelling in a general direction. The amplitude profile of any line cutting this wave would be a cosine. Three special directions are the normal to the wavefront, where the wavelength is $1/k$, the x-axis, where the wavelength is $1/u$ and the y-axis, where it is $1/v$. $k^2 = u^2 + v^2$ (figure 11.10).

Any function $f(x, y)$ can be made up of a weighted summation of such waves. In three dimensions visualizing the wave is more difficult, but a cosine-varying wave of pressure is an appropriate image. In this case the three spatial frequencies u, v, w are associated with the three axes x, y, z and again $k^2 = u^2 + v^2 + w^2$ where k is the frequency along the normal to the wavefront.

$f(x, y)$ is said to be a function in *real space*, whereas $F(u, v)$ is a function in *frequency space*. These terms simply refer to the axes of these spaces. This section will be seen to be important when we come to spatial filtering of images in section 14.4 of Chapter 14.

Amplitude and phase

A second way to write the expansion of $f(x, y)$ given above is to define an angle θ such that

$$F_R(u, v) = |F(u, v)| \cos \theta(u, v)$$

$$F_I(u, v) = |F(u, v)| \sin \theta(u, v)$$

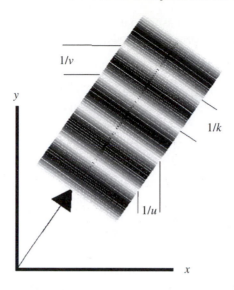

Figure 11.10. *A two-dimensional cosine wave.*

where

$$|F(u, v)| = \sqrt{F_R(u, v)^2 + F_I(u, v)^2}.$$

Then

$$f(x, y) = \iint |F(u, v)| \cos \theta (u, v) \cos 2\pi (ux + vy) + |F(u, v)| \sin \theta (u, v) \sin 2\pi (ux + vy) \, du \, dv$$

which reduces to

$$f(x, y) = \iint |F(u, v)| \cos(2\pi (ux + vy) - \theta (u, v)) \, du \, dv. \qquad (11.7)$$

$|F(u, v)|$ is called the modulus of the Fourier transform and $\theta (u, v)$ is the phase of the transform. Displaying $F(u, v)$ can pose problems because it is a complex function. The usual practice is to display $|F(u, v)|$ and ignore the phase. This can give insights into the way image information is distributed in frequency space, although in practice most of the structural information in an image would appear to be in the phase component.

From the above definitions we can write $F(u, v)$ as

$$F(u, v) = |F(u, v)| e^{-j\theta (u,v)}$$

11.7.2. The shifting property

The Fourier transform is linear. Another useful property is the following. Suppose we shift the function $f(x, y)$ by x_0 and y_0. Then,

$$\iint f(x - x_0, y - y_0) e^{-2\pi j(ux + vy)} \, dx \, dy$$

$$= e^{-2\pi j(ux_0 + vy_0)} \iint f(x - x_0, y - y_0) e^{-2\pi j((x - x_0)u + (y - y_0)v)} \, dx \, dy$$

$$= e^{-2\pi j(ux_0 + vy_0)} F(u, v).$$

The FT of a shifted image is the FT of the unshifted image multiplied by a complex exponential term. As we saw above the FT could be written in terms of the modulus and in this notation the shifted form is

$$F(u, v)\, e^{-2\pi j(ux_0+vy_0)} = |F(u, v)|\, e^{-2\pi j(ux_0+vy_0)-j\theta(u,v)}.$$

The modulus is unchanged by shifting the function $f(x, y)$ and this result explains a somewhat cavalier attitude found in books on image processing to specifying the origin of a function when computing the Fourier transform.

11.7.3.   The Fourier transform of two simple functions

The Fourier transform of a function can be determined as long as the function fulfils some conditions. It must not contain any infinite discontinuities and its integral must be finite. Computing the Fourier transform of an arbitrary function is not always possible and most applications require numerical methods. However, it is possible to work out an analytical solution for some cases. Perhaps the most interesting is the Fourier transform of the Gaussian function. In two dimensions this is

$$\mathrm{FT}\left(\frac{1}{2\pi d^2}\, e^{-(x^2+y^2)/2d^2}\right) = e^{-(2\pi d)^2(u^2+v^2)/2}$$

and this is useful as a model of the Fourier transform of a point spread function.

The delta function would appear to violate the conditions for allowing an FT to be made, but we can still derive the FT of this function by taking the limit of the Gaussian function. As d goes to zero the FT becomes $e^{-0} = 1$. The FT of the delta function (or rather the modulus of the FT) has unit amplitude for all frequencies. Conversely, the FT of the function with unit amplitude over all space is a delta function.

11.7.4.   The convolution equation

If the imaging process can be represented by a convolution integral then it can be shown (see section 13.6.3) that the Fourier transforms of the object function ($F(u, v)$), the image function ($G(u, v)$) and the point spread function ($H(u, v)$) are related through the simple relationship

$$G(u, v) = F(u, v)\, H(u, v).$$

This is a remarkable result. It is only valid if $h(x, y)$ is position independent. Similar results hold for the three-dimensional case and the one-dimensional case.

The three transforms are in general all complex. If $h(x, y)$ is symmetric about its centre, for example if it can be represented by a Gaussian function, then the imaginary part of $H(u, v)$ is zero for all values of u and v, so we can write

$$G(u, v) = F(u, v)\, H_R(u, v)$$

where H_R is real valued. To appreciate what is happening we expand $g(x, y)$ in the form

$$g(x, y) = \iint (G_R(u, v) \cos 2\pi (ux + vy) + G_I(u, v) \sin 2\pi (ux + vy))\, du\, dv$$

and expand the $G(u, v)$ terms

$$g(x, y) = \iint (H_R(u, v)\, F_R(u, v) \cos 2\pi (ux + vy) + H_R(u, v)\, F_I(u, v) \sin 2\pi (ux + vy))\, du\, dv.$$

This equation shows that $g(x, y)$ has been obtained by decomposing $f(x, y)$ into its constituent cosine and sine terms, multiplying each of these by the appropriate weighting factor ($H_R(u, v)$) and adding the new cosine and sine terms back together again. The imaging system independently modifies the amplitudes of the cosine and sine terms in $f(x, y)$ to form the image.

If we work through the appropriate mathematics in detail we can also show that the moduli of $G(u, v)$, $F(u, v)$ and $H(u, v)$ are related by

$$|G(u, v)| = |F(u, v)||H(u, v)|.$$

For this reason $|H(u, v)|$ is known as the modulation transfer function (MTF) of the imaging system.

11.7.5. Image restoration

The Fourier transform approach to describing the imaging process gives a very graphic description of this process. However, there is a more powerful reason for the interest in this description, which we will look at again in Chapter 14. In general, in the imaging process we know the function $g(x, y)$, the image, and we know the PSF, $h(x, y)$, but do not know the object function $f(x, y)$. However, by dividing both sides of the Fourier representation of the imaging equation by $H(u, v)$ we have

$$F(u, v) = \frac{G(u, v)}{H(u, v)}$$

which, in principle, gives us the Fourier transform of the object function and hence this function itself. This result can be used in image processing and this will be described in Chapter 14.

11.8. IMAGE RECONSTRUCTION FROM PROFILES

One of the major developments in medical imaging over the past two decades has been the development of techniques for constructing images representing slices through three-dimensional objects. These techniques are called tomography (tomos = slice) and are based on the idea that an object may be constructed from *projections* of the object. That this was possible was first shown by Radon in 1917, but the method was not formally developed until the end of the 1960s.

11.8.1. Back-projection: the Radon transform

The term *tomography* usually refers to the reconstruction of two-dimensional slices through three-dimensional objects. The images we are dealing with in this case are therefore two dimensional, although we can and often do go on to construct three-dimensional images by stacking slices. This means that we can approach image reconstruction by considering how we would construct a two-dimensional slice.

The object function $f(x, y)$ we are considering is a two-dimensional slice through a three-dimensional object. We can form a one-dimensional *profile* $p(y)$ by integrating the function in the x direction,

$$p(y) = \int f(x, y) \, dx$$

where the integration is from boundary to boundary. In an x-ray CT system this integration is effectively performed by the x-ray beam, but these specific details will not concern us here.

We formally integrate in an arbitrary direction θ by writing

$$p(s, \theta) = \int\int f(x, y) \, \delta(x \cos \theta + y \sin \theta - s) \, dx \, dy \tag{11.8}$$

where the δ function is non-zero only along the line given by $x \cos \theta + y \sin \theta = s$ (figure 11.11).

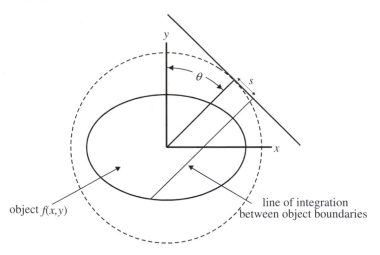

Figure 11.11. *The geometry of the Radon transform.*

For each value of θ, $p(s, \theta)$ is a projection of the object function in the direction defined by θ. The above integral is the Radon transform mapping $f(x, y)$ to $p(s, \theta)$. The function $p(s, \theta)$ is often called the *sinogram* of $f(x, y)$ (see figure 11.12(b)). Radon showed that given $p(s, \theta)$ it was possible to derive $f(x, y)$.

The key concept in reconstructing $f(x, y)$ from $p(s, \theta)$ is the idea of back-projection. To back-project a profile, say the profile collected for angle θ, we replicate the value $p(s, \theta)$ at all points along the direction

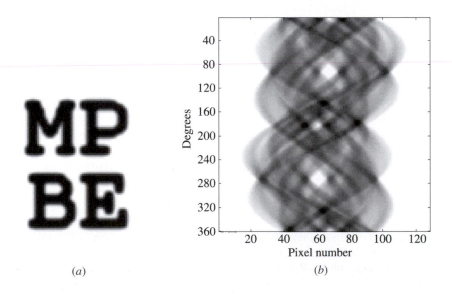

Figure 11.12. *(a) An image and (b) the Radon projection or sinogram of this image. The wavy sine like patterns explain the use of the term sinogram.*

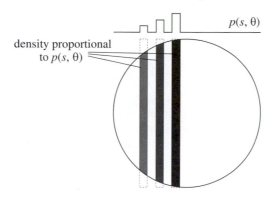

Figure 11.13. *Back-projection of a profile.*

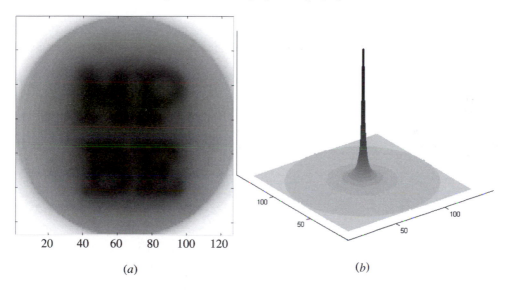

(a)　　　　　　　　　　　　　　　　　(b)

Figure 11.14. *(a) Image produced by back-projecting the sinogram given in figure 11.12(b); (b) the response to a point object in the middle of the field of view.*

normal to the profile for this angle. This is formally represented by the equation

$$b_\theta(x, y) = \int p(s, \theta)\, \delta(x \cos \theta + y \sin \theta - s)\, \mathrm{d}s.$$

Again the δ function selects out a line in space along which the appropriate value is replicated. Figure 11.13 shows how the back-projection of a profile looks in practice.

We can do this for the profile collected at each value of θ and sum all the images together. Figure 11.14(a) shows the image produced by back-projecting the sinogram given in figure 11.12(b). This is clearly very blurred. The reason why this is the case can be seen if we look at the image that would be produced by back-projecting the profiles produced from a point object at the centre of the field of view; this is shown in figure 11.14(b). Although correctly positioned at the centre of the field of view, this is clearly a smoothing

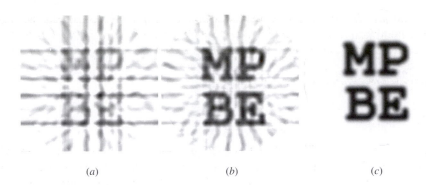

(a)   (b)   (c)

Figure 11.15. *The image reconstructed from the sinogram of figure 11.12(b). (a) The image reconstructed from the profiles collected every 30°; (b) using profiles spaced at 10°; (c) profiles spaced at 2°. The importance of taking enough profiles is apparent from these figures.*

function. A little thought should convince you that the same shape of function will be produced wherever the point object is, but centred on the position of the point object.

If we now recognize that this image is a point spread function then we can immediately write a general convolution equation relating a general object to the image we would get if we collected profiles from that object and back-projected them. The equation is

$$g(x, y) = \iint f(\xi, \eta)\, h_b(x - \xi, y - \eta)\, d\xi\, d\eta$$

where h_b is the PSF described above. We have already seen that this equation can be solved if the Fourier transforms of the function in it are known. Therefore,

$$F(u, v) = \frac{G(u, v)}{H_b(u, v)}.$$

We can compute $G(u, v)$ by back-projecting the profiles and computing the FT of the resulting image. We can compute $H_b(u, v)$ by doing the same for a point object and so we can compute $F(u, v)$ and from this the object function f. It turns out that $H_b(u, v) = \sqrt{u^2 + v^2}$ and so

$$F(u, v) = G(u, v)\sqrt{u^2 + v^2}.$$

This shows that f can be reconstructed from its profiles and gives us a way of doing it. It is not the best way. If we remember that the whole process is linear and that g is constructed from a sum of individual back-projections, then we can consider taking each individual back-projection and applying the above filter to it before adding all the back-projections together.

The Fourier transform of the profile $p(s, \theta)$ taken over s is given by

$$P(u, \theta) = \int p(s, \theta)\, e^{-2\pi j u s}\, ds$$

and it can be shown that if a new profile is formed by computing

$$q(s, \theta) = \int |u|\, P(u, \theta)\, e^{+2\pi j u s}\, du$$

(filtering $P(u, \theta)$ with $|u|$, the 'ramp' filter), then back-projecting $q(s, \theta)$ reconstructs the image $f(x, y)$ without any further filtering. This is known as 'filtered back-projection' and, in various forms, represents a key method of reconstructing tomographic and hence three-dimensional images. As profiles are collected sequentially they can be filtered and back-projected while further profiles are being collected, leading to a very efficient approach to image reconstruction.

11.9.   SAMPLING THEORY

So far we have concentrated on images and objects as continuous functions. However in practice images may need to be represented as discrete arrays of numbers. This is especially the case with digital imaging and tomography. The image is represented as a discrete array by sampling the image intensity at points on a grid and using the derived array of numbers as a description of the image. Since the regions between the points on the grid are not sampled we conclude that some information may be lost in this process. The aim of sampling theory is to determine how close together the grid points have to be for no information, or for insignificant amounts of information, to be lost. Fourier transform theory comes to the rescue here.

11.9.1.   Sampling on a grid

We assume a continuous image function $g(x, y)$ which is sampled by multiplying by a grid of regularly spaced δ functions

$$g_s(x, y) = \sum_{i,j=-\infty}^{\infty} g(x, y)\,\delta(x - i\,\Delta x, y - j\,\Delta y)$$

where Δx and Δy are the spacing between grid points. The Fourier transform of an array of grid points is itself an array of grid points. In Fourier space this product becomes a convolution, so

$$G_s(u, v) = G(u, v) \otimes \sum_{i,j=-\infty}^{\infty} \delta(u - i\,\Delta u, v - j\,\Delta v)$$

where

$$\Delta u = \frac{1}{\Delta x} \quad \text{and} \quad \Delta v = \frac{1}{\Delta y}.$$

The effect of this convolution is to produce an array of replications of $G(u, v)$ throughout frequency space. These will overlap unless the maximum frequencies (u, v) for which $G(u, v)$ has a significant amplitude are less than $\Delta u/2$ and $\Delta v/2$. These limiting frequencies are the Nyquist frequencies in the u and v directions. If, as often happens, $\Delta x = \Delta y$ then we can speak of a single Nyquist frequency.

The effects of sampling can be illustrated with a simple one-dimensional example. Consider the one-dimensional case

$$g(x) = \int G(u) \cos(2\pi u - \phi(u))\,\mathrm{d}u$$

The value of $G(u) \to 0$ as $u \to \infty$.

Consider the highest value of u for which $G(u)$ has significant amplitude (e.g. is greater than noise). The cosine wave for this u is shown in figure 11.16(a).

This cosine is now sampled at various equally spaced points. In figure 11.16(b) the wave is sampled with sufficient frequency for the wave to be reconstructed uniquely from the points. In figure 11.16(c) the wave is just sampled sufficiently for it to be reconstructed. In figure 11.16(d) however the sampling is so far

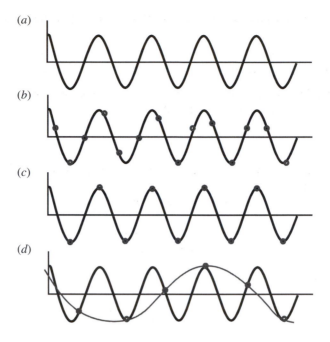

Figure 11.16. *(a) The cosine wave; (b) oversampled; (c) just adequately sampled; (d) undersampled.*

apart that an alternative interpretation of the sample values is possible. Sampling is inadequate. The samples appear to describe a cosine wave of lower frequency.

Suppose a wave of this lower frequency is really present as well. Then the sampling process adds the contribution of the high-frequency wave to the low-frequency wave and thus alters the amplitude of the low-frequency wave. This is known as *aliasing*.

If $u_{\lim}$ is the limiting (Nyquist) frequency then

$$u_{\lim} = \frac{1}{2\Delta x}.$$

The same argument applies for the y direction,

$$v_{\lim} = \frac{1}{2\Delta y}.$$

Usually $\Delta x = \Delta y$.

11.9.2. Interpolating the image

Suppose we have a sampled image. We now represent the image on a finer grid with the grid values between the known data samples set to zero. If we now compute the FT of this resampled image we will end up with a Fourier transform of replications as outlined above. By setting all of the frequency space to zero except for the central replication we have restored the Fourier transform to that of the original unsampled image. If we take the Fourier transform of this we end up with the original unsampled function. This seems miraculous. However, it does depend on the transform $G(u, v)$ going to zero inside the space bounded by the Nyquist frequency. If this is not the case then the replicated transforms will overlap. Frequency components close to the Nyquist frequency will be corrupted by components beyond the Nyquist frequency, i.e. aliasing will have occurred.

11.9.3. Calculating the sampling distance

As an example of calculating the appropriate sample frequency, let us assume we have an imaging system with a Gaussian PSF of parameter d. The FT of this PSF is

$$e^{-(2\pi d)^2(u^2+v^2)/2}.$$

Assume this is effectively zero when it reaches a value of 0.001. Then in the u direction ($v = 0$) the limiting value of u for which this holds is

$$\log_e(1000) = \tfrac{1}{2}(2\pi d)^2 u_{\text{Nyq}}^2$$

i.e.

$$u_{\text{Nyq}} = \frac{1}{d}\sqrt{\frac{\log_e(1000)}{2\pi^2}}$$

and so the sampling distance is

$$\Delta x = \frac{1}{2u_{\text{Nyq}}} = d\sqrt{\frac{\pi^2}{2\log_e(1000)}}.$$

The FWHM was given by $2.36d$ so that

$$\Delta x = \frac{\text{FWHM}}{2.8}.$$

Now since $G(u, v) \le H(u, v)$, it follows that the sampling distance selected in the above manner is probably acceptable. A rule of thumb therefore is that to avoid aliasing the sampling distance should be less than one-third of the FWHM of the imaging system.

11.10. PROBLEMS

11.10.1. Short questions

a Do we perceive 2D or 3D images?
b What is meant by the 'object distribution' in imaging?
c What is meant by an imaging system being described as 'linear'?
d What is the ideal point spread function for a high-quality image?
e What is the point sensitivity function?
f Why is a PSF a good method of describing image quality?
g What term describes the fact that straight lines may not be preserved in an imaging process?
h What is the FWHM?
i If two points in an image are 0.5 FWHM apart will they be clearly separated to the eye?
j What is meant by stationary noise in an image?
k Give an example of non-stationary noise in medical imaging.
l An image is a blurred and noisy representation of an object. Conceptually, is the noise added to the image before or after blurring?
m What is convolved with what in the convolution equation as applied to imaging?

n   Say in words what the advantage is of the Fourier transform in dealing with the imaging equation.
o   A Fourier transform can be expressed in real and imaginary terms. What is the alternative representation?
p   What is tomography?
q   What is an image profile?
r   What is the Nyquist frequency?
s   What is an appropriate Nyquist frequency to avoid aliasing if the FWHM is 6 mm?
t   What is a sinogram?
u   What is aliasing in an image?

11.10.2.   Longer questions (answers are given to some of the questions)

Question 11.10.2.1

Describe what is meant by an imaging system being linear and indicate why this is a desired property of an imaging system.

Describe carefully the difference between a nonlinear imaging system and a position-dependent imaging system.

Distinguish between the nonlinearity of a nonlinear system and spatial nonlinearity.

If the image produced by a linear imaging system is recorded on film the system becomes nonlinear. Show with a simple example why this is formally so. In the case of film the nonlinearity can in most cases be removed by post-processing of the image. Give an example of when this is not the case.

Answer

An imaging system is linear when the sum of the images separately produced from two different objects is the same as the image produced by imaging the two objects together. In the limit any object can be considered to be made up of a set of independent point objects. If the system is linear then the response of the imaging system for all objects can be obtained from knowledge of the response to a set of point objects. If the system is nonlinear, i.e. the images of point objects interfere with each other, then the only way to discover what the image of an object will look like is to image that object.

In a position-dependent system the image of a point object may vary in shape from point to point. However, the imaging system can still be linear in the sense of the above paragraph. Conversely, it is possible for the image of a point object to be the same shape wherever the point object is placed, but for the images of objects to interfere with each other. In this case the system is position independent but nonlinear.

If the object and the image are the same size dimensionally we might expect that the position of the image of a point object and the position of the point object would be the same. However, this need not be the case, as the image position could be shifted relative to the object position. This means that the image of a straight line need not be straight. The system is position dependent, but can still be linear. However, this effect is known as spatial nonlinearity.

The response of a film is nonlinear. This means that the darkness of the film is not linearly related to the image intensity falling on it. Suppose we have two objects which for simplicity we can make uniform. Then the brightness of the first object is A and that of the second object is B. If we image then independently the first image would have a brightness $f(A)$ and the second image a brightness $f(B)$. If we imaged then together we would have a brightness $f(A + B)$ and, in general, for film

$$f(A + B) < f(A) + f(B).$$

However, if we know f then we can convert the measured brightness $f(A+B)$ to $A+B$ by using the inverse function f^{-1}. We can do the same for $f(A)$ and $f(B)$ so we can make the system linear. This would not be possible if $f(A+B)$ saturated the film.

Question 11.10.2.2

An object consists of a two-dimensional cosine wave (see figure 11.10) of wavelength L and of amplitude A combined with a uniform background of amplitude B such that $B > A$. The object is imaged using an imaging system with a PSF of $1/(2\pi d^2) e^{-(x^2+y^2)/2d^2}$. The result is an image consisting of a 2D cosine wave of amplitude C on a background B. What is the amplitude C of this wave and what is its wavelength?

The imaging device has a FWHM of 5 mm. What is the wavelength of the object cosine wave for which the amplitude is reduced in the image to one-tenth of its value in the object?

Answer

The FT of the Gaussian PSF is given by $H(u, v) = e^{-(2\pi d)^2(u^2+v^2)/2}$. Assume that the cosine wave lies along the x-axis. Then $v = 0$ and $u = 1/L$. The value of $H(u, v)$ at this frequency is $e^{-(2\pi d)^2/(2L^2)}$ and therefore the amplitude of the cosine wave in the image is $C = Ae^{-(2\pi d)^2/(2L^2)}$. The wavelength is unchanged at $u = 1/L$. If the FWHM = 5 mm the value of d is $5/2.36 = 2.119$ mm. Then $0.1 = e^{-88.632/L^2}$ and

$$L = \sqrt{\frac{88.632}{\log_e(10)}} = 6.2 \text{ mm}.$$

Question 11.10.2.3

Back-projection of the profiles produced from a point object produces an image similar to figure 11.14(*b*). In this case, although the image is centred at the position of the object it is blurred. If the profiles are filtered with a 'ramp' filter before back-projection the image of a point object is itself a point object with no blurring. What feature of the filtered profiles makes this possible? Sketch what you think the filtered profiles might look like.

Answer

The filtered profiles have a central peak with negative side lobes. When back-projected these lobes exactly cancel out the positive tails of the original back-projected image to produce a blur-free result.

Question 11.10.2.4

A CT imaging system has a FWHM of 2 mm. How small do the image pixels have to be to avoid aliasing problems? For a field of view of radius 25 cm how many pixels (N_s) are required along each side of the image?

Answer

The pixels have to be of the order of one third of the FWHM. This means they must be 0.6 mm in size or less. The number of pixels along each side of the image is therefore $500/0.6 = 830$ pixels. In practice 1024 would be used.

Question 11.10.2.5

For CT scanning the angular sampling needs to be taken into account as well as the spatial sampling. Spatial sampling considerations determine the number of samples along a profile. A CT imaging system has a FWHM of 2 mm and a radius of field of view of 25 cm. Assuming that the image sampling has to be adequate out to the edge of the field of view, and that a translate–rotate configuration is being used, estimate the angular sampling required and from this how many profiles (N_p) will be needed.

The total number of data points collected will be of the order of $N_p N_s$. This is generally larger than $N_s N_s$, i.e. we appear to need to oversample the image. Why is this so?

Answer

For adequate angular sampling the sampling angle $\Delta\theta$ must be such that $R\Delta\theta$ is less than or equal to the pixel size at the periphery of the field of view. The radius of the field of view is R. Then if the pixel size is p, $\Delta\theta = p/R$ rad. The number of profiles is therefore $2\pi/\Delta\theta = 2\pi R/p$, although for an x-ray system only half of these are needed since opposing profiles give the same data. For $p = 0.6$ mm and $R = 250$ mm this gives 1309 projections. The total number of data points is therefore $1309 \times 830 = 1\,086\,467$. The number of pixels in the image, recognizing that the image is bounded by a circle of radius R, is $830 \times 830 \times \pi/4 = 541\,061$. There are more data values than pixel values so the image is overdetermined. This arises because if the angular sampling is such that the peripheral parts of the image are adequately sampled, the inner parts are oversampled. The number of profiles could be reduced by almost one-half and the image still be overdetermined, but at the risk of undersampling the periphery.

Question 11.10.2.6

Isotropic resolution means that the PSF has circular (in 2D) or spherical (in 3D) symmetry. 3D images can be created by stacking a set of 2D tomographic images on top of each other. To avoid undersampling the image planes need to be sufficiently close together. Would you expect the resolution associated with such a 3D image to be isotropic and if not, why not?

Answer

The resolution is unlikely to be isotropic in 3D. Even if the angular sampling is sufficient for the PSF in the slice plane to be isotropic, this PSF is the result of an image reconstruction process and in practice is unlikely to have the same shape as the PSF in the axial direction, which is derived from a simple sampling process. For example, in practice the PSF in the slice plane may well have small negative side lobes, whereas this should not be the case for the PSF in the axial direction.

Question 11.10.2.7

Outline the history and discuss the application to medical imaging of the Radon transform.

Answers to short questions

a   The image on the retina is 2D, but the brain is able to extract depth information and thus constructs a 3D image.
b   The 'object distribution' is the distribution within the body of the tissue parameter to be imaged.
c   An imaging system where the image of two or more objects together is the same as the sum of the images produced from each of the objects in turn.

d   A delta function is the ideal point spread function. The image is a perfect representation of the object.

e   The point sensitivity function represents the total signal collected from a point object. It is the integral of the PSF.

f   The PSF describes the image that results from a point object. The more closely the PSF describes a point the better is the imaging system. A very diffuse PSF will result in poor image quality.

g   Spatial linearity.

h   FWHM is the full width at half the maximum amplitude of the point response function.

i   No.

j   Stationary noise is noise that has the same mean, variance and covariance everywhere in the image.

k   The noise in a radioisotope image has a variance which is proportional to the count density and thus varies from point to point. It is non-stationary.

l   The noise is added after blurring.

m   The PSF is convolved with the object distribution to give the image

n   The Fourier transform allows the convolution integral to be reduced to the multiplication of two functions—the Fourier transforms of the PSF and the object distribution.

o   The alternative to the real and imaginary description is that in amplitude and phase.

p   Tomography is the imaging of 2D slices from a 3D object.

q   A profile in a 2D imager is a line integral in the x direction as a function of y.

r   The Nyquist frequency is the minimum spatial frequency at which we can sample image data in order to preserve the highest spatial frequencies present in the object distribution.

s   An appropriate Nyquist frequency is 0.25 mm^{-1}.

t   A sinogram is the function produced by performing a line integral through the object distribution over many directions.

u   Aliasing is the cause of image artefacts as a result of inadequate data sampling. High spatial frequency features can appear as low spatial frequency, objects.

BIBLIOGRAPHY

Macovski A 1983 *Medical Imaging Systems* (Englewood Cliffs, NJ: Prentice-Hall)

Radon J 1917 Uber die Bestimmung von Funktionen durch ihre Integralwerte längs gewisser Mannigfaltigkeiten *Ber. Verh. Saechs. Wiss. Leipzig Math. Phys. Kl.* **69** 262–77

Webb S (ed) 1988 *The Physics of Medical Imaging* (Bristol: Hilger)

CHAPTER 12

IMAGE PRODUCTION

12.1. INTRODUCTION AND OBJECTIVES

Medical imaging techniques have been used since the discovery of x-rays by Röentgen in 1895. Techniques other than simple planar radiographs have become available during the last 50 years. These include, in historical order, radioisotope (NM) images, ultrasound (US) images, computed tomography (CT) images and magnetic resonance (MR) images. In particular, the past two decades have seen fairly dramatic developments in all these technologies. Although x-ray planar images and planar isotope images still form the backbone of radiology and nuclear medicine the trend is steadily towards the production of tomographic images, images of slices through a three-dimensional object, or even full three-dimensional images. Ultrasound images have always been tomographic images, as have MR images. The invention of computed tomography allowed tomographic x-ray images to be generated, as did the application of similar techniques to radioisotope imaging, although in fact the generation of radioisotope tomographic images predated CT by a few years.

For a century a medical image has meant a film of some sort. The next decade will see the progressive replacement of films with digital images on a monitor. This is already well under way for the modalities listed above, and planar x-ray is in hot pursuit. All the imaging methods listed above form digital images. This allows the images to be manipulated in a variety of ways, from simple display manipulation to the extraction of numerical information to, possibly, automatic interpretation. The introduction of digital technology into image production has also changed the way in which some of the older imaging devices work. In this chapter we shall look at the way images are formed by the different imaging devices, concentrating on current approaches rather than historical ones and on principles rather than a detailed technical discussion.

At the end of the chapter you should be able to:

- Appreciate that there are very many ways in which to image a part of the body.
- Explain the principles of the gamma camera, SPET and PET.
- Understand how Doppler and echo information can be combined in an ultrasound image.
- Describe what a pulse sequence consists of in magnetic resonance imaging.
- Appreciate how the pattern of currents imposed in electrical impedance tomography (EIT) dictates how images might be reconstructed.
- Understand the distinction between anatomical and functional imaging.

This chapter is not heavily mathematical. In it we attempt to present the principles of the various imaging techniques. You do not have to read this chapter as a whole but can jump to the particular modality in which you are interested. However, you should have read Chapters 5 and 6 before looking at section 12.2 on radionuclide imaging. Similarly, you need to have read Chapter 7 as an introduction to section 12.3 on ultrasonic imaging. Chapter 8 could be a useful introduction to sections 12.4 and 12.6 on MRI and EIT, respectively.

Medical imaging as a subject has grown enormously over the past few decades and as a result this chapter is long. However, we cannot hope to be comprehensive. Our intention is to present the basic ideas involved in the various imaging methods. What they all have in common is the objective of showing the spatial distribution of a particular property of tissue.

12.2. RADIONUCLIDE IMAGING

The principles behind the use of radioisotopes in diagnosis are outlined in Chapter 6. Imaging the distribution of a radiopharmaceutical in a patient requires an imaging device. At one time the spatial distribution of a radiopharmaceutical in the human body was determined by moving a γ-ray detector in a rectilinear raster or scan over the region of interest. The activity, i.e. the number of γ-rays being detected per second, in a small, approximately cylindrical, volume below the detector was measured at each point and these measurements were used to build up an image of the activity distribution in the patient. Mechanical scanning instruments (rectilinear scanners) have now been completely replaced by non-scanning imaging systems called gamma cameras. However, it is still fairly common to call images produced by gamma cameras, radionuclide images, 'scans' although no scanning motion is necessary. In more recent times gamma cameras have been used to produce three-dimensional images. For all images produced using radionuclides which emit single photons the gamma camera needs to scan completely around the patient. Also images of the whole-body distribution of a radiopharmaceutical can be obtained by scanning a gamma camera detector along the body of the patient, so in an informal way the use of the term 'scan' still applies to some images.

12.2.1. The gamma camera

The principal radionuclide imaging system currently available is the gamma camera. Radionuclide imaging using a gamma camera is also referred to as scintigraphy. The principle of the gamma camera was described by Anger about 40 years ago. Anger's camera formed images using analogue circuitry. Modern gamma cameras make extensive use of digital circuitry, but the underlying principles are still close to those of the original camera. However, the intervening years have seen steady improvements in the resolution of gamma cameras and improvements in linearity and uniformity, concepts described below. The method of image formation depends on whether the radioisotope being imaged emits a single γ-ray, such as ^{99m}Tc, or two opposed photons via the emission of positrons, such as those from ^{11}C. Most images are formed from radioisotopes of the former type so we shall concentrate on these, but the case of positron emitters will be described in section 12.2.6 on PET.

Single-photon imaging

The gamma camera consists of two major components, the collimator and the photon localization system (figure 12.1(a)). The collimator consists of a large block of absorbing material, usually lead, through which a set of parallel holes has been driven. The thickness of the collimator is typically 50 mm or so, with holes of a few mm diameter. For a 400 mm field of view there are therefore many thousands of holes. γ-rays emitted from the patient can only pass through a collimator hole if they are travelling parallel or almost parallel to the axes of the holes, otherwise they strike the side of the hole and are absorbed. The radiation passing through each hole represents a measure, modified by attenuation within the patient, of the activity directly below the hole. In principle, a small scintillation detector could be placed at the end of each collimator hole away from the patient. A map of all the detector outputs would constitute an image of the activity distribution in the patient. However, this is not currently practical, although solid-state detectors offer the possibility of doing this in the future. Instead a large single crystal of scintillation material (thallium-doped NaI) is placed behind the collimator (see section 5.6.3 for a description of scintillation counters). This crystal may be 10 mm thick and 400 mm or more in linear dimensions. γ-rays passing through the collimator cause scintillations in the crystal. The key feature of the gamma camera is how the positions of these scintillations are determined without the need for a vast array of detectors.

Photon localization

In order to detect and localize a scintillation in the crystal the crystal is bonded onto an array of photomultiplier (PM) tubes. Although the number of tubes used has increased dramatically from the seven used in the original gamma camera the number of tubes is far less than the number of holes in the collimator. Typically between 60 and 90 will be used in a modern camera. When a scintillation occurs in the crystal the light emitted from the scintillation spreads out and is detected by several PM tubes. The scintillation does not have to occur directly below a tube for the tube to register light from it.

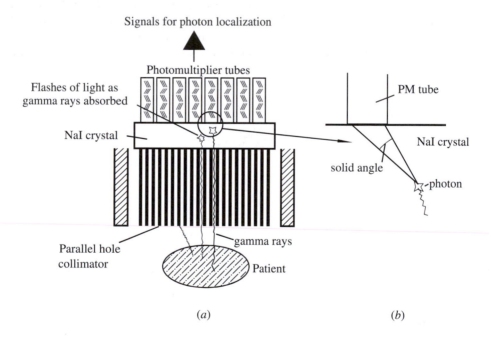

(a) $\qquad\qquad\qquad\qquad\qquad\qquad\qquad\qquad$ (b)

Figure 12.1. *(a) Photomultipliers bonded onto the crystal; (b) enlarged view. Light from a scintillation reaching a photomultiplier tube.*

Each tube will have a spatial sensitivity associated with it, a measure of the amount of light reaching it from a scintillation occurring at a particular distance from the centre of the tube. Assume the centre of the tube is placed at a point (x_i, y_i) of a coordinate system. A scintillation occurs in the crystal at a point (x_s, y_s). The amount of light reaching the photomultiplier will depend on the distance between the scintillation and the tube and is largely a function of the solid angle subtended by the sensitive face of the PM tube when viewed from the scintillation (figure 12.1(*b*)). It will also depend on how the front face of the crystal reflects light and on the relative refractive indices of the scintillator material and any intermediate materials between the crystal and the PM tube. The magnitude of the output signal can be represented by a function $L(x_s - x_i, y_s - y_i)$. The shape of L will depend on all the above factors but, in general, will decrease in amplitude as $x_s - x_i$ and $y_s - y_i$ increase, and ideally will only be a function of the radial distance from the centre of the PM tube, especially for tubes placed well away from the edge of the crystal. We would expect it to be a maximum at the centre of the tube. Near the edge of the crystal this symmetry will be lost.

If we get a large signal from the tube at point (x_i, y_i) when a scintillation occurs at (x_s, y_s) we can take this as evidence that the scintillation has occurred near this position, i.e. (x_s, y_s) is close to (x_i, y_i). If the signal is weak we can take this as evidence that the signal has occurred some distance away. An estimate of the position of the scintillation is given by computing the centre of gravity of the PM tube outputs. If N is the number of tubes then

$$x_s = \frac{\sum_1^N L_i x_i}{\sum_1^N L_i} \qquad y_s = \frac{\sum_1^N L_i y_i}{\sum_1^N L_i}.$$

As the density of PM tubes increases, and assuming that each tube has the same L function, it is not too difficult to show that (x_s, y_s) becomes an unbiased estimate of the position of the scintillations. For a realistic array of PM tubes this is not the case and the value of (x_s, y_s) deviates from the true value in a position-dependent way. If L is radially symmetric then these estimates will be unbiased for scintillations directly under the centre of a tube and for other positions of symmetry within the array (e.g. at points equidistant from three tubes in an hexagonal array) provided edge effects can be ignored. Elsewhere the position estimates will be biased. For typical L, estimates will be shifted towards the positions of the nearest PM tubes (figure 12.2). The degree to which this occurs will depend on the shape of L.

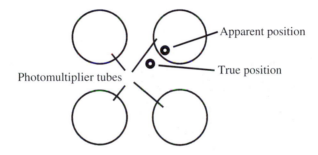

Figure 12.2. *The movement of the apparent position of a scintillation towards the centre of a PM tube.*

A consequence of this sort of error is that the image of a straight-line object will not appear straight. This represents a failure of spatial linearity.

Early gamma cameras computed the summations given above using resistor or capacitor arrays. Significant effort was put into the design of the coupling between the crystal and the PM tubes (the light pipe) to shape L so that the systematic errors were minimized. The values of the resistor array were tuned at the edges to minimize edge effects. Even when this component of the system was optimized it was important to ensure that the magnitude of L did not vary as a function of time, and great effort was put into ensuring the stability of the gain of the PM tubes and tuning the system to optimum performance.

Intrinsic resolution

As noted in Chapter 5, even if a scintillation detector absorbs γ-rays of a constant energy the apparent energy varies from event to event because of the variation in the number of photoelectrons created at the photocathode of the PM tube. The detector has a limited energy resolution. The same situation occurs in the gamma camera. Even if photons of a constant energy are absorbed at a fixed point in the crystal, the magnitude of the signals generated by the PM tubes will vary from scintillation to scintillation. This means that the calculated value of the position of the scintillation will vary from scintillation to scintillation. This uncertainty in the position of the scintillation represents a loss of resolution in the image. The accuracy with which the position of a scintillation can be determined is known as the 'intrinsic' resolution of the gamma camera. It should be below 4 mm (FWHM; see section 11.4.1) for a modern gamma camera imaging 140 keV γ-rays but is dependent on the energy of the γ-rays being detected. It generally improves as the energy of the γ-ray increases, because the energy resolution increases, but this potential gain is offset by other effects.

If a PM tube is far away from the scintillation then its output will be small. Intrinsic resolution may be improved by dropping the contribution of this tube from the position summations since it will only add noise. This may make spatial linearity worse, but this can now be corrected using digital techniques.

Early cameras attempted to optimize performance and minimize the systematic position errors by the design of the crystal and PM tube assembly and the analogue summation circuitry. The introduction of digital circuitry has allowed position errors to be significantly reduced in magnitude. To deal with this problem the apparent position (x_a, y_a) of a scintillation as measured by the gamma camera is compared to the known position (x_s, y_s) of the scintillation, as determined from the known position of a beam of γ-rays directed onto the crystal. The apparent position is determined to sufficient accuracy by averaging over many events, so that the effects of intrinsic resolution become unimportant. For each apparent position the true position is stored in a computer memory. When a scintillation is detected its position coordinates are used to address the appropriate memory location and the true position extracted. Provided the systematic errors are not too great, so that there is a unique relationship between the apparent and true positions, systematic positions errors can be largely eliminated.

12.2.2.   Energy discrimination

The principal interaction between γ-rays of the energies used in radionuclide imaging and human tissue is the Compton interaction (see section 5.2.2 and figure 5.5(b)). The γ-ray flux emerging from the patient is rich in γ-rays of lower energy than the primary emission. The point of origin of an unscattered γ-ray can be localized to a limited volume of tissue by the collimator. This is no longer true for a scattered ray and so such rays, if detected by the gamma camera, will degrade the quality of the image. The summation term $\sum_{i=1}^{N} L_i$ is a measure of the energy of the detected γ-ray (within the limits of the energy resolution of the camera). This term can be used to reject γ-rays which appear to have energy lower than the unscattered energy, using pulse height analysis, and so scattered radiation can in principle be rejected. Regrettably, for the most common isotope used in imaging, ^{99m}Tc, if a reasonable percentage of unscattered radiation is to be accepted then the width of window required will also accept a proportion of the scattered radiation. For example, if an acceptance window of $\pm 10\%$ of the photopeak energy is used then the lower limit will represent 126 keV, and this corresponds to a scattering angle of 53°. This situation could be improved if the energy resolution could be improved, but this seems unlikely with existing technology. Solid-state detectors have much better energy resolution, but solid-state gamma cameras seem a long way off. Techniques have been developed for removing the effects of scatter, but these do increase the noise levels in the images and have not yet found widespread acceptance.

The summation term $\sum_{1}^{N} L_i$ is also a function of position. In practice, the apparent energy of the photopeak will appear lower for scintillations between PM tubes than for scintillations under a tube. An energy acceptance window optimized for scintillations below a tube would be rather too high for scintillations between tubes. If the window is widened to allow for the detection of more photopeak events, then increased scattered radiation will be accepted.

This problem is also dealt with by digital techniques. The energy spectrum for a beam of monoenergetic γ-rays can be measured at each point on the crystal. The measured value of the photopeak, and the optimum value of the upper and lower window levels can be determined at each point and stored in computer memory. When an event is detected the position of the scintillation is used to identify the appropriate location in memory and return the window levels. These are then used to reject or accept the γ-ray. Such techniques require high-speed digital computation. Modern gamma cameras have incorporated digital techniques right down to the level of computing the position coordinates using directly digitized PM signals.

Dead time

In spite of the use of arrays of PM tubes, the gamma camera is still a serial device and can only process one γ-ray at a time. If two γ-rays produce a scintillation within the resolving time of the crystal then this will be seen as a single flash of light. The position computation would produce a result positioned between the scintillations. Fortunately, the apparent energy of the event is up to twice the energy of a single γ-ray so most times it gets rejected unless the two γ-rays are scattered events whose sum energy falls within the photopeak. A γ-ray which undergoes a Compton interaction in the crystal and is then absorbed some distance away can also appear as a single event. For this to happen it would have to be scattered into the plane of the crystal. This sequence of interactions would also be detected as a single event. Fortunately at the energy of ^{99m}Tc most of the interactions within the crystal are photopeak events so this problem is not critical.

The duration of a scintillation within the crystal is of the order of a microsecond and scintillations spaced further apart in time than this can be distinguished in principle. Once detected the scintillation data has to be processed and this can take up to several microseconds. While this is happening the gamma camera should ideally ignore further events. This period is known as the 'dead time'. Suppose the processing of a scintillation takes τ microseconds, then during this time no further events can be detected. If the rate of arrival of γ-rays is low then the probability of a γ-ray arriving during the dead time is low. As the count rate increases the probability of a γ-ray arriving during this time and being lost increases. The relationship between true count rate c_t and observed count rate c_o is

$$c_o = c_t e^{-c_t \tau}$$

where τ is the dead time. As the count rate increases a greater fractional number of counts is lost and the plot of c_o against c_t demonstrates a maximum value (see figure 12.3). Beyond this the count rate falls even though the true number of events is increasing. A gamma camera should never be operated in this count-rate range.

Even scattered events which are ultimately going to be rejected have to be processed so that the fraction of counts lost actually depends on the total γ-ray flux reaching the crystal rather than the γ-ray flux being accepted within the photopeak. Although the formula given above would appear to allow estimation of the fraction of γ-ray events lost, an estimate based on the observed photopeak count will generally be an underestimate. It is important to appreciate that, for a gamma camera with a single processing channel, the fractional loss of counts is the same across the whole of the gamma camera, irrespective of the local count density. Low count rate regions lose an identical fraction of counts as high count rate regions. The fractional loss is determined by the total γ-ray flux integrated over the whole of the camera detector.

12.2.3. Collimation

The crystal and PM tube assembly detects and identifies the position of a γ-ray in the crystal. The distribution of γ-rays must represent an image of the distribution of radioactivity within the object being imaged and this image needs to be formed. In an optical camera the image is formed on the detector (the film) using a lens, but no γ-ray lens exists so another method must be used. Optical images can be formed without a lens using a

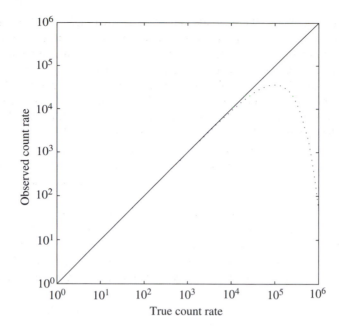

Figure 12.3. *The relationship (dotted curve) between the true count rate (c_t) and the observed count rate (c_o). τ is 10 μs. The full line shows the result for zero dead time.*

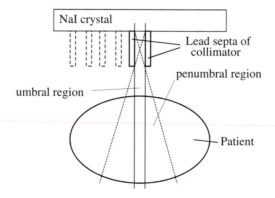

Figure 12.4. *The patient volume viewed by a single collimator hole. From the umbral region the whole of the back hole of the collimator can be seen. From the penumbral region only a fraction of this hole can be seen.*

small hole (the 'pinhole camera') and this was the method used by the earliest gamma cameras. However, this was not a particularly sensitive method and most images are formed using a parallel hole collimator, although as the pinhole collimator can produce magnified images it is still in use for imaging small organs.

Consider a single hole in the collimator (figure 12.4). Viewed from the detector, through this hole only a small piece of the radioactive object can be seen, so the portion of the detector directly behind the hole is only sensitive to γ-rays emitted by this limited region of the object. γ-rays from other regions of the object, if they enter the hole, will strike the wall of the hole and, provided this wall is thick enough, will be absorbed. The portion of the detector behind each hole is only sensitive to the portion of the object in front of that hole. The pattern of γ-rays reaching the crystal is therefore a map or image of the distribution of radioactivity in the object. Clearly, the extent of the object seen by each hole will depend on the diameter of the hole and also on its length. The resolution of the image projected onto the detector will depend on these dimensions.

As the diameter of the hole is made smaller or the hole is made longer the fraction of γ-rays which are able to get through the hole will decrease. In the limit only γ-rays travelling in a direction parallel to the axis of the hole will reach the detector. These γ-rays will have been emitted from points along a line through the object, so the signal detected at each point will represent a planar projection of the activity distribution in the object being imaged. Each point in the object would contribute γ-rays to one point on the detector. A practical collimator produces an approximate planar projection, modified by attenuation of γ-rays within the object, of the object onto the detector.

The parallel hole collimator works by excluding γ-rays moving in inappropriate directions. Unfortunately in the limiting case described above all the γ-rays will have been rejected. If the holes are of finite size then γ-rays not travelling exactly along the axis of the collimator will be accepted and this means that γ-rays from points within the object can pass through more than one hole (see figure 12.5). This means that it is not possible to exactly identify the line along which a γ-ray was emitted. γ-rays emitted from closely adjacent points in the object cannot be distinguished. In other words, as the hole dimensions change to accept more γ-rays the resolution gets worse. However, accepting more γ-rays represents an increase in sensitivity, so there is clearly a trade-off between sensitivity and resolution.

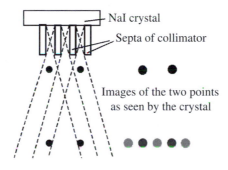

Figure 12.5. *The decreasing resolution of the collimator with distance. The two objects appear as two objects when close to the collimator but as blurred objects when further away.*

Considering the imaging due to the collimator alone, this is clearly not a position-independent process in the sense of Chapter 11 (section 11.4). The image falling on the detector will be made up of small discrete hole-shaped patches. However, the boundaries between these patches are usually blurred by the intrinsic resolution of the detector and the image usually appears as continuous rather than discrete.

There are lead walls between the holes called septa. These need to be thick enough to stop γ-rays entering one hole at an angle from passing through to an adjacent hole. For the γ-rays of ^{99m}Tc with energy 140 keV this is easily achieved with lead. In fact, it is difficult to make septa thin enough for penetration to be a problem. With higher-energy γ-rays the septa have to be increased in thickness and the collimator becomes less efficient because some detector area is shielded by the septa.

Sensitivity

An estimate of the sensitivity of a collimator can be computed from simple geometry. A useful measure is the sensitivity of the collimator per unit area of collimator to a plane source of activity placed in front of it. It is easily shown from geometric considerations that the fraction (f) of γ-rays emitted per unit area of source which get through the collimator is given by

$$f = k_s \frac{a^2}{t^2}$$

where a is the hole diameter, t is the collimator thickness and k_s is a constant that depends on the spacing between the holes, the shape of the holes (square or circular) and the way they are packed together (square or hexagonal). The value of k_s is of the order of 0.05. If there is significant penetration of the septa the above result will be affected, i.e. k_s is a function of γ-ray energy.

Suppose the plane source is moving away from the hole. Provided the source extends sufficiently so that the edge of the source can be ignored, the fractional acceptance remains constant. In other words, the count rate from the source is the same irrespective of how far away it is from the hole. This is because the effect of reduced count rate due to the inverse square law as the source is moved away is exactly compensated for by the increase in source area seen by the detector. Sensitivity increases with hole diameter and decreases with hole length.

Collimator resolution

We can use this result to estimate the resolution or FWHM of the collimator. The actual PSF (see section 11.3.1) is a position dependent function with a shape dependent on the distance from the collimator face. If we approximate it by a Gaussian function it can be shown that the FWHM at a distance z from the collimator face is given by

$$\text{FWHM}_c = k_r a \frac{(t + z)}{t}$$

where k_r is a constant of value of the order of unity. This resolution is proportional to the diameter of the hole, but also depends on the distance z in front of the collimator. It increases with z. For the highest resolution with a given collimator it is important that the patient is imaged as close to the collimator as possible.

Comparing the equations for sensitivity and FWHM_c it can be seen that improving resolution can realistically only be achieved by sacrificing sensitivity and vice versa. Usually a range of collimators are provided to enable this trade-off to be made in different ways for different imaging requirements.

Total resolution

The overall resolution is given by combining the collimator resolution and the intrinsic resolution of the detector system. The overall FWHM_t is given to a good approximation by

$$\text{FWHM}_t = \sqrt{\text{FWHM}_i^2 + \text{FWHM}_c^2}$$

where FWHM_i is the intrinsic resolution of the collimator. Clearly, there is little benefit in reducing the collimator resolution much below that of the detector, as sensitivity is lost without any marked improvement in overall resolution.

12.2.4.   Image display

All modern gamma cameras collect the image in digital form. The x and y signals produced by the detector when a γ-ray is absorbed are digitized and the two binary numbers thus produced used to identify a memory location. The contents of this location are incremented by one. At the end of the data acquisition the contents of this location represent the number of γ-rays detected in a small area of the detector. Typically for 'static' images the x and y signals are each divided into 256 discrete values and so the image is effectively divided into a grid of 256×256 squares or pixels. Use of fewer pixels than this may degrade image quality. For dynamic studies there may be fewer pixels per image. To display the image the numerical value stored in each memory location is converted to a brightness value on a monitor at the appropriate location. This monitor may be photographed to produce a 'hard-copy' image for the clinician or physicist to inspect. However, the trend is for interpretation to be made directly from the screen.

Between 200 000 and 500 000 γ-ray events are usually acquired for a static image. The time required to do this will depend on the activity injected, the radioisotope used, the collimator and the detector and other factors such as the physique of the patient. In general, acquisition will take at least tens of seconds and it is important that the patient remains still within this time frame. The number of events collected in dynamic images, as well as depending on the above, will also depend on the rate at which the underlying distribution of tracer is changing. This generally means that fewer counts can be collected per image than for a static study and so use of fewer pixels than 256×256 will not significantly degrade image quality.

12.2.5. *Single-photon emission tomography (SPET)*

The gamma camera can be used to produce three-dimensional images. Data are collected by rotating a gamma camera around a patient (figure 12.6).

This produces a sequence of images. In order to produce an image of a slice through the patient a profile is taken from each image (see figure 12.7). These slices are taken to be projections of the activity distribution in the sense of Chapter 11 (section 11.8). The filtered back-projection reconstruction algorithm can be used to reconstruct images of the radioactivity distribution through the patient.

The slice computation can be performed for all profiles in the image. In practice, this means that a three-dimensional image is always produced. For the reconstruction algorithm to be correct there should

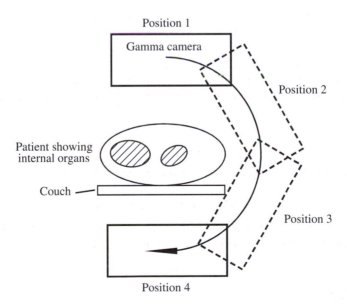

Figure 12.6. *The rotation of the gamma camera detector around the patient allows 3D images to be produced.*

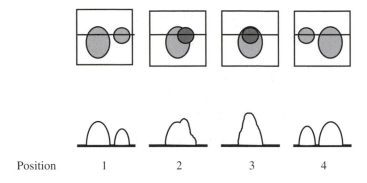

Position   1   2   3   4

Figure 12.7. *Activity profiles taken at the four positions of figure 12.6.*

be no attenuation of the γ-rays and the resolution of the camera should be independent of distance from the collimator and there should be no scattered radiation. None of these conditions are true. In spite of this the filtered back-projection reconstruction algorithm works well, although the resulting images are not quantitatively accurate. In principle, provided the attenuation coefficient is the same throughout the object, the reconstruction algorithm can be modified to produce correct images in the presence of attenuation and non-constant resolution, but not generally in the presence of scatter. Usually scatter and resolution effects are ignored and the effects of attenuation approximately corrected using a sensitivity matrix.

SPET imaging is now common for cardiac imaging and imaging of brain perfusion. With improvements in detector performance, especially in terms of spatial linearity, uniformity of sensitivity and features such as automatic body-contour following, which maximizes resolution by ensuring that the detector remains as close to the body as possible and improved methods of attenuation correction, SPET is likely to extend into other applications, especially if quantitative estimates of organ functions are needed.

12.2.6. *Positron emission tomography (PET)*

When a positron is emitted from a nucleus it almost immediately, within 1 or 2 mm in tissue, annihilates with an electron to form two 511 keV γ-rays. For all practical purposes these move off in exactly opposite directions. If they are detected simultaneously (in coincidence) by two detectors it follows that the point of origin of the γ-rays lies along the line joining the two detectors. This means that an image can be formed without collimation. Imaging systems which use this property of positron emission have been available for some time. They consist of rings of detectors (see figure 12.8). Coincidence events can be detected between any pair of detectors. It can easily be shown that the data collected can be sorted into a form comparable with a conventional SPET data set and a similar algorithm used for reconstruction, so an image of a slice can be formed. Alternatively a direct reconstruction algorithm can be used. Multiple rings of detectors can be used. This arrangement is, in principle, capable of using all the photons emitted in the plane of the detectors so that the utilization of photons can be much better than for SPET. However, this benefit is offset by the fact that detection of the photons is harder because of their relatively high energy and both must be detected for a coincidence to be registered. If P is the probability of a photon being detected when it enters a crystal, the probability of both being detected is proportional to P^2 so it is important for P to be large, otherwise the sensitivity will be low.

If two unrelated γ-rays enter a pair of detectors at the same time then a false coincidence will be detected. The probability of a false or random coincidence will depend on the count rate and will increase proportionately as the true coincidence count rate increases. However, the rate of random coincidences between a pair of

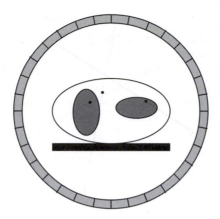

Figure 12.8. *A PET detector ring configuration. The patient is shown on the couch with a complete circle of detectors around.*

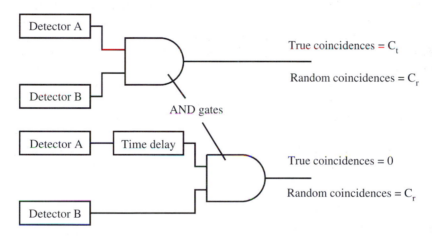

Figure 12.9. *Separation of true from random coincidences.*

detectors can be estimated by inserting a delay time after one of the detectors (see figure 12.9). The number of true coincidences will fall to zero but the number of random coincidences will be unaffected (assuming the count rate is not changing with time).

Use of multiple rings of detectors along one axis means that multiple slices can be formed. If coincidences are allowed between rings then potential photon utilization is very high. In the limit data are available for a full three-dimensional reconstruction rather than multiple-slice reconstruction. As no mechanical motion is required it is possible to generate dynamic data. The volume of image data generated can be very large, but the potential for accurate quantitative measurement, following attenuation correction which can be performed using transmission data derived from measurements using an external source, is high. A serious problem with full 3D imaging is that the amount of scattered radiation detected is high and estimates of the amount detected need to be produced and subtracted from the image data if quantitative accuracy is to be preserved.

Physiologically PET is interesting because several isotopes of physiological significance, carbon, oxygen and nitrogen, have positron-emitting isotopes (^{11}C, ^{15}O, ^{13}N). In addition, fluoro-deoxy-glucose (^{18}FDG), an analogue of glucose, may be used to investigate glucose metabolism. All these isotopes are generated with a cyclotron and, with the possible exception of FDG, need to be generated close to the point of application to the patient because their half-lives are short. Positron ring detector imaging systems are very expensive, as is a cyclotron, and are therefore not very common. They form invaluable research devices but their cost effectiveness for routine clinical use is more questionable.

Recently, several gamma camera manufacturers have produced prototype systems which use two gamma camera detectors in coincidence mode (figure 12.10).

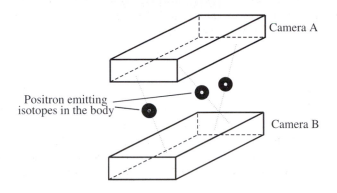

Figure 12.10. *A hybrid PET/SPET system.*

These detector pairs rotate around the patient as with SPET systems. They are likely to be significantly less expensive (25–30% of the cost) of a full ring PET detector system. Their disadvantages are that the coincidence sensitivity is low compared to a ring system because the detector crystals are thinner and the relatively low rate (compared to ring systems) at which the detectors can process γ-ray events limits the amount of tracer that can be used. ^{18}FDG has a physical half-life of 100 min and therefore it could be produced centrally within a reasonably sized conurbation and delivered as required. It has uses in cardiology, neurology and oncology. When not being used in PET mode the imager, with added collimators, can be used as a conventional dual-headed SPET system. The cost effectiveness of these systems for routine clinical use looks much more attractive than ring detector systems.

12.3. ULTRASONIC IMAGING

The basic physics of sound and the generation of ultrasound were described in Chapter 7. Ultrasound is sound at too high a frequency for the human ear to detect. This definition applies to any sound above 20 kHz, but the frequency of sound used in medical imaging is well above this, usually in the range 1–10 MHz. There are two main uses for ultrasound. The first is to produce images of the interior of the human body and the second is to provide a means of measuring blood flow in the arteries and veins (see Chapter 19, section 19.7). In the first case pulse–echo techniques are used to produce images. In the second case measurement of the Doppler shift within a single-frequency ultrasound beam is used to measure the velocity within blood vessels. More recently it has become possible to combine these techniques and image blood flow within vessels and this latter development will be discussed in this chapter (see section 12.3.6).

12.3.1. Pulse–echo techniques

Pulse–echo techniques use the same technique as sonar or radar. A pulse of ultrasonic energy is emitted from a transmitter. As the pulse moves through the medium, in this case the body, objects within the body reflect some of the energy back towards the transmitter. The time taken for the echo to return is a measure of the distance of the object. If the ultrasound is transmitted into the body, the interfaces between the different structures in the body will produce echoes at different times. If the wave velocity is known then these times can be translated into distances from the transmitter. Provided the position and orientation of the transmitter and receiver are known, and the echoes are used to intensity modulate a display, a two-dimensional map or image of the structures within the body can be obtained. If the velocity of sound is c m s^{-1} and the time taken for the echo to return is t s, then the distance of the reflecting object is

$$d = \frac{2c}{t}.$$

The ultrasound frequency will be altered if it is reflected from moving structures. This is the Doppler effect. The most common use of this effect is the measurement of blood velocity using a continuously transmitted beam of ultrasound. This is dealt with in Chapter 19 (section 19.7).

Sounds are the result of the transmission of mechanical (pressure) vibrations through a medium. High frequencies are required because the resolution achievable will depend upon the wavelength of the ultrasound and this decreases as the frequency increases. The relationship between the frequency, the wavelength and the velocity of sound is given by

$$\lambda = \frac{c}{f}.$$

The velocity of sound in water at the frequencies of interest in imaging is approximately 1500 m s^{-1}. At 1 MHz the wavelength is 1.5 mm. Objects smaller than this cannot be resolved so that if higher resolution is required the frequency must be increased. In practice, at higher frequencies absorption in tissue limits the resolution which can be achieved. The velocity of sound is fairly constant within the body, but does vary slightly in tissues. A constant velocity is needed for imaging since the calculation of the distance to a reflecting object is based on the assumption that the velocity is known. A 2.5% change in velocity (muscle and soft tissue) over 40 mm can result in a 1 mm positioning error and this is of the order of the potential resolution.

12.3.2. Ultrasound generation

Ultrasound energy is generated by a transducer (a transducer is a device for converting energy from one form into another) which converts electrical energy into ultrasonic energy. The physics of this subject was introduced in Chapter 7 (section 7.3). Piezoelectric materials such as quartz and tourmaline change their shape when subjected to an electric field. The converse effect is also obtained: mechanical deformation of the crystal produces an electric field across it which can be measured. Piezoelectric crystals are therefore suitable both for transmitting and receiving transducers. In practice, at the frequencies which are used for medical diagnosis, an artificial material lead titanium zirconate (PZT) is used. A crystal of a particular thickness has a particular resonant frequency and this determines the frequency at which the transducer can be used. A typical lead titanium zirconate transducer operating at its fundamental resonant frequency would be half a wavelength thick (about 2 mm thick for a 1 MHz transducer).

Figure 12.11 shows the construction of a typical transducer. Electrodes are evaporated or sputtered onto each face of the crystal. A voltage pulse applied between the electrodes will cause the crystal to 'ring' at its resonant frequency. The decay of the oscillations, and hence the shape of the emitted ultrasound pulse, is determined by the damping material. If the ultrasound is to be transmitted as pulses, it is usual to use the same transducer as both transmitter and receiver.

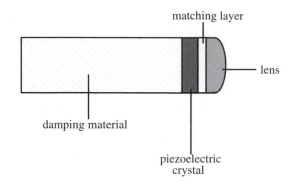

Figure 12.11. *The construction of an ultrasonic transducer.*

The ultrasound beam

The transducer can be thought of as a vibrating piston which generates a pressure wave which then propagates into the body. The physics of this was covered in section 7.3.1 and it was shown that there is a complex pattern close to the transducer (the Fresnel zone), where interference patterns occur, but a much simpler pattern in the far-field (Fraunhofer) zone. Detailed analysis shows that in the near field, the beam energy is confined to a cylinder which has approximately the same diameter as the transducer. The length of this region z_{nf} is given by

$$z_{nf} = \frac{r^2}{\lambda} - \frac{\lambda}{4}$$

where r is the radius of the transducer and λ is the wavelength. For a 2 MHz transducer of 2 cm diameter, the length of the near field in soft tissue is 13 cm.

In the far field, the beam is simpler, with a central high-intensity area which decreases in size with distance, and an annular outer area which increases in size with distance. A detailed analysis shows that there are side lobes around the main beam. The central beam diverges slowly, with an angle of divergence δ given by

$$\sin \delta = \frac{0.61\lambda}{r}.$$

For the same 2 MHz transducer, $\delta = 2.7°$. The far field is usually only of interest when precise information about the amplitude of the reflected echoes is required (for instance, when the size and shape of the reflected echoes is used to characterize the type of tissue).

The sensitivity distribution of the transducer when used as a receiver is identical in form to the distribution of power in the ultrasound beam when used as a transmitter. This is because the wave equations describing the behaviour of the transducer are the same whether the wave is travelling away or towards the transducer.

Pulse shape

The above results are for continuous ultrasonic emission at a single frequency. However, the transducer is operated in pulsed mode. In this mode a brief voltage pulse is applied across the transmitting crystal and the crystal rings at its resonant frequency in the same way that a bell rings when struck with a hammer. The backing material behind the crystal absorbs energy and damps the oscillation. For illustrative purposes the shape of the pulse can be represented by a cosine wave with an envelope of Gaussian form. If this envelope is given by $A(t)$:

$$A(t) = A_0 e^{-t^2/2r^2}$$

with a FWHM equal to $2.36r$, then the returning pulses need to be this distance apart for them to be recognized as two discrete pulses. The frequency spectrum of this pulse is given by

$$A(\omega) = e^{-(\omega-\omega_0)^2 r^2/2}$$

where ω_0 is the resonant frequency of the crystal and it can be seen that as the width of the pulse decreases, improving resolution along the axis of the transducer, the frequency spectrum widens. If the output of the transducer is less damped the pulse duration is longer. For each frequency in the pulse spectrum there is a transmitted power distribution in the near field and as the shape of this distribution is dependent on the frequency, the zeros and maxima for different frequencies will not coincide. This means that the power distribution in the near field is smoother than shown in figure 7.2. For the same reason the sensitivity distribution within the beam is also smoother.

Range and lateral resolution

Resolution is a measure of the ability of the system to separate objects which are close together. The range resolution is the smallest separation of two objects that can be distinguished along the axis of the ultrasound beam. In practice, the range resolution is roughly twice the wavelength of the ultrasound. For a 1 MHz transducer, which gives a wavelength of 1.54 mm in soft tissue, the range resolution is about 3 mm. The range resolution can obviously be improved by increasing the transmitted frequency, but this will also increase the absorption of the ultrasound.

The lateral resolution is the smallest separation of two objects that can be distinguished at right angles to the axis of the beam, and is roughly equal to the diameter of the beam so that the lateral resolution is worse than the range resolution. The diameter of the transducer can be decreased with increasing frequency, so that the lateral resolution also improves at higher frequencies. In practice, lateral resolution can also be improved by focusing the beam.

The interaction of ultrasound with the human body is complex. The more important effects are reflection and refraction, absorption, attenuation and scattering. In addition, movement within the ultrasound beam will cause a change in frequency of the ultrasound (the Doppler effect).

12.3.3. Tissue interaction with ultrasound

The propagation velocity of ultrasound depends on the density and compressibility of the material. The ultrasound velocity is higher in harder materials. The speed of sound is given by

$$c = \sqrt{\frac{B}{\rho}}$$

where B is the elastic modulus (stress/strain) and ρ is the density of the material.

Ultrasound propagates as waves through a medium and therefore undergoes reflection at interfaces within the medium. The property which determines how ultrasound is reflected and refracted is the acoustic impedance. This is the product of density and ultrasound velocity,

$$Z = \rho c.$$

Table 7.1 gave the acoustic impedance and the ultrasound velocity for air, water, some soft tissues and bone. Section 7.4 described the reflection and refraction at boundaries of differing acoustic impedance in some detail. The fraction of ultrasound energy reflected at an interface between two materials with different acoustic impedances Z_1 and Z_2 is given by

$$R = \frac{(Z_1 - Z_2)^2}{(Z_1 + Z_2)^2}.$$

The energy transmitted is given by

$$T = \frac{4Z_1 Z_2}{(Z_1 + Z_2)^2}.$$

For two tissues, or tissue and water, the fraction reflected is small and most energy continues through the interface between the two materials. This is important as it means reflections can be obtained from interfaces further away from the transmitter. In the case of an interface between tissue and bone, although ultrasound will pass through bone, invariably the ultrasound arrives at the bone from soft tissue and the large difference in acoustic impedance means that much of the energy is reflected and comparatively little is transmitted, so the bone acts as a barrier to the ultrasound wave. This is why it is very difficult to image through bone.

For the interface between soft tissue and air the fraction of reflected energy is also high and therefore it is not possible to image through an air cavity. Ultrasound is transmitted into the body from a transmitter placed on the surface of the body. Invariably there will be an air gap between the front surface of the transmitter and the skin and this would normally prevent the ultrasound energy from entering the body. If the air is replaced by a gel or oil with similar acoustic impedance to tissue then the ultrasound can enter the body. This process of increasing transmission by using materials of matching impedance is known as coupling.

What are the timings involved? If an ultrasound pulse is propagated from the skin, through the soft tissue to a muscle, and the echo arrives 20 μs later, how large is the echo and how far away is the muscle? If the acoustic impedances are inserted in equation (7.5), it will be found that the amplitude of the reflected signal is 2.1% of the incident amplitude. The velocity of sound in soft tissue is 1540 m s^{-1}, so that the distance to the muscle and back to the transducer is $1540 \times 20 \times 10^{-6}$ m $= 3.08$ cm, so that the muscle is 1.54 cm below the skin. This assumes, of course, that the interface is a flat smooth surface and that the ultrasound pulse is reflected directly back to the transducer.

The transmitted ultrasound beam will be refracted at an interface, that is the incident angle θ_i will not be the same as the refracted angle θ_t and the ultrasound beam will deviate from a straight line. This is analogous to the refraction of light, and the bending of the ultrasound beam can be found from Snell's law (see section 3.5.1):

$$\frac{\sin \theta_i}{\sin \theta_t} = \frac{c_1}{c_2}$$

where c_1 and c_2 are the velocities of the ultrasound in the two media. For incident angles of less than 30° the deviation at most interfaces will be less than 2°. At a soft-tissue–bone interface, the deviation will be 20°. For most cases this will lead to only a small degradation of the image, but may give serious problems for tissue–bone interfaces. However, it is worth noting that a 2° deviation will cause the apparent position of a reflecting point at a depth 5 cm from the refracting interface to be displaced by 1.5 mm, again an amount equivalent to the expected resolution.

The above considerations assume that the interface is flat. Many interfaces are not smooth and will therefore give diffuse reflection of the ultrasound. As the ultrasound will be scattered in all directions, only a small proportion will be intercepted by the transducer, and the apparent size of the echo will be smaller.

Scattering, absorption and attenuation of ultrasound

Scattering

Specular reflection (analogous to the reflection of light by a mirror) of the ultrasonic energy will take place when the interface is smooth over an area which is several times as great as the ultrasound wavelength, λ. When the features of the surface are about the same size as the wavelength, diffuse reflection will occur. If the scatterers are very small compared with the wavelength, then Rayleigh scattering will take place, in which the incident energy is scattered uniformly in all directions. Obviously, with this form of scattering, very little energy will be reflected back to the transducer. Red blood cells are about 8–9 μm in diameter and produce Rayleigh scattering.

Absorption

The transmission of mechanical vibrations through a medium absorbs energy from the ultrasound beam. This occurs through a variety of mechanisms but essentially arises because the changes in kinetic and potential energy, caused by the passage of the pressure wave through the medium, are not always reversible and so some of the energy is lost as heat. This is especially important at higher frequencies.

Attenuation

Loss of signal produced by reflecting interfaces between the transmitter and the reflecting object and by scattering and absorption of the signal will reduce the strength of the detected signal. The divergence of the ultrasound beam will also reduce the received energy. The received echoes used to form a typical abdominal scan may be 70 dB below the level of the transmitted signal, and the signals from moving red blood cells may be 100–120 dB below the transmitted signal. These effects taken together represent beam attenuation.

Attenuation follows an exponential law. The beam intensity at depth x is given by

$$I = I_0 e^{-\alpha x}$$

where α, the attenuation coefficient, consists largely of an absorption plus scattering component. I_0 is the unattenuated beam intensity. The attenuation coefficient is a function of frequency. Although the relationship is complex, α is roughly proportional to frequency. The decibel attenuation coefficient μ is defined as $\mu = 4.3\alpha$. Some approximate values of μ were given in table 7.2.

A pulse of ultrasound contains a range of frequencies and as the pulse propagates through tissue these will be differentially attenuated, leading to a net reduction of the average frequency which will lead to a loss in resolution and a change in the apparent attenuation.

It is necessary to compensate for the increasing attenuation of the signals with distance. The receiving amplifier compensates for the increase in attenuation of the echo when it has travelled through a greater depth of tissue. The amplifier gain is increased with time after the transmitted pulse. This process is termed *time–gain control* (TGC) and is illustrated in figure 12.12. By adjustment of the TGC characteristic a uniform image can be obtained.

Power outputs

Ultrasound is transmitted through tissue as a mechanical vibration. If the power levels are sufficiently high then mechanical damage can arise as a result of thermal effects and cavitation. There is a very large literature

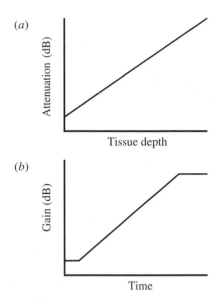

Figure 12.12. *(a) The attenuation of ultrasound with distance in tissue and (b) the time–gain profile used to compensate for this attenuation.*

on the possible damage to tissue when exposed to ultrasound. However, there is no clear consensus as to what levels should be considered as safe. Doppler systems with a continuous power output of up to 0.5 mW mm^{-2} are used. Imaging systems use peak powers up to 500 mW mm^{-2} with pulses of about 1 μs in duration.

12.3.4. Transducer arrays

All imaging uses pulsed emission of ultrasound. If a single beam of ultrasound is directed into the body, echoes will be returned from reflectors and scatterers in the beam. If the intensity of the returning signal is plotted against time a graph of the position of reflecting sources along the direction of the beam will be obtained. This is traditionally called an A scan. The repetition rate of the pulses must be sufficiently slow for the echoes from the furthest interface in the body to have time to return to the transducer. At 1 kHz in soft tissue, there is time for echoes to return from an interface 77 cm away from the probe, which is more than adequate. At 10 kHz, the distance is reduced to 7.7 cm, which would be insufficient in abdominal scanning. This will limit the rate at which an image can be formed.

The time axis of an A scan be converted to distance if the velocity of sound is known. If the direction of the beam is changed systematically the information from successive A scans can be used to build up an image, by plotting the intensity distribution along each beam as a function of beam position and direction. Each beam position relative to previous and future beam positions needs to be known.

Early imaging systems mounted the transducer on a mechanical arm so that the position and orientation of the transducer could be measured at all times. Scanning of the transducer was performed manually. In a development of this the transducer was scanned mechanically and systems of this form can still be found. However, although some mechanical scanners have a very high performance the trend is to form images using arrays of transducers and this is the approach to image formation that will be described here. Future developments will be based on transducer arrays so it is important to appreciate how such systems work.

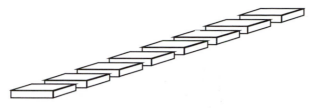

Figure 12.13. *An array of transducers.*

The basic form of the transducer is a linear array of small flat piezoelectric transducers placed side by side (figure 12.13). There may typically be over 100 of these. The length of each elemental transducer is substantially larger than the width. The width of each elemental transducer is small enough for the sound wave it produces to be an approximation to a cylindrical wave. Suppose all the transducers are excited simultaneously in phase. Then the cylindrical waves generated add together to produce a plane wave moving away from the transducer. If a subset of adjacent transducers is excited together then they produce a wave equivalent to a single transducer of the same overall shape. For example, suppose the total length of the array is 100 mm, the width is 10 mm and there are 100 crystals. By exciting a block of 10 adjacent transducers a 10 mm square transducer is simulated. By suitable selection of adjacent groups of transducers it is possible to sweep this 'equivalent' transducer along the length of the linear array, thus simulating mechanical movement of a single transducer.

Focused arrays

Consider a small array of transducers (figure 12.14). Suppose the transducers at the ends of this small array are excited before the transducers at the centre. The sound wave propagating away from the transducers will no longer be plane, but will curve in towards the central axis. If we assign a variable delay to each of the transducers so that the moment at which each transducer is excited can be controlled, then by suitable selection of the delay times we can shape the wavefront as we wish.

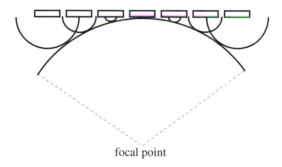

focal point

Figure 12.14. *A focused beam produced by phased excitation of the transducers.*

In the above array the outer transducers are excited earlier than the central transducers and by choice of suitable excitation times the advancing wavefront can be made to take any desired shape. In particular, we can generate a circular wavefront which comes to a focus at some point in front of the transducer array. The different excitation times are produced by generating a single excitation signal and then delaying its arrival at the transducers in a systematic way. In the above case the delay is maximum for the central transducer and zero for the transducers at the end of the array.

The advantage of focusing the transmitted beam in this way is that the width of the beam can be reduced and the resolution improved relative to a parallel beam. Associated with a particular pattern of delays will be a focal point and resolution will be maximized at the focal point. Elsewhere the resolution will be less good. In principle, we can obtain good resolution over the whole of the depth of field. We can select only those pulses emitted within the focal area by rejecting pulses which arrive too early or too late relative to the time of flight of a signal reflected from the focal point. When we have received these we can change the focus by changing our delays and repeat the sequence. In this way the signals we receive are always from or near to a focal region. In practice, this approach is not used because it slows down data collection, because we are discarding most echoes and increasing the deposition of ultrasonic power in the patient.

One thing we can do, however, is to select a fixed focus for transmission but vary the focus for reception. The focusing strategy works for reception as well as for transmission. Each transducer in the transducer array receives a signal and these signals are added together within the receiver. If we delay the signals from the central transducer relative to the end transducers we can ensure that the signals received from a source at the focal point are in phase. After transmission with some fixed focus the delays are set to place the focal point near the transducer. As time proceeds, and any signals received are being reflected from deeper and deeper in the patient the focal point is moved away from the transducer. At any time we know the depth where the pulses being received at that time are coming from and can focus on this point. In this way maximum reception resolution can be achieved over the whole of the range of the transducer. To build up an image the group of active transducers is scanned along the array of transducers (figure 12.15).

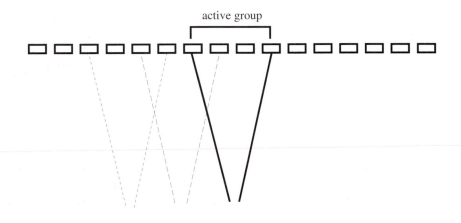

Figure 12.15. *Scanning an active group of transducers.*

Curved arrays

In the form given above the volume of patient imaged is limited by the length of the transducer array, which therefore needs to be of reasonable length. In order to produce a larger field of view, particularly at depth, the array can be curved (figure 12.16).

This allows a physically small array to scan a relatively large volume. There is clearly potential for some loss of resolution at depth compared to a flat array but careful design of the array and focusing strategies can minimize this.

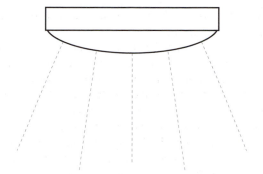

Figure 12.16. *A curved array of transducers. By producing diverging beams a large field of view can be obtained.*

Phased arrays

A second approach to increasing the field of view at depth is the phased array. Consider the flat array of transducers again. If the delays are identical the wave generated moves away in a direction normal to the plane of the transducer array. However, if we impose a linear delay (figure 12.17) on the transducers a plane wave is still generated, but this time the plane wave is at an angle to the normal to the transducer array.

Figure 12.17. *The phased array. By delaying the activation of the elements increasingly from left to right the ultrasound beam can be steered to the left.*

By adjusting the magnitude of the delays, in addition to any focusing delays, the transducer array can be constrained to image in a direction away from the direction normal to the array. Appropriate adjustment of the delays can sweep this direction from side to side, thus expanding the width of tissue imaged at a given depth.

With this arrangement (figure 12.18) all points in the extended volume are visited at least once, with some points being viewed from a range of angles.

The transducer in this form images a slice of tissue where the thickness of the slice is determined by the width of the transducer. Focusing to narrow this plane can be achieved using a cylindrical ultrasonic lens, but resolution normal to the imaging plane will tend to be worse than within the plane.

Timings

Signals from a depth of 15 cm, which we can conveniently take to be the maximum range, return in 200 μs, so this is the time that we must wait at each point along the transducer array. If we have 100 positions along the array, then the total time to make a complete sweep is 20 ms, which means we can generate 40 images per second. A transducer array can therefore generate real-time images. This is of major importance as real-time

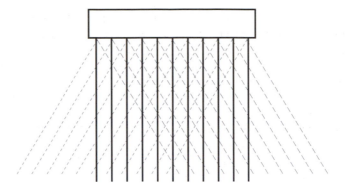

Figure 12.18. *Scanning from left to right and back with the phased array enables a volume of tissue to be imaged.*

images are often much easier to understand than static ones. Because movement can be seen it is relatively easy to identify cardiac- and respiratory-related components within an image.

12.3.5. Applications

Imaging the foetus (see figure 12.19) is probably still the most widely used application of ultrasound in medicine. It is not possible to image the foetus during the first few weeks of pregnancy, because it is hidden

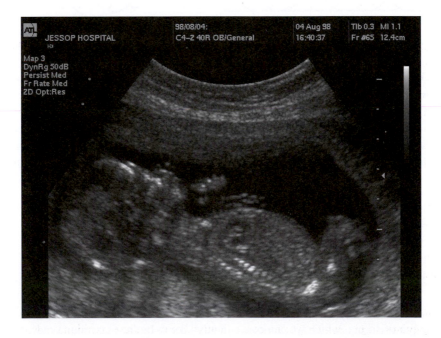

Figure 12.19. *A foetal image obtained using ultrasound. The heart, spine and other features are clearly seen. Image kindly provided by Dr S B Sherriff.*

by the pelvis, although it is possible to scan the uterus through the bladder, if it is full of urine. Multiple pregnancies can be diagnosed and the location of the placenta and the presentation of the foetus can be seen.

Foetal maturity can be assessed by measuring the biparietal diameter of the foetal head, that is, the distance between the parietal bones of the skull. Several scans of the foetal head are made, to establish the plane of the maximum diameter, and electronic callipers are then used to measure the biparietal diameter. The velocity of ultrasound in the foetal head is usually taken as about 1600 m s^{-1}. This value was established by comparing ultrasound measurements immediately before birth with calliper measurements immediately after birth. The positioning of the markers on the scan image is usually such that the measurement includes transit through part of the skull, which has a much greater ultrasound velocity than the brain. This may explain why measured values of the velocity in the foetal brain are somewhat lower than 1600 m s^{-1}. At typical frequencies of 2–3 MHz, an accuracy of biparietal measurement of better than 1 mm is unlikely. The most reliable estimation of gestation can be made during weeks 14–20, when the foetus is growing most rapidly, but useful predictions can be made up to 30 weeks. After 30 weeks, the spread of ages corresponding to a particular head size is too great for useful prediction.

There are very many other applications of ultrasonic imaging, particularly in abdominal and cardiac imaging. In many cases the images combine Doppler and pulse–echo imaging so that vascular and cardiac images can be superimposed on the anatomical detail given by the pulse–echo image. This gives a combined anatomical and functional image.

12.3.6. Doppler imaging

The use of continuous wave Doppler for the measurement of blood flow is described in some detail in Chapter 19. If the ultrasound beam of frequency f_0 is reflected or scattered back from an object moving at velocity v_b in the direction of the beam then the shift in frequency is given by

$$f_d = 2f_0 \frac{v_b}{c}$$

where c is the velocity of sound. If the velocity vector is at an angle θ to the direction of the beam then v_b is the component of the velocity in the direction of the beam $v \cos(\theta)$.

The shift in frequency is extracted from the returning signal by multiplying the signal by the emitted signal. Suppose the transmitted signal is at frequency f_0 and the returning signal is at frequency $f_0 + f_d$. If we multiply the two signals together we obtain

$$\cos 2\pi f_0 t \cos 2\pi (f_0 + f_d)t$$

which can be rewritten as

$$(\cos 2\pi (2f_0 + f_d)t + \cos 2\pi f_d t)/2.$$

The first term is at a very high frequency and can be filtered out using a low-pass filter to reveal the shift frequency. We can also multiply the returning signal by the sine of the transmitted signal:

$$\sin 2\pi f_0 t \cos 2\pi (f_0 + f_d)t$$

which can be rewritten as

$$(\sin 2\pi (2f_0 + f_d)t + \sin 2\pi f_d t)/2.$$

If the blood is moving towards the transmitter then f_d will be positive, but if it is moving away then f_d will be negative. Simply measuring $\cos 2\pi f_d t$ cannot distinguish between these two signals. Knowing $\sin 2\pi f_d t$ as well allows us to determine the sign of f_d.

Continuous wave Doppler provides an integrated value of the velocities contained within the beam. However, using pulsed techniques it is possible to localize the velocity values along the beam. Time-gating of pulse signals allows selection of a portion of the beam. Analysis of the frequency shifts associated with this spatial location allows the velocities associated with this region to be determined. Not surprisingly, using pulsed signals rather than a continuous beam makes things more complex.

As with imaging we can sample the signal coming from a particular depth by only accepting signals returning at a particular time. If pulses are emitted every T seconds we have a stream of pulses returning with this time spacing. In effect, the continuous signals $\cos 2\pi f_\mathrm{d} t$ and $\sin 2\pi f_\mathrm{d} t$ are sampled every T seconds. From these data we need to extract the frequency f_d.

In fact, from a small volume of tissue there will be a spectrum of frequencies since there will be a range of velocities. The 'cosine' and 'sine' signals can be generalized to two functions $x(t)$ and $y(t)$ and these can be combined into a complex signal $z(t) = x(t) + \mathrm{j} y(t)$. The mean frequency ω_m representing the mean velocity and the spread of frequencies σ, called the turbulence indicator, can be computed from the autocorrelation function of this complex signal. We will not give the details here but note that it has to be done very quickly if real-time images are to be produced. The accuracy of these values will depend on the number of pulses over which the calculation is performed. The more pulses included the more accurate the results will be, but the longer the calculation will take and the slower the response to changes in flow.

There is an upper limit to the velocities that can be detected. The velocity-related signals, $\cos 2\pi f_\mathrm{d} t$ and $\sin 2\pi f_\mathrm{d} t$, are sampled at intervals of T seconds. Sampling theory confirms that frequencies above $1/2T$ will be aliased into lower frequencies. However, T cannot be shorter than the time required for the return of a signal from the maximum depth that can be imaged. Compromises have to be made to ensure accuracy of velocity measurement against resolution.

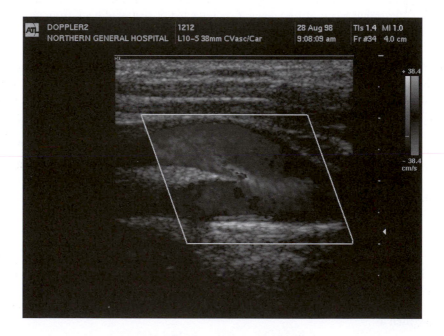

Figure 12.20. *An ultrasound image of the carotid artery in the neck. The bifurcation (from the right) into the internal and external carotid vessels is clearly seen. By superimposing the Doppler signal the direction of the blood flow can be seen and small areas of turbulence can be identified. The velocity scale is shown in the top right-hand corner of the figure. Image kindly provided by Dr S B Sherriff. This figure appears in colour in the plate section.*

Since the direction of flow is available, images are displayed usually with red representing flow towards the transducer and blue representing flow away from the transducer. The turbulence signal is sometimes added as a green component to the signal. The colour information gives an immediate visual impression of the flow patterns within the vessel and can visually identify regions of turbulent flow within the vessel being imaged. As with continuous wave Doppler the velocity of the blood can only be determined if the angle of insonation is known accurately, although this is in principle easier to determine since an image of the anatomy is simultaneously available. Figure 12.20 shows a combined Doppler and pulse–echo image of the bifurcation of the carotid artery in the neck. It illustrates well the value of combining both anatomical and functional imaging.

12.4. MAGNETIC RESONANCE IMAGING

Magnetic resonance imaging (MRI) has become an important imaging technique over the past few years. It appears to be a safe procedure (see Chapter 8 for a discussion of the possible biological effects of electromagnetic fields) and it can produce images with high contrast between different tissues. In addition, modifications to the way data are collected allow flow in blood vessels to be visualized. This makes it a very flexible technique and its uses are likely to continue to grow. However, the physics of MR image formation is less straightforward than some of the more familiar techniques of imaging. Magnetic resonance imaging is able to measure the density of protons in the body as a function of position. It can also be used to measure properties which reflect the environment of the protons. For example, it can produce images which distinguish between protons in fat and protons in water. The key to understanding MRI is to understand the way the protons behave in a magnetic field.

12.4.1. The nuclear magnetic moment

The nucleus of many atoms can be thought of as having a spin and as a consequence of this (the nucleus is effectively a rotating electric charge) acts as a small magnet. The nucleus therefore has a magnetic moment. The size of this moment depends on the nature of the nucleus. Nuclei with even numbers of protons and even numbers of neutrons do not have a spin and so do not have a magnetic moment. An example is ^{12}C. Nuclei with even numbers of protons but odd numbers of neutrons have a spin with an integer value. Nuclei with an odd number of protons have a spin which is half-integral. The best known example is ^{1}H with a spin of $\frac{1}{2}$, but other values are possible with other nuclei. Most MRI imaging consists of imaging the distribution of protons in tissue, and we shall concentrate on this. Although other elements can be imaged in principle, in practice the abundance of hydrogen in tissue means that it is the easiest nucleus to image.

12.4.2. Precession in the presence of a magnetic field

The magnetic moment has both a magnitude and a direction. If the proton is placed in a magnetic field there will be a couple on the proton attempting to make it line up with the field. However, because the proton is spinning and has angular momentum like a spinning top the protons, rather than lining up with the field, will precess around the field direction (figure 12.21). There is a characteristic frequency associated with this, which is given by

$$\omega_0 = -\gamma B_0$$

where γ is the gyromagnetic ratio for the proton (the ratio of the magnetic moment and the angular momentum) and B_0 is the magnetic field strength. ω_0 is known as the Larmor frequency. It is proportional to B_0.

The gyromagnetic ratio for protons is 42.6 MHz T^{-1}, i.e. in a field of 1 T (tesla) the protons will precess at a frequency of 42.6 MHz. The effect will also take place in the Earth's magnetic field. This is about 50 μT so the corresponding frequency of precession is about 2 kHz.

Figure 12.21. *Precession of a proton around the field B_0.*

In the absence of a magnetic field there are equal numbers of protons oriented in all directions. However, if a magnetic field is applied there will be a tendency for the protons to orientate in the direction of the field. The fraction which line up with the magnetic field (the fractional excess) is given by

$$\Delta n = \frac{\gamma h B_0}{2\pi kT}$$

where k is Boltzmann's constant and h is Planck's constant. Δn is small, of the order of 3×10^{-6} for a field of 1 T (10 000 G) at room temperature. Although the magnetic moments of the individual protons are still largely randomly oriented there is now a net magnetic moment m formed from the vector sum of all the individual moments. This magnetization has a spin associated with it. The tissue sample will now have a net magnetization in the direction of the field proportional to the field strength.

The fact that Δn increases with field strength is one reason why MR imaging is carried out at high magnetic field strengths. The signals which arise from the precessing protons are much easier to record at the high field strengths because the fractional excess is larger.

To understand what happens when we apply a radio-frequency field to the precessing protons we need to define some appropriate coordinate systems. The laboratory coordinate system has the z-axis along the direction of the field B_0. The x- and y-axes are at right angles to this. We will also be using a rotating coordinate system in which we rotate about the z-axis with an angular velocity ω (figure 12.22). The rotating axes normal to the z-axis are x' and y'.

There are two effects due to the effect of applying a magnetic field. The first is that the protons are precessing around the z-axis with frequency ω_0. The second is that there is an induced magnetization in the

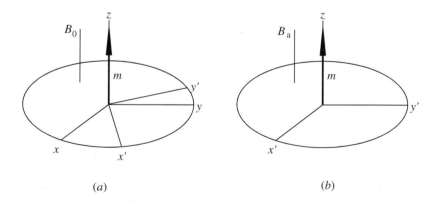

Figure 12.22. *The laboratory (a) and rotating (b) coordinate systems.*

tissue sample. Suppose we now move to the rotating coordinate system which rotates with angular velocity ω. In this coordinate system the protons precess with a frequency $\omega_0 - \omega$. As far as the rotating observer is concerned this loss of angular velocity can only be explained by a reduction in the applied field. In fact, it is fairly easily shown that the applied field appears to be given by $B_a = B_0 + \omega/\gamma$ and this goes to zero when $\omega = \omega_0$.

Rotation of the magnetization

Suppose we now move from the laboratory axes to the coordinate system rotating with angular velocity ω_0. In this coordinate system the field B_0 will appear to have disappeared. We now apply a field B_1 in the direction of the x'-axis in the rotating coordinate system (see figure 12.23). Under these conditions, while the field B_1 is present, all the protons will precess about the x'-axis with angular velocity

$$\omega_1 = -\gamma B_1.$$

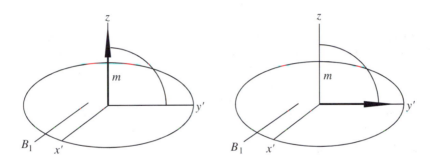

Figure 12.23. *Precession about a field along the x'-axis.*

If the field is applied for a time $t = 1/8\pi\gamma B_1$ individual protons will rotate through $90°$ and therefore the net magnetization vector m, initially in the direction of the z-axis, will have rotated into the plane transverse to the main field, i.e. along the y'-axis. Returning to the stationary coordinate system, the magnetization vector now rotates in the transverse plane at the precession frequency ω_0 (see figure 12.24).

In order to apply a constant field along the x'-axis in the rotating frame we have to apply a rotating field in the stationary frame. To do this we apply a radio-frequency field at the precession frequency. This can be resolved into two fields rotating in opposite directions,

$$B \cos \omega_0 t = \tfrac{1}{2}\left(B\mathrm{e}^{\mathrm{j}\omega_0 t} + B\mathrm{e}^{-\mathrm{j}\omega_0 t}\right)$$

In the rotating coordinate frame this becomes

$$B = \tfrac{1}{2}\left(B + B\mathrm{e}^{-\mathrm{j}2\omega_0 t}\right).$$

The component rotating against the precession can be ignored since in the rotating coordinate system it appears to have a frequency of twice the Larmor frequency and so oscillates too fast to affect the orientation of the magnetization. The field rotating with the precession is the one we want. The application of an appropriate radio-frequency signal has flipped the magnetization into the (x, y)-plane. This is an example of resonance. If any other frequency is used then the strength of the residual field which has not gone to zero at that frequency, ensures that the magnetization is not moved significantly from the z-axis.

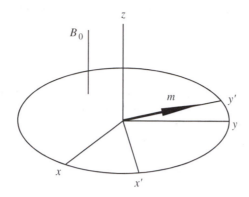

Figure 12.24. *Rotation of the magnetization vector in the laboratory coordinate system.*

After a radio-frequency pulse of sufficient length is applied the magnetization is in the plane at right angles to the main field. As the magnetization has been rotated through 90° this pulse is known as a 90° pulse. If the pulse is too short rotation will be through a lesser angle. The magnetization now rotating in the transverse (x, y)-plane will induce a current in a coil perpendicular to the plane. The frequency of this induced current will be the precession frequency of the protons. After a 90° pulse the magnetization along the main axis is reduced to zero.

Saturation

Let us consider the state of the individual protons rather than the net magnetization. In the presence of the magnetic field protons can be excited to the higher level by the application of radio-frequency radiation as shown above. But if a proton is already in the higher state the radiation can trigger a return to the lower level with the emission of radiation. If equal numbers of protons are in each of the levels then the chance of a proton in the lower state being excited to the upper state is equal to that of a proton in the upper state being stimulated to emit radiation and drop to the lower state. Under these conditions there can be no net change either way and the tissue is said to be saturated.

12.4.3. T_1 and T_2 relaxations

T_1 relaxation

The protons do not precess in the (x, y)-plane for ever. Energy has been supplied to the protons to rotate them into the transverse plane. This energy can be released by stimulation with a field oscillating at ω_0 and this frequency is contained, along with others, in the fluctuating magnetic fields produced by molecules surrounding the proton. Energy is lost in this way by the precessing protons and they realign with B_0. The transverse axial magnetization realigns with the z-axis. The axial magnetization cannot be observed directly, but the rotating transverse magnetization can be seen to decay. The reappearance of the axial magnetization follows an exponential relationship with a characteristic time constant conventionally called T_1,

$$m(t) = m_z(1 - e^{-t/T_1}).$$

The rate at which this occurs depends on the way the protons are shielded from the fluctuating fields by surrounding electrons. Much of the contrast in MRI imaging represents the different behaviour between protons in fat and protons in water. Protons in fat-rich tissues lose their transverse magnetization faster than those in low-fat tissues.

T_2 relaxation

As the axial magnetization increases the transverse magnetization decreases. However, in general, the rates are not the same. A second effect of local fluctuations is that the net field B_0 around the protons will differ from proton to proton. This means that the protons will not all precess at quite the same angular velocity. Although after the 90° pulse all the protons set out in phase with a magnetic moment of m in the transverse plane, the variation in precession rates means that after an interval of time they drift out of step, with some ahead of the average and some behind (see figure 12.25). The vector sum of the individual moments steadily decreases from m towards zero. The random fluctuations in this process ensure that this loss of coherence is irreversible and eventually results in the transverse component of magnetization falling to zero.

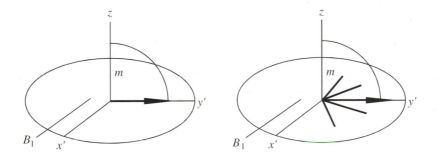

Figure 12.25. *Loss of coherence in the rotating transverse magnetization.*

This can happen before the axial magnetization is fully restored. The decay of transverse magnetization is also exponential and has a characteristic decay time called T_2,

$$m(t) = m_{xy}e^{-t/T_2}.$$

In fact, the rate of dephasing is speeded up by other fixed factors, typically inhomogeneities in the main field, and this means that rather than observing T_2 a composite decay factor is observed, usually called T_2^*.

Following the application of a 90° pulse the signal generated by the rotating magnetization vector can be measured. The resulting signal is called the free induction decay signal or FID; it can be digitized. The Fourier transform of this signal shows a peak at the precession frequency.

12.4.4.   *The saturation recovery pulse sequence*

Suppose we now excite a sample of material with the appropriate 90° pulse and then follow the decay of the transverse magnetization. The axial magnetization starts off as zero. The transverse magnetization at time 0 is $m_{xy}(0)$. This will decay with half-time T_2^*, but as we have seen this is not the rate at which axial magnetization is restored. If we wait a time *TR* which is significantly longer than T_2 so that the FID has fallen to zero but significantly less than T_1, the axial magnetization will only have been partly restored, say to a value $m_z(1 - \exp(-TR/T_1))$. If we then apply a second 90° pulse then the maximum transverse magnetization we can generate is equal to this latter value. This sequence of events is summarized in table 12.1.

Table 12.1. *The saturation recovery pulse sequence.*

Time	Axial magnetization	Transverse magnetization
Before 90° pulse	m_z	0
After 90° pulse	**0**	$m_{xy} = m_z$
TR before 90° pulse	$m_z(1 - \exp(-T/T_1))$	0
TR after 90° pulse	**0**	$m_{xy} = m_z(1 - \exp(-TR/T_1))$

The magnitude of the final signal is dependent on the magnetization m_z, the value of T_1 and the time *TR*. Clearly, if we have two tissues a and b with the same m_z, but different T_1, the ratio of signals following the above sequence will be given by $(1 - \exp(-TR/T_{a1}))/(1 - \exp(-TR/T_{b1}))$, which is a function of *TR*. The deviation from unity of the ratio of the two signals is largest for $TR = 0$ and the ratio moves towards unity as *TR* increases. Suppose we have two tissue samples with equal magnetization values and subject each tissue sample to a sequence of pulses separated by a repetition time *TR* and, once the sequence had been established, measured the transverse magnetization immediately after the pulse was applied. With a long repetition time (long enough for the axial magnetization to have recovered) the signals from both samples will be the same. If a short repetition time is used, shorter than either T_1, the ratio of the signals will be in inverse proportion to the T_1 values. So by a suitable choice of *TR* we can produce high or low contrast between tissues. If the magnetizations of the tissues are not the same, a more likely situation, then in the first case the ratio of signals is in proportion to the proton density and in the second case, although still dominated by the T_1 values, it is modified by the proton density thus producing T_1-weighted signals

The sequence of pulses and their associated timings are simply called a *pulse sequence*.

The inversion recovery sequence

A variation of this sequence is to apply a longer pulse which completely inverts the direction of the magnetization. This is called a 180° pulse (figure 12.26).

After this pulse there is no transverse magnetization. The size if the axial magnetization is measured at some time T_1 by the application of a 90° pulse.

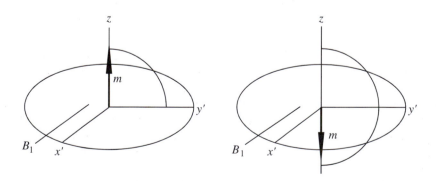

Figure 12.26. *Application of a 180° pulse.*

12.4.5.   The spin–echo pulse sequence

The most common sequences are spin–echo sequences (figure 12.27). Suppose we apply a 90° pulse to a sample and then measure the FID. The decay of the FID depends on the value of T_2^* and this is largely dependent on the local field inhomogeneities, which are not specifically dependent on tissue relaxation properties. The component of dephasing caused by random field fluctuations in the local fields is irreversible. However, the component caused by fixed inhomogeneities can be reversed. This is done by applying a second transverse pulse at a time conventionally called $TE/2$. This has twice the duration of the 90° pulse and precesses the protons in the rotating transverse plane through 180°.

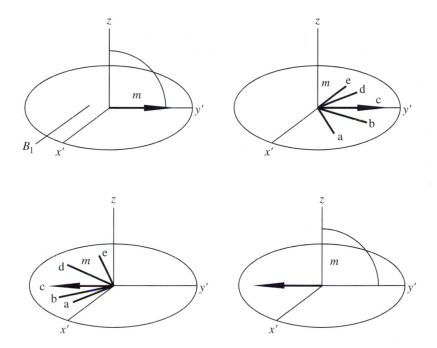

Figure 12.27. *Spin–echo sequence: the individual protons* (a, b, c, d, e) *precess at slightly different rates because of inhomogeneities in the main field caused both by fluctuations in the field generated by neighbouring molecules and non-uniformity of the main field. These protons are flipped through 180° by a suitable pulse which reverses their position in the 'race' and after a suitable interval of time they are in line again.*

The effect of this is that protons which were in the lead are now trailing by an amount equal to their lead. After a period of time equal to the time delay between the 90° pulse and the 180° pulse, i.e. a further time $TE/2$ from the 180° pulse, the rapidly precessing protons catch up with the slowly precessing photons and all are now in step. The transverse signal is measured at this time. This does not mean that the magnetization in the transverse plane is back to the value it was before immediately after excitation since both spin–lattice (T_1) energy transfer and spin–spin (T_2) transfer has occurred. It does mean that the effect of field inhomogeneities has been removed. After a time TR from the initial 90° pulse a further pulse 90° pulse is given, starting the whole sequence again.

Spin–echo techniques are capable of giving signals which are weighted in a variety of ways. If TE is kept short, then little T_2 decay has occurred and the signals from the different tissues are only weakly affected by their T_2 values. As in saturation recovery if TR is short then the tissues will not have recovered their saturation values and the relative signal size will be dominated by their T_1 values. Conversely, if *TE* time is long then the relative signal sizes will be strongly affected by T_2 values and if *TR* is long enough for saturation to have largely recovered then different T_1 values will not strongly affect the relative signal sizes. In both cases the signal size will be proportional to the proton density, so signals from these different pulse sequences are said to be T_1- and T_2-weighted signals rather then T_1 and T_2 signals. Producing a signal which is only dependent on T_1 or T_2 is more complex. A short *TE* and long *TR* produces a measurement which is dominated by proton density.

Many other sequences are possible but cannot be covered in this short account. Some values of T_1 and T_2 are given in table 12.2.

Table 12.2. *Typical values for T_1 and T_2 for the brain tissue and fat.*

Tissue	T_1 (ms)	T_2 (ms)
White matter	600–800	60
Grey matter	800–1000	80
Fat	150	50

12.4.6.   Localization: gradients and slice selection

If a uniform magnetic field is applied to a large sample of tissue, then the precession frequency will be the same for all parts of the sample. The size of the returning signal will reflect the total number of protons in the sample. This is clearly no use for imaging. In order to localize the signal to a particular point in space, the applied field is made non-uniform. Suppose we apply an extra non-uniform field $G_z(z)$, a field which generates a gradient in the z direction so that the steady field is now given by

$$B(z) = B_0 + G_z(z).$$

The precession frequency will now be a function of position along the z-axis. The z-axis is normally directed along the long axis of the patient. If we now apply a single frequency, or a narrow band of frequencies, to the object then only protons where z is such that the applied frequency is the resonant frequency for the field at that point will absorb and re-emit energy. The combination of the gradient and the frequency bandwidth of the radio-frequency pulse define a two-dimensional slice normal to the z-axis in which protons are excited (see figure 12.28). No other slices contribute to the emitted signal. The slice thickness Δz is given by

$$\Delta z = \gamma G_z \Delta \omega$$

where $\Delta\omega$ is the bandwidth of the radio-frequency pulse.

12.4.7.   Frequency and phase encoding

Frequency encoding: the x direction

Let us assume for simplicity that we have used a single frequency to select a slice at a point along the z-axis. The coordinates within this plane are x and y. Once excitation has ceased the induced transverse magnetization in this slice produces a FID signal. The z gradient field can now be removed since it has done its work. Before G_z is removed the protons in the slice are precessing with frequency

$$\omega = -\gamma B(z).$$

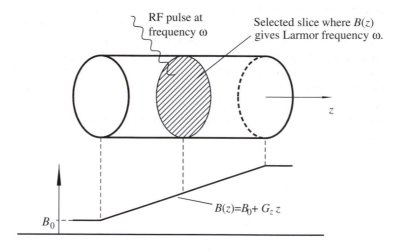

Figure 12.28. *Selection of a slice in the z direction.*

When the gradient is removed this frequency reverts to

$$\omega_0 = -\gamma B_0.$$

Only the protons in the slice have a net magnetization in the transverse plane but at this stage are all generating a transverse magnetization of the same frequency. In order to determine the contributions of protons at different positions in the slice we need to force these protons to precess at different frequencies. Let us consider first how we can select the signal from protons located in strips parallel to the y-axis of the slice. We do this by applying a second gradient, this time along the x-axis (figure 12.29) Then

$$B(x) = B_0 + G_x x.$$

The previously excited protons now increase or decrease their precession frequency depending on whether their local field has been increased or decreased by the x gradient field. The transverse magnetization

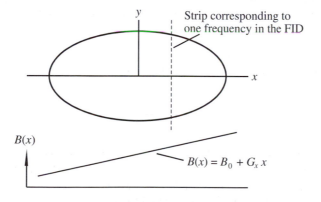

Figure 12.29. *FID selection in the x direction.*

at a point x now precesses with a frequency of $\omega(x) = -\gamma(B_0 + G_x x)$ and the FID now contains a range of frequencies, one for each x value. All the protons with the same x value, i.e. all those lying within a strip centred on x, will precess with the same frequency. By performing a Fourier transform of the FID and picking out the frequency $\omega(x)$ we have a signal dependent on the total transverse magnetization in a strip along the y-axis at position x. The Fourier transform of the FID is, in fact, a profile of the transverse magnetization in the x direction, integrated along the y-axis. This process is called frequency encoding.

This signal now constitutes a line integral in the y direction, the classic profile required by the computed tomography reconstruction algorithm. We could also set up a gradient in the y direction and measure the profile in that direction or indeed, by judicious combination of gradients in both x and y directions, generate a gradient, and hence a profile, at any direction in the xy plane. This allows us to develop a data set from which an image could be reconstructed using filtered back-projection (see Chapter 11, section 11.8.1). This could and has been done.

Phase encoding: the y direction

However, the preferred and most common method of image reconstruction uses a different approach. Let us return to the situation where the protons are excited and this time apply a gradient in the y direction (see figure 12.30).

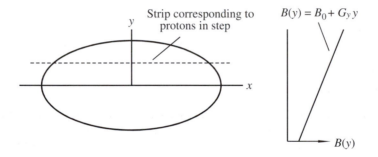

Figure 12.30. *FIDS selection in the y direction.*

As before, protons for which the field is increased increase their angular velocity of precession, whereas those for which the field is decreased reduce their angular velocity. Initially all components of the transverse magnetization along the y-axis (for all x) were in step, but the gradient field causes the components of magnetization at different y positions to become progressively out of step, or out of phase. After a period of time the gradient is removed. Let us suppose that, before the gradient field is applied, we have a rotating magnetization generating a signal,

$$m_{xy}(x, y)\, e^{+j\omega_0 t}$$

we now apply a gradient in the y direction so that the precessing frequency at y becomes

$$\omega_0 + \omega(y) = -\gamma(B_0 + G_y y).$$

The transverse magnetization now precesses at a different rate,

$$m_{xy}(x, y)\, e^{+j(\omega_0 + \omega(y))t}.$$

The net advance in signal at the level y compared to $y = 0$ at which $\omega(y) = 0$ after the y gradient has been applied for τ seconds is given by

$$m_{xy}(x, y)\, e^{+j\omega_0\tau} = m_{xy}(x, y)\, e^{-j\gamma G_y\tau}$$

and the integrated signal from the slice is given by

$$\iint m_{xy}(x, y)\, e^{-j\gamma G_y\tau}\, dx\, dy = \iint m_{xy}(x, y)\, e^{-jk_y y}\, dx\, dy$$

where $k_y = \gamma G_y\tau$. This formula is clearly a Fourier transform in the y direction so we can write it as

$$\int m_{xy}(x, k_y)\, dx.$$

As it stands the signal is integrated over the x direction. However, we already know how to differentiate signals from different points along the x direction by applying a gradient along that direction. The FID measured under these circumstances is the Fourier transform of the magnetization in the x direction, so the FID measured following prior application of a y gradient is $M(k_x, k_y)$, the components of the Fourier transform of $m_{xy}(x, y)$ in the k_x direction for one value $k_y = \gamma G_y\tau$. We can select other values of k_y by changing G_y or τ (usually the former) and measuring another FID. In this way we can determine the whole of $M(k_x, k_y)$ and hence $m_{xy}(x, y)$. This method of extracting a line in frequency space is called phase encoding.

12.4.8. The FID and resolution

The free induction decay (FID) signal for a spin–echo sequence looks like figure 12.31. As noted above the Fourier transform of this function represents the spatial distribution of magnetization in the frequency-encoded direction (the x direction in the discussion above). For a single point object the FT of the FID is the point spread function in this direction. If the whole of the FID is transformed the width of the PSF is dependent on the decay of the FID, i.e. T_2^* and the strength of the x gradient. Substituting various values of these parameters suggests that resolutions of the order of 0.1 mm should be achievable. In practice, the FID is noisy and this means that the tails of the FID are lost in noise. It is pointless to transform these so in practice a truncated FID is used. The effect of this is to increase the width of the FT and hence degrade the spatial resolution. Resolution is therefore dependent on noise, and since signal size can be improved by increasing the strength of the main field this is an important way of increasing image quality.

The width of the FT of the FID is the width of the image. This width is determined by the sampling frequency of the FID. Doubling the frequency with which the FID is sampled will double the field size. If the sampling frequency is inadequate the image will show aliasing effects.

Noise

The calculation of noise levels is complex. Principal sources of noise are 'thermal' noise in the receiver coils and sample noise where the sample appears as a resistance due to the presence of eddy currents. The signal-to-noise ratio (SNR) is proportional to the volume of the element. If we shrink (in the 2D case) the size of the pixel dimensionally by half the SNR decreases by 4. Note that we cannot recover the SNR for the original pixel size by pixel averaging. We could only get a twofold improvement. This is unlike radionuclide imaging. In addition, the SNR is proportional to the concentration of resonant nuclei.

The dependence on field strength is complex. If the coil noise dominates then SNR will increase as (approximately) $B_0^{7/4}$, if the sample noise dominates the increase is as B_0.

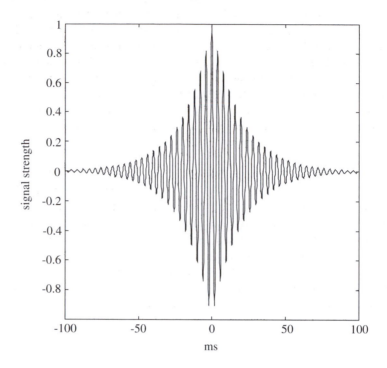

Figure 12.31. *A FID signal from a spin–echo sequence.*

12.4.9.   *Imaging and multiple slicing*

A typical sequence of events for spin–echo imaging is therefore the following.

- Set the z gradient and excite with a narrow-band 90° pulse to select a slice in the z direction.
- Apply a gradient G_y in the y direction for a time τ to phase encode the magnetizations.
- After time $TE/2$ apply a 180° pulse to rephase the FID and after another period $TE/2$ apply a gradient in the x direction to frequency encode the magnetizations and digitize the FID.
- After a time TR repeat the process, but with a different G_y.
- Continue this sequence until all the data are collected.

The above steps will collect data for one slice. It is useful to realize that, as far as the gradients are concerned, there is nothing special about the x, y and z directions. For example, slice selection could equally be achieved by applying a gradient in the x direction, phase encoding in the y direction and frequency encoding in the z direction. This would directly select and image a slice parallel to the axis of the main field rather than transverse to it. Combinations of gradients can select and image slices at any orientation.

Multiple slices

If sensitivity is not to be lost it is important that sufficient time elapses between excitation for the axial magnetization to have returned to its initial level. This means that *TR* must be sufficiently long for this to have occurred. In the period following *TE* the system is not preparing or recording data. During this period it is quite possible to select another slice, for example, at a different *z* value, and take data from this. Provided this new slice and the previous slice are sufficiently far apart, then the magnetization in the new slice will not have been affected by the excitation of the first slice (because of slice selection). Therefore, in the period during which magnetization is recovering in the first slice, data can be collected from the new slice without any interference from the previous slice. Typically the sequence of data acquisition might be to select a sequence of slices and measure data for one value of k_y in each slice in sequence, and then return to the first slice and repeat these steps for a second value of k_y and so on. It is possible to collect as many slices in the same time for one slice as *TE* will go into *TR*.

In the examples given above the image data are assembled sequentially. The basic time to collect the data is related to the pulse repetition time *TR* and the number of rows in *k* space. The value of *TR* varies according to the pulse sequences being used, but short values are typically 500 ms and long values 2000 ms. Total imaging times for a slice are measured in minutes. This can include multiple slices, but clearly the patient must lie still over this period of time if artefacts are to be avoided. *TE* values will vary typically from 10 to 60 ms depending on the pulse sequence used. Fast imaging sequences are available, for example, echo-planar imaging, but these are difficult to implement and generally are less sensitive than the more conventional approaches. There are a variety of other pulse sequences available each of which have their advantages. These cannot be covered here but are dealt with in much more detail in other publications.

Magnets and coils

This discussion is only the briefest introduction to MR imaging. We have not covered the practical aspects of building an MR scanner.

How do you produce the static magnetic fields of several tesla that are required for MR imaging? The larger fields require superconducting magnets but air-cored coils are used in lower-cost, lower-resolution systems. We showed in section 12.4.2 that the fractional excess increases with field strength so that larger signals are obtained at higher fields. This enables higher-quality images to be obtained in a shorter time. MR magnets are very expensive devices. One of the reasons for this is that very high field uniformity is required over the whole of the imaging field.

What are the power levels required for the radio-frequency pulses used in the various pulse sequences we have described? The power levels are actually very high, although their duration is such that the power deposited in the patient is not sufficient to cause dangerous temperature rises. Section 8.6 of Chapter 8 considered some of the effects of exposing tissue to radio-frequency fields.

How are the signals produced by the precessing protons detected in order to be used to measure the FID? Coils of various designs are used. These may be part of the main part of the MR scanner or they may be 'surface coils' placed close to the patient. The principle is simply the induction of a potential in the coil when placed in a changing magnetic field. Section 9.4.3 of Chapter 9 considered the problems of measuring small magnetic fields using a coil as the detector. One of the problems of measuring small signals is that of interference from external sources such as radio transmitters and other hospital equipment. The subject of interference and how to reduce its significance was described in sections 9.4.4 and 10.4. Considerable effort goes into the preparation of a site before an MR system is installed and a major part of this is the reduction of interference.

MR systems have improved dramatically since their first introduction. Resolutions down to levels of a few tens of microns are now possible. These improvements have arisen as a result of improvements in both the hardware and software of the scanners. More uniform magnetic fields and lower-noise systems for recording the FID have played a major part, but there have also been major improvements in the pulse sequences and image reconstruction algorithms.

12.5.   CT IMAGING

The origins of x-ray computed tomography imaging date to the early 1970s. It was the not the first tomographic imaging technique to gain widespread application. B-mode ultrasonic imaging was in use prior to the introduction of CT imaging. Neither was it the first to use the idea of image reconstruction from profiles, since a version of radionuclide tomography preceded the introduction of CT imaging. However, it was the first to produce section images of high quality and opened the way to an explosion of tomographic and three-dimensional imaging techniques. The success of CT depended largely on the development of a fast and accurate image reconstruction algorithm. This, in turn, generated a general interest in reconstruction algorithms some of which promise to improve the quality of tomographic imaging in areas other than x-ray computed tomography.

12.5.1.   *Absorption of x-rays*

The mechanism of absorption of x-rays is well documented (see section 5.2). For mono-energetic radiation a beam of x-rays is attenuated in an exponential fashion as it passes through matter. For the case of a homogeneous medium, i.e. one with constant linear attenuation coefficient μ, the transmitted intensity is given by

$$I = I_0 e^{-\mu L}$$

where L is the thickness of the object. If μ is a function of position then, using the geometry shown in figure 12.32

$$I(x, z) = I_0 e^{-\int \mu(x,y,z)\,dy}.$$

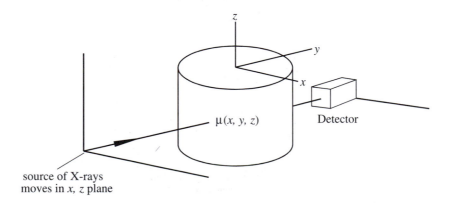

Figure 12.32. *Collection of profile data.*

If logarithms are taken of both sides of this equation

$$\log_e(I_0) - \log_e(I(x, z)) = \int \mu(x, y, z)\, dy.$$

The right-hand side of this result is a line integral through the linear attenuation profile. If the source and detector are scanned across the object and the above transform performed on the data the result is a profile representing a projection of the linear attenuation coefficient in the sense of Chapter 11 (section 11.8). Given a full set of these profiles the filtered back-projection algorithm can be used to reconstruct the distribution of the linear attenuation coefficient within the patient being imaged.

12.5.2.   *Data collection*

Early generation CT imaging systems physically scanned a source and a detector across the subject. When a profile had been collected the gantry holding the source and detector was rotated through typically 1° and the process repeated. Given that the scanning motion took 1 s, 3 min were required for a complete scan, which meant that the patient had to remain still within this time. CT images are very sensitive to motion of the object during data collection, and it is critical that the patient does not move during the investigation. The first investigations concentrated on imaging the head and it was possible to constrain head motion mechanically. However, this was not a useful solution for imaging the chest and abdomen, where there was significant internal movement due to respiration.

Current CT imaging systems use a ring of detectors (figure 12.33) and a rotating x-ray source to generate data. The x-ray beam produced is often referred to as a fan-beam. The source used is, in principle, no different from conventional x-ray sources, although there are some design requirements related to the relatively high power outputs required. The detectors must have a stable sensitivity. Solid-state detectors are typically a cadmium tungstate scintillator coupled to photodiodes and have high efficiency and a large dynamic range. An alternative is to use detectors filled with xenon under pressure. Ionization in the gas allows current to flow. These detectors have good stability but lower efficiency.

The scanning time is determined physically by the time taken to rotate the source around the subject and this can be reduced to a few seconds.

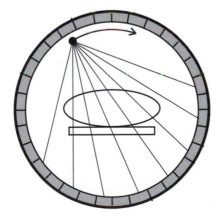

Figure 12.33. *Arrangement of source and detectors in a fourth-generation system. The x-ray source produces a fan-beam that is rotated around the body shown on the couch. There are detectors in a complete ring.*

At some time in the scan cycle each detector receives an unattenuated beam of x-rays. Assuming the x-ray source is constant in intensity this signal can be used to calibrate each detector during the scan. This ensures that there are no artefacts produced due to varying sensitivity between detectors.

12.5.3.   Image reconstruction

The data collected from a CT scanner of the configuration shown in figure 12.33 are not in parallel beam form. The profile data generated at each position of the source are not in the form required for Radon's transform (see section 11.8.1) and therefore the method described in Chapter 11 does not apply. However, a reconstruction algorithm does exist for the transform represented by the above geometry. The filtering function used on the profiles is different from the simple ramp filter. In addition, a weighting is applied depending on the position of every point from the x-ray source. However, the general principles are the same. We can show that fan-beam data acquisition is equivalent to parallel beam data acquisition by demonstrating that the fan-beam data can be resolved into parallel beam form as illustrated in figure 12.34.

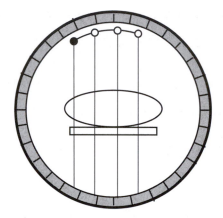

Figure 12.34. *An equivalent parallel beam configuration for a fan-beam system. The four selected positions of source and detector give parallel collection.*

For each position of the source there is a beam direction parallel to the first position. By selecting the appropriate detector for each beam position it is possible to generate a set of parallel beam data. With the above geometry the spacing of these data will not be uniform but this can be corrected by interpolation. In this way the fan-beam data can be transformed into parallel beam data and therefore the two imaging geometries are equivalent.

12.5.4.   Beam hardening

A primary assumption behind the image reconstruction algorithm used for CT imaging is that profile generation is a linear process (see section 11.2.2). In fact, this is not strictly the case. The x-ray source does not emit photons at a single energy but over a spectrum of energies, with a maximum energy equal to the voltage applied to the x-ray tube, and an average energy somewhat lower than this. The subject of x-ray spectra is dealt with in Chapter 21 and illustrated in figure 21.4.

Low-energy x-rays are absorbed more readily than higher-energy x-rays so as the beam passes through the patient or object being imaged the low-energy photons are preferentially absorbed and the average energy of the beam increases. This effect is called beam hardening. A practical effect of this is that the image of an object of uniform attenuation will appear to have a lower (higher energy) value of μ in the centre than at the periphery. The reason for this is that a beam passing through the centre of the object will be hardened more than a beam passing through the periphery and therefore the apparent uniform value of μ this attenuation represents will be lower than for an unhardened beam going through the edge regions.

Although the data are inconsistent an image can still be formed, however the apparent average μ for regions in the centre of the object is lower than for regions near the edges. This effect can be corrected. For soft tissues, including water, the linear attenuation coefficient for the energies typically used in x-ray CT is proportional to density, so that for any energy

$$\mu(E) = k(E)d$$

where $k(E)$ is a function of photon energy but is independent of the tissue and d is the density of the tissue. Then we can write

$$\mu(E) = \frac{d}{d_w}\mu_w(E)$$

where $\mu_w(E)$ and d_w are the values for water. For any photon energy we can write

$$I(x, z, E) = I_0(E)\,e^{-\int \mu(x,y,z,E)\,dy}$$

and

$$I(x, z, E) = I_0(E)\,e^{-(\mu_w/d_w)\int d(x,y,z,E)\,dy}.$$

The integral is now over density and is the same for all energies. Taking logarithms of both sides we have

$$\frac{\log_e(I_0(E)) - \log_e(I(x, z, E))}{k(E)} = \int d(x, y, z, E)\,dy.$$

Integrating over all energies gives

$$\int \frac{\log_e(I_0(E)) - \log_e(I(x, z, E))}{k(E)}\,dE = \int d(x, y, z)\,dy.$$

The right-hand side of this equation is simply an integration over density. There will exist a thickness of water which has exactly the same integrated density as given by $\int d(x, y, z)\,dy$ and which will attenuate the x-ray beam in exactly the same way.

For a given x-ray tube voltage we measure the ratio of input intensity to transmitted intensity for a variety of thicknesses of water. When we pass the same beam through a block of tissue we determine the ratio of input to transmitted intensity and look up the thickness of the equivalent amount of water. This is then the integral $\int d(x, y, z)\,dy$ we require for image reconstruction. Provided the x-ray spectrum is the same in both cases the effect of beam hardening can be eliminated. Unfortunately, this correction cannot be used to correct beam hardening by bony structures as the relationship between the attenuation coefficients as a function of energy does not hold between bone and water.

12.5.5. Spiral CT

The CT scanner collects data for a single slice. The limitation of this approach is that the position of the appropriate slice needs to be known. Usually several slices need to be taken to adequately cover the volume of interest. To produce multiple slices the patient is moved relative to the detector ring along the axis of the scanner. Normally the space between slices is larger than the pixel size within a slice. A 3D image can be produced by taking multiple slices but unless care is taken to ensure that the step along the axis of the scanner is the same as the pixel size the voxels will not be cubic. Scanning and stepping is not particularly efficient.

An alternative and automatic way of generating data for a 3D image is to move the patient continuously though the detector ring. The source effectively traces a spiral around the patient, hence the name of the technique. As continuous rotation of the source many times around the patient is required it is necessary to utilize slip-ring technology to transfer power to the x-ray source. Some modifications to the reconstruction algorithm are required to deal with the spiral scan path. Scanners of this sort allow useful volume (3D) images to be generated.

As for MR imaging we have not considered the practicalities of putting together a CT scanner. It is not a trivial task. The mean x-ray energy used in CT scanning is around 90 keV and the beam is generated using a rotating anode tube rated at perhaps 30 kW. The scan time may be about 5 s and doses of the order of 0.01 Gy can be delivered to the patient. Considerable effort has gone into improving the efficiency of detection of the x-ray beam so that the dose to the patient can be minimized. CT scanners are still developing and improvements in image quality being achieved. Some texts for further reading are given in the bibliography at the end of the chapter.

12.6.   ELECTRICAL IMPEDANCE TOMOGRAPHY (EIT)

12.6.1.   Introduction and Ohm's law

Human tissue conducts electricity quite well. This explains why we can receive an electric shock. In Chapter 8 (section 8.3) we described some models for the process of electrical conduction in tissue. In section 8.5 comparisons were made between tissues and table 8.4 gave the resistivities of a range of tissues. It can be seen that there is a wide range of resistivities and for this reason considerable effort has been expended to attempt to produce images of tissue impedance. The techniques developed for electrical impedance tomography (EIT) are still at the research stage and the technique is not an established medical imaging modality. However, we will consider it briefly in this chapter as it is an active area for research and it also illustrates techniques for imaging a nonlinear process.

The aim of electrical impedance tomography is to produce images of the distribution of electrical impedance within the human body. For simple conductors we can speak of either the resistance or the conductance, where conductance is the inverse of resistance and vice versa. If the conductor is complex we speak of the impedance and its inverse the admittance. The specific resistance of a material is the resistance between opposite faces of a unit cube of the material. This is usually called the resistivity with its inverse the conductivity. Although not yet in common use the corresponding terms for the case of complex conductors would be impedivity and admittivity. Most theoretical formulations of electrical impedance tomography are formulated in terms of admittance rather than impedance and it is common to use the term conductivity rather than admittivity, it being understood that in this context conductivity is complex, i.e. it has both a real and an imaginary part. We shall use this terminology in this section.

The forward problem

For volume conductors the relationship between the current vector J within the conductor, the (complex) conductivity σ and the electric field vector E is given by

$$J = \sigma E.$$

If the material is isotropic then E and J are parallel. If the material is not isotropic then this is generally not the case. We will assume isotropic conductivity, although some tissues are anisotropic. Every point within a conducting object through which electric current is flowing has an associated electrical potential ϕ. E is the local gradient of potential, usually written as $\nabla\phi$. If there are no sources or sinks of current within the object then the divergence of J, written as ∇J, is zero. So we can write

$$\nabla(\sigma \nabla \phi) = 0.$$

We can place electrodes on the surface of the object and apply a pattern of current to the object through these electrodes. The only constraint is that the total current passed into the object equals the total current extracted from the object. Suppose we apply a known current pattern through an object with conductivity distribution σ. ϕ within the object must satisfy the above equation. If we know the conductivity distribution σ, it is possible to calculate the distribution of ϕ within the object and on the surface of the object for a given current pattern applied to the surface of the object. This is known as the forward problem. For some simple cases it is possible to do this analytically. However, for most cases this is not possible and a numerical solution must be used. The technique used to do this is the finite-element method (see section 1.9.7). The conductivity is divided into small volume elements and it is usually assumed that the conductivity value within each element is uniform. It is then possible to solve numerically the last equation and calculate the potential distribution within the object and on the surface of the object. The principal reason for doing this is to predict the surface voltage measurements obtained for a real object whose internal conductivity distribution is known. The accuracy with which this can be done will depend on the number of volume elements and their shape and distribution, how accurately the boundary shape of the object is known and how well the electrodes are modelled.

The inverse problem

We cannot, in general, measure ϕ within the volume of the object but we can measure it on the surface of the object. Since ϕ on the surface of the object is determined by the distribution of conductivity within the object and the pattern of current applied to the surface of the object, an obvious question is: can we use knowledge of the applied current pattern and the measured surface voltage values to determine the distribution of conductivity within the object? This is known as the inverse problem. Provided the voltage is known everywhere on the surface and sufficient different current patterns are applied the answer is yes, provided the conductivity is isotropic. For the anisotropic case the answer is no. The best that can be achieved is that the conductivity can be isolated to a class of possible conductivities. This is a rather esoteric result which in practice is ignored.

The above arguments have implicitly assumed that the current is DC. In practice, we want to use alternating current because direct current can produce polarization effects at the electrodes which can corrupt measurements and can stimulate responses in the nervous system. Accurate measurement of alternating voltages is easier than DC voltages and in any case we want to see how conductivity changes with frequency. Provided the wavelength of the alternating current (i.e. the velocity of light divided by the frequency) is much larger than the size of the object, we can treat alternating currents as though they are DC currents. This is called the quasi-static case. The frequencies used in EIT are usually less than 1 MHz and the quasi-static assumption is a reasonable one.

12.6.2. Image reconstruction

In order to fix our ideas consider the situation where we have a conducting object with four electrodes attached to the surface of the object as shown in figure 12.35. Unit current is passed between two of the electrodes (the drive pair) and the voltage between the other two (the receive pair) is measured. Then if we represent this measurement by g_{rd} it can be shown that its value is given by

$$g_{rd} = \iiint \sigma(x, y, z) \nabla \phi \, \nabla \psi \, dx \, dy \, dz$$

where $\nabla \phi$ is the potential gradient at the point x, y, z produced by unit current being passed between the drive electrodes and $\nabla \psi$ is the field at this point which would be produced if unit current were passed between

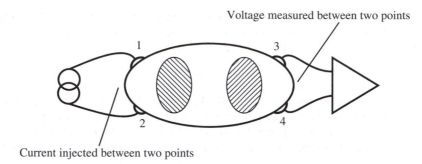

Figure 12.35. *Current is injected between electrodes 1 and 2 and the resulting potential is measured between electrodes 3 and 4. The ratio of voltage to current is a transfer impedance.*

the receive electrodes. Note that if we exchange the drive and receive pairs we obtain the same measured value. This is the principle of reciprocity. If we can reduce the conductivity distribution to a set of M volume elements such that the conductivity σ_m within each element can be assumed to be uniform then we can write this integration as a summation

$$g_{\mathrm{rd}} = \sum_{m=1}^{M} \sigma_m (\nabla \phi_{\mathrm{d}} \cdot \nabla \psi_{\mathrm{r}})_m.$$

We can now consider many combinations of the drive and receive pairs, say N such combinations. If we replace rd by n we can write

$$g_n = \sum_{m=1}^{M} \sigma_m ((\nabla \phi \cdot \nabla \psi)_n)_m = \sum_{m=1}^{M} \sigma_m A_{nm}$$

or in matrix form

$$\boldsymbol{g} = A\boldsymbol{\sigma}.$$

Although this has been illustrated using the idea of a pair of electrodes to pass current and a pair to measure the voltage, we could have considered other configurations. For example, the current could have been applied through many electrodes, forming a current pattern distributed across the surface of the object. For this case there will still be a potential at each point in the volume produced by the applied current, so the equations are still valid.

One thing which makes EIT images harder to reconstruct than some other tomographic methods is that $\nabla \phi$ and $\nabla \psi$, and hence the matrix A, depend on the conductivity distribution. The last equation should really be written as

$$\boldsymbol{g} = A(\sigma)\,\boldsymbol{\sigma}$$

which makes it difficult to solve since the answer (σ) needs to be known before A can be determined and σ is just what we do not know. The reconstruction problem, determining σ from g, is nonlinear and the practical consequence of this is that an iterative process needs to be used to reconstruct the image. Geselowitz obtained an important result which has bearing on this. He showed that if the conductivity changed by an amount $\Delta\sigma$ then the change in voltage measurement is given by

$$\Delta g_{\mathrm{rd}} = -\iiint \Delta \sigma(x, y, z) \, \nabla \phi_{\mathrm{b}} \, \nabla \psi_a \, \mathrm{d}x \, \mathrm{d}y \, \mathrm{d}z$$

where $\nabla\phi_b$ is the gradient before the conductivity change occurred and $\nabla\psi_a$ is the gradient after the change occurred. Provided that the change is small this equation can be linearized by replacing $\nabla\psi_a$ with the value $\nabla\psi_b$ before the change occurred. The matrix formulation of this is now

$$\Delta g = S\Delta\sigma$$

where S is known as the sensitivity matrix. We are trying to determine the conductivity distribution given the measured surface voltages. In order to do this we start off with an assumed conductivity distribution σ_0. This is usually taken to be uniform and equal to the average conductivity within the object. The matrix S_0 is calculated using finite-element modelling (FEM) for this uniform case and for the pattern of drive and receive electrodes used. The values of the data vector g_0 are also calculated using FEM. g_0 is subtracted from the actual measurements g_m to form the vector Δg. The matrix S_0 is inverted and the vector $\Delta\sigma$ formed using

$$\Delta\sigma = S_0^{-1}\Delta g.$$

A new conductivity estimate is then formed from

$$\sigma_1 = \sigma_0 + \Delta\sigma.$$

A new sensitivity matrix S_1 and a new data vector g_1 are formed using σ_1. This new data vector should be closer to g_m than g_0. A new Δg and hence $\Delta\sigma$ is calculated and added to the current estimate of conductivity. This process is repeated until the differences between the measured data and the computed data are within the measurement error. The required distribution of conductivity within the object is the distribution used to calculate the final data vector. Although there is no guarantee that convergence will occur, in practice this algorithm works quite well under many circumstances and converges to the correct conductivity distribution.

The forward problem predicts the voltage distribution on the surface given the conductivity distribution. This problem is very ill-posed. In this case this means that large changes in conductivity may only produce small changes in the surface voltages. Conversely, small errors in determining the surface voltages can produce large errors in the estimate of the conductivity distribution. These errors arise for two reasons. The first is measurement noise. In EIT the noise arises primarily in the instrumentation used to make the measurements. It is possible to reduce the noise levels down to about 0.1% of the signal with careful design, but thermal noise then represents a fundamental limit to what can be achieved. The second source of error is that in the reconstruction algorithm it is necessary to calculate at the kth iteration the vector g_k, and subtract it from g_m. If g_k is in error then so is Δg and the calculated conductivity may well diverge from the true value. In fact, when the current conductivity is correct the computed value must equal the measured value to an accuracy at least as good as the noise. This is very difficult to achieve for a variety of reasons. One important reason with medical imaging is that accurate calculation requires accurate knowledge of the boundary shape of the object in three dimensions and accurate knowledge of the electrode positions. For example, the size of the signal measured between two electrodes will depend in part on the distance between them. Changing this by 1% will typically result in a change in magnitude of the measurement by this amount. If the measured values are to be predicted to this accuracy then determination of the electrode positions must be made to this accuracy. This is difficult for *in vivo* measurements.

The fact that we have had to consider the above problem in three dimensions illustrates the biggest problem in EIT. In x-ray CT the value of each point in a profile is simply a line integral along the path of the x-ray beam. However, in EIT every value is a volume integral as current spreads throughout the object.

Differential imaging

Suppose that we only implement one step of the iterative algorithm proposed in the above section. Then

$$\Delta\sigma = S_0^{-1}(g_m - g_0).$$

However, instead of calculating g_0 we measure it before some change in conductivity occurs and then measure g_m after a change occurs. $\Delta\sigma$ is then the change in conductivity between these two states. In general, S_0 is clearly not the correct matrix to use. Suppose now we construct a diagonal matrix G whose diagonal elements are the average values of the corresponding elements of g_m and g_0, i.e.

$$G_{ii} = \tfrac{1}{2}(g_{mi} + g_{0i})$$

$$G_{ij} = 0 \qquad i \neq j.$$

Then the above reconstruction equation can be written as

$$\Delta\sigma = S_0^{-1} G G^{-1}(g_m - g_0)$$

$$\Delta\sigma = (G^{-1}S_0)^{-1} G^{-1}(g_m - g_0)$$

$$\Delta\sigma = F^{-1}\left(2\frac{g_m - g_0}{g_m + g_0}\right).$$

Whereas the values of the elements of the vector $g_m - g_0$ will be sensitive to the exact placing and spacing of the electrodes and the shape of the object, we expect the vector $(g_m - g_0)/(g_m + g_0)$ to be less sensitive to these things as ideally they do not change between the two measurements, whereas the conductivity does change. Similarly, we anticipate that the matrix F is less sensitive to the values of the placing and spacing of the electrodes and the shape of the object and the exact conductivity distribution than S. Experimentally this appears to be the case to such an extent that the last equation has been used to construct images of the changes in conductivity within the human body, and to date is the only reconstruction method which has been able to do this reliably. F is usually constructed from a simple model of the object. Typically the conductivity is assumed to be uniform, the boundary circular and the electrodes equally spaced around the boundary.

Strictly speaking F can only be inverted if it is square which is only the case if the number of measured values is equal to the number of elements of conductivity. In practice, this is rarely the case and F^{-1} must be replaced by F^+, where F^+ is known as a pseudo-inverse of F. In fact, because F is generally hard to invert (it is ill-conditioned) the pseudo-inverse is required even if F is square. Construction of the best pseudo-inverse to use depends on several factors, including the noise on the data and these issues are well covered in the appropriate literature.

12.6.3. Data collection

Current is applied to the object being imaged through electrodes placed on the surface of the object (figure 12.35). Surface measurements of voltage are made using electrodes applied to the surface of the object. The reconstruction theory given above has made no mention of which current patterns are optimum. Most EIT systems apply current between a pair of electrodes at a time (a bipolar drive configuration). However, there is no reason why current cannot be applied and extracted through many electrodes simultaneously. In fact, it can be shown that for any conductivity distribution there is a set of optimum current patterns in the sense that, all other things being equal, the signal-to-noise ratio is maximized when this set is used. In some simple cases the optimum set can be computed analytically. For the two-dimensional case of any radially symmetric distribution of conductivity the optimum current patterns are a set of sine and cosine functions. If there are P electrodes equally spaced around the circular boundary of the object at positions θ_p, then the current through the pth electrode has magnitude $I = I_0 \cos n\theta_p$ or $I = I_0 \sin n\theta_p$, where n can take values from 1 to $P/2$.

Amongst the possible bipolar patterns the most commonly used is the adjacent drive configuration. Current is passed between sets of electrodes which are adjacent in space. For example, with N electrodes placed around the boundary of a two-dimensional object current is passed in turn between the N possible

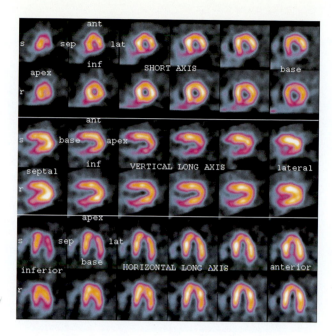

Figure 6.13. *Slices normal to the long axis of the left ventricle. A SPET myocardial examination. The three-dimensional image of the left ventricle has been displayed as slices normal to the long axis of the ventricle and slices in two orthogonal planes to this. The top row in each of the three sections is a stress image and the second row a rest image of the same subject. There is a region of decreased perfusion in the inferior wall of the heart in the stress image which is perfused in the rest image, indicating ischaemic heart disease in this territory.*

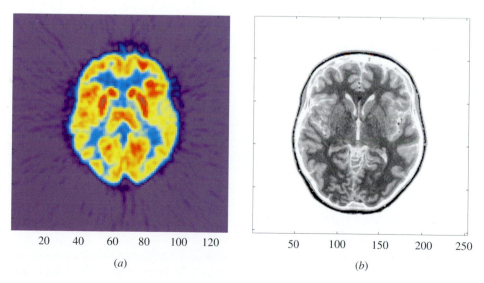

Figure 11.1. *(a) A PET image of the brain of a normal subject. This is an image of function, since the colour is related to the uptake of tracer (^{18}FDG) in the brain tissues and this is a measure of physiological activity (glucose metabolism). (b) An MR image of approximately the same slice. This shows the anatomy of the brain. Both these images are images of the same object, but imaging a different aspect of the brain.*

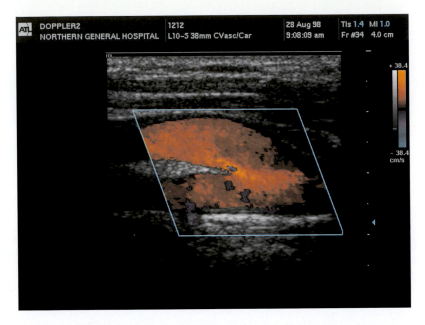

Figure 12.20. *An ultrasound image of the carotid artery in the neck. The bifurcation (from the right) into the internal and external carotid vessels is clearly seen. By superimposing the Doppler signal the direction of the blood flow can be seen and small areas of turbulence can be identified. The velocity scale is shown in the top right-hand corner of the figure. Image kindly provided by Dr S B Sherriff.*

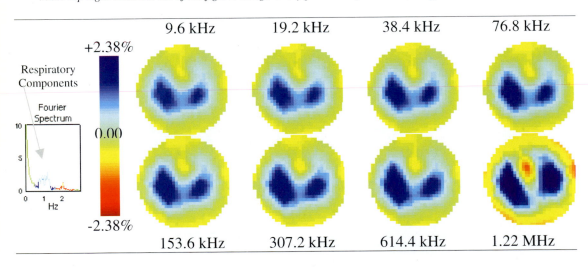

Figure 12.36. *By extracting a band of frequencies (marked in blue on the left) prior to image reconstruction, EIT images can be obtained during tidal breathing in neonates. The eight images correspond to data collected at eight different frequencies of applied current. It can be seen that there is an increase in the resistivity change of the lungs with increasing frequency. In all these images anterior is shown at the top and the left of the body appears on the right of the image. This is the conventional orientation for medical images. (The image is taken from the PhD thesis of A Hampshire.)*

adjacent pairs. In fact, one of these pairs is redundant. Suppose we pass current between electrodes 2 and 3, and make voltage measurements, then between 3 and 4, and make measurements, and then between 4 and 5 and so on up to passing current between N and 1, and then add all the sets of voltage measurements together. The principle of superposition ensures that the resulting voltage set is identical to that measured when current is passed between electrodes 1 and 2.

Although distributed voltage measurements can be made the most common procedure is to measure the voltage difference between adjacent electrodes. For each applied current N such measurements can be made. Using a similar argument to the above, it is easily shown that one of these measurements is redundant. In addition, if we interchange drive and receive electrode pairs we get the same result because of the reciprocity principle. From these considerations it is easily shown that from N electrodes it is possible to obtain a maximum of $N(N-1)/2$ independent measurements. As any bipolar or distributed configuration of drive and receive patterns can be synthesized from the adjacent drive configurations it is not possible to do better than this. It is possible to do worse.

The above result assumes that it is possible to make a useful voltage measurement on an electrode through which current is flowing. This is actually unreliable because of the variability of electrode impedances. For an adjacent electrode configuration where the same electrodes cannot be used for driving and receiving the number of available measurements then reduces to $N(N-3)/2$. For a 16-electrode configuration 104 independent measurements can be made. This represents a fairly coarse resolution image as, of course, the number of independent image pixels cannot be greater than the number of independent measurements. Increasing the number of electrodes will increase the number of independent measurements. For 32 electrodes the number increases to 464. However, more detailed theoretical analysis of the problem suggests that this increase in measurements cannot always be translated into a comparable increase in image resolution, especially towards the centre of the object.

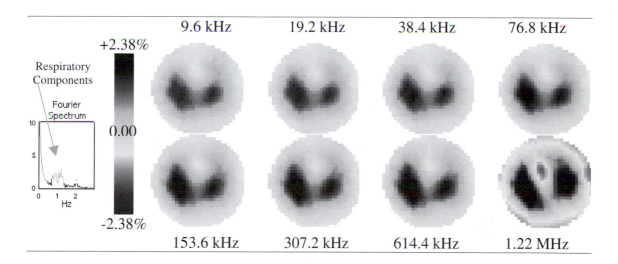

Figure 12.36. *By extracting a band of frequencies (marked in blue on the left) prior to image reconstruction, EIT images can be obtained during tidal breathing in neonates. The eight images correspond to data collected at eight different frequencies of applied current. It can be seen that there is an increase in the resistivity change of the lungs with increasing frequency. In all these images the anterior is shown at the top and the left of the body appears on the right of the image. This is the conventional orientation for medical images. (The image is taken from the PhD thesis of A Hampshire.) This figure appears in colour in the plate section.*

Figure 12.36 shows a set of lung images obtained from 16 electrodes placed around the thorax of a neonate. The images are differential images with the reference taken at expiration and the data at inspiration. Reference and data have been extracted from a record of tidal breathing using Fourier analysis. Only the frequency components corresponding to respiration at around 1 Hz have been used to reconstruct the images. The lungs can be seen quite clearly but the spatial resolution is poor.

12.6.4.   *Multi-frequency and 3D imaging*

It is possible to use the idea of differential imaging to observe changes of tissue impedance with frequency. We showed in Chapter 8 that tissue impedance is complex and that the magnitude of the impedance will fall with frequency. By using measurements made at one frequency as a reference and then using measurements made at other frequencies as data, images showing how impedance changes with frequency can be obtained. The impedance spectrum for each pixel in the image can then be fitted to a Cole model (see section 8.3.2), and hence images showing the spatial distribution of the Cole parameters can be determined. This has been shown to be an important innovation and allows quantitative measurements of tissue properties to be imaged.

Three-dimensional imaging

All the results developed so far relate to imaging of either two- or three-dimensional objects. In practice, most objects of interest are three dimensional. However, most images produced so far have attempted to treat the object as a series of two-dimensional slices by collecting data from electrodes placed around the borders of a plane through the object. Unfortunately, unlike other three-dimensional imaging methods such as x-ray CT, a slice cannot be simply defined in this way in EIT. Electric current injected into a three-dimensional object cannot be confined to a plane, but flows above and below the plane defined by the electrodes. This means that conductivity distributions in these regions contribute to the measured signals. In order to eliminate this problem data collection and image reconstruction must be considered to be a full three-dimensional problem. This can be achieved by placing electrodes over the full surface of the objects. Typically this is done by placing electrodes in a series of planes around the object. Any pair of electrodes, both within and between planes, can be used for applying current and any pair used for making measurements. Reconstruction, although involving more measurements and larger sensitivity matrices, is identical in principle to the two-dimensional case.

EIT is a fast and inexpensive method of imaging tissue properties. Electrode application, especially for three-dimensional imaging, is still a problem and collection of data of adequate quality and reliability can be difficult. Some areas of clinical application have been explored. These include the measurement of lung water and of gastric emptying. However, the technique is still a subject for research.

12.7.   PROBLEMS

12.7.1.   *Short questions*

a   What is a planar image?
b   What is a tomographic image?
c   What does multi-modality mean in the context of medical imaging?
d   What is a radiopharmaceutical?
e   Name the isotope most commonly used in scintigraphy/gamma camera imaging.
f   Approximately how many PM tubes are used in the imaging head of a modern Anger gamma camera?
g   Of what is the collimator of a gamma camera made?

h   What limitation does 'dead time' impose?
i   Is 0.5 mm a typical intrinsic resolution for a gamma camera?
j   Which scintillation crystal is typically used in an Anger gamma camera?
k   Why are positron emitters used in PET?
l   Does ultrasound travel faster or slower in bone than in soft tissue?
m   About how long does it take an ultrasound echo to travel across the trunk?
n   How does an ultrasound array transducer focus ultrasound at a point in the body?
o   Can an ultrasound phased array be used to direct the beam to the side?
p   What limits the maximum velocity of blood that can be measured in a Doppler imaging system?
q   What is the approximate value of the Larmor frequency for protons in a magnetic field of strength 1 T.
r   What resonates in magnetic resonance imaging?
s   Is 100 μs a typical value for a T_2 time constant in tissue?
t   What effect might cause the centre of an x-ray CT image of a uniform object appear to have a lower value of absorption coefficient μ than the periphery?
u   What is meant by 'the forward problem' in electrical impedance tomography?
v   What is meant by 'reciprocity' in EIT?
w   What spirals in spiral CT?
x   What is the basis of magnetic resonance imaging? How is the position of the resonating protons found?
y   What limits the best spatial resolution that can be obtained in any imaging system?
z   How are the protons caused to precess in an MRI system?

12.7.2.   Longer questions (answers are given to some of the questions)

Question 12.7.2.1

With the aid of diagrams, describe the gamma camera collimator, its function and the design parameters that are critical to its performance.

Sketch the following collimator types:

(a)   Plane parallel collimator.
(b)   Diverging collimator.
(c)   Pinhole collimator.

Describe how nuclear imaging can be used in the diagnosis of kidney disease.

Answer

No γ-ray lens exists, so a collimator is used to produce an image on a scintillation crystal of a distribution of radioactivity. The collimator consists of a matrix of holes in a lead plate several centimetres thick. This sits on top of a large NaI crystal.

Only γ-rays that are coaxial with the holes will reach the crystal—this reduces scatter.
The parameters which are critical to collimator performance are:

•   hole diameter;
•   hole length;
•   hole spacing (septal thickness).

High-energy imaging requires thick septa to stop off-axis γ-rays. This reduces the hole count, and therefore reduces sensitivity.

Longer holes produce better collimation and better resolution, but also reduce sensitivity. Hole diameter affects resolution (point response function).

Plane parallel collimator

This is a general purpose collimator. It has a magnification of unity. High- and low-energy versions exist. This type of collimator is used for bone scanning.

Diverging collimator

This has a wide field of view. The magnification is less than one. It has the disadvantage of causing some image distortion.

Pinhole collimator

This consists of a small diverging hole at the apex of a lead cone. The magnification is greater than one. Again it causes some image distortion. This type of collimator is used for thyroid scanning.

Kidney imaging

Dynamic imaging can be used to investigate renal function. Consecutive images are acquired over a period of time following the injection of a radiopharmaceutical. A Tc-based tracer is injected intravenously and imaging is carried out after 30 min. The back of the patient gives the best view of the kidneys and urinary system. 70×20 s images are often taken. It is possible to assess kidney drainage by drawing a region of interest around the kidneys and monitoring radiopharmaceutical uptake with time. Abnormalities may appear as an asymmetry between the left and right kidneys. Clearance rates can be measured and compared with normal values.

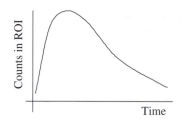

Figure 12.37. *A region of interest (ROI) drawn on the kidney image can be used to give a curve showing the uptake and clearance of a radiopharmaceutical by the kidneys.*

Question 12.7.2.2

An MRI system operates at a magnetic field strength of 1 T. Compute the resonant frequency of hydrogen nuclei.

For a spin–echo sequence the functional form of the FID obtained from a point object at position x along the field gradient axis is given by

$$\mathrm{FID}(t) = \cos\omega(x)t\, e^{-|t|/T_2^*}.$$

It will look like

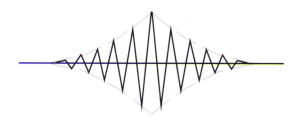

Figure 12.38. *Free induction decay from a spin-echo sequence.*

where the frequency $\omega(x)$ is a function of position because of the gradient G_x. Compute the Fourier transform of this function. This function is a point spread function because it describes the spatial response to a point object. Calculate the FWHM of this PSF in terms of the spatial coordinate x. Is the PSF position independent?

Answer

The Fourier transform of the FID is given by

$$f(\omega) = \frac{2}{T_2^*} \frac{1}{\frac{1}{(T_2^*)^2} + (\omega - \omega(x))^2}$$

and hence the FWHM of this is given by $\Delta\omega = 2/T_2^*$. Since $\omega(x) = -\gamma(G_x x + B_0)$, $\Delta\omega = \gamma G_x \Delta x$ and hence $\Delta x = 2/(\gamma G_x T_2^*)$, which is the spatial FWHM. This FWHM is position dependent as it depends on T_2^* which is position dependent. In practice the resolution is limited by noise which reduces this dependence substantially.

Question 12.7.2.3

Explain what is meant by the 'resonant frequency' of the piezoelectric crystal used to make an ultrasound transducer. How is the resonant frequency related to the thickness of the crystal?

What are the two characteristic properties of an ultrasound pulse that are determined by the Q of a transducer and how does the Q relate to the frequency spectrum? Is the Q kept high or low in ultrasonic imaging and how is this achieved?

A phased array can be used to scan through an angle. This is achieved by delaying the activation of the transducers in the array relative to each other. For a phased array of length L derive a formula for the delay required for the transducer at point X along the array (origin at the centre of the array) when a beam of angle θ to the normal to the array is being generated.

For a phased array of length 10 cm calculate the relative delay between the end transducers when a beam travelling at 30° to the normal to the array is being produced.

Answer

The resonant frequency is the dominant frequency at which a crystal vibrates when it has been excited by a short electrical pulse. This is usually the mode of vibration in which the opposite faces of the crystal oscillate inwards and outwards. The frequency of the ultrasound will be given by the velocity of sound in the crystal divided by twice the crystal thickness.

The Q of a transducer determines the range of frequencies contained within a pulse and the rate at which the pulse will decay. A high Q will give a narrow range of frequencies and a slow decay. Q can be measured by dividing the peak intensity in the spectrum by the FWHM of the peak.

A low Q is required for imaging so that the pulse is short. This gives the best time resolution and hence image resolution. It is achieved by placing an absorber on the back face of the transducer.

If the thickness of the crystal is t then the longest resonant wavelength is $\lambda = t/2$. If the velocity of the wave in the crystal is c then the frequency of the resonant wave is $f = c/\lambda$ and so $f = 2c/t$. The resonant frequency is inversely proportional to the thickness of the crystal.

The spatial delay d at the point x along the phased array is given by $d = x \sin \theta$ and for a velocity of sound c this corresponds to a time $t(x) = t = (x/c) \sin \theta$.

The relative time difference for $L = 10$ cm, $\theta = 30°$, $c = 150\,000$ cm s^{-1} is 33.3 μs.

Question 12.7.2.4

Describe the principles of four different medical imaging modalities and explain why you might categorize them as either anatomical or functional modalities.

Question 12.7.2.5

Discuss the principles involved in imaging the electrical resistivity of tissue and explain why this technique is unlikely to allow high spatial resolution images to be produced.

Answer

It can be shown that provided the conductivity is isotropic, the conductivity distribution of an object can be deduced by applying an independent set of current patterns through the surface of the object and measuring the voltage distribution generated on the surface of the object. Unfortunately, if current is applied between a pair of electrodes it does not flow directly between them but spreads out. This means that the value of the voltage developed on any electrode is affected by the conductivity throughout the object, rather than just along a line as is the case for CT. This 'blurring' of the signal (the voltages measured on the electrodes) relative to the object (the conductivity distribution) means that the resolution of the image is relatively poor. Although, in principle, this could be corrected by signal processing the presence of noise on the data prevents this.

In practice, the resolution of an EIT image, as with all images, is also limited by the number of data points collected. This number does rise as the square of the number of electrodes. In practice, the previously mentioned effects mean that in 2D using more than 16 electrodes gains little in the way of image improvement.

Question 12.7.2.6

In spiral CT the patient is moved through the scanner. This means that the data collected during a rotation of the x-ray source are not strictly consistent. Can you suggest how the collected data might be modified to extract consistent data? What are the assumptions behind your answer?

Answer

The CT system consists of a series of detectors. At each rotation a detector returns to the same position relative to the imaging system but the patient has moved along the axis of the imaging system, the z-axis. Associated with each repeat of the detector position is a z-axis value. The set of these values will be different for each detector and position, but will be known. The time sequence of data collected by a detector at a particular angular position can therefore be interpolated to a standard set of z positions, and these interpolated consistent sets of data can be used to produce images. The principal assumption behind this approach is that the data in the time sequence represent a rapid enough sampling (in space) for interpolation to be accurate. Failure to achieve this will result in aliasing effects.

Question 12.7.2.7

You wish to examine, using MRI, a group of patients with heart disease. You feel that it would be prudent to monitor heart rate. What methods are available to you and what problems can you identify?

Answers to short questions

a   A planar image is a 2D image formed by transmission through a 3D object.
b   Tomographic images are slice images through a 3D object.
c   Multi-modality imaging means the use of more than one type of image to investigate part of the body, e.g. ultrasound and CT used together.
d   A radiopharmaceutical is a substance to which a radioactive label has been attached in order to be able to trace its position within the body.
e   The isotope of technetium ^{99m}Tc is probably the most commonly used.
f   60–90 photomultiplier tubes are often used in gamma cameras.
g   Collimators are made of lead.
h   Dead time limits the maximum count rate that can be dealt with by a gamma camera.
i   No, a more typical intrinsic resolution would be 3–5 mm.
j   Sodium iodide doped with thallium is the crystal usually employed in a gamma camera.
k   Positron emitters annihilate to produce two γ-rays that move directly away from each other. If both γ-rays can be detected then the line along which the positron was emitted is known. This can be use to produce a PET image.
l   Ultrasound travels about three times faster in bone than in soft tissue.

m   Ultrasound travels at about 1.5 mm μs^{-1}. If a typical trunk is 30 cm across then the ultrasound will take 200 μs to travel through it.

n   If the elements of an ultrasound array are pulsed at different times then a focused beam can be produced. The peripheral elements must be pulsed first and the central elements last.

o   Yes, an ultrasound beam can be steered sideways by delaying the pulses applied to the elements on the side to which the beam is to be steered.

p   The time taken for an echo to travel to the blood vessel and back to the transducer limits the maximum Doppler frequency that can be measured because of aliasing.

q   The Larmor frequency for protons in a field of 1 T is 42.5 MHz.

r   It is usually protons that resonate in the high static magnetic field used in magnetic resonance imaging.

s   No, 100 ms is a more likely value for T_2.

t   Beam hardening can make μ appear less in the centre of an image.

u   The forward problem in EIT is that of calculating the potentials on the surface of the body, when an electrical current is applied, when the conductivity distribution within the body is known.

v   Reciprocity is used to describe the fact that if a current between one pair of electrodes on the body causes a potential between a second pair of electrodes, then the pairs of electrodes can be interchanged without changing the measured potential.

w   The point where the x-ray beam enters the body spirals around and down the body in spiral CT.

x   The position of the resonating protons is found by applying gradients of a magnetic field so that the frequency of precession then uniquely identifies the position of the proton.

y   The limitation on spatial resolution is determined by the number of independent measurements that can be made from the object. You cannot have more pixels in the image than the number of measurements made.

z   Protons are caused to precess by applying a pulse of radio-frequency energy.

BIBLIOGRAPHY

Boone K, Barber D C and Brown B H 1997 Review: imaging with electricity: report of the European Concerted Action on Impedance Tomography *J. Med. Eng. Technol.* **21** 201–32

Hampshire A R 1998 Neonatal impedance tomographic spectroscopy *PhD Thesis* University of Sheffield

Hendrick W R, Hykes D A and Starchman D E 1995 *Ultrasound Physics and Instrumentation* 3rd edn (St Louis, MO: Mosby)

Hounsfield G N 1973 Computerised transverse axial scanning (tomography) *Br. J. Radiol.* **46** 1016–51

Lerski R A (ed) 1985 *Physical Principles and Clinical Applications of Nuclear Magnetic Resonance* (London: IPSM)

Mansfield P and Pykett I L 1978 Biological and medical imaging by NMR *J. Magn. Reson.* **29** 355–73

McDicken W N 1991 *Diagnostic Ultrasonics—Principles and Use of Instruments* 3rd edn (Edinburgh: Churchill Livingstone)

Ridyard J N A 1991 Computerised axial tomography *Biological and Biomedical Measurement Systems* ed P A Payne (Oxford: Pergamon)

Sharp P F, Dendy P P and Keyes W I (eds) 1985 *Radionuclide Imaging Techniques* (New York: Academic)

Shirley I M, Blackwell R J and Cusick G 1978 *A User's Guide to Diagnostic Ultrasound* (Tunbridge Wells, UK: Pilman Medical)

Webb S (ed) 1988 *The Physics of Medical Imaging* (Bristol: Hilger)

CHAPTER 13

MATHEMATICAL AND STATISTICAL TECHNIQUES

13.1. INTRODUCTION AND OBJECTIVES

Any investigation of the performance of the human machine requires the taking of measurements and the interpretation of the resulting data. Many of the chapters of this book deal with the problems of measurement of signals in the human environment, and the purpose of this chapter is to introduce some of the techniques used for the classification, description and interpretation of measured signals. The primary motives for the development of the signal analysis methods presented in this chapter are:

- The qualitative and quantitative description of signals.
- The description of the inherent variability of most biological signals.
- The understanding of the effects of discrete and limited-time sampling of signals.
- The statistical comparison of signals.
- The removal of noise from signals.
- The interpolation or extrapolation of signals to provide predictions of behaviour.
- The identification of underlying processes responsible for the signals.

This chapter does not relate directly to any one of the other chapters in the book. It is relevant to most areas of medical physics and biomedical engineering. It does contain mathematics as the subject is essentially mathematical but the non-mathematical reader should be able to use the text to gain some understanding of the principles of the various techniques. Test your understanding with the list of short questions given in section 13.7.1. The section on applied statistics (13.5) is not intended to be a short course on statistics. There are many books on statistics so no purpose would be served by duplication. Section 13.5 aims to act as a refresher by reintroducing the concepts of statistics and highlights some aspects that are most relevant to the biomedical scientist.

Readers are referred to Lynn (1989) for a good introduction to the processing of signals. Some parts of this chapter follow the thinking presented by Lynn.

13.1.1. Signal classification

We shall classify signals according to their attributes as follows:

Dimensionality

In general, we measure a quantity at a point in space at a point in time. It might therefore be assumed that the quantity under investigation must be described as a function of four parameters—three length coordinates and a time coordinate. Often we will be interested only in a subset of these parameters. For example, an x-ray image might be intended to convey data about a slice through the body at a particular time, in which case the intensity of the image varies with two spatial parameters, whereas an electrocardiogram measures a quantity that is assumed to be position independent as it changes with time. In this chapter we shall concentrate on the simple data-against-time signal, although the techniques that are described apply equally to more general data. We might regard such a signal as one dimensional (its amplitude varies only as a function of time), and we can draw it on two-dimensional graph paper or a computer screen.

Periodicity

A periodic signal is one that repeats itself after some interval. An aperiodic signal never repeats.

Continuity

Continuity is used in two senses. In the conventional sense a function is continuous if it does not have jumps in value. Sometimes signals in the body are assumed to be 'on or off ': a good example is the transmission of electrical pulses down a nerve fibre. These signals are assumed to take a value of zero or one at any particular instant, and might be regarded as discontinuous. An alternative use of the notion of continuity is that which arises when a continuous function is sampled only at discrete points in time. The samples form a discrete signal, with a value that is defined only at the sampling points. At all other points the signal is undefined.

Determinism

A deterministic signal is one that can be expressed in terms of a mathematical equation. Such a signal has a predictable value at all times providing its underlying equation is known. A non-deterministic signal is one that cannot be expressed in this way. It might nevertheless be possible to describe some of the properties of the signal. For example, although its amplitude at any time is unknown, it might be possible to define the mean amplitude over a period of time and to define the probability that the amplitude lies within a particular range at any one point in time. Often a signal has both deterministic and non-deterministic elements: there is an underlying physical process that can be described mathematically, but random elements are superimposed on the signal.

13.1.2. Signal description

The appropriate descriptors used to measure and to compare signals depend to a large degree on the classification of the signal. We might be able to describe a deterministic signal very accurately and concisely in terms of a mathematical equation, and this in itself could tell us all we need to know. Often, however, the equation will not in itself promote an understanding of the signal, and we might wish to derive descriptors that give us a more intuitive feel for the nature of the signal. We may, for example, want to know what the average value of the signal is, and what sort of variation it exhibits. These measures are referred to as statistical descriptors of the signal. We may also want to know something about the frequency domain structure of the signal, i.e.

whether it consists of high- or low-frequency components, and what sort of energy levels are present in each of the components.

For non-deterministic signals no exact mathematical description is available. It becomes necessary to talk in terms of the probability of the signal lying within a particular range at a particular time. Nevertheless the statistical and frequency domain descriptors might still be very useful: although we cannot say exactly what value the signal will take at any time it might be enough to understand its fundamental form in more general terms.

13.2. USEFUL PRELIMINARIES: SOME PROPERTIES OF TRIGONOMETRIC FUNCTIONS

Many of the processes in the body are regular in that they have a fundamental and underlying periodicity. It is natural for us to describe such processes in terms of the trigonometrical sine and cosine functions, since these are the periodic functions with which we are most familiar. We shall see that these functions are, in fact, quite versatile and that we can also use combinations of them to represent aperiodic functions. The trigonometric functions have some special properties and attributes that make them particularly appropriate for the description of signals, and we shall investigate some of them in this section.

13.2.1. Sinusoidal waveform: frequency, amplitude and phase

The general equation describing the waveform illustrated in figure 13.1 is

$$f(t) = A \sin(\omega t + \phi).$$

In this graph the parameters ω and ϕ have been set to 1 and A is set to 3. These three parameters tell us everything about a sinusoidal wave. The frequency of the sinusoid, ω, dictates how many wavelengths will occur in a unit time. A sinusoid that repeats at an interval of T seconds (the period of the wave) has a frequency of $1/T$ cycles per second (hertz). The frequency parameter ω is measured in radians per second (rad s^{-1}), and is therefore equal to $2\pi/T$. The amplitude of the wave, A, is a measure of its magnitude. The sinusoid is a simple sine wave shifted along the time axis. The phase parameter ϕ is a measure of the shift. All possible forms of the signal can be described by a shift in the range $-\pi \leq \phi < \pi$. If the shift is zero the wave is a pure sine wave, a shift of $\pi/2$ gives a cosine wave, and shifts of $-\pi$ and $-\pi/2$ give inverted sine and cosine waves, respectively. The time shift corresponding to the phase shift lies in the range $-T/2$ to $+T/2$.

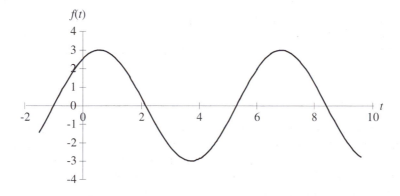

Figure 13.1. *General sinusoidal waveform.*

We can readily show that

$$a \cos \omega t + b \sin \omega t = A \sin(\omega t + \phi)$$

where

$$A = \sqrt{a^2 + b^2} \qquad \phi = \tan^{-1}(a/b) \text{ rad s}^{-1}.$$

This means that a linear sum of sine and cosines of the same frequency can be expressed as a sine wave with a phase shift. Conversely, any sinusoidal waveform at a given frequency can be expressed as a linear sum of sine and cosine components. This property of the trigonometric functions is most useful when we develop Fourier series representations of arbitrary time functions. Often we are interested only in the amplitude of a wave, and we might ignore the phase shift. In this context the general sinusoidal expression, which contains only one amplitude parameter, is more useful than a sum of sine and cosine terms.

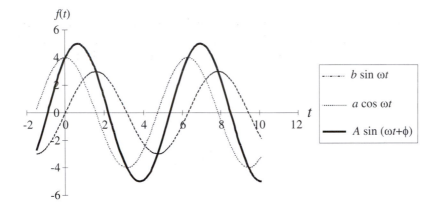

Figure 13.2. *The dotted sinusoids add to give the sinusoid plotted as a full curve.*

For the example illustrated in figure 13.2 $a = 4$ and $b = 3$. The sum of the cosine and sine waves is a sinusoid with amplitude $A = 5$ and shift $\phi = 0.93$ rad. The wave illustrated has a frequency of 1 rad s^{-1}, and so the time shift is $-\phi/\omega = -0.93$ s. Any integer multiple of 2π can be added to the shift in radians, and so it might alternatively be expressed as a positive shift of 5.35 s.

13.2.2.   Orthogonality of sines, cosines and their harmonics

A pair of functions $f_m(t)$ and $f_n(t)$ are said to be orthogonal over a range t_1 to t_2 if

$$\int_{t_1}^{t_2} f_n(t) f_m(t) \, dt = 0 \qquad \text{for} \quad m \neq n.$$

This is true over the interval from 0 to 2π of sine and cosine functions and of their harmonics. Thus,

$$\int_0^{2\pi} \cos n\theta \sin m\theta \, d\theta = 0 \qquad \text{for all } m \text{ and } n$$

$$\int_0^{2\pi} \cos n\theta \cos m\theta \, d\theta = \int_0^{2\pi} \sin n\theta \sin m\theta \, d\theta = 0 \qquad \text{for } m \neq n.$$

The sine and cosine functions and their harmonics are therefore orthogonal over the range 0 to 2π. The integral of the product of the functions with themselves is

$$\int_0^{2\pi} \cos^2 n\theta \, d\theta = \int_0^{2\pi} \sin^2 n\theta \, d\theta = 2\pi \qquad (m = n).$$

Functions for which the above integral yields unity are said to be orthonormal. The sine and cosine functions are orthonormal if multiplied by $1/\sqrt{2\pi}$.

    We shall meet integrals of sine and cosine products again and again as we learn to represent and to manipulate data describing physiological systems. The orthogonality of the sine and cosine functions and their harmonics is fundamental to the efficient description and processing of such data.

13.2.3.   *Complex (exponential) form of trigonometric functions*

The description of a general sinusoidal wave as a linear sum of a sine and cosine component, together with the orthogonality of these two components, suggests the representation of the sinusoid as a vector with sine and cosine as bases and with their amplitudes as coefficients. Graphically the sinusoid is represented as shown in figure 13.3.

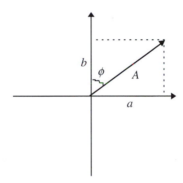

Figure 13.3. *The horizontal axis is the amplitude of the cosine component and the vertical axis that of the sine component. From the geometry of the system we can deduce that the length (magnitude) of the vector is the amplitude and the angle from the vertical is the phase of the general sinusoidal representation.*

    This graph suggests that we might represent a general sinusoid of a particular frequency as a complex number. The horizontal axis, corresponding to the cosine component, would be the real part and the vertical axis, corresponding to the sine component, would be the imaginary part.

    The complex forms of the trigonometric series are written in terms of the Euler identity, which is expressed as follows:

$$e^{j\theta} = \cos\theta + j\sin\theta.$$

Based on this identity any general sinusoid at a particular frequency can be expressed as complex numbers multiplied by $\exp(\pm j\theta)$.

$$A\sin(\omega t + \phi) = a\cos\omega t + b\sin\omega t = c_+ e^{j\omega t} + c_- e^{-j\omega t}$$

where

$$c_+ = \tfrac{1}{2}(a - jb) \qquad c_- = \tfrac{1}{2}(a + jb).$$

Obviously a pure cosine wave has entirely real coefficients and a pure sine wave has entirely imaginary coefficients. Like the parent functions, the complex representation of the cosine is symmetrical ($c_+ = c_-$) and that of the sine function is anti-symmetrical ($c_+ = -c_-$).

13.3.   REPRESENTATION OF DETERMINISTIC SIGNALS

13.3.1.   Curve fitting

The first step in the description of a deterministic signal is to derive a mathematical equation that represents it. Once the equation is known it is possible to perform all manner of manipulations to derive alternative measures and descriptions appropriate for many different purposes. The description of a signal by an equation is essentially a curve-fitting process. If a signal is sampled at discrete intervals, its value is known only at the sampling times. The first equations that we met at school were polynomials and this is the form that many people would first think of to describe data. It is possible to fit an nth-order polynomial through $(n + 1)$ points, and therefore to describe any signal sampled at a finite number of points by a polynomial. The lowest-order polynomial that fits the data is unique, but there are infinitely many higher-order polynomials that could be chosen: for example, although only one straight line (first-order polynomial) can be fitted between two points, any number of parabolics, cubics, etc could also be chosen. Similarly, there are infinitely many other mathematical equations that could be chosen to represent the discretely sampled data. When we choose an equation to use, it would make sense to choose one that is convenient to manipulate to produce any other descriptors of the data that we might like to derive. For the moment we will gloss over the problem that our chosen equation contains much more information (about times in between samples and before and after sampling) than does our original recording.

For continuous signals, with effectively infinitely many sampling points, the problem is reversed. There is now only one mathematical equation that fits the data precisely. It is most unlikely that any convenient and manipulable equation will describe a real measured process. The problem now is to choose an equation that fits the data well enough for all practical purposes, and exhibits the required mathematical properties. The problem of how well the original signal is represented by the equation is not addressed directly in this chapter, although some of the numerical examples will give an insight into the levels of error that might be anticipated.

When we first look at a signal we will sometimes have an immediate notion of the type of mathematical function that will describe the data. All physicists will be familiar with the general shapes of simple polynomial, trigonometric and exponential functions. When the data become more complicated it might not be apparent which, if any, of these forms is most appropriate. There are generalized forms of polynomial that can be used to represent any function to any desired level of accuracy over a given interval. The basic form of such polynomials is that of the Taylor series expansion of a function about a point. There are three fundamental problems with the Taylor series representation of a function: it is not intuitive, its range might be limited and it is not particularly easy to manipulate.

Many physiological signals exhibit a marked periodicity. An obvious example is the electrocardiogram. Such signals can be described elegantly as a Fourier series. The Fourier series is a form of trigonometric series, containing sine and cosine terms. It was first developed by Jean-Baptiste Joseph Fourier, who published the basic principles as part of a work on heat conduction in Paris in 1822. The power and range of application of the methodology was immediately recognized, and it has become one of the most important mathematical tools in the armoury of the physicist and engineer.

13.3.2. Periodic signals and the Fourier series

Our first motivation for the development of the Fourier series is to express a periodic function in a manner which is both concise and easy to manipulate. We shall soon find that this form of representation also yields directly some interesting properties of the function that will be valuable to us in their own right. The mathematical functions that spring immediately to mind as building blocks for any periodic function are the trigonometric ones of sine and cosine. A great advantage of the trigonometric functions is that they are very easy to integrate and to differentiate. Fourier showed that all periodic functions, a typical example of which is illustrated in figure 13.4, could be expressed as a combination of such terms.

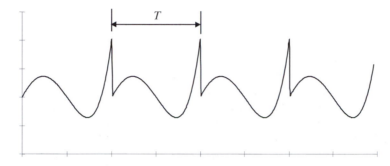

Figure 13.4. *A general periodic function of time with a period T.*

Fourier proposed that a function with a period T could be represented by a trigonometric series of the following form:

$$f(t) = a_0 + \sum_{n=1}^{\infty} \left(a_n \cos n\omega t + b_n \sin n\omega t \right)$$

where $\omega = 2\pi/T$ is the frequency of repetition of the function (in rad s^{-1}).

For any known periodic function $f(t)$ the coefficients in the series can be calculated from the following equations:

$$a_n = \frac{2}{T} \int_c^{c+T} f(t) \cos n\omega t \, dt \qquad a_0 = \frac{1}{T} \int_c^{c+T} f(t) \, dt \qquad b_n = \frac{2}{T} \int_c^{c+T} f(t) \sin n\omega t \, dt.$$

The starting point for the integrations (the lower limit, c) is arbitrary. Any starting point must, of course, produce the same value for the integral of the function over one period.

This series representation of the function is called its Fourier series, and the coefficients of the trigonometric terms are called the Fourier coefficients. The equations from which the coefficients are evaluated are called the Euler formulae. The Euler formulae can be developed from the original Fourier series by multiplying the series by, for example, $\cos(n\omega t)$ and integrating the result over the period of the function. The summation of integrals looks difficult, but an analytical inspection of each of the terms reveals that all but one of the multipliers of the coefficients integrate to zero. What is left is the Euler formula. This is just one example of the way in which the manipulation of Fourier series tends to produce results that can be expressed elegantly and concisely.

The Fourier series of the function has some important properties:

- The frequency of the first sine and cosine terms is that of the function.
- The increments of frequency between the terms as n increases are equal to the frequency of the function.

Another way of looking at these properties is to note that:

●   The period of the first sine and cosine terms is that of the function.
●   Each of the terms in the series represents an integer number of sine or cosine waves fitted into the period of the function.

This seems intuitively satisfactory because any trigonometric terms other than these integer multiples of frequency would return values that do not exhibit the periodicity of the function. Proof of the completeness and uniqueness of the Fourier series is beyond the scope of this text. It is enough for the present that the basic attributes of the series seem reasonable from a physical standpoint.

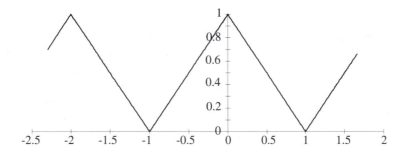

Figure 13.5. *A triangular periodic signal.*

Example: Fourier series of a triangular wave

The triangular wave illustrated in figure 13.5 has a period, T, of 2 s. The function can be written as a mathematical equation as follows:

$$f(t) = t - k \qquad k \le t \le k+1 \quad k \text{ odd}$$
$$= -t + k + 1 \qquad k \le t \le k+1 \quad k \text{ even}.$$

The Fourier coefficients can be written down from the Euler formulae,

$$a_0 = \frac{1}{T} \int_c^{c+T} f(t)\, \mathrm{d}t = \frac{1}{2}$$

$$a_n = \frac{2}{T} \int_c^{c+T} f(t) \cos n\omega t\, \mathrm{d}t = \int_0^T f(t) \cos n\pi t\, \mathrm{d}t = \frac{2}{n^2\pi^2}(1 - (-1)^n)$$

$$b_n = \frac{2}{T} \int_c^{c+T} f(t) \sin n\omega t\, \mathrm{d}t = \int_0^T f(t) \sin n\pi t\, \mathrm{d}t = 0.$$

The Fourier series representing this triangular wave is therefore

$$f(t) = \frac{1}{2} + \sum_{n=1,3,5,\dots}^{\infty} \frac{4}{n^2\pi^2} \cos n\pi t.$$

Writing out the terms in the series:

$$f(t) = \frac{1}{2} + \frac{4}{\pi^2}\left(\cos \pi t + \frac{\cos 3\pi t}{9} + \frac{\cos 5\pi t}{25} + \frac{\cos 7\pi t}{49} + s \right).$$

The question immediately arises as to how many terms will need to be retained to return a value of the function to within a given accuracy. The usefulness of the Fourier series would be greatly restricted if a great number of terms (or even all of them) had to be retained. In practice it is often possible to represent most functions to an acceptable level of accuracy using relatively few terms. The actual number of terms required will depend on the function and on the accuracy desired. The more distorted the periodic function is from the shape of a simple sine wave, the more terms will usually be required in the truncated Fourier series representation. The reader might experiment by plotting the Fourier series representation of the triangular wave at various truncations.

Even and odd functions

In the preceding example it was found that the coefficients of the sine terms in the Fourier series were all zero. This makes intuitive sense since the triangular wave under consideration is symmetrical about the vertical axis ($t = 0$); an attribute that it shares with a cosine wave but not with a sine wave which is anti-symmetrical. Any function that is symmetrical is said to be even and can always be represented purely in cosine terms: conversely an anti-symmetrical function is said to be odd and can always be represented purely in sine terms. Clearly the knowledge that a function is odd or even can save a lot of work in the unnecessary evaluation of (potentially long) integrals that are going to come to zero anyway!

Amplitude and shift

The terms in the Fourier series at each harmonic of the fundamental frequency have been described as a linear combination of cosines and sines. As discussed in section 13.2.1, it is always possible to describe such a combination in terms of a general sinusoid,

$$f_n(t) = a_n \cos n\omega t + b_n \sin n\omega t = A_n \sin(n\omega t + \phi_n)$$

where A_n is the amplitude of the combined wave and ϕ_n is proportional to the shift of the wave in time relative to a sine wave. This form of representation is often used, and it is common practice in many physics and engineering applications to report only the amplitude of the composite harmonic and to ignore the shift.

Time and frequency domain representations of a periodic function

A periodic function, $f(t)$, that varies with time can be written as an equation with time as a variable parameter. The Fourier series is one form of the equation that might be used. The function could also be plotted against time in the form of a graph. Either expression of the function, as a graph against time or as an equation in time, is called a *time domain representation*. The Fourier series offers an alternative representation of the function in terms of the frequency domain. Instead of plotting the function against time, a histogram could be constructed with the horizontal axis as frequency and the vertical axis as the amplitude of each frequency. This would be termed a *frequency domain representation*. A frequency domain representation of the triangular wave considered earlier is illustrated in figure 13.6.

This representation gives an immediate understanding of the relative amplitudes of the frequency components. The function considered in this example is even, and so there are only coefficients of cosine terms. More generally there will be sine coefficients too, and both would be represented on the histogram. There are other representations in the frequency domain that might be useful, such as the amplitude and phase of the equivalent sinusoid. A common parameter that is used is the power of each frequency component (see section 13.3.4).

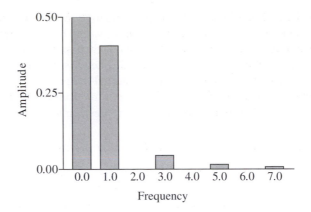

Figure 13.6. *Frequency domain representation of a triangular wave. Amplitude versus frequency, truncated at the seventh harmonic ($7\omega t$).*

- It is important to recognize that the time and frequency domain representations are two sides of the same coin. The information that they contain is identical, but sometimes one will be more convenient than the other.
- There are two senses in which the frequency domain representation of the function is not closed. Firstly, the sequence extends to infinity: infinitely many integer multiples (harmonics) of the fundamental frequency might be included. Secondly, the interval between the bars of the histogram might be filled with infinitely many subdivisions: for example if the fundamental frequency is halved there will be one bar between each of the existing pairs, and so on. It has been shown that it is inappropriate to subdivide the frequencies for periodic functions since only harmonics of the fundamental frequency will exhibit the required periodicity. We shall pursue this further when we look at the Fourier representation of aperiodic data.

Complex (exponential) form of the Fourier series

It has already been stated that the Fourier series is one of the most powerful tools in the armoury of the physicist. Once a concept becomes familiar, it is always tempting to find 'shorthand' ways to express it, mainly to save writing it out in tedious detail every time it is used. Sometimes the shorthand expressions can have merit in their own right. Such is the case with the complex form of the Fourier series when used to represent real functions (or real data). The complex form of the Fourier series will now be investigated, but the reader should be clear that it is only a mathematical convenience. Any understanding of the principles behind the trigonometric series already discussed applies equally well to the complex form. The complex form contains no more and no less information than the trigonometric form, and it is always possible to translate between the two.

Using the Euler identity (section 13.2.3) the Fourier series can be written in complex form as follows:

$$f(t) = \sum_{n=-\infty}^{\infty} \left(c_n \, e^{jn\omega t} \right) \qquad \text{where} \qquad c_n = \frac{1}{T} \int_c^{c+T} f(t) \, e^{-jn\omega t} \, dt.$$

The coefficients of the series, c_n, are complex. When real functions are represented in this way the real parts of the complex coefficients are related to the cosine coefficients and the imaginary parts to the sine coefficients of the trigonometric form. The relationships between the coefficients are given in section 13.2.3.

In the frequency domain the real and imaginary components might be plotted separately, or a composite measure such as the amplitude of the complex number might be plotted. It is important to recognize that, in the complex form, the frequency must be allowed to take both positive and negative values—otherwise it would not be possible to reconstitute the original sine and cosine terms. This presents no real difficulty mathematically, but does not have a ready physical interpretation.

Clearly the complex representation of the Fourier series is more concise than the trigonometric form. We shall see later that it also has other useful attributes.

13.3.3.   Aperiodic functions, the Fourier integral and the Fourier transform

It has been demonstrated that a periodic function can be expressed in terms of a Fourier series. An obvious question is whether the principle can be extended to represent an aperiodic function. Suppose, for example, that a function described a triangular blip around zero time, but was zero at all other times (see figure 13.7). Would it be possible to write a Fourier series to represent this function?

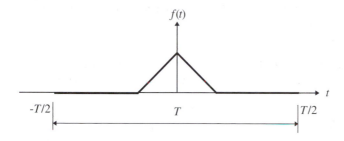

Figure 13.7. *A triangular blip signal placed symmetrically about time zero.*

The function can be represented over any interval T as shown in figure 13.7, and if we assume that the function repeated outside this interval then it can be represented as a Fourier series. Clearly we require a Fourier series in which the period T is extended to infinity. What are the consequences of such an extension for the Fourier series representation? As $T \to \infty$:

- the fundamental frequency, ω, becomes infinitesimally small, and therefore
- the interval between frequencies in the Fourier series becomes infinitesimally small, i.e. all frequencies are present in the Fourier series. This means that the frequency domain representation will now be a continuous curve rather than a discrete histogram.

Intuitively it might be expected that the Fourier series must in some sense become an integral, because in the limit all frequencies are present and instead of summing over discrete frequencies the summation becomes one over a continuous range of frequencies. We shall not attempt a formal description of the process by which a Fourier series becomes an integral but it can be shown that, subject to certain mathematical constraints on the function, an aperiodic function can be represented as follows:

$$f(t) = \frac{1}{\pi} \int_0^\infty \left(A(\omega) \cos \omega t + B(\omega) \sin \omega t \right) d\omega$$

where

$$A(\omega) = \int_{-\infty}^\infty f(t) \cos \omega t \, dt \qquad \text{and} \qquad B(\omega) = \int_{-\infty}^\infty f(t) \sin \omega t \, dt.$$

This form of representation is known as the Fourier integral.

Complex form of the Fourier integral

The Fourier integral, like the Fourier series, can be expressed in complex form. The complex form of the Fourier integral is very important in medical physics and biomedical engineering: it forms the basis for a mathematical description of measured physiological signals, and it has certain properties which make it particularly useful for data manipulation in applications such as image processing (see section 11.7). In complex form, the Fourier integral representation of a function can be expressed as follows:

$$f(t) = \frac{1}{2\pi} \int_{-\infty}^{\infty} F(\omega)\, e^{j\omega t}\, d\omega \qquad \text{where} \quad F(\omega) = \int_{-\infty}^{\infty} f(t)\, e^{-j\omega t}\, dt.$$

Frequency domain representation and the Fourier transform

The frequency domain representation of an aperiodic function can be presented in a similar manner to that of the periodic function. The difference is that the function $F(\omega)$ replaces the discrete coefficients of the Fourier series, thus yielding a continuous curve. The function $F(\omega)$ is a frequency domain representation of the function $f(t)$ in the time domain: $F(\omega)$ is referred to as the Fourier transform of $f(t)$. In some texts the terms Fourier integral and Fourier transform are used interchangeably and the function $F(\omega)$ is referred to as the Fourier spectrum of $f(t)$. In the context of our investigation the Fourier transform switches information from the time domain to the frequency domain. It will be shown that the converse is also true and that essentially the same transformation switches frequency domain information back into the time domain, but first it is time for a numerical example.

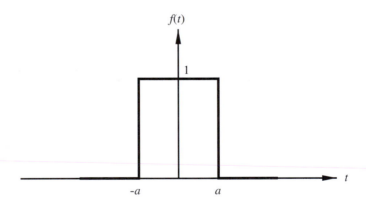

Figure 13.8. *An isolated square pulse.*

Example: Fourier integral and Fourier transform of an isolated square pulse

The single impulse illustrated in figure 13.8 has a duration of $2a$ seconds, but its period is infinite (it is never repeated). It can be defined as a mathematical equation as follows:

$$f(t) = 1 \qquad |t| < a$$
$$= 0 \qquad |t| > a.$$

A discontinuous function like this can present some difficulties at the points of discontinuity: for now the value of the function at the points $t = a$ and $-a$ will not be considered. The function as described above has a value of 1 at all points inside the interval and of zero at all points outside it.

Using the definitions of the Fourier integral:

$$F(\omega) = \int_{-\infty}^{\infty} f(t)\,e^{-j\omega t}\,dt = \int_{-a}^{a} e^{-j\omega t}\,dt = -\frac{1}{j\omega}\left[e^{-j\omega t}\right]_{-a}^{a} = \frac{2\sin\omega a}{\omega}$$

$$f(t) = \frac{1}{2\pi}\int_{-\infty}^{\infty} F(\omega)\,e^{j\omega t}\,d\omega = \frac{1}{2\pi}\int_{-\infty}^{\infty} \frac{2\sin\omega a}{\omega}\,e^{j\omega t}\,d\omega = \begin{cases} 1 & |t| < a \\ \frac{1}{2} & |t| = a \\ 0 & |t| > a. \end{cases}$$

The evaluation of the integral to evaluate $f(t)$ is not trivial. There are tables of integrals in the literature to assist in these cases. As a point of interest, the Fourier integral representation of the function does actually return a value for it at the point of discontinuity: the value returned is the average of the values on either side of the discontinuity.

The frequency domain representation of the function can now be written down. For the Fourier integral the function $F(\omega)$ replaces the discrete coefficients of the Fourier series, thus yielding a continuous curve. The frequency domain representation of the isolated square pulse centred at the origin is illustrated in figure 13.9.

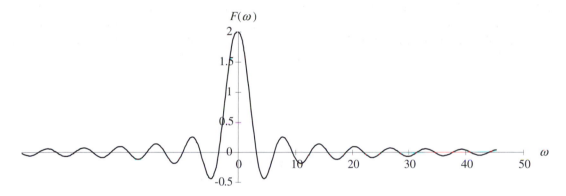

Figure 13.9. *Frequency domain representation of an isolated square pulse.*

The function considered in this example is even, and so the coefficient function of the Fourier integral is real. In the general case the coefficient function will be complex, reflecting both sine and cosine terms in the makeup of the function to be represented.

Symmetry of the Fourier transform

Inspection of the relationships between $f(t)$ and its Fourier transform $F(\omega)$ reveals that the transformations in both directions are very similar. The transformations are the same apart from the sign of the exponent and the factor $1/2\pi$ outside the integral, arising from the definition of ω in rad s^{-1}. In some texts the factor $\sqrt{1/2\pi}$ is taken into $F(\omega)$: we saw in section 13.2.2 that the sine and cosine functions are orthonormal when multiplied by $\sqrt{1/2\pi}$.

By taking the Fourier transform of the Fourier transform of a function you actually get the complex conjugate of the function. Many mathematical software packages offer an inverse Fourier transform operator. It was demonstrated in the preceding example that the Fourier transform of a square impulse centred at the origin in the time domain is a function of the form $\sin(\omega a)/\omega$ in the frequency domain. A function of form $\sin(t)/t$ in the frequency domain will transform to a square impulse in the time domain as shown in figure 13.10.

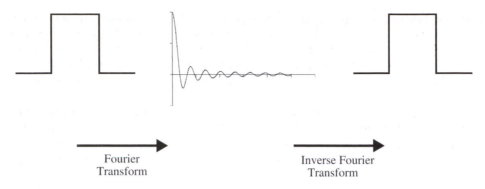

Figure 13.10. *Illustrates the symmetry of the Fourier transform process.*

Practical use of the Fourier transform

It has already been mentioned that the value of the Fourier transform, apart from giving an immediate appreciation of the frequency makeup of a function, is in its mathematical properties. We shall find that it is a most valuable tool in linear signal processing because operations that are potentially difficult in the time domain turn out to be relatively straightforward in the frequency domain. Several applications of the Fourier transform are found in the chapters of this book: in particular, extensive use of the three-dimensional Fourier transform is made in the chapters on imaging.

Before leaving this section on the Fourier transform, we might usefully catalogue some of its attributes.

- The transform of a periodic function is discrete in the frequency domain, and conversely that of an aperiodic function is continuous.
- The transform of a sinusoid is two discrete values in the frequency domain (one at $+\omega$ and one at $-\omega$). The two values are a complex conjugate pair.

It is valuable for the reader to work through examples to calculate the frequency domain representation of various functions, but in practice the Fourier transforms of many commonly occurring functions are readily found in the literature.

13.3.4.   *Statistical descriptors of signals*

Mean and variance (power)

We have seen how a signal can be described as a Fourier series or as a Fourier integral depending on its periodicity. The time and frequency domain representations contain a lot of detailed information about the signal. In this section we shall look at some alternative measures of the signal that can give us a simple appreciation of its overall magnitude and the magnitude of the oscillation at each frequency.

The mean of a signal is a measure of its time average. Formally, for any deterministic signal described by a function $f(t)$:

$$\text{mean} \qquad \mu = \frac{1}{T} \int_0^T f(t)\, \mathrm{d}t.$$

The variance of a signal is a measure of the deviation of the signal from its mean. Formally,

$$\text{variance (power)} \qquad \sigma^2 = \frac{1}{T} \int_0^T (f(t) - \mu)^2\, \mathrm{d}t.$$

The terms used to describe these properties of the signal are familiar to anyone with a basic knowledge of statistics. The variance is the square of the standard deviation of the signal. Analysis of a simple AC resistive circuit will reveal that the average power consumed by the resistor is equal to the variance of the signal. Because much of our understanding of signals is based on Fourier series representations, essentially a linear superposition of AC terms, the term power has come to be used interchangeably with variance to describe this statistical property.

The mean and power of any signal over any interval T can, in principle, be computed from the above equations providing that the function $f(t)$ is known. In practice we might not be able to perform the integration in closed form and recourse might be made to numerical integration. For a particular signal sample over a finite time each statistical descriptor is just a scalar quantity. These descriptors offer two very simple measures of the signal.

We know that any deterministic signal can be represented as a Fourier series or as a Fourier integral, and we might choose to use statistical measures of each of the frequency components in place of (or to complement) the amplitude and phase information. The mean of a sine wave over a full period is, of course, zero. For a sine wave monitored for a long time relative to its period the mean approaches zero. The power in each of the frequency components of a signal can be calculated from the basic definition. We have seen that each frequency component can be written as a sinusoid, and

$$\text{power in sinusoid} = \frac{1}{T} \int_{-T/2}^{T/2} (A \sin(\omega t + \phi))^2 \, \mathrm{d}t = \frac{A^2}{2}.$$

A signal is often described in the frequency domain in terms of the power coefficients. Note that the phase information is lost in this description.

13.3.5. *Power spectral density*

We know that we can decompose any signal into its Fourier components, and we have now developed the concept of the power that is associated with each component. We might choose to describe each component in the frequency domain in terms of the average power at that frequency instead of the amplitude. Such a representation is called the *power spectral density* of the signal.

13.3.6. *Autocorrelation function*

The statistical descriptors of the signal give no information about frequency or phase. Frequency information is available from a quantity called the autocorrelation function. The autocorrelation function, r_{xx}, is constructed by multiplying a signal by a time-shifted version of itself. If the equation of the signal is $f(t)$, and an arbitrary time shift, τ, is imposed

$$r_{xx}(\tau) = \int_{-\infty}^{\infty} f(t) \, f(t + \tau) \, \mathrm{d}t.$$

The principle behind the autocorrelation function is a simple one. If we multiply a signal by a time-shifted version of itself then the resulting product will be large when the signals line up and not so large when they are out of phase. The signals must always be in phase when the time shift is zero. If they ever come back into phase again the autocorrelation function will be large again, and the time at which this occurs is the period of the signal. Hence the computation of the autocorrelation function might help us to identify periodicities that are obscured by the form of the mathematical representation of a deterministic signal. It will obviously be even more useful for a non-deterministic signal when we do not have a mathematical description to begin with.

The principle is clearly illustrated by looking at the autocorrelation function of a sinusoid. When the time shift is small the product in the integral is often large and its average is large. When the shift is greater the product is generally smaller and indeed is often negative. The autocorrelation function therefore reduces with increased time offset. The worst alignment occurs when the waves are 180° out of phase, when the autocorrelation function is at a negative peak. The graphs illustrated in figure 13.11 have been constructed by forming the autocorrelation function for the sinusoid from its basic definition.

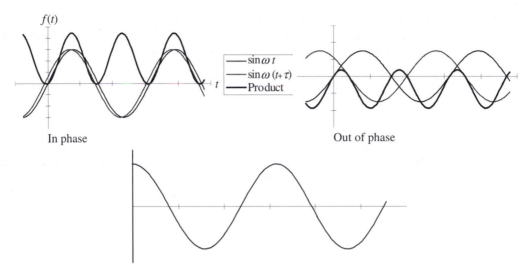

Figure 13.11. *The upper traces show how the autocorrelation function (r_{xx}) of a sinusoid is derived. The lower trace shows r_{xx} as a function of the delay τ.*

The periodicity of the autocorrelation function reveals, not surprisingly, the periodicity of the sinusoid. It also looks like the autocorrelation function of our arbitrary sinusoid might be a pure cosine wave. We can prove this by evaluating it in closed form. The autocorrelation function of a sinusoid is

$$r_{xx}(\tau) = \int_{-\infty}^{\infty} A \sin(\omega t + \phi) \, A \sin(\omega(t + \tau) + \phi) \, dt = \tfrac{1}{2} A^2 \cos \omega \tau.$$

Hence we see that:

- The autocorrelation function of any sinusoid is a pure cosine wave of the same frequency.
- Its maximum occurs at time zero (and at intervals determined by the period of the wave).
- All phase information is lost.
- The amplitude of the autocorrelation function is the power in the sinusoid.

A corollary of the above is that the autocorrelation function of any signal must be a maximum at time zero, since it can be expressed as a linear combination of sinusoids. If the signal is periodic the sinusoids are harmonics of the fundamental one and its autocorrelation function will be periodic.

In the same way that we can construct a frequency domain representation of the signal by taking its Fourier transform, we can construct that of the autocorrelation function. We know that the Fourier transform of a pure cosine wave is a symmetrical pair of discrete pulses, each of the magnitude of the amplitude of the wave, at plus and minus ω. The Fourier transform of the autocorrelation function of a sinusoid is therefore symmetrical, and its amplitude is the average power in the sinusoid. Writing the Fourier transform of $r(t)$ as $P(\omega)$, for the sinusoid

$$P_{xx}(\omega) = \int_{-\infty}^{\infty} r_{xx}(t) \, e^{-j\omega t} \, dt = \tfrac{1}{2} A^2.$$

We should recognize this function as the power of the sinusoid. For a general signal the equivalent would be the power spectral density (section 13.3.5).

- The Fourier transform of the autocorrelation function of a signal is the power spectral density of the signal.
- The autocorrelation function is a time domain representation of the signal that preserves its frequency information. The frequency domain equivalent is the power spectral density. Both of these measures preserve frequency information from the signal but lose phase information.

13.4.   DISCRETE OR SAMPLED DATA

13.4.1.   *Functional description*

The discussion in section 13.3 has focused on the representation of a continuous function as a Fourier series or as a Fourier integral. In practice in a digital environment the function is sampled at discrete intervals, and furthermore it is sampled only for a limited period. The function is of known magnitude only at the sampling times, and is unknown at all other times. A function sampled at an interval T is illustrated in figure 13.12.

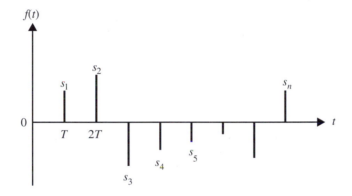

Figure 13.12. *Function sampled at discrete intervals in time.*

If this discrete function is to be manipulated, the first requirement is for a mathematical description. Perhaps the obvious way to define the function is to fit a curve that passes through each of the sampling points, giving a continuous time representation. This does, however, contain a lot more information than the original sampled signal. There are infinitely many functions that could be chosen to satisfy the constraints: each would return different values at any intermediate or extrapolated times, and only one of them represents the true function that is being sampled. One possible representation of the function $f(t)$ is written in terms of a unit impulse function as follows:

$$f(t) = s_0 \, \mathbf{1}(t) + s_1 \, \mathbf{1}(t - T) + s_2 \, \mathbf{1}(t - 2T) + \cdots = \sum_{k=0}^{n} s_k \, \mathbf{1}(t - kT)$$

where

$$\mathbf{1}(t) = \begin{cases} 1 & \text{if} \quad t = 0 \\ 0 & \text{if} \quad |t| > 0. \end{cases}$$

This representation of $f(t)$ returns the appropriate value, s_k, at each of the sampling times and a value of zero everywhere else. From this point of view it might be regarded as a most appropriate representation of a discrete signal. A useful way to look at this representation is to regard the point value of the function, s_k, as a coefficient multiplying the unit impulse function at that point. The function is then just a linear combination of unit impulse functions occurring at different points in time.

13.4.2.   The delta function and its Fourier transform

The Fourier transform of a discrete signal has the same range of application as that of a continuous function. Adopting the representation of the function suggested in the preceding paragraph, and noting that the discrete values of the function are simply coefficients, the Fourier transform of the signal is just a linear combination of the transforms of unit impulses occurring at different points in time. The definition of the Fourier transform requires the continuous integral of a function over all time, and since the unit impulse function is non-zero at only a single point, having no width in time, the integral would be zero. It is appropriate to define the delta function (see figure 13.13) representing a pulse of infinite height but infinitesimal duration centred at a point in time, with the product of the height times the duration equal to unity.

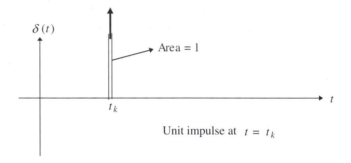

Figure 13.13. *A delta function at time t_k.*

Formally,

$$\int_{-\infty}^{\infty} \delta(t - t_k)\, \mathrm{d}t = 1.$$

An understanding of the nature of the delta function can be gained by consideration of a problem in mechanics. When two spheres collide energy is transferred from one to the other. The transfer of energy can occur only when the balls are in contact, and the duration of contact will depend on the elastic properties of the material. Soft or flexible materials such as rubber will be in contact for a long time, and hard or stiff materials such as ivory (old billiard balls) or steel for a relatively short time. The energy that is transferred is a function of the force between the balls and of its duration, but in practice both of these quantities will be very difficult to measure and indeed the force will certainly vary during the time of contact. The total energy transferred is an integral quantity involving force and time and is relatively easy to measure: perhaps by measuring the speeds of the balls before and after impact. In the limit, as the material becomes undeformable, the force of contact tends towards infinity and the duration becomes infinitesimal, but nevertheless the total energy transfer is measurable as before. The delta function represents the limit of this impact process.

Consider now the product of a function, f, with the delta function, δ, integrated over all time:

$$\int_{-\infty}^{\infty} f(t)\, \delta(t - t_k)\, \mathrm{d}t.$$

The function $f(t)$ does not change over the very short interval of the pulse, and it assumes the value $f(t_k)$ on this interval. It can therefore be taken outside the integral which, from the definition of the delta function, then returns a value of unity. The integral effectively 'sifts out' the value of the function at the sampling points and is known as the sifting integral,

$$\int_{-\infty}^{\infty} f(t)\,\delta(t - t_k)\,\mathrm{d}t = f(t_k).$$

It is now apparent that the integral of the delta function behaves in the manner that was prescribed for the unit impulse function in the chosen representation of the discrete signal. The signal can thus be expressed in terms of the delta function and the discrete measured values of the signal.

The Fourier transform of the delta function is easy to evaluate based on the property of the sifting integral: $f(t) = e^{-j\omega t}$ is just a function like any other, and

$$F(\delta(t - t_k)) = \int_{-\infty}^{\infty} e^{-j\omega t}\,\delta(t - t_k)\,\mathrm{d}t = e^{-j\omega t_k}.$$

The amplitude of $e^{-j\omega t}$ is unity for any value of ω, and the frequency domain representation of the impulse function is therefore a straight line parallel to the frequency axis at a height of one. All frequencies are present in the impulse, and all have equal magnitude. The frequency components are time-shifted relative to each other, and the shift is proportional to the frequency.

It should be noted that the delta function is represented by a discrete 'blip' in the time domain, but is continuous in the frequency domain (see figure 13.14).

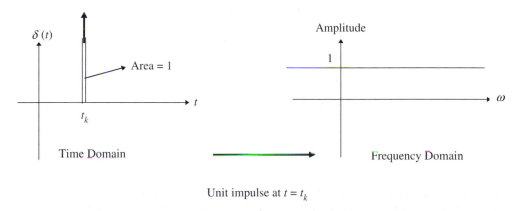

Unit impulse at $t = t_k$

Figure 13.14. *A delta function shown in the time domain on the left and the frequency domain on the right.*

13.4.3. *Discrete Fourier transform of an aperiodic signal*

The representation of the discrete signal in terms of the delta function presents no difficulties when a signal is sampled for a limited period of time: outside the sampling period the function is assumed to be zero, and it is therefore regarded as aperiodic. The Fourier transform of the discrete signal sampled at uniform intervals T starting at time zero is:

$$F(f(t)) = f_0\,e^0 + f_1\,e^{-j\omega T} + f_2\,e^{-j2\omega T} + \cdots = \sum_{k=0}^{n} f_k\,e^{-jk\omega T}.$$

Remembering that the complex exponential term is just a real cosine and imaginary sine series, three important attributes of the Fourier transform of the discrete series can be noted.

- Although the aperiodic-sampled signal is discrete in the time domain, its Fourier transform is, in general, continuous in the frequency domain.
- Like that of the continuous aperiodic function the Fourier transform of the discrete signal is, in general, complex. For even signals it is real and for odd signals it is imaginary.
- The function $\exp(jk\omega T)$ is periodic with a period of $kT/2\pi$. The period of the fundamental frequency is $T/2\pi$, and all other frequencies are harmonics of the fundamental one. The Fourier transform of the discrete signal is therefore periodic with a period $T/2\pi$.

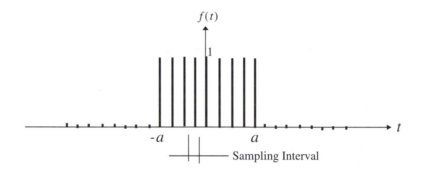

Figure 13.15. *A square wave sampled at discrete points.*

Example: discrete Fourier transform of a sampled square pulse

The signal illustrated in figure 13.15 can be expressed as a linear combination of unit impulses:

$$f(t) = 1 \times \mathbf{1}(t+a) + 1 \times \mathbf{1}(t+a-T) + 1 \times \mathbf{1}(t+a-2T) + \cdots = \sum_{k=0}^{2a/T} \mathbf{1}(t+a-kT).$$

Its Fourier transform is

$$F(f(t)) = \sum_{k=0}^{2a/T} e^{-j\omega(kT-a)} = e^{j\omega a} \sum_{k=0}^{2a/T} e^{-jk\omega T}.$$

For the signal indicated in the diagram, if a is taken as 1 s, there are nine samples equally spaced over the interval of 2 s, and the sampling interval is $T = 0.25$ s. The fundamental frequency of the Fourier transform is therefore 8π rad s^{-1}. The frequency domain representation of the sampled signal is illustrated in figure 13.16.

Clearly there is no point in calculating explicitly the transform for frequencies outside the range $0 \le \omega \le 2\pi/T$ because the function just repeats. Furthermore, the cosine function is symmetrical about the origin, and its value in the interval $\pi/T \le \omega \le 2\pi/T$ corresponds to that in the interval $-\pi/T \le \omega \le 0$. This implies that it is enough to study the transform over the interval $0 \le \omega \le \pi/T$: the remainder of the function can be constructed by reflection. The validity of this argument for the even signal is demonstrated by figure 3.16. For arbitrary signals the cosine terms will behave as shown and the sine terms will be reflected about both vertical and horizontal axes. In mathematical terms, the transform at $\omega = (\pi/T + \psi)$ is the complex conjugate of that at $\omega = (\pi/T - \psi)$. A corollary of the above is that all known information about

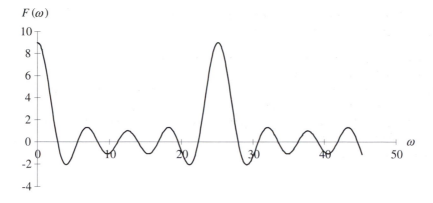

Figure 13.16. *Frequency domain representation of a discretely sampled isolated square pulse.*

the signal can be extracted from the Fourier transform in the frequency range $0 \le \omega \le \pi/T$. Information in the frequency domain outside this range is redundant.

In summary, for a signal sampled at discrete intervals:

- the sampling interval T determines the range of frequencies that describe the signal, and
- the useful range, over which there is no redundant information, is $0 \le \omega \le \pi/T$.

13.4.4. The effect of a finite-sampling time

The Fourier transform of the discrete signal is continuous because it is assumed that sampling occurs over the whole time that the signal is non-zero, or that outside the sampling time the signal is zero. In practice the signal is sampled for a finite time, and the value that it might take outside this time is unknown. Instead of assuming that the function is zero outside the sampling time it might be assumed that it is periodic, repeating at the period of the sample. The Fourier transform would then be discrete, like that of a continuous periodic function.

If n samples are taken to represent a signal and it is assumed that thereafter the signal repeats, then the sampling time is nT seconds. This implies that the fundamental frequency is $2\pi/nT$ rad s^{-1}, and the Fourier transform would contain just this frequency and its harmonics. For example, for a signal sampled for a period of 10 s, the fundamental frequency is $\pi/5$ rad s^{-1} and harmonics with frequencies of $2\pi/5$, $3\pi/5$, $4\pi/5$, etc would be present in the Fourier transform. If the sampling interval were 0.1 s then the maximum frequency component containing useful information would be 10π rad s^{-1}, and 50 discrete frequencies would lie within the useful range. Since each frequency component is described by two coefficients (one real and one imaginary), 100 coefficients would be calculated from the 100 data points sampled. This is obviously consistent with expectations, and it is not surprising that frequency information outside the specified range yields nothing additional. To obtain information about intermediate frequencies it would be necessary to sample for a longer period: reducing the sampling interval increases the frequency range but does not interpolate on the frequency axis. It should be noted that the continuous frequency plot produced earlier for the square impulse is based on the implicit assumption that the signal is zero for all time outside the interval $-a < t < a$.

In summary:

- a finite-sampling time produces a fixed resolution on the frequency axis. Finer resolution requires a longer sampling time.

13.4.5. Statistical measures of a discrete signal

The statistical measures that we developed for continuous signals are readily applied to a discrete signal, and indeed we might well be more familiar with the discrete versions giving us the mean and variance,

$$\mu = \frac{1}{n} \sum_{k=0}^{n} f_k \qquad \sigma^2 = \frac{1}{n} \sum_{k=0}^{n} (f_k - \mu)^2.$$

In the next section we review some of the basic ideas involved in statistics.

13.5. APPLIED STATISTICS

The aim of this section is to refresh the readers mind about the basic concepts of statistics and to highlight some aspects that are particularly relevant to the biomedical scientist. There are very many books on statistics and the assumption is made that the reader has access to these.

Background

In the *Hutchinson Concise Dictionary of Science* statistics is defined as 'that branch of mathematics concerned with the collection and interpretation of data'. The word 'statistics' has its origins in the collection of information for the State. Around the 17th Century taxes were raised to finance military operations. Knowledge of an individual's taxable assets was required before taxes could be levied. 'State information' infiltrated many affairs of Government as it still does today. The 17th Century also saw the birth of a new branch of mathematics called probability theory. It had its roots in the gaming houses of Europe. Within a hundred years of their inception the approximation of 'State-istics' by models based on probability theory had become a reality. Armed with this information predictions could be made and inferences drawn from the collected data. Scientists of that era recognized the potential importance of this new science within their own fields and developed the subject further. Victor Barnett, a Professor of Statistics at Sheffield University, has summed up the discipline succinctly: *statistics is 'the study of how information should be employed to reflect on, and give guidance for action in, a practical situation involving uncertainty'.*

Two intertwined approaches to the subject have evolved: 'pure' statistics and 'applied' statistics. In general, pure statisticians are concerned with the mathematical rules and structure of the discipline, while the applied statistician will usually try to apply the rules to a specific area of interest. The latter approach is adopted in this section and applied to problems in medical physics and biomedical engineering.

Statistical techniques are used to:

- describe and summarize data;
- test relationships between data sets;
- test differences between data sets.

Statistical techniques and the scientific method are inextricably connected. An experimental design cannot be complete without some thought being given to the statistical techniques that are going to be used. All experimental results have uncertainty attached to them.

A variable is some property with respect to which individuals, objects, medical images or scans, etc differ in some ascertainable way. A random variable has uncertainty associated with it. Nominal variables like pain or exertion are qualitative. Ordinal variables are some form of ranking or ordering. Nominal variables are often coded for in ordinal form. For example, patients with peripheral vascular disease of the lower limb might be asked to grade the level of pain they experience when walking on a treadmill on a scale from 1 to 4. In measurement, discrete variables have fixed numerical values with no intermediates. The heart beat, a photon or an action potential would be examples of discrete variables. Continuous variables have no missing intermediate values. A chart recorder or oscilloscope gives a continuous display. In many circumstances the measurement device displays a discrete (digital) value of a continuous (analogue) variable.

13.5.1.   Data patterns and frequency distributions

In many cases the clinical scientist is faced with a vast array of numerical data which at first sight appears to be incomprehensible. For example, a large number of values of blood glucose concentration obtained from a biosensor during diabetic monitoring. Some semblance of order can be extracted from the numerical chaos by constructing a frequency distribution. The difference between the smallest and largest values of the data set (the range) is divided into a number (usually determined by the size of the data set) of discrete intervals and the number of data points that fall in a particular interval is recorded as a frequency (number in interval/total number). Figure 13.17 gives an example of a frequency distribution obtained from 400 blood glucose samples. A normal level of blood glucose is 5 mmol l^{-1}, so the values shown are above the normal range.

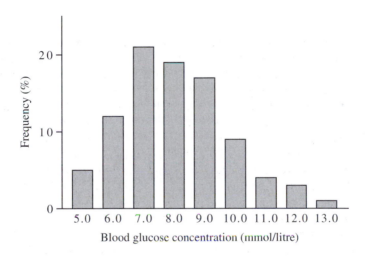

Figure 13.17. *Frequency distribution obtained from 400 blood glucose samples.*

Data summary

The data can always be summarized in terms of four parameters:

- A measure of central tendency.
- A measure of dispersion.
- A measure of skewness.
- A measure of kurtosis.

Central tendency says something about the 'average value' or 'representative value' of the data set, dispersion indicates the 'spread' of the data about the 'average', skew gives an indication of the symmetry of the data pattern, while kurtosis describes the convexity or 'peakedness' of the distribution.

There are four recognized measures of central tendency:

- Arithmetic mean.
- Geometric mean.
- Median.
- Mode.

The arithmetic mean uses all the data and is the 'average' value. Geometric mean is the 'average' value of transformed variables, usually a logarithmic transformation that is used for asymmetrical distributions. The median is a single data point, the 'middle' piece of data after it has been ranked in ascending order. The mode is the mid-point of the interval in which the highest frequency occurs. Strictly speaking, not all of the data are used to determine the median and mode.

Symmetrical and asymmetrical distributions

All measures of central tendency are the same for a symmetrical distribution. The median, mean and mode are all equal. There are two types of asymmetrical distribution:

- Right-hand or positive skew.
- Left hand or negative skew.

Positively skewed distributions occur frequently in nature when negative values are impossible. In this case the mode < the median < the arithmetic mean. Negatively skewed distributions are not so common. Here the arithmetic mean < the median < the mode. Asymmetry can be removed to a certain extent by a logarithmic transformation. By taking the logarithm of the variable under investigation the horizontal axis is 'telescoped' and the distribution becomes more symmetrical. The arithmetic mean of the logarithms of the variable is the geometric mean and is a 'new' measure of central tendency.

13.5.2.   Data dispersion: standard deviation

All experimental data show a degree of 'spread' about the measure of central tendency. Some measure that describes this dispersion or variation quantitatively is required. The range of the data, from the maximum to the minimum value, is one possible candidate but it only uses two pieces of data. An average deviation obtained from averaging the difference between each data point and the location of central tendency is another possibility. Unfortunately, in symmetrical distributions there will be as many positive as negative deviations and so they will cancel one another out 'on average'. The dispersion would be zero. Taking absolute values by forming the modulus of the deviation is a third alternative, but this is mathematically clumsy. These deliberations suggest that the square of the deviation should be used to remove the negative signs that troubled us with the symmetric distribution. The sum of the squares of the deviations divided by the number of data points is the mean square deviation or variance. To return to the original units the square root is taken. So this is the root mean square (rms) deviation or the standard deviation.

Coefficient of variation

The size of the standard deviation is a good indicator of the degree of dispersion but it is also influenced by the magnitude of the numbers being used. For example, a standard deviation of 1 when the arithmetic mean is 10 suggests more 'spread' than a standard deviation of 1 when the mean is 100. To overcome this problem the standard deviation is normalized by dividing by the mean and multiplied by 100 and then expressed as a percentage to obtain the coefficient of variation. In the first case the coefficient of variation is 10%, while the second has a coefficient of variation of only 1%.

13.5.3. *Probability and distributions*

There are two types of probability: subjective and objective probability. The former is based on 'fuzzy logic' derived from past experiences and we use it to make decisions such as when to cross the road or overtake when driving a car. The latter is based on the frequency concept of probability and can be constructed from studying idealized models. One of the favourites used by statisticians is the unbiased coin. There are two possible outcomes: a head or a tail. Each time the coin is thrown the chance or probability of throwing a head (or a tail) is $\frac{1}{2}$.

Random and stochastic sequences

In a random sequence the result obtained from one trial has no influence on the results of subsequent trials. If you toss a coin and it comes down heads that result does not influence what you will get the next time you toss the coin. In the same way the numbers drawn in the national lottery one week have no bearing on the numbers that will turn up the following week! In a stochastic sequence, derived from the Greek word *stochos* meaning target, the events appear to be random but there is a deterministic outcome. Brownian motion of molecules in diffusion is an example of a 'stochastic sequence'.

In an experiment or trial the total number of possible outcomes is called the sample space. For example, if a single coin is thrown four times in succession, at each throw two possibilities only (a head or a tail) can occur. Therefore with four consecutive throws, each with two possible outcomes, the totality of outcomes in the trial is $2^4 = 16$. HHHT would be one possible outcome, TTHT would be another. To make sure you understand what is happening it is worth writing out all 16 possibilities. If the total number of outcomes is 16 then the probability of each possibility is $1/16$. Therefore for the random sequence the sum of the probabilities of all possible outcomes in the sample space is unity (1). This an important general rule: probabilities can never be greater than 1 and one of the possible outcomes must occur each time the experiment is performed. The outcome of a (random) trial is called an event. For example, in the coin tossing trial four consecutive heads, HHHH, could be called event A. Then we would use the notation probability of event A occurring, $P(A) = 1/16 = 0.0625$.

Conditional probability

Conditional probability is most easily understood by considering an example. Anaemia can be associated with two different red cell characteristics: a low red cell concentration or, in the case of haemoglobin deficiency, small size. In both cases the oxygen-carrying capacity of the blood is compromised. In some cases the cells are smaller than normal and in addition the concentration is abnormally low. For example, in a sample of size 1000 there were 50 subjects with a low count, event A, and 20 with small red cells, event B. Five of these patients had both a low count and the cells were small. In the complete sample the proportion of subjects with small cells is $20/1000 = 0.02$, $P(B) = 0.02$, whereas the proportion in the subsample that had a low count is $5/50 = 0.1$. The conditional probability of event B knowing that A has already occurred, $P(B/A) = 0.1$. The probabilities are different if we already know the person has a low count!

Statistical independence

All the events in the sample space must be mutually exclusive, there must be no overlapping, like the subjects with both a low count and small cells, for statistical independence. This characteristic is an important criterion when scientific hypotheses are tested statistically and, unfortunately, is often ignored, especially by clinicians, in the scientific literature.

Probability: rules of multiplication

For two statistically independent events A and B, the probability that A and B occur is the product of the probability of A and the probability of B, $P(A \text{ and } B) = P(A) \times P(B)$. For example, in the coin throwing experiment, the probability of obtaining four heads on two consecutive occasions, event A is 4H and event B is 4H, is $P(A \text{ and } B) = 0.0625 \times 0.0625 = 0.003\,906\,25$ or 0.0039 rounded to four decimal places.

In the absence of statistical independence the rules change. Some account must be taken of the fact that events A and B are not mutually exclusive. With statistical dependence the multiplication rule becomes $P(A \text{ and } B) = P(A) \times P(B/A)$, where $P(B/A)$ is the probability of event B knowing that A has already occurred.

Probability: rules of addition

For two statistically independent events A and B, the probability that A or B occurs is the sum of the probability of A and the probability of B, $P(A \text{ or } B) = P(A) + P(B)$. In the coin throwing experiment, if 4H is event A and 4T is event B then the probability that four consecutive heads or four consecutive tails are thrown is $P(A \text{ or } B) = 0.0625 + 0.0625 = 0.125$.

Again with statistical dependence some account of the 'overlap' between two events A and B must be included in the rule of addition. In this case $P(A \text{ or } B) = P(A) + P(B) - P(A \text{ and } B)$, where the probability of A and B occurring has been defined above.

Probability distributions

Probability distributions are used for decision making and drawing inferences about populations from sampling procedures. We defined a random variable earlier as a variable with uncertainty associated with it. We will use the symbol x for the random variable in the following discussion. Consider the coin throwing experiment and let x denote the number of heads that can occur when a coin is thrown four times. Then the variable, x, can take the values 0, 1, 2, 3 and 4, and we can construct a probability histogram based on the outcomes (see table 13.1).

Table 13.1. *Analysis of the result of throwing a coin four times.*

Value of x (the number of heads)	Combinations of four throws
0	TTTT
1	HTTT, THTT, TTHT, TTTH
2	HHTT, TTHH, HTTH, THHT, HTHT, THTH
3	HHHT, HHTH, HTHH, THHH
4	HHHH

Each outcome has a probability of 0.0625 so the histogram looks like figure 13.18.

Frequency histograms obtained from sample data are an approximation to the underlying probability distribution. In fact, frequency = probability × sample size.

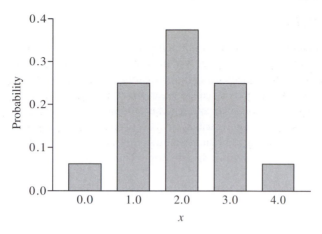

Figure 13.18. *A probability distribution.*

There are very many frequency distributions that have been applied to medical data. We will consider a few of them briefly.

Binomial distribution

This is a counting distribution based on two possible outcomes, success and failure. If the probability of success is p then the probability of failure is $q = 1 - p$.

For n trials, the arithmetic mean of the binomial distribution is np and the standard deviation is $\sqrt{npq}$.

Poisson distribution

This is also a counting distribution, derived from the binomial distribution, but based on a large number of trials, n is large, and a small probability of success. In these circumstances the mean of the distribution is np and the standard deviation is $\sqrt{npq} \cong \sqrt{np}$ because $q = (1 - p)$ is approximately unity. The process of radioactive decay is described by Poisson statistics (see Chapter 6, section 6.4.1).

From binomial to Gaussian distribution

The binomial distribution is based on a discrete (count) random variable. If the probability of success is a different size to the probability of failure and the number of trials, n, is small then the distribution will be asymmetrical. If, however the number of trials is large the distribution becomes more symmetrical no matter what value p takes. With a large number of trials the binomial distribution approximates to a Gaussian distribution. The Gaussian distribution is based on a continuous random variable and can be summarized by two statistical parameters: the arithmetic mean and the standard deviation (SD). It is often called the normal distribution because purely random errors in experimental measurement 'normally' have a Gaussian distribution. It is named after Carl Friedrich Gauss who developed the mathematics that describes the distribution. The area between ± 1 SD is equal to 68% of the total area under the curve describing the Gaussian distribution, while ± 2 SD equals 95% of the area. Virtually all the area (99%) under the curve lies between ± 3 SD.

Standard Gaussian distribution

A continuous random variable, x, can take an infinite number of values with infinitesimally small probability. However, the probability that x lies in some specified interval, for example the probability that $34 < x < 45$, is usually required. This probability, remembering that probability can never be greater than 1, is related directly to the area under the distribution curve. Because an arithmetic mean and a standard deviation can take an infinite number of values and any Gaussian curve can be summarized by these two parameters, there must be an infinite number of possible Gaussian distributions. To overcome this problem all Gaussian distributions are standardized with respect to an arithmetic mean of 0 and a standard deviation of 1 and converted into a single distribution, the standard Gaussian distribution. The transformation to a standard normal random variable is

$$z = (x - \text{mean})/\text{SD}.$$

With this formulation all those values of z that are greater than 1.96 (approximately 2 SD above the mean of zero for the standardized distribution) are associated with a probability—the area under the distribution—that is less than 0.025 or in percentage terms less than 2.5% of the total area under the curve. Similarly, because the mean is at zero, all those values of z that are less than -1.96 (approximately 2 SD below the mean of zero) are also associated with a probability of less than 0.025 or 2.5%. This means that any value of a random variable, x, assuming it has a Gaussian distribution, that transforms into a value of $z > 1.96$ or $z < -1.96$ for the standard distribution, has a probability of occurring which is less than 0.05. If z is known then the value of x that corresponds to this value of z can be obtained by a back transformation

$$x = (z\,\text{SD}) + \text{mean}.$$

13.5.4. *Sources of variation*

There are three major sources of variation:

- Biological variation.
- Error variation in experimental measurement.
- Sampling variation.

It is usually impossible to separate these sources of variation. Clinically, decisions are made to investigate subjects that lie outside the 'norm', but what is normal? When the Gaussian distribution is called 'normal' it is often confused with the concepts of conforming to standard, usual, or typical values. Usually, there is no underlying reason why biological or physical data should conform to a Gaussian distribution. However, if test data happen to be Gaussian or close to Gaussian then the opportunity should be taken to use the properties of this distribution and apply the associated statistical tests. However, if the distribution is not 'normal' then other tests should be used. It may be best to consider the range of the data from minimum to maximum.

Sensitivity and specificity

Sensitivity is the ability to detect those subjects with the feature tested for. It is equal to $1 - \beta$. If the β error is set at 5% then the sensitivity of the test will be 95%. The higher the sensitivity, the lower the number of false negatives. Specificity is the ability to distinguish those subjects not having the feature tested for. It is equal to $1 - \alpha$. If the α error is set at 5% then the specificity of the test is 95%. For example, a parathyroid scan is used to detect tumours. The feature being tested for is the presence of a tumour. If the scan and its interpretation by an observer correctly identifies subjects who have not got a tumour 95 times out of each 100 scans, then its specificity is 95%. If a tumour is picked out by a 'hot spot' and at operation a tumour is found 95 times out of a 100 then the sensitivity of the test is 95%. A test with 100% sensitivity and 100% specificity would never throw up false negatives or false positives. In practice this never happens,

$$\text{Sensitivity} = 1 - \text{the fraction of false negatives}$$
$$\text{Specificity} = 1 - \text{the fraction of false positives.}$$

The 95% paradox

If the α error of each test in a battery of statistically independent tests is set at 5% then the probability of being 'abnormal' is increased as the number of tests in the battery increases. The probability rules indicate that for two events A and B that $P(\text{A and B}) = P(\text{A}) \times P(\text{B})$. So for two tests the probability that at least one test value falls outside the reference range is $1 - (0.95 \times 0.95) = 0.10$. When the number of tests rises to five the probability of 'abnormality' increases to 0.23. This is a very important conclusion. If we are looking for a particular result, perhaps an association between cancer and coffee, then if we keep making different trials we actually increase the probability of getting a false answer.

13.5.5.   Relationships between variables

Suppose that changes in a variable, y, depend on the manipulation of a second variable, x. The mathematical rule linking y to x is written as $y = F(x)$, where F is some function. The variable, y, is the dependent variable or effect and the variable, x, is the independent variable or cause. Two statistical models can be developed:

- The simple linear regression model.
- The bivariate regression model.

The mathematics is the same for both models and relies on a straight-line fit to the data by minimizing the square of the difference between the experimental value of y for a specified value of x and the value of y predicted from the straight line for the same values of x. However, in the former case the values of x have no uncertainty associated with them, the regression of height with age in children, for example, while in the latter case both variables arise from Gaussian distributions, the regression of height against weight, for example.

Correlation

Correlation research studies the degree of the relationship between two variables (it could be more in a multi-variate model). A measure of the degree of the relationship is the correlation coefficient, r. A scatterplot of the data gives some idea of the correlation between two variables. If the data are condensed in a circular cloud the data will be weakly correlated and r will be close to zero. An elliptical cloud indicates some correlation and in perfect correlation the data collapses onto a straight line. For positive correlation $r = 1$ and for negative correlation $r = -1$. Perfect positive or negative correlation is a mathematical idealization and rarely occurs in practice. The correlation coefficient, r, is a sample statistic. It may be that the sample chosen is representative of a larger group, the population, from which the sample was drawn. In this case the results can be generalized to the population and the population correlation coefficient can be inferred from the sample statistic.

Partial correlation

It is possible that an apparent association is introduced by a third hidden variable. This may be age, for example. Blood pressure tends to increase with age while aerobic capacity decreases with age. It may appear that there is a negative correlation between blood pressure and aerobic capacity. However, when allowances are made for changes with age the spurious correlation is partialed out. In fact, there may be minimal correlation between blood pressure and aerobic capacity. The possibility of partial correlation is another very important conclusion.

13.5.6. *Properties of population statistic estimators*

All population estimations are assessed against four criteria:

- Consistency, increases as sample size approaches the population size.
- Efficiency, equivalent to precision in measurement.
- Unbiasedness, equivalent to accuracy in measurement.
- Sufficiency, a measure of how much of the information in the sample data is used.

The sample arithmetic mean is a consistent, unbiased, efficient and sufficient estimator of the population mean. In contrast, the standard deviation is biased for small samples. If the value of n, the sample size, is replaced by $n - 1$ in the denominator of the calculation for standard deviation the bias is removed in small samples.

Standard error

Suppose a sample is drawn from a population. The sample will have a mean associated with it. Now return the sample to the population and repeat the process. Another sample will be generated with a different sample mean. If the sampling is repeated a large number of times and a sample mean is calculated each time, then a distribution of sample means will be generated. However, each sample has its own standard deviation so that this sample statistic will also have a distribution with a mean value (of the sample SDs) and a standard deviation (of the sample SDs). These concepts lead to a standard deviation of standard deviations and is confusing! Thus standard deviations of sample statistics like the mean and SD are called standard errors to distinguish them from the SD of a single sample.

When the sampling statistic is the mean then the standard deviation of the distribution of sample means is the standard error about the mean (SEM). In practice, the SEM is calculated from a single sample of size n with a standard deviation SD by assuming, in the absence of any other information, that SD is the best estimate of the population standard deviation. Then the standard error about the mean, the standard deviation of the distribution of sample means, can be estimated:

$$SEM = SD/\sqrt{n}$$

13.5.7. *Confidence intervals*

The distribution of sample means is Gaussian and the SD of this distribution of sample means is the SEM. However, ± 2 SD of a Gaussian distribution covers 95% of the data. Hence $x \pm 2$ SEM covers 95% of the sampling distribution, the distribution of sample means. This interval is called the 95% confidence interval. There is a probability of 0.95 that the confidence interval contains the population mean. In other words, if the population is sampled 20 times then it is expected that the 95% confidence interval associated with 19 of these samples will contain the population mean.

A question that frequently arises in statistical analysis is 'Does the difference that is observed between two sample means arise from sampling variation alone (are they different samples drawn from the same population) or is the difference so large that some alternative explanation is required (are they samples that have been drawn from different populations?)' The answer is straightforward if the sample size is large and the underlying population is Gaussian. Confidence limits, usually 95% confidence limits, can be constructed from the two samples based on the SEM. This construction identifies the level of the α error and β error in a similar way to the specificity and the sensitivity in clinical tests. Suppose we call the two samples A and B, consider B as data collected from a 'test' group and A as data collected from a 'control' group. If the mean of sample B lies outside the confidence interval of sample A the difference will be significant in a statistical sense, but not necessarily in a practical sense. However, if the mean of sample A lies outside the confidence interval of sample B then this difference may be meaningful in a practical sense. Setting practical levels for differences in sample means is linked to the power of the statistical test. The power of the test is given by $1 - \beta$. The greater the power of the test the more meaningful it is. A significant difference can always be obtained by increasing the size of the sample because there is less sampling error, but the power of the test always indicates whether this difference is meaningful, because any difference which the investigator feels is of practical significance can be decided before the data are collected.

Problems arise when samples are small or the data are clearly skewed in larger samples. Asymmetrical data can be transformed logarithmically and then treated as above or non-parametric statistics (see later) can be used. As sample size, n, decreases the SEM estimation error increases because SEM is calculated from $S/\sqrt{n}$, where S is the sample SD. This problem is solved by introducing the 't' statistic. The t statistic is the ratio of a random variable that is normally distributed with zero mean, to an independent estimate of its SD (of sample means) based on the number of degrees of freedom. As the number of degrees of freedom, in this case one less than the sample size, increases the distribution defining t tends towards a Gaussian distribution. This distribution is symmetrical and bell-shaped but 'spreads out' more as the degrees of freedom decrease. The numerator of t is the difference between the sample and population mean, while the denominator is the error variance and reflects the dispersion of the sample data. In the limit as the sample size approaches that of the population, the sample mean tends towards to the population mean and the SD of the sample means (SEM) gets smaller and smaller because n gets larger and larger so that true variance/error variance $= 1$.

Paired comparisons

In experimental situations the samples may not be statistically independent, especially in studies in which the same variable is measured and then repeated in the same subjects. In these cases the difference between the paired values will be statistically independent and 't' statistics should be used on the paired differences.

Significance and meaningfulness of correlation

Just like the 't' statistic the correlation coefficient, r, is also a sample statistic. The significance and meaning-fulness of r based on samples drawn from a Gaussian population can be investigated. The larger the sample size the more likely it is that the r statistic will be significant. So it is possible to have the silly situation where r is quite small, close to zero, the scatter plot is a circular cloud and yet the correlation is significant. It happens all the time in the medical literature. The criterion used to assess the meaningfulness of r is the coefficient of determination, r^2. This statistic is the portion of the total variance in one measure that can be accounted for by the variance in the other measure. For example, in a sample of 100 blood pressure and obesity measurements with $r = 0.43$, the degrees of freedom is $100 - 2 = 98$, the critical level for significance $r_{98;5\%} = 0.195$, so statistically the correlation between blood pressure and obesity is significant. However, $r^2 = 0.185$ so that only about 18% of the variance in blood pressure is explained by the variance in obesity. A massive 82% is unexplained and arises from other sources.

13.5.8.   Non-parametric statistics

When using sampling statistics such as t and r assumptions are made about the underlying populations from which the samples are drawn. These are:

- The populations are Gaussian.
- There is homogeneity of variance, i.e. the populations have the same variance.

Although these strict mathematical assumptions can often be relaxed, to a certain extent, in practical situations and the statistical methods remain robust, techniques have been developed where no assumptions are made about the underlying populations. The sample statistics are then distribution free and the techniques are said to be non-parametric. Parametric statistics are more efficient, in the sense that they use all of the data, but non-parametric techniques can always be used when parametric assumptions cannot be made.

Chi-squared test

Data are often classified into categories, smoker or non-smoker, age, gender, for example. A question that is often asked is 'Are the number of cases in each category different from that expected on the basis of chance?' A contingency table can be constructed to answer such questions. The example in table 13.2 is taken from a study of myocardial infarction (MI) patients that have been divided into two subgroups (SG1 and SG2) based on the size of their platelets, small circulating blood cells that take part in blood clotting. We wondered whether different sites of heart muscle damage were more prevalent in one group compared to the other. The numbers in the table are the experimental results for 64 patients; the numbers in brackets are those expected purely by chance. These expected figures can be calculated by determining the proportion of patients in each subgroup, 39/64 for SG1 and 25/64 for SG2, and then multiplying this proportion by the number in each MI site category. For example, there are 30 patients with anterior infarction so the expected value in MISG1 would be $39 \times 30/64 = 18.28$. Rounded to the nearest integer, because you cannot have a third of a person, this gives 18 for the expected value.

Table 13.2. *64 patients with myocardial infarction analysed as two groups depending upon platelet size.*

	Anterior infarction	Inferior infarction	Subendo-cardial infarction	Combined anterior and superior infarction	Totals
MI SG1	16(18)	10(11)	8(6)	5(4)	39
MI SG2	14(12)	8(7)	1(3)	2(3)	25
	30	18	9	7	64

Chi-squared is evaluated by considering the sum of squares of the difference between the observed (O) and expected values (E) normalized with respect to the expected value (E).

$$\chi^2 = \sum[(O - E)^2/E].$$

In general, the total number of degrees of freedom will be given by $(r - 1)(c - 1)$ where r is the number of rows and c is the number of columns. In this case the degrees of freedom will be $3 \times 1 = 3$. From appropriate tables showing the distribution of chi-squared, the critical value at the 5% level is 7.81, whereas $\chi^2 = 4.35$. There is no statistical reason to believe that the site of infarction is influenced by platelet size. Chi-squared can also be used to assess the authenticity of curve fitting, between experimental and theoretical data, but it is influenced by the number of data points. As the number of data points increases the probability that chi-squared will indicate a statistical difference between two curves is increased.

Rank correlation

Instead of using the parametric correlation measure, r, which assumes a bivariate Gaussian distribution, the data can be ranked, the difference between the ranks calculated and a non-parametric measure, named after Spearman, can be calculated.

Difference between variables

A non-parametric equivalent of the 't' statistic can be applied to small groups of data. The Mann–Whitney U statistic only relies on ranking the data and makes no assumption about the underlying population from which the sample was drawn. When the data are 'paired' and statistical independence is lost, as in 'before' and 'after' studies, a different approach must be adopted. The Wilcoxon matched-pairs signed-ranks statistic is used to assess the 'paired' data.

13.6. LINEAR SIGNAL PROCESSING

Up to this point we have looked at ways in which we can describe a signal, either as a mathematical equation or in terms of statistical descriptors. We shall now turn our attention to the problem of manipulation of signals. There are two facets of the problem that will be of particular interest:

- Can we process a signal to enhance its fundamental characteristics and/or to reduce noise?
- Can we quantify the effects of an unavoidable processing of a signal and thus in some way recover the unprocessed signal?

A signal processor receives an input signal f_{in} and outputs a signal f_{out} (see figure 13.19). As before we shall concentrate on signals that are functions of time only, but the techniques developed have wider application.

Figure 13.19. *Diagrammatic representation of the processing of a signal.*

In the most general case there will be no restriction on the performance of the signal processor. However, we will focus on linear signal processing systems: nonlinear systems fall outside the scope of this book. The properties of a linear system are summarized below.

- The output from two signals applied together is the sum of the output from each applied individually. This is a basic characteristic of any linear system.
- Scaling the amplitude of an input scales the amplitude of the output by the same factor. This follows from the first property described above.
- The output from a signal can contain only frequencies of the input signal. No new frequencies can be generated in a linear system. This property is not immediately obvious, but follows from the form of solutions of the linear differential equations that describe analytically the performance of a linear processor.

We have shown that we can take the basic building block of a signal to be a sinusoidal wave of frequency ω, phase ϕ and amplitude A. Any signal can be considered to be a linear sum of such components, and the

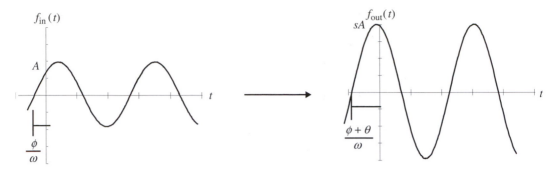

Figure 13.20. *Action of a linear processor on a sinusoid.*

output from a linear signal processor will be the sum of the output from each of the sinusoidal components. For a wave of frequency ω, the amplitude scaling factor is σ_ω and the phase shift is θ_ω (see figure 13.20). Note that the scaling and shifting of the sinusoid will, in general, be frequency dependent even for a linear system, and in principle we should be able to make use of the scaling to amplify or to attenuate particular frequencies.

13.6.1.   *Characteristics of the processor: response to the unit impulse*

A fundamental premise of linear signal processing is that we should like to be able to describe the processor in such a way that we can predict its effect on any input signal and thus deduce the output signal. Obviously if we are to do this we need to be able to characterize the performance of the processor in some way. In principle, the input signal might contain sinusoidal components of any frequency, and so in order to characterize the processor it would seem to be necessary to understand its response to all frequencies. At first sight this might seem to be a rather onerous requirement, but in practice we can make profitable use of a function that we have already met, the unit impulse function. We have seen that the spectrum of the Fourier transform of the unit impulse is unity, i.e. that it contains all frequencies and that all have unit magnitude. This suggests that if we can quantify the response of the processor to a unit impulse it will tell us something about its response to every frequency.

   The response of a linear processor to a unit impulse can be considered to be a descriptive characteristic of the processor. Suppose, for example, that the response is that illustrated in figure 13.21.

   For the processor characterized by this particular response function the peak output occurs after about one-third of a second, and thereafter the response decays until it is negligible after 0.3 s. The function $\mathbf{1}_{\text{out}}(t)$ characterizes the processor in the time domain, and is known as its impulse response.

13.6.2.   *Output from a general signal: the convolution integral*

We have already seen (section 13.4.1) that we can think about a discrete signal as a weighted sum of unit impulses,

$$f(t) = f_0\,\mathbf{1}(t) + f_1\,\mathbf{1}(t-T) + f_2\,\mathbf{1}(t-2T) + \cdots = \sum_{k=0}^{n} f_k\,\mathbf{1}(t-kT).$$

This is a useful starting point for the description of the output generated by a linear processor for a general input signal. Suppose that the processor is characterized by the response $\mathbf{1}_{\text{out}}(t)$. Each pulse initiates an output response that is equal to the magnitude of the pulse times the response to the unit impulse. At a time

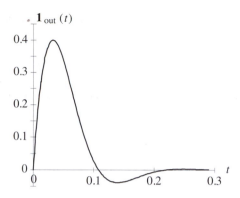

Figure 13.21. *Example of the impulse response of a linear processor.*

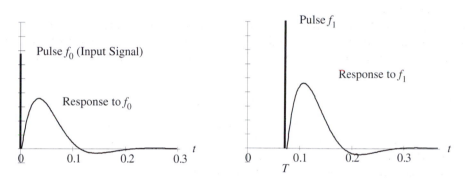

Figure 13.22. *The temporal response to successive pulses f_0 and f_1.*

t after commencement the output from the processor due to the first pulse is $f_0 \mathbf{1}_{\text{out}}(t)$. After an interval of T s another pulse hits the processor. The responses to each pulse are illustrated in figure 13.22.

In the interval $T \leq t < 2T$ the output will be the sum of the two:

$$f_{\text{out}}(t) = f_0 \mathbf{1}_{\text{out}}(t) + f_1 \mathbf{1}_{\text{out}}(t - T).$$

The procedure is readily extended to give the response on any interval, $nT \leq t < (n+1)T$:

$$f_{\text{out}}(t) = \sum_{k=0}^{n} f_k \mathbf{1}_{\text{out}}(t - kT).$$

For a continuous signal the sampling period is infinitesimally small and the value of the input signal is $f_{\text{in}}(t)$. The output from the processor (the limit of the expression derived above) is

$$f_{\text{out}}(t) = \int_0^\infty f_{\text{in}}(\tau) \mathbf{1}_{\text{out}}(t - \tau) \, d\tau.$$

This expression is called the convolution integral. In our derivation it defines the output from a linear signal processor in terms of the input signal and the characteristic response of the processor to a unit impulse.

It should be noted that this expression is correct providing that no input pulses were presented to the processor before commencement of monitoring of the output. If such signals existed, the output would continue to include their effect in addition to that of any new signals. If we suddenly start to monitor the output from a signal that has been going on for some time, the lower limit of the integral should be changed to minus infinity. Furthermore, the unit impulse response must be zero for any negative time (the response cannot start before the impulse hits), and so the value of the convolution integral for any τ greater than t is zero. We can therefore replace the limits of the integral as follows:

$$f_{out}(t) = \int_{-\infty}^{\infty} f_{in}(\tau)\, \mathbf{1}_{out}(t - \tau)\, d\tau.$$

Although we have developed the principles of the convolution integral with respect to time signals, it might be apparent that there will be widespread application in the field of image processing. All imaging modalities 'process' the object under scrutiny in some way to produce an image output. Generally the processing will be a spreading effect so that a point object in real space produces an image over a finite region in imaging space. This is similar to the decaying response to the unit impulse function illustrated above. An understanding of the nature of the processing inherent in the imaging modality should, in principle, permit a reconstruction in which the spreading process can be removed, or at least moderated. This subject is considered further in section 14.4 of Chapter 14 on image processing.

Graphical implementation of the convolution integral

An understanding of the underlying process of the convolution integral can be promoted by the consideration of a graphical implementation of the procedure as presented by Lynn (see figure 13.23). A clue to the graphical implementation can be obtained by looking at the terms in the integral. The input signal f_{in} at time τ seconds is multiplied by the value of the impulse response at time $(t - \tau)$ seconds. This suggests that we must be able to plot both functions on the same graph if we displace the impulse response by t seconds (moving its starting point from the origin to the time, t, at which the output signal is to be computed), and then reflect it about the vertical axis (because of the negative sign of τ). The product of the two curves can be formed, and the convolution integral is the area under this product curve at any time t.

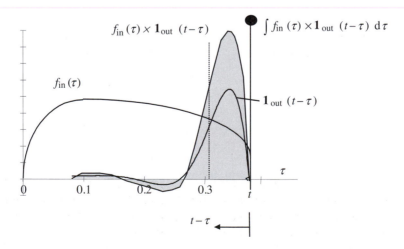

Figure 13.23. *Graphical illustration of the convolution integral (as presented by Lynn (1989)).*

Essentially the superposed graphs reflect a simple observation: the response to the input signal that occurred at time τ has, at time t, been developing for $(t - \tau)$ seconds. Examining the above graph, we can see that the input signal and the shifted and reflected unit response function take appropriate values at time τ. The processes of the formation of the product and then the integral are self-evident.

13.6.3.   *Signal processing in the frequency domain: the convolution theorem*

The convolution integral defines the output from the linear processor as an integral of a time function. The actual evaluation of the integral is rarely straightforward, and often the function will not be integrable into a closed-form solution. Although the graphical interpretation described above might assist with the understanding of the process, it does nothing to assist with the evaluation of the integral. We can, however, take advantage of the properties of the Fourier transform by performing the integration in the frequency domain.

The Fourier transform of the output signal is

$$F_{\text{out}}(\omega) = \int_{-\infty}^{\infty} \left(\int_{-\infty}^{\infty} f_{\text{in}}(\tau)\, \mathbf{1}_{\text{out}}(t - \tau)\, \mathrm{d}\tau \right) \mathrm{e}^{-\mathrm{j}\omega t}\, \mathrm{d}t.$$

We can multiply the exponential term into the inner integral so that we have a double integral of a product of three terms. The order of integration does not matter, and we can choose to integrate first with respect to t,

$$F_{\text{out}}(\omega) = \int_{-\infty}^{\infty} \int_{-\infty}^{\infty} f_{\text{in}}(\tau)\, \mathbf{1}_{\text{out}}(t - \tau)\, \mathrm{e}^{-\mathrm{j}\omega t}\, \mathrm{d}t\, \mathrm{d}\tau.$$

The input function is independent of t, and so it can be taken outside that integral,

$$F_{\text{out}}(\omega) = \int_{-\infty}^{\infty} \left(\int_{-\infty}^{\infty} \mathbf{1}_{\text{out}}(t - \tau)\, \mathrm{e}^{-\mathrm{j}\omega t}\, \mathrm{d}t \right) f_{\text{in}}(\tau)\, \mathrm{d}\tau.$$

Changing the variable of integration in the bracket to $t' = t - \tau$:

$$F_{\text{out}}(\omega) = \int_{-\infty}^{\infty} \left(\left(\int_{-\infty}^{\infty} \mathbf{1}_{\text{out}}(t')\, \mathrm{e}^{-\mathrm{j}\omega t'}\, \mathrm{d}t' \right) \mathrm{e}^{-\mathrm{j}\omega \tau} \right) f_{\text{in}}(\tau)\, \mathrm{d}\tau.$$

The term in the inner bracket is, by definition, the Fourier transform of the response to the unit impulse, $\mathbf{1}_{\text{out}}(\omega)$. In this equation it is independent of the parameter τ, and can be taken outside its integral,

$$F_{\text{out}}(\omega) = \mathbf{1}_{\text{out}}(\omega) \int_{-\infty}^{\infty} \mathrm{e}^{-\mathrm{j}\omega \tau}\, f_{\text{in}}(\tau)\, \mathrm{d}\tau.$$

The remaining integral is the Fourier transform of the input signal. Finally then,

$$F_{\text{out}}(\omega) = \mathbf{1}_{\text{out}}(\omega)\, F_{\text{in}}(\omega).$$

This is a very important result. It is known as the convolution theorem, and it tells us that the convolution of two functions in the time domain is equivalent to the multiplication of their Fourier transforms in the frequency domain. We have made no use of any special properties of the functions under consideration (indeed we do not know whether they have any), and so the result is completely general.

The Fourier transform, $\mathbf{1}_{\text{out}}(\omega)$, of the impulse response is an alternative (and equally valid) characteristic description of a linear processor. It characterizes the processor in the frequency domain, and is known as the frequency response. We should note that it is generally complex, representing a scaling and a phase shift of each frequency component of the signal. We can now envisage a general frequency domain procedure for the evaluation of the response to a general input signal.

- Form the Fourier transform of the input signal.
- Multiply by the frequency response of the signal processor. The resulting product is the frequency domain representation of the input signal.
- If required take the Fourier transform of the product to recover the time domain representation of the input signal.

The relative ease of multiplication compared to explicit evaluation of the convolution integral makes this procedure attractive for many signal processing applications. It is particularly useful when the primary interest is in the frequency components of the output rather than in the actual time signal, and the last step above is not required.

13.7.   PROBLEMS

13.7.1.   *Short questions*

a   Would you consider a nerve action potential as a continuous or discontinuous signal?
b   Is the ECG/EKG a periodic signal?
c   Is the equation for a straight line a polynomial?
d   Does a triangular waveform have finite harmonics at all multiples of the fundamental frequency?
e   What is the result of carrying out a Fourier transform on a rectangular impulse in time?
f   Is the variance of a data set equal to the square root of the standard deviation?
g   How are power and amplitude related?
h   What is an autocorrelation function and where is its maximum value?
i   Is an EMG signal periodic?
j   What conclusion could you draw if the mean, mode and median values of a distribution were all the same?
k   What type of statistical distribution describes radioactive decay?
l   If the number of false positives in a test for a disease is 12 out of 48 results then what is the sensitivity of the test?
m   If we carried out 10 different tests on a group of patients then would we change the chance of getting an abnormal result compared with just making one test?
n   If the SD is 6 and $n = 64$ then what is the SEM?
o   What does a Mann–Whitney U statistic depend upon?
p   Rank the following data: 0.15, 0.33, 0.05, 0.8 and 0.2 .
q   Is the χ^2 test a parametric test?
r   What is the application of non-parametric tests?
s   What is the convolution integral?
t   What do you get if you multiply the Fourier transform of a signal by the frequency response of a system?

13.7.2.   Longer questions (answers are given to some of the questions)

Question 13.7.2.1

Figure 13.24 shows 5 s of an ECG/EKG recording. Table 13.3 gives the amplitude of the signal during the first second of the recording, digitized at intervals of 25 ms.

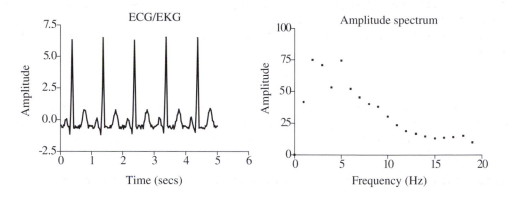

Figure 13.24. *An ECG/EKG measured over 5 s (left) and the associated Fourier amplitude spectrum (right). The first second of data are tabulated (table 13.3).*

Table 13.3. *Tabulated data for figure 13.24.*

Time (ms)	Amplitude	Time (ms)	Amplitude
0	−0.4460	500	−0.5829
25	−0.5991	525	−0.6495
50	−0.6252	550	−0.5080
75	−0.4324	575	−0.6869
100	−0.4713	600	−0.4347
125	−0.0339	625	−0.5665
150	−0.0079	650	−0.3210
175	0.0626	675	−0.0238
200	−0.3675	700	0.5087
225	−0.5461	725	0.7614
250	−0.5351	750	0.7505
275	−1.1387	775	0.7070
300	0.4340	800	0.3653
325	2.4017	825	−0.0457
350	6.3149	850	−0.2246
375	2.2933	875	−0.6473
400	−0.6434	900	−0.4424
425	−0.6719	925	−0.5100
450	−0.4954	950	−0.4602
475	−0.6382	975	−0.5227

The amplitude spectrum shown was produced by carrying out a Fourier transform on the 5 s of recording. Use what signal analysis software is available to you to carry out a Fourier analysis of the tabulated ECG/EKG (table 13.3) and compare your results with the given amplitude spectrum.

Explain why the maximum component of the amplitude spectrum is not at the frequency of the heartbeat.

Would you expect the mean amplitude of the ECG/EKG to be zero?

Question 13.7.2.2

Figure 13.25 shows the autocorrelation function (r) of the ECG/EKG presented in figure 13.24. The heartrate is about 60 bpm. Explain how this function has been calculated.

What can be deduced by comparing the amplitude of the function at time zero with that for a time delay of 1 s? What is the likely cause for the reduced amplitude of the second peak?

What function would result if the Fourier transform of the autocorrelation function was calculated?

Answer

See section 13.3.6. The reduced amplitude of the second peak is almost certainly caused by variations in the R–R interval time of the ECG/EKG. A completely regular heart rate would give a peak equal in amplitude to that at time zero. The heart rate is of course never completely regular.

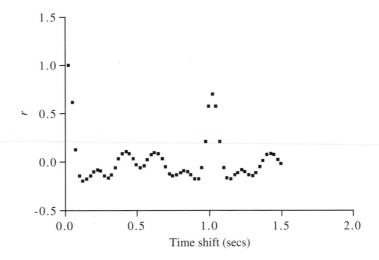

Figure 13.25. *Autocorrelation function of the ECG/EKG shown in figure 13.24.*

Question 13.7.2.3

Some clinical trial data are given to you. Table 13.4 gives the results of measuring the heights of 412 patients in the clinical trial. The results have been put into bins 4 cm wide and the tabulated value is the mid-point of the interval.

Table 13.4.

Height (cm)	Number of patients
134	5
138	10
142	40
146	55
150	45
154	60
158	100
162	50
166	25
170	10
174	11
178	1

Calculate the mean, median and mode for the distribution. What conclusion can you draw about the symmetry of the distribution? Why are the three calculated parameters not the same? What further information would you request from the supplier of the data?

Answer

Mean	154.4
Median	~153
Mode	158

The distribution is skewed. When plotted it appears there may be two distributions. You should ask for the sex of the patients and any other information that may explain the presence of two populations.

Question 13.7.2.4

Measurements are made on a group of subjects during a period of sleep. It is found that the probability of measuring a heart rate of less than 50 bpm is 0.03. In the same subjects a pulse oximeter is used to measure oxygen saturation Po_2 and it is found that the probability of measuring a value of Po_2 below 83% is 0.04. If the two measurements are statistically independent then what should be the probability of finding both a low heart rate and low oxygen saturation at the same time?

If you actually find the probability of both the low heart rate and low oxygen saturation occurring at the same time to be 0.025 then what conclusion would you draw?

Answer

The combined probability if the two measurements are independent would be 0.0012.
    If the probability found was 0.025 then the conclusion would be that heart rate and oxygen saturation measurements are not statistically independent. This would not be a surprising finding as the two measurements have a physiological link.

Question 13.7.2.5

A new method of detecting premalignant changes in the lower oesophagus is being developed and an index is derived from the measurements made on the tissue *in vivo*. In the clinical trial a tissue biopsy is taken so that the true state of the tissues can be determined by histology. The results for both normal and premalignant tissues are presented in figure 13.26 and table 13.5.

Calculate both the sensitivity and specificity of the new technique as a method of detecting premalignant tissue changes. Note any assumptions that you make.

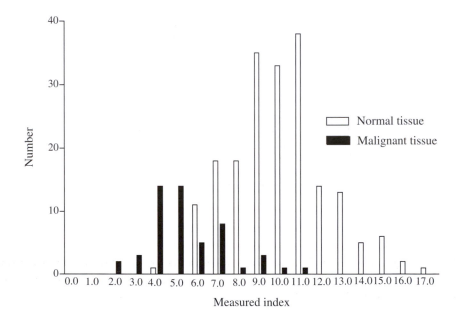

Figure 13.26. *The distribution of measurements made on samples of normal and malignant tissue.*

Answer

Assume that we put the division between normals and abnormals at an index value of 6.5.

   The number of false negatives in the malignant group is 14 from a total of 52. The fraction of false negatives is 0.269 and hence the sensitivity is 0.731.

   The number of false positives in the normal group is 22 from a total of 195. The fraction of false positives is 0.112 and hence the specificity is 0.888.

   We could put the division between normals and abnormals at any point we wish and this will change the values of sensitivity and specificity. The division may be chosen to give a particular probability of false negatives or positives. In this particular case it would probably be desirable to minimize the number of false negatives in the malignant group.

Table 13.5. *The data illustrated in figure 13.26.*

Measured index	Number of samples	
	Normal tissue	Malignant tissue
0	0	0
1	0	0
2	0	2
3	0	3
4	1	14
5	0	14
6	11	5
7	18	8
8	18	1
9	35	3
10	33	1
11	38	1
12	14	0
13	13	0
14	5	0
15	6	0
16	2	0
17	1	0

Question 13.7.2.6

The ECG/EKG signal given in figure 13.24 is passed through a first-order low-pass filter with a time constant of 16 ms. Convolve the transfer function of the filter with the given spectrum of the ECG/EKG to give the amplitude spectrum to be expected after the signal has been low-pass filtered.

Answer

The transfer function of the low-pass filter constructed from a resistance R and capacitance C is given by

$$\frac{-j/\omega C}{(R - j/\omega C)}.$$

The magnitude of this can be found to be

$$\frac{1}{(1 + R^2\omega^2C^2)^{1/2}}.$$

This can be used to determine the effect of the filter at the frequencies given in figure 13.24. The time constant CR of 16 ms will give a 3 dB attenuation at 10 Hz.

Answers to short questions

a   A nerve action potential should probably be considered as discontinuous as it moves very rapidly between the two states of polarization and depolarization.

b   The ECG/EKG is periodic, although the R–R interval is not strictly constant.

c   Yes, a straight line is a first-order polynomial.

d   No, a triangular waveform has a fundamental and odd harmonics.

e   You obtain a frequency spectrum of the form $\sin(t)/t$ if you carry out a Fourier transform on a rectangular impulse.

f   No, the variance is equal to the square of the standard deviation.

g   Power is proportional to the square of amplitude.

h   The autocorrelation function is produced by integrating the product of a signal and a time-shifted version of the signal. The maximum value is at zero time delay.

i   An EMG signal is not periodic. It is the summation of many muscle action potentials which are asynchronous.

j   The distribution must be symmetrical if the mean, mode and median are the same.

k   A Poisson distribution is used to describe radioactive decay.

l   The sensitivity would be $1 - 12/48 = 0.75$.

m   Yes, we would increase the chance of getting an abnormal result if we made 10 tests instead of just one.

n   0.75. The SEM $= \mathrm{SD}/\sqrt{n}$.

o   The Mann–Whitney U statistic depends upon ranking the data.

p   0.05, 0.15, 0.2, 0.33 and 0.8.

q   No, the χ^2 test is a non-parametric test.

r   Non-parametric tests are used when no assumptions can be made about the population distribution of the data.

s   The convolution integral gives the output of a system in terms of the input and the characteristic response of the system to a unit impulse.

t   If you multiply the FT of a signal by the frequency response of a system then you get the FT of the output from the system.

BIBLIOGRAPHY

Downie N M and Heath R W 1974 *Basic Statistical Methods* (New York: Harper and Row)

Leach C 1979 *Introduction to Statistics. A Non-Parametric Approach for the Social Sciences* (New York: Wiley)

Lynn P A 1989 *An Introduction to the Analysis and Processing of Signals* (London: MacMillan)

Lynn P A and Fuerst W 1998 *Introductory Digital Signal Processing with Computer Applications* (New York: Wiley)

Marple S L 1987 *Digital Spectral Analysis, with Applications* (Englewood Cliffs, NJ: Prentice Hall)

Moroney M J 1978 *Facts from Figures* (Harmondsworth: Penguin)

Neave H R 1978 *Statistics Tables* (London: Allen and Unwin)

Proakis J G and Manolakis D G 1996 *Introduction to Digital Signal Processing: Principles, Algorithms and Applications* (Englewood Cliffs, NJ: Prentice Hill)

Reichman W J 1978 *Use and Abuse of Statistics* (Harmondsworth: Penguin)

Siegel S and Castellan N J 1988 *Nonparametric Statistics* (New York: McGraw-Hill)

CHAPTER 14

IMAGE PROCESSING AND ANALYSIS

14.1. INTRODUCTION AND OBJECTIVES

Many of the medical images which are currently produced are digital in nature. In some cases, such as MRI and CT, the production of the images is intrinsically digital, as they are the result of computer processing of collected data. In other cases, such as radioisotope images, they can be either digital or directly recorded onto film, although the latter approach is almost extinct in practice. The last remaining bastion of direct film (analogue) images is conventional planar x-ray imaging. However, digital imaging is becoming more common and film is likely to become obsolete, with all images being acquired digitally and viewed from a screen rather than a film.

Some of the objectives of this chapter are to address issues such as:

- How can we make the best use of an image which we have acquired from the body?
- Can we 'optimize' an image?
- How do we deal with images obtained by different methods, but from the same body segment?
- Can we obtain quantitative information from an image?

At the end of this chapter you should be able to:

- Assess the requirements for image storage.
- Adjust the intensity balance of an image.
- Know how to enhance edges in an image.
- Know how to smooth images and appreciate the gains and losses.

This chapter follows Chapters 11 and 12 on image formation and image production. It is the last of the chapters that address the basic physics of a subject rather than the immediate practical problems of an application. Masters level students in medical physics and engineering should have no difficulty with the mathematics of this chapter. The undergraduate and general reader may prefer to skip some of the mathematics but we suggest you still read as much as possible and certainly sufficient to gain an understanding of the principles of various types of image manipulation.

14.2. DIGITAL IMAGES

In the abstract any image is a continuous spatial distribution of intensity. Mathematically it can be represented as a continuous real function $f(x, y)$. If the image is a true colour image then the value at each point is a vector of three values, since any colour can be represented by the magnitudes of three primary colours (usually red, green and blue). Practically all images are bounded in intensity and in space. There is a finite maximum intensity in the image and a finite minimum intensity which for most raw images, i.e. images not processed in any way, is always greater than or equal to zero.

14.2.1. Image storage

If the continuous image is to be stored in a computer it must be reduced to a finite set of numerical values. Typically a two-dimensional image is made up of M rows each of N elements, conventionally known as *pixels*. A three-dimensional image is made up of K slices or planes each containing M rows each of N elements. The elements in this case are *voxels*. Pixels are invariably square. The equivalent may not be true for voxels, although it is very convenient if they are cubic. In practice for MR and CT images the distance between slices, the slice thickness, may be larger than the dimensions of the pixel within a slice. Often interpolation is used to ensure that the voxels are cubes. Each voxel has a set of three integer coordinates n, m, k where each index runs from 0, so k runs from 0 to $K - 1$ (or possibly from 1 to K depending upon the convention used). In the rest of this chapter the term *image element*, or simply *element* will be used when either pixel or voxel is appropriate.

   The issue of how many image elements are needed if no information is to be lost in converting the collected data into an image is covered in sections 11.9 and 13.4. Once the size of the image element is determined, a value can be assigned to each element which represents either the integrated or average image intensity across the element, or a sampled value at the centre of the element. In either case information will be lost if the element size is too large. In the majority of cases the element size is usually adequate. However, the slice thickness may not always satisfy the requirements of sampling theory.

   Images are stored usually as one image element per computer memory location. The intensity measured is usually converted to an integer before storage if it is not already in integer form. For any image the intensity values will range from the minimum value in the image to the maximum value. Typical ranges are 0–255 (8 bits or 1 byte), 0–65 535 (16 bits or 2 bytes) or less commonly 0–4294 967 295 (32 bits). After processing the image may contain elements with negative values. If these are to be preserved then the range from zero to the maximum value is effectively cut by half, since half the range must be preserved for negative values. Which of these ranges is appropriate depends on the type of image. For radionuclide images the output is automatically integer (since the number of γ-rays binned into a pixel is always a whole number) and this is the value stored. It will be necessary to choose the number of bytes per pixel or voxel to be sufficient to ensure that the value of the element with maximum counts can be stored. However, if the image has a large number of elements considerable space can be saved if only 1 byte is assigned to each element. If the maximum intensity exceeds 255 when converted to an integer then the intensity values will need to be scaled. If the maximum intensity in the image is V_{max} and the intensity in the ith element is V_i (positive or zero only) then the scaled value is

$$v_i = 255 \frac{V_i}{V_{max}}.$$

Element values now lie in the range 0–255. There will inevitably be some truncation errors associated with this process and scaling irreversibly removes information from the image. The principal justification for scaling is to reduce storage space, both in memory and on disk or other storage devices. These requirements are less critical with up-to-date computer systems and there is no longer any real justification for using scaling, with the possible exception of high-resolution planar x-ray images. All medical images can be comfortably stored at 2 bytes per image element without any significant loss in performance. If data storage on disk or elsewhere is a problem then image compression techniques are now commonly available which can save significant amounts of storage. These are not appropriate for storage in memory where unrestricted access to image elements is required, but computer memory is relatively inexpensive.

Images can also be stored in floating point format. Typically 4 bytes per image element are required. This apparently frees us from having to know the maximum and minimum when displaying the image. However, for a given number of bits per element the precision of floating point is always less than an integer of the same number of bits since in floating point format some bits must be assigned to the exponent.

14.2.2. *Image size*

M and N can be any size, but it is common to make them a power of 2 and often to make them the same value. Sizes range from 64 for low count density radionuclide images to 2048 or larger for planar x-ray images. K may often not be a power of 2.

An image of 2048×2048 elements will contain 4.19×10^6 elements and 64 Mbits of data. If this formed part of a video image sequence then the data rate would be several hundred megabits per second. It is interesting to compare this figure with the data rates used in television broadcasting. The data rate output from a modern TV studio is 166 Mbit s^{-1} for a very high-quality video signal. However, the data rate used in high-definition television is only 24 Mbit s^{-1} and that used in standard television and video is only about 5 Mbit s^{-1}. The difference is explained largely by the compression techniques that are used in order to reduce the required data transfer rates. By taking advantage of the fact that our eyes can only detect limited changes in intensity and that most picture elements in a video do not change from frame to frame, very large reductions can be made in the required data rates. We will confine our interests to still images in this chapter but we should not forget that sequences of images will often be used and that there is an enormous literature on techniques of image data compression.

Images are usually stored with elements in consecutive memory locations. In the case of a three-dimensional image if the starting address in memory is A then the address of the element with coordinates n, m, k is given by

$$\text{address (image}(n, m, k)) = A + n + m \times N + k \times N \times M.$$

In other words, the first row of the first slice is stored first, then the second row and so on. There is nothing special about this, but this is the way most computer languages store arrays.

14.3. IMAGE DISPLAY

Once the image is in computer memory it is necessary to display it in a form suitable for inspection by a human observer. This can only be done conveniently for 2D images so we shall assume that a 2D image is being displayed. The numerical value in memory is converted to a voltage value which in turn is used to control the brightness of a small element on a display screen. The relationship between the numerical value and the voltage value is usually linear, since a digital-to-analogue converter (DAC) is used to generate the voltage. However, the relationship between the voltage and the brightness on the screen may well not be linear.

The appearance of the image will be determined by the (generally nonlinear) relationship between the numerical image value and the brightness of the corresponding image element on the screen. This relationship can usually be controlled to a certain extent by altering the brightness and contrast controls of the display monitor. However, it is usually best to keep these fixed and if necessary control the appearance of the image by altering the numerical values in the image. This process is known as display mapping.

When displayed, image values V directly generate display brightness values B (see figure 14.1). Under these conditions image value V will always be associated with a particular brightness. Suppose we want the image value V to be associated with a different brightness B. There will be a value D that gives the brightness we want. If we replace each occurrence of V in the image with D, then when the image is displayed, each image element with the original value V will be displayed at brightness B.

To achieve a particular relationship or *mapping* between image values and brightness values we need to define a function which maps V values to D values. Suppose we assume that the image values V span the

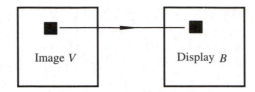

Figure 14.1. *Mapping from the image held in memory to the display.*

range from 0.0 to 1.0. In practice the values will span a range of integer values but we can always convert to the 0–1 range by dividing by the maximum image value (provided the image is converted to a floating point format!). A display mapping is a function f which converts or maps each image value V to a new value D.

$$D = f(V).$$

We choose f so that D also lies in the range 0.0–1.0. The display of an image now becomes a two-stage process. We first map the values of the image to a new set of values D (a new image) using f and then display these values (see figure 14.2).

Clearly, the choice of f allows us to generate almost any brightness relationship between the original image and the display.

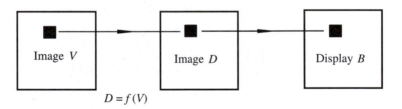

Figure 14.2. *Mapping from the image to the display via a display mapping function.*

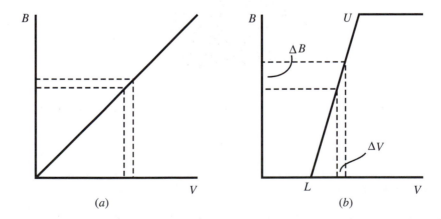

Figure 14.3. *(a) A linear display mapping; (b) a nonlinear display to increase contrast.*

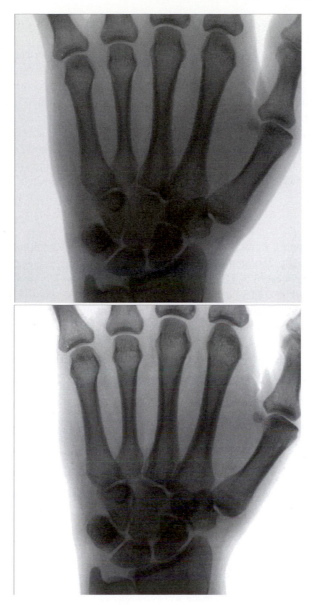

Figure 14.4. *Contrast enhancement. The lower image has the contrast enhanced for the upper 75% of the intensity range.*

14.3.1. Display mappings

In order to distinguish adjacent features in the displayed image they must have a different brightness. The human eye requires a certain change in relative brightness between two adjacent regions, typically 2%, before this change can be seen. This means that changes in the image intensity which are smaller than this cannot be perceived unless they can be made to generate a visible brightness change on the display. Small changes can be made visible using an appropriate display mapping.

Suppose that the brightness of the screen is proportional to the numerical value input to the display. This relationship can be represented by a linear display mapping as shown in figure 14.3(a). However, if we now apply a mapping as in figure 14.3(b), then the brightness change ΔB associated with the change in image value ΔV is now larger than in the previous case. This means that the change is more visible. However, if the intensity change occurred in a part of the image with intensity less than L or greater than U it would not be seen at all. The price paid for being able to visualize the intensity change as shown above is that changes elsewhere may not be seen. Many imaging systems allow the values of L and U to be chosen interactively. Figure 14.4 shows an x-ray image of the hand modified by a display mapping which increases the contrast of the upper 75% of the intensity range.

14.3.2. Lookup tables

The application of the mapping function is usually performed using a lookup table (see figure 14.5). This is an array of values, typically 256 elements long. In this case the image is assumed to be displayed with (integer) intensity values between 0 and 255. The value of each image element in turn is used as an address to an element of this array. The content of this address is the value to be sent to the display device. In fact, it is common for three values to be stored at this location, representing the intensity of the red, green and blue components of the colour to be displayed (if they are equal in value a grey scale is produced). The image, scaled to span the intensity range 0–255, is transferred into a display buffer. Hardware transfer of the image values to the display then proceeds via the lookup table. Modification of the appearance of the image can then be achieved by altering the contents of the lookup table rather than the image values themselves. Changing the lookup table is much faster than changing the image.

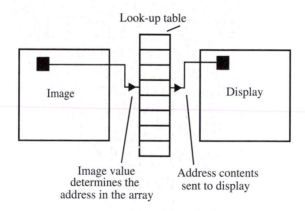

Figure 14.5. *A lookup table.*

14.3.3. Optimal image mappings

The eye responds to relative changes in intensity of the order of 2%. This means that a larger absolute change of intensity is required if the change occurs in a bright part of the image than if it appears in a dim part. In order to provide equal detectability of an absolute change in intensity it is necessary to have a mapping function which increases in steepness as V increases. The brightness on the screen is given by b and if $\Delta b/b$

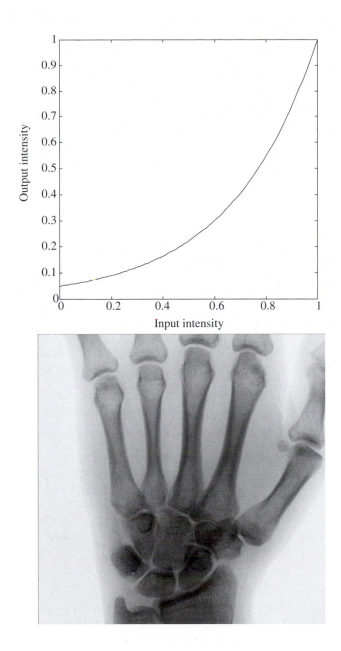

Figure 14.6. *The exponential display mapping shown at the top gives the image of the hand at the bottom.*

is to be constant then the relationship between the image value V and the brightness value b should be given by

$$b = Be^{kV}.$$

The values of B and k are given by specifying the slope of the relationship for some value of V and the value of b at some value of V. For example, if we set a gradient of 3 at $V = 1$ and set $b = 1$ at $V = 1$ this gives $k = 3$ and $B = 1/e^3$. This gives the mapping and the image shown in figure 14.6.

  This last result assumes that the screen brightness is proportional to the intensity of the output of the mapping. If this is not the case than a different mapping will be required.

14.3.4. Histogram equalization

Even if the nonlinear relationship between perceived brightness and image intensity is corrected it does not necessarily mean that the display is being used to best advantage. The histogram of an image is a plot of the number of pixels in the image against the pixel value. Consider a histogram of the form shown in figure 14.7(*a*)).

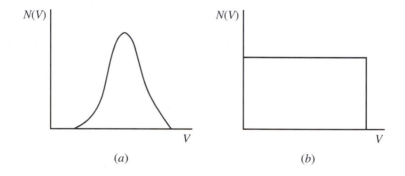

(*a*)   (*b*)

Figure 14.7. (a) A typical image histogram. N(V) is the number of pixels holding value V. (b) An intensity-equalized histogram.

  Clearly in this case most pixels have a value fairly close to the peak of this curve. This means that there are relatively few bright pixels and relatively few dark pixels. If this image is displayed then the bright and dark ranges of the display will be wasted, since few pixels will be displayed with these brightness values. On the other hand, if the histogram is of the form shown in figure 14.7(*b*), then each brightness value gets used equally, since there are equal numbers of pixels at each image intensity. The aim of histogram equalization is to find an intensity mapping which converts a non-uniform histogram into a uniform histogram. The appropriate mapping turns out to be

$$D(V) = \int_0^V N(s)\, \mathrm{d}s.$$

For the histogram of figure 14.7(*a*) the mapping would look approximately as shown in figure 14.8.

  This makes sense. For those parts of the intensity range where there are many pixels the contrast is stretched. For regions where there are few pixels it is compressed. This mapping assumes that in effect all pixels are of equal importance. If the pixels of interest are, in fact, in the low or high brightness range then this mapping might make significant changes less visible.

14.4. IMAGE PROCESSING

We have seen above that, if we have the image in digital form, we can modify the appearance of the image by manipulating the values of the image elements. In this case each image element is modified in a way which is independent of its neighbours and is only dependent on the value of the image element. However, there are

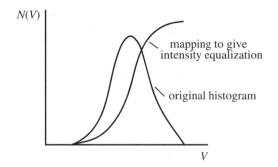

Figure 14.8. *The display mapping to produce a histogram-equalized image.*

many other ways in which we may want to modify the image. All medical images are noisy and blurred so that we may want to reduce the effects of noise or blur. We may want to enhance the appearance of certain features in the image, such as the edges of structures. Image processing allows us to do these things. For reasons of convenience image processing can be divided into image smoothing, image restoration and image enhancement, although all of these activities use basically similar techniques.

14.4.1. Image smoothing

Most images are noisy. The noise arises as part of the imaging process and if it is visible can obscure significant features in the image. The amplitude of noise can be reduced by averaging over several adjacent pixels. Although all the following theory applies to three-dimensional images as well as two-dimensional ones, for convenience we shall concentrate on the two-dimensional case. A pixel and its immediate neighbours can be represented as shown in figure 14.9.

$i-1, j+1$	$i, j+1$	$i+1, j+1$
$i-1, j$	i, j	$i+1, j$
$i-1, j-1$	$i, j-1$	$i+1, j-1$

Figure 14.9. *Coordinates of a pixel and its neighbours.*

We can form the average of the values of the pixel and its neighbours. The averaging process may be represented as a set of weights in a 3×3 array as shown in figure 14.10. This array is called an image filter. The use of the term filter is historical and derives from the use of electrical 'filtering' circuits to remove noise from time-varying electrical signals.

1/9	1/9	1/9
1/9	1/9	1/9
1/9	1/9	1/9

Figure 14.10. *The equal-weights* 3×3 *image filter.*

1	1	1	1	1	1
1	1	1	3	3	3
1	2	2	3	3	3
1	2	2	2	3	3
1	1	2	2	3	3
2	2	2	2	2	2

(*a*)

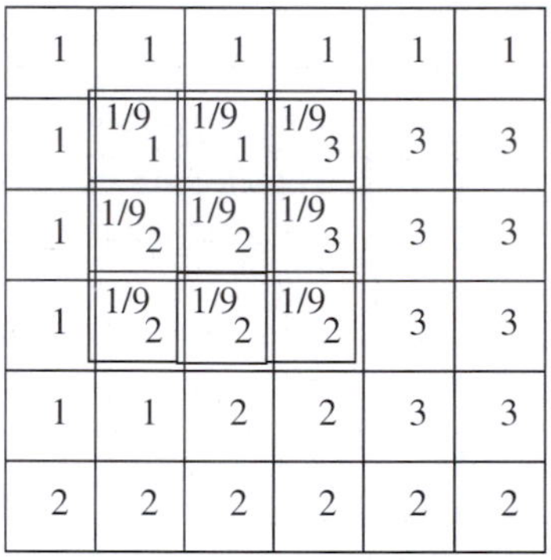

(*b*)

Figure 14.11. *(a) An array of pixel values: a simple image. (b) Averaging over the local neighbourhood using an array of weights.*

Now, imagine we have an image as in figure 14.11(*a*). To compute an average around a pixel in this image we place the grid of weights over the image, centred on that pixel. The average value in this case is the sum of the pixel values lying under the 3×3 grid divided by 9. This value can be placed in the corresponding pixel of an initially empty image as illustrated in figure 14.12.

The 3×3 grid is then moved to the next position and the process repeated, and then continued until all the pixels have been visited as shown in figure 14.13.

In the simple case above the weights are all the same. However, we can generalize this process to a weighted average. If the value of the (i, j)th pixel in the original image is f_{ij} and the value in the (k, m)th element of the 3×3 grid is w_{km} (rather than simply $\frac{1}{9}$) then the value of the (i, j)th pixel in the output image is given by

$$g_{ij} = \sum_{m=-1}^{1} \sum_{k=-1}^{1} w_{km} f_{i+k, j+m}.$$

Note that conventionally the weights w_{km} sum to unity. This is to ensure that the total intensity in image f (the original image) is the same as the total intensity in image g (the 'filtered' image).

1	1	1	1	1	1
1	1/9 (1)	1/9 (1)	1/9 (3)	3	3
1	1/9 (2)	1/9 (2)	1/9 (3)	3	3
1	1/9 (2)	1/9 (2)	1/9 (2)	3	3
1	1	2	2	3	3
2	2	2	2	2	2

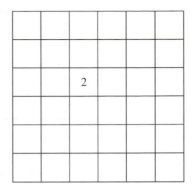

Figure 14.12. *Building a new image from the averaged values. The values on the left, multiplied by the filter weights, are entered into the corresponding pixel of the empty image on the right.*

1	1	1	1	1	1
1	1	1/9 (1)	1/9 (3)	1/9 (3)	3
1	2	1/9 (2)	1/9 (3)	1/9 (3)	3
1	2	1/9 (2)	1/9 (2)	1/9 (3)	3
1	1	2	2	3	3
2	2	2	2	2	2

Image *f*

1.22	1.67	2.00	2.33
1.44	2.00	2.44	2.89
1.89	2.00	2.45	2.77
1.67	1.89	2.22	2.44

Image *g*

Figure 14.13. *By moving the filter over the image on the left the filtered image is produced as shown on the right.*

Suppose that image f is noisy. For simplicity we assume that the variance of the noise on each pixel is σ^2 and that the noise on each pixel is independent of its neighbours. Then the variance of the pixel formed from a weighted average of its neighbours is given by

$$\sigma_g^2 = \sum_{m=-1}^{1} \sum_{k=-1}^{1} w_{km}^2 \sigma_f^2.$$

For the equal-weights filter $\sigma_g^2 = \sigma_f^2/9$ so that the amplitude of the noise is reduced threefold. However, after filtering with this filter the noise on neighbouring pixels is no longer independent since common pixels were used to form these filtered pixel values.

A popular alternative set of values or weights is that shown in figure 14.14.

In this case the reduction in the variance of the noise after averaging is given by $9\sigma_f^2/64$. This is not as great as for the case where the weights are all equal (which is $9\sigma_f^2/81$). However, consider the effect of these two different filters on a step in intensity as illustrated in figure 14.15.

1/16	2/16	1/16
2/16	4/16	2/16
1/16	2/16	1/16

Figure 14.14. *The 421 filter.*

0	0	0	100	100	100
0	0	0	100	100	100
0	0	0	100	100	100
0	0	0	100	100	100
0	0	0	100	100	100
0	0	0	100	100	100

(a)

0	0	33.3	66.6	100	100
0	0	33.3	66.6	100	100
0	0	33.3	66.6	100	100
0	0	33.3	66.6	100	100
0	0	33.3	66.6	100	100
0	0	33.3	66.6	100	100

(b)

0	0	25	75	100	100
0	0	25	75	100	100
0	0	25	75	100	100
0	0	25	75	100	100
0	0	25	75	100	100
0	0	25	75	100	100

(c)

Figure 14.15. *(a) The original image; (b) filtered with the equal-weights filter; (c) filtered with the 421 filter.*

In the original image the edge is sharp. After processing with the equal-weights filter the edge is blurred. The same is true for the 421 filter but less so. Generally this means that the more the filter reduces the amplitude of noise, the more the edges and fine structures in the image are blurred. This is a general rule, and attempting to maximize noise reduction, while minimizing blurring, is the aim which drives much of image processing. It is possible to develop filters which are optimal in the sense that they minimize both noise and blurring as far as possible, but this is beyond the scope of this chapter. Figure 14.16(*a*) shows a copy of the hand image to which noise has been added. Figure 14.16(*b*) shows the same image smoothed with the 421 filter. The noise has been reduced, but the image is now blurred.

Clearly many weighting schemes can be devised. The pattern of weights is not simply confined to 3×3 grids but can span many more pixels. Multiple passes of the same or different filters can also be used. For example, if g is formed by filtering f with the 421 filter and then g is also filtered with the 421 filter, the net effect is the same as filtering f with a 5×5 filter (equivalent to filtering the 421 filter with itself). Each pass of the filter reduces noise further, but further blurs the image.

The summation equation above is a discrete version of the convolution equation introduced in Chapters 11 and 13 (sections 11.7.4 and 13.6.2). Just as convolution can be implemented through the use of Fourier transforms so discrete convolution can be, and often is, implemented through the use of discrete Fourier transform methods. This can often be faster than the direct discrete convolution because very efficient methods

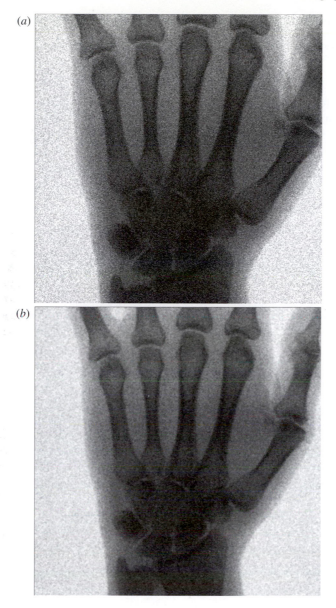

Figure 14.16. *(a) A noisy version of the hand image; (b) image (a) smoothed with the 421 filter.*

of computing the discrete Fourier transform exist (the fast Fourier transform or FFT). The use of Fourier transform techniques requires the filter to be the same at all points on the image, but within this framework optimum filters have been developed.

Nonlinear filters: the median filter

The filters described above are linear filters. This means that

$$(a + b) \otimes c = a \otimes c + b \otimes c$$

where $\otimes$ is the symbol for convolution. Attempts have been made to improve on the performance of linear filters by using nonlinear methods. The most well known, but also one of the simplest, nonlinear filters is the median filter. The equal-weights filter computes the average or mean value of the pixels spanned by the filter. The median filter computes the median value. Consider the two 3×3 regions in the step intensity image of figure 14.17.

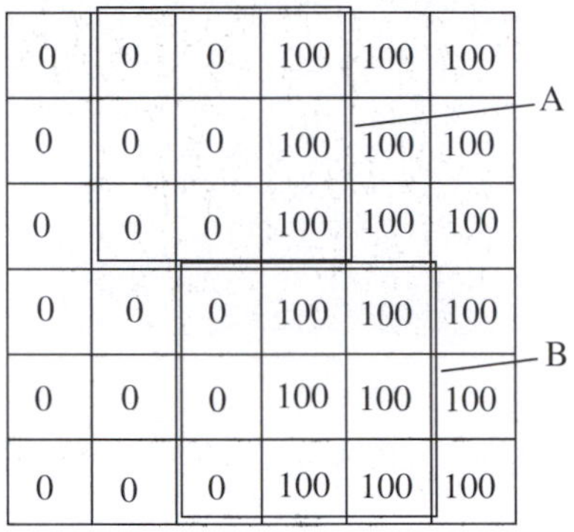

0	0	0	100	100	100
0	0	0	100	100	100
0	0	0	100	100	100
0	0	0	100	100	100
0	0	0	100	100	100
0	0	0	100	100	100

Figure 14.17. *Two filter positions shown on a step intensity change.*

The pixel values for position A are 0, 0, 100, 0, 0, 100, 0, 0, 100 and the median value is calculated by ranking these in magnitude, i.e. 0, 0, 0, 0, 0, 0, 100, 100, 100 and then selecting the middle value, 0 in this case. For position B the ranked values are 0, 0, 0, 100, 100, 100, 100, 100, 100 and the middle value is 100. Using the median filter the step edge in the above image is preserved. When scanned over a uniform but noisy region the filter reduces the level of noise, since it selects the middle of nine noisy values and so reduces the magnitude of the noise fluctuations. The reduction in noise is not as great as for the averaging filter, but is nevertheless quite good. Although the median filter can produce some artefacts it can generally preserve edges while simultaneously reducing noise. Figure 14.18 shows the noisy hand image filtered with a median filter.

14.4.2. Image restoration

We saw in Chapter 11 (section 11.7.4) that the convolution equation mathematically related the image formed to the object and the point spread function of the imaging device. We also saw that if the convolution equation is expressed in terms of the Fourier transform of the constituent functions there was a direct way of obtaining the Fourier transform of the object, given the Fourier transform of the image and the point spread function. The form of this relationship was given as

$$F(u, v) = \frac{G(u, v)}{H(u, v)}$$

where $G(u, v)$ is the Fourier transform of the image, $H(u, v)$ is the Fourier transform of the point spread function and $F(u, v)$ is the Fourier transform of the object being imaged. There is also a three-dimensional

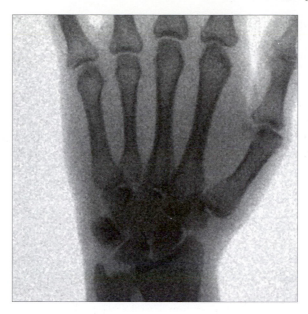

Figure 14.18. *The noisy hand image of figure 14.16(a) filtered with a 3 × 3 median image.*

form of this equation. This equation is important because it implies that the resolution degradation caused by the imaging device can be reversed by processing the image after it has been formed. This process is known as image restoration. This equation has caused a great deal of excitement in its time because it implied that the effects of poor resolution, such as those found in gamma cameras, could be reversed. Unfortunately, deeper study showed that this goal could not be achieved and image restoration has generally fallen out of favour in medical imaging.

There are two reasons why image restoration runs into difficulties. The first is that $H(u, v)$ may go to zero for some values of u and v. This means that $G(u, v)$ is also zero (because $G(u, v) = F(u, v)H(u, v)$) and so $F(u, v)$ is undetermined.

The second, and more serious, reason is that all images are corrupted by noise. The Fourier transform of a real noisy image can be written as

$$G_n(u, v) = G(u, v) + N(u, v).$$

It can be shown that, although $N(u, v)$ fluctuates in amplitude randomly across frequency space, it has the same average amplitude for all u, v. On the other hand, $H(u, v)$ and $G(u, v)$ fall in amplitude with increasing frequency. Dividing both sides of the above equation by $H(u, v)$ gives

$$\frac{G_n(u, v)}{H(u, v)} = \frac{G(u, v)}{H(u, v)} + \frac{N(u, v)}{H(u, v)}.$$

The first term on the right-hand side of this equation behaves well (except where $H(u, v) = 0$); its value is $F(u, v)$. However, the second term behaves badly because the amplitude of $N(u, v)$ remains constant (on average) as u and v increase, while $H(u, v)$ decreases in amplitude. This ratio becomes very large as u and v increase and soon dominates. The noise in the image is amplified and can completely mask the restored image term $G(u, v)/H(u, v)$ unless the magnitude of the noise is very small. In effect, once the amplitude of $G(u, v)$ falls below that of $N(u, v)$ this signal ($G(u, v)$) is lost and cannot be recovered. A considerable amount of research has been devoted to trying to recover as much of $G(u, v)$ (and hence $F(u, v)$) as possible, while not amplifying the noise in the image. It is difficult to improve resolution this way by even a factor of two, especially for medical images which are rather noisy, and only limited success has been reported. If high resolution is required it is usually better to try and collect the data at high resolution (for example by using a higher-resolution collimator on a gamma camera) even if sensitivity is sacrificed than to collect a lower resolution image at higher sensitivity and attempt to restore the image.

The restoration process outlined above is a linear technique. However, we sometimes have additional information about the restored image which is not specifically taken into account in this restoration process. For example, in general there should not be any regions of negative intensity in a restored image. It may also be known that the image intensity is zero outside some region, a condition which is true for many cross-sectional images. If such constraints can be incorporated into the restoration process then improvements in image restoration beyond that achievable using simple linear methods can be obtained. It is generally true that the more constraints that can be imposed on the solution the better the results. However, computational complexity increases and such methods have not found widespread use.

14.4.3.   Image enhancement

There is ample evidence that the human eye and brain analyse a visual scene in terms of boundaries between objects in the scene and that an important indication of the presence of a boundary at a point in the image is a significant change in image brightness at that point. The number of light receptors (rods and cones) in the human eye substantially outweighs the number of nerve fibres in the optic nerve which takes signals from the eye to the brain (see Chapter 3). Detailed examination of the interconnections within the retina shows that receptors are grouped together into patterns which produce strong signals in response to particular patterns of light falling on the retina. In particular, there appear to be groupings which respond in a particularly strong way to spatial gradients of intensity. In fact, the retina appears fairly unresponsive to uniform patterns of illumination compared to intensity gradients. In a very simplistic way it appears that the signals transmitted to the brain correspond to an image representing the visual scene convolved with a Laplacian function (if f is the image then the Laplacian filtered image is given by $\partial^2 f / \partial x^2 + \partial^2 f / \partial y^2$), or rather a set of Laplacian functions of differing smoothness, rather than the visual scene itself. Given the preoccupation of the human visual system with intensity gradients it seems appropriate to process an image to be presented to a human observer in such a way that intensity gradients are enhanced. This is the basis of image enhancement techniques.

Unlike image restoration or optimal image smoothing there is no firm theoretical basis for image enhancement. We know that when we smooth an image the fine detail, including details of intensity changes at object boundaries, is blurred. In frequency space this corresponds to attenuating the amplitude of the high spatial frequency components, while preserving the amplitude of low-frequency components. If we subtract on a pixel by pixel basis the smoothed version of an image from the unsmoothed version then we are achieving the opposite effect, attenuating the low spatial frequencies and enhancing the high spatial frequencies, and by implication the intensity gradients in the image. In general, this process can be written as

$$e(x, y) = f(x, y) - A f(x, y) \otimes s(x, y)$$

where $s(x, y)$ is a smoothing function, A is a scaling factor, usually less than 1, and $e(x, y)$ is the enhanced image. There are clearly many degrees of freedom here in the choice of $s(x, y)$ and A. Since this process will also enhance noise it may be necessary to smooth $e(x, y)$ with an additional smoothing filter.

Figure 14.19 shows an enhanced version of the hand image. Note that the fine detail in the bones now shows more clearly. In this case s is the 421 filter and $A = 0.95$. One interesting observation is that, whatever

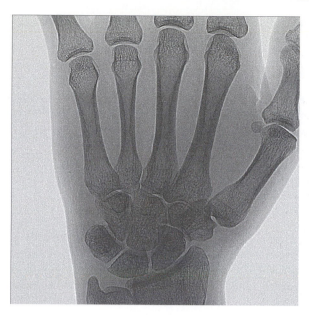

Figure 14.19. *An enhanced version of the hand image of figure 14.4.*

the histogram of f, the histogram of e, with $A = 1$, tends to be a Gaussian function about zero intensity. Such images display well with histogram equalization.

14.5.   IMAGE ANALYSIS

Image analysis is largely about extracting numerical information or objective descriptions of the image contents from the image. This is an enormous subject and we can only scratch at the surface. Much of image analysis consists of empirical approaches to specific problems. While this can produce useful results in particular situations, the absence of any significant general theory on image analysis makes systematic progress difficult. Nevertheless, some basic techniques and principles have emerged.

14.5.1.   Image segmentation

A central preoccupation in image analysis is image segmentation, the delineation of identifiable objects within the image. For example, we may want to find and delineate cells on a histological image, or identify the ribs in a chest x-ray. There are a variety of methods for image segmentation. Many of them reduce to identifying sets of image elements with a common property, for example that the intensity of the element is above a certain value, or identifying borders or edges between different structures on the image. Once we have identified the elements belonging to an object we are then in a position to extract descriptions of the object, for example its volume and dimensions, its shape, its average intensity and so on. Many medical images are two-dimensional representations of three-dimensional objects and this can make image segmentation more difficult than it needs to be, or even invalid. Three-dimensional images of three-dimensional objects are preferred.

The simplest and potentially the most powerful method of identifying the limits of an object is manual outlining. This is the method most able to deal with complex images but is slow and demonstrates inter-observer variability, especially when the images are blurred. Ideally we would like to get away from the need for manual delineation of objects. It is only really practical for two-dimensional images. Delineation of object boundaries in three dimensions is far too tedious for routine analysis.

14.5.2.   *Intensity segmentation*

The simplest computational method of segmenting an image is intensity thresholding. The assumption behind thresholding is that the image elements within the desired object have an intensity value different from the image elements outside the object. We select a threshold T.

Then if P_i is element intensity;

$$P_i > T \longrightarrow \text{object}$$
$$P_i \leq T \longrightarrow \text{not object}$$

for a bright object (the reverse for a dark object). T may be selected manually, or by automatic or semi-automatic methods. One way of selecting a useful T is to inspect the intensity histogram. If several objects are present the choice of several values of T may allow segmentation of multiple objects as shown in figure 14.20(*a*).

In this simple case selecting values of T in the gaps between the peaks would separate the objects from each other and the background. In practice a histogram of the form shown in figure 14.20(*b*) is more likely

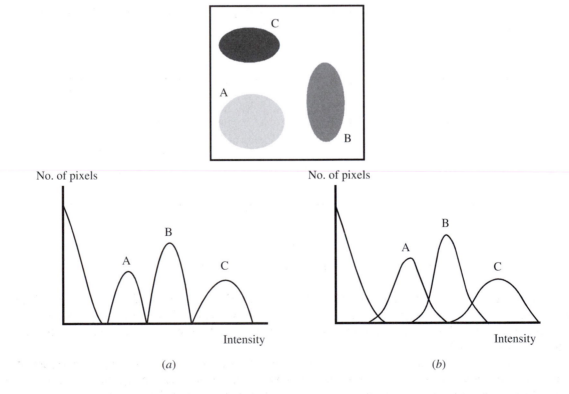

Figure 14.20. *An image (on the right) of three objects with different intensities and below: (a) the intensity histogram showing the separation of the objects. (b) A more likely intensity histogram with overlap of the regions.*

to be found and clearly the regions cannot be separated so neatly by thresholding. Using a local histogram can sometimes solve this problem. Two-dimensional images often suffer from overlapping structures and thresholding will generally not perform well under these circumstances.

14.5.3. Edge detection

For thresholding to work reliably the image needs to obey rather strict conditions. In practice we know that we can visually segment an image which cannot be reliably segmented by thresholding. What we often appear to do visually is detect regions of high-intensity gradient and identify them as object boundaries. This can be done numerically by first generating an image of intensity gradients and then processing this image. The gradient image is given in two dimensions by

$$g'(x, y) = \left[\left(\frac{\partial g}{\partial x} \right)^2 + \left(\frac{\partial g}{\partial y} \right)^2 \right]^{1/2}$$

and in three dimensions by

$$g'(x, y, z) = \left[\left(\frac{\partial g}{\partial x} \right)^2 + \left(\frac{\partial g}{\partial y} \right)^2 + \left(\frac{\partial g}{\partial z} \right)^2 \right]^{1/2}.$$

This image has large values at points of high-intensity gradient in the original image and low values at regions of low gradients as shown in figure 14.21. A threshold is applied to this image to identify edge points and produce an edge image. This edge image is a binary image.

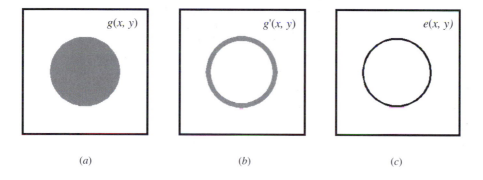

(a) (b) (c)

Figure 14.21. *An image (a), the gradient image (b) and the edge image (c).*

In practice the edge may be broken because the amplitude of the intensity gradient is not uniform all the way round the object. Lowering the threshold may produce many small irrelevant edges, whereas a choice of too high a threshold will produce a severely broken edge (figure 14.22). The edge points need to be connected to form a closed boundary.

This is not easy and there is no robust general method for doing this. A useful approach, especially if a model is available for the shape of the object, is to fit that model to the available edge points. For example, if the object boundary can be modelled by an ellipse the available edge points can be used to derive the parameters of the ellipse which can then be used to fill in the missing values. Models based on the Fourier transform and other sets of orthogonal functions can also be used. A particularly elegant approach in two dimensions is to treat the coordinates of each point on the edge as a complex number (see figure 14.23).

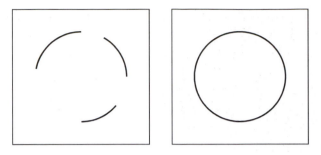

Figure 14.22. *A 'broken' edge image and the ideal closed version.*

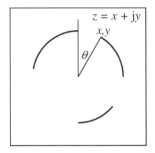

Figure 14.23. *The complex representation of the edge points.*

Then,

$$z(\theta) = x(\theta) + \mathrm{j}y(\theta)$$

The one-dimensional Fourier transform of this is

$$Z(u) = \int z(\theta)\,\mathrm{e}^{-2\pi\mathrm{j}u\theta}\,\mathrm{d}\theta.$$

The inverse transform is

$$z(\theta) = \int Z(u)\,\mathrm{e}^{2\pi\mathrm{j}u\theta}\,\mathrm{d}u$$

and by computing estimates of $Z(u)$ from the available points, the complete boundary can be interpolated. A smooth boundary can be ensured by only using a few low-order components of $Z(u)$.

Once a complete closed boundary has been obtained it is easy to determine whether a pixel is inside or outside the boundary by determining whether there is a route to the edge of the image which does not cross the boundary. If this is the case the pixel or voxel lies outside the object. Figure 14.24 puts the problem of identifying edges in an image into perspective.

14.5.4. Region growing

An alternative to edge detection is region growing. In these techniques a pixel is chosen and its neighbours examined. If these have a similar value to the first pixel they are included in the same region. Any pixels which are not sufficiently similar are not included in the region, which is grown until no further pixels can be added. In fact, any test of similarity can be used although the most common one is to keep a running total of

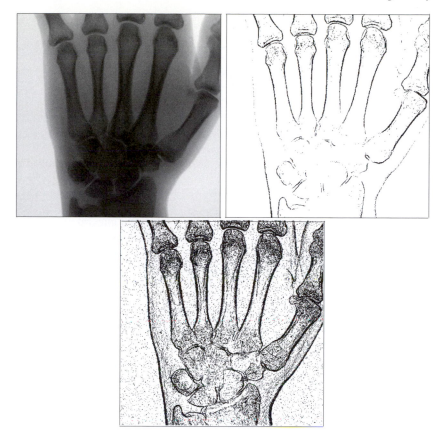

Figure 14.24. *The hand image and two edge images produced using different thresholds.*

the mean and standard deviation of the pixel intensities within the growing region and accept or reject pixels based on whether their values differ significantly from the current mean. A starting pixel is needed which may have to be supplied interactively. One way round this is to progressively divide the image into quarters, sixteenths, etc until the pixel values within each segment become (statistically) uniform. These can then form starting points for region growing.

Region growing requires that within regions the image is in some sense uniform. This is more likely to be the case for three-dimensional images than two-dimensional images. For example, the hand image has some reasonably well defined regions (the bones) but the intensity values within these regions are hardly uniform.

14.5.5. Calculation of object intensity and the partial volume effect

Once an object has been extracted parameters such as position, area, linear dimensions, integrated intensity and others can be determined. These can be used to classify or identify the extracted regions or volumes.

In functional imaging, especially radionuclide studies, integration of the total activity within a structure can be an important measurement. Unfortunately the image of the structure is invariably blurred. This means that even if the boundary of the object is known, estimation of the total intensity within the object will be incorrect. Consider the simple one-dimensional example illustrated in figure 14.25.

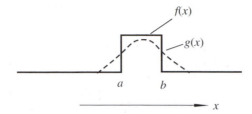

Figure 14.25. *The partial volume effect showing how the integral under the filtered function g(x) is less than under the original f(x).*

The image $g(x)$ is a smoothed version of the object $f(x)$. The total integral under f is the same as the integral under g. However, in practice the integral under g is computed within a region defining the boundary of f and in this case

$$\int_a^b g(x)\,\mathrm{d}x < \int_a^b f(x)\,\mathrm{d}x.$$

This loss of intensity is known as the partial volume effect and the loss of intensity increases as the dimensions of f approach the resolution of the imaging system. It may be possible to estimate the loss of counts from the object from a simple model in which case a correction can be made. Accuracy is further lost if there is a second structure close to f since intensity from this will also spill over into f.

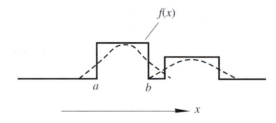

Figure 14.26. *The partial volume effect with adjacent structures.*

The only principled solution to this problem is image restoration, although we have argued that this is not very effective if the data are noisy. Again, if the underlying distribution can be modelled in some way it may be possible to estimate the magnitude of the partial volume effect.

14.5.6.   *Regions of interest and dynamic studies*

There are many situations, especially in nuclear medicine but also in other imaging modalities, where the significant thing is the way the intensity in a structure varies as a function of time. A dynamic study consists of a sequence of images (see figure 14.27) such that the distribution of intensity within the images changes from image to image.

The way the intensity varies within the structure of interest can be used to determine its physiological function. A common way of determining the time-varying functions associated with each image structure is to use regions of interest (ROIs). In this method regions are drawn around the structures of interest (see figure 14.28) and the (usually integrated) intensity in each image of the sequence within each region is determined. This will produce a time–activity curve (TAC) for that region. If the images are two-dimensional

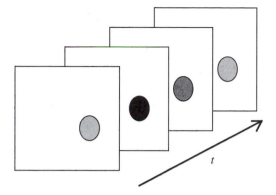

Figure 14.27. *An idealized dynamic study.*

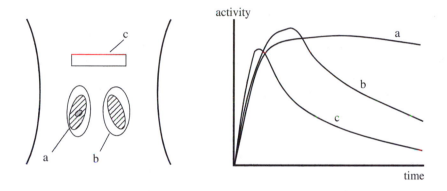

Figure 14.28. *Regions of interest analysis. Three regions of interest (ROIs) give three time–activity curves. Region c could be used to give the 'background' activity.*

the intensity distribution at each point will contain contributions from several structures, which in general can overlap. As far as possible it is usually assumed that structures do not overlap. However, in almost all cases as well as a contribution from the structure of interest the TAC will contain a contribution from tissues lying above and below the structure, the 'background' contribution. In order to eliminate this contribution the usual approach is to draw a region over a part of the image which does not contain the structure, i.e. a background region, and then subtract the TAC associated with this region from the TAC derived from the region over the structure of interest. The amplitude of the background TAC will need to be scaled to account for the different areas of the two regions. Account must also be taken of the fact that the intensity of the signal in the background region may be different from the intensity of the background signal in the region over the structure. Even if the intensity of the signal per unit volume of background tissue is the same for both regions the structure will almost certainly displace some of the background tissue, so the background signal will be smaller for the region over the structure than for the background region. A subtler problem is that the time variation of the signal may be different in the two regions because the background tissues are not quite the same. There is no general solution to these problems, although partial solutions have been found in some cases.

Once the TACs have been extracted appropriate parameters can be extracted from them, either on an empirical basis or based on a physiological model, to aid diagnosis. Region-of-interest analysis is most commonly used with radionuclide images since these are the most widely used functional images.

14.5.7. Factor analysis

Isolating the TAC associated with an individual structure is difficult with region-of-interest analysis for the reasons outlined above. In two dimensions overlap is possible. Estimating the intensity in region A of figure 14.29 is made difficult by the fact that it overlaps region C as well as containing a contribution from the background B. In principle, apart from the partial volume effect, there is no overlap in three dimensions and quantitation of three-dimensional images is more accurate.

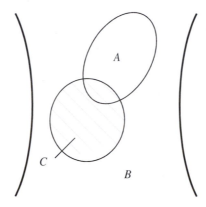

Figure 14.29. *Overlapping objects in a planar image.*

One way of isolating the TACs associated with each of the structures in a dynamic study is factor analysis. Suppose we have N homogeneous structures within the object. Homogeneous means in this context that the intensity varies in the same way with time at each point in the structure. For the nth structure the variation with time is given by $a_n(t)$. The variation of intensity with time for the image element at (x, y) is given by

$$p(x, y, t) = \sum_{n=1}^{N} A_n(x, y) \, a_n(t)$$

where $A_n(x, y)$ is the contribution from the curve $a_n(t)$ for the image element at position (x, y). $a_n(t)$ is called the nth factor curve and $A_n(x, y)$ the nth factor image. $p(x, y, t)$ represents the change in activity with time in the pixel at (x, y). Given that N is usually fairly small (<5) we would like to be able to extract A_n and $a_n(t)$ from the data, but unfortunately A_n and a_n are not unique. In order to find suitable sets of A and a we need to apply some constraints. We can find a set of N orthogonal functions, the principal components, and their corresponding factor images which satisfy these equations and this is often a first step in analysing this problem. However, these functions rarely correspond to any of the a_n, which are, in practice, never orthogonal to each other.

More useful constraints which have been used include the fact that both A_n and a_n cannot physically be negative, i.e. if the factor images correspond to real structures within the object then they will have zero amplitude over much of the image, and that a_n can be modelled by an appropriate function. None of these approaches has provided a universal solution to the general problem of finding the factors, but some useful results for special cases have been published, especially with three-dimensional dynamic studies using PET.

14.6. IMAGE REGISTRATION

The use of three-dimensional imaging is increasing. In particular, the availability of reliable and accurate SPET imaging systems (see section 12.2 of Chapter 12) is resulting in an increased use of this modality for routine imaging. Three-dimensional cardiac imaging is now commonplace as is brain imaging, but the use of SPET is being extended into other areas. SPET images are images of function with relatively poor resolution. On the other hand, CT images and MR images are still primarily images of anatomy but have relatively high resolution. It seems likely that there are important benefits to be obtained if the anatomical information from a CT or MR image can be mapped onto a functional image such as a SPET image. In addition, an important precursor to the automatic analysis of three-dimensional images is likely to be the ability to transform an image to a standard position and shape so that standard regions of interest or other approaches can be used. Both these applications require the ability to transform a pair of images so that they are in registration.

Given two images the principal requirement for registering the images is that a correspondence between the coordinates of corresponding points in the two images can be established. This requires a coordinate transfer function (CTF), which is a set of functions which map pixel values in one image to those in the other (see figure 14.30). Suppose we have two images $f(x, y, z)$, the fixed image, and $m(x, y, z)$, the image to be moved into registration with f. Then we can define three functions

$$u = a(x, y, z)$$
$$v = b(x, y, z)$$
$$w = c(x, y, z).$$

For each point (x, y, z) in the fixed image these functions give the coordinates (u, v, w) of the corresponding pixel in the moved image. Given these functions, the moved image is registered in the following way.

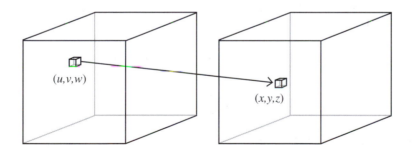

Figure 14.30. *Mapping one image to another with a coordinate transfer function.*

We start off with an empty image (g). For each point (x, y, z) in g we calculate the corresponding point in $m(u, v, w)$ from the CTF and extract the value of the voxel at this point. This value is placed at (x, y, z) in g. Since (u, v, w) will, in general, not lie at a voxel centre in m we will need to interpolate the voxel value from surrounding voxels. These operations are repeated for all the voxels in g. The resulting image is a version of m registered to f.

In order to do this we need to know the CTF. In the absence of any specific information it is usual to model the CTF as a polynomial in x, y and z. For registration which only involves image translation the CTF has the form

$$u = a_{11} + x$$
$$v = a_{21} + y$$
$$w = a_{31} + z.$$

The next simplest general form is

$$u = a_{11} + a_{12}x + a_{13}y + a_{14}z$$
$$v = a_{21} + a_{22}x + a_{23}y + a_{24}z$$
$$w = a_{31} + a_{32}x + a_{33}y + a_{34}z$$

which is known as the affine transform. These equations can be written in matrix form as

$$
\begin{vmatrix} 1 \\ u \\ v \\ w \end{vmatrix}
=
\begin{vmatrix}
1 & 0 & 0 & 0 \\
a_{11} & a_{12} & a_{13} & a_{14} \\
a_{21} & a_{22} & a_{23} & a_{24} \\
a_{31} & a_{32} & a_{33} & a_{34}
\end{vmatrix}
\begin{vmatrix} 1 \\ x \\ y \\ z \end{vmatrix}.
$$

In fact, the affine transform is not quite the next simplest form. It contains 12 parameters and therefore has 12 degrees of freedom. Many of the applications of image registration involve registering images of the same patient taken using different modalities. In these circumstances, assuming that the images have the same voxel dimensions which can always be achieved in principle, registration will only involve translation plus rotation about the three image axes, in total only six degrees of freedom. For this case the CTF can be written as

$$
\begin{vmatrix} 1 \\ u \\ v \\ w \end{vmatrix}
=
\begin{vmatrix}
1 & 0 & 0 & 0 \\
a_{11} & \cos\beta\cos\alpha & \cos\beta\sin\alpha & \sin\beta \\
a_{21} & -\cos\gamma\sin\alpha - \sin\gamma\sin\beta\cos\alpha & \cos\gamma\cos\alpha - \sin\gamma\sin\beta\sin\alpha & \sin\gamma\cos\beta \\
a_{31} & \sin\gamma\sin\alpha - \cos\gamma\sin\beta\cos\alpha & -\sin\gamma\cos\alpha - \cos\gamma\sin\beta\sin\alpha & \cos\gamma\cos\beta
\end{vmatrix}
\begin{vmatrix} 1 \\ x \\ y \\ z \end{vmatrix}
$$

where α, β and γ are the angles of rotation about the three axes. The difference between this CTF and the full affine transform can be visualized in terms of registering two tetrahedra. The last CTF allows two rigid tetrahedra of identical shape to be registered. The full affine transform allows any two tetrahedra to be registered. Higher-order polynomial CTFs can be defined.

The key to registering two images is to find the parameters of the CTF. The most direct method is to use landmarks. Suppose that we can identify corresponding points on both images. These points might well be the location of prominent anatomical features. Then for each feature we obtain two sets of coordinates. By hypothesis these sets are related by a CTF of the proposed form. If we have sufficient sets of coordinates, typically at least twice the number of degrees of freedom, we can insert them into the equations of the CTF and solve the resulting overdetermined set of equations for the parameters of the CTF. In the case of the affine transform, the CTF is linear in its parameters and a direct solution is possible. For the rigid transform this is not the case, but a solution is nevertheless possible. If natural landmarks do not exist, artificial ones can be used. For example, to register an MR image to a SPET image markers which are both radioactive and provide a strong MR signal can be fixed to the surface of the patient being imaged. These will be visible in both images and provided they are appropriately positioned can be used as landmarks. Caution must be exercised to ensure that landmarks form an independent set. For example, if three landmarks are placed in a line, only the coordinates of two of these are independent under an affine transform.

The use of landmarks is not always convenient or possible. Accurate identification of anatomical landmarks in SPET images is notoriously difficult. An alternative approach is to use a function minimization approach. At any stage in the registration process a measure of the quality of the registration of the two images can be computed. The aim is to find a set of parameters which minimize (or maximize) this measure. This is a classic function minimization problem and a variety of algorithms exist for implementing it, given a way of calculating the quality measure from the parameters.

There are three problems with this method.

- The first is that of finding a suitable quality measure. If the images are from the same modality then a measure such as the sum of the squares of the voxel differences may be appropriate. If the images are from different modalities then a more complex measure may be required. One important approach is to use a two-dimensional histogram. Consider registering two images, say a SPET image and an MR image as shown in figure 14.31. We assume that the images consist of regions (e.g. white matter, grey matter, CSF) of fairly uniform intensity. The actual relative intensities associated with these regions will not be the same for both images. For example, if the SPET image is an image of perfusion then grey matter will be on average four times as bright as white matter, because perfusion is much higher for the former, whereas the brightness of grey matter on the MR image may be less than the white matter.

Assume that the images are in registration. For each voxel position there will be two intensity values, one from the SPET image and one from the MR image. These values can be plotted as points on a two-dimensional plot or 2D histogram. The points associated with each region will cluster around some average value. If the images are not in registration these regions will be much more diffuse since they will contain points drawn from more than one region. An appropriate quality measure is therefore one which is a maximum or minimum when the regions in the two-dimensional histogram are as compact as possible. One measure which has been proposed is that of mutual entropy, which does appear to capture the required histogram properties.

Another approach uses image segmentation techniques to define the border of the object to be registered, in most cases the brain. The two surfaces are registered using a quality measure based on the normal distance between the two surfaces.

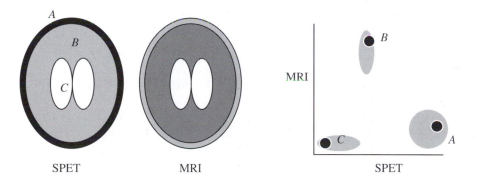

Figure 14.31. *Image registration using a two-dimensional histogram.*

- The second problem is a computational one. Function minimization procedures need to evaluate how the quality-of-fit measure varies as the parameter varies. For each change in parameter value the quality measure needs to be recomputed, which will require a computation over the whole of the images. This is time consuming. The number of times this needs to be done will depend very strongly on the number of parameters. In practice most applications to date have used the six-degree-of-freedom transform. Moving beyond this has generally proved computationally too expensive.

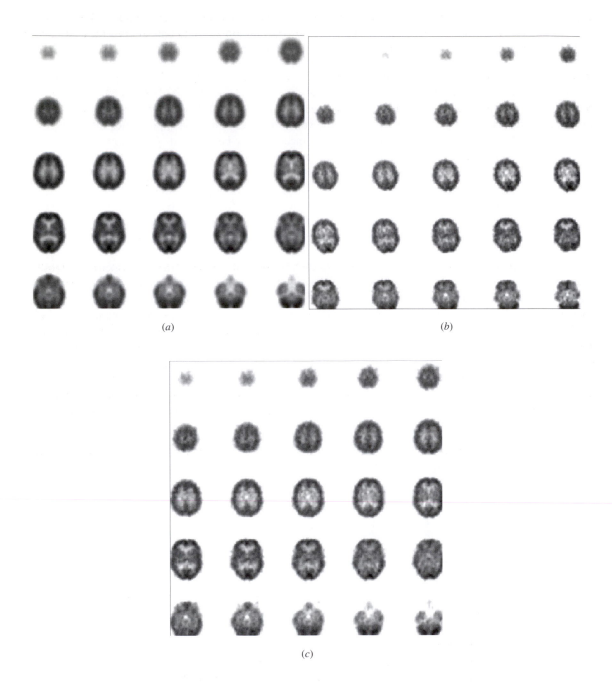

Figure 14.32. *(a) A standard* ^{99m} *Tc-HMPAO brain image. This image is the average of six registered normal images. (b) A normal brain image. This shows significant rotation. (c) The image of (b) registered to the image (a). The rotation has been corrected and the linear dimensions of the image scaled to match the standard image.*

• The third problem is that function minimization techniques may become stuck in a local minimum of the quality measure and be unable to find the global minimum. Techniques do exist for dealing with this problem but increase the computational requirements further.

If the images to be registered are from the same modality and are reasonably similar a method exists for very efficiently computing the parameters of the CTF, provided the CTF is linear in these parameters. This can be illustrated with the affine transform. It can be shown that if the displacement between the two images is small, for each voxel the relationship

$$f(x, y, z) - m(x, y, z) = (a_{11} + (a_{12} - 1)x + a_{13}y + a_{14}z)\frac{\partial f}{\partial x}$$

$$+ (a_{21} + a_{22}x + (a_{23} - 1)y + a_{24}z)\frac{\partial f}{\partial y}$$

$$+ (a_{31} + a_{32}x + a_{33}y + (a_{34} - 1)z)\frac{\partial f}{\partial z}$$

holds. There are many voxels in the image and the corresponding set of equations represents a substantially overdetermined set of simultaneous equations. The a_{ij} can be determined directly from this set. If the initial displacement between the images is not small it is still possible to achieve a solution by iteration. Experimentally, provided the initial displacement is not too large, the process converges on the correct solution in a few iterations. Unlike the function minimization methods, all the parameters are estimated simultaneously. This makes it possible to consider higher-order transforms. The limit to transform complexity is set by issues of whether the image data are accurate enough to support such a transform rather than the time taken to compute the parameters. This transform has been used to routinely register SPET brain images and cardiac images to standard images, i.e. to a standard position and size both for display and analysis purposes. Figure 14.32 shows a SPET brain scan before and after registration to a standard brain image, defined as the average of a set of previously registered images.

14.7. PROBLEMS

14.7.1. Short questions

a   How does a pixel relate to a voxel?
b   What is a display mapping function?
c   What is the disadvantage of using a nonlinear image mapping to improve image contrast?
d   What is an image look-up table?
e   What is the purpose of histogram equalization?
f   Why do the weights in an image filter usually sum to unity?
g   What penalty is paid by applying a filter to reduce noise in an image?
h   What is the particular advantage of using a median filter to reduce image noise?
i   What is the basis of image enhancement?
j   How does image intensity segmentation work?
k   What is the most difficult problem in using edge detection to segment objects in an image?
l   Does the partial volume effect increase or decrease the measured volume of an object?
m   What is meant by image registration?
n   How many pixels might there be in a typical digitized x-ray image?

o   What are the main advantages of using an image look-up table?
p   What is the best reduction in noise amplitude that can be achieved using a 4×4 image filter?
q   Why is image restoration usually too difficult?
r   What is a typical information transfer rate for a compressed video signal?
s   Can you get negative values in an intensity image?
t   Why is accurate background subtraction difficult?
u   What image changes can be achieved using an affine mapping?

14.7.2.   Longer questions (answers are given to some of the questions)

Question 14.7.2.1

The 421 filter has the set of weights shown in figure 14.14. Smoothing an image with this filter and then smoothing the filtered image with this filter again (i.e. smoothing twice) is equivalent to smoothing with a filter produced by smoothing the 421 filter with itself. Calculate the weights of this equivalent filter.

The 421 filter is a 3×3 filter. How big is the new filter?

Calculate how many multiplications and additions are required to apply the 421 filter twice and the new filter once. Which is the most efficient in terms of the number of operations.

Many extended, i.e. larger than 3×3, filters could be synthesized in this way. However, a penalty is paid for this efficiency. What is it?

Answer

The weights are

1/256	4/256	6/256	4/256	1/256
4/256	16/256	24/256	16/256	4/256
6/256	24/256	36/256	24/256	6/256
4/256	16/256	24/256	16/256	4/256
1/256	4/256	6/256	4/256	1/256

The new filter is 5×5.

To apply a 3×3 filter requires nine multiplications and eight additions. To apply it twice requires 34 operations. The equivalent 5×5 filter requires 25 multiplications and 24 additions $= 49$ operations. Two smooths are more efficient.

A 5×5 filter has 25 assignable weights. With two passes of two possibly different 3×3 filters there are only 10 possible weights that can be assigned. Therefore some 5×5 filters cannot be generated by using two 3×3 filters.

Question 14.7.2.2

Show that the 421 filter (figure 14.14) can be decomposed into the operation of smoothing each row in the image with a one-dimensional filter with weights $[1/4, 2/4, 1/4]$ and then filtering each column of the resulting image with the same 1D filter.

Extend this concept to producing a three-dimensional equivalent of the 2D 421 filter. What are the values of the weights for this filter?

Can the uniform filter (figure 14.13) be synthesized in a similar manner?

Answer

The filters

1/4	2/4	1/4

and

1/4
2/4
1/4

when convolved, produce the 421 filter. The weights for the equivalent 3D filter are:

1/64	2/64	1/64
2/64	4/64	2/64
1/64	2/64	1/64

2/64	4/64	2/64
4/64	8/64	4/64
2/64	4/64	2/64

1/64	2/64	1/64
2/64	4/64	2/64
1/64	2/64	1/64

The uniform filter can be synthesized in a similar manner.

Question 14.7.2.3

In probability theory a consequence of the central limit theorem is that if a distribution function is repeatedly convolved with itself the resulting function will become closer and closer to a normal (Gaussian) distribution. If the variance of the distribution function is σ^2, the variance of the function produced by convolving N copies of the function is $N\sigma^2$. This result applies to repeated smoothing with a filter. Calculate the variance of the 421 filter assuming the pixel size is p mm, and hence calculate the FWHM of the equivalent filter produced by N smoothings with the 421 filter.

How many smoothings with the 421 filter are required to synthesize a Gaussian filter of FWHM 20 mm if the pixel size is 2 mm?

Repeat these calculation for the 3D case. Assume cubic voxels.

Answer

The variance of the 421 filter is p^2. Variance is defined for a centrosymmetric function as

$$\sigma^2 = \frac{\int (x^2 + y^2) P(x, y) \, dx \, dy}{\int P(x, y) \, dx \, dy}$$

and in the discrete case we replace the integral with summations.

The FWHM for N smoothings of the filter is given by

$$\text{FWHM} = 2.36\sqrt{N}\,p$$

and so for a FWHM of 20 mm and a p of 2 mm N is 18.

For 3D the variance is $1.5p^2$ and so the number of smooths in this case is 12.

Question 14.7.2.4

The 2D Laplacian of an image is

$$\frac{\partial^2 g}{\partial x^2} + \frac{\partial^2 g}{\partial y^2}.$$

The Laplacian of a 3D image is

$$\frac{\partial^2 g}{\partial x^2} + \frac{\partial^2 g}{\partial y^2} + \frac{\partial^2 g}{\partial z^2}.$$

The gradient $\partial g/\partial x$ of an image in the x direction can be approximated by filtering the rows of the image with the weights $[-1, 1]$. The gradient in the y direction can be generated by applying this filter along the columns of the image and for 3D images the gradient in the z direction can be derived similarly. Using these filters derive the set of weights for a digital 2D and 3D Laplacian filter.

Answer

Filtering twice with the $[-1, 1]$ filter produces an effective filter $[1, -2, 1]$. The Laplacian is constructed by adding this filter to a similar one in the column direction. The 2D Laplacian is therefore:

0	1	0
1	−4	1
0	1	0

The equivalent filter in 3D is

0	0	0
0	1	0
0	0	0

0	1	0
1	−6	1
0	1	0

0	0	0
0	1	0
0	0	0

Question 14.7.2.5

The partial volume effect. A small object can be represented by a two-dimensional Gaussian

$$f(x, y) = \frac{A}{2\pi d^2} e^{-(x^2+y^2)/2d^2}.$$

The integral under this object is A. An edge detection program selects the position of the edge as the circular contour of maximum gradient. Calculate the radius of this contour. Compute the integral of the object intensity within this region. What fraction of the true object intensity (A) is contained within this region?

Answer

The object is circularly symmetric so the contour of maximum gradient will be a circle centred on the peak of the object. It follows that the position of the maximum gradient can be found by finding the position of the maximum gradient along one of the axes, i.e. along the x-axis with $y = 0$,

$$\frac{\partial f}{\partial x} = -\frac{x}{d^2} \frac{A}{2\pi d^2} e^{-(x^2+y^2)/2d^2}$$

and the second differential by

$$\frac{\partial^2 f}{\partial x^2} = -\frac{1}{d^2} \frac{A}{2\pi d^2} e^{-(x^2+y^2)/2d^2} + \frac{x^2}{d^4} \frac{A}{2\pi d^2} e^{-(x^2+y^2)/2d^2}.$$

The gradient is a maximum when this is zero which is obtained when $x = d$, so this is the point (radius) of maximum gradient.

The volume of the object inside this region relative to the whole volume under the object is given by

$$v = \frac{\int_0^d r e^{-r^2/2d^2} \, \mathrm{d}r}{\int_0^\infty r e^{-r^2/2d^2} \, \mathrm{d}r} = e^{-1/2} = 0.607$$

which is the fraction of the true object intensity within the contour.

Question 14.7.2.6

A patient is L cm thick. A 'kidney' lies D cm under the skin surface and is uniformly T cm thick with an area A. The uptake of radio-tracer in the non-organ tissues is B units cm^{-3} and K units cm^{-3} in the organ. The attenuation coefficient of the emitted γ-ray is μ cm^{-1}. The sensitivity of the gamma camera viewing the organ is S.

    The counts over the kidney are measured in an appropriate region covering the kidney and the counts in a similar area of background. The background is subtracted and the result taken as a measure of the kidney uptake.

    For values $L = 20$ cm, $D = 5$ cm, $T = 1$ cm, $A = 10$ cm^2, $\mu = 0.12$ cm^{-1}, $K = 50$ units cm^3 and $B = 10$ units cm^3, by how much does the measured background-corrected value of organ uptake differ from the value that would have been obtained in the absence of attenuation and background activity, i.e. the 'true' value. The effects of scatter and resolution should be ignored for this exercise.

A second identical kidney does not overlap the first kidney but is at a depth of 7 cm. The activity in this kidney is calculated in the same way as the first kidney. The ratio of uptakes is calculated. How much does this ratio differ from unity?

Answer

The counts from the background region per unit time are given by

$$C_\mathrm{B} = \frac{BSA}{\mu}[1 - e^{-\mu L}]$$

and the counts from the kidney region as

$$C_K = \frac{KSA}{\mu}[1 - e^{-\mu T}]e^{-\mu D} + \frac{BSA}{\mu}[1 - e^{-\mu L}] - \frac{BSA}{\mu}[1 - e^{-\mu T}]e^{-\mu D}.$$

The last term is a correction for the background tissues displaced by the kidney.

The background-corrected counts are therefore

$$C_K = \frac{(K - B)SA}{\mu}[1 - e^{-\mu T}]e^{-\mu D}$$

unless an explicit correction is made to the background counts for the thickness of the kidney.

The 'true' counts are $C = KSAT$ and so the ratio of observed to true is

$$R = \frac{(K - B)}{KT\mu}[1 - e^{-\mu T}]e^{-\mu D}.$$

If we can assume μT is fairly small then this becomes

$$R = \frac{(K - B)}{K}e^{-\mu D}.$$

For the values given $R = 0.44$.

The ratio between this and a second identical kidney at a different depth is

$$R = e^{-\mu(D_1 - D_2)}$$

and this is 1.27. The depth difference has produced an apparent 27% difference in uptake in the two kidneys.

Answers to short questions

a   A pixel is an image element in 2D. A voxel is a volume element in a 3D image.

b   A display mapping function is a function that converts each image value to a new value in order to improve an image.

c   For pixel values outside the expanded range, nonlinear image mapping to improve contrast may reduce image quality.

d   An image look-up table is an array that can be addressed by the value of each image element such that the value in the array is displayed rather than the image value itself.

e   Histogram equalization is used to transform the distribution of pixel values so that they are spread uniformly. The objective is to improve image quality.

f   By arranging for the weights in an image filter to sum to unity the total intensity of an image is preserved after filtering.

g   A noise-reducing filter will also cause blurring of the image.

h   A median filter preserves edges in an image whilst also reducing noise.

i   Image enhancement enhances edges or intensity gradients to improve an image.

j   Image segmentation identifies regions of interest in an image by selecting different ranges of element intensity.

k   Edges can be detected in an image but connecting the edges to form closed boundaries is very difficult.

l   The partial volume effect reduces the measured volume of an object.

m   Image registration is used to align two images of the same object by mapping the coordinates of the first image to the coordinates of the second.

n   There are likely to be at least 4×10^6 pixels in a digitized x-ray image.

o   Use of an image look-up table is fast and it leaves the original image data unchanged.

p   A 4×4 image filter could reduce the noise amplitude by a factor of four if the noise is random and an equal-weights filter is used.

q   Image restoration is often impossible because of image noise and the fact that the Fourier transform of the PSF can contain zeros.

r   A compressed video signal has an information transfer rate of typically 6 Mbit s^{-1}.

s   Yes, you can get negative values in an image. They arise because imaging systems such as x-ray CT have performed calculations on the measured data before the image stage.

t   It is difficult to accurately determine the scaling factor by which the background term should be multiplied before subtraction.

u   The affine transform can be used to rotate the image and scale its linear dimensions.

BIBLIOGRAPHY

Banks S 1990 *Signal Processing, Image Processing and Pattern Recognition* (Englewood Cliffs, NJ: Prentice-Hall)

Glasbey C A and Horgan G W 1995 *Image Analysis for the Biological Sciences* (New York: Wiley)

Teuber J 1992 *Digital Image Processing* (Englewood Cliffs, NJ: Prentice-Hall)

Webb S (ed) 1988 *The Physics of Medical Imaging* (Bristol: Hilger)

CHAPTER 15

AUDIOLOGY

15.1. INTRODUCTION AND OBJECTIVES

Hearing impairment is very common and arises as a result of factors such as disease, physical accident, exposure to high-intensity sounds, and the process of ageing. Surgical procedures can correct some defects of the middle ear, cochlear implants can be used in some cases of profound hearing loss, and hearing aids can help to overcome defects within the inner ear and the neural pathways, but the hearing loss has to be measured before treatment is planned. Audiometry, or the measurement of hearing loss, forms the major part of this chapter, which concludes with a section on hearing aids.

Both technicians and graduate scientists are employed in hearing test clinics to carry out investigations, some of which are simple and others quite complex. Many of the techniques, such as pure-tone audiometry, are well established and routine whereas others, such as evoked-response audiometry, and oto-acoustic emission testing are still developing.

Some of the questions we hope to answer in this chapter are:

- What determines the range of frequencies that we can hear?
- Can hearing be assessed objectively?
- What determines the shape of an auditory evoked response?
- What limits the assistance that a hearing aid can give to the wearer?

When you have finished this chapter, you should be aware of:

- The steps involved in the process of hearing.
- What defects in hearing can occur.
- The essentials of pure-tone audiometry.
- What is meant by the acoustic impedance of the middle ear?
- Several techniques for the objective assessment of hearing loss.
- The main components of a hearing aid.

This chapter is practical rather than theoretical. Chapter 3 (section 3.4) covered the physics of sound and the mathematics of wave propagation. The content of this chapter should be accessible to all our readers who can test their comprehension by trying to answer the set of questions at the end.

15.2.   HEARING FUNCTION AND SOUND PROPERTIES

Some of the anatomy and the physics of the sense of hearing were explained in Chapter 3 (section 3.4.4). In this section we will briefly review and expand on some aspects of these topics. The ear is a transducer which is connected to the brain via the eighth cranial nerve. Sound travels down the external ear canal and causes the eardrum (the tympanic membrane) to vibrate. The eardrum is oval in shape, has a maximum width of about 8 mm and is about 0.1 mm in thickness. The vibrations of the membrane, which have an amplitude approximately equal to one atomic radius (10^{-10} m) at sound intensities close to the threshold of hearing, cause the small bones within the middle ear to vibrate. These bones, or ossicles, transmit the vibrations to a membrane that covers the entrance to the fluid-filled inner ear. The ossicles appear to be pivoted in a manner which makes them insensitive to vibrations of the skull but able to magnify the forces applied to the eardrum, by a factor of about 1.6. The difference in area of the tympanic membrane and the window into the inner ear gives a much larger increase in transmitted pressure. These magnified changes in force and pressure act within the cochlea of the middle ear to transduce the vibratory energy into electrical energy. There is no widely accepted theory to explain how the cochlea generates the pattern of nerve impulses which we interpret as sound. Some of the possible mechanisms were described in Chapter 3.

15.2.1.   Anatomy

Figure 15.1 shows the anatomy of the ear. The middle ear is air filled and communicates through the Eustachian tube and the pharynx with the throat. If these pathways are open, then both sides of the eardrum are at the same pressure, so that the ear cannot sense a very slowly changing pressure, as there will be no differential across the eardrum. Infections of the middle ear or of the pharynx can block the pathways. This results in an abnormal and unpleasant sensitivity to the pressure changes which can occur in aircraft and cars during rapid changes in altitude.

The vestibular apparatus is part of the inner ear but is not a sound transducer. It consists of three fluid-filled semi-circular canals which are set at almost 90° to each other. Any movement of the head will cause fluid to move in the canals which contain hair cells to sense this movement. The hair cells generate nerve impulses which the brain uses, together with input from vision and muscle proprioception, to maintain our sense of balance. The sense of balance is a very important sense, but one of which we are not normally aware. The dizziness which results from a pirouette is partially caused by movement of fluid in the semi-circular canals.

Theories of hearing

Certainly the most complex part of the ear is the cochlea. It is a tube about 35 mm long and coiled to form two and a half turns. The tube is divided along its length by the basilar membrane. When the ossicles, responding to a sound stimulus, move the oval window, the resultant fluid disturbance passes along the cochlea. During its passage, the disturbance distorts the basilar membrane, on whose surface there are thousands of sensitive hair cells which transform the distortion into nerve impulses. High-frequency sounds only disturb the basilar membrane close to the oval window, whereas lower frequencies are transmitted over the whole length of the cochlea. A 3 kHz sound only causes disturbances about halfway down the cochlea, whereas a 50 Hz sound disturbs the whole of the basilar membrane.

In order to perceive a sound we need information on both the intensity and the frequency components of that sound. In general, the body senses an increased intensity of sensation by increasing the frequency of nerve impulses which are generated. If we first touch something which is hot and then another object which is cold, the frequency of nerve impulses generated by the end organs in the skin will change. One theory of hearing is that different parts of the cochlea respond to different frequencies of sound, and that the number

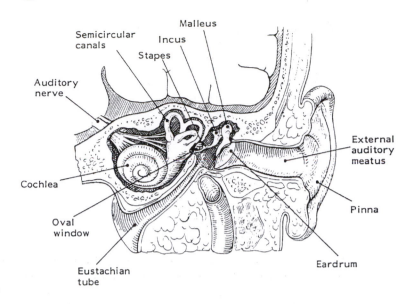

Figure 15.1. *Anatomy of the ear.*

of nerve impulses produced by a particular part of the cochlea is determined by the intensity of sound at that frequency. The intensity of sound changes the frequency of nerve impulses and the frequency of the sound corresponds to a particular spatial position within the cochlea. This simple theory can be partially supported by experiments which show that the frequency of auditory nerve potentials, recorded using microelectrodes in an animal, change with the intensity of a sound. It can also be shown that particular hair cells in the cochlea respond to particular sound frequencies. Figure 15.2 shows how the sensitivity of one hair cell changes with the frequency of a sound.

This theory of hearing has the virtue of simplicity, but unfortunately it is much too simple. When sound intensity increases, animal experiments have shown that not only does the number of nerve impulses generated by a particular cochlear nerve fibre increase, but also more nerve fibres are stimulated. It has also been shown that the frequency of a sound changes both the number of active nerve fibres and their frequency of discharge. Section 3.4.5 of Chapter 3 followed this theory of hearing in a little more depth.

Electrodes can be implanted within the cochlea of totally deaf patients whose nerve pathways are still complete. Sound is converted to an electrical signal within the implant which then stimulates the cochlea via the implanted electrodes. Most implants use several electrodes which are driven by different frequency components of the sound. This is an attempt to use the simplest theory of hearing where position along the cochlea corresponds to different sound frequencies. Patients with a cochlear implant can normally perceive sound although this is much less clear than a functioning ear.

In summary therefore, it would seem that both the frequency and the amplitude of a sound affect the number of nerve impulses initiated by particular hair cells within the cochlea, together with the number of hair cells initiating impulses. Whilst very little is known about how the cochlea performs this transformation, it is probably true to say that even less is known about how the brain interprets the signals.

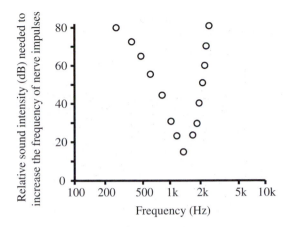

Figure 15.2. *The graph shows how the sensitivity of a single auditory nerve fibre in the cat changes with the frequency of the sound stimulus. The minimum of the curve at 1.5 kHz corresponds to a maximum sensitivity. Different nerve fibres exhibit their maximum sensitivity at different frequencies. (Redrawn from S Y S Kiang 1965 Discharge patterns of single fibres in the cat's auditory nerve Res. Mon. No 35 (Cambridge, MA: MIT Press). Reprinted by permission of MIT Press.)*

15.2.2. Sound waves

Chapter 3 (section 3.4.2) covered this in some detail. It is important to understand what sound is and how it can be measured before you attempt to measure a patient's ability to hear. Any vibration can be transmitted through the air as one molecule disturbs its immediate neighbours, so that the disturbance is propagated in the same way as ripples in a pool when the surface is disturbed. In a pool of water the disturbance results in vertical movements of the water, but sound disturbances are longitudinal movements. The cone of a loudspeaker will either push the air in front of it, and so increase the air pressure, or move backwards to reduce the air pressure. These pressure changes will propagate away from the loudspeaker as a series of alternate increases and decreases in air pressure (figure 15.3).

The simplest way to represent these longitudinal pressure changes is as a sinusoidal signal:

$$p = p_0 \sin 2\pi f t$$

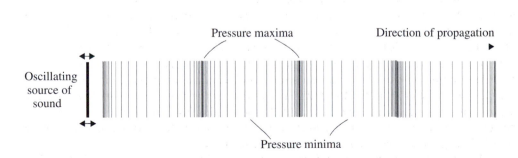

Figure 15.3. *The longitudinal vibrations produced in air as a result of an oscillating source. The density of the vertical lines represents the magnitude of the air pressure.*

where p_0 is the amplitude of the pressure changes, f is the frequency of the regular changes in pressure, and t is time. This is the simplest sound which is referred to as a 'pure tone'. All sounds which we hear can be represented as a mixture of many pure tones ranging in frequency from about 20 Hz to 20 kHz.

The basic concept of frequency analysis was introduced in Chapter 13 where it was shown that a complex signal can be seen as a combination of many sine wave components. In figure 15.4 the sound 'O' has been analysed into its frequency components. Most of the components of the sound 'O' are in the range 200–600 Hz, but it can be seen that there are many other components in the sound.

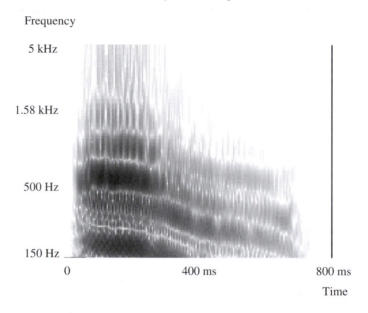

Figure 15.4. *A frequency analysis of the sound 'O'. The density of the trace represents the amplitude of the frequency components which change during the sound.*

15.2.3.   *Basic properties: dB scales*

For the pure tone $p = p_0 \sin 2\pi f t$, p_0 is a measure of the loudness of the sound. The units in which pressure is measured are newtons per square metre (N m^{-2}) or Pascals (Pa).

The lowest sound pressure which the human ear can detect is about 20 μPa, and the highest before the sound becomes painful is about 100 Pa. This represents a range of more than a million to one, and so for convenience a logarithmic scale is used. The response of the ear also appears to be approximately logarithmic in that equal fractional increases in sound pressure are perceived as approximately equal increases in loudness. There are, therefore, good reasons to adopt a logarithmic scale. The logarithmic scale which is used adopts 20 μPa (20 $\times$ 10^{-6} N m^{-2}) as a reference pressure, p_0, so that the sound pressure, p, is expressed as $\log_{10}(p/p_0)$. Two further modifications are made to this expression. Firstly, because sound power (often called intensity) is a better measure of loudness than amplitude, p^2 is used instead of p, and secondly, because the basic logarithmic unit, called the bel, is large it is divided into 10 decibels. The final definition is therefore

$$\text{sound pressure level} = 10 \log_{10} \left[\frac{p}{p_0} \right]^2 = 20 \log_{10} \left[\frac{p}{p_0} \right] \text{dB.}$$

The sound pressure will change from cycle to cycle of a sound so that the average value of p is normally measured. The average value is shown as $\bar{p}$.

If the average sound pressure is 2 Pa then the sound pressure level (SPL) will be

$$\text{SPL} = 20 \log_{10} \frac{2}{20 \times 10^{-6}} = 100 \text{ dB}.$$

On this scale the minimum sound we can hear is about 0 dB and the maximum before the sound becomes painful is about 134 dB. A 10 dB increase in SPL corresponds to about a doubling in the subjective loudness of the sound. This scale of measurement has become widely used and is very often referred to as the sound level rather than the sound pressure level. In summary, the unit used is the bel, defined as a tenfold increase in sound power or intensity. Therefore,

$$\text{SPL (in decibels)} = 10 \log_{10} \frac{I}{I_0}$$

or

$$\text{SPL (in decibels)} = 20 \log_{10} \frac{p}{p_0}$$

where I is the intensity, p power and $I_0 = p_0^2$ the reference intensity.

15.2.4. Basic properties: transmission of sound

The transmission of sound through air is a very complex subject that was introduced in section 3.4.2. In order to deal with transmission over long distances we would need to consider temperature, humidity and the pressure of the air. Winds and temperature gradients would also need to be taken into account. In clinical audiology the sounds considered usually have only short distances to travel and so the detailed properties of the air can be neglected. However, some factors still need to be taken into account.

The *velocity* of sound in air at normal temperature and pressure is 340 ms^{-1}. This is a relatively slow velocity and can cause quite significant delays in evoked response measurements. If the sound source is placed 0.34 m from the ear then it will take 1 ms to cover this distance; as the time from stimulus to neural response in the technique of cochleography (see section 15.5.4) may only be 2 ms, an additional 1 ms delay is quite significant.

The ear is capable of hearing pure tones over the frequency range of about 20 Hz–20 kHz. The transmission of sound changes quite markedly within this range. The higher the frequency of a sound the more it behaves like a light ray which can only travel in a straight line. This effect is exploited by bats which use frequencies as high as 200 kHz to locate objects by their echoes. It would be impossible to use low frequencies for direction finding because the sound would be scattered around objects rather than be reflected. Whilst bats appear to have the most highly developed direction finding system, many other animals are also able to hear very high frequencies: whales, a number of rodents and moths are all able to hear sounds with frequencies beyond 100 kHz.

The reason for the different behaviour of high- and low-frequency sounds is connected with their *wavelength*. This is the distance which the sound travels in one cycle of the pure tone:

$$\text{wavelength} = \text{velocity}/\text{frequency}.$$

A 35 Hz pure tone has a wavelength of about 10 m, and a 9 kHz pure tone has a wavelength of about 37 mm.

If the wavelength is large compared to the size of an object in the path of the sound, then the sound will be diffracted around the object. If the wavelength is small then the sound may be reflected by the object but will not be scattered around it. This is why low-frequency sounds appear to be able to travel around corners; an effect which is easily appreciated by listening to a band as it disappears around a corner, when the high sounds fade away but the drums continue to be heard. High-frequency sounds are used in direction finding because the sounds are reflected by objects and not diffracted around them.

15.2.5. *Sound pressure level measurement*

The simplest possible sound pressure level meter need only consist of four components: a microphone to transduce sound pressure to an electrical signal; an amplifier to increase the size of the electrical signal; a rectifier to convert the alternating electrical waveform to a DC signal and some form of meter or digital display (see figure 15.5). If the sound level meter is to record the sound level in decibels then either additional circuitry is necessary to convert from a linear to a logarithmic scale, or the logarithm of intensity is calculated for display.

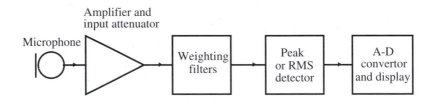

Figure 15.5. *The major components within a sound level meter.*

This type of system is said to have a linear response because it responds equally to all frequencies of sound. Many commercial sound level meters have a 'linear' switch position so that a true sound level measurement can be made. Unfortunately the human ear does not have the same sensitivity at all frequencies so that a sound pressure level of 20 dB at 2 kHz can be heard, whilst a sound of the same level but at 50 Hz cannot be heard. If the minimum sound pressure level which can be heard by a group of normal subjects is found over a range of pure-tone frequencies, then the graph shown as the lower curve in figure 15.6 is obtained. This curve is called the normal threshold of hearing and it shows that the ear is at its most sensitive between about 500 Hz and 5 kHz, which is where most of the information in human speech is contained.

The change in sensitivity of the human ear with frequency can be presented in another way; this is in terms of an *equal loudness contour.*

A pure tone of, say 40 dB at 1 kHz, is first presented and the normal subject is then asked to control the sound level of a sequence of other frequencies such that all the sounds have the same subjective loudness. This variation of sound level to give the same subjective loudness is plotted as a function of frequency to give an equal loudness contour. These measurements have been made on a very large number of normal subjects and the resulting average curves have been agreed internationally. The curve for a sound of level 40 dB at 1 kHz is shown in figure 15.6 (upper curve). Similar, although not identical, curves are obtained at different intensities; these show that, at very low intensities, the ear is even less sensitive both to very low and very high frequencies.

The equal loudness contour is used to alter the frequency response of sound level meters such that a sound which reads 40 dB will have the same subjective loudness at any frequency. This modification of the frequency response is called a dBA response and it is the one which is used for most measurements of noise level. By definition the dBA response is most accurate when recording sound levels around 40 dB. Other responses (dBB and dBC) have been developed for use at higher sound levels, but these responses are not often used.

The normal sound level meter will include, therefore, in addition to the components shown in figure 15.5, a filter, with a response which is the inverse of the upper curve shown in figure 15.6. This dBA weighting is shown in figure 15.7.

To make a measurement of *background noise level* the following procedure should be followed. Place the meter securely where the measurement is to be made, making sure that there are no objects close to the

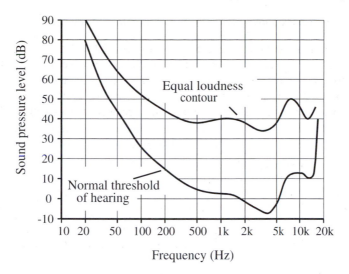

Figure 15.6. *This graph shows how the normal threshold of hearing (lower curve) depends upon the frequency of a pure tone. The curve is an average for many normal subjects (redrawn from values given in ISO389-7, see the bibliography). The upper curve is an equal loudness contour. The curve was obtained as an average for many normal subjects and it joins together points which correspond to sounds that give the same subjective loudness. An SPL of 40 dB at 1 kHz was taken as the reference.*

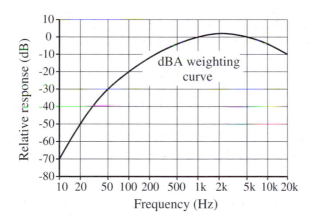

Figure 15.7. *The dBA weighting curve which compensates for the changes in sensitivity of the ear as shown in figure 15.6.*

microphone. Switch ON the meter and select the dBA position. Move the 'range' switch until the meter reading is on-scale.

Some sound level meters include a slow/fast response switch. In the 'slow' position the meter response is slowed by using a low-pass filter which will make the meter less subject to sudden changes in sound level. The slow position allows a more accurate measurement of the average noise level to be obtained than the fast position.

15.2.6.   Normal sound levels

Table 15.1 gives the sound pressure levels both in pascals and in decibels, corresponding to nine circumstances.

Damage to the ear occurs immediately for sound levels of about 160 dB. Normal atmospheric pressure is about 10^5 Pa and 160 dB is 2×10^3 Pa so that damage occurs at about 0.02 atm. The threshold of hearing is the other extreme. This pressure represents 2×10^{-10} atm; if we were to measure this pressure with a mercury manometer then the mercury level would only change by 1.5×10^{-10} m.

The range of sound levels which are encountered in normal living is very wide, although there has been increasing pressure in recent years to limit the maximum sound levels to which people are exposed. There is no international agreement on standards for occupational exposure but most of the developed countries have adopted a limit of 90 dB for continuous exposure over a normal 8 h working day, with higher levels allowed for short periods of time. In some countries the level is set below 90 dB.

In a room where hearing tests are carried out, the background noise level should not be greater than 40 dBA and a level below 30 dBA is preferred. Lower noise levels are needed if 'free field' testing is to be carried out (see ISO 8253-2 for more detailed guidance). The use of sound-reducing material in the walls, floor and ceiling of the audiology test room is often necessary. Noise-reducing headsets are a cheap way of reducing the background noise level for a patient.

Table 15.1. *Nine typical sound pressure levels, expressed on the dBA scale.*

Sound pressure ($\mathrm{N\,m^{-2}} = \mathrm{Pa}$)	Sound pressure level (dBA)	Circumstances
2×10^3	160	Mechanical damage to the ear perhaps caused by an explosion
2×10^2	140	Pain threshold, aircraft at take-off
2×10	120	Very loud music, discomfort, hearing loss after prolonged exposure
2×10^0	100	Factory noise, near pneumatic drill
2×10^{-1}	80	School classroom, loud radio, inside a car
2×10^{-2}	60	Level of normal speech
2×10^{-3}	40	Average living room
2×10^{-4}	20	Very quiet room
2×10^{-5}	0	Threshold of hearing

15.3.   BASIC MEASUREMENTS OF EAR FUNCTION

A measure of speech comprehension is the most desirable feature of a hearing test. Tests are used in which speech is presented to the subject at a range of intensities and their ability to understand is recorded. Speech audiometry is a valuable test of hearing, although the results depend not only on the hearing ability of the subject but also upon their ability to comprehend the language which is used. Sounds other than speech are also used: a tuning fork can be used by a trained person to assess hearing quite accurately. Sources of sound such as rattles are often used to test a child's hearing: the sound level required to distract the child can be used as evidence of their having heard a sound.

In this section an account is given of some commonly used hearing tests. In pure-tone audiometry a range of pure tones is produced and the subject is asked whether they can hear the sounds. Middle-ear impedance audiometry is another type of hearing test which enables an objective measurement to be made of the mechanical function of the middle ear. Otoacoustic emissions are a further objective measure of hearing, but the origin of these emissions is not well understood.

15.3.1. *Pure-tone audiometry: air conduction*

The pure-tone audiometer is an instrument which produces sounds, in the form of pure tones, which can be varied both in frequency and intensity. They are presented to the patient either through headphones for air conduction measurements, or through a bone conductor for bone conduction measurements. In the test situation the patient is instructed to listen carefully and respond to every sound. This response may be to raise a finger or to press a button; if the patient is a child then they may be asked to respond by moving bricks or some other toy when they hear the sounds. The threshold level is said to be the minimum intensity at which the tone can be heard on at least 50% of its presentations.

The audiometer contains an oscillator which produces a sinusoidal waveform. The frequency of this sine wave can be changed, and the minimum available frequencies are 250, 500, 1000, 2000, 4000 and 8000 Hz. The output from the oscillator is taken to an audio amplifier and then into an attenuator, which may be either stepped or continuously variable. A standard range would be from -10 to 120 dB. The output from the attenuator is taken to the headphones. The input connection to the amplifier can be interrupted by a switch which allows the sound to be presented as bursts of a pure tone. This is the most common way of presenting sounds for manual audiometry.

A loud sound is presented to the patient and the intensity reduced slowly until they can no longer hear the sound. The threshold found by this method will not be the same as that which is found if the sound intensity is increased slowly from zero to the point where it can first be heard. For this reason it is important that a consistent test procedure be adopted. There is no universal agreement on the procedure to be adopted in pure-tone audiometry, but the following is a widely used system. The sounds are initially presented in decreasing intensity and then both upward and downward changes are made close to the threshold to determine the level at which 50% of the sounds are heard.

Procedure for routine air conduction pure-tone audiometry

Place the headphones comfortably on the patient, making sure that the red phone is over the right ear. Spectacles can be most uncomfortable when headphones are worn and are therefore best removed.

Start at a level of 50 dB and 1000 Hz.

Present tones of about 2 s in duration with varying intervals (1–3 s).

If the tone is heard, then reduce the level in 10 dB steps until it is no longer heard. If the starting tone is not heard, then raise the level in 20 dB steps until it is heard, and then descend in 10 dB steps.

From the first level at which the tone is not heard, first raise the level in 5 dB steps until it is heard, then down in 10 dB steps until it is not heard, then up again in 5 dB steps. This enables two ascending threshold measurements to be made.

After testing at 1000 Hz proceed to 2000, 4000 and 8000 Hz. Repeat the reading at 1000 Hz and then make measurements at 500, 250 and 125 Hz.

Great care must be taken to vary the interval between the tones in order to detect where incorrect responses are given.

15.3.2.   *Pure-tone audiometry: bone conduction*

Instead of presenting the sound vibrations through headphones a vibrator can be attached over the mastoid bone behind the ear. The vibrator is usually attached by a sprung band passing over the head. Sounds presented by this means bypass the eardrum and middle ear and are able to stimulate the inner ear directly. A patient with disease of the middle ear, such that sounds are attenuated in passing through the middle ear, may have a raised threshold to sound presented through headphones but a normal threshold to sound presented through the bone conductor.

The procedure for making a threshold determination through a bone conductor is the same as that which was described for air conduction. The results of both air and bone conduction threshold measurements are presented graphically as shown in figure 15.8. Different symbols are used for the right and left ears and also for air and bone conduction thresholds.

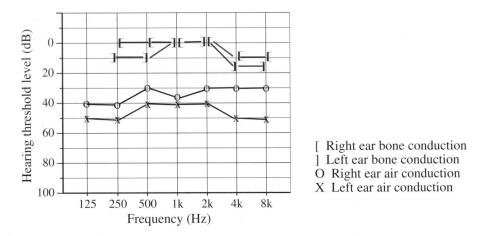

[Right ear bone conduction
] Left ear bone conduction
O Right ear air conduction
X Left ear air conduction

Figure 15.8.   *This pure-tone audiogram shows the variations in hearing level for both air- and bone-conducted sounds.   The patient has normal bone conduction thresholds but a 40 dB loss for air conduction.*

15.3.3.   *Masking*

A hearing loss which only affects one ear is called unilateral and a loss to both ears, but of different degrees, is called asymmetrical. If a patient has a much greater hearing loss in one ear than the other then it is possible that sounds presented to the poor ear may be heard in the good ear. 40 dB is the minimum reduction in intensity of a sound presented to one ear but heard in the other ear. If the difference in pure-tone thresholds between the two ears is greater than 40 dB then special techniques have to be used in testing.

In order to obtain a 'true' threshold when testing the poor ear, a masking noise is presented to the good ear to prevent cross-over. The masking noise of choice is narrow-band noise; this is a random noise such as the hiss which is produced by a high-gain audio amplifier, but filtered to present a one-third octave band of noise centred on the test frequency.

The criteria to assess when masking is needed, are:

- where the difference between left and right unmasked air conduction thresholds is 40 dB or more; and
- where the unmasked bone conduction threshold is at least 10 dB better than the worst air conduction threshold.

This is necessary because sounds are conducted through the skull with very little loss of intensity; a sound presented through the mastoid bone on one side of the head can be heard at the same intensity on the other side.

Procedure for masking

Measure the threshold of the masking signal in the non-test ear.

Present the tone to the poor ear at unmasked threshold level.

Introduce narrow-band masking into the good ear at the masking signal threshold.

Now present the tone to the poor ear again:

1   If the patient still hears the tone then increase the masking level to the good ear in 5 dB steps up to a maximum of 30 dB above threshold. If the tone is still heard then this is considered to be the true threshold for the poor ear.
2   If the patient does not hear the tone then increase the intensity of the tone presented to the poor ear in 5 dB steps until it is heard. Then proceed as in 1.

The test is not considered satisfactory until the tone in the poor ear can be heard for an increase of 30 dB in the masking to the good ear. (See the reference to 'Recommendations for masking in pure-tone threshold audiometry', in the bibliography.)

15.3.4.   Accuracy of measurement

Pure-tone audiometry gives a measure of hearing threshold over a range of sound frequencies. However, the measurement is a subjective one because it depends upon the co-operation of the patient and their ability to decide when a sound can be heard. Hearing threshold will vary amongst a group of normal people; it can also change from day to day and is affected by exposure to loud sounds. For these reasons a range of -10 to $+15$ dB is normally allowed before a threshold measurement is considered to be abnormal.

Very many factors can contribute to inaccuracies in measurement but only a few can be mentioned here. These factors can arise either from the equipment or from the operator.

Pure-tone audiometry equipment should be calibrated at least once a year using an artificial ear; this is a model of an ear with a microphone included so that the actual sound level produced by headphones can be measured. In addition, a routine weekly test of an audiometer should be made by the operator by testing his or her own hearing. If the threshold readings change by more than 5 dB and there is no reason for their hearing to have been affected, then the audiometer is probably at fault and should be recalibrated.

There are many ways in which the operator can obtain inaccurate results. Switch positions or displays can be misread or the threshold plotted incorrectly on the audiogram. Correct placement of the earphones or the bone conductor is very important; if the earphone is not placed directly over the ear canal significant errors can arise.

In addition to disease many other factors can change hearing thresholds. Aspirin, some antibiotics, and menstruation are just three factors which, it has been claimed, can cause changes. The common cold can cause the Eustachian tubes to become partially blocked and this will change the threshold. An audiologist must be alert to these factors which might explain an abnormal hearing threshold.

Some explanation of how hearing defects can be diagnosed from the audiogram is given in section 15.4.

15.3.5. Middle-ear impedance audiometry: tympanometry

This is a technique for measuring the integrity of the conduction between the eardrum and the oval window to the inner ear by measuring the acoustic impedance of the eardrum (see figure 15.1). The primary function of the middle ear is that of an impedance matching system, designed to ensure that the energy of the sound wave is transmitted smoothly (with minimum reflection) from the air in the outer ear to the fluid in the inner ear. Middle-ear impedance audiometry (tympanometry) is a technique for measuring the integrity of this transmission system. If the middle ear is defective (whether due to a mechanical defect or physical inflammation) then the impedance matching might be lost and most of the energy of an applied sound will be absorbed or reflected. The acoustic impedance (see section 3.4.2) of the eardrum and middle ear is analogous to an electrical impedance. If the ear has a low impedance then an applied sound will be transmitted with very little absorption or reflection. If the middle ear is inflamed then the impedance may be high and most of an applied sound will be absorbed or reflected.

Electrical impedance is measured by applying a potential, V, across the impedance and recording the current, I, which flows. Then,

$$\text{impedance} = V/I.$$

The analogy in acoustics is that an alternating pressure is applied to the impedance and the resulting airflow is recorded,

$$\text{acoustic impedance} = \frac{\text{pressure}}{\text{flow}} = \frac{\text{pressure}}{\text{velocity} \times \text{area}} \left(\frac{\text{N m}^{-2}}{\text{m s}^{-1}\, \text{m}^2} \right).$$

The acoustic impedance is measured in acoustic ohms which have the units of N s m^{-5}. Figure 15.9 shows how sound can be applied as an alternating pressure to a volume whose acoustic impedance is to be measured for a given constant flow. The sound pressure at the entrance to the volume will be proportional to the acoustic impedance. If the acoustic impedance doubles, then the sound pressure level at the entrance to the volume will double. Now the volume flow provided by the loudspeaker can elicit two responses: the pressure within the ear canal might rise or the eardrum might move. In practice both of these responses will occur.

The wavelength of the sound used for the test is large compared with the length of the ear canal, so that there will be approximately uniform pressure in the ear canal. The frequency normally used is in the range 200–220 Hz, which corresponds to a wavelength in air of 1.7 m.

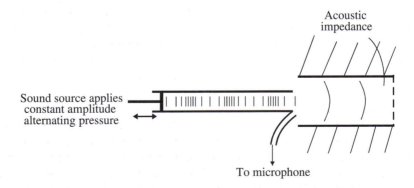

Figure 15.9. *Sound presented to a cavity which will have a certain acoustic impedance. The relative magnitude of the absorption and reflection of the sound determines the intensity of sound which is measured by the microphone.*

There are several designs of equipment which can be used to measure the acoustic impedance from the external auditory meatus. It is difficult to separate the impedance of the ear canal from that of the tympanic membrane and the middle ear. A complete analysis is outside the scope of this book. The measurement of acoustic impedance is widely used but, in most cases, only relative values of impedance are measured. The rest of this short section will be devoted to a qualitative description of the technique.

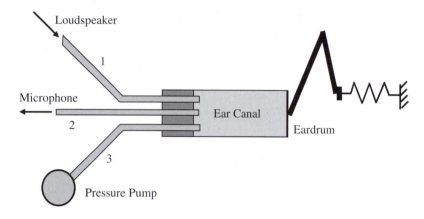

Figure 15.10. *A system for measuring the acoustic impedance of the eardrum and middle ear.*

A probe containing three tubes is introduced into the external ear canal; the tip of the probe is in the form of a plug which makes an airtight seal with the walls of the ear canal. The first of the tubes is connected to a sound source and the second to an air pump which enables the static pressure between the probe and the tympanic membrane to be controlled. The third tube is connected to a microphone which feeds an amplifier and recorder (figure 15.10). Under normal circumstances the pressure in the middle ear is equal to atmospheric pressure, the Eustachian tube having the function of equating middle-ear pressure to that close to the pharynx. If for any reason there is a pressure difference across the tympanic membrane then this stress will increase the stiffness of the membrane and hence its impedance. The system shown in figure 15.10 can be used to apply a positive pressure to the tympanic membrane and thus increase its impedance. The sound applied down the coupling tube will be reflected from the tympanic membrane back into the microphone tube. If the positive pressure is now reduced, then less sound will be reflected until a minimum impedance is reached when the pressure on both sides of the eardrum is the same. If the pressure is further reduced to a negative value then the impedance will rise again. In figure 15.11, the output from the impedance meter has been plotted as a graph of impedance versus the pressure applied to the eardrum. This is the result for a normal ear which has a well-defined minimum impedance when the applied pressure is zero.

Most impedance meters used clinically are calibrated by measuring the impedance of a known volume within the range 0.2–4.0 ml. The larger the volume, the smaller will be the impedance. However, if the reciprocal of impedance is used instead of impedance then there is a linear relationship with volume. The reciprocal of impedance is compliance, which is analogous to conductance in electrical terms. (Strictly, the reciprocal of impedance is admittance, which depends on the frequency of the applied pressure, but at the frequencies used in clinical impedance meters compliance and admittance are equal.) In the case of the ear a floppy eardrum will have a high compliance and a taut eardrum a low compliance. If compliance is used instead of impedance then the impedance curve of figure 15.11 can be replotted with the vertical axis calibrated as an equivalent volume (figure 15.11). This type of display is called a tympanogram and is widely used. Some examples of how otitis media (fluid in the middle ear) or a perforated eardrum affect the shape of the tympanogram curve are given in section 15.4.2.

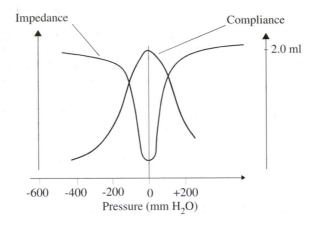

Figure 15.11. *Acoustic impedance plotted as a function of the pressure applied to an eardrum. The equivalent compliance is plotted as a tympanogram.*

Acoustic impedance measurements can be made at frequencies ranging from 100 Hz to several kilohertz and both the amplitude and phase of the impedance can be recorded. In the routine clinic a frequency of about 220 Hz is often used. The impedance is usually recorded as the applied steady pressure is changed from +200 mm of water pressure to −400 mm of water (+2 kPa to −6 kPa). In a normal ear, the minimum impedance is usually found between +100 and −100 mm of H_2O.

Stapedius reflex

There are two muscles within the middle ear: the tensor tympani and the stapedius. These muscles respond to acoustic stimulation. A loud sound introduced into one ear normally provokes bilateral contraction of the stapedius muscle. The muscle acts on the ossicles to stiffen the tympanic membrane. The intensity of sound normally required to cause this reflex is about 80 dB above threshold. The increase in stiffness of the eardrum changes the impedance of the ear. Observation of the impedance change resulting from the stapedius reflex contraction can be of some value in assessing hearing threshold.

15.3.6.   Measurement of oto-acoustic emissions

The ear may be thought of as a transducer which converts sounds into a stream of nerve impulses. The way in which this is carried out is still not clear but it appears to be an active process. Part of the evidence for this is the observation that the ear actually produces sounds. Spontaneous emissions can be recorded in about 50% of normally hearing subjects although they are at very low levels and need very quiet conditions and very careful recording in order to be observed.

It was found by Kemp in 1978 that evoked oto-acoustic emissions could be recorded. Oto-acoustic emissions can be defined as *acoustic energy produced by the cochlea and recorded in the outer ear canal*. By applying a short acoustic stimulus to the ear a microphone placed within the ear canal can then be used to record an emitted sound. It can be shown that the emissions are not simply passive acoustic reflections of the sound; they occur over a period of about 20 ms following the stimulus, which is much too long a period for simple acoustic reflection; the response is related to the stimulus in a nonlinear manner, whereas a simple reflection would be linear, and the emissions cannot be recorded from ears with damaged outer hair cells.

There has been considerable experimental work carried out since 1978 to determine the origin of oto-acoustic emissions and the conclusion is that the origin is related to the function of structures related to the outer hair cells within the cochlea. Because these emissions can be recorded relatively easily and they are sensitive to small amounts of hearing loss the technique has been developed as a means of testing the function of the cochlea objectively. Most studies show that oto-acoustic emissions cannot be recorded if the hearing loss is greater than 30 dB.

Figure 15.12 shows the acoustic emissions recorded from a normal subject following a stimulus with an intensity which was varied over the range 35–80 dB SPL. It can be seen that the response appears over a period of about 10 ms and has a maximum amplitude of about 500 μPa, i.e. 28 dB SPL. Now in a test subject the background noise level in the ear canal, even in a very quiet environment, may be about 30 dBA SPL because of the noise produced by blood flow, breathing, muscle contractions and joint movements in the head. In order to record oto-acoustic emission signals filtering and then averaging must be used. In a typical recording the stimulus will be repeated at 50 pps and 500 or more signals will be averaged.

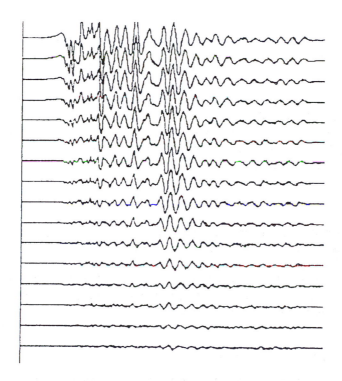

Figure 15.12. *Oto-acoustic emissions recorded in a normal adult for stimuli of intensity varied over the range 35–80 dB SPL (bottom to top of diagram). Time runs from left to right on a linear scale from 0 to 20 ms. The vertical scale is the same for all traces and the largest response is about 500 μPa (SPL 28 dB). It can be seen that the responses change with stimulus intensity but not in a linear fashion. In each case two average signals are presented so that the consistency of the response can be assessed. (From Grandori et al 1994 Advances in Otoacoustic Emissions, Commission of the European Communities, ed F Grandori.)*

Oto-acoustic emissions are usually recorded as the response to a transient stimulus. However, they can also be recorded either during the frequency sweep of a low-level tone stimulus or by recording the distortion products produced when two continuous sine waves of different frequencies are applied.

The acoustic emissions are recorded from a probe containing both a miniature loudspeaker and a microphone. This probe must be well sealed in the ear canal otherwise low-frequency signals will be reduced in amplitude and there will be a consequent change in the shape of the oto-acoustic emissions It is possible to use the microphone to monitor the stimulus as a means of checking the probe seal. The technique of measuring oto-acoustic emissions is still relatively new and is still developing. However, it may be a useful screening technique and is being investigated for use both in adults and also in neonates (Stevens *et al* 1991), where an objective measurement at an early age can be particularly important.

15.4. HEARING DEFECTS

Clinical audiology is a complex subject which cannot be covered in a brief section. The information given here is only intended to cover the more common hearing defects and illustrate the application of pure-tone audiometry and impedance techniques. Pure-tone audiometry is used to help in the diagnosis of ear pathology and also to help in the planning of treatment. The next three sections describe the main causes of hearing loss.

15.4.1.   *Changes with age*

There is a progressive deterioration in hearing after the age of 30 years. The process is given the title of presbycusis, but it is only evident at frequencies above 1 kHz and the degree of loss is very variable. The following are approximate average figures for loss of hearing at 3 kHz:

> 5 dB at age 40
> 10 dB at age 50
> 20 dB at age 70.

It is quite possible that presbycusis may be caused, in part, by exposure to damaging levels of sound. There is some evidence that people in remote societies who are not exposed to very loud sounds do not suffer hearing loss in old age. Noises with intensities above 90 dB cause temporary changes in the threshold of hearing and prolonged exposure will cause a permanent defect. Noise-induced deafness usually causes a drop in the pure-tone audiogram at around 4 kHz, even though the noise which gave rise to the damage was at lower frequencies. This dip in the pure-tone audiogram at 4 kHz can help in the difficult problem of distinguishing noise-induced deafness from presbycusis, which gives a progressive fall in the pure-tone audiogram at high frequencies.

15.4.2.   *Conductive loss*

Any defect which interrupts the transmission of sound from the ear canal, via the ossicles, to the oval window of the inner ear is termed a conductive defect. The ossicles are the three bones, i.e. the malleus, the incus and the stapes, which conduct sound from the tympanic membrane to the oval window (see figure 15.1). Many conditions can give rise to conductive hearing loss; the most common are wax in the ear canal, damage to the tympanic membrane, malfunctions of the ossicular chain as a result of physical damage, and middle-ear disease. Respiratory infections can result in infections and a build-up of fluid in the middle ear, i.e. otitis media, which will interfere with sound transmission.

A conductive hearing loss will result in reduced air conduction hearing thresholds but bone conduction readings will be normal. In a person with normal hearing, air and bone conduction thresholds will not differ by more than 10 dB. Figure 15.13 shows the pure-tone audiogram for a person with a conductive hearing loss. The bone conduction thresholds are 40 dB lower than the air conduction thresholds.

Conductive hearing loss will usually change the acoustic impedance of the ear and therefore the tympanogram may be helpful in diagnosis. Figures 15.14 and 15.15 show the effects of otitis media and a

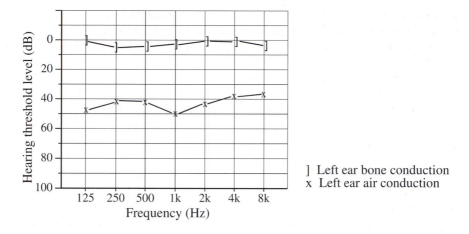

Figure 15.13. *Pure-tone audiogram from a person with a conductive hearing loss.*

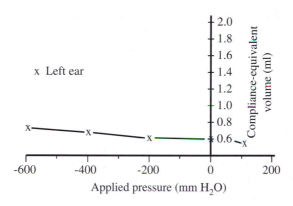

Figure 15.14. *Tympanogram in a patient with otitis media (fluid in the middle ear).*

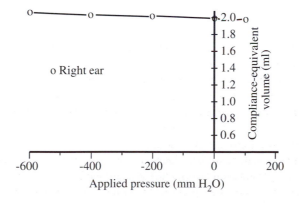

Figure 15.15. *Tympanogram in a patient with a perforated eardrum.*

perforated tympanic membrane, respectively; in both cases the normal sharp peak at zero pressure is absent. Fluid in the middle ear causes most of the incident sound to be reflected from the eardrum and so a high acoustic impedance is found, whereas a perforated eardrum allows the sound to pass unimpeded and so the acoustic impedance is low.

15.4.3. *Sensory neural loss*

Defects within the inner ear, in the transmission along the nerve pathways, or in perception by the brain are termed sensory neural defects. There are many possible causes of these defects: rubella (i.e. German measles) in the mother can result in a congenital cochlea deafness in the child; a viral infection or vascular accident within the inner ear can cause sudden deafness; tumours can occur and compress the eighth nerve or cause damage in the brainstem. The damage which results from long-term exposure to high-intensity sounds is sensory neural damage.

Sensory neural loss should affect both air and bone conduction thresholds equally; bone and air conduction thresholds should be within ±10 dB of each other at all frequencies. Sensory neural loss often occurs progressively at higher frequencies as shown in the audiogram of figure 15.16. It is possible for both sensory neural and conductive deafness to occur together and give rise to a separation of air and bone conduction thresholds in addition to an increased threshold at higher frequencies.

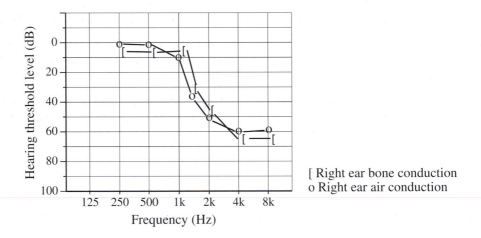

[Right ear bone conduction
o Right ear air conduction

Figure 15.16. *Sensory neural loss often occurs progressively at higher frequencies as shown in this pure-tone audiogram.*

15.5.   EVOKED RESPONSES: ELECTRIC RESPONSE AUDIOMETRY

In order for us to perceive a sound it must be conducted to the inner ear where the cochlea produces a corresponding pattern of nerve impulses. The impulses are carried by about 50 000 nerve fibres, each of which can carry a few hundred impulses each second. When the nerve impulses reach the brain they activate a particular pattern of neurones and we interpret this as sound. All the events from the cochlea onwards are electrical events and will give rise to current flow within the body. For every sound presented to the ears there will be a corresponding electrical evoked response resulting from the activity of the cochlea, the cochlear nerve, and the brain. These evoked responses can be recorded and used to investigate a person's hearing.

One of the advantages of recording an evoked electrical potential is that it is an objective measurement of hearing and does not depend upon a voluntary response from the subject. This is particularly important when making hearing tests on mentally retarded or psychiatric patients, testing babies and children under the age of three and in adults with a vested interest in the results of the test, e.g. those claiming compensation for industrial hearing loss.

The use of evoked potential measurements allows a more complete picture to be obtained of hearing loss and some of the following techniques are available in most large audiology departments. Four types of responses will be described, all of which result from a click or tone-burst stimulus. All the evoked responses are small and require an averager to be used in order to reduce the effect of background noise. The detail given in the next sections is sufficient to introduce the techniques, but is not sufficient to enable all aspects of the procedures to be carried out. Chapter 10 described the technique of evoked-response measurement.

15.5.1. Slow vertex cortical response

This is a response which is thought to be generated by the cortex; it can be recorded by placing a surface electrode on the vertex of the head (figure 15.17). The response has a latency of 50–300 ms, i.e. it appears within this period after the sound has been presented to the ear, and the amplitude of the response is about 10 μV, which is less than the amplitude of the background EEG signal. However, by averaging about 32 responses, the signal can be recorded. The slow evoked response is shown in figure 15.18.

Typical control settings for the equipment and notes on the recording technique are as follows:

Electrodes. Chlorided silver discs are attached to the skin. The positive input is connected to the electrode placed on the vertex of the head; the negative input to the electrode placed over the mastoid bone behind the ear; the ground or reference electrode is placed on the forehead.

Stimulus. The stimulus may be applied as a free-field sound generated by a loudspeaker, or applied through headphones. The sound may be a filtered tone burst with a duration of 300 ms, a rise time of 10 ms and a frequency of 1 kHz. The sound is repeated once every 2 s.

Averager sweep time:	1.0 s.
Averager number of sweeps:	32.

The slow vertex response is reliable and it is possible to obtain a response for stimulus intensities down to only 10 dB above the subjective threshold. It can be used to measure responses for stimuli of different frequencies. The major disadvantage of the technique is that the test may take one and a half hours to perform. Each average of 32 stimuli will take more than 1 min and repeat averages are required to check the consistency of the response at each stimulus intensity level. A co-operative and relaxed patient is necessary otherwise noise generated by electrode movements and EMG signals will obscure the responses. Electrical interference can also obscure signals and, for this reason, the test is usually performed in a screened room.

15.5.2. Auditory brainstem response

Using exactly the same electrode positions as for the vertex recording, another evoked response can be found with a much shorter latency than that shown in figure 15.18. The response occurs within 10 ms of the stimulus and is thought to originate from the brainstem. The latency of the start of the response is consistent with a delay of about 1 ms from the sound source to the ear, 2 ms within the cochlea and a conduction time of a few milliseconds from the cochlea to the brainstem. The amplitude is less than one-tenth that of the slow vertex response so that many more signals need to be averaged in order to extract the signal from the background EEG and the electronic noise.

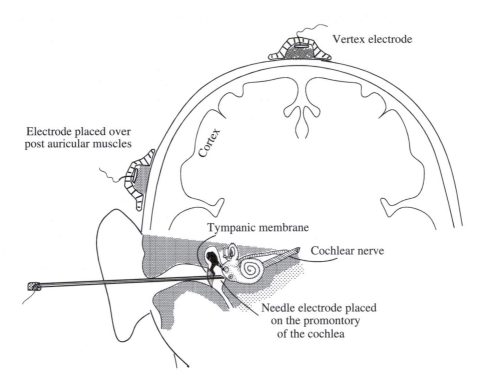

Figure 15.17. *Diagram showing sites for recording evoked responses. Electrodes for vertex, myogenic reflex and cochlear evoked responses are all shown, but only the active electrode is shown in each case.*

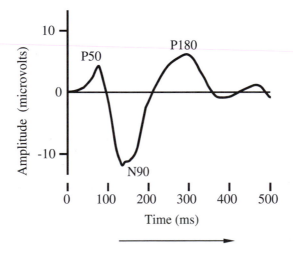

Figure 15.18. *Typical slow vertex response recorded following a stimulus 80 dB above threshold. This response is an average following 32 stimuli.*

Typical control settings for the equipment and details of the recording technique are as follows:

Electrodes. Chlorided silver discs are attached in the same positions as for recording the slow vertex response.

Stimulus. A sound field stimulus from a loudspeaker or headphones may be used. Because the evoked response is very small, great care is necessary in designing the equipment to eliminate direct pick-up of the stimulus by the recording electrodes and leads (see figure 15.19).

The sound may be a short tone burst of approximately 5 ms duration or a click stimulus can be used. No response is observed for tone bursts of frequency less than 1 kHz.

Averager sweep time: 20 ms.
Averager number of sweeps: 2000.

Use of this early brainstem response can be used to record evoked responses to within 10 dB of subjective threshold. The disadvantage of this test, in comparison with the cortical evoked response, is that it only gives the response at a low level in the brain and cannot be used to measure the sensitivity to different frequencies of auditory stimulus. At least half an hour is required for the test and, because the response is small, great care is needed in the attachment of electrodes. In order to reduce background noise and electrical interference a screened room is needed for this test.

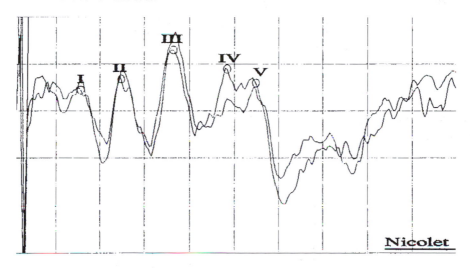

Figure 15.19. *Typical brainstem evoked responses recorded following a click stimulus 80 dB above subjective threshold. The signals are the average responses following 3000 stimuli presented at 11.1 s⁻¹. The time scale is 1 ms per division and the vertical scale 0.12 µV per division. The five marked peaks have latencies of 1.56, 2.48, 3.64, 4.84 and 5.48 ms. (Data kindly provided by Dr J C Stevens, Sheffield.)*

15.5.3. *Myogenic response*

Myogenic means that the signal arises from a muscle. Many animals produce a 'startle' response to a sudden sound such as a handclap. This response is reflex in origin; the sound stimulus causes nerve action potentials to pass to the brainstem which then initiates other nerve action potentials which may cause a muscle twitch. There are tiny superficial muscles behind our ears and these post-auricular muscles twitch slightly in response to a sudden sound; whilst the twitch is not sufficiently strong to produce a noticeable twitch of the ears, it can be recorded electrically as an evoked response.

The post-auricular myogenic (PAM) response can be recorded from electrodes placed behind the ear (see figure 15.17) and the amplitude of the signals may be up to several tens of microvolts.

Typical control settings for the equipment and details of the recording technique are as follows:

Electrodes. Chlorided silver discs are attached to the skin. The active (positive) electrode is placed over the post-auricular muscles at the base of the mastoid bone. The position of the reference (negative) electrode is not critical but is often placed in front of the ear. The ground or common electrode is placed on the forehead.

Stimulus. Again a sound field stimulus from a loudspeaker or headphones may be used. The stimulus parameters are similar to those for the brainstem evoked-response test.

Averager sweep time: 40 ms.
Averager number of sweeps: 100–500.

The PAM response is not a reliable response; in about 20% of normal subjects no response can be obtained. Another disadvantage is that the response threshold can vary by more than 40 dB in normal subjects, from about 10 to 50 dB. It is not widely used.

15.5.4. *Trans-tympanic electrocochleography*

Cochleography is a technique for recording an evoked response from the cochlea and the cochlear nerve. In order to record this response an electrode has to be placed within a few millimetres of the cochlea; a needle electrode can be placed through the tympanic membrane so that its tip lies on the promontory of the cochlea and close to the base of the incus (figure 15.17). This is obviously an invasive measurement technique which may need to be performed under anaesthetic, but no permanent damage is caused to the eardrum. The latency of the response is less than any of the other evoked potentials because the signal originates directly from the cochlea (figure 15.20). For a high-intensity stimulus the response is several microvolts in amplitude and can be seen without averaging but if the smaller response to a lower stimulus level is to be recorded, then a signal averager is necessary.

Typical control settings for the equipment and details of the recording technique are as follows:

Electrodes. A fine stainless steel needle electrode is placed through the eardrum; a microscope is used to control this procedure. The shaft of the needle is insulated so that only the tip is exposed and able to record the evoked response. The needle is sterilized before use and the wire attached to the needle is supported by a small frame which surrounds the ear. The negative electrode is a chlorided silver surface electrode which is placed over the mastoid bone. The ground or reference electrode is placed on the forehead.

Stimulus. The frame which supports the needle electrode is also used to support an earphone which produces tone bursts or clicks with the same parameters as are used in the brainstem evoked-response test.

Averager sweep time: 10 ms.
Averager number of sweeps: 200.

When carried out carefully, electrocochleography is a reliable test which enables cochlear function to be measured for sounds of intensity close to subjective threshold. It can also be used to differentiate between conductive and sensory neural hearing loss. The latency and amplitude of the signal change with stimulus level is very different in the two types of hearing loss. The disadvantage of the test is that it is invasive and can only be justified where the information to be obtained is essential to the management of the patient.

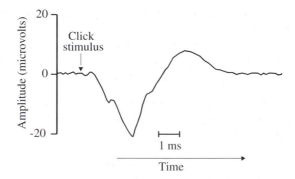

Figure 15.20. *Typical evoked response obtained during an electrocochleography test. The stimulus was presented 256 times at 80 dB above subjective threshold.*

15.6. HEARING AIDS

Hearing aids were first introduced in the 1930s, but were cumbersome devices and produced very poor sound quality. These aids used carbon granule microphones and amplifiers. Carbon granules change their electrical resistance when they are subject to pressure changes and so they can be used to modify an electrical current as sound changes the pressure applied to a diaphragm. Miniature valves superseded the carbon granule amplifiers and piezoelectric microphones replaced the carbon granule microphones. In the 1950s, transistors were introduced and currently most aids use integrated circuit amplifiers. Future aids are likely to be all-digital devices. Piezoelectric microphones are still used, although ceramic materials are used as the piezoelectric element. A piezoelectric material has a crystal structure such that, when pressure is applied, shared electric charges are redistributed and so a potential is produced across the material. The diaphragm of a ceramic microphone is directly connected to the piezoelectric ceramic so that movement of the diaphragm gives a proportional potential difference across the material.

The need for a hearing aid is usually assessed on the basis of hearing tests, and table 15.2 gives some basic information on the classification of auditory handicap.

15.6.1. *Microphones and receivers*

Microphones and receivers are both transducers; the first converting from sound to electrical energy and the second vice versa. The 'receiver' is rather inappropriately named because it is the earpiece which actually produces the amplified sound, although it does allow the patient to receive the sound. In current hearing aids the microphone and the receiver are the largest components, with the exception of the battery. Ceramic microphones are the most commonly used but magnetic types are also in use; the magnetic type consists of a diaphragm connected to a ferromagnetic armature which is within the magnetic field produced by a coil. Movement of the diaphragm causes the armature to move and thus induces a potential in the coil. Most receivers are also magnetic types that use the current through the coil to move a metal core attached to a diaphragm.

The coupling between the receiver and the ear canal is very important as it modifies the frequency response of the aid. Old aids of the body-worn type had a separate receiver placed directly over the ear canal but aids which are worn behind the ear contain both the microphone and receiver so that the sound has to be conducted to the ear canal through a short plastic tube. Many aids are now placed within the ear and the receiver makes acoustic connection very close to the eardrum.

Table 15.2. *Some of the problems which are associated with different levels of hearing loss.*

Average hearing loss	Speech understanding	Psychological implications	Need for a hearing aid
25 dB HL	Slight handicap; difficulty only with faint speech	Child may show a slight verbal deficit	Occasional use
35 dB HL	Mild handicap; frequent difficulty with normal speech	Child may be educationally retarded. Social problems begin in adults	Common need for hearing aid
50 dB HL	Marked handicap; difficulty even with loud speech	Emotional, social and educational problems more pronounced	The area of most satisfaction from a hearing aid
65 dB HL	Severe handicap; may understand shouted speech but other clues needed	Pronounced educational retardation in children. Considerable social problems	Hearing aids are of benefit—the extent depends upon many factors
85 dB	Extreme handicap; usually no understanding of speech	Pronounced educational retardation in children. Considerable social problems	Lip reading and voice quality may be helped by hearing aid

15.6.2.   *Electronics and signal processing*

The three most important factors in the specification of the performance of a hearing aid are: gain, frequency response and maximum output.

Gain can be varied by using the volume control and, in many aids, a range of 0–60 dB (1–1000) is provided. The maximum possible gain is usually limited by acoustic feedback from the receiver to the microphone which will cause a howl or oscillations if the gain is increased too far.

Frequency response should, ideally, cover the whole audio bandwidth but in practice, the performance of the receiver and the microphone limit the bandwidth. A typical frequency response for an aid is shown in figure 15.21.

Maximum output may be the most important part of the specification. A normal person might hear sounds with intensities ranging from 0 to about 90 dB. If these sounds are to be amplified by 60 dB then the range of intensities to be produced by the aid should be 60–150 dB. It is very difficult to produce sound at a level of 150 dB without distortion; many aids will only produce about 110 dB and the very best aids 140 dB. However, in most cases the maximum output required is much less than 150 dB.

The maximum output is limited both by the receiver and the electrical power available. The power supply for a hearing aid has to be provided from a battery. The current consumption in the quiescent state, i.e. no sound, may be about 100 μA and in a noisy environment 5 mA. If the battery is to last for more than a week of normal use then the battery capacity must be greater than about 60 mA h, e.g. for 100 h usage made up of 10 h at 5 mA and 90 h at 100 μA would require 59 mA h.

The trend in hearing aid design is to make them more versatile and adaptable to the hearing loss of a particular patient. Several computer programmes are available which will select a hearing aid appropriate to patient audiometric characteristics. The enormous recent advances in electronics have enabled many improvements to be made in hearing aid design. Not only can aids be made smaller but also sophisticated

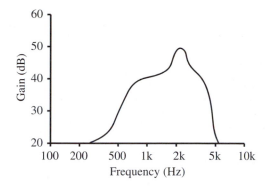

Figure 15.21. *Representative frequency response of a hearing aid.*

signal processing can be included. Digitized hearing aids have been developed to maximize high-frequency gain, filter out non-speech signals and exactly mirror a particular hearing loss. One way to enable a wide range of input sounds to be accommodated with a more limited output range is to use sound compression. Sound compression increases the range of sounds which can be amplified. Complex frequency responses can be provided and matched to the hearing loss of a particular patient. This field is changing so rapidly that the reader is advised to study manufacturers' literature as well as more detailed texts.

15.6.3. *Types of aids*

The range of hearing aid types is very wide but they can be classified according to where they are worn. Five major categories are:

1   Body worn.
2   Behind the ear (BTE).
3   In the ear (ITE).
4   In the canal (ITC).
5   Completely in the canal (CIC)

Body-worn aids can be relatively large which enables high-quality components and large batteries to be used. For these reasons the body-worn aid usually gives the largest 'maximum output' and the best sound quality. However, behind-the-ear and within-the-ear aids are usually more acceptable than the body-worn aids because they are more convenient and more socially acceptable. Body-worn aids are now rarely used, except where very high output is required.

The BTE aid hangs behind the ear with a tube connecting to the canal via an ear mould. The relatively large size allows complex circuitry to be included, but they are more visible than an ITE aid. The fact that the microphone is outside the pinna means that no natural ear resonance effects are obtained.

The ITE aids are popular and are often custom-made to fit the ear. Quite high gain is possible without feedback if the aid fits well into the ear. The aids are quite visible, however, and the microphone can be subject to wind noise. The maximum output is lower than that of a BTE aid.

The ITC aids sit in the concha portion of the ear and extend down into the ear canal. They have the advantage of being less visible than other aids, a reduced effect of wind noise and the ability to use natural resonances within the ear. Disadvantages include short battery life and an increased chance of feedback at higher gains.

CIC aids sit entirely within the canal and are almost completely invisible to casual view. However, the constructional demands are high and performance may be limited. Battery life is short and control of the device difficult. Some ear canals are not sufficiently large to accommodate them.

The total performance of a hearing aid is determined by the microphone characteristics, amplifier characteristics, receiver/ear mould characteristics and the way in which these elements might interact. The ear mould is the plastic plug which is made to fit a particular ear and the sound-conducting tube which connects the aid to the mould. The acoustic properties of the plastic mould are relatively unimportant but it must make a tight seal to the walls of the ear canal. An analysis of the physics involved in the performance of the coupling is difficult and not appropriate to this introductory text.

15.6.4.   Cochlear implants

Cochlear implants convert sound signals into electrical currents which are used to stimulate auditory nerve cells via electrodes placed within the cochlea. They are fundamentally different from normal acoustic hearing aids in that they replace the function of the hair cells within the inner ear. Implants are only used in patients who are profoundly deaf, whose hearing loss exceeds 105 dB. This is a small patient group representing less than 1% of the population.

These devices sample the sound and then use a processed version to electrically stimulate the auditory nerve within the cochlea. The microphone and signal processing are carried out within a body-worn unit. The output is then inductively coupled to the implant via a transmitter coil which is often held in place over the mastoid by magnets within both the transmitter coil and the implant. Implants allow for the perception of sound when the middle and inner ear are damaged but the auditory nerve is intact. Cochlea implantation is expensive and hence there is rigorous selection of patients in order to maximize the success of an implantation. The selection takes into account motivation, expectations, emotional state, age, intelligence and general fitness, as well as the expected benefit to hearing. Alessandro Volta in about 1800 was the first person to apply an electrical stimulus to the ear. Systematic research into electrical stimulation of the auditory nerve was carried out in the 1960s and 1970s, but self-contained implants were first developed in the 1980s both in Melbourne, Australia and in the USA.

Principles of operation

There are about 50 000 fibres in the auditory nerve and each of them is normally sensitive to a narrow range of frequencies over a dynamic range of intensity of about 30–40 dB. The intensity of sound is coded by the rate at which cells fire and the number of cells which are excited. Sounds of different frequencies are attenuated by different amounts as they are transmitted into the cochlea. It appears that the brain uses the spatial distribution of cells stimulated within the cochlea to determine the frequency of a sound and the temporal distribution of cell impulses to determine the intensity of a sound. This last sentence is oversimplified but it is the basis upon which cochlea implants have been designed.

In a cochlear implant either a single electrode or an array of electrodes are usually implanted, via a hole drilled close to the round window, inside the scala tympani, which places the electrodes close to the auditory nerve cells. Figure 15.22 shows an array of electrodes that can be placed within the cochlea. The electrodes are typically made from platinum–iridium and up to 22 electrodes are used. In some cases the electrodes are in pairs (bipolar) and an electrical stimulus is applied between the two, but in other cases the stimulus is applied between one electrode (the cathode) placed close to the nerve cells and an indifferent electrode placed further away. The principles of neural electrical stimulation are covered in sections 10.2 and 16.5. It is very important that tissue damage cannot occur as a result of the electrical stimuli applied via the electrodes. It is generally accepted that the charge density per pulse should be below 0.2 μC mm^{-2} to minimize the risk of long-term damage.

If a single pulse has an amplitude of 0.5 mA and a duration of 100 μs then the charge per pulse is 0.05 μC. If the electrode is a square of sides 0.5 mm then the charge density per pulse will be 0.2 μC mm^{-2}.

In order to limit the current which can flow a constant current as distinct from a constant voltage stimulus is usually used. The constant current stimulus also has the advantage that the threshold to stimulation does not depend upon electrode impedance and so will not change with time as the electrode interface changes. In addition to limiting the charge which can be delivered it is important to avoid electrode polarization and the release of toxic products from electrochemical reactions by preventing any DC current flow. This is usually done by using a charge-balanced bipolar stimulus with both positive and negative phases.

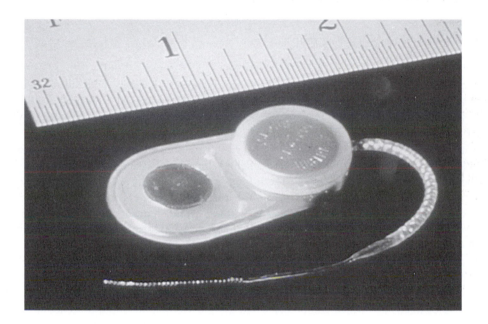

Figure 15.22. *A cochlear implant. The array of electrodes which are passed through the round window and placed within the scala tympani can be seen. The scale shown is in inches. (Courtesy of Cochlear Ltd.)*

Sound coding

Whilst most implants make some attempt to mimic normal cochlear function there is nonetheless a very wide range of strategies. Systems are generally multi-channel, and processing techniques range from mapping the frequency of the stimulus to electrode location in the cochlea to more complex coding strategies. These might code the stimuli as pulses, square waves or amplitude-modulated high-frequency carriers. All the implants seek to include the frequency information contained in human speech which is mainly between 300 and 3000 Hz. The square wave and pulse types are usually designed to track voice pitch or first formant. The carrier types amplitude modulate a carrier at about 15 kHz with a filtered audio signal.

Cochlear implants are still developing rapidly and a standard method of sound coding and impulse delivery has not yet emerged.

Performance and future

For adults, cochlear implants provide their users with significant benefits even though they do not replace the function of a normal ear. They help the wearer to recognize environmental sounds such as bells, calls and knocks at the door. They can also help improve lip-reading and the user's speech is likely to be more intelligible as they can monitor their own voice. For children, the pre- as well as postlingually deaf are now considered for implantation. Many children have learnt language through the use of an implant.

Patients are carefully selected and an implant programme includes a long period of training, learning and assessment after the implant procedure.

15.6.5.   *Sensory substitution aids*

Cochlear implants do not enhance what remains of an impaired sense of hearing. They substitute one sensation for another in that sound is used to electrically stimulate the auditory nerve. Nonetheless, the sensation is perceived as coming from the ears. Many attempts have been made to use senses other than hearing to carry sound information. Research groups have tried converting a sound into a visual stimulus, an electrical stimulus or a vibratory stimulus. Some of these sensory substitution aids have reached the point of clinical application as part of the range of aids which might help the profoundly deaf.

The best developed of these sensory substitution aids is the vibrotactile aid. These devices convert sound into a modulated vibration which can then be used to vibrate a finger or elsewhere on the body. Such aids can certainly be used to give the wearer some awareness of environmental sounds. However, the information carrying capacity of such an aid is very limited and is almost certainly insufficient to convey normal speech. The range of frequencies which can be usefully used in a tactile aid is between 20 and 500 Hz as below 20 Hz individual cycles are perceived and above 500 Hz sensitivity falls rapidly. The percentage change in frequency which can be identified is about 25% which compares poorly with the ear where a change of about 0.3% can be heard.

The range of intensities which can be perceived using a vibrotactile stimulus is about 55 dB and the smallest detectable percentage change is about 2 dB. The 55 dB range is very much less than the range of the ear but the detectable change in intensity is similar to that of the ear. Temporal resolution is often measured as the ability to detect gaps or periods of silence and values of about 10 ms are quoted for vibrotactile input.

From this evidence it is clear that the tactile sense can be used for communication but it will have a much smaller information transfer capacity than the auditory system.

Information transfer rate can be defined as follows:

$$\text{Information transfer rate} = \text{the number of binary bits of information which can be communicated each second.}$$

The information transfer rate required for speech transmission is usually considered to be about 20 bits s^{-1}. The number of bits which can be transmitted using a single-channel vibrotactile aid is at best about 4 bits s^{-1}. It is possible that by using several channels applied to the fingers for example that 20 bits s^{-1} might be communicated but this has not yet been proven.

Vibrotactile sensory substitution aids have been compared with cochlear implants (Stevens 1996) and it has been concluded that a tactile aid may offer advantages over a cochlear implant for certain categories of patient. In particular, those who are prelingually deaf may obtain limited benefit from a cochlear implant.

15.7.   PRACTICAL EXPERIMENT

15.7.1.   *Pure-tone audiometry used to show temporary hearing threshold shifts*

Objectives

- To obtain practice in recording a pure-tone audiogram.
- To observe the effect on hearing threshold of exposure to high-intensity sounds.

Equipment

A pure-tone audiometer with headphones, facilities for narrow-band noise masking and an attenuation in 2 dB steps.

Method

You may use either yourself or a volunteer as the subject for this experiment. The experiment involves recording a pure-tone audiogram, then applying narrow-band noise to an ear at 90 dB and then repeating the pure-tone audiogram. Sections 15.2.5 and 15.3.1 should be read before carrying out this experiment.

1   Use the procedure described in section 15.3.1 to obtain a pure-tone audiogram, taking threshold measurements over the range 500 Hz to 8 kHz. Test both ears.
2   Now apply a narrow-band masking sound at 90 dB to the right ear, with a centre frequency of 1 kHz. Make sure that the sound is not increased above 90 dB for long periods whilst adjustments are made. Apply this sound for a period of 10 min.
3   Immediately following step 2 repeat step 1, testing first the right ear and then the left.
4   Wait for a further period of 20 min and then repeat step 1 yet again.

Results and conclusions

Plot the three pure-tone audiograms on separate graph sheets.

Were there any changes in the hearing thresholds after exposure to the 90 dB noise? Did changes occur at some particular frequencies? (It may help if you compare the average threshold change for 500 Hz, 1 kHz and 2 kHz with that for 4, 6 and 8 kHz.) Did the hearing thresholds return to their original values within 20 min of exposure? Would you anticipate any long-term effects of this exposure to noise at 90 dB?

15.8.   PROBLEMS

15.8.1.   *Short questions*

a   Where, in the ear, is sound converted into neural signals?
b   What part do the semi-circular canals play in hearing?
c   Do low or high frequencies penetrate furthest into the cochlea?
d   How long would sound take to get from a loudspeaker to your ear 3 m away?
e   Which sounds are diffracted best round corners?
f   Why is the dBA scale for measuring sound levels nonlinear with frequency?
g   At about what frequency is the ear most sensitive?
h   What is distraction testing?
i   Why is 'masking' used in hearing tests?
j   Should a hearing loss of 10 dB be a cause for concern?

k   Would fluid in the ear increase or decrease the acoustic impedance of the ear?
l   Do your ears produce sounds?
m   What is a typical latency for a stimulated oto-acoustic emission?
n   What is presbycusis?
o   What is meant by conductive hearing loss?
p   About how many fibres are there in the auditory nerves?
q   About how large is the slow vertex response to a high-level sound stimulus?
r   Could you average 1024 slow vertex responses in 32 s?
s   How close to subjective threshold can an auditory brainstem response be obtained?
t   What is the receiver in a hearing aid?
u   What factor limits the maximum possible gain of a hearing aid?
v   Where are the electrodes placed in a cochlea implant?
w   About how many binary bits per second are required to convey a speech signal?
x   How can the electrocochleogram be used to measure the nature of hearing loss?
y   The amplitude of the auditory brainstem reponse is approximately 1 μV and the background electroen-
       cephalogram is about 10 μV. Explain how you might predict the number of averages required to obtain
       a reasonable signal-to-noise ratio.
z   What are the advantages in terms of measuring hearing of using the cortical evoked response as opposed
       to the auditory brainstem response?

15.8.2.   Longer questions (answers are given to some of the questions)

Question 15.8.2.1

A person listening to music produced by a pair of headphones is causing complaints from a student who is
trying to work nearby. Estimate the SPL in dB at a distance of 1.6 m from the headphones. Assume that a
SPL of 100 dB at 1 kHz is produced by the headphones to the listener and that the headphones themselves
have an equivalent absorption thickness of 15 mm with an absorption of 20 dB cm^{-1}. The headphones are
approximately hemispherical with a radius of 5 cm. State any further assumptions which you need to make
in arriving at your answer. Is the student's complaint justified?

Answer

The headphones themselves will give an absorption of $1.5 \times 20 = 30$ dB. Therefore the SPL close to the
headphones will be $100 - 30 = 70$ dB.
       If we assume that the headphones radiate omnidirectionally then the SPL will fall off as an inverse square
law, i.e. at 6 dB for each doubling of distance. At the surface of the headphones the equivalent distance is 5 cm
if we assume that the whole surface of the headphones radiates. The SPL at 160 cm will be 5×6 dB $= 30$ dB
less than at the surface, i.e. 40 dB. A sound level of 40 dB is low so the complaint is probably not justified
although even a quiet sound can be distracting.

Question 15.8.2.2

An alarm produces a sound with most energy at a frequency of about 3 kHz. The SPL is measured as 93 dB on
a linear scale at a distance of 0.5 m from the alarm. Estimate the SPL in dBA at 1 m using the data provided
in figure 15.7. Also estimate the maximum distance at which the alarm might be heard. List any assumptions
you have to make.

Answer

The linear and dBA scales will be the same at 1 kHz, but at 3 kHz, using the curve of figure 15.7, the dBA reading will be about 3 dB more, i.e. 96 dBA at 0.5 m.

If the alarm is omnidirectional then we can assume the SPL falls off as an inverse square law at 6 dB for each doubling of distance. The SPL at 1 m will be 96–90, i.e. 90 dBA. To fall off to the threshold of hearing the distance will have to double $90/6 = 15$ times. $2^{15} = 32\,768$ so that the alarm might be heard at a distance of 32 km. In practice this will be an overestimate as we have assumed that no absorption of the sound takes place and that no obstacles fall between the alarm and the listener.

Question 15.8.2.3

Explain the concept of 'equal loudness contours' to a novice.

Question 15.8.2.4

Write a brief description of the electrocochleogram, the auditory brainstem response and the slow vertex response as recorded from the human subject. Explain as far as possible the reasons for the amplitude, latency and shape of the waveforms obtained. Explain the choices to be made of electrode position, filters and averaging required.

Question 15.8.2.5

A patient's pure-tone audiogram shows a difference between the air and bone conduction thresholds of 40 dB and this is independent of frequency.

Explain the meaning of the above statement and say what you might infer about the likely cause of the hearing loss. What other measurements might help to clarify the situation?

Answer

The threshold to hearing has been measured by two separate means, one by air conduction and the other by bone conduction. In a normal ear the two thresholds should be the same. The fact that there is a difference of 40 dB means that there is a significant loss. The bone conduction method bypasses the middle ear, whereas the air conduction method is affected by transmission through the middle ear. There is almost certainly a conductive hearing loss in this patient. A further investigation might be a test to measure the middle-ear impedance, as this would assess the function of the eardrum and middle ear.

Answers to short questions

a   Sound in converted to neural signals in the inner ear or cochlea.
b   None. The semi-circular canals are concerned with balance.
c   Low frequencies penetrate further than high into the cochlea.
d   Sound would take about 9 ms to travel 3 m.
e   Low-frequency sounds get diffracted around corners.
f   The dBA scale is nonlinear with frequency in order to compensate for the nonlinear response of the ear.
g   The ear is most sensitive in the range 0.5–5 kHz.
h   Distraction testing is often used to test hearing in children. It involves assessing what level of sound will distract a child from a task or from play.
i   Masking is used to prevent sound presented to one ear being heard by the other.

j   No, a hearing loss of 10 dB is within the measurement error of a hearing test.

k   Fluid in the ear would increase the acoustic impedance of the ear.

l   Yes, ears actually produce sounds—acoustic emissions.

m   5–10 ms is a typical latency for a stimulated oto-acoustic emission.

n   Presbycusis is the process of hearing deterioration with age.

o   Conductive hearing loss is caused by defects in the middle ear that conducts sound to the inner ear.

p   There may be about 50 000 fibres in the auditory nerves.

q   The slow vertex response may be 10 μV in response to a large sound.

r   No, the slow vertex response may last for 1 s so 1024 responses would require about 15 min to record.

s   An auditory brainstem response can be recorded within about 10 dB of subjective threshold.

t   The receiver is the sound generator in a hearing aid.

u   Feedback from receiver to microphone and the maximum output level usually limit the maximum gain that can be obtained from a hearing aid.

v   The electrodes are placed in the cochlea in a cochlear implant.

w   About 20 bits s^{-1} are required to convey a speech signal.

x   The change in latency and amplitude of the electrocochleogram with stimulus level can be used to differentiate between conductive and sensory neural hearing loss.

y   On the assumption that the EEG is a random signal the noise reduction on averaging will be proportional to $\sqrt{n}$, where n is the number of averages. For a signal-to-noise ratio of 5 an improvement of 50 is required and hence an average of 2500 responses will be required.

z   The cortical evoked potential allows hearing thresholds to be measured at different frequencies. It also allows hearing to be measured at a higher point in the auditory pathway than does the brainstem response.

BIBLIOGRAPHY

Cooper H (ed) 1991 *Cochlear Implants: a Practical Guide* (London: Whurr)

Flanagan J L 1972 *Speech Analysis: Synthesis and Perception* (Berlin: Springer)

Hassall J R and Zaveri K 1979 *Acoustic Noise Measurement* (Denmark: Bruel and Kjaer)

Haughton P M 1980 *Physical Principles of Audiology* (Bristol: Hilger)

International Standards Organisation 1993 *ISO 389-7* Acoustics—reference zero for the calibration of audiometric equipment—reference threshold of hearing under free-field and diffuse-field listening conditions

International Standards Organisation 1992 *ISO 8253-2* Thresholds of hearing using sound field audiometry with pure-tone and narrow-band test signals

Katz J (ed) 1985 *Handbook of Clinical Audiology* 3rd edn (Baltimore, MD: Williams Wilkins)

Keidel W D 1979 Neurophysiological requirements for implanted cochlear prostheses *Acta Oto-Laryngol.* **87** 163

Kemp D T 1978 Stimulated acoustic emissions from within the human auditory system *J. Acoust. Soc. Am.* **64** 1386–139

Pickles J O 1988 *An Introduction to the Physiology of Hearing* 2nd edn (New York: Academic)

Recommendations for masking in pure-tone audiometry 1986 *Br. J. Audiol.* **20** 307–14

Schindler R A and Merzenich M M (eds) 1985 *Cochlear Implants* (New York: Raven)

Simmons F B, Mongeon C J, Lewis W R and Huntington D A 1964 Electrical stimulation of the acoustic nerve and inferior colliculus in man *Arch. Oto-Laryngol.* **79** 559

Stevens J C, Webb H D, Hutchinson J, Connell J, Smith M F and Buffin J T 1991 Evaluation of click-evoked oto-acoustic emissions in the newborn *Brit. J. Audiol.* **25** 11–14

Stevens J C 1996 Tactile aids *Scott–Brown's Otolaryngology: Adult Audiology* (Butterworth-Heinemann)

Summers I (ed) 1992 *Tactile Aids for the Hearing Impaired* (London: Whurr)

CHAPTER 16

ELECTROPHYSIOLOGY

16.1. INTRODUCTION AND OBJECTIVES: SOURCES OF BIOLOGICAL POTENTIALS

Neurology is the branch of medicine dealing with all aspects of the nervous system. At least two types of electrophysiological measurement are usually made in a department of neurology; these are electroencephalography (EEG) and electromyography (EMG). EEG measurements can help in the diagnosis of epilepsy and are also useful in the investigation of brain tumours and accidental damage to the brain. EMG is usually taken to include both the recording of electrical signals from muscle and also the measurement of neural function using techniques such as nerve conduction measurement. EMG measurement is used in the diagnosis of muscle disease such as muscular dystrophy and also in the investigation of nerve damage which has resulted either from disease or physical injury.

Cardiology deals with all aspects of cardiac function. The heart is a single muscle which produces an electrical signal when the muscle depolarizes. The measurement of this electrical signal (the electrocardiograph) is one of the oldest medical techniques. The ECG/EKG was first recorded by Einthoven in the Netherlands and Waller in England in 1895.

In this chapter the origin of some bioelectric signals and the methods of recording these electrical signals from the body will be described. The origin of almost every electrical potential which arises within the body is a semi-permeable membrane. A single nerve consists of a cylindrical semi-permeable membrane surrounding an electrically conducting centre or axon. The membrane is called semi-permeable because it is partially permeable to ions such as potassium (K^+) and sodium (Na^+), which can pass more freely in one direction through the membrane than the other. The result of these electrical properties of the membrane is that a potential of approximately 0.1 V is generated across the membrane. Changes in potentials of this type are the origin of signals such as the EEG, EMG and ECG/EKG.

The objectives of this chapter are to address questions such as:

- Why can an ECG/EKG be recorded from the limbs?
- What factors affect the speed at which nerves conduct impulses?
- What determines the shape of bioelectric signals?
- What determines the frequency content of a bioelectric signal?

When you have finished this chapter, you should be aware of:

- The organization of the nervous system.
- How information is transferred to and from the brain.
- The origin of bioelectric signals.
- The essentials of EEG recording.
- The essentials of EMG recording.

There is very little mathematics in this chapter which should therefore be accessible to all of our intended readers. It does not assume much prior knowledge, but this chapter does follow logically from Chapters 8–10.

16.1.1. The nervous system

It is not appropriate to give a detailed introduction to the anatomy and physiology of the nervous system in this text. There are many introductory texts, written primarily for medical and nursing staff, that are adequate for students of medical engineering and physiological measurement. This and the following three sections explain the organization of the nervous system, and also introduce some of the terminology which is needed for an understanding of electroencephalography and electromyography.

The brain, nerves and muscles are the major components of the nervous system. The brain is supplied with information along sensory or afferent nerves which are affected by sensations such as heat, touch and pain. On the basis of the information received, the brain can make decisions and pass instructions down the motor or efferent nerves to produce an effect by causing muscles to contract (figure 16.1). The human nervous system is a highly parallel system in that there are many nerves in a single nerve trunk. For example, a mixed peripheral nerve might contain 30 000 nerve fibres.

The junctions between nerves are called *synapses* and communication across these junctions is by diffusion of the chemical acetylcholine. The chemical transmitter involved in smooth muscle activity is noradrenaline. It may seem surprising that a chemical transmitter can cause such rapid action but the gaps involved at synapses are very narrow. If a gap of 1 μm is to be traversed in 1 ms then the speed of diffusion required is only 1 mm s^{-1}.

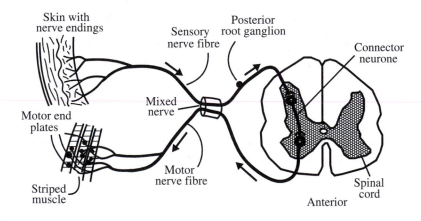

Figure 16.1. *The organization of the nerve pathways from the spinal cord to the periphery.*

The synapses allow reflex loops to arise via the spinal column. H reflexes arise by feedback from the sensory to the motor pathways. Impulses from the brain can inhibit reflex loops within the spinal cord. Motor control also involves feedback from the proprioceptive fibres which sense the changes in length of muscle fibres. The spinal cord contains 31 spinal nerve trunks (eight cervical, 12 thoracic, five lumbar, five sacral, one coccyx and 12 cranial). The posterior roots are sensory and the anterior ones motor. The motor and sensory fibres are not separated in more peripheral nerve trunks.

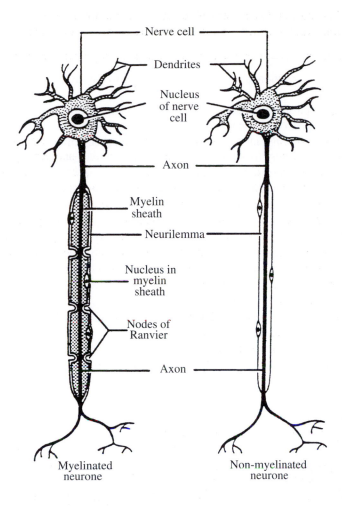

Figure 16.2. *Two neurones. The myelinated neurone will conduct impulses much more rapidly than the non-myelinated neurone. Information interchange is through the dendrites. Action potentials transmitted along an axon cause a muscle fibre to twitch or carry sensory information centrally.*

The basic component of both the brain and nerves is the neurone. There are many forms of neurone but all consist of a cell body, dendrites which radiate from the cell body rather like the tentacles of an octopus, and an axon which is a long cylindrical structure arising from the cell body. Figure 16.2 gives a diagram of two such neurones. It is simplest to consider the dendrites as the means of information input to the cell, and the axon as the channel for the output of information. The axon allows a cell to operate over a long distance, whereas the dendrites enable short-distance interactions with other cells. The cell body of the neurone may be within the brain or within the spinal cord and the nerve axon might supply a muscle or pass impulses up to the brain. The brain itself is a collection of neurones which can interact electrically via the dendrites and axons and so functions in a similar manner to an electronic circuit.

A neurone is enclosed by a membrane which has strange properties in terms of ionic permeability such that the inside of the neurone is normally about 50–100 mV negative with respect to the outside. An impulse is conducted along a nerve axon by means of a propagating wavefront of depolarization followed by repolarization. During depolarization the inside of the neurone becomes slightly positive with respect to the outside. This process of depolarization is caused by Na^+ passing to the inside of the neurone and K^+ passing to the outside of the neurone.

Nerves can conduct impulses in either the normal (orthodromic) or the opposite (antidromic) direction. The impulses are typically 1 ms in duration and the frequency of impulses in a single fibre may vary from zero up to about 200 pps. Neurones are not all the same size, but a typical cell body has a diameter of 100 μm and the axon may be up to 1 m long with a diameter of 15 μm. A large nerve trunk will contain many nerve fibres which are axons. The ulnar nerve runs down the arm and is very superficial at the elbow where it may be knocked and cause a characteristic feeling of pins and needles; this nerve trunk looks rather like a thick piece of string and contains about 20 000 fibres in an overall diameter of a few millimetres. The optic nerve within the head contains even more fibres and the brain which it supplies is estimated to contain approximately 10^9 neurones.

16.1.2.   *Neural communication*

The body is completely controlled by electronic impulses, and the electrical signals which the brain, nerves and muscles generate are not the result of activity but the cause of it. If we make an analogy between the brain and a computer then we have to consider the brain as a digital and not an analogue computer. The signals which travel down nerves are pulses of electricity whose repetition frequency changes, but whose amplitude is constant. If we wish to inform the brain that a more intense pain has been received then it is not the amplitude of the electrical pulses which changes but their frequency.

Coding of sensory information along nerves is by frequency modulation (FM) of the train of pulses carried by the nerve; the more intense the sensation the higher the frequency of nerve impulses. The normal frequency of impulses passing along a single sensory or afferent nerve fibre may be 1 pulse per second (pps), but if the pressure sensor which supplies the nerve senses a high pressure applied to the skin, the frequency of impulses will be increased to perhaps 100 pps. The relation between intensity of sensation and the frequency of nerve impulses is approximately logarithmic (figure 16.3). It is interesting to speculate as to why a logarithmic relation has evolved. The relation between sensory input $s(t)$ and frequency of nerve impulses $f(t)$ could be a linear one. However, we need to consider the nature of the sensory inputs. The sounds that we can hear have a dynamic range of at least $10^6/1$, i.e. 120 dB. The eye is sensitive to a similarly wide range of intensities. If $f(t) = k s(t)$ and the maximum frequency of impulses which a nerve can handle is 100 pps, then the minimum sensory input will correspond to a frequency of 10^{-4} pps. This is clearly impractical, as the interval between the nerve pulses is so long (almost 3 h) that the sensory input would have changed before we perceived the change.

By evolving a logarithmic relation such as

$$f(t) = a \log(s(t)) + b$$

a wide dynamic range can be handled.

If the constants a and b are 20 and 5, respectively then for an input $s(t)$ which ranges from 1 to 10^6 the output $f(t)$ will change from 5–125 pps. We have compressed a dynamic range of $10^6/1$ down to one of 25/1.

There is a price to pay for the logarithmic compression which is the inability to distinguish a large number of amplitude steps. If the number of discriminable steps in intensity of stimulus between threshold and pain is say 120, then there are only 20 per decade available if the total dynamic range is of six decades.

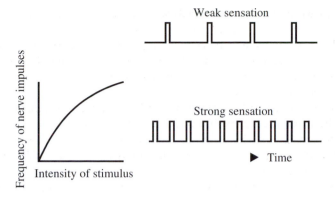

Figure 16.3. *An increase in intensity of stimulus causes an increase in the frequency of nerve impulses. These pulses are shown on the right and the relation between intensity of sensation and pulse frequency on the left.*

20 steps per decade means that each step is given by

$$(x)^{20} = 10 \quad \text{i.e.} \quad 20 \log(x) = 1 \quad \text{and} \quad x = 10^{0.05} = 1.122.$$

The intensity needs to change by 12.2% in order to be discriminated.

Superficially it seems unlikely that the control of our muscles is digital because we make smooth graded actions rather than the twitches which would result from a digital system. However, the system is digital and to increase the contraction of a muscle, it is the frequency of impulses travelling down the efferent nerve which is increased. A smooth contraction is obtained because a muscle consists of many muscle fibres which do not twitch simultaneously, with the result that the integrated effect of the twitches is a smooth contraction. This concept is illustrated in figure 16.4 and explained further in section 16.1.5.

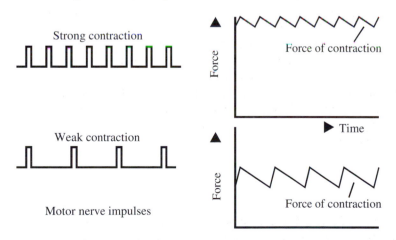

Figure 16.4. *The force of contraction of a single motor unit (group of muscle fibres supplied by one nerve fibre) is shown for different frequencies of nerve impulse.*

16.1.3. The interface between ionic conductors: Nernst equation

Before we consider nerve potentials in more detail we will look at what happens when two ionic conductors are separated by a membrane. Consider the situation shown in figure 16.5, where a volume of saline is enclosed within a thin membrane and placed in a volume of de-ionized water. The saline will contain both Cl^- (anions) and Na^+ (cations) ions and these will be able to diffuse through the membrane which is semi-permeable. We will assume that the membrane is impervious to the Na^+ ions. If we consider just the Cl^- ions then as they pass through the membrane they will produce a negative potential outside the membrane. As this potential increases Cl^- ions will be repelled and so eventually a point will be reached at which equilibrium is established, with the electrostatic force of repulsion balanced by the force produced by the diffusion gradient across the membrane.

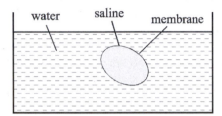

Figure 16.5. *A volume of saline contained within a semi-permeable membrane and suspended in a volume of de-ionized water.*

The *Nernst equation* relates the transmembrane potential V, measured with respect to the outside of the membrane, to the concentration of cations as follows:

$$V = \frac{RT}{Z_p F} \log_e \frac{C_o}{C_i} \cong \frac{58}{Z_p} \log_{10} \frac{C_o}{C_i} (mV) \tag{16.1}$$

where C_i and C_o are the concentrations of the Cl^- ions inside and outside the membrane bag, respectively, and Z_p is the valence (the number of electrons added or removed to ionize the atom) of the particular cation considered. R is the gas constant, F the Faraday and T the absolute temperature. At room temperature the second part of equation (16.1) gives the approximate transmembrane potential in millivolts. Typical values for the concentrations of the three major ions found in neural tissue are given in table 16.1. We can use these to calculate the equivalent Nernst potentials by using equation (16.1). It is not simple to determine the transmembrane potential when several ions are present. In the resting state nerve membranes are almost impervious to Na^+ ions so that these do not contribute to the transmembrane potential, but this is not the case during the generation of a nerve action potential when the Na^+ ions do make a contribution. However, it can be seen from table 16.1 that we can normally expect the inside of a nerve in the resting state to be negative on the inside with respect to the outside.

This process is important to neural function as it explains the potential generated between the inside and outside of a neurone, but it is also important in understanding the generation of a potential across an electrode where ions and electrons may diffuse across the junction.

Table 16.1. *Typical ionic concentrations between the inside and outside of a large nerve axon.*

Ion	Intracellular concentration (mM)	Extracellular concentration (mM)	Nernst potential inside wrt outside
K^+	400	20	-75 mV
Na^+	50	450	$+55$ mV
Cl^-	40	550	-66 mV

16.1.4. *Membranes and nerve conduction*

The nerve impulses referred to in section 16.1.2 have an amplitude of approximately 0.1 V and a duration of 1 ms. Their amplitude is measured between the inside and the outside of the nerve fibre and the impulses can travel along the nerve fibre at a speed of about 50 m s^{-1}.

A single nerve fibre consists of a cylindrical semi-permeable membrane which surrounds the axon of a neurone. The properties of the membrane normally give rise to a high potassium ion concentration and low sodium ion concentration inside the nerve fibre, which results in a potential of about -80 mV between the inside and the outside of the fibre. The nerve is said to be polarized. The membrane potential in the polarized state is called the resting potential, which is maintained until some kind of disturbance upsets the equilibrium. Measurement of the resting potential is made with respect to the potential of the surrounding extracellular body fluids as shown in figure 16.6.

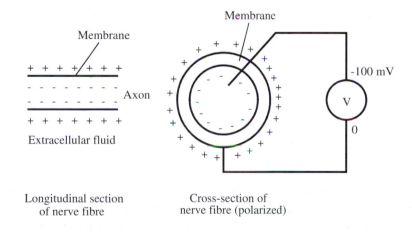

Figure 16.6. *A resting neurone.*

When a section of the nerve membrane is excited, either by the flow of ionic current or by an externally supplied stimulus, the membrane characteristics change and begin to allow sodium ions to enter and potassium ions to leave the nerve axon. This causes the transmembrane potential to change which, in turn, causes further changes in the properties of the membrane. We can make an electrical analogy by saying that the membrane resistance depends upon the voltage across it. The result is an avalanche effect rather like the effect of positive feedback in an electronic circuit. This process is called depolarization and it results in the inside of the nerve becoming positive with respect to the outside; the process of depolarization is the beginning of a nerve action potential. Depolarization is not a permanent state because the properties of the semi-permeable membrane

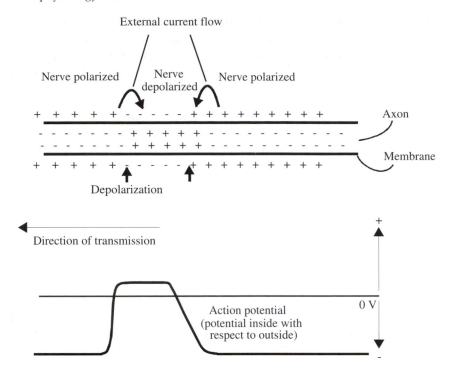

Figure 16.7. *The upper part shows a single nerve fibre and an action potential transmitted from right to left. The lower part shows the potential across the nerve membrane at all points along the fibre.*

change with time so that, after a short time, the nerve fibre reverts to the polarized state. Figure 16.7 shows how the transmembrane potential changes along a nerve fibre which is first depolarized and then repolarized. What is shown is a single nerve action potential lasting for about 1 ms.

A nerve action potential (NAP) is the impulse of depolarization followed by repolarization that travels along a nerve. Transmission can be in either direction. Muscle fibres can also transmit action potentials which result in a contraction of the muscle. Muscle action potentials (MAPs) are described in the next section.

Because a nerve fibre is immersed in a conducting fluid, ionic currents will flow around it from the polarized to the depolarized parts. These external currents are very important because they are the only external evidence that an action potential is present; it is these external currents which give rise to the bioelectric signals which can be recorded. For example, the heart gives rise to external currents of approximately 100 μA when it is active and it is these currents which give rise to the ECG.

External current flow around a nerve fibre is also responsible for the transmission of an action potential along the nerve. The external current flow at the point of depolarization disturbs the transmembrane potential further along the fibre and this causes depolarization to spread. An action potential is transmitted along a fibre with a speed of a few metres per second. It can be shown experimentally that external current flow is essential to the transmission of action potentials, by removing a single nerve fibre from the surrounding extracellular fluid; under these conditions an action potential is not transmitted. In figure 16.2 the nerve axon on the left is shown surrounded by bands of myelin. *Myelin* is an electrical insulator which prevents current flowing from the nerve axon into the extracellular fluid, and if it were continuous along the nerve then no action potentials would be possible. However, the myelin is actually in bands with areas called the *nodes of Ranvier* between the bands. External current can flow from the nodes of Ranvier and the effect of the myelin is to speed up the transmission of nerve action potentials which jump from one node to the next. This process is called saltatory conduction and it allows action potentials to be transmitted at about ten times the speed which fibres without myelin conduct impulses. The fast nerve fibres which supply our muscles are myelinated fibres, whereas the slow fibres used to transmit pain sensation are slow non-myelinated fibres.

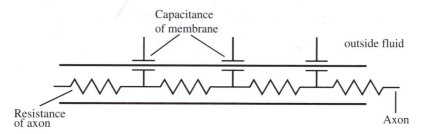

Figure 16.8. *Axonal resistance and capacitance control impulse transmission.*

The speed of transmission of a nerve impulse is actually determined by the capacitance of the membrane and myelin which separate the axon from the outside fluid, and the resistance of the axon (figure 16.8). Any resistance and capacitance have a time constant which controls the rate at which the potential across the capacitance can change. A cylindrical membrane surrounding a neural axon will have a capacitance C, associated with the area and dielectric constant ε of the neural membrane, and an axonal longitudinal resistance R, determined by the size of the axon and the resistivity ρ of the axonal fluid. If the length of the axon is L and the diameter D, then we can determine R and C as follows:

$$R = \frac{4\rho L}{\pi D^2} \quad \text{and} \quad C = \varepsilon \pi D L$$

so that the time constant of the membrane is

$$CR = \rho\varepsilon \frac{4L\pi D}{\pi D^2} = \rho\varepsilon \frac{4L^2}{D}. \tag{16.2}$$

The velocity of conduction will be inversely related to the time constant per unit of length and hence we can expect velocity to be directly related to D, the diameter of the axon.

A typical nerve membrane has a capacitance of 1 μF cm^{-2} (10^4 μF m^{-2}). If the axon has a diameter of 10 μm and is 10 mm long then:

$$\text{capacitance of membrane} = \text{membrane area} \times 10^4 \ \mu\text{F}$$
$$= 2\pi \times 5(\mu\text{m}) \times 10(\text{mm}) \times 10^4(\mu\text{F})$$
$$\cong 3 \times 10^{-3} \mu\text{F}$$
$$\text{resistance of the axon} = \frac{\rho \times \text{length}}{\pi \times 25 \times 10^{-12}} \cong 1 \times 3 \times 10^8 \ \Omega$$

where ρ is the resistivity of the fluid in the axon (a value of 1 Ω m has been assumed).

Therefore, the time constant of the membrane is

$$\text{time constant} = 3 \times 10^{-9} \times 1.3 \times 10^8 \cong 0.4 \text{ s}$$

This is quite a long time constant and it controls the speed at which a nerve impulse can be transmitted. Bear in mind that this is the time constant of a 10 mm length of axon. A shorter length will have a shorter time constant in proportion to L^2. Also a myelinated fibre will, of course, conduct more rapidly because the time constant is reduced still further. More will be said about nerve conduction velocities in section 16.5.1.

16.1.5.   Muscle action potentials

All the muscles in the body produce electrical signals which also control contractions. Muscles are subdivided into smooth and striated, which are also called involuntary and voluntary types; smooth muscle looks smooth under an optical microscope and it contracts without the need for conscious control, but striated muscle looks striped under the microscope and requires voluntarily produced nerve signals before it will contract. The muscle from which our intestines are made and the muscle in the walls of blood vessels is smooth muscle, but the muscles which move our limbs are of the striated type. This subdivision of muscles is an oversimplification because there are muscles such as those of respiration which are striated and yet not normally voluntarily controlled, and also some smooth muscle which can be partially controlled consciously; however, it is a useful subdivision for our purposes.

The way in which smooth muscle contractions are controlled by electrical changes is not well understood, even though it has been the subject of considerable research in recent years. Smooth muscle is intrinsically active; it can produce electrical signals without any neural or hormonal trigger. Electrical changes can be recorded, e.g. the electrogastrogram, which is the electrical signal produced by the stomach, but their uses are only in research and therefore they will not be considered further in this book.

The end of a nerve axon within the spine may make contact with either the cell body or dendrites of another neurone; this contact is called a synapse. A synapse is a junction and it is these junctions which allow neurones to influence one another. Where a motor nerve joins a striated muscle there is a special type of junction called a motor end plate, which allows a NAP from the nerve to initiate a MAP and subsequently a twitch in the muscle fibre. Striated muscle fibres are similar to nerve fibres in that they are cylinders of semi-permeable membrane which can transmit an action potential at speeds of a few metres per second; the speed is slower than in a motor nerve, even though the fibre diameter may be as large as 100 μm, because there is no myelin around the muscle fibre. Associated with the muscle fibres are longitudinal molecules of actin and myosin which are attached in a way such that they can slide over one another. When an action potential travels down the muscle fibres it is followed by a contraction as the molecules of actin and myosin move over one another. The release of calcium is an important step in the chain reaction between electrical and mechanical changes.

Exactly how this happens is not well understood but it is true that an action potential that might last for 5–10 ms, is followed by a single twitch from the muscle fibres lasting between 50 and 100 ms. A muscle fibre does not contract on its own because fibres are grouped into motor units. A single motor nerve splits into many terminal fibres, each of which supplies a muscle fibre and the muscle fibres supplied by one motor nerve are called a motor unit. Figure 16.9 illustrates the process of muscle contraction and the following list gives the steps which lead to a muscle twitch.

- A nerve impulse (NAP) is initiated by the brain and travels down an axon within the spinal cord.
- The nerve impulse will cross a synapse within the spinal cord and initiate an action potential, which then travels down a motor nerve at a speed up to about 100 m s^{-1}.
- The motor nerve may branch a few times along its course but, a few millimetres before it reaches the muscle, it branches into many terminal nerve fibres. Each of these fibres supplies one muscle fibre.
- A junction of a terminal nerve fibre with a muscle fibre is called a motor end plate where a chemical called acetylcholine is released. Acetylcholine diffuses across the gap to the muscle fibre and causes it to be depolarized.
- Depolarization of the muscle fibre gives rise to a conducted action potential (MAP) which travels in both directions from the end plate. This change in transmembrane potential is about 100 mV in amplitude and it travels along the muscle fibre at about 1–5 m s^{-1}.

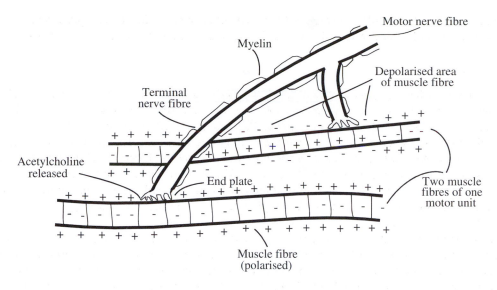

Figure 16.9. *Two striated muscle fibres which are supplied from one motor nerve fibre.*

- Following the action potential, a single contraction of the fibre takes place over a period of about 100 ms.
- One motor nerve supplies all the muscle fibres in one motor unit. In some very small muscles, e.g. the extraocular muscles, there may be only five fibres in a unit but large muscles may contain 1000 fibres in a unit, and be made up of several hundred units. The tension developed by one motor unit is a few grams.
- The action potentials and resulting twitches are asynchronous, i.e. not simultaneous, so that a smooth muscular contraction is obtained.

16.1.6.   Volume conductor effects

In this section we will determine the potential distribution which we can expect to find around an active neurone in tissue. We showed in section 9.2.3 that if we inject a current I at a point on the skin surface then the integral form of Poisson's equation gives us the potential ϕ at distance r as

$$\phi = \frac{1}{4\pi\sigma} \int \frac{I_v}{r} \, dV \tag{16.3}$$

where σ is the scalar conductivity of the medium, I_v is the current density and the integration is over the volume V. If we fix the potential at infinity as zero, and consider the tissue to be homogeneous with resistivity ρ and infinite in extent, then we can perform the integration with radial distance to obtain

$$V(r) = I \int_r^\infty \frac{\rho}{4\pi r^2} \, dr = \frac{\rho I}{4\pi r}. \tag{16.4}$$

This shows us how the potential changes with radial distance from a point source of current I. We can now extend this to consider the case of a cylindrical nerve fibre within a uniform conducting medium. This is illustrated in figure 16.10.

If the current which emerges from an element dx of the membrane is $I_m \, dx$ then

$$V(r) = \frac{\rho I_m \, dx}{4\pi r}. \tag{16.5}$$

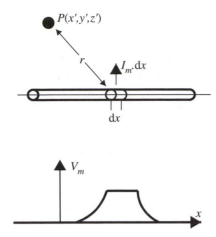

Figure 16.10. *A cylindrical nerve fibre set within an infinite-volume conductor and extending in the positive x direction. We calculate the potential at a point P where the coordinates are (x', y', z'). The current emerging from the element dx is of magnitude $I_m\,dx$ and is related to the transmembrane potential V_m.*

If the element I_m is located at (x, y, z) then we can calculate the contribution made to the potential field at a point $P(x', y', z')$ as shown in figure 16.10 as follows:

$$r = [(x - x')^2 + (y - y')^2 + (z - z')^2]^{1/2} \tag{16.6}$$

and we can integrate along the fibre to obtain the superimposed potential at point P. We will place the fibre along the x-axis such that $y = z = 0$.

$$\phi(x, y, z) = \int \frac{\rho I_m(x)}{4\pi[(x - x')^2 + y'^2 + z'^2]^{1/2}}\,dx. \tag{16.7}$$

In order to determine the potential distribution we need to know I_m. The transmembrane potential V_m was shown in figure 16.10. We can relate this to I_m as follows using the notation of figure 16.11:

$$I_m(x) = I_i - I_o = \frac{[V(x) - V(x - dx)]a}{\rho\,dx} - \frac{[V(x + dx) - V(x)]a}{\rho\,dx} = \frac{2a}{\rho}\frac{d^2V}{dx^2} \tag{16.8}$$

where a is the cross-sectional area of the fibre and ρ the resistivity of the axoplasm.

If we know the form of the transmembrane potential V_m we can use equation (16.8) to calculate I_m and then equation (16.7) to determine ϕ at any point.

This is illustrated in figure 16.12 where we have calculated ϕ as a function of x for three different values of y, with z taken as zero. V_m was defined over the time 1–280 and ϕ was calculated for $y = 30, 40$ and 50. V_m was defined as $(x - 120)^2$ for $120 \le x < 130$, 100 for $130 \le x < 150$, $(160 - x)^2$ for $150 \le x < 160$ and zero elsewhere. The double differentiation for equation (16.8) and integration for equation (16.7) were performed numerically—be careful to handle the gaps in the segments properly.

The form of the potential shown in figure 16.12 is interesting. It shows that we can expect a triphasic waveform to be recorded as an action potential is propagated past a recording electrode. It also shows that the shape of the signal will depend upon the distance between the nerve and the electrode. Not only the size of the signal changes with distance, but the timing of the successive peaks also changes. We expect the

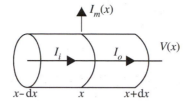

Figure 16.11. *The current I_m emerging from an element of a nerve fibre is related to the transmembrane potential V.*

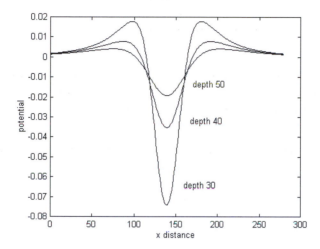

Figure 16.12. *The potential field around a nerve at different perpendicular distances from the nerve.*

frequency content of the signal to change with the distance between electrode and the nerve. This is shown in figure 16.13, where the Fourier transforms of the three signals shown in figure 16.12 are plotted. It can be seen that as the distance between nerve and recording electrode decreases the amplitude of the high-frequency components rises. It is found in practice that, as electrodes are moved closer to the source of a bioelectric signal, the bandwidth of the measured signal increases.

One further comment can be made about the waveforms shown in figure 16.12. As the action potential approaches an electrode the potential is first positive, then negative and finally positive again. The positive potential is shown as an upwards deflection. However, many recordings of NAPs and MAPs show a negative potential as an upwards deflection. This can be confusing and must be remembered when making EMG or nerve conduction measurements.

16.2.   THE ECG/EKG AND ITS DETECTION AND ANALYSIS

The ECG/EKG can be recorded from electrodes attached to the skin because the body is a good conductor of electricity. Electrical changes which take place within the body give rise to currents in the whole of the body which we can regard as a volume conductor. This is a very useful effect as it allows us to record electrical events from the body surface; however, it is a very difficult effect to understand because the electrical changes

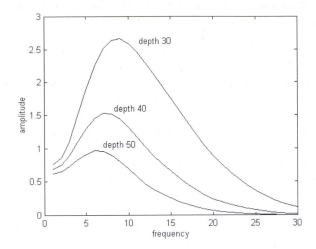

Figure 16.13. *This shows the frequency transforms of the three signals given in figure 16.12. Note the increase in high-frequency components close to the nerve axon.*

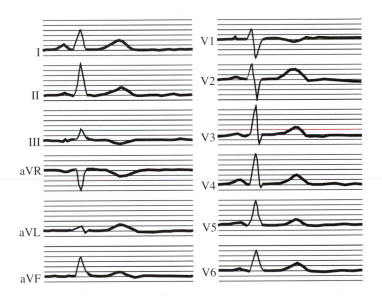

Figure 16.14. *An ECG/EKG corresponding to a single cardiac contraction but recorded from 12 different recording electrode sites. It can be seen that both the amplitude and shape of the ECG/EKG depend upon the position of the recording electrodes.*

on the surface of the body are related in quite a complex way to the source which gives rise to them. An analogy may be helpful. If a lighthouse simply produced a flashing light, then the flashes would be seen simultaneously by any observer; however, if the lighthouse produces a rotating beam of light, then the timing of the flashes will depend upon where the observer stands. Looking at an electrical source within the body is analogous to looking at a combination of the two types of lighthouse because the source is changing in intensity and also moving, as action potentials spread along semi-permeable membranes. The result can be seen in figure 16.14 which shows the ECG/EKG recorded from many combinations of surface electrodes. Not only does the size of the ECG/EKG change with recording site, but also the shape of the signals changes.

This change in ECG/EKG signal shape with distance arises in the same way as the changes in nerve action potentials which we considered in section 16.1.5. We can make another observation if we consider again the field around a single neurone which was illustrated in figure 16.6. Because all points on the surface of the neurone are at the same potential there are no circulating currents and therefore no potential changes in the volume conductor. This is an important point. *An inactive source, even though it is polarized, does not cause any potential change in the surrounding tissue.* We cannot detect it at a distance.

Similarly, if a neurone is completely depolarized, then all parts of its surface will be at the same potential and there will be no potential change in the surrounding tissue. However, at the point where the semi-permeable membrane becomes depolarized and also at the point where it becomes repolarized, the surface of the neurone is not all at the same potential and therefore external currents will flow. The second important point is therefore: *potential changes in the tissue surrounding a neurone only occur at the points where the transmembrane potentials are changing.*

This observation largely explains the shape of the ECG/EKG (figure 16.15), which only demonstrates when the activity of the heart is changing. The P wave corresponds to initial contraction of the atria, the QRS complex corresponds to relaxation of the atria and initiation of ventricular contraction and the T wave corresponds to ventricular relaxation. Between these points, where the heart is either fully contracted or relaxed, there are no potential changes in the ECG/EKG.

A precise derivation of the relation between surface recordings and transmembrane potentials is difficult and well outside the scope of this book.

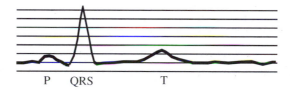

Figure 16.15. *The normal electrocardiogram.*

16.2.1.   *Characteristics of the ECG/EKG*

Transmission of the electrical impulse

During the normal heartbeat, the electrical impulse from the sino-atrial node spreads through the ordinary (non-auto-rhythmic) myocardial cells of the right atrium and is conducted rapidly to the left atrium along a specialized bundle of fibres, so that the contraction of the two atria takes place together. The AV node and the fibres leading to the bundle branches are the only conducting connection between the atria and the ventricles. The AV node delays the excitation by about 100 ms, to give time for the atria to contract completely, and the impulse then spreads rapidly down the specialized conducting fibres of the bundle branches and through the myocardial cells of the ventricles. This ensures that the whole of the muscle of each ventricle contracts almost simultaneously.

Skeletal muscle has a short absolute refractory period of 1–2 ms following contraction, during which the membrane is completely insensitive to a stimulus. Cardiac muscle has an absolute refractory period of about 250 ms, starting after depolarization of the membrane. This is almost as long as the contraction and is an important safeguard as the muscle will always relax before contracting again, thus ensuring that the heart will continue to act as an effective pump, even if stimuli are arriving at many times the normal rate. The absolute refractory period is followed by a relative refractory period during repolarization, during which a larger than normal stimulus is needed to initiate depolarization. A premature beat during this period (or an external electrical stimulus) can cause ventricular fibrillation.

The electrocardiogram

The electrocardiogram recorded from the right arm and the left leg has a characteristic shape shown in figure 16.15. The start of the P wave is the beginning of depolarization at the SA node. The wave of depolarization takes about 30 ms to arrive at the AV node. There is now a delay in conduction of about 90 ms to allow the ventricles to fill. The repolarization of the atria, which causes them to relax, results in a signal of opposite sign to the P wave. This may be visible as a depression of the QRS complex or may be masked by the QRS complex. After the conduction delay at the AV node, the His–Purkinje cells are depolarized, giving rise to a small signal which is usually too small to be visible on the surface. The conduction through the His–Purkinje system takes about 40 ms, and the depolarization and contraction of the ventricles then begins, giving rise to the QRS complex. Finally, repolarization of the ventricles takes place. This is both slower than the depolarization and takes a different path, so that the resulting T wave is of lower amplitude and longer duration than the QRS wave, but has the same polarity.

The mechanical events in the heart give rise to characteristic heart sounds, which can be heard through a stethoscope or can be recorded using a microphone on the chest wall. The first sound is low pitched and is associated with the closure of the atrio-ventricular valves as the ventricles start to contract. The second, a high-pitched sound, is associated with the closure of the aortic and pulmonary valves as the ventricles relax. Other sounds are usually the result of heart disease.

16.2.2.   The electrocardiographic planes

The heart can be thought of as a generator of electrical signals that is enclosed in a volume conductor—the body. Under normal circumstances we do not have access to the surface of the heart and must measure the electrical signals at the surface of the body. The body and the heart are three dimensional, and the electrical signals recorded from the skin will vary depending on the position of the electrodes. Diagnosis relies on comparing the ECG/EKG from different people, so some standardization of electrode position is needed. This is done by imagining three planes through the body (figure 16.16).

The electrodes are placed at standard positions on the planes. The frontal plane is vertical and runs from left to right. The sagittal plane is also vertical but is at right angles to the frontal plane, so it runs from front to back. The transverse plane is horizontal and at right angles both to the frontal and the sagittal plane.

ECG/EKG monitoring, which simply checks whether the heart is beating or not, uses electrodes placed in the frontal plane. To diagnose malfunctions of the heart, the ECG/EKG is recorded from both the frontal and the transverse plane. The sagittal plane is little used because it requires an electrode to be placed behind the heart—often in the oesophagus.

The frontal plane ECG/EKG: the classical limb leads

The ECG is described in terms of a vector: the cardiac vector. The electrical activity of the heart can be described by the movement of an electrical dipole which consists of a positive charge and a negative charge separated by a variable distance. The cardiac vector is the line joining the two charges. To fully describe the cardiac vector, its magnitude and direction must be known. The electrical activity of the heart does not consist of two moving charges, but the electric field which is the result of the depolarization and repolarization of the cardiac muscle can be represented by the simple model of a charged dipole.

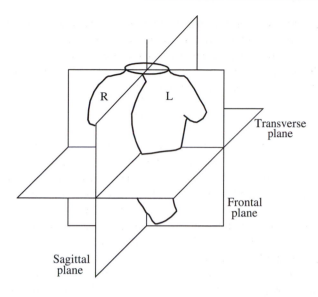

Figure 16.16. *The electrocardiographic planes.*

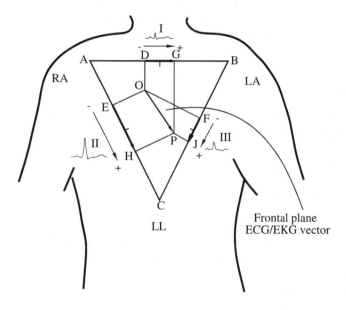

Figure 16.17. *Einthoven's triangle.*

In the physical sciences, a vector is usually described by its length in two directions at right angles (e.g. the *x*- and *y*-axes on a graph). With the frontal plane ECG/EKG, it is usual to describe the cardiac vector by its length in three directions at 60° to each other. The resulting triangle (figure 16.17) is known as Einthoven's triangle, and the three points of the triangle represent the right arm (RA), the left arm (LA) and the left leg (LL). Because the body is an electrical volume conductor, any point on the arm, from the shoulder down to the fingers, is electrically equivalent, and recording from the left leg is electrically equivalent to recording from anywhere on the lower torso.

The three possible combinations of the three electrode sites are called leads I, II and III, and convention stipulates which is the positive electrode in each case:

> Lead I   RA (−) to LA (+)
>
> Lead II   RA (−) to LL (+)
>
> Lead III   LA (−) to LL (+).

If the amplitude of the signals in the three leads is measured at any time during the cardiac cycle and plotted on the Einthoven triangle, the direction and amplitude of the cardiac vector can be found. In practice, 'cardiac vector' refers to the direction and amplitude of the cardiac vector at the peak of the R wave.

The use of Einthoven's triangle assumes that the human torso is homogeneous and triangular. This, of course, is not true but it is ignored in practice as the interpretation of the ECG/EKG is empirical and based on the correlations between the shape of the ECG/EKG and known disorders of the heart.

In order to work out the direction of the cardiac vector, recordings from leads I, II and III must be made. Draw an equilateral triangle ABC (figure 16.17) and mark the centre point of each side. Measure the height of the R wave on the same ECG/EKG complex for each of leads I, II and III. This is taken as the algebraic sum of the R and S waves, i.e. measure from the lowest point on the S wave to the highest point on the R wave. Note whether this is positive or negative. Using a suitable scale (e.g. 5 cm = 1 mV), draw each of the R wave amplitudes in the correct direction along the appropriate side of the triangle (DG, EH, FJ). Place the centre of the R wave vector at the centre of the side of the triangle. Draw in the perpendiculars from each end of the vectors (DO, EO and FO; HP, GP and JP). The point of intersection, O, is the beginning of the cardiac vector, and the point of intersection, P, is the end. Draw in the cardiac vector OP. In practice, the measurements will not be perfect. The three lines will not meet at a point P, but will form a small triangle, within which is the end of the cardiac vector.

The normal cardiac vector direction depends on age and body build. The direction of lead I, from right to left, is taken as $0°$. (Remember that we are looking from the front of the body, so that this runs from left to right on the diagram.) In young children, the axis is vertically downwards at $+90°$. During adolescence, the axis shifts to the left. A tall thin adult will have a relatively upright axis, whereas a short stocky adult might have an axis between $0°$ and $-30°$. An axis between $-30°$ and $-180°$ is referred to as left-axis deviation, and an axis between $+90°$ and $+180°$ is referred to as right-axis deviation.

The frontal plane ECG/EKG: augmented limb leads

Leads I, II and III are referred to as bipolar leads, because the measured signal is the difference in potential between two electrodes. Unipolar measurements are made by recording the potential at one electrode with respect to the average of the other two potentials. These are referred to as aVR, aVL and aVF (augmented vector right, left, and foot). The combinations are:

> aVR   (LA + LL)/2(−)   wrt RA(+)
>
> aVL   (RA + LL)/2(−)   wrt LA(+)
>
> aVF   (RA + LA)/2(−)   wrt LL(+).

The three unipolar leads have a direct vector relationship to the bipolar leads:

$$aVR = -(I + II)/2$$
$$aVL = (I - III)/2$$
$$aVF = (II + III)/2.$$

They are, in fact, the projection of the frontal plane cardiac vector onto three axes which are rotated 30° to the left from the Einthoven triangle. The direction and size of the cardiac vector can obviously be determined from the unipolar lead recordings in the same way as from the bipolar lead recordings.

The transverse plane ECG/EKG

The transverse plane ECG/EKG is recorded unipolarly with respect to an indifferent electrode formed by summing the signals from the left and right arms and the left leg (LA + RA + LL). Six electrodes are usually used, labelled V1 to V6. The electrodes are placed close to the heart and their position is more critical than the position of the frontal plane electrodes. They are placed on a line running round the chest from right of the midline to beneath the left axilla (figure 16.18). V1 and V2 are placed in the fourth intercostal space immediately to the right and left of the sternum. V3 is placed halfway between V2 and V4, and V4 is placed in the fifth intercostal space directly below the middle of the left clavicle. V4, V5 and V6 all lie on the same horizontal line, with V5 directly below the anterior axillary line (the front edge of the armpit), and V6 directly below the mid-axillary line (the mid-point of the armpit). The electrical signals recorded from the transverse plane electrodes are also shown in figure 16.18.

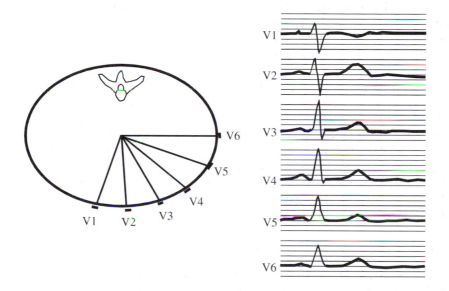

Figure 16.18. *The position of the chest electrodes for leads V1–6. Typical waveforms are also shown.*

The sagittal plane ECG/EKG

The sagittal plane ECG is rarely recorded. The indifferent electrode is again formed by the summation of the signals from the right and left arms and the left leg, and the active electrode is placed behind the heart. This is done using an oesophageal electrode, consisting of an electrode at the end of a catheter. The catheter is placed through the nose and down the oesophagus, until the electrode lies in the same horizontal plane as the heart.

16.2.3.   Recording the ECG/EKG

For diagnostic purposes, the ECG/EKG is recorded on paper, with either the six frontal planes only or the six frontal planes and the six transverse plane electrodes. For monitoring purposes, the ECG/EKG is displayed on a VDU screen, though provision may be made for recording abnormal stretches of the ECG either to paper or to memory. If a long-term record of the ECG/EKG of the patient's normal work is required, a miniature recorder will be used. This is described in section 16.2.4.

It is commonplace nowadays for a patient to have an in-dwelling cardiac catheter, either for recording the ECG/EKG and pressure from within the heart, or for pacing using an external pacemaker. The presence of a direct electrical connection to the heart greatly increases the danger of small electrical leakage currents causing fibrillation. If the patient has an in-dwelling cardiac catheter, then ECG/EKG equipment which is isolated from earth must be used—this applies to equipment connected either to the cardiac catheter or to surface electrodes. All modern ECG equipment is isolated from earth.

Monitoring or recording equipment must be capable of measuring the ECG/EKG, which has an amplitude of about 1 mV, whilst rejecting the interfering common-mode signal due to the presence of the 50/60 Hz mains supply. The subject of amplifier design and interference rejection was dealt with in Chapters 9 (section 9.3.3) and 10 (section 10.3).

Electrodes and filters

Electrodes are dealt with in detail in section 9.2. Two types of electrode are commonly used for ECG recording. A six- or 12-lead ECG/EKG recording will only take a few minutes at the most, but may be done on a very large number of patients each day. Plate electrodes are used with either saline-soaked pads or gel pads which are held on to the arms and legs with rubber straps. Suction electrodes are used for the transverse plane electrodes. For long-term monitoring, where the ECG/EKG may be recorded continuously for several days, disposable silver–silver chloride electrodes are used. These have a flexible backing material and use an electrode jelly formulated to give a minimum of skin reaction. Plate electrodes should not be used for long-term recording, as any corrosion of the plate can give rise to unpleasant skin reactions. With good skin preparation, the impedance of the electrodes will be less than 10 kΩ, so that an amplifier input impedance of 1 MΩ is adequate. In practice, the electrodes will not have exactly the same impedance. The electrode impedance will act as a potential divider with the common-mode input impedance of the amplifier. To achieve 80 dB common-mode rejection with a 10 kΩ difference in impedance between the electrodes, a common-mode input impedance of 100 MΩ is required (see section 10.3.3).

Interference on the ECG/EKG is often caused by the electrical signals from any muscles which happen to lie between the electrodes. The majority of the EMG spectrum lies above the frequency range required for recording the ECG/EKG, so that most of the EMG interference can be removed by suitable filtering of the signal. For long-term monitoring, the electrodes are placed on the chest so that the signals from the arm and leg muscles are eliminated. The bandwidth needed for diagnosis (which requires accurate reproduction of the waveshape) is 100 Hz, whilst 40 Hz is adequate for monitoring. The lowest frequency of interest in the ECG/EKG is at the repetition rate, which is not normally lower than about 1 Hz (60 b.p.m.). However, because the waveshape is important, a high-pass filter at 1 Hz cannot be used, because the distortion due to the phase shift of the filter will be unacceptable. The usual solution is to reduce the centre frequency of the high-pass filter until there is no significant phase shift at the lowest frequency of interest. A low-frequency 3 dB point of 0.05 or 0.1 Hz is usually used. The introduction of digital filters is enabling the 0.1 Hz cut off to be increased so that improved baseline stability is obtained without risk of waveform distortion.

The electrocardiograph

The electrocardiograph usually records the ECG/EKG on paper with 1 mm divisions in both directions and every fifth line emphasized. The standard paper speed is 25 mm s^{-1} (400 ms cm^{-1}), with a sensitivity of 10 mm mV^{-1} (1 mV cm^{-1}). An historical oddity is that the amplitude of the various parts of the ECG/EKG is quoted in millimetres (assuming the standard calibration). As there is nothing fundamental about the calibration, it would be more logical to quote the amplitude in millivolts.

The standard ECG/EKG for inclusion in the patient's notes is recorded using a portable electrocardiograph, which may record one- or three-lead positions simultaneously. The three-lead machines may switch between the leads automatically to give a fixed length of recording from each lead, ready for mounting in the notes. If the lead switching is to be done manually, the recording for each lead would be continued until 5 or 10 s of record has been recorded free from artefacts.

First of all, the patient should be encouraged to relax. Taking an ECG/EKG may be routine for the technician, but it is not routine for the patient, who may think that something in the test is going to hurt or they may be apprehensive about the results. The skin should be cleaned gently and the electrodes applied. For an automatic recorder, all 12 electrodes will have to be applied. For a one-channel recording, the three electrodes will be moved to the appropriate sites between each recording. Check that there is no mains interference—it may be necessary to earth yourself by touching the machine or the electrocardiograph may have to be moved to reduce the interference. If the patient is relaxed, there should be no EMG interference.

16.2.4. Ambulatory ECG/EKG monitoring

The traditional method of studying the ECG/EKG of a patient with suspected abnormalities that are not visible on a standard ECG/EKG recording is to confine the patient to bed for a few days with an ECG/EKG monitor connected, and tell a nurse to look for abnormalities on the ECG/EKG. Automatic arrhythmia detectors are also available which will do the nurse's job without fatigue, but this is still a very expensive method of diagnosis, and it may not be successful because many ECG/EKG abnormalities occur as a result of stress during the normal working day. Monitoring the ECG/EKG of patients during their normal working day is both cheaper and more effective. The monitoring is usually done using a small digital recorder.

The heart contracts about 100 000 times in 24 h. If a 24 h long recording were replayed onto an ECG/EKG recorder with a paper speed of 25 mm s^{-1}, the record would be 1.26 km long. Some form of automatic analysis is obviously needed, and this is usually performed by a special purpose computer. First of all, the R wave must be detected reliably. Most of the energy in the R wave lies between 10 and 30 Hz. The ECG is therefore passed through a bandpass filter and full wave rectified (because the R wave may have either polarity) to give a trigger signal. The R–R interval can be measured, and alarm limits set for low and fast heart rates (bradycardia and tachycardia). More sophisticated analyses can be performed. The results of the analysis can be made available as trend plots or as histograms, and the analyser will write abnormal sections of ECG on a chart recorder for visual analysis by a cardiologist. One person can analyse about 50 24 h recordings per week using an automatic analyser.

16.3. ELECTROENCEPHALOGRAPHIC (EEG) SIGNALS

The EEG technician's major role is to provide the medical specialist with a faithful recording of cerebral electrical activity, but in order to do this the technician must have an understanding of both the recording equipment and the characteristics of the EEG and its source. Electroencephalograph simply means a graph of the electrical changes from the *enkephalos* (Greek for brain).

The EEG arises from the neuronal potentials of the brain but, of course, the signals are reduced and diffused by the bone, muscle and skin which lie between the recording electrodes and the brain. There is a technique called electrocorticography (ECoG) where electrodes are placed directly on the cortex during surgery, but this is not a routine technique. The advantage of ECoG is that the electrodes only record from an area of the cortex of about 2 mm diameter, whereas scalp electrodes record from an area about 20 mm in diameter.

16.3.1. Signal sizes and electrodes

The EEG is one of the most difficult bioelectric signals to record because it is very small; this probably explains why the ECG/EKG was first recorded in about 1895, but the EEG was not recorded until 1929. There were simply no methods of recording signals as small as the EEG in the first decades of the 20th Century.

   The normal EEG has an amplitude between 10 and 300 μV and a frequency content between 0.5 and 40 Hz. If electrodes are applied perfectly and the very best amplifier is used, there will still be a background noise of about 2 μV p–p, which is significant if the EEG is only 10 μV in size. Every care must be taken to reduce interference and to eliminate artefacts, such as those which patient movement can produce, if a good EEG is to be recorded.

   The best electrodes are Ag–AgCl discs which can be attached to the scalp with collodion. The scalp must be degreased with alcohol or ether and abraded before the electrode is held in place; collodion is run round the edge of the electrode and allowed to dry. Electrolyte jelly is then injected through a hole in the back of the disc electrode to form a stable scalp contact.

   In a routine EEG clinic it is normally much too time consuming to apply many disc electrodes with collodion, which has to be removed with acetone after the test, and so electrode skullcaps are often used. The skullcap is an elastic frame which can be used to hold saline pad electrodes in place. The electrodes are a chlorided silver core with a saline loaded (10 g l^{-1}) cotton material around the core. These electrodes can be attached quickly and give good results.

   Electrodes can be placed all over the scalp and different combinations used for recording. The most commonly used electrode placement system is the 10–20 system, so named because electrode spacing is based on intervals of 10% and 20% of the distance between specified points on the head. These points are the nasion and inion (the root of the nose and the external occipital protuberance at the back of the head), and the right and left pre-auricular points (the depressions felt in front of the upper part of the ear opening). The 10–20 system is shown in figure 16.19. In this diagram the letters correspond to anatomical areas of the brain as follows: O, occipital; P, parietal; C, central; F, frontal; FP, frontal pole; T, temporal; and A, auricular. Nineteen electrodes are used in the 10–20 system.

16.3.2. Equipment and normal settings

An EEG machine is basically a set of differential amplifiers and recorders. The distinguishing features are that there is usually a minimum of eight channels—and in many cases 16 or more channels—on the recorder; and there may be provision for 44 input electrode connections. The eight-channel machines are normally portable types. The amplifier outputs are usually digitized so that a computer can be used for further analysis and display.

   Sixteen differential amplifiers will have a total of 32 input connections plus one earth connection. The input selector switches allow the correct combination of electrodes to be connected to the differential amplifiers; on some machines every electrode may be selected separately, but in others a complete combination (called a montage) can be selected by one switch. If the electrodes are selected individually then it must be remembered that each differential amplifier has both a non-inverting and an inverting input. These + and − inputs usually correspond to white and black wires, respectively, and must be connected correctly.

   There are internationally agreed 'normal' or 'standard' settings for an EEG recording; these are listed below:

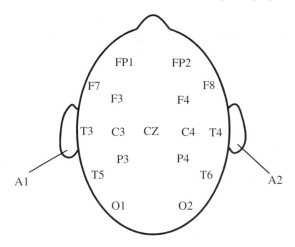

Figure 16.19. *The 10–20 system of electrode placement.*

Chart speed. Speeds of 15, 30 and 60 mm s^{-1} are usually provided but 30 mm s^{-1} is the standard setting.

Gain setting. Switched settings are usually given but 100 μV cm^{-1} is the standard for routine recording.

Time constant. The low-frequency response of an EEG recorder is usually quoted as a time constant (TC) and not as a -3 dB point. 0.3 s is the standard time constant: it corresponds to a -3 dB point of 0.53 Hz.

Filters. The high-frequency response of an EEG recorder is quoted as a -3 dB point. 75 Hz is the standard setting but other values such as 15, 30 and 45 Hz are available to reduce interference which cannot be eliminated by other means.

A calibration facility is included so that the gain settings can be checked. The calibration allows a signal of say 100 μV to be introduced at the inputs of the differential amplifiers. This type of calibration does not check that the electrodes have been carefully applied and are performing correctly. Many machines include an electrode impedance test circuit which allows every electrode to be tested; an impedance below 10 kΩ is necessary for the best recording. Some machines also include a facility called a biological test whereby one electrode on the body is driven with a standard test signal; this test signal should appear equally on all channels if all the electrodes and amplifiers are functioning correctly.

16.3.3. *Normal EEG signals*

It is not possible in this short section to describe the 'normal EEG'. What we can do is to outline a normal recording procedure and to give one example of an EEG tracing.

A complete EEG test will take about 30 min and it be essential that the test is conducted in a quiet environment. The room must be both acoustically quiet and electrically quiet if interference is not to be troublesome. A location which is remote from sources of interference such as operating theatres and physiotherapy departments is best. Only one person should normally be in the room with the patient: a bell call system can always be used to bring rapid help if required. Resuscitation equipment, including oxygen, should always be on hand. In some cases, for example, young children, it may be necessary to have two people present in the room. Application and testing of electrodes may take about 10 min, after which the patient is asked to relax for the duration of the test. In order to record the EEG during a range of states the patient is first asked to relax with their eyes closed for 5–10 min; a further shorter recording is then made with the

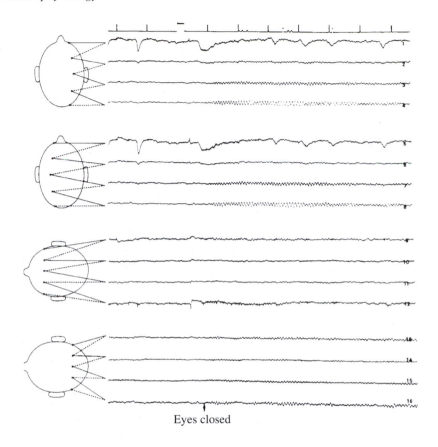

Eyes closed

Figure 16.20. *A 16-channel EEG recorded from a normal subject. Time runs from left to right and, at the top, one second marker blips are shown; the amplitude of these blips corresponds to 75 µV. When the eyes are closed an artefact can be seen on channels 1 and 5, and a regular rhythm within the alpha range (8–13 Hz) appears in several channels. (Courtesy of Dr J A Jarratt, Department of Neurology, Royal Hallamshire Hospital, Sheffield.)*

eyes open. Following this the patient is asked to hyperventilate (breathe as fast as possible) for about 3 min. Hyperventilation is a form of stimulation to the brain as oxygen levels are increased and carbon dioxide levels decreased; another form of stimulation which can make EEG abnormalities more obvious is a flashing light.

Flashes at a repetition frequency of 15 s^{-1} are often used as this can precipitate abnormal rhythms in patients suffering from epilepsy.

Figure 16.20 shows a normal 16-channel EEG recording. The waveform varies greatly with the location of the electrodes on the scalp. There are, however, certain characteristics which can be related to epilepsy, seizures and a number of other clinical conditions.

Epilepsy can give rise to large-amplitude spike and wave activity and localized brain lesions may give distinctive, large-amplitude, slow waves. An alert and wide-awake normal person usually displays an unsynchronized high-frequency EEG, whereas if the eyes are closed a large amount of rhythmic activity in the frequency range 8–13 Hz is produced. As the person begins to fall asleep, the amplitude and frequency of the waveforms decrease. Many more 'normal' patterns can be described

Problems and artefacts

There are very many practical problems which arise when recording an EEG. Some of the problems are associated with the equipment, but other problems originate from the patient, whose co-operation must be obtained if spurious signals from movement of the eyes or muscles are to be avoided. The list which follows includes some of the more common causes of artefacts on an EEG trace: electrode artefacts are usually the most troublesome.

Eye potentials. There is a potential of several millivolts between the back and front of the eyes. This potential gives rise to current flow through the tissues surrounding the eyes and this current will change as the eyes move. The effect can be seen as large deflections of the EEG trace when the eyes are moving.

ECG. The ECG is not usually a major problem in EEG recording, but if the recording electrodes are spaced a long way apart an ECG will be recorded and seen as sharp regular deflections on the recording. The artefact which results if the patient has an implanted cardiac pacemaker is very large and cannot be removed.

Electrode artefacts. If the patient moves or the wires leading to the electrodes are disturbed, then the electro-chemical equilibrium underneath the electrodes will be changed and so potential changes will occur. Another effect can occur if the patient is perspiring as this will also disturb the electrochemical equilibrium under the electrodes and give rise to quite large potential changes. These changes are usually slow baseline changes on the EEG.

There are very many more sources of spurious signals on the EEG trace. Ways in which electrical interference can arise were described in Chapter 10 (section 10.4). It has even been suggested that problems have arisen from dental fillings, where an electrical discharge between different metallic fillings gave rise to artefacts in the EEG. The EEG technician must always be on guard for possible sources of interference.

Particular EEG patterns have been associated with many conditions such as cerebral tumours, epilepsy, haematomas, concussion and vascular lesions. However, no attempt will be made to describe these patterns here. Analysis of EEG signals is not easy because it is difficult to describe the signals. The ECG/EKG can be described in simple terms because there are only about five major components to the waveform, but the EEG is a much more complex signal. The various frequency ranges of the EEG have been arbitrarily assigned Greek letter designations to help describe waveforms. Electroencephalographers do not agree on the exact ranges, but most classify the frequency bands as follows: below 3 Hz, delta rhythm; from 3–7 Hz, theta rhythm; from 8–13 Hz, alpha rhythm; and from 14 Hz upwards, beta rhythm.

Most humans develop EEG patterns in the alpha range when they are relaxed with their eyes closed. The alpha rhythm seems to be the idling frequency of the brain and as soon as the person becomes alert or starts thinking the alpha rhythm disappears. This is the rhythm which is used in biofeedback systems where the subject learns to relax by controlling their own alpha rhythm.

Very many attempts have been made to analyse the EEG using computers, in order to help clinical interpretation. Many EEG machines include a frequency analyser which presents the frequency components of the EEG on the same chart as the EEG. Currently none of the methods of analysis have been found useful routinely and so they will not be described here.

Many EEG departments make EEG evoked response measurements in addition to the background EEG.

16.4. ELECTROMYOGRAPHIC (EMG) SIGNALS

An electromyograph is an instrument for recording the electrical activity of nerves and muscles. *Electro* refers to the electricity, *myo* means muscle and the graph means that the signal is written down. The electrical signals can be taken from the body either by placing needle electrodes in the muscle or by attaching surface electrodes over the muscle. Needle electrodes are used where the clinician wants to investigate neuromuscular disease by looking at the shape of the electromyogram. He may also listen to the signals by playing them through a loudspeaker, as the ear can detect subtle differences between normal and abnormal EMG signals. Surface electrodes are only used where the overall activity of a muscle is to be recorded; they may be used for clinical or physiological research but are not used for diagnosing muscle disease. Both surface electrodes and needle electrodes only detect the potentials which arise from the circulating currents surrounding an active muscle fibre, and do not enable transmembrane potentials to be recorded. Nerves and muscles produce electrical activity when they are working voluntarily, but it is also possible to use an electrical stimulator to cause a muscle to contract and the electrical signal then produced is called an evoked potential. This is the basis for nerve conduction measurements, which allow the speed at which nerves conduct electrical impulses to be measured. This technique can be used to diagnose some neurological diseases and the principles of the method are explained in section 16.5.1.

Needle electrode measurements are almost always performed and interpreted by clinical neurologists, although both technical and scientific assistance may be required for the more sophisticated procedures. Nerve conduction measurements can be made as an unambiguous physiological measurement which is then interpreted either by medically or technically qualified staff.

16.4.1.   Signal sizes and electrodes

The functional unit of a muscle is one motor unit but, as the muscle fibres which make up the unit may be spread through much of the cross-section of the muscle, it is impossible to record an EMG from just one unit. If a concentric needle electrode, of the type shown in figure 9.3, is placed in a weakly contracting muscle, then the EMG obtained will appear as in figure 16.21. Each of the large spike deflections is the summation of the muscle action potentials from the fibres of the motor unit which are closest to the tip of the needle electrode.

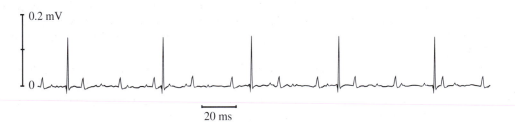

Figure 16.21. *An EMG recorded via a concentric needle electrode in a weakly contracting striated muscle.*

Remember that an upwards deflection represents a negative potential. The largest spikes all come from the same motor unit which is firing repetitively every 50 ms, but many other smaller spikes can be seen from fibres which are further away from the needle. The signal shown in figure 16.21 is a normal needle electrode EMG. The signal has a maximum amplitude of about 500 μV and the frequency content extends from about 10 Hz to 5 kHz. If the strength of contraction of the muscle is increased, then more motor units fire and the spike repetition frequency is increased, but the frequency content of the signal will not change significantly.

If a more localized recording is required then a bipolar concentric needle electrode can be used (figure 16.22). With a monopolar concentric needle electrode the signal is recorded from between the tip of the needle and the shaft; with a bipolar needle, the signal is that which appears between the two exposed faces of platinum at the needle tip, and the shaft is used as the earth or reference electrode. A bipolar needle only records from the tissue within about 1 mm of the tip and the signals obtained are smaller and also have a higher-frequency content than concentric needle recordings.

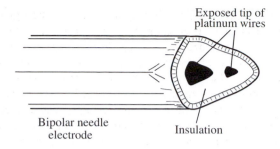

Figure 16.22. *The tip of a bipolar needle electrode.*

Any surface electrode placed over an active muscle can be used to record an EMG. If one electrode is placed over a muscle and the other electrode is placed several centimetres away, then EMG signals will be obtained from all the muscles lying between the electrodes. A more specific recording can be made if smaller electrodes are used and the separation is reduced to a few millimetres, but if the separation is reduced below about 4 mm then the amplitude of the signal falls rapidly. A very convenient and cheap electrode can be made from metal foil which can be cut to the desired size, will conform to the contours of the body and can be attached with adhesive tape. The skin must of course be cleaned and abraded before electrolyte jelly is applied and the electrode attached.

It is not possible, using surface electrodes, to record an EMG from just one muscle without interference from other muscles lying nearby. Even if two small electrodes are placed on the forearm, the EMG obtained will arise from many muscles. Localized recordings can only be made from needle electrodes, but these are uncomfortable and cannot be left in place for long periods of time. There is a type of electrode called a 'fine wire electrode' which can be left in a muscle for long periods: a wire is passed down the centre of a hypodermic needle which is then withdrawn, leaving the wire within the muscle. This can give an excellent long-term EMG recording.

The high-frequency content of surface electrode EMG signals is less than that from needle electrodes because of the volume conductor effects which were described in section 16.1.6. The recording amplifier should have a bandwidth from 10 to 1000 Hz. The amplitude of the signals depends upon the relative position of the electrodes and the muscle, but signals up to about 2 mV are typical.

16.4.2. *EMG equipment*

An EMG machine can be used to record both voluntary signals and evoked potentials. The amplitude of the signals will range from less than 1 μV up to 10 mV; the smallest signals are those produced by nerves and recorded from surface electrodes, whereas the largest are those evoked potentials from large muscles. Figure 16.23 gives a block diagram of an EMG machine.

The pre-amplifier will be a differential amplifier with the following typical specification.

Amplification	100
Input impedance	10 MΩ
Noise with input shorted	2 μV p–p
Common-mode rejection ratio	80 dB
Bandwidth (-3 dB points)	10 Hz–10 kHz

The output from the pre-amplifier is taken to the main amplifier and then to the A–D converter and host computer. The signal is also usually taken to a loudspeaker as EMG signals fall within the audio band and the

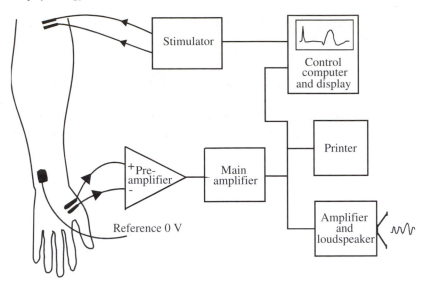

Figure 16.23. *A block diagram of an electromyograph.*

ear is very sensitive to subtle distinctions between signals. The remaining component of the EMG machine is the stimulator which is used for nerve conduction measurements. This will be considered in some more detail in section 16.5.

Equipment testing

The most common fault in all electrophysiological equipment is broken leads. Plugs and leads must be inspected regularly. Surface electrodes will have three connections to the differential amplifier: two inputs and an earth connection. A check on the operation of an EMG amplifier can be made by shorting together the three input connections and setting the amplifier gain to maximum. This may give a display of 10 μV per division on the screen, and if the amplifier is operating correctly a band of noise will be seen on the trace. By increasing the volume control on the loudspeaker amplifier the noise should be heard as a random broad-band signal. The simplest way to check the stimulator is to hold both output connections in *one* hand and to increase the output control slowly. A shock should be felt at an output of about 60 V.

16.4.3.   Normal and abnormal signals

Clinical electromyography using needle electrodes consists of inserting the sterilized needle into a muscle and then recording a voluntary EMG pattern from several points within the muscle. Samples are taken at several points because a diseased muscle may contain both normal and abnormal fibres. The neurologist will usually listen to the EMG signal, which sounds like intermittent gunfire. The patient will normally be asked to make only a mild contraction of the muscle so that individual spikes can be identified. When a strong contraction is made, a complete interference pattern is obtained sounding rather like an audience clapping. Not all muscles give the same sound although the difference between muscles of the same size is not great. An extreme case is the signals which can be recorded from a fine needle placed in the small muscles which move the eyes; these muscles are very small and the spikes obtained are of very short duration.

An individual action potential or spike when viewed on a screen has a total duration of a few milliseconds and usually contains only two or three deflections. In a myopathic muscle the action potentials are often smaller, may have more than three phases, and are of shorter duration than normal signals.

It is very difficult to distinguish individual spike potentials from a surface electrode recording. The amplitude of the EMG waveform is the instantaneous sum of all the action potentials generated at any given time. Because these action potentials occur in both positive and negative directions at a given pair of electrodes, they sometimes add and sometimes cancel. Thus the EMG pattern appears very much like a random noise waveform with the energy of the signal a function of the amount of muscle activity (figure 16.24).

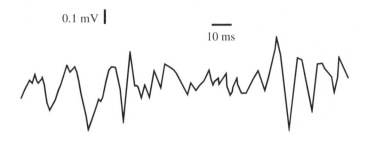

Figure 16.24. *An EMG recorded from surface electrodes.*

Signal analysis and clinical uses

Electromyography is used:

- in the diagnosis of neuromuscular disorders;
- as a measure of relaxation in the application of biofeedback techniques;
- as an index of muscle activity in physiological studies such as gait analysis.

There are very many clinical uses of electromyography, but it must be said that electromyography is really an extension of the classical methods of clinical examination, and each patient must be studied as an independent exercise in neurology. The skill of the electromyographer is as much in the planning of the examination as in its performance and interpretation. It seems improbable that electromyography will ever become a routine test performed by a technician under remote supervision.

Having made the point of the last paragraph it is of interest to outline very briefly the areas where clinical electromyography is useful. Following damage to a nerve, EMG signals give characteristic patterns of denervation which allow a prediction to be made about recovery. Damaged nerves may recover over periods as long as several years. EMG patterns characteristic of denervation include spontaneous activity such as small fibrillation potentials of short duration, instead of normal voluntarily produced spike potentials. Central neurogenic lesions such as motor neurone disease, poliomyelitis, and also spinal cord compression, cause characteristic EMG patterns which include large spike potentials with many deflections, synchronized motor unit activity, and some spontaneous electrical activity. Various inherited myopathies such as muscular dystrophy also give characteristic EMG patterns where the spike potentials are small, look ragged and contain more high-frequency components than the normal EMG.

Many methods of signal analysis have been tried to quantify EMG patterns; some depend upon measuring the frequency content of the signals. These can be of some use in quantifying the EMG and they have been shown to be helpful in identifying the carriers of muscular dystrophy, but they are not yet applied routinely.

16.5.   NEURAL STIMULATION

Because the body is a good conductor of electricity, and because our nerves and muscles function electrically, we would expect to see physiological effects when current is applied to the body. These effects can be a source of hazard but they can also be utilized both for the diagnosis and treatment of disease.

Chapter 8 dealt with the physiological effects of electricity. The three most important physiological effects of electricity were shown to be electrolysis, neural stimulation and heating. Neural stimulation was dealt with in some detail in section 10.2.1, but we will consider two applications of neural stimulation in this section.

16.5.1.   *Nerve conduction measurement*

An evoked response can be recorded from a nerve following electrical stimulation. The response has the form shown in figure 16.12 and, by measuring the time between the stimulus and the response, a conduction time and velocity can be determined. The average nerve conduction velocity is about 50 m s^{-1} or 112.5 mph. This velocity is actually very slow and certainly far slower than the speed at which electrical signals travel down a length of wire; this speed approaches the speed of light. Just how slow a nerve conduction velocity is may be appreciated by thinking about what happens to the nerve impulses which would arise from your feet, if you jumped off a very high building. If the building were 200 m high then you would hit the ground at about 63 m s^{-1} (we have neglected air resistance in making this calculation) which is a higher velocity than that at which the nerve signals would travel towards your brain, so that you would not feel your feet hit the ground. However, the slow conduction velocity of nerves does not appear to handicap the human body and it does make measurement of the velocity very easy. 50 m s^{-1} is the conduction velocity of myelinated nerve fibres, but there are also non-myelinated fibres which transmit signals much more slowly. At birth, myelination is not complete and this is one of the reasons why nerve conduction velocities increase over the first few years of life. Because the velocity is changing rapidly before birth a premature baby will have a slower velocity than a full term child; indeed, it is possible to determine the gestational age of a child at birth by measuring the nerve conduction velocity. In figure 16.25 the conduction velocity of the posterior tibial nerve is shown for groups of children up to the age of one year.

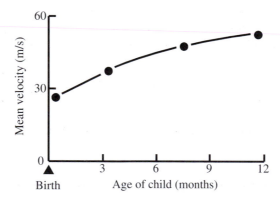

Figure 16.25. *Nerve conduction velocity in the posterior tibial nerve is shown as a function of age. The points shown are mean values for a group of normal children. From B H Brown, R W Porter and G E Whittaker (1967) Nerve conduction measurements in Spina Bifida Cystica, 11th Annual Report of the Society for Research into Hydrocephalus (Carshalton).*

Measurement of motor nerve conduction velocity

Following an electrical stimulus to a nerve, an action potential will travel down the nerve, along the terminal nerve fibres, across the neuromuscular junction and then cause a muscle action potential to spread along the muscle fibres. All these processes take some time and yet only the total time can be determined. This total time is called the 'latency'. However, by stimulating the nerve at two points and recording the difference in latency between the two we can determine the conduction time between the points of stimulation. Figure 16.26 shows where electrodes may be placed and equipment attached to make a measurement from the ulnar nerve, and figure 16.27 shows the results that should be obtained.

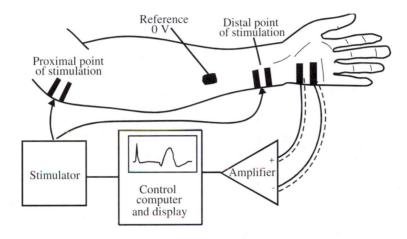

Figure 16.26. *The system used to record an ulnar nerve conduction velocity.*

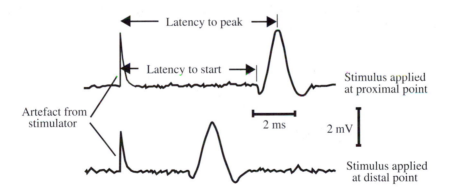

Figure 16.27. *Recordings made following stimulation of a nerve trunk at two points.*

Proximal means the point closest to the point of attachment of the limb to the body, and distal refers to the point which is further away; distance is measured between the stimulation cathodes because the action potential is initiated underneath the cathode. Latency is measured to the start of the muscle action potential by most workers, but to the major negative peak (upwards deflection) by some workers. There are established routines in every neurology department and 'normal' values will have been obtained for the particular method of recording.

One particular regime is now listed for recording a median nerve motor conduction velocity. The median nerve is one of the three major nerves which supply the arm; the other two are the ulnar and the radial. Figure 16.28 shows the major nerve trunks of the upper and lower limbs.

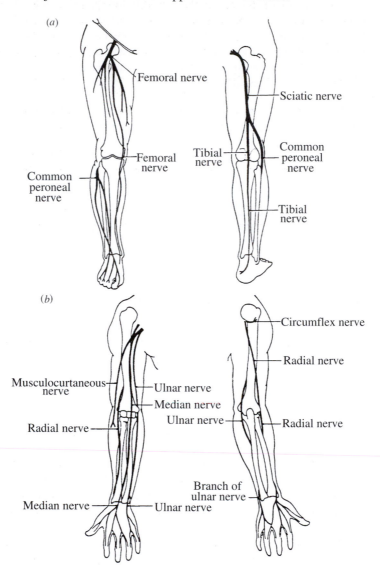

Figure 16.28. *(a) The major nerve trunks of the legs; (b) the major nerve trunks of the arms.*

Make sure that the patient is warm. Nerve conduction velocity decreases by several per cent for each centigrade degree drop in body temperature.

Prepare the skin by abrasion over the antecubital fossa, the wrist and the thenar muscles.

Make up three pairs of metal foil electrodes on adhesive tape, apply a small amount of electrode jelly to these and then attach them to the skin.

Set the EMG machine amplifier to 2 mV per division, the filters to a bandwidth of at least 10 Hz to 5 kHz and the time base to 2 ms per division. Check that the stimulator output is set to zero.

Connect the distal pair of electrodes to the stimulator with the cathode closest to the hand. Connect the amplifier to the electrodes on the hand, with the non-inverting input to the most distal electrodes; this will give a positive deflection for a negative signal underneath the first recording electrode. The leads to the electrodes should be secured to the arm, so that movement of the arm does not cause electrode movement artefact.

Prepare a skin site on the forearm and attach the earth or reference electrode. The patient may be lying down with the arm resting on the bed. Alternatively, they can sit on a chair and rest their arm on a convenient support about 20 cm above the level of the chair. It is important that the patient be asked to relax, because tension and the resultant continuous EMG activity make measurement difficult.

Obtain a continuous trace on the screen and check that no 50 Hz interference is present. If interference is present, first check that the electrodes have been applied well before investigating the many other possible causes of the interference.

Set the stimulator to give 100 μs width pulses at 1 pps. Increase the stimulus amplitude progressively until the EMG response shows no further increase in amplitude. You then have a supramaximal response.

Measure the latency of the response.

Turn the stimulator output down to zero.

Now apply the stimulus to the upper pair of electrodes and repeat the previous three steps to obtain a value for the distal latency.

Disconnect the stimulator and amplifier and then record the distance between the two stimulation cathodes.

Calculate the conduction velocity.

Measurement of sensory nerve conduction velocity

The EMG signals obtained when a motor nerve is stimulated are several millivolts in amplitude and therefore relatively easy to record. When a sensory nerve is stimulated, the only signal obtained is that which arises from the circulating currents surrounding the active nerve, and this signal is very small. The signals are nearly always less than 10 μV and a signal averager is required to reduce the background noise (see section 10.3.4).

There is one situation where a relatively large sensory nerve action potential is obtained and this can be useful for demonstration and teaching purposes. Following stimulation of the median nerve at the wrist, nerve impulses will be conducted in both directions along the nerve fibres; conduction up the sensory nerves in the normal direction is termed orthodromic conduction and conduction in the opposite direction is called antidromic conduction. Antidromic conduction gives rise to quite a large nerve action potential if recording electrodes are placed around the base of the index finger and a stimulus applied to the median nerve at the wrist. This signal can be recorded without a signal averager and should have a latency of about 5 ms to the peak of the action potential.

Accurate measurement of sensory nerve conduction velocities is often difficult because two convenient sites for stimulation of the same nerve are not available. For this reason only the latency may be measured and this value compared with a 'normal' range of values; alternatively an arbitrary reference point may be found on the nerve action potential waveform and the latency to this point used to calculate a 'velocity', using only a single nerve action potential recording.

The equipment used for signal averaging has already been described in the sections on evoked response measurement in Chapter 10. No further details will be given here.

Possible measurements and clinical value

If great care is taken, then both sensory and motor conduction velocity measurements can be made from most of the major nerves in the body. Needle electrodes are needed in order to obtain sufficiently large signals in some cases. The following short list is of the stretches of nerve from which recordings can be made easily using only surface electrodes.

Motor measurements. Arm, median and ulnar nerves—axilla to elbow to wrist to hand; Leg, posterior tibial—behind the knee to the ankle to the foot; anterior tibial - head of the fibula to the ankle to the foot.

Sensory measurements. Arm, median nerve index finger to wrist; ulnar nerve—little finger to wrist.

When used in conjunction with EMG recording, nerve conduction measurement is an extremely useful diagnostic tool. The major use of nerve conduction measurements in isolation is in the diagnosis and investigation of entrapment neuropathies. The most common of these is 'carpal tunnel syndrome' where compression of the median nerve at the wrist causes an increased latency for both motor and sensory signals. Compression within the tarsal tunnel at the ankle can also be investigated by making MCV measurements. A further use of these measurements is in cases of nerve lesions, e.g. damage to the ulnar nerve at the elbow; stimulation above and below the site of the suspected lesion may show slowing across the lesion.

One final point should be made, that nerve conduction velocity is actually the conduction velocity of many nerve fibres not all of which will have the same conduction velocity. There is a range of conduction velocities and there are methods of measuring this spectrum, but these methods are not yet widely used.

16.6.   PROBLEMS

16.6.1.   *Short questions*

a   Which ion is actively transported out of a resting cell?
b   What is the normal resting potential across a human nerve cell: -153, -70 or -15 mV?
c   What type of relationship is there between the intensity of sensation and pulse repetition rate in the human nervous system?
d   The largest component of the ECG/EKG is the R wave. To what mechanical activity of the heart does the R wave correspond?
e   There is a delay in the electrical conduction system of the heart between the upper and lower chambers (the atria and ventricles). What is the function of this delay?
f   What is the name of the insulating material that surrounds many human nerves?
g   What are nerve membranes made from?
h   What is meant by saying that smooth muscle is intrinsically active?
i   Will a monopolar recording of the potential produced in a volume conductor by a single wavefront of depolarization have a biphasic or triphasic shape?
j   What advantage does myelination of a nerve axon confer on a nerve cell?
k   What amplitude and frequency range would you expect for electrical signals generated within the body?
l   How long would a typical nerve action potential take to travel from the brain to the feet?
m   What is the approximate amplitude of the ECG/EKG when recorded from the skin?
n   What event does the T wave of the ECG/EKG correspond to?
o   What is the physiological function of acetylcholine?
p   Are the fastest conducting myelinated nerve fibres larger or smaller than the slowest fibres?

q   Give an example of smooth muscle.

r   What is bradycardia?

s   Describe briefly the 10–20 electrode system used for recording the electrical activity of the brain and give the number of electrodes used.

t   Describe briefly one clinical application of the EEG.

u   Why does the shape of the ECG depend upon where the electrodes are placed upon the body?

v   What is the name of the conduction path within the heart which carries signals from the atria to the ventricles?

w   Why is the sagittal plane rarely used in electrocardiography?

x   Why was the first EEG recording not made until the late 1920s when the first ECG recording was made around 1895?

y   What is the name of the wave in the EEG which is associated with relaxation?

z   Why would a bipolar needle electrode be used in clinical electromyography?

16.6.2.   Longer questions (answers are given to some of the questions)

Question 16.6.2.1

It is known that information transmitted along nerves is coded by the frequency of nerve impulses. An increasing strength of sensation is coded as an increase in frequency of the impulses. The maximum pulse repetition frequency is about 100 pps. The range of intensities of sound which are encountered are at least $10^6/1$. Explain why this makes a linear relation between intensity of sensation and nerve impulse frequency modulation impractical.

If a stimulus $s(t)$ gives rise to a frequency of nerve impulses given by $f(t)$, where $f(t) = a \log(s(t)) + b$, then calculate a and b if a range of stimuli of intensities ranging from 1 to 10^6 are to give nerve impulses which are frequency modulated between 2 and 100 pps.

If the percentage change in intensity of a stimulus which can just be perceived is 10% then calculate the number of steps in intensity of stimulus which can just be discriminated between the threshold of sensation and pain.

Question 16.6.2.2

If the nodes of Ranvier in a myelinated nerve fibre, of diameter 10 μm, are spaced at intervals of 5 mm and form a gap of exposed membrane 100 μm wide around the fibre then calculate an approximate value for the speed of conduction of an action potential along the fibre. Assume that the action potential hops from node to node with a time delay in each case equal to two time constants of the nodal membrane. Also assume that the fibre membrane has a capacitance of 1 μF cm^{-2} and the axon a resistivity of 1 Ω m.

Answer

62.5 m s^{-1}.

Question 16.5.2.3

Would you expect an EKG/ECG recorded from a cardiac catheter to have a wider or lower bandwidth than a signal recorded from the classical limb lead II position? Give a reasoned answer.

Question 16.5.2.4

Using the model given in section 16.1.5 determine how the amplitude of a recorded nerve action potential can be expected to fall-off with the depth of the nerve below the recording electrodes.

Question 16.5.2.5

Compare the information transfer rate of a single nerve fibre to that of a modem interface to a personal computer. The modem can handle 19 600 bytes per second. The nerve can carry impulses of 1 ms duration at frequencies up to 200 pps. Note any assumptions you need to make.

How does the human body increase the information transfer rate?

Answer

If we regard each impulse as one bit of information then the nerve can handle 200 bits s^{-1}, whereas a typical modem can handle 19 600 bytes (8-bit words), i.e. 156 800 bits s^{-1}. The modem is almost 800 times faster than the human nerve fibre.

We assume that the system is purely digital and uses frequency modulation.

The body compensates for the low information transfer rate by having many thousands of nerve fibres which operate in parallel.

Answers to short questions

a Na^+ is normally actively expelled from cells in the resting state.
b -70 mV is the normal resting potential across the membrane of a human nerve cell.
c   There is an approximately logarithmic relationship between the intensity of a sensation and the frequency of pulses generated in the nervous system.
d   The R wave of the ECG/EKG corresponds to ventricular contraction.
e   The delay between the atria and ventricles is to allow time for ventricular filling.
f   Myelin is the insulating material that surrounds many nerve axons.
g   Nerve membranes are made from lipids.
h   Smooth muscle is said to be intrinsically active because it can produce electrical signals spontaneously without a neural or hormonal trigger.
i   A monopolar recording from a single wavefront will be biphasic in shape. If both depolarization and repolarization are considered then the signal is triphasic.
j   Myelin allows an action potential to jump from node to node and hence speeds up conduction along a nerve axon.
k   A frequency range of 0.01 Hz to 5 kHz and an amplitude range of 1 μV to 1 mV are typical of the electrical signals generated within the body.
l   For a typical velocity of 50 m s^{-1} and height of 1.6 m it would take 32 ms for a signal to travel from the brain to the feet.
m   1 mV is a typical amplitude for the ECG/EKG.
n   The T wave corresponds to ventricular repolarization.
o   Acetylcholine is a neurotransmitter. It acts as the messenger between a nerve and a muscle.
p   The fastest nerve fibres have the largest diameter.
q   The gut and blood vessels are examples of smooth muscle.
r   Bradycardia is an abnormally low heart rate.

s   A diagram of the 10–20 system should show the reference points and indicate the 10% and 20% divisions between points. 19 electrodes are used.

t   Applications of EEG measurements could include the identification of space-occupying lesions, the investigation of epilepsy and various psychological investigations.

u   The shape of an ECG/EKG depends upon electrode positions because the origin of the signals are moving wavefronts and not a single source within the heart.

v   The bundle of HIS or atrio-ventricular bundle carries electrical signals between the atria and the ventricles of the heart.

w   The sagittal plane is rarely used in ECG/EKG recording because an electrode must be placed behind the heart, normally in the oesophagus

x   ECG/EKG signals were recorded before the EEG because the amplitude of the EEG is much smaller than that of the ECG. Equipment able to record microvolt signals did not exist at the beginning of the 20th Century.

y   Alpha waves of the EEG are associated with relaxation.

z   A bipolar needle electrode might be used to measure the neural activity from a small volume of tissue.

BIBLIOGRAPHY

Barber C, Brown B H and Smallwood R H 1984 *Dictionary of Physiological Measurement* (Lancaster: MTP)

Bowsher D 1979 *Introduction to the Anatomy and Physiology of the Nervous System* (Oxford: Blackwell)

Brazier M A B 1968 *The Electrical Activity of the Nervous System* (Baltimore, MD: Williams and Wilkins)

Fleming J S 1979 *Interpreting the Electrocardiograph* (London: Update)

Hector M L 1980 *ECG Recording* (London: Butterworths)

Katz B 1966 *Nerve Muscle and Synapse* (New York: McGraw-Hill)

Longmore D 1971 *The Heart* (London: Weidenfield and Nicolson)

Marieb E N 1991 *Human Anatomy and Physiology* 3rd edn (Menlo Park, CA: Benjamin-Cummings).

Noble D 1975 *The Initiation of the Heart Beat* (Oxford: Clarendon)

CHAPTER 17

RESPIRATORY FUNCTION

17.1. INTRODUCTION AND OBJECTIVES

The whole purpose of the respiratory system, which includes the nose, pharynx, trachea, bronchi and lungs, is to allow the intake of fresh air into the lungs, where oxygen is absorbed and carbon dioxide discharged. The process by which the air enters the lungs is called *ventilation* and the method by which the blood fills the blood vessels in the lungs is called *perfusion*.

Diseases such as bronchitis, asthma and pneumoconiosis affect the process of ventilation and physiological measurements allow the degree of impairment to be assessed. Other conditions, such as pulmonary embolism, can affect the process of lung perfusion and again measurements are needed to assess this condition. Most larger hospitals have respiratory function or pulmonary function laboratories where these measurements can be made. Very often these laboratories are attached to cardiology, anaesthetic or medical physics departments. This chapter covers the basic principles of the techniques found in most respiratory function laboratories with the exception of the techniques of blood gas analysis. Some information on transducers for blood gas analysis is given in section 9.5.3, but readers are referred to the more specialized texts listed in the bibliography for wider coverage of this subject.

Respiratory function does not involve the application of just one area of physics in the way that nuclear medicine is the application of nuclear physics in medicine. Respiratory function is basically a clinical area where the scientist with some knowledge of many aspects of physics and engineering can help the clinician in the diagnosis of disease. By describing, in quantitative terms, how well a person is breathing, it is possible to form an initial evaluation of a person who complains of breathlessness, to follow the course of a disease during treatment, and to make a pre-operative evaluation of a surgical patient with a high risk of respiratory complications. Whilst in the respiratory function laboratory, a technician in training should receive some instruction in the following basic techniques: peak air flow measurement, forced expiratory volume measurement, flow/volume curve recording, determination of lung residual capacity, and use of an ergometer and observation of respiratory changes during exercise.

The physics which you will need in order to understand the techniques of respiratory function testing concerns electronics and the physics of gases. Some of the electronics and its application to instrumentation is covered in this book, but the basic physics of gases is not covered; for example, an understanding of what is meant by partial pressure and the relationship between pressure and volume is assumed. Respiratory function is a large and complex subject. This chapter should be seen as an introduction to the topic. Questions we hope to answer include:

- What are the best indicators of the effectiveness of respiration?
- What tissue properties change with respiration?
- What determines normal respiratory rate?

When you have finished this chapter, you should be aware of:

• The parameters most commonly used to describe ventilation and some typical values.
• Several methods of measuring ventilation.
• A method of measuring gas exchange within the lungs.
• Some methods of monitoring respiratory effort and oxygenation.

There are three major sections to this chapter: firstly, the basic physiology of respiration is explained; secondly, the terminology which you will find used in the laboratory is covered; and thirdly, some of the methods of measuring respiration are explained, including the principles of the instrumentation involved. A final section is included on respiratory monitoring found outside the respiratory function laboratory in areas such as the intensive care unit, special baby care units and in research situations.

17.2. RESPIRATORY PHYSIOLOGY

When air is inhaled (inspiration) the lungs expand and when it is exhaled (expiration) they contract. After passing through the nose or the mouth the cleaned and warmed air enters the pharynx. Just below the level of the lower jaw the pharynx divides to form two passages, one of which leads to the stomach and the other to the lungs. The passage to the lungs is called the trachea, which is a muscular tube bound by semi-circles of cartilage and lined with a mucous membrane. It divides into the two bronchi, each of which enters a lung and within the lungs the bronchi divide many times to produce the tiny bronchioli which terminate in the alveoli. The alveoli are like tiny balloons and it is here that the exchange of gases between blood and air takes place.

Blood is carried, from the right side of the heart, to the lungs through the pulmonary artery. This artery is one of the great blood vessels from the heart and it divides into two branches serving the two lungs. In the lungs, the blood vessels divide repeatedly until they become the tiny blood vessels called capillaries which are interwoven with the alveoli, being separated by only a small amount of tissue. Oxygen is breathed into the alveoli where it passes through the pulmonary membrane and enters the blood. Carbon dioxide and water within the blood pass in the reverse direction.

17.3. LUNG CAPACITY AND VENTILATION

The lungs are shaped rather like two upright cones within the rib cage and they have the consistency of a sponge. The heart lies in front of the left lung. Each of the lungs is divided into lobes which are enclosed by the pleural membranes. There are three lobes in the right lung and two in the left.

The lungs are not emptied completely each time air is exhaled, indeed when a person is resting only about 0.5 l of air may be exhaled and 4 l stay in the lungs. The air which is actually exhaled and then replaced is called the tidal air and its volume the 'tidal volume'. By taking a very deep breath the tidal volume can be increased, but some of the 4 l will still remain: this volume is called the 'residual volume'. This residual volume is composed of the air in the mouth and trachea as well as that which remains in the lungs. Yet another term which is used is the 'vital capacity', which is the maximum volume of air which can be expired following the deepest possible breath. The measured values of these parameters are a little arbitrary, of course, as they depend upon the effort made by the patient.

Another measurement of respiratory performance is the flow rate of air into and out of the lungs. Both the upper airways and the lungs themselves exhibit resistance to airflow and this airways' resistance is changed by many respiratory diseases. By analogy to an electrical circuit, airways' resistance is calculated by dividing pressure (equivalent to voltage) by flow (equivalent to current).

The total amount of air entering the lungs each minute is also determined by the respiration rate. The normal respiration rate for a baby is about 60 min^{-1} but, by the age of one year, this has fallen to 30–40 min^{-1} and, in an adult, 12–15 min^{-1} is a normal respiration rate. In general, respiration rate increases as body weight decreases so that an elephant might breath about 4 min^{-1} and a rat 200 min^{-1}. There are good physical reasons for this as it can be shown that there is an optimum respiratory rate which requires minimum effort, depending upon the size of the animal. Both elastic and inertial forces are involved in expanding the lungs and these forces can balance at a certain resonant frequency when minimum work is expended.

If we represent the inertial component by a mass m and the elastic component by a coefficient E as illustrated in figure 17.1, then we can write down the following equation to represent the balance of forces:

$$m\ddot{x} + Ex = 0$$

which has the general solution

$$x = a \sin\left[\sqrt{\frac{E}{m}}t + b\right]$$

and the resonant frequency is given by

$$\text{frequency} = \frac{1}{2\pi}\sqrt{\frac{E}{m}}.$$

Hence, we might expect the normal respiratory frequency to be inversely proportional to the square root of body mass. Figure 17.1 gives a plot of $1/\sqrt{m}$ against frequency for the elephant, rat, adult and neonate using the frequencies given above and masses of 3000, 0.3, 70 and 3 kg, respectively.

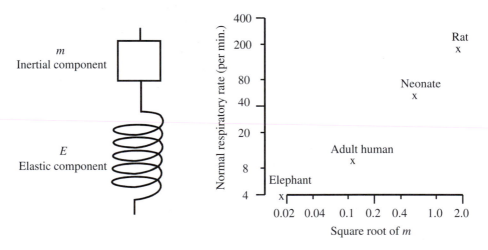

Figure 17.1. *On the left an analogy is given for the inertial (m) and elastic (E) forces acting on the lungs. On the right a graph is given of $1/\sqrt{m}$ against normal respiratory frequency for the elephant, rat, adult and neonate using frequencies of 4, 300, 12 and 60 min^{-1}, and masses of 3000, 0.3, 70 and 3 kg, respectively.*

The respiration rate is not completely regular and changes with exercise, during talking and according to the environment. The control centre for breathing is in the medulla of the brain and the major influence on this is the CO_2 level in the blood. The medulla, through its influence on the nerves supplying the respiratory muscles and the diaphragm, can increase or decrease respiration rate and tidal volume. Even a slight increase in the amount of CO_2 stimulates the respiratory centres to increase the rate and depth of breathing. A reduction in arterial oxygen content has a similar effect. For this reason measurements of arterial O_2 and CO_2 are extremely useful measures of respiratory function.

Figure 17.2 is included so that you can appreciate the complexity of the way the body controls respiration. A well-equipped research laboratory would be able to investigate all parts of the control system.

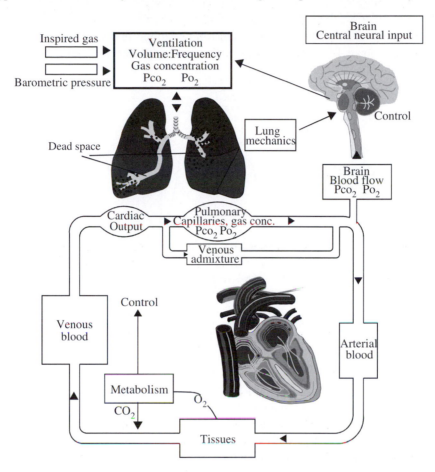

Figure 17.2. *This figure shows in diagrammatic form the various factors which influence the respiratory system.*

17.3.1.   Terminology

Breathing is not easily described because it is such a complicated process. What actually matters is the efficiency with which oxygen is transferred from the air to the blood and carbon dioxide is removed, but this is very difficult to measure unless continuous samples of air and blood can be taken.

In clinical practice, breathing is described by very many measurements which need to be carefully defined if normal ranges are to be established. A number of measurements can be made directly from a recording of the lung volume changes which occur when the patient is asked to carry out set procedures. Some of these measurements are illustrated in figure 17.3.

Vital capacity (VC). This is the maximum volume of air that can be expired after a maximum inspiration. Normal values for young men are about 5 l. The values are less in women (about 3.5 l) and decrease with age. (Note, some laboratories use inspired VC. This is the volume that can be inspired after a maximum expiration.)

Residual volume (RV). This is the volume of air remaining in the lungs after a maximum expiration. Normal values for young men are about 1.2 l. This volume increases with age and is slightly lower in women.

Forced expiratory volume (FEV$_1$). This is the volume of air expired in 1 s following full inspiration. For young men the FEV$_1$ is about 4 l. The range of normal values is different for men and women and changes with height, weight and age.

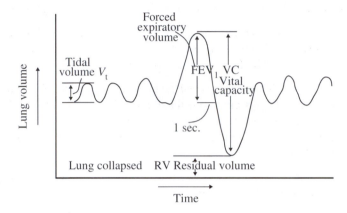

Figure 17.3. *This graph, of lung volume versus time, shows the changes which occur during normal tidal breathing followed by maximal inspiration and then by forced maximal expiration. Four parameters which describe the graph are marked.*

Peak expiratory flow rate (peak-flow rate) (PEFR). This is the maximum flow rate during forced expiration following full inspiration. This is a very commonly used measurement which, in normal adult males, gives a value of about 7 l s^{-1}. However, it must again be borne in mind when interpreting results that the normal values are different for men and women and also change with age.

Tidal volume (V_t). This is the volume of air inspired or expired at each breath. This can be at any level of breathing activity. Quite obviously, normal values depend upon the effort of the subject. For a normal adult at rest V_t is about 300 ml. Note the very large difference between this and the VC. Only a small fraction of the lung capacity is used when resting.

The five measurements defined so far are all measurements made from a single breath. There are many more measurements which can be made, but these five are the most commonly used. Ventilation of the lungs depends upon how rapidly the person breathes and the depth of each breath. The total ventilation can be defined in many ways but only two measurements will be defined here.

Maximum voluntary ventilation (MVV). This is the volume of air inspired per minute during maximum voluntary hyperventilation. The patient is asked to breathe at 50 min^{-1} as deeply as possible for 15 s. The total volume of inspired air is measured and multiplied by four to give the MVV. Normal values for men are about 150 l min^{-1}, but values change with age and body size. There is no standard definition of MVV, with the result that some laboratories use a different breathing rate and make their measurements over a longer or shorter time. Normal values must be established for each particular method of testing.

Alveolar ventilation (VA). This is the volume of air entering the alveoli in 1 min. VA is important because all the gas exchange takes place in the alveoli. However, VA cannot be measured directly but it can be calculated from other measurements and is normally about 80% of the inspired air.

All the measurements given so far can be obtained from a record of lung volume against time. The six definitions which follow are much more difficult to implement. They are given here to help you understand the language used in the respiratory function laboratory, and some of the definitions will be referred to later in the chapter.

Minute volume. This is the volume of blood passing through the lungs in 1 min. This is obviously an important measurement as the exchange of oxygen in the lungs will depend upon both the flow of air and the flow of blood through them. In a normal resting adult the blood flow through the lungs is about 5 l min^{-1}.

Ventilation: perfusion ratio. This is the ratio of the alveolar ventilation to the blood flow through the lungs. In resting adults, the alveolar ventilation is about 4 l min^{-1} and the lung blood flow 5 l min^{-1} so that the ventilation:perfusion ratio is about 0.8. The ratio increases during exercise.

Arterial oxygen pressure (PAO$_2$). This is the partial pressure of oxygen dissolved in the arterial blood. The partial pressure of a component of a gas mixture is the pressure it would exert if it alone occupied the whole volume of the mixture. To measure the partial pressure of gases in a liquid such as blood, the liquid is allowed to equilibrate with the surrounding gas and the pressure measurement made in this gas. Normal values for adults are 80–100 mmHg (10.6–13.3 kPa).

Alveolar oxygen pressure (PaO$_2$). This is the partial pressure of oxygen in the air present in the alveoli. Normal values are 95–105 mmHg (12.6–14.0 kPa).

These last two definitions are particularly important because it is the difference between alveolar and arterial pressures which causes oxygen to diffuse from the alveolar air into the blood. Unfortunately, it is impossible to measure either of these pressures directly because we cannot gain access to the alveoli. However, it is possible to measure the oxygen pressure in arterial blood. The alveolar oxygen pressure can be calculated indirectly: knowing that the partial pressure of oxygen in the atmosphere is 160 mmHg, and by making allowance for the increase in temperature when air is inspired and the fact that water vapour displaces about 7% of the air in the lungs, the alveolar oxygen pressure can be found.

Compliance of the lung (C). This is the expansibility of the lungs expressed as the volume change per unit pressure change. It is a measure of the effort needed to expand the lungs. Normal values are about 200 ml cmH$_2$O^{-1} (2 l kPa^{-1}).

Airways' resistance. This is a measure of the resistance to airflow in the airways expressed as the air pressure divided by the flow. Normal adult values are about 2 cmH$_2$O l^{-1} s (0.2 kPa l^{-1} s).

17.4. MEASUREMENT OF GAS FLOW AND VOLUME

Air is surprisingly heavy and should therefore be relatively easy to measure. A room of $4 \times 4 \times 3$ m^3 contains about 62 kg of air. Even so, it is quite difficult to measure the flow of air into and out of the lungs. Several instruments with rather elaborate names have been devised to measure airflow and volume and it is easy to become confused when people talk of pneumographs, pneumotachographs, spirometers, respirometers and plethysmographs. The origin of these words is actually quite simple.

Pneumo comes from Greek and means lungs, so that the pneumograph is an instrument which gives a graph of lung function; this is actually a graph of lung volume against time. *Tacho* is also a Greek word and means speed so that the pneumotachograph produces a graph of the speed of airflow into or out of the lungs. *Spiro* is the Latin word meaning breath so that the spirometer is an instrument which measures breath. The measurement is usually of volume. *Pletho* is another Greek word meaning fullness so that the plethysmograph is an instrument which gives a graph of fullness. The body plethysmograph gives a measurement of lung fullness.

17.4.1.   The spirometer and pneumotachograph

A spirometer can use the same principle as a commercial gas holder. An inverted cylindrical container floats in a liquid and the volume of gas in the container can be calculated from the vertical displacement of the cylinder.

The patient's nose is closed with a clip and, following maximum inspiration, he is asked to breathe out as much as possible into the spirometer. The volume read on the calibrated scale is the vital capacity. Actually a correction factor has to be applied to allow for the fact that the air cools and thus contracts when it passes from the patient into the spirometer. A thermometer is usually built into the spirometer and the correction factor is found from scientific tables (e.g. J E Cotes *Lung Function*, see the bibliography). The volume measured by the spirometer is less than the volume of the air in the lungs and the correction factor is of the order of 10%.

The basic instrument shown in figure 17.4 was used for many years to measure vital capacity. It was superseded by devices that do not require a liquid support for the chamber and can be used at various orientations. For example, a wedge-shaped chamber and bellows which expand to an angle proportional to the volume of air can be used.

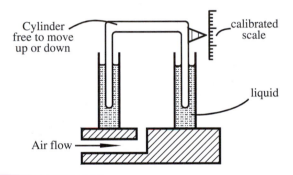

Figure 17.4. *Principle of the spirometer. In practice a counterweight is used to reduce the force needed to move the cylinder.*

There are many types of spirometer but they all consist basically of a chamber which can expand in proportion to the volume of air contained. The maximum volume is usually about 8 l, which is sufficient to cope with the largest lung volumes.

If the pointer attached to the spirometer chamber is used to record position on a chart, then the instrument is called a spirograph and it can then be used to measure FEV_1 and V_t. However, the instrument still has a basic disadvantage in that, because it is closed system, it cannot be used to measure ventilation over several breaths. The basic spirograph can only be used to make measurements from a single breath. This problem can be overcome by supplying oxygen to the chamber at the same rate as the patient is absorbing the gas, and also by absorbing the carbon dioxide and water vapour in the exhaled air as shown in figure 17.5. Instruments used to measure residual volume, RV, by helium dilution, and transfer factor, TL, by carbon monoxide absorption, utilize this technique, and thus allow the patient to breathe continuously into the spirograph.

The spirograph measures flow volume. In order to measure flow rate a *pneumotachograph* is used. The pneumotachograph measures flow rate by means of a transducer through which the patient breathes. The air

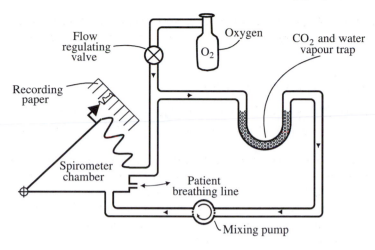

Figure 17.5. *System which allows a patient to breathe continuously into a spirograph.*

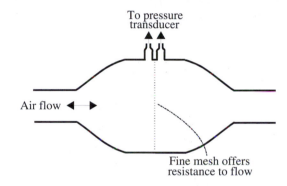

Figure 17.6. *A pneumotachograph transducer. The pressure difference across the fine mesh is proportional to flow velocity.*

passes through a fine mesh which offers a small resistance to flow, with the result that there will be a pressure drop across the mesh in proportion to the flow rate. The mesh is chosen such that a flow of about $10 \, \mathrm{l \, s^{-1}}$ will give a pressure drop of a few mmHg. This pressure can be measured with a sensitive pressure transducer and so a direct measure of flow rate obtained.

Figure 17.7 shows a block diagram of the complete pneumotachograph. It consists of the flow head or transducer which is connected by two fine tubes to the pressure transducer (see section 18.4). This is a differential pressure transducer because it measures the difference between the two pressures P_1 and P_2. The transducer output is amplified and the output voltage displayed both on a meter and on a chart recorder. The most important characteristic of the amplifier and transducer is that they must have very low drift, because the pressure applied to the pressure transducer will only be a few mmH_2O (0.01 kPa).

In addition to flow rate, the instrument also calculates volume by integrating the flow signal. By this means the same type of volume-against-time graph which the spirograph produces is also made available, but with the advantage that the patient can continue to breathe fresh air through the transducer whilst the measurements are taken. This volume output from the pneumotachograph can be used to measure MVV, in

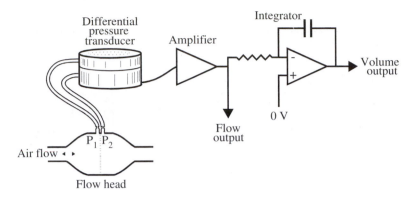

Figure 17.7. *Block diagram of a pneumotachograph.*

addition to the other measurements which can be made with a spirometer.

It might seem that the pneumotachograph is the ideal instrument for making ventilation measurements but, in practice, the instrument has some disadvantages. The major one is that the sensitivity to expired and inspired air is not the same, with the result that the volume trace does not have a stable baseline. The difference in sensitivity is due to the fact that the expired air is warmer and contains more water vapour than the inspired air. Some instruments try to correct for these factors by heating the air and the mesh in the flow head. However, it is impossible to correct the errors completely. It must also be remembered that, because the body is removing oxygen from the air and exhaling carbon dioxide, the actual volumes of inspired and expired air are not actually equal. To overcome this problem some pneumotachographs incorporate an electronic switch which resets the integrator at the end of each inspiration so that the volume of each inspiration can be measured from a stable baseline.

Figure 17.8 shows the outputs available from the pneumotachograph during a period of tidal breathing. On the left of the volume trace baseline drift can be seen. On the right, the integrator is reset on each cycle of expiration which enables the inspired volume to be recorded.

For routine monitoring purposes the pneumotachograph can be calibrated by using a large plastic syringe. A 400 ml capacity syringe can be moved in and out at a normal respiratory rate in order to calibrate the volume trace.

17.4.2. *Body plethysmography*

The pneumotachograph which has just been described is more likely to be used in a research situation than in routine patient measurement. The same is true for the body plethysmograph. We will show how it can be used to measure residual volume.

A body plethysmograph consists of a box in which the patient sits and breathes through a hole in the side. Each time the patient inhales volume V, the total volume of their body increases and so the pressure in the box increases. If the initial pressure in the box, P_b, and the air volume, V_b, are known, then the reduction in volume, V, can be calculated from the pressure rise, p, during inspiration,

$$P_b V_b = (P_b + p)(V_b - V)$$

therefore

$$V = V_b \frac{p}{P_b + p}. \tag{17.1}$$

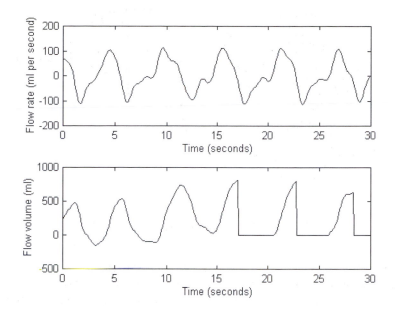

Figure 17.8. *The outputs of flow (upper trace) and volume (lower trace) from the pneumotachograph obtained during a period of tidal breathing. On the left of the volume trace baseline drift can be seen. On the right, the integrator is reset on each cycle.*

The system can be calibrated by injecting a known volume, from which V_b may be calculated.

If a pressure transducer is connected close to the mouthpiece then both the pressure of air entering the lungs, P_m, and the lung volume changes can be measured and then used to calculate residual volume.

- The patient is asked to breathe out as much as possible and P_{m1} is measured at the end of expiration. We want to find RV, the residual volume of the lungs.
- The pressure in the box, P_b, is noted.
- The patient is asked to attempt to breathe in deeply and, at the end of inspiration, the mouth pressure P_{m2} and the rise in box pressure, p, are noted. However, at the moment the patient attempts to inspire, a shutter valve (see figure 17.9) is operated to close the breathing tube so that no air enters the system. The air within the patient's lungs expands when they try to breathe, but no air is inspired.

Because no air enters the system, the volume of the patient's lungs increases and the lung pressure decreases when they attempt to inspire. Under these circumstances the product of pressure and volume will be constant if the temperature has not changed. The temperature of the air in the lungs will be constant and so this is a reasonable assumption.

Therefore,

$$P_{m1} RV = P_{m2}(RV + V)$$

where V is the increase in volume of the lung. This can be calculated from equation (17.1). Therefore,

$$RV = \frac{P_{m2} V}{P_{m1} - P_{m2}}.$$

(17.2)

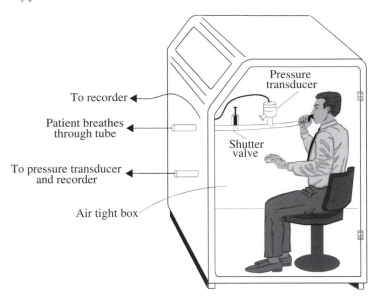

Figure 17.9. *Basic system of a body plethysmograph.*

The principle of the body plethysmograph is simple and it can give accurate measurements of lung volume. However, there are many practical difficulties such as those of sealing the box and measuring the very small pressure changes which appear in it.

The system is suitable for research purposes but not for routine patient measurements where problems such as the claustrophobic effect of being sealed in an airtight box also need to be considered.

17.4.3. Rotameters and peak-flow meters

A simple way of measuring flow is to insert a small turbine into the airflow. The turbine will behave rather like a windmill and the rate of revolution will increase with increasing flow. The turbine must have a very small inertia and it must be free to turn even at very low flow rates. In practice, it is not possible to make a perfect turbine, with the result that, at very low flow rates, the blades do not move and, at very high flow rates, the turbine tends to run at a constant speed. However, errors can be kept to about 10% at normal flow rates and instruments based upon this principle are used routinely. Air entering the transducer emerges from a series of tangential holes at the back of the turbine blades so that the meter only responds to flow in one direction. The number of turns of the turbine is therefore proportional either just to inspired air or just to expired air. The number of rotations of the turbine are counted by mounting a light source and phototransistor on either side of one of the blades. The flashes of light are counted electronically and the output calibrated directly in litres of air. This type of instrument is often called a rotameter or a respirometer and is widely used for routine respiratory function tests. There is another simple mechanical device which is widely used: the *peak-flow meter*, see figure 17.10. Peak-flow meters are instruments used to measure peak expiratory flow rate (PEFR). PEFR can be measured with a pneumotachograph, but the peak-flow meter is a much simpler instrument which can be carried around and used in a busy clinic or on the ward.

The patient expires as forcefully as possible into the flowmeter which balances the pressure exerted by the flow of air past a movable vane against a spiral spring. The peak-flow meter consists of a cylinder about 15 cm diameter and 4 cm deep with a nozzle through which the patient blows. There is an annular gap around

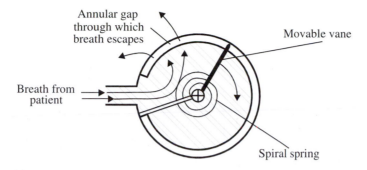

Figure 17.10. *A peak-flow meter.*

the edge of the cylinder so that the air can escape. The harder the patient blows the further the movable vane rotates so that more of the annular gap is exposed to allow the air to escape. There is a ratchet to hold the vane in the maximum position reached so that the peak reading can be made after the patient has blown through the nozzle. Either a washable plastic mouthpiece or a disposable cardboard mouthpiece are used to avoid the spread of infection between patients. A normal adult male should give a reading of about $7\,l\,s^{-1}$ but, of course, account must be taken of the sex, weight, age and height of the patient when interpreting a result. It is normal practice to take an average of at least three measurements of peak-flow rate to reduce the variability caused by patient effort.

17.4.4. Residual volume measurement by dilution

The most commonly used method of measuring total lung volume is by dilution. A known quantity of a substance is added to the system and from the measured dilution of the substance the volume of the system can be calculated.

If quantity, q, is added to the system of volume v, and the concentration after dilution is c, then

$$cv = q.$$

Obviously none of the quantity, q, must be lost and so an inert gas such as helium is used. Helium is not absorbed by the lungs and so the final concentration after a known amount of gas has been added will be solely determined by the volume of the lungs.

The principle of the technique is simple, but the equipment to actually make the measurement is quite complicated. It is shown in figure 17.11.

Once the helium has been added, the patient must be allowed to breathe for several minutes during which the helium can become thoroughly mixed with the air in their lungs. During this period they must breathe into a spirograph to which oxygen is added at the rate which the lungs are absorbing the gas (typically $300\,ml\,min^{-1}$). The carbon dioxide and water vapour exhaled must also be absorbed.

The following is a typical experimental sequence:

- Height, weight, age and sex data are taken, the patient is seated and a nose clip is attached. There is no helium in the system which has been flushed with air.
- The patient is connected to the system as in figure 17.11, with an oxygen supply of $300\,ml\,min^{-1}$ and the blower operating both to mix the gases and to pass them through the CO_2 and water vapour trap.
- The spirograph should give a tracing of tidal volume. The oxygen supply is adjusted so that the total system volume does not increase or decrease. If the supply is too great then the spirograph tracing will have a rising baseline, and if too little the baseline will fall.

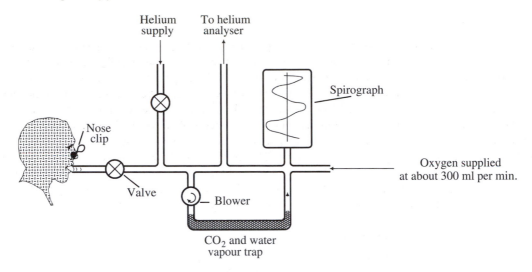

Figure 17.11. *System for making lung volume measurements by helium dilution.*

- When the correct supply of oxygen has been found the valve close to the patient's mouth is closed and the patient removed from the spirograph. The oxygen supply is closed.
- Sufficient helium is now introduced into the system to give a concentration of about 15% as read on the helium analyser. This reading should be constant and will be used to calculate the total amount of helium present. The amount will be the product of the concentration and the known volume, V_s, of the system.
- The patient is asked to make a maximal expiration and at this point is again connected to the spirograph system. The oxygen is reconnected at the rate determined above. The patient breathes normally into the system for 2–3 min, during which the helium should dilute to a point of equilibrium.
- The final value of helium concentration is measured and the patient is disconnected from the system.

The following measurements have been made:

$$V_s = \text{the system volume}$$
$$C_i = \text{the initial concentration of He}$$
$$C_f = \text{the final concentration of He.}$$

The total quantity of helium present at the beginning and the end of the test is the same and therefore

$$(RV + V_s)\, C_f = V_s C_i$$
$$RV \text{ (residual volume)} = V_s(C_i - C_f)/C_f.$$

Using this technique the residual volume can be determined to an accuracy of about 10%. Reproducibility of the measurement depends greatly upon the care which the operator takes, particularly in what he asks the patient to do, and in the care he takes to assess the correct end-expiration point at which the patient is connected to the system.

The helium analyser is usually based upon a measurement of the thermal conductivity of the gas. The gas is passed through a cell and the transfer of heat from a heater to a temperature sensor is measured. This type of measurement is not specific to a particular gas and so corrections have to be applied for the amount of other gases, such as oxygen and nitrogen, which are present.

17.4.5. Flow volume curves

All the measurements described so far have been expressed as a single number. A single number such as the PEFR or the FEV_1 is easy to handle and ranges of normal values can be established. However, a lot of information has been rejected by taking only one value. Current developments in respiratory function testing are towards the use of more information in order to give better discrimination between normal people and different clinical groups. One example of this is the use of flow/volume curves for single-breath analysis.

The pneumotachograph will give simultaneous measurements of airflow and total air volume. If these two measurements are plotted as X against Y, then a graph of flow against volume will be obtained. The result for a single expiration is given in figure 17.12. Obviously the vital capacity and the peak-flow reading can be made from this graph. However, the actual shape of the curve contains a lot of information about the ventilation capacity of the particular subject. It is possible for the vital capacity and the peak flow to be identical in two subjects and yet the shape of the two curves may be very different.

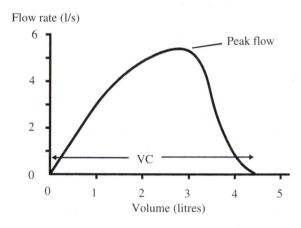

Figure 17.12. *A flow/volume curve obtained during a single expiration.*

17.4.6. Transfer factor analysis

What really matters when a person is unable to breathe properly is that oxygen fails to enter the blood. The technique of transfer factor analysis is used to give a measurement which quantifies the transfer of oxygen from the air to the blood.

It is possible to measure the concentration of oxygen in both expired and inspired air and so obtain a measurement of oxygen transfer across the alveolar membranes. Unfortunately oxygen can pass both ways through the membranes and it is therefore impossible to make accurate transfer measurements. However, carbon monoxide will also pass from air through the alveolar membrane, but once it reaches the blood it is much more readily absorbed by the haemoglobin. Therefore, by measuring the rate at which CO is absorbed by the lungs, a consistent indicator of both ventilation and perfusion can be obtained (figure 17.13).

The measurement system is similar to that used for helium dilution. A known amount of CO is added to the system into which the patient breathes and the rate at which the concentration of CO decreases is measured.

The CO analyser is usually based upon a measurement of the absorption of infrared light by the gas. Gases such as CO and CO_2 absorb infrared energy at a characteristic wavelength and so an analyser can be constructed which is specifically sensitive to one gas. A heated wire is usually used to emit infrared energy

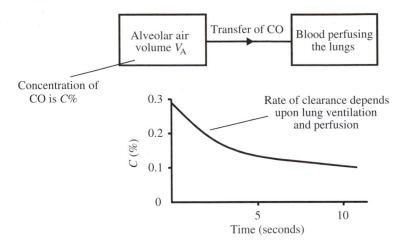

Figure 17.13. *Measurement of the transfer of CO from the lungs to the blood gives a good indicator of both ventilation and perfusion. The greater the transfer of CO the more rapid will be the fall in concentration in a closed system into which the patient breathes.*

which is passed through an optical cell containing the gas. The transfer of heat is detected with some type of temperature sensor.

The concentration of CO used is less than 1%, as higher concentrations could be a danger to the patient and would affect the breathing. This concentration of the gas is usually provided from a gas cylinder already containing the mixture with oxygen and nitrogen. In addition to the CO, the mixture contains about 15% helium; this is necessary in order to correct for the initial dilution of the CO by the air in the lungs.

The measurement sequence is:

- After maximum expiration, the patient breathes a measured volume of gas containing known concentrations of CO and He.
- The patient holds their breath for 10 s during which the CO is partially absorbed.
- The final concentrations of CO and He are measured.

The transfer factor can then be calculated from equation (17.6), the derivation of which will now be shown.

The patient inspires volume V_i of gas mixture, containing a carbon monoxide concentration of COI and helium concentration of HEI.

V_i is mixed with the volume of air already in the lungs, RV, and, after 10 s, the expired air contains a concentration, COF, of carbon monoxide, and concentration, HEF, of helium. Now,

$$V_i \, \text{HEI} = (\text{RV} + V_i) \, \text{HEF}.$$

$(\text{RV} + V_i)$ is the total volume, which is often called the effective alveolar volume. We will call it V_A

$$V_i \, \text{HEI} = V_A \, \text{HEF}. \tag{17.3}$$

The initial concentration of CO in the lungs will be

$$\frac{\text{COI} \, V_i}{V_A} = \frac{\text{COI} \, \text{HEF}}{\text{HEI}}.$$

This will decay as the CO is absorbed. The fall with time, t, will be exponential because the rate at which CO is absorbed will be proportional to the concentration of the gas. Therefore if the concentration is $C(t)$,

$$C(t) = \frac{\text{COI HEF e}^{-kt}}{\text{HEI}}.$$

The volume of CO present at any time, t, is

$$V_A\, C(t) = V_A \frac{\text{COI HEF e}^{-kt}}{\text{HEI}}. \tag{17.4}$$

In this equation, k is the rate constant that determines how rapidly the CO is absorbed.

The rate of loss of CO is given by differentiating this equation

$$V_A \frac{\text{d}C(t)}{\text{d}t} = -k V_A C(t). \tag{17.5}$$

From equation (17.4) we know that COF on expiration is

$$\text{COF} = \frac{\text{COI HEF e}^{-kt}}{\text{HEI}}.$$

Taking logarithms

$$\log_e \frac{\text{COI HEF}}{\text{COF HEI}} = kt.$$

Substituting into equation (17.5) we get the rate of loss of CO as

$$-V_A C(t) \frac{1}{t} \log_e \frac{\text{COI HEF}}{\text{COF HEI}}.$$

The negative sign indicates a loss of CO.

This is a formula for the transfer factor. However, the result is usually expressed as a fraction of the partial pressure of CO. This is proportional to $C(t)$ and so the following formula is obtained:

$$\text{transfer factor} = \frac{V_A 160}{t} \log \frac{\text{COI HEF}}{\text{COF HEI}}. \tag{17.6}$$

V_A can be found from equation (17.3). 160 is a constant which is correct if V_A is in litres, t in seconds and the concentrations in percentage form. The transfer factor has the units ml of CO absorbed per minute for each mmHg of the CO partial pressure. A typical normal value is 30 ml min^{-1} mmHg^{-1}.

17.5. RESPIRATORY MONITORING

Monitoring simply means 'listening in' in the way that you might monitor a telephone conversation. During an operation the anaesthetist often wishes to monitor the patient's cardiac activity or respiration. In an intensive care unit following surgery, both heart rate and respiration may be monitored and, in addition, perhaps arterial and venous pressures. Young babies, particularly those born prematurely, often have heart rate, respiration and temperature monitored continuously. They are particularly prone to produce periods of apnoea (cessation of breathing), and so apnoea alarms are used on special baby care units.

The level of respiratory monitoring given in these situations is much simpler than that which is possible in a respiratory function unit. Whilst it is possible to connect a patient undergoing surgery to a pneumotachograph it is not feasible in most other situations. What is required is continuous monitoring of breathing using a technique which interferes least with the patient. The techniques used for monitoring ventilation range from those, such as impedance plethysmography, which give a reasonably accurate continuous measurement of lung volume, to movement detectors which only give an indirect measurement of respiration. The major purpose of respiration is of course to transport oxygen into the body. Blood gas analysis can provide data on the effectiveness of oxygen transport, but it is not a convenient method of monitoring as it involves gaining access to the blood. However, some indirect methods of monitoring do exist and we continue this chapter with an explanation of the technique of pulse oximetry which depends upon the optical absorption properties of blood.

17.5.1. Pulse oximetry

Traditionally the concentration of O_2 and CO_2 in the blood are determined by withdrawing a small volume of blood, either via a catheter or an arterial catheter. The gas concentrations can then be measured by various spectrometry or electrochemical techniques. However, withdrawal of blood carries some risks, it is certainly invasive and can only be applied intermittently. Errors can arise in the blood sampling and very often the time delay between sampling and availability of the result reduces the usefulness of the measurement. Continuous monitoring can offer major advantages in the management of critically ill patients.

Skin is normally fairly impervious to blood gases but not completely so and the diffusion through the skin can be increased dramatically by heating the skin. This observation has given rise to many techniques of transcutaneous gas analysis. A heated probe is put on the skin and various electrochemical techniques can be used to measure the gases which diffuse through the epidermis of the skin from the capillaries in the dermis. The techniques have been found to work much better in neonates than in adults because diffusion through the skin is much higher in neonates. Non-invasive blood gas monitoring has not replaced invasive blood gas analysis, but the development of transcutaneous techniques has been adopted in neonatal intensive care situations.

Pulse oximetry is a transcutaneous technique for measuring oxygen saturation but it does not depend upon gas diffusion through the skin. It depends upon the observation that blood has a characteristic red or blue colour depending on the concentration of oxygen in the blood.

Oxygen transport

Most of the oxygen in blood is transported by attachment to haemoglobin (Hb) molecules to form oxy-haemoglobin (HbO_2) which is carried inside the red blood cells. A very small amount of oxygen is carried in the water of the plasma. The *oxygen saturation* So_2 is the ratio of the oxyhaemoglobin to the total amount of haemoglobin. It is usually expressed as a percentage and in normal arterial blood is about 98%—in venous blood it is about 75%. So_2 depends upon the partial pressure of oxygen (Po_2) and the two are related by the sigmoid-shaped oxyhaemoglobin dissociation curve which appears in many texts. So_2 and Po_2 are related by the dissociation curve, but care must be taken as the curve depends upon factors such as blood temperature and pH.

Optical properties of blood

Arterial blood appears red and venous blood more blue because haemoglobin and oxyhaemoglobin have different optical absorption spectra. The attenuation of electromagnetic radiation by a sample can be described by the *Beer–Lambert* law in the form

$$I = I_0 10^{-\alpha d} \tag{17.7}$$

where I_0 is the intensity incident on the sample, I the exit intensity and d the thickness of the sample. α is the linear absorption coefficient which will depend upon the concentration of the absorber within the sample. α will in general also depend upon the wavelength of the radiation which is being transmitted through the sample. Strictly speaking, the Beer–Lambert law only applies if the radiation is monochromatic, the radiation is collimated and the sample contains no scattering particles. These assumptions are usually not true for measurements made on blood but the law is found to apply fairly well in practise.

Figure 17.14 shows how the absorption coefficient α varies for haemoglobin and oxyhaemoglobin over the wavelengths 600–1000 nm. 600 nm is in the visible red part of the spectrum and 1000 nm is in the infrared. It can be seen that at some wavelengths, such as 660 nm, there is a large difference between the absorption coefficients for Hb and HbO$_2$, whereas at about 800 nm the absorption coefficient is about the same for both Hb and HbO$_2$ (this is called the isobestic wavelength). By making measurements at two different wavelengths it is possible to determine the oxygen saturation So$_2$.

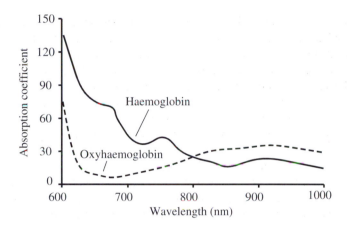

Figure 17.14. *Sketch of the optical absorption spectra for haemoglobin (Hb) and oxyhaemoglobin (HbO$_2$) in the red to near infrared part of the spectrum. The linear absorption coefficient has the units of mmol^{-1} l m^{-1}.*

From equation (17.7) we can obtain

$$\alpha d = \log_{10} \frac{I_0}{I}. \tag{17.8}$$

By measuring I_0 and I we can determine αd but only α will be a function of wavelength. If we make measurements at wavelengths λ_1 and λ_2 then we can obtain

at λ_1: $\alpha_{\lambda 1} = A(\text{cHb} + k\,\text{cHbO}_2)$ (17.9)

and at λ_2: $\alpha_{\lambda 2} = B(\text{cHb} + \text{cHbO}_2)$ (17.10)

where cHb and cHbO$_2$ are the concentrations of haemoglobin and oxyhaemoglobin, respectively, and A, k and B are constants determined by the relative absorption coefficients for Hb and HbO$_2$ at the two wavelengths.

The total concentration cH will be given by cHb + cHbO$_2$.

Substituting into the two previous equations

$$\alpha_{\lambda 1} = A(cH - cHbO_2 + k\,cHbO_2) = A(cH - (1-k)\,cHbO_2)$$

$$\alpha_{\lambda 2} = B\,cH$$

$$\frac{\alpha_{\lambda 1}}{\alpha_{\lambda 2}} = \frac{A}{B}\left(1 - \frac{(1-k)\,cHbO_2}{cH}\right) = \frac{A}{B}(1 - (1-k)\,So_2)$$

where So_2 is the oxygen saturation $cHbO_2/cH$.

Therefore,

$$So_2 = \left(1 - \frac{B}{A}\frac{\alpha_{\lambda 1}}{\alpha_{\lambda 2}}\right)\frac{1}{(1-k)}. \tag{17.11}$$

Now, if we know the approximate overall attenuation, we can determine $\alpha_{\lambda 1}/\alpha_{\lambda 2}$ using equation (17.8) and the measured intensities at the two wavelengths. B/A and k are simply ratios of absorption values which can be determined from curves such as shown in figure 17.14. Thus we can determine the oxygen saturation.

General form of spectroscopy

In the previous section we made measurements at just two wavelengths to identify the two concentrations of Hb and HbO$_2$, but the principle can be extended to a large number of wavelengths and absorbers with different absorptions. We can generalize equation (17.7) and obtain

$$I = I_0(10^{-(dc_1\alpha_1)} \times 10^{-(dc_2\alpha_2)} \ldots 10^{-(dc_n\alpha_n)})$$

and

$$\frac{1}{d}\log_{10}\left(\frac{I_0}{I}\right) = R = c_1\alpha_1 + c_2\alpha_2 + \cdots + c_n\alpha_n. \tag{17.12}$$

This is one equation in n unknowns, the values of c. By making measurements at $(n-1)$ other wavelengths we can produce a set of equations of the form of equation (17.12),

$$R_1 = c_1\alpha_{11} + c_2\alpha_{12} + c_3\alpha_{13} + \cdots + c_n\alpha_{1n}$$

$$\vdots \tag{17.13}$$

$$R_n = c_1\alpha_{n1} + c_2\alpha_{n2} + c_3\alpha_{n3} + \cdots + c_n\alpha_{nn}$$

where the subscript on R and the first digit of α refers to the wavelength. These equations can be solved by matrix inversion and hence the set of concentrations c_n can be determined. Thus we have a general method by which, if we know the absorption spectra of a range of substances in a mixture, we can measure the absorption of the mixture over a range of wavelengths and hence calculate the relative concentrations. However, it must be born in mind that we will need to make measurements at as many wavelengths as we have components in the mixture. Also the choice of wavelengths will be important otherwise the discrimination between components will be lost in the measurement errors.

Practical measurement

In early *ear oximetry* optical transmission measurements were made through the pinna of the ear over a range of wavelengths. The problem in simply using equation (17.11) was in taking account of the components other than blood which are in the ear. To avoid this problem, instead of detailed measurements of the absorption values α_n, a set of values determined empirically from measurements made on a set of normal subjects, where So_2 was measured independently, were used.

Pulse oximetry uses a different approach to the problem of light absorption and scattering from other tissues in the ear. On cardiac systole the volume of blood in the ear increases slightly and this causes an increase in optical absorption by the ear. Now this increase is caused solely by arterial blood and hence if this increase of absorption is measured at two wavelengths then equation (17.11) can be used to determine S_{O_2}. The values of α are determined empirically by measuring the optical absorptions associated with independent measurements of arterial oxygen saturation.

The sensor of a pulse oximeter usually consists of two LEDs at red and infrared wavelengths (often 660 and 805 nm) and a photodetector. Most systems place the transmitter and detectors on opposite sides of the earlobe, finger or toe to measure the absorption on transmission, but reflectance systems have also been developed. Typical recorded waveforms are shown in figure 17.15. The cardiac-related changes are usually 1–5% of the baseline transmitted signal. If the arterial inflow was all of HbO_2 then the amplitude of the cardiac pulse would be much reduced at the longer wavelength, but if the inflow was of Hb, then the amplitude would be higher at the longer wavelength.

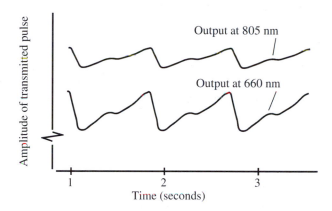

Figure 17.15. *This shows typical optical transmission waveforms at the two wavelengths 660 nm and 805 nm. The pulsatile waveform is caused by the inflow of arterial blood on cardiac systole. The amplitude of the pulse is less at 805 nm because HbO_2 has a higher absorption coefficient at this wavelength. The break in the vertical scale is to show that the pulsatile component is very much less than the total transmitted intensity.*

Pulse oximetry is a simple and usually reliable technique which is now in widespread clinical use. The cardiac-related change in optical absorption is relatively easy to record but problems can arise if there is hyperthermia or vascular disease which reduce the size of the arterial systolic pulse. You can test the operation of a pulse oximeter by holding your breath for about 30 s; this should cause the S_{O_2} to fall from about 98% to 80%.

17.5.2. Impedance pneumography

Impedance plethysmography means the measurement of fullness by first measuring electrical impedance. The technique can be used to monitor respiration and when used for this purpose it is often called *impedance pneumography*. In this technique two surface electrodes are placed on either side of the thorax and the electrical impedance between them is measured. It is found that the impedance rises on inspiration and falls on expiration. The changes are about 1 Ω on a base impedance of about 200 Ω and by recording the changes a continuous indication of lung ventilation is obtained. The method is widely used as a method of monitoring breathing. We will firstly consider what might be the origin of these changes and then describe how such measurements can be made. When we breathe, both the shape of the thorax and the impedance of lung tissue change because of the increase in air content. Both these changes can affect the transthoracic impedance.

Changes in shape of the thorax

You should remember that a short fat electrical conductor will have a lower resistance than a long thin conductor. Even if we use a constant volume of copper, the electrical resistance will depend upon the shape of the conductor (see section 20.2.2 on wire strain gauges). The human body can be considered as an electrical conductor and, if its shape changes, then the electrical impedance will change. Because the shape of the chest changes as we breathe, it is reasonable to expect that impedance will change.

   Calculating the electrical impedance of a complex shape such as the thorax is very difficult. We will simply consider a cylindrical conductor (see figure 17.16) of resistivity ρ, length l and cross-sectional area a and consider the impedance between the faces. The impedance Z will be given by $Z = \rho l/a$.

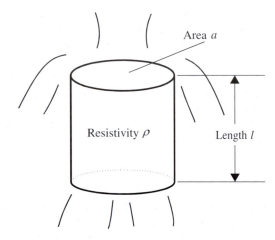

Figure 17.16. *Cylindrical model of the thorax.*

   If we take the cylinder as an approximation to the thorax then we can consider what happens when air enters the thorax and so the volume V changes. This situation is considered in detail in Chapter 19 (section 19.6.2) and the conclusion is that the impedance will not change if both area a and length l change together. However, if the change is only in length then impedance Z will *rise* in proportion to volume change dV. If the change is only in cross-sectional area then impedance will *fall* in proportion to dV.

   This is a disappointing conclusion as it means that unless we know exactly how a change in lung volume distorts the shape of the thorax then we cannot relate a measured impedance change to an associated change in lung volume. Fortunately changes in the volume of the thorax as a result of respiration are found to be less important in determining thoracic impedance than the changes in impedance of lung tissue. Nonetheless the affects of geometry are not insignificant and we should therefore consider impedance pneumography as a qualitative rather than quantitative method unless we can control for changes in geometry.

Changes in the impedance of lung tissue

When air enters the lungs it increases the resistivity of the tissue. Electrical current has to flow around the air-filled alveoli and hence the resistivity rises with increasing air content. Table 17.1 lists typical values for

Table 17.1. *Typical resistivity values for some tissues.*

Tissue	Resistivity (Ω m)
Blood	1.6
Muscle: transverse	>4
longitudinal	1.5–2.5
Fat	16
Lung	10–20 (expiration to inspiration)
Bone	>100

the resistivity of a range of human tissues. Lung tissue has a relatively high resistivity and the resistivity approximately doubles from maximum expiration to maximum inspiration.

When an impedance measurement is made across the thorax the lungs make a major contribution to the measurement and the changes with lung ventilation are large. The average resistivity of a limb is about 2 Ω m, whereas because of the high resistivity and large volume of the lungs the average for the thorax is about 5 Ω m.

Calculation of the expected changes in resistivity as air enters the alveoli is a difficult problem which we cannot cover here. However, it can be assumed that lung tissue resistivity will rise with increasing air content and this is the main contribution to the impedance pneumograph signal.

Before we leave the subject of impedance pneumography we will consider what frequency we should use when making a transthoracic impedance measurement.

Impedance changes with frequency

In the preceding sections the words 'resistance' and 'impedance' were used rather indiscriminately. If two electrodes are attached to the skin, then the resistance between them can be measured by applying a DC voltage, measuring the current that flows and then applying Ohm's law. If an AC voltage is applied, then we can measure the impedance between the electrodes, which will be found to decrease as the frequency of the AC voltage is increased. Figure 17.17 shows the results which might be obtained. The fact that both tissue and electrodes have a complex impedance was discussed in sections 8.3–8.5 and 9.2, respectively.

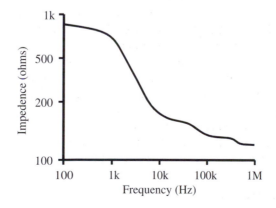

Figure 17.17. *The electrical impedance measured between a pair of skin electrodes attached to the thorax.*

A very similar change of impedance to that shown in figure 17.17 is found from an equivalent circuit of two resistors and a capacitor (figure 17.18). At very low frequencies the capacitor C has a very high impedance and so the total resistance is approximately equal to R; at very high frequencies the capacitor has a low impedance and so the total impedance is equal to S in parallel with R. More complex models are described in Chapter 8 and indeed are needed to explain the high-frequency inflections seen in the graph of figure 17.17

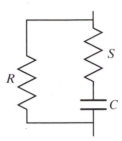

Figure 17.18. *An equivalent circuit which exhibits similar impedance changes to those shown in figure 17.17.*

S, R and C can be shown to depend very much upon the electrodes and how these are applied, whereas the impedance of the tissues is less significant. The impedance of the tissues actually gives rise to the small changes with frequency which are seen at the frequencies above about 10 kHz in figure 17.17. If we wish our impedance pneumograph to have maximum sensitivity to tissue impedance and minimum sensitivity to electrode impedance then we should make our measurement at a frequency well above 10 kHz. 100 kHz is often used and there is good evidence that even higher frequencies would be better still. The actual change in impedance with ventilation depends on where the electrodes are placed across the chest and the build of the person. A greater percentage change is found in small people than in large people.

The equipment

Most impedance pneumographs measure the impedance at about 100 kHz. This frequency is too high to stimulate nerves and so there is no hazard attached to the technique. The IEC safety standards permit a current of 10 mA at a frequency of 100 kHz; the currents used in impedance plethysmography are usually less than 1 mA.

A pair of electrodes placed 10 cm below the armpits might give an impedance of 200 Ω which will fluctuate by 1 Ω as the person breathes. We can construct equipment to measure these changes relatively easily. The necessary electronics are:

- A 100 kHz oscillator.
- A current generator circuit which will pass a constant amplitude alternating current between the electrodes.
- A peak detector circuit. This will give a DC voltage equal to the amplitude of the voltage developed between the electrodes. As the current is constant this voltage will be directly proportional to the electrode impedance.
- An AC amplifier which will reject the baseline impedance between the electrodes and record just the changes caused by respiration.

Impedance plethysmography is the most popular method of measuring respiration over long periods of time, but it must be remembered that, because it is not measuring lung volume directly, it can give false results. If the patient moves their body or arms then artefacts are produced. Movement of the electrodes will change their impedance slightly and so give a false deflection on the trace. The amplitude of the impedance change is also changed by the position of the patient. Simply by moving from a sitting position to a supine position the impedance fluctuations may change in amplitude by 50%.

Many patient monitoring systems include both ECG/EKG and respiratory monitoring and, in this case, it is possible to use the same electrodes for both measurements. The impedance plethysmograph operates at 100 kHz, whereas the components of the ECG/EKG are below 100 Hz so that the two need not interfere with each other.

17.5.3. Movement detectors

The impedance plethysmograph gives an indirect measurement of lung volume. There are many other methods of making an indirect measurement of lung volume. Many of these methods rely upon detecting chest wall movements. *Strain gauge plethysmography* uses a band placed round the chest and a strain gauge to record the changes in tension of the band as the chest wall moves. The word 'pneumography' is often used to describe a system consisting of a length of large diameter tubing which is tied around the chest. The tubing is connected to a pressure transducer, which records the fluctuations in pressure as the chest expands and thus changes the volume of the air filling the tubing. There are many other methods, but none are used very widely and so do not justify much study.

Many methods have been devised for measuring breathing patterns in babies. One method uses an air mattress on which the child lies. There are sections to the mattress and, as the baby breathes, air flows from one section to another. This flow of air is detected by means of a thermistor through which a current is passed. The air will cool the thermistor and so change its resistance. A second method uses a pressure-sensitive pad which is placed underneath the cot mattress. Respiratory movements produce regular pressure changes on the pad, and these alter the capacitance between electrode plates incorporated into the pad. The capacitance changes can be detected by applying an alternating voltage across the pad. As the impedance of the pad will be inversely proportional to capacitance the current which passes will increase as the capacitance increases.

If these types of movement detector are set up carefully underneath a young baby they can give a good record of breathing. However, they are subject to artefact as they cannot distinguish between breathing and the baby waving its arms around or somebody knocking the cot.

17.5.4. Normal breathing patterns

The greatest difficulty in the use of almost every physiological measurement technique is in deciding what is a 'normal' result. Normal people are not identical so that even a simple measurement such as temperature will give a range of values for normal subjects.

A great deal of work has been done to define normal results for the various respiratory function tests. The ways in which the results of respiratory function tests are presented to the doctor are explained in section 17.6.2, and this includes an assessment of whether the results are normal or abnormal. However, normal values have not been established for long-term breathing patterns. People do not breathe completely regularly and it is actually very difficult to describe the record that is obtained if breathing is recorded over a period of several hours.

Patterns such as that shown in figure 17.19 may be analysed using a computer and features such as rate, variability in rate and periods of apnoea extracted but there are no clear definitions of what features are 'normal'. Figure 17.20 shows the results of measuring respiration rate in 67 infants and plotting the results against the age of the infants. You can see that respiratory rate decreases with age. The regression line showing how the average rate changes with age is drawn, and also the two lines which are two standard deviations outside this average. There is only a 1 in 20 chance that a normal child will have a respiratory rate

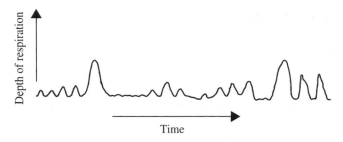

Figure 17.19. *The breathing pattern recorded from a baby over a period of 1 min.*

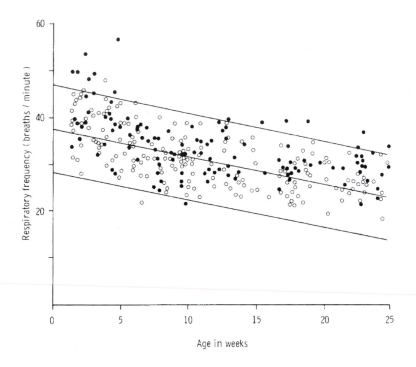

Figure 17.20. *Change of respiratory frequency with age in 67 infants (some infants had measurements made at monthly intervals). The full dots are from infants shown to be at high risk of sudden infant death (cot death). (From C I Franks et al 1980 Respiratory patterns and risk of sudden unexpected death in infancy, Arch. Disease Childhood **55** 595–9.)*

outside the limits of these two lines. This is the type of information which can be used to analyse records and assess the probability that the records are normal or abnormal.

17.6.   PROBLEMS AND EXERCISES

17.6.1.   *Short questions*

a   What is meant by lung perfusion?
b   What is a *normal* adult resting respiration rate?
c   Give one important reason why small animals breath more rapidly than we do.
d   What does the term 'residual volume' mean?
e   Is the oesophagus or the bronchus the way into the lungs?
f   State the two main functions of the lungs.
g   Is a peak expiratory flow rate (PEFR) of $5 \, \mathrm{l \, s^{-1}}$ likely to be a normal figure?
h   Is the flow of air through the lungs very much greater than the flow of blood?
i   If quantity q of an indicator is added to a sealed system and the mixed concentration is c then what is the volume of the system?
j   What is FEV_1?
k   What is a whole-body plethysmograph?
l   Which two gases are used in transfer factor analysis?
m   What *pulse* is referred to in the title *pulse oximetry*?
n   What are the units in which airways' resistance is measured?
o   Why is pulse oximetry more accurate in babies than in adults?
p   What is the main reason why the trace of air volume recorded using a differential pressure pneumo-tachograph will drift upwards or downwards with time?
q   About what fraction of lung volume do we use during normal tidal breathing?
r   What is PAO_2?
s   What does the *Beer–Lambert* law describe?
t   Is 1 kHz a normal operating frequency for an impedance pneumograph?
u   What is apnoea?
v   Does the impedance pneumograph show an increase or a decrease of impedance during inspiration?
w   Does $1 \, \mathrm{m^3}$ of air weigh about 10 g, 100 g or 1000 g?
x   Do blood gases diffuse through skin?

17.6.2.   *Reporting respiratory function tests*

Objective

To calculate commonly measured respiratory parameters and to present the results in statistical terms.

Method

You will need to refer to sections 17.3.1, 17.4.4 and 17.4.6 in order to complete this exercise.
   A request for respiratory function tests may appear on the type of form shown in table 17.2 which also allows for recording the results. Seven measurements are required. These are:

FVC	Forced vital capacity
FEV_1	Forced expiratory volume at 1 s
PEFR	Peak expiratory flow rate
TL	Ventilation perfusion transfer factor
TLC	Total lung capacity
FRC	Functional residual capacity
RV	Residual volume

Table 17.2. *A request form for a range of respiratory function tests.*

Clinical information:		Name:			
		Address:			
		Unit No.:			
		Consultant:	Hospital:	Ward:	
	Date of test:	Age: 31			
Test No. Ref.:		Height (m): 1.75	Weight (kg): 72	Sex: M	

	FVC (*l*)	FEV₁ (*l*)	PEFR (*l/min*)	TL (*ml/min/mmHg*)	TLC (*l*)	FRC (*l*)	RV (*l*)
Measured			6.2		5.0	2.7	
Predicted							
% Predicted							

You will find all of these measurements explained earlier in the chapter with one exception. The functional residual capacity is the volume of air in the lungs at the resting expiratory level during normal breathing. The measured value is compared with the value predicted for a patient of the given age, sex, height and weight. The comparison is made by expressing the measured value as a percentage of the predicted value.

Some of the measurements are already marked. Use the following data to complete the request card (figure 17.21).

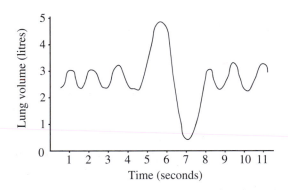

Figure 17.21. *This is the spirograph record for the patient.*

During the helium test the patient was asked to make a maximal expiration and they were then connected to the spirograph system for 2–3 min. These results were obtained:

initial concentration of helium   14.6%
final concentration of helium   10.5%.

The volume of the spirometer and the dead space of the connecting tubing were determined as 8 litres.

During the carbon monoxide transfer test the following results were obtained:

Initial concentration of CO	0.28%
Concentration of CO on expiration after 10 s	0.10%
Initial concentration of He	14%
Final concentration of He	8.0%
Volume of gas mixture inspired	3.81

Results

In order to complete the form giving the expected values you will have to consult tables of normal values (see, for example, J H Cotes, *Lung Function*). Are the results which you have obtained normal or abnormal?

17.6.3. Use of peak-flow meter

Objective

To determine the reproducibility of peak-flow readings and to compare a group of normal subjects before and after exercise.

Equipment

A moving vane peak-flow meter or a pneumotachograph.

Method

- Use yourself as a subject and take 30 peak-flow readings at 30 s intervals.
- Calculate the mean and standard deviation for your group of 30 readings. Also plot a graph showing how your peak-flow readings changed with time.
- Now obtain the co-operation of a group of normal subjects. Try to find at least five people.
- Take three flow readings for each subject at intervals of 15 s. Then ask the subjects to exercise for a period of 3 min. You can ask them to step up and down from a step at a fixed rate. Repeat your three peak-flow readings after the exercise.
- Calculate the mean and standard deviation of your readings for all the subjects; firstly before the exercise and then after exercise.

Results

Can you obtain answers to the following questions:

Would you be able to measure a fall in your own PEFR of 15%?

Is it best to take just one reading of PEFR or to take an average of several readings?

Did your results change with time? Can you suggest a statistical test that would enable you to prove a change with time?

What was the effect of exercise on your normal subjects?

17.6.4. Pulse oximeter

A pulse oximeter makes optical transmission measurements through the ear at wavelengths of 660 and 805 nm. The amplitude of the cardiac-related changes are found to be 150 and 50 mV, respectively. Calculate the oxygen saturation on the assumption that 805 nm is the isobestic point, that the ratio of the absorption coefficients for haemoglobin at 660 and 805 nm, respectively, is 3 and that the ratio of the absorption coefficients at 660 nm for haemoglobin and oxyhaemoglobin, respectively, is 10:1. Also assume that the effective total absorption across the ear lobe at 660 nm is 90% of the incident intensity.

Notes on the answer

Call 660 nm λ_1 and 805 nm λ_2.
We can use equation (17.8) to determine αd as $\log 10 = 1$.
We can then find α_2 as $(1 + \log(150/50)) = 1.477$.
We can then use equation (17.11) where $B/A = \frac{1}{3}$ and $k = \frac{1}{10}$ to give the oxygen saturation as 86.3%.

Answers to short questions

a   Perfusion of the lungs is the process by which blood flows through the capillaries around the alveoli.

b   A normal adult resting respiration rate is about 12 min^{-1}.

c   Small animals breath more rapidly than we do because the balance of elastic and inertial forces means that they expend less energy by breathing rapidly.

d   Residual volume is the air remaining in the lungs of a subject who has expired as far as possible.

e   The bronchii are the tubes which enter the lungs.

f   The purpose of the lungs is to transfer oxygen from the air into the blood and to discharge carbon dioxide in the reverse direction.

g   Yes, $5 \, \text{l} \, \text{s}^{-1}$ is a typical PEFR.

h   No, the flows of blood and air through the lungs are very similar.

i   Volume $= q/v$.

j FEV_1 is the volume of air that can be expired in 1 s following full inspiration.

k   A whole-body plethysmograph is an airtight box inside which a subject can sit but breath through a tube connected to the outside of the box. It enables accurate measurements of airflow and volume to be made.

l   Carbon monoxide and helium are used in making transfer factor analysis measurements.

m   Pulse oximetry uses the change in optical transmission during the cardiac arterial pulse to determine oxygen saturation of the blood.

n   Airways resistance is pressure/flow rate and the units are kPa l^{-1} s.

o   Pulse oximetry works better in neonates than in adults because there is less absorption in the tissues between the capillary bed and the skin surface. The cardiac pulse can be measured with a higher signal-to-noise ratio.

p   Pneumotachographs drift because they have a different sensitivity to inspired and expired air. The difference is caused by the different temperatures, densities and humidities of the two flows.

q   We only use between 5% and 10% of total lung volume during tidal breathing.

r PAO_2 is the arterial oxygen pressure: the partial pressure of oxygen dissolved in the arterial blood.

s   The Beer–Lambert law relates the attenuation of light to the thickness of tissue traversed and an absorption coefficient.

t   No, most impedance pneumographs operate between 20 and 200 kHz.

u   Apnoea is a period where breathing ceases. Both adults and neonates stop breathing from time to time.

v   Impedance increases with inspiration.

w   The density of air is just over 1 kg m^{-3}.

x   Blood gases do diffuse through skin. The diffusion increases with skin temperature.

BIBLIOGRAPHY

Cotes J E 1979 *Lung Function* 4th edn (Oxford: Blackwell)

Mendelson Y 1988 Blood gas measurement, transcutaneous *Encyclopaedia of Medical Devices and Instrumentation* ed J G Webster (New York: Wiley) pp 448–60

Petrini M F, Coleman T G and Hall J E 1986 Basic quantitative concepts of lung function *Mathematical Methods in Medicine* pt 2 ed D Indram and R F Bloch (Chichester: Wiley)

Petrini M F 1988 Pulmonary function testing *Encyclopaedia of Medical Devices and Instrumentation* ed J G Webster (New York: Wiley) pp 2379–95

Webster J G (ed) 1997 *Design of Pulse Oximeters* (Bristol: IOP Publishing)

CHAPTER 18

PRESSURE MEASUREMENT

18.1. INTRODUCTION AND OBJECTIVES

The measurement of pressure and flow form a central part of many physiological measurements. The measurement techniques are less straightforward than they appear at first sight, because of the need to be minimally invasive. In this chapter, we examine both invasive and non-invasive measurement of pressure, with one example of each considered in detail. Most blood flow measurements are indirect, with the flow being calculated on the basis of a set of assumptions which are satisfied to a varying extent. An understanding of the importance of these assumptions is an essential part of any indirect measurement technique.

Questions we hope to answer include:

- What pressures might we find in the body?
- Is there such a thing as a 'normal' blood pressure?
- How can we measure the pressure inside the eye?
- What are the problems in using a fluid-filled catheter to measure blood pressures?
- Can we measure pressures accurately without being invasive?

In Chapter 2 we considered the basic physics of pressures and you should read section 2.2 before reading this chapter. In this chapter we set out to consider how to measure pressure. We look at the problems and the sources of inaccuracy. Section 18.5 is mathematical and deals with the measurement of pressure at the end of a fluid-filled catheter. It is an important subject but the general reader can skip the mathematics. Nonetheless they should use the short questions at the end to test their comprehension of the principles.

18.2. PRESSURE

Pressure is defined as force per unit area, and the SI unit of pressure is therefore newtons per square metre ($N\,m^{-2}$). This definition reminds us that forces can be determined by measuring the pressure over a known area. An example is the urethral pressure profile, in which the force exerted by the urethral sphincter (muscles exert forces) is measured by a fluid pressure.

The SI unit of pressure is not obviously related to physiological pressures, and a pedantic insistence on using $N\,m^{-2}$ or Pa can obscure the practical details of pressure measurement. As an example, bladder pressure is traditionally measured in cmH_2O, and thus the size of the measurement error due to vertical displacement of a transducer with a fluid-filled catheter is immediately obvious. Table 18.1 lists the relationship between the different units that are commonly encountered.

Table 18.1. *Pressure unit conversion.*

1.0 kPa	1.0 kN m^{-2}	10.0 mb
1.0 kPa	7.5 mmHg	10.0 mb
1.0 mb	100.0 Pa	0.75 mmHg
1.0 mmHg	133.32 Pa	1.33 mb
1.0 cmH$_2$O	0.74 mmHg	98.07 Pa
1.0 dyn cm^{-2}	0.1 Pa	
1.0 kg cm^{-2}	98.07 kPa	
1.0 atm	101.3 kPa	

18.2.1. *Physiological pressures*

Blood pressure is possibly the most important pressure that is commonly measured. Perfusion of oxygen and nutrients depends on pressure gradients, and the blood pressure is an important indicator of the condition of the cardiovascular system. The normal ranges for arterial and venous blood pressure are well documented, and examples of these are given in tables 18.2 and 18.3 (abridged from information in Documenta Geigy).

Table 18.2. *Arterial blood pressure (mmHg) versus age. Pressures are (mean $\pm$ standard deviation).*

	Systolic		Diastolic	
Age (years)	Men	Women	Men	Women
1	96 $\pm$ 15	95 $\pm$ 12	66 $\pm$ 12	65 $\pm$ 15
10	103 $\pm$ 6	103 $\pm$ 7	69 $\pm$ 6	70 $\pm$ 6
20	123 $\pm$ 14	116 $\pm$ 12	76 $\pm$ 10	72 $\pm$ 10
30	126 $\pm$ 14	120 $\pm$ 14	79 $\pm$ 10	75 $\pm$ 11
40	129 $\pm$ 15	127 $\pm$ 17	81 $\pm$ 10	80 $\pm$ 11
50	135 $\pm$ 19	137 $\pm$ 21	83 $\pm$ 11	84 $\pm$ 12
60	142 $\pm$ 21	144 $\pm$ 22	85 $\pm$ 12	85 $\pm$ 13
70	145 $\pm$ 26	159 $\pm$ 26	82 $\pm$ 15	85 $\pm$ 15
80	145 $\pm$ 26	157 $\pm$ 28	82 $\pm$ 10	83 $\pm$ 13
90	145 $\pm$ 23	150 $\pm$ 24	78 $\pm$ 12	79 $\pm$ 12

Table 18.3. *Venous blood pressure (mmHg) measured at the median basilic vein at the elbow at the level of the tri-cuspid valve.*

	Mean	Range
Children: 3–5 years	3.4	2.2–4.6
5–10 years	4.3	2.4–5.4
Adults: men	7.4	3.7–10.3
women	6.9	4.4–9.4

The definition of a 'normal range' for blood pressure is not easy, because the definition of a normal group would involve the exclusion of individuals with raised blood pressure, which is a circular definition. Different definitions are: normal range of systolic or diastolic pressure = mean $\pm$ 1.28 $\times$ standard deviation; lower limit of hypertension = mean + 2 $\times$ standard deviation; the systolic and diastolic limits for hypertension are 140–150 and 90 mmHg, respectively.

Bladder pressure is commonly measured during urodynamic investigations of bladder function. The normal basal pressure due to the muscular wall of the bladder (the detrusor muscle) is a few cmH$_2$O, with a normal voiding pressure of a few tens of cmH$_2$O. In a normal bladder the resting pressure is almost independent of the volume in the bladder, and there are no pressure peaks except at voiding. An increase in baseline pressure or contractions is pathological. For instance, neurological problems may cause detrusor instability, with fluctuations in bladder pressure sufficient to cause leakage. In patients with severe obstruction (e.g. in paraplegics with dysinergia, the simultaneous contraction of the external urethral sphincter and the detrusor muscle) the peak pressures may reach 200 cmH$_2$O and cause reflux of urine up the ureters to the kidneys. Chronic reflux will destroy the kidneys.

Raised intracranial pressure may occur in infants or adults with hydrocephalus, in patients with a space-occupying lesion within the brain, or following traumatic head injury, and will lead to irreversible brain damage. In the case of hydrocephalus, a valve is inserted to replace the non-functional or absent drainage mechanism for cerebro-spinal fluid. In other cases, it may be necessary to surgically increase the drainage until the underlying cause is treated.

Glaucoma—raised intraocular pressure—causes vision disturbances and ultimately blindness. Chronic glaucoma may not be noticed until the optic nerve is permanently damaged. The cause may be obstruction of the drainage of aqueous fluid from the anterior compartment of the eye, which can be treated by a simple surgical operation. We will return to the measurement of intraocular pressure.

Bed sores are ulcers which develop as a result of excessive pressures being applied to the skin for an extended period of time. The blood supply to the skin is from the subcutaneous tissue. Any continuous pressure which obstructs the blood supply for more than a few hours will lead to tissue necrosis and a pressure sore. This is particularly likely in patients with impaired peripheral blood supply or damaged peripheral nerves (as a result of diabetes, leprosy or spinal injury). With the exception of the soles of the feet, the skin cannot sustain pressures of more than 40 mmHg.

Constraints on the measuring system

The limitations of the pressure measuring system are no different from those encountered in an industrial or laboratory system—sensitivity, thermal effects, frequency response, repeatability. The measurement of physiological variables introduces other constraints—safety and sterilization are the most important. The measuring process may alter the pressure being measured. The anterior chamber of the eye has a very small volume and low compliance, so inserting a measuring device would inevitably change the pressure. In contrast, the bladder has a large volume and the pressure is essentially independent of volume, so that the measuring device will have a negligible effect on the pressure.

It may not be possible to place the transducer at the measuring site, and the frequency response of the system external to the transducer may then be important. We will return to this problem. Long-term monitoring may result in ingress of physiological fluids into the transducer, with consequent safety and performance problems.

18.3. NON-INVASIVE MEASUREMENT

An invasive measure of the pressure is often neither feasible nor justifiable, so numerous non-invasive techniques have been developed. The commonest is the sphygmomanometer for measuring arterial blood pressure.

The sphygmomanometer consists of an inflatable bag attached to a cuff which can be fastened around the arm, a pump, a pressure gauge and a pressure release valve. Arterial blood pressure does not usually vary greatly with posture, but blood pressure readings are nevertheless normally made with the patient seated. The inflatable bag is placed on the inner surface of the upper arm, overlying the brachial artery, and is held in place by wrapping the cuff around the arm. If the pressure in the cuff is raised above the systolic pressure, the artery will be completely compressed and there will be no pulse and no blood flow downstream from the cuff. If the pressure is released, the pulse will return when the cuff pressure falls below the systolic pressure. If a stethoscope is placed over the artery, faint tapping sounds (the Korotkoff sounds) will be heard. As the pressure in the cuff falls to diastolic pressure, the sounds will become muffled. The standard procedure is to increase the pressure to about 30 mmHg above the point at which the pulse disappears, and then release the pressure at about 3 mmHg s^{-1}. The pressure at which the Korotkoff sounds are first heard is the systolic pressure, and the pressure at which the sounds become muffled is the diastolic pressure. After the diastolic pressure reading has been taken, the pressure is rapidly reduced to zero to restore the blood flow.

Several automated sphygmomanometers are available for non-invasive monitoring of blood pressure in intensive care units. The cuff is inflated at regular intervals by a motor driven pump, and the Korotkoff sounds are detected by a microphone. The signal from the microphone is suitably filtered to give a signal denoting the onset and muffling of the sounds. The cuff pressure, measured by a transducer, is recorded on an event recorder. In a fluid-filled system, the pressure transducer must be at the same level as the pressure which is being measured, so that the weight of the fluid column does not affect the reading. However, as the cuff is pressurized using air, which has a negligible weight, the height of the pressure transducer is not important. These instruments work adequately, but have difficulty in distinguishing between genuine Korotkoff sounds and sounds caused by patient movement.

Many other non-invasive methods of measuring blood pressure have been described in the literature, but none of them are now used clinically. Non-invasive methods are routinely used to measure the pressure within the eye (the intraocular pressure), and are described below. A similar technique can be used to measure intracranial pressure through the anterior fontanelle in infants.

18.3.1. *Measurement of intraocular pressure*

As an example of non-invasive pressure measurement, we will examine the measurement of intraocular (within the eye) pressure. Glaucoma (raised intraocular pressure) causes vision disturbances and ultimately blindness. Chronic glaucoma may not be noticed until the optic nerve is permanently damaged. The cause may be obstruction of the drainage of aqueous fluid from the anterior compartment of the eye. It is obviously important to know the normal range of intraocular pressure, and to be able to measure it non-traumatically. Unfortunately, the only definition of 'normal' for intraocular pressure is the pressure that does not lead to glaucomatous damage of the optic nerve head.

Numerous large studies of intraocular pressure have been made. The normal pressure is about 15 mmHg, with a standard deviation of about 3 mmHg. Figure 18.1 shows the interpretation of the results as two populations (non-glaucoma and glaucoma) which overlap considerably. The standard deviation should also be treated with caution, as the distributions are skewed. The figures suggest a normal range of approximately 10–20 mmHg. The intraocular pressure is influenced by a number of factors (blood pressure, drugs, posture, blinking, etc) which are discussed in standard ophthalmology texts.

The problem, as far as we are concerned, is how we measure intraocular pressure non-invasively. The eye is a fluid-filled globe, and we can measure the pressure within by relating the deformation of the globe to an externally applied force. Tonometers (devices for measuring the pressure within the eye) are of two types: indentation and applanation (flattening). We will consider only the applanation type.

Applanation tonometers rely on the relationship between wall tension and pressure in an elastic sphere. The surface tension γ is related to the pressure difference ΔP across the curved wall by $\Delta P = 2\gamma/r$, where r

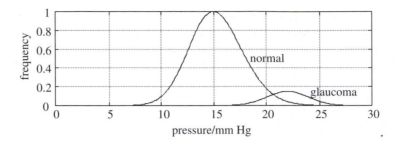

Figure 18.1. *Distribution of intraocular pressure for normal and glaucoma groups.*

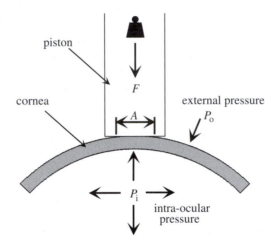

Figure 18.2. *Applanation tonometer applied to the cornea.*

is the radius. If the wall of the sphere is locally flattened (i.e. $r = \infty$), the pressure difference ΔP will be zero, and the pressure within the sphere can be related to the force applied to the wall by *pressure = force/area*. Ophthalmologists give this latter expression the grandiose name of the Maklakov–Fick or Imbert–Fick law, but it is no more than the standard definition of pressure.

The assumptions are that the eye is a spherical shell, perfectly flexible, has an infinitely thin wall, and that there are no surface tension effects due to fluids in contact with either surface of the wall. In practice, the eye violates all of these assumptions (see section 3.5.2 for the anatomy of the eye). The cornea has a central thickness of approximately 0.55 mm (figure 18.2), so that the outer contact area is less than the internal flattening. The wet surface gives a surface tension acting on the plunger, and a finite force is required to bend the cornea. Surface tension has a component acting in the same direction as the applied force, which is opposed by the bending force. In the literature, it is stated that it has been found empirically that these forces balance out when the contact area is 3.06 mm diameter. Given all the assumptions about the technique, and the wide range of intraocular pressures, this precision seems inappropriate. The resulting volume displacement is about 0.5 μl, compared to 180 μl for the volume of the anterior chamber and 60 μl for the posterior chamber, so that the applanation has a negligible effect on the intraocular pressure.

The Goldmann-type applanation tonometer applies a force to the surface of the cornea, and uses an optical system to determine when the applanated area is 3.06 mm in diameter. The cornea is anaesthetized with a topical anaesthetic, and the tear film is stained with sodium fluorescein. The meniscus surrounding the applanated area is visible because of the fluorescence, and is split into two semi-circles by biprisms. The geometry is such that the inner edges of the semi-circles touch when the area applanated is correct. The intraocular pressure can then be measured directly from a scale on the tonometer. The thickness and curvature of the cornea can produce errors in the measurement.

Tonometers are now available that use a puff of air to deform the surface of the cornea. By measuring the velocity and volume of the air and the resulting deformation of the corneal surface it is possible to provide a reasonably accurate measurement of intraocular pressure in a very non-invasive manner.

18.4.   INVASIVE MEASUREMENT: PRESSURE TRANSDUCERS

In general, greater accuracy will be achieved if the measuring device is inserted into the measurement volume, that is, if the measurement is invasive. The pressure transducer may be connected to the measurement volume by means of an in-dwelling air- or fluid-filled catheter, or may be mounted on the end of a catheter and inserted into the measurement volume. Intracranial pressure may be measured by placing a catheter in the ventricles, or by sub- or extradural transducers. The method of choice will be determined by cost, accuracy, frequency response, movement artefact and the time period over which monitoring will take place.

Placing the transducer itself within the measurement volume will give the best frequency response; freedom from movement artefact; no change in baseline pressure due to the patient moving relative to an external transducer; and will not require infusion of the catheter to maintain patency. However, catheter-tip transducers are expensive and fragile, and do not take kindly to repeated sterilization. If used for extended periods they may suffer from ingress of physiological fluids, and it is difficult to check the zero-pressure output of an in-dwelling transducer. Catheter-tip transducers which have only the diaphragm at the tip, with the diaphragm deflection measured externally through a fibre optic link, may overcome some of these problems.

The theoretically less desirable external transducer, with its fluid-filled catheter connected to the measurement volume, enables the technical problems of pressure measurement to be separated from the physiological problems. The problem of repeated sterilization destroying the delicate transducer has been overcome by physically separating the fluid from the transducer, so the transducer only requires, at most, surface decontamination. This has been achieved by including a flexible membrane in the transducer dome, which is in close apposition to the measuring diaphragm, and transmits the fluid pressure. The domes are sold as sterile disposable units. Indeed, this philosophy has been pursued to such an extent that sterile disposable transducers are commonly used.

Similar systems can be used for most fluid pressure measurements. The fluid-filled catheter system does suffer from movement artefacts and from apparent pressure changes if the position of the transducer changes relative to the measurement volume. Air-filled catheters overcome this problem, but are clearly limited to applications where air embolus is not a problem. One such area is the measurement of bladder function in ambulatory patients.

Intracranial pressure can be measured with an in-dwelling catheter, though this may provide a route for infection and cause trauma if any force is exerted on the catheter. Transducers are available which fit into a burr-hole, so that they are rigidly connected to the skull, thus reducing the risk of trauma. The actual pressure measurement may be made sub-durally, or, less invasively, by the transducer diaphragm being in contact with the external surface of the dura mater so that the dura mater remains intact.

Transducers

The most common type of pressure transducer consists of a diaphragm, one side of which is open to the atmosphere and the other connected to the pressure which is to be measured. Pressure causes a proportional displacement of the diaphragm which can be measured in many ways. The most common method is to use *strain gauges* (see section 20.2.2). A strain gauge is a device which measures deformation or strain. A single crystal of silicon with a small amount of impurity will have an electrical resistance which changes with strain. If a silicon strain gauge is attached to the diaphragm of the pressure transducer then its resistance will change with the pressure applied to the diaphragm. Unfortunately the resistance of the silicon will also change with temperature, but the effect of this change may be eliminated by attaching four strain gauges to the diaphragm (figure 18.3).

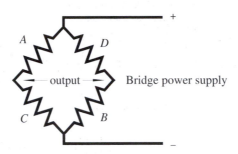

Figure 18.3. *The four arms of the resistive bridge are attached to the diaphragm of the pressure transducer.*

These strain gauges then form the four arms of a resistance bridge. By placing two of the strain gauges tangentially and close to the centre of the diaphragm, where the strain is positive and hence their resistance will increase with applied pressure, and the other two radially and close to the periphery, where the strain is negative and hence their resistance will decrease with applied pressure, the resistance bridge will be unbalanced by a change in pressure. However, if the temperature changes, all four resistances will change by the same percentage and this will not change the output from the resistance bridge. In figure 18.3 the two strain gauges, A and B, might increase their resistance with pressure, whereas C and D will decrease.

We can calculate how the bridge output will change as follows. In the general case, the resistance of one of the strain gauges will be $f(S, T)$, where f is a functional relationship of the strain S and temperature T. However, for small changes we can assume that R_A is given by

$$R_A = R_0 + kS + cT \tag{18.1}$$

where R_0 is the resistance when the strain is zero and k and c are constants. Because the same current flows through arms A and C we can determine the potential at the junction of A and C as V_{AC} when the bridge power supply is $+V$ and $-V$,

$$V_{AC} = V\frac{(C - A)}{(C + A)}.$$

We can perform a similar calculation for the potential at the junction of D and C, and thus obtain the output of the bridge as

$$\text{output} = V\frac{(C - A)}{(C + A)} - V\frac{(B - D)}{(D + B)}. \tag{18.2}$$

It can be seen that the output will be zero when all the bridge arms have the same resistance. It can also be seen that the output will be zero if the temperature changes because the same change will occur in all four arms of the bridge. However, if a strain occurs then we can substitute from equation (18.2), taking into account the fact that the strain will be in opposite directions in arms A, B and D, C,

$$\text{output} = V\frac{2kS}{R_0 + cT}. \tag{18.3}$$

The output is thus mainly determined by the strain and the temperature dependence is reduced to a second-order effect.

The complete transducer may have a dome, which can be unscrewed for cleaning and sterilization but it more likely that the transducer will now be a disposable type. Figure 18.4 shows how the transducer can be connected for an arterial pressure measurement. There are usually two connections to the transducer so that saline can be flushed through and all the air removed or a constant flow system can be connected as shown.

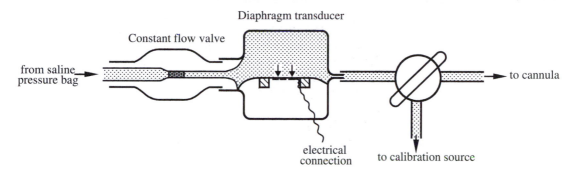

Figure 18.4. *A disposable pressure transducer is shown connected to an arterial catheter/cannula. An additional line is used to continuously flush the system with a slow flow of saline at typically 3 ml h⁻¹. The tap enables the transducer to be connected to a calibration pressure whilst the arterial catheter is isolated.*

If air bubbles remain then false pressure readings will be obtained and the frequency response of the transducer will be reduced. The reason for this is that a rapid rise in pressure will compress the air bubble instead of moving the diaphragm of the transducer. A transducer connected to a source of pressure by a fluid-filled catheter is a complex system which will usually have a resonant frequency. This system is considered in some detail in the next section, but we will consider very briefly here the effect of pressure transducer diaphragm compliance. The pressure transducer works by measuring the deformation of a diaphragm when a pressure is applied. If a pressure change ΔP causes the diaphragm to move such that a volume ΔV of fluid is displaced, then the ratio $\Delta V / \Delta P$ is called the compliance C_p of the transducer. The stiffer the diaphragm the smaller the compliance of the transducer.

Now if we make the simplest assumptions which are that the catheter connection to the transducer is completely rigid, viscous forces are negligible and the transducer diaphragm is not elastic then the acceleration of the diaphragm will be given by

$$\text{acceleration} = \frac{\Delta P\, a}{m}$$

where a is the cross-sectional area of the catheter and m is the mass of the fluid in the catheter. The distance Δs which the diaphragm has to move to displace the volume ΔV is given by

$$\Delta s = \frac{C_p \Delta p}{a}$$

and hence we can use Newton's laws of motion to calculate the time Δt for the displacement to occur as

$$\Delta t = \sqrt{\frac{2\Delta s}{\text{accel.}}} = \sqrt{\frac{2C_p l \rho}{a}} \qquad (18.4)$$

where l is the length of the catheter and ρ is the density of the fluid in the catheter. Therefore, we can conclude that the response time of the transducer will increase with the compliance of the transducer and the length of the catheter. It will increase in inverse proportion to the square root of the diameter of the catheter. In fact, if viscous forces are taken into account the dependence upon catheter diameter is even more important. To keep the response time short the catheter length should be short, the diameter large and the compliance small.

A syringe can be used to pass saline through the pressure lines and the transducer. In addition to allowing bubbles to be removed, the syringe can also be used to ensure that the catheters, i.e. the pressure lines, are not blocked. Electrical connections to the transducer will have at least four wires: two power supply connections to the bridge and two output connections. The output connections drive a differential amplifier whose common or reference connection will be to one of the power supply wires. The sensitivity of the transducer will usually be quoted as X μV V^{-1} mmHg^{-1}. This means that for each volt applied to the transducer bridge there will be an output of X μV for each mmHg pressure applied to the transducer. Many transducers are calibrated to have a standard sensitivity of 50 μV. Obviously the differential amplifier must be able to handle, without distortion, a voltage equal to the product of the maximum applied pressure, X, and the power supply voltage.

In selecting a transducer for a particular application there are very many factors to be considered. A wide variety of transducers are available, ranging from large, general-purpose types to miniature, catheter-tip types for arterial pressure measurement. These miniature types have the advantage that no fluid-filled catheter is required and so many problems are avoided. Transducers used for routine intensive care monitoring are now often disposable so that sterilization problems are avoided.

The manufacturer's specification of a transducer will normally include at least the following:

Sensitivity	μV V^{-1} mmHg^{-1}
Maximum applied bridge voltage	volts
Maximum applied pressure	mmHg or kPa
Operating pressure range	mmHg or kPa
Temperature drift	mmHg °C^{-1}
Nonlinearity and hysteresis	% of FSD (full-scale deflection)

The temperature drift is important in the measurement of small pressures such as bladder or airways' pressure. In some cases the drift can be greater than the pressure to be measured. Gain changes are not very important as they can be taken into account in calibration, but nonlinearity, hysteresis and zero shifts are very difficult to take into account.

A typical practical procedure for the use of a pressure transducer is:

- Make the electrical connection between the transducer and the measurement equipment.
- Connect the pressure lines, with the exception of the patient connection.
- Fill the system with saline and remove all the air bubbles; it may be necessary to tap and tilt the transducer to dislodge all the bubbles.
- Open the transducer to the atmosphere using the three-way tap.
- Zero the bridge and amplifier so that zero output corresponds to atmospheric pressure.
- Close the tap to atmosphere and now make the patient connection.
- Flush a small volume of saline through the system and check that blood is not entering the system. The system should now be ready for measurement.

Figure 18.5 gives an interesting example of the use of a fluid-filled catheter system to measure the venous pressure at the ankle. The recording system used disposable pressure transducers and an output connected

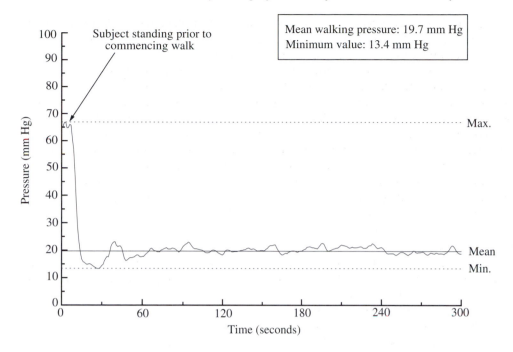

Figure 18.5. *Venous pressure at the ankle measured using a saline-filled catheter and disposable pressure transducer. The pressure when standing is determined by hydrostatic pressure but falls rapidly during walking (Data kindly provided by E Matthews who carried out the study as a Masters project at the University of Sheffield.)*

to a data logger to record the long-term changes in venous pressure. It can be seen that when standing the venous pressure is determined by the hydrostatic pressure of the blood in the vascular system. However, walking caused the pressure to drop dramatically. The muscular activity of walking compresses the venous system and forces blood through the venous valves back towards the heart. This reduces the pressure within the veins. This figure illustrates the importance of walking to the maintenance of blood flow and low venous pressures.

18.5. DYNAMIC PERFORMANCE OF TRANSDUCER–CATHETER SYSTEM

The use of a catheter in conjunction with a transducer is the most common method of measuring blood pressure and other pressures such as the pressure within the bladder. The addition of a flexible fluid-filled catheter to the relatively rigid transducer introduces severe problems associated with the dynamic performance of the system. We will use a simple lumped parameter model, analogous to an electrical *RLC* circuit, to explore the dynamic performance problems (this follows an exposition given in Webster (ed) 1998 *Medical Instrumentation: Application and Design* 3rd edn (New York: Wiley) pp 296–300).

The model is shown in figure 18.6. A pressure change at the end of the catheter will cause fluid to move through the catheter and displace the diaphragm of the transducer. The displacement of the diaphragm is proportional to the applied pressure. The hydraulic properties that are important are inertance, resistance and compliance, the electrical analogues of which are inductance, resistance and capacitance. These properties

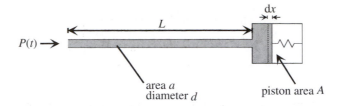

Figure 18.6. *Mechanical model of transducer and catheter.*

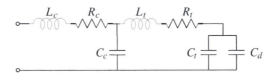

Figure 18.7. *Electrical model of transducer and catheter.*

are due to the inertia, friction and elasticity within the system. Figure 18.7 shows the resistance, inertance and compliance per unit length of the catheter (with subscript c), and for the transducer (with subscript t) and the compliance of the diaphragm (with subscript d).

In practice, for a reasonably stiff catheter and bubble-free fluid, the compliance of the catheter and friction and inertia in the transducer can be neglected, giving the simplified model in figure 18.8. If we introduce a bubble of air (with high compliance) into the catheter, we can model it by adding an additional capacitance C_b in parallel with C_d (figure 18.9). We will now determine the relationship of R, L and C to the physical properties of the system.

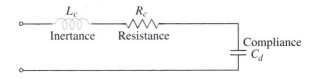

Figure 18.8. *Simplified electrical model of transducer and catheter.*

The resistance R_c is due to shear forces within the liquid flowing through the catheter, and is given by

$$R_c = \frac{\Delta P}{F} \quad (\text{Pa s m}^{-3})$$

where ΔP is the pressure difference (Pa) and F is the flow rate $(\text{m}^3 \text{ s}^{-1})$.

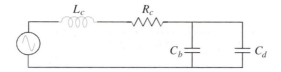

Figure 18.9. *Electrical model of transducer and catheter with inserted bubble (C_b).*

Poiseuille's law relates the pressure differential and flow rate to the viscosity and size of the tube:

$$\frac{F}{\Delta P} = \frac{\pi r^4}{8\eta L}$$

i.e.

$$R_c = \frac{8\eta L}{\pi r^4}.$$

The inertance L_c is the ratio of the pressure differential to rate of change of flow:

$$L_c = \frac{\Delta P}{\mathrm{d}F/\mathrm{d}t} = \frac{\Delta P}{a A} \quad (\text{Pa s m}^{-3})$$

where a is the acceleration and A is the cross-sectional area of the catheter.

By making use of *force = pressure × area = mass × acceleration*, and $m = \rho L A$, we find

$$L_c = \frac{\rho L}{\pi r^2} \quad \text{where } \rho \text{ is the liquid density.}$$

The compliance C_d of the transducer diaphragm is defined as

$$C_d = \frac{\Delta V}{\Delta P} = \frac{1}{E_d} \quad (\text{m}^3 \text{ Pa}^{-1})$$

where E_d is the volume modulus of elasticity of the transducer diaphragm.

The compliance C_b of a bubble is given by

$$C_b = \left.\frac{\Delta V}{\Delta P}\right|_{\text{air}} \times \text{(volume of air bubble)}.$$

If we now apply a sinusoidal input voltage v_i to the model (analogous to a varying pressure), we can calculate the output voltage across C_d (analogous to the measured pressure). We do this by summing the voltages around the circuit to give

$$L_c C_d \frac{\mathrm{d}^2 v_0}{\mathrm{d}t^2} + R_c C_d \frac{\mathrm{d}v_0}{\mathrm{d}t} + v_0 = v_i.$$

We can rewrite this as

$$\frac{\mathrm{d}^2 v_0}{\mathrm{d}t^2} + \frac{R_c}{L_c}\frac{\mathrm{d}v_0}{\mathrm{d}t} + \frac{1}{L_c C_d} v_0 = \frac{1}{L_c C_d} v_i$$

and compare this to the standard form

$$\ddot{x} + 2\omega_0\zeta\dot{x} + \omega_0^2 x = y$$

where the damping factor $\zeta = (R_c/2)(C_d/L_c)^{1/2}$ and the natural frequency $\omega_0 = (L_c C_d)^{-1/2}$.

By substitution, we can relate the natural frequency $f_0 = \omega_0/2\pi$ and the damping factor ζ to the physical parameters of the system:

$$f_0 = \frac{r}{2}\left(\frac{1}{\pi\rho L}\frac{\Delta P}{\Delta V}\right)^{1/2} \quad \text{and} \quad \zeta = \frac{4\eta}{r^3}\left(\frac{L(\Delta V/\Delta P)}{\pi\rho}\right)^{1/2}.$$

We can now calculate the frequency response of a real transducer and catheter system. Webster (see the bibliography) gives values of the constants for air and water:

η	Water at 20 °C	1.0×10^{-3} Pa s
η	Water at 37 °C	0.7×10^{-3} Pa s
η	Air at 20 °C	0.018×10^{-3} Pa s
ρ	Air at 20 °C	1.21 kg m^3
$\Delta V/\Delta P$	Water	0.53×10^{-9} m^5 N^{-1} m^{-3}
$\Delta V/\Delta P$	Air	1.02×10^{-5} m^5 N^{-1} m^{-3}
E_d	Typical transducer	0.5×10^{15} N m^{-5}

Example

A pressure transducer is connected to a 2 m long stiff catheter of internal radius 0.5 mm. The system is filled with water at room temperature. Calculate the natural frequency and damping factor with and without a 5 mm long air bubble in the catheter.

Figure 18.10 shows the frequency response in the two cases. Interestingly, the bubble improves the damping of the system (i.e. it is closer to critical damping where $\zeta = 1$), but the resonant peak in the frequency response is sufficiently low that it will have a significant effect on the measured pulse shape for an arterial pressure signal. The requirements for good frequency response and optimum damping are conflicting, and the only practical solution is to dispense with the fluid-filled catheter and place the transducer at the measurement site.

As the frequency response of second-order systems is so important, it is worth pursuing this a little further. The transfer function for this system is given by

$$A(\mathrm{j}\omega) = \frac{1}{(\mathrm{j}\omega/\omega_0)^2 + (2\zeta\,\mathrm{j}\omega/\omega_0) + 1}$$

so the amplitude and phase are given by

amplitude: $\quad A = \dfrac{1}{\sqrt{\left\{\left[1 - (\omega/\omega_0)^2\right]^2 + 4\zeta^2\omega^2/\omega_0^2\right\}}}$

phase: $\quad \phi = \arctan\left(\dfrac{2\zeta}{\omega/\omega_0 - \omega_0/\omega}\right).$

The figures (figures 18.11 and 18.12) show the frequency and step response for values of damping from 0.1 (under-damped—oscillatory step response and large peak in frequency response) to 1.0 (critically damped).

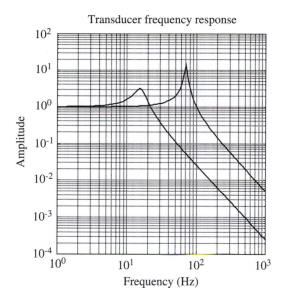

Figure 18.10. *Frequency response for transducer and catheter. The lower frequency response is with a bubble in the catheter.*

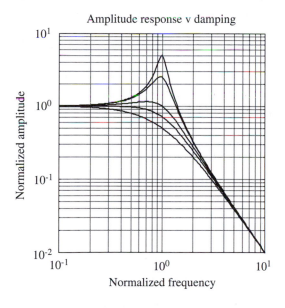

Figure 18.11. *Frequency response with different damping—lower damping gives a higher resonant peak.*

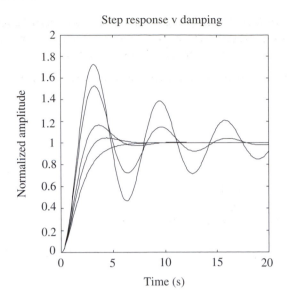

Figure 18.12. *Step response with different damping—lower damping gives more overshoot.*

Figure 18.11 shows the effect of varying the damping on the amplitude response of an ideal system, and figure 18.12 shows the effect of damping on the response to a step change in the input.

The response of the system to a step input has three forms for the three cases of over-, under- and critical damping.

Under-damped, $0 < \zeta < 1$:

$$x(t) = 1 - \left[\frac{\exp(-\zeta \omega_0 t)}{\sqrt{(1 - \zeta^2)}} \right] \sin \left(\omega_0 t \sqrt{(1 - \zeta^2)} + \tan^{-1} \left[\frac{\sqrt{(1 - \zeta^2)}}{\zeta} \right] \right).$$

Critically damped, $\zeta = 1$:

$$x(t) = 1 - (1 + t/T) \exp(-t/T) \qquad \text{where} \quad T = 1/\omega_0.$$

Over-damped, $\zeta > 1$:

$$x(t) = 1 + \frac{1}{\alpha^2 - 1} \left[\exp(-\alpha t/T) - \alpha^2 \exp(-t/\alpha T) \right]$$

where $\zeta = (1 + \alpha^2)/2\alpha$.

18.5.1. Kinetic energy error

The blood in an artery is not stationary, and this must be taken into account when measuring the pressure. If the end of the catheter (or the diaphragm of a catheter-tip transducer) faces upstream, the fluid column in the catheter will stop the blood which impinges on it. The pressure recorded will therefore be the sum of the lateral pressure and the converted kinetic energy of the bloodstream,

$$\text{kinetic energy/unit volume} = \rho v^2/2.$$

For $v = 1$ m s^{-1}, the pressure is 3.76 mmHg at $\rho = 1.0$. This is an error of a few per cent at normal arterial pressures. To avoid this error, pressures are measured using a catheter with the aperture in the side wall of the catheter, so that there is no normal component of blood velocity.

18.6. PROBLEMS

18.6.1. Short questions

a   What is 10 mmHg pressure in kPa?

b   Do men or women have the higher systolic blood pressure on average?

c   What pressure in the bladder might initiate voiding?

d   What is hydrocephalus?

e   Is a pressure of 70 mmHg applied to the skin over an extended period likely to cause any detrimental effects?

f   Why is it difficult to measure the pressure in a small volume low compliance system such as the anterior chamber of the eye?

g   Does posture affect intraocular pressure?

h   What is a tonometer?

i   What assumptions does the applanation tonometer technique make?

j   What effect do air bubbles have on the frequency response of a fluid-filled catheter pressure measurement system?

k   What are Korotkoff sounds?

l   Is high compliance good or bad in a pressure transducer?

m   Is 15 mmHg a normal intraocular pressure?

n   How will increasing the diameter of a catheter affect the resonant frequency and damping of a pressure measurement system?

o   Will a catheter pressure system underestimate or overestimate the true arterial pressure if the catheter is facing downstream?

p   Why is there normally a slow flow of saline down a pressure measuring catheter?

q   Why are there four strain gauges on the diaphragm of a pressure transducer?

r   To what type of amplifier is the electrical output of a pressure transducer connected?

s   What output will a transducer of sensitivity 20 μV V^{-1} mmHg^{-1} give if a pressure of 100 mmHg is applied and the bridge voltage is 5 V?

18.6.2. Longer questions (answers are given in some cases)

Question 18.6.2.1

Distinguish between invasive and non-invasive methods of blood pressure measurement, including in your discussion their relative disadvantages and sources of error from the clinical point of view.

Question 18.6.2.2

Sketch the physical arrangement and the approximate electrical equivalent circuit of a liquid-filled catheter using an external diaphragm as a sensor. Indicate clearly the correspondence between the mechanical and electrical counterparts in the diagram.

Why is it important to measure the dynamic as well as the static calibration of such a system? Draw and explain the typical frequency response of this system. Indicate how this response is modified and the electrical equivalent circuit altered when the following are included:

- a bubble
- a constriction
- a leak.

Question 18.6.2.3

We wish to make some basic calculations in the design of a tonometer to measure the pressure inside the eye by using air deformation of the cornea. Calculate the pressure on the cornea if 100 ml s^{-1} of air is directed onto an area of 10 mm^2 (3.3 mm × 3.3 mm) of the cornea. Would this exert sufficient pressure to flatten the surface of the cornea? (Assume that the density of air is 1.3 kg m^{-3} and use figure 18.1 to estimate normal eye pressure.)

Answer

The velocity of the air stream will be 100 ml/10^{-1} cm^2 = 10 m s^{-1}.

The mass of air directed at the eye will be 10^{-4} × 1.3 kg s^{-1}.

The force on the eye will equal the rate of change of momentum of the air. If we assume the air is deflected sideways then

$$\text{force} = 10 \text{ m s}^{-1} \times 1.3 \times 10^{-4} \text{ kg s}^{-1} = 1.3 \times 10^{-3} \text{ kg m s}^{-2}.$$

Pressure over the 10 mm^2 will be given by

$$\text{pressure} = \text{force/area} = 130 \text{ Pa}.$$

Now a normal intraocular pressure is about 15 mmHg (see figure 18.1) which is approximately 2000 Pa.

The air stream will not flatten the cornea, but it is sufficient to produce a measurable movement of the surface.

Answers to short questions

a   1.33 kPa is equivalent to 10 mmHg pressure.
b   There is no significant difference in the systolic BP of men and women.
c   A few tens of cmH$_2$O pressure can initiate voiding of the bladder.
d   Hydrocephalus is an increased intracranial pressure, usually caused by blockage of the exit for cere-brospinal fluid (CSF).
e   Yes, it is usually considered that pressure above 40 mmHg applied to the skin can cause long-term tissue damage.
f   In order to measure pressure a volume change has to be produced. In a small-volume low-compliance system (e.g. the anterior chamber of the eye), this volume change will cause a significant change in the pressure to be measured.
g   Yes, posture does affect intraocular pressure.
h   A tonometer is a device used to measure intraocular pressure by deforming the front surface of the cornea.
i   The applanation tonometry technique has to assume that the eye is spherical, flexible, has a very thin wall and that surface tension is negligible.
j   Air bubbles will reduce the frequency response of a fluid-filled catheter pressure measurement system.

k   Korotkoff sounds can be heard using a stethoscope during blood pressure measurement. They disappear when cuff pressure falls below the diastolic pressure.

l   High compliance is bad in a transducer as it will reduce the response time.

m   Yes, 15 mmHg is a normal intraocular pressure.

n   Increasing catheter size will increase the resonant frequency and decrease damping.

o   If the catheter is facing downstream then the system will underestimate the true pressure.

p   Continuous flushing is used to stop blockage of a pressure measurement catheter.

q   Four strain gauges are used to reduce thermal drift in a pressure transducer.

r   A differential amplifier is connected to a pressure transducer.

s   10 mV.

BIBLIOGRAPHY

Documenta Geigy 1972 *Scientific Tables* 7th edn (Basle: Ciba–Geigy)

McDonald D A 1974 *Blood Flow in Arteries* (London: Arnold)

Webster J G (ed) 1998 *Medical Instrumentation: Application and Design* 3rd edn (New York: Wiley)

CHAPTER 19

BLOOD FLOW MEASUREMENT

19.1. INTRODUCTION AND OBJECTIVES

If blood circulation is to be maintained, and tissues are to be perfused with oxygen, then the correct pressures must be maintained within the vascular system. We considered pressure measurements in chapter 18. However, if oxygen and nutrients are to reach tissues then a flow of blood must be maintained. Pressure does not necessarily result in flow.

Blood flow measurements are made at very many points within the cardiovascular system. Cardiac output is often measured as an index of cardiac performance, flow through an arterial graft may be used to ensure that a graft has been successfully inserted during surgery, or the blood flow in peripheral arteries and veins may be measured in the assessment of vascular disease.

A very large number of methods have been used in an attempt to quantify blood flow in organs and vessels in the intact body. Venous occlusion plethysmography is the nearest approach to an absolute technique. All other techniques are based on a set of assumptions, of varying validity, and need to be calibrated in some way in order to give quantitative results. On examination, it is found that there are actually only a limited number of underlying mathematical techniques, which have been used with a wide variety of indicators of blood flow. We will examine these basic techniques and describe briefly the different applications, before moving on to methods for measuring blood velocity. A good general reference is Woodcock (1975).

Some examples of the questions we hope to answer in this chapter are:

- What is a typical arterial blood flow?
- Is venous flow pulsatile?
- Is arterial flow always in the same direction?
- What are your options if you want to measure blood flow non-invasively?

At the end of the chapter you should be aware of:

- Several methods of measuring blood flow by adding an indicator to the blood.
- The relation between thermal conductivity and blood flow.
- The signal analysis techniques required to obtain a Doppler signal proportional to blood velocity.
- The principles of various methods of plethysmography.

This chapter is quite mathematical in parts but we would urge our readers not to avoid the whole chapter. Look for the conclusions at the end of each section. We have tried to state the conclusions which have resulted from the mathematics. The short questions at the end of the chapter are based largely upon these conclusions and set out to test your comprehension of the principles of blood flow measurement.

What are typical values for some blood flow measurements? Cardiac output is typically about $6 \, l \, min^{-1}$ which corresponds to a stroke volume (the volume expelled by the left ventricle on cardiac systole) of 100 ml if the heart rate is 60 beats per minute. This 100 ml is distributed throughout the body with about 20 ml going to the kidneys, 17 ml to the muscles and 12 ml to the brain. If we know the sizes and numbers of the blood vessels then we can calculate the blood velocities. This is relatively easy for the main arteries, because there are only a few of them, and we conclude that the peak velocities might be of the order of $1\ \mathrm{m\ s^{-1}}$. As the blood vessels branch and get smaller the average velocities fall and in the capillary beds the red blood cells (erythrocytes) move through in single file and very slowly. In the major veins the velocities involved are a few centimetres per second.

Cardiac output per minute is similar to the total blood volume so that we can expect the whole of our blood to get around our bodies at least once each minute. This fact can be important in some techniques, as a label on the blood will recirculate well within a minute.

19.2.   INDICATOR DILUTION TECHNIQUES

We will start by considering methods that rely on indicators to measure the blood flow. The indicator could be a dye, an isotope, a gas—anything in fact which is transported by the blood, which is acceptable to the body, and the concentration of which can be measured. The underlying mathematics of the techniques is independent of the indicator.

Indicator methods are usually sub-divided into dilution methods (based on indicators which remain in the bloodstream but are diluted by mixing), and transport or clearance methods (based on the blood volume which is required to deliver or remove the indicator from an organ or tissue). We will consider dilution techniques first.

Stewart introduced dilution techniques for the study of blood volume in the heart and lungs in 1897. Hamilton and his co-workers, in 1932, looked at cardiac output and developed the use of mean circulation time to determine the volume of the vascular bed. Meier and Zierler (1954) examined the requirements for the theory to be valid. The theory which follows assumes that the system has a single input and output vessel, has a volume V, and the blood enters and leaves the system at a constant rate F (figure 19.1). The system can take any form between the input and the output. Branching vessels will introduce a spectrum of transit times between input and output. Other assumptions are that the system has stationarity (the distribution of transit times within the system does not change); the indicator must have the same distribution of transit times as the blood; there is no recirculation of the indicator; and the fluid in the system is eventually eliminated, i.e. there is no storage of indicator within the system.

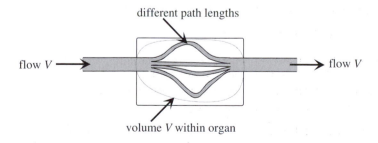

Figure 19.1. *The principle of the indicator dilution method. The indicator is injected on the left and the resulting concentration of indicator is measured on the right.*

19.2.1. *Bolus injection*

Two techniques are used to measure a steady blood flow—bolus injection (a single rapid injection), and continuous injection at a constant rate. Figure 19.2 shows the variation in the concentration of an indicator downstream of the injection site. For a bolus injection, the concentration will rise to a peak, and then decay. The second smaller peak is due to recirculation.

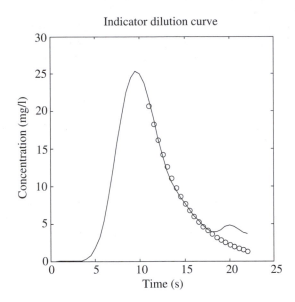

Figure 19.2. *Concentration of a marker after a bolus injection. The small peak occurring at about 20 s is caused by recirculation of the marker. The circles indicate an exponential curve fitted to remove the effect of recirculation.*

Consider the mathematics that describes the concentration versus time curve. Suppose X units of indicator are injected at $t = 0$, and $c(t)$ is the concentration at the sampling site at time t. If F is the constant flow in the system, then in a small interval δ an amount $F\,c(t)\,\delta$ will leave the system. Eventually, all the indicator leaves the system, therefore

$$X = F \int_0^\infty c(t)\,\mathrm{d}t$$

i.e. the flow is given by

$$F = \frac{X}{\int_0^\infty c(t)\,\mathrm{d}t}. \tag{19.1}$$

The integral term is the area under the concentration–time curve for the bolus injection, so the flow can be calculated from the measured area and the quantity of indicator injected. This is the *Stewart–Hamilton equation*.

We are often interested in the volume of blood contained within a particular organ or tissue. We can show that this is given by the (constant) flow rate multiplied by the mean transit time for the indicator:

$$V = F\bar{t}. \tag{19.2}$$

In terms of the observed function $c(t)$ this is

$$\bar{t} = \frac{\int_0^\infty t\, c(t)\, dt}{\int_0^\infty c(t)\, dt}. \tag{19.3}$$

This requires no assumptions to be made about the route of the indicator through the organ. In other words, we do not need to know anything about the blood vessels within the volume of tissue—we need only measure the output concentration with time.

There is one important problem which is that very often the indicator is not eliminated during one passage through the vascular system. Recirculation will affect the accuracy. To take recirculation into account the concentration versus time curve is extrapolated to zero assuming that the decay in the curve is exponential. An exponential decay is assumed (figure 19.2), giving

$$c(t) = c_{max} e^{-kt}$$

where k and c_{max} can be determined. The circles in figure 19.2 denote an exponential curve extrapolation to remove the effect of recirculation.

19.2.2. Constant rate injection

If we inject the marker continuously instead of as a bolus then the measured concentration will rise and approach a steady value asymptotically. However, recirculation is quite difficult to take into account. If the marker is not excreted from the system then the concentration will simply rise continuously with time. We will assume that the marker is removed from the system.

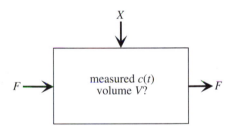

Figure 19.3. *Quantity X per unit time injected into the system.*

In this case (figure 19.3), the indicator is infused at a rate of X per unit time, and the concentration $c(t)$ rises asymptotically to C_{max}, i.e. the rate of change of concentration is given by

$$\frac{dc(t)}{dt} = \frac{F}{V}(C_{max} - c(t)). \tag{19.4}$$

When this equilibrium is reached, the inflow of indicator is balanced by the amount removed by the blood flow, giving

$$X/F = C_{max} \tag{19.5}$$

and the volume V of the system is given by

$$V = \frac{F}{C_{max}} \int_0^\infty (C_{max} - c(t))\, dt. \tag{19.6}$$

The integral term is the area between the curve of $c(t)$ versus time and C_{max}, i.e. it is the area lying above the $c(t)$ curve and below C_{max}.

Time-varying blood flow or infusion rate

Blood flow is often pulsatile and even in the venous system will change with time. The instantaneous flow rate at time t is given by

$$F(t) = X(t)/c(t).$$

If the infusion rate X is constant, and the flow F varies, the flow volume V in the interval t_1 to t_2 is given by

$$V = \int_{t_1}^{t_2} F(t)\,dt = X \int_{t_1}^{t_2} \frac{1}{c(t)}\,dt.$$

The mean volume flow during the interval t_1 to t_2 is therefore given by

$$\bar{V} = X\overline{[1/c(t)]}. \tag{19.7}$$

This is the *variable flow equation*. If the infusion rate X varies, but the flow rate F is constant, then the instantaneous infusion rate X is

$$X(t) = Fc(t)$$

$$\therefore \int_{t_1}^{t_2} X(t)\,dt = F \int_{t_1}^{t_2} c(t)\,dt \tag{19.8}$$

$$\text{i.e.} \quad F = \bar{X}/\bar{c}(t)$$

Equation (19.8) is the *constant flow equation*.

In practice, the instantaneous concentration of the indicator will vary if the flow rate varies, e.g. throughout the cardiac cycle. Applying the constant flow equation will underestimate the true flow by an amount which depends on both the size and waveshape of the fluctuations.

19.2.3. Errors in dilution techniques

We have already considered the error that can arise from recirculation. There are a number of other possible errors described in the literature. A straight blood vessel with well-developed laminar flow will have a parabolic velocity profile, i.e. the velocity at a distance r from the central axis will be given by

$$V_r = V_{max}\left(1 - r^2/R^2\right) \tag{19.9}$$

where R is the vessel radius and V_{max} is the velocity along the vessel axis. A bolus injection into the vessel will result in the indicator being dispersed in a parabola along the catheter. The result is that the mean spatial concentration across the vessel cross-section is constant, but the mean flux of indicator varies across the vessel. If a photoelectric detector is used, which illuminates the whole cross-section of the vessel, then the spatial average of the indicator is measured.

A major error occurs when flow is variable. The flow value calculated for cardiac output using a short bolus injection varies greatly depending on when the injection takes place relative to the cardiac cycle. A constant infusion will give a valid measure of average flow.

The theory of dilution techniques demands that the indicator is representative of the flow being measured. Blood is an inhomogeneous suspension, in which the cells travel faster than the plasma. In general, therefore, it is not possible to measure the flow of the individual components with a single indicator. Dyes tend to measure plasma flow, whereas labelled red cells will obviously measure cell flow.

19.2.4. *Cardiac output measurement*

The dye dilution technique can be used to measure the cardiac output. A small quantity of dye is injected into the venous bloodstream close to the heart through a catheter. Arterial blood is drawn off through another catheter by a motorized syringe, and the dye concentration is measured. Indocyanine green (ICG) is most commonly used. It has a relatively low toxicity, does not persist for long in the bloodstream and has a maximum optical absorption at 805 nm. At this wavelength, the absorption of haemoglobin is independent of oxygenation, so that the absorption is only a function of concentration (see section 17.5.1 on pulse oximetry).

The same principle can be used relatively non-invasively by injecting a bolus of radioactivity into an arm vein, and measuring the activity from a detector (a collimated scintillation detector or a gamma camera) placed over the heart. A non-diffusible tracer is required, as the whole of the administered dose must pass through the heart, and the external count rate can only be related to blood concentration if the tracer does not leave the circulation. Commonly used radioactive tracers are ^{99m}Tc-HSA (human serum albumin) or ^{99m}Tc labelled red cells.

We will return to the measurement of cardiac output in section 19.4.3 where we consider the use of thermal dilution to make this measurement.

19.3. INDICATOR TRANSPORT TECHNIQUES

Indicator transport techniques look at the blood volume required for the delivery and removal of an indicator. Fick suggested in 1870 that cardiac output could be measured from the known concentrations of oxygen or carbon dioxide in arterial and venous blood, and the uptake of oxygen or release of carbon dioxide in the lungs. This is an application of the concept that, in a time interval dt the amount Q_i of a substance entering a region must be equal to the quantity Q_s stored plus the quantity Q_m metabolized plus the amount Q_o leaving the region as illustrated in figure 19.4.

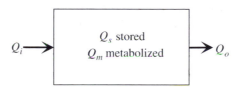

Figure 19.4. *A compartment showing the input of indicator Q_i and output Q_o*

Expressed mathematically,

$$\frac{Q_i}{dt} = \frac{Q_s}{dt} + \frac{Q_m}{dt} + \frac{Q_o}{dt}.$$

In a steady state, oxygen is metabolized but does not accumulate and the blood supply is the only route for supply and removal. C_A and C_V are the constant arterial and venous concentrations, and the blood flow is F, then

$$F(C_A - C_V) = Q_m/dt \qquad (19.10)$$

which is the *Fick equation* (units (ml s^{-1}) (quantity/ml) = (quantity/s)).

If the concentration of oxygen or carbon dioxide in arterial and mixed venous blood is known, and in addition the rate of inhalation of oxygen or exhalation of carbon dioxide is known, then we can use the Fick equation to calculate the cardiac output. The method works, but it is invasive in that blood gas concentrations have to be measured fairly close to the lungs. The rate of inhalation or exhalation of the gases can be measured using a spirometer (see section 17.4.1).

Types of indicator

Many different tracers have been used in circulation studies. Biologically inert indicators were particularly important for the early development of tracer studies—nitrogen, ethylene and acetylene have been used to measure cardiac output; nitrous oxide has been used to measure cerebral blood flow and coronary blood flow. The use of gamma-emitting indicators removes the need for blood sampling in order to measure the indicator concentration. Hydrogen is also biologically inert and freely diffusible, and can be measured *in situ* using polarography, in which the current flow due to the oxidation of the hydrogen at a platinum electrode is measured.

Transport indicators can be divided into three groups: freely diffusible, restricted diffusibility, and selective. Freely diffusible indicators are small lipophilic molecules which can cross the capillary endothelium over its entire surface area. Examples are ^{85}Kr, ^{79}Kr, ^{133}Xe and ^{125}I or ^{131}I iodo-antipyridine. In most normal tissues there is complete equilibrium with venous capillary blood, so the rate of removal depends only on capillary blood flow.

Indicators of restricted diffusibility are small hydrophilic molecules or ions. Examples are $^{24}Na^+$, $^{42}K^+$ and ^{131}I. The rate of removal is mainly determined by capillary permeability.

Selective indicators are mainly hydrophilic molecules of about 350 nm diameter, including colloids such as $^{131}Iodinated$ human serum albumin (IHSA) and colloidal ^{198}Au. Examples of their action are the removal of ortho-iodohippuric acid (Hippuran) by the kidneys and of indocyanine green by the liver.

19.3.1.   Selective indicators

Para-aminohippuric acid (PAH) is selectively removed by the kidneys and excreted in the urine. There is no accumulation or metabolism in the kidneys, but some PAH appears in the venous drainage from the kidneys (i.e. the extraction rate is not 100%). If Q_u is the amount of indicator appearing in the urine in time dt, then

$$\frac{Q_u}{dt} = F(C_A - C_V).$$

The renal blood flow can thus be calculated from a knowledge of the arterial and venous concentrations and the rate of appearance of the indicator in the urine. There are numerous assumptions:

- The arterial concentration is the same as the concentration in the blood supply to the tissue or organ, and the venous concentration is the same as the concentration in the blood leaving the organ.
- The indicator is only removed by the organ or tissue in question.
- The indicator is not produced, metabolized or altered by the organ.
- There is no storage of the indicator.
- The indicator is not removed by the lymphatic drainage.
- The indicator is not returned to the venous supply downstream of the sampling point.
- The excretion of the indicator does not change the blood flow.
- There is no indicator in 'dead space', e.g. in a residual volume in the urine.

19.3.2.   Inert indicators

An inert gas introduced into the blood is neither stored nor metabolized. The blood is distributed throughout the body and diffusion of the gas occurs across the capillary membranes. Equilibrium is reached when the inert gas tension in the arterial and venous blood and the tissue is the same. As the gas is not metabolized the

Q_m term is zero:

$$\frac{dQ_i}{dt} = F(C_A - C_V)$$

where Q_i is the quantity taken up by the tissue.

Kety showed that measurement of Q_i is not necessary, provided that the venous blood concentration is in equilibrium with the tissue. If $C(t)$ is the tissue concentration at time t, then

$$\frac{dC(t)}{dt} = \frac{1}{V}\left(\frac{dQ_i}{dt}\right) = \frac{F}{V}(C_A - C_V)$$

where V is the organ or tissue volume. The concentration in the organ is therefore

$$C(t) = \frac{F}{V}\int_0^t (C_A - C_V)\,dt.$$

In general, the venous concentration will not be the same as the concentration in the tissue. At equilibrium $C(t) = \alpha C_V$, where α is the specific partition coefficient for the indicator between blood and tissue. The partition coefficient gives the relative concentration of the indicator in different tissues at equilibrium. For Kr, the partition coefficient is 1.0 for muscle and 5.0 for adipose (fatty) tissue; for Xe the values are 0.7 and 10.0, respectively. This therefore gives

$$\frac{F}{V} = \frac{\alpha C_V}{\int_0^t (C_A - C_V)\,dt}. \tag{19.11}$$

The flow per unit volume of the tissue or organ can thus be found from a knowledge of the partition coefficient and the arterial and venous concentrations of the indicator. The partition coefficient can be a major source of error as it varies greatly from tissue to tissue and may change with time.

Techniques based on these equations are called clearance techniques. They can be used for measuring blood flow without venous sampling in, for instance, the brain, liver and kidneys.

19.3.3.   Isotope techniques for brain blood flow

Brain blood flow can be measured with the gamma-emitting radionuclide ^{133}Xe, which can be administered either by injecting the gas dissolved in saline into the carotid artery, or by inhalation.

After injection, the activity versus time curve from the cerebral hemispheres is recorded for about 15 min. After the first passage through the brain, the activity passes through the left heart to the lungs. Xenon is ten times more soluble in air than in blood, so most of it is exhaled, i.e. C_A falls rapidly to zero. Inserting $C_A = 0$ in equation (19.11),

$$\frac{F}{V} = \frac{\alpha C_V}{\int_0^t C_V\,dt}.$$

Note that it is not necessary to know the concentration of Xe in the blood—the right-hand side of the equation is a ratio of the concentration to the integral of concentration, which is identical to the ratio of activity to its integral.

The ^{133}Xe can also by administered by breathing on a closed circuit containing Xe for 5 min. Uptake and subsequent clearance is monitored over a 40 min period. This is completely atraumatic, but requires the use of deconvolution techniques (see section 13.6) to obtain the curve shape that would have resulted from a single injection of ^{133}Xe—each breath will have delivered a separate bolus. There is also time for the Xe to diffuse into extracerebral tissues, which also complicates the analysis.

Positron emitters have been used for the study of regional cerebral blood flow, using positron-emission tomography (PET). This stimulated the development of gamma-emitting radiopharmaceuticals that could be used with a tomographic camera (SPET, single-photon emission computed tomography) for regional cerebral blood flow imaging.

19.3.4.   Local clearance methods

Tissue blood flow can be measured by injecting the tracer directly into the tissue. In this case, the arterial concentration is zero, so

$$\frac{\mathrm{d}C(t)}{\mathrm{d}t} = -\frac{F}{V}C_{\mathrm{V}}.$$

With a partition coefficient α, we can show that

$$C_{\mathrm{V}} = C(t)/\alpha.$$

The clearance rate is given by

$$\frac{\mathrm{d}C(t)}{\mathrm{d}t} = kC(t)$$

where k is the fractional clearance rate. Eliminating $\mathrm{d}C/\mathrm{d}t$ and C_{V} from the first equation gives

$$F/V = -k\alpha \qquad (19.12)$$

As k is the rate constant of an exponential decay process, it can be found from the slope of the log(activity) versus time curve.

Measurement is performed by injecting a small quantity of saline (0.1 ml) containing dissolved ^{133}Xe. The technique has been used in muscle, skin, myocardium and the eye.

19.4.   THERMAL TECHNIQUES

Some of the earliest techniques for measuring blood flow were thermal. They include calorimetry (the measurement of heat flow), thermal dilution, resistance thermometry, measurement of thermal conductivity and temperature measurement. We will look at the principles of some of these techniques, but only examine the measurement of tissue blood flow in any detail.

19.4.1.   Thin-film flowmeters

The basic idea of a thin-film flowmeter—sometimes called a thin-film anemometer—is to heat a very small probe to a fixed temperature above that of blood and then to measure the heat loss. Increased flow past the probe will increase the heat loss. The theoretical treatment of the heat loss from the heated film to the flowing blood is difficult. It has been shown that

$$\frac{I^2 R}{\Delta T} = A\frac{k\sigma^{1/3}}{\eta^{1/2}}U^{1/2} + B$$

where $I^2 R$ is the power applied to the film, ΔT is the temperature difference between the film and the fluid, U is the fluid velocity, A and B are constants, k is the thermal conductivity of the fluid, σ is the Prandtl number of the fluid and η is the kinematic viscosity. The value of $k\sigma^{1/3}/\eta^{1/2}$ is nearly constant for water, saline and plasma, so that the probe (figure 19.5) can be calibrated by measuring I^2 versus $U^{1/2}$ at a constant temperature difference in water.

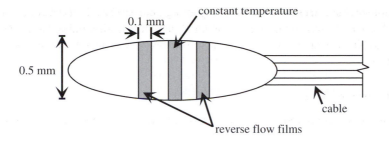

Figure 19.5. *Diagram of a thin-film flowmeter probe.*

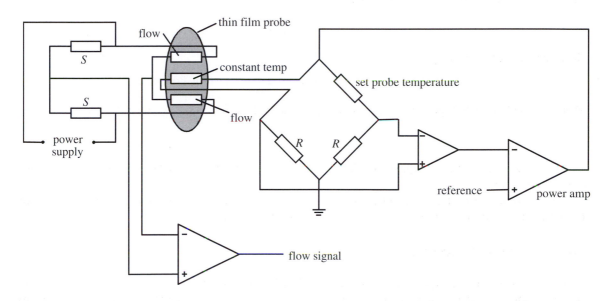

Figure 19.6. *Circuit that maintains the probe at a constant temperature and measures the power required to do so.*

The sensing element uses a material with a high thermal coefficient of resistivity such as platinum, which is deposited on a rigid substrate, such as glass. Typical surface areas are about 10^{-3}–10^{-2} cm^2, with a resistance of a few ohms. The film is placed in one arm of a Wheatstone bridge. The output of the bridge is a measure of the change in film resistance due to temperature change. The system is operated in a constant temperature mode, in which the output voltage controls the power supply to the bridge. If the film resistance falls as a result of an increase in flow, the bridge output increases, the voltage across the bridge increases, and the temperature (and thus resistance) of the film increases until the bridge is balanced. This relies on the thermal mass of the film being much less than that of the other resistors in the bridge, i.e. the temperature, and hence the resistance, of the other resistors is constant. The measured output is the voltage applied to the bridge. As in all such feedback control systems, an error voltage is needed to drive the power amplifier, but this can in principle be made negligibly small by increasing the loop gain of the system (figure 19.6).

A single film cannot resolve directional flow—the heat loss is the same for flow in either direction. A modification of the probe uses two films R_1 and R_2 placed on either side of a heated film R_0, which is maintained at a constant temperature (1–5 °C above blood temperature). The blood is warmed slightly in passing over the heated film, causing a change in resistance of the downstream film. This gives an output signal from the bridge, the polarity of which indicates the direction of flow (figure 19.6).

Thin-film devices measure velocity, not volume flow. Their small size (the spatial resolution can be < 0.1 mm) makes them well suited to making measurements of the flow profile in blood vessels (Belhouse *et al* 1972). They have also been used to study turbulence.

19.4.2.   Thermistor flowmeters

Thermistors are small semiconductor devices with a negative temperature coefficient of resistance (typically −4% °C^{-1}). For temperature measurement, the current through the device is kept very low to avoid self-heating. For flow measurement, the thermistor is maintained at an elevated temperature, as in the thin-film device. Early thermistor flowmeters used constant current heating. This has the disadvantage that the thermistor temperature approaches the blood temperature asymptotically as flow increases, i.e. the sensitivity is lower at high flow rates. At zero flow, the constant heat input will cause the temperature to increase for a considerable time, so that there is no clear indication of the cessation of flow.

The preferred method is to use the same principle as the thin-film flowmeter, with the thermistor maintained at a constant temperature difference from the blood (typically 1 °C above blood temperature). Thermistor flowmeters are larger than thin-film flowmeters, have a poor frequency response as a result of their thermal mass and cannot resolve the direction of flow. They are, however, very much cheaper than thin-film devices, as they are a common electronic component.

It should be noted that it is assumed that the only cause of a temperature change is a change in blood velocity. In practice, the probe is maintained at a constant temperature, not at a constant differential temperature with respect to stationary blood. A change in blood temperature would therefore appear as an apparent change in flow.

19.4.3.   Thermal dilution

Thermal dilution is a variation on the dye dilution technique and can be used for repeated measurements. Cardiac output (see also section 19.2.4) is measured by injecting a cold bolus of saline into the superior vena cava or the right atrium, and detecting the resulting temperature change (see figure 19.7) in the pulmonary artery or aorta. The technique is sometimes referred to as the injection of coolth. The coolth is the thermal energy of the cool saline. The measurements can be made using a single multi-lumen catheter. A temperature measuring thermistor is in the tip of the catheter, and the injection orifice 30 cm distal to the tip. The catheter is passed through the venous system into the right atrium, through the right ventricle and into the pulmonary artery. When the measuring thermistor is in the pulmonary artery, the injection lumen is in the right atrium.

It is assumed that there is complete mixing of the saline indicator and blood, and that the injection does not alter the flow rate. If the indicator temperature is T_i and the blood temperature T_B, then the heat gained by the indicator is

$$H = V_i S_i (T_f - T_i) \rho_i \quad \left(\text{units:} \quad J = m^3 \, J \, kg^{-1} \, K^{-1} \, K \, kg \, m^{-3} \right)$$

where V_i is the volume, S_i the specific heat and ρ_i the density of the cold saline, and T_f is the final temperature of the blood/indicator mixture. If the injection and equilibration were instantaneous, T_f would be the temperature immediately after injection. In practice, T_f is found as the mean of the thermal dilution curve. This heat gain is at the expense of a heat loss from the blood:

$$H = V_B S_B (T_B - T_f) \rho_B$$

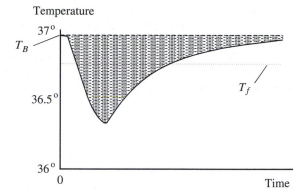

Figure 19.7. *Thermal dilution curve showing the temperature change measured in the pulmonary artery following injection of cool saline into the right atrium.*

where V_B is the total volume of blood in which the saline is diluted, i.e. the blood volume which passes the temperature sensor between the beginning and end of the dilution curve. If the flow rate (of blood + saline) is F, and the time between the beginning and end of the dilution curve is t, then the volume passing the sensor is Ft, and the blood volume is

$$V_B = Ft - V_i.$$

Assuming no loss of heat,

$$V_B S_B (T_B - T_f)\rho_B = V_i S_i (T_f - T_i)\rho_i$$

therefore

$$(Ft - V_i) S_B (T_B - T_f)\rho_B = V_i S_i (T_f - T_i)\rho_i$$

thus

$$F = \frac{V_i}{t}\left[1 + \frac{S_i(T_f - T_i)\rho_i}{S_B(T_B - T_f)\rho_B}\right]. \tag{19.13}$$

So in order to calculate the flow F we need to know:

- The volume (V_i) and temperature (T_i) of the injected saline.
- The specific heat (S_B), density (ρ_B) and temperature T_B of the blood.
- T_f, which we calculate by integrating the area under the thermal dilution curve (see figure 19.7) and dividing by t.

Saline indicator volumes of 3.5–10 ml and temperatures of 13 °C are quoted (compare this with the stroke volume of 70 ml and blood temperature of 37 °C). Errors can occur as a result of unwanted heat exchange. The first injection is usually discarded, as the injection catheter will have been warmed by the blood. In subsequent injections, heat exchange may occur through the wall of the catheter, which will lower the blood temperature and raise the saline temperature. The effect of this can be reduced by using a double-walled catheter and using a thermistor at the end of the catheter to measure the actual injectate temperature. Saline remaining in the catheter will also exchange heat with the blood after the injection.

Recirculation does occur, but can be allowed for by extrapolation. The blood temperature returns rapidly to normal, so that repeat measurements are possible. With dye dilution, the accumulation of dye precludes repeat measurement. Thermal dilution is a widely used method of measuring cardiac output, but it can be subject to quite large errors. The temperature change is only 0.2–0.5 °C, and the spontaneous fluctuations in the temperature of blood are of the order of ±0.03 °C. The signal-to-noise ratio is low. However, the fluctuations will be averaged over the time period of the dilution curve and repeat measurements can be taken (Forrester *et al* 1972).

19.4.4.   *Thermal conductivity methods*

If a heat source is surrounded by an infinite-thermal conductor, a temperature difference ΔT will be established, where

$$P = 4\pi R \,\Delta T\, \beta.$$

R is the radius of the sphere across which the temperature difference is measured, β is the thermal conductivity and P is the power input. This equation can be used to measure the thermal conductivity of a volume of tissue. The thermal conductivity will appear to increase if blood flow increases. However, the problem is that the thermal conductivity will not be zero even when blood flow is zero. A typical value for the thermal conductivity of skin and subcutaneous tissue is 0.3 W m^{-1} K^{-1}. In tissue, heat is removed by blood flow as well as by the thermal conductivity of the tissue. The technique has been used to estimate local blood flow in the brain, the gut and the skin. In practice, it is not possible to distinguish between a change in blood flow and a change in thermal conductivity (due to, for instance, a change in the water content of the tissue).

The method has been applied in two ways: either apply constant power to the heater and measure the resulting temperature gradient or measure the power required to maintain a constant temperature difference between the probe and the tissue (Holti and Mitchell 1979). The advantage of the latter method is that the vasodilatation caused by the increased tissue temperature should not be a function of the rate of heat removal.

19.4.5.   *Thermography*

All objects at a temperature greater than 0 K emit radiation, the spectrum of which depends on the surface temperature T_S and the emissivity of the surface. The Stefan–Boltzmann law gives the heat transferred from a surface to the environment at temperature T as

$$E = \sigma\left(T_S^4 - T^4\right) \quad \text{W m}^{-2}$$

where σ is the Stefan–Boltzmann constant. $\sigma = 5.6705 \times 10^{-8}$ W m^{-2} K^{-4}. An image of the surface temperature of the body can be formed by a scanning system which measures the emitted radiation. Surface temperature is a reflection of the underlying blood flow, so that the images can be used to assess skin blood flow. However, the skin blood flow is part of the thermo-regulatory system, and the images will therefore also depend on the extent of thermal equilibration between the body and the surrounding environment. This process may take a considerable length of time (half an hour or more), particularly if the subject has come from a cold environment, and the skin blood flow is therefore reduced.

Thermography can produce nice images showing body surface temperature but it cannot be regarded as a quantitative method of measuring blood flow. By comparing left and right sides of the body abnormalities can sometimes be detected, but in general it is too difficult to control the environment for quantitative indications of blood flow to be measured.

19.5.   ELECTROMAGNETIC FLOWMETERS

The use of magnetic fields to measure blood flow is based on Faraday's law of magnetic induction (figure 19.8). A voltage is induced in a conductor which is moving in a magnetic field. Blood is an electrical conductor (see table 8.4 in section 8.5.1), so that a voltage will be developed when blood flows through a magnetic field. The magnetic field, the direction of motion and the induced voltage are mutually at right angles as shown in figure 19.8. The induced voltage is proportional to the velocity of the conductor (i.e. the blood velocity), the strength of the magnetic field and the length of the conductor at right angles to the direction of motion (i.e. the diameter of the blood vessel). Put mathematically the induced voltage E is given by

$$E = DBV$$

where D is the dimension of the conductor in the direction of the induced voltage, B is the magnetic field and V is the velocity. This assumes that the magnetic field is uniform and of infinite extent and, for the case of a conducting fluid, that the flow is axially symmetrical.

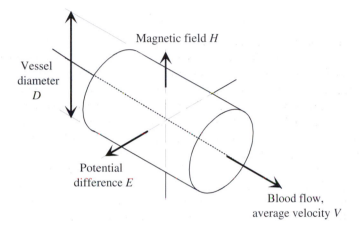

Figure 19.8. *The principle of the electromagnetic flowmeter.*

We can put some values into the above equation for a typical blood vessel in order to estimate the likely size of E. For a 10 mm diameter blood vessel placed in a magnetic field of 0.025 T the induced voltage will be 250 μV for a blood velocity of 1 m s^{-1}. 250 μV is quite a small signal and may be difficult to measure. Faraday did not suggest his principle of magnetic induction could be applied to the measurement of blood flow, but he did suggest that it could be used to measure the flow of water in the river Thames. The Earth provides a magnetic field so there should be a potential induced across the river in proportion to the water velocity. Unfortunately, attempts to try to measure this potential failed simply because the voltage was too small for contemporary instruments to measure (see problem 19.8.2.5).

Early electromagnetic blood flowmeters used DC excitation, i.e. a steady magnetic field was applied across the flow. The induced potential was measured using small electrodes placed in contact with the vessel wall. Now if the blood flow is steady then the induced voltage will also be steady. With DC excitation it was found that the flow signal of about 100 μV could not be distinguished from potentials caused by electrode polarization (see section 9.2.2). For this reason flowmeters now use AC excitation. By using an AC magnetic field the induced signal proportional to blood flow will also be an AC signal and this can be separated from electrode polarization potentials.

If the electromagnet (see figure 19.9) is excited by an alternating current then the flow signal will also be an alternating one. However, there is a problem with AC excitation which is that the changing magnetic field will induce an AC voltage between the electrodes, even in the absence of any blood flow. This induced voltage is referred to as the transformer signal and will be proportional to the rate of change of the magnetic field.

For an AC excitation flowmeter with a sinusoidal magnetic field, the induced voltage is given by

$$E = DB(V \sin \omega t + k \cos \omega t)$$

where $B \sin \omega t$ is the magnetic field. The $k \cos \omega t$ term is due to the transformer effect, and is the voltage induced in the electrical circuit through the electrodes. In principle, sampling the signal when $\cos \omega t = 0$ will give the flow signal (figure 19.10). In practice there is also capacitive coupling between drive and signal, so that the minimum of the transformer signal is not at 90° to the drive signal. This problem can be circumvented by using a square wave excitation. Remembering that the induced voltage $e \propto \mathrm{d}B/\mathrm{d}t$, there will be no transformer voltage at the maxima and minima of the drive voltage.

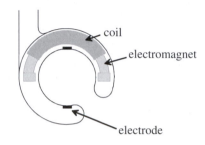

Figure 19.9. *Electromagnetic flowmeter probe.*

Probes are of three types, perivascular (i.e. fitted around the outside of intact blood vessels, figure 19.9), cannula (the blood vessel is cut and the probe inserted in-line with the vessel), and intravascular (the probe is mounted on a catheter). The electromagnet can be iron or air cored and a typical excitation frequency is 500 Hz. The magnet is necessarily small, so that the magnetic field is non-uniform, particularly in the direction of flow. This results in the induced voltage decreasing with distance from the electrodes, leading to circulating currents in the blood. The measured voltage is consequently reduced. In addition, the sensitivity to velocity can be shown to vary across the vessel. For point electrodes, the sensitivity varies from 1.0 on the axis to 0.5 on the vessel wall adjacent to the magnet, and approaches infinity near the electrodes. This sensitivity can be made more uniform by increasing the size of the electrodes and by modifying the magnetic field shape (Wyatt 1972).

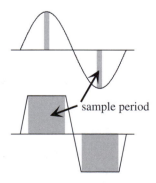

Figure 19.10. *Sine wave and square wave excitation showing the sampling time during which $\mathrm{d}B/\mathrm{d}t = 0$.*

The electromagnetic flowmeter actually measures velocity, not volume flow. To determine flow, the internal diameter of the vessel and the flow profile must be known. In practice, it would be assumed that the flow was laminar, and the probe would be calibrated on a test rig for laminar flow. The vessel wall is (indeed, it must be!) electrically conducting, which applies a shunt impedance to the induced voltage. For an average artery, with a ratio of inside-to-outside diameter of 0.85, the sensitivity is reduced by about 7%. If the outside of the blood vessel were wet, this would further reduce the voltage across the electrodes by about 10–15%.

19.6. PLETHYSMOGRAPHY

Arterial blood flow is usually pulsatile, whereas venous flow is steady. A direct consequence of this is that the volume of a body segment will change during the cardiac cycle. Blood vessels and the surrounding tissues have of course to be elastic in order for the volume to change. In principle, blood flow can be determined by measuring the rate of change of volume of a body segment. A plethysmograph measures changes in volume, and was first described in the early 17th Century for the measurement of the contraction of isolated muscle. Numerous different methods have been used to measure changes in volume, including air or water displacement and strain gauge plethysmography, impedance measurement, capacitive transducers and photoelectric transducers.

If you try to hold you arm in a horizontal position you will find that the fingers move up and down with the cardiac cycle. One reason for this is that on systole the arm volume increases by a few millilitres and hence is heavier. A second reason for the change is the inertial forces that result from the pulsatile direction of arterial blood flow. Separating the two effects is not easy, but it is an illustration of the fact that as our heart beats there are consequent volume changes in parts of the body.

The classical method of measuring the change in volume is to place the limb or digit in question within a sealed chamber. The chamber is filled with either air or water, and the change in volume is measured by means of a miniature spirometer for air (see section 17.4.1) or the movement of the water level in a glass tube. Figure 19.11 shows a typical waveform obtained from a plethysmograph. The volume of the body segment increases during cardiac systole because arterial inflow is greater than venous outflow but decreases during diastole as venous flow is then greater than arterial flow.

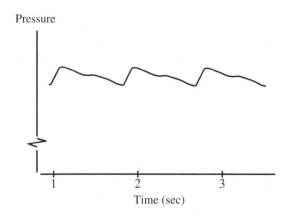

Figure 19.11. *Pulsatile changes in pressure observed in a container sealed around a body segment.*

Calculating blood flow from traces such as that shown in figure 19.11 is difficult. One reason is that the volume changes result from the net blood flow, i.e. the difference between arterial and venous flows, and so cannot be used to determine absolute arterial or venous blood flow. There are other problems, such as those with the measurement system. An air-filled system is sensitive to changes in temperature of the air enclosed. A water-filled system imposes a hydrostatic pressure on the tissue, which may affect local blood supply. If the temperature of the water is such that there is a net heat flow to or from the tissue, this will alter the degree of vasodilatation, thus altering the blood flow.

We will examine only the use of plethysmography for the measurement of blood flow, but the principles can be applied to the measurement of other volume changes. The whole-body plethysmograph, for instance, is used to measure changes in lung volume (see chapter 17, section 17.4.2).

19.6.1.   *Venous occlusion plethysmography*

A limb or organ will swell if the venous return is occluded, due to the inflow of arterial blood (see figure 19.12). Venous flow can be occluded by using a pressure cuff that is inflated well above venous pressure, but below systolic arterial pressure. The volume change will give a measure of the blood volume of the limb or organ, and the rate of change of volume gives a measure of the arterial blood flow. This is the basis of venous occlusion plethysmography. It avoids the problem of volume changes resulting from the net difference between arterial and venous flows by reducing the venous flow to zero. However, the technique does have to make some assumptions. The assumptions underlying the technique are that occlusion of the venous return does not affect the arterial supply, that the resulting increase in venous pressure does not reduce the arterial flow rate, that the swelling of the tissue in question is equal to the increase in the blood volume and that the venous occlusion is only applied for a limited period.

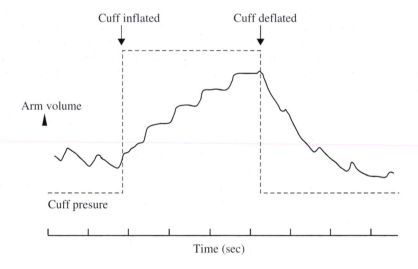

Figure 19.12. *Volume changes of an arm when the venous outflow is occluded.*

Venous occlusion plethysmography is one of the better techniques for measuring blood flow in that flow and volume can be well related. However, it is certainly invasive and it can only be applied to accessible body segments such as the limbs. Another disadvantage is that it cannot be reapplied too frequently.

Various workers have studied the validity of these assumptions. A cuff pressure of about 50 mmHg is sufficient to occlude the venous return, and no change in arterial inflow is found until the venous pressure has risen 4–12 mmHg above the resting value.

19.6.2.   Strain gauge and impedance plethysmographs

An index of the change in volume of a limb can be found by measuring the change in circumference. The mercury-in-rubber strain gauge consists of a column of mercury contained in a silastic tube. This is wrapped around the limb. As the limb expands, the tube is stretched, the column of mercury becomes longer and thinner and its resistance therefore increases. A typical gauge has a bore of 0.8 mm and a wall thickness of 0.5 mm. It is left as an exercise for the reader to show that

$$dV/V = dZ/Z \qquad (19.14)$$

i.e. the fractional volume change is directly proportional to the fractional resistance change.

Other systems can be used to measure limb or trunk circumference. A strain gauge can be used with an elasticated band or a large corrugated tube connected to a pressure transducer. Care must be taken in the use of all these techniques to avoid movement artefacts.

Impedance plethysmograph

As an alternative to measuring volume directly it is possible to relate the electrical impedance of a body segment to the volume of the segment.

Any change in volume of the limb or digit in question will alter the impedance according to the usual relationship $Z = \rho L/A$, where L is the length, A the cross-sectional area and the segment is assumed to have a uniform resistivity ρ. In practice the blood and surrounding tissue have different resistivities, so that the change in impedance will be a function of both the altered dimensions of the segment and of the change in blood volume. An additional effect is due to the impedance of the blood, which changes with velocity. The resistivity changes with velocity arise because erythrocytes tend to align themselves with the blood flow and this results in transverse resistivity rising and longitudinal resistivity falling as blood flow increases. The effect is an interesting one although it is a relatively small contribution to the signals obtained in impedance plethysmography.

The usual measurement technique is to inject a current at 100 kHz and 1 mA into one pair of electrodes, and measure the resulting voltage at a second pair (figure 19.13). The standard four-electrode technique for measuring resistivity (see section 8.9.2) eliminates the effect of electrode impedance. It has been claimed that the impedance plethysmograph signal is an artefact due to changes in electrode impedance caused by movement, but this is ruled out by the four-electrode technique.

It is instructive to consider the change in impedance resulting from a change in tissue dimensions alone, as this illustrates the effect of the assumptions that are made. If we take a segment of length L and cross-sectional area A, then the impedance Z is given by

$$Z = \rho L/A = \rho L^2/V \qquad \text{where} \quad V = LA.$$

We could make three different assumptions about the expansion of this segment:

- it is radial only;
- axial only; or
- expands isotropically.

Taking the first two cases, and finding the change in impedance with volume

$$\left[\frac{\delta Z}{\delta V}\right]_L = -\frac{Z}{V} \qquad \text{and} \qquad \left[\frac{\delta Z}{\delta V}\right]_A = \frac{Z}{V}. \qquad (19.15)$$

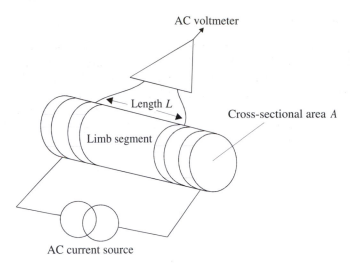

Figure 19.13. *Principle of the electrical impedance plethysmograph.*

The first two assumptions give the same *magnitude* of change, but *opposite* sign! A more reasonable assumption is that the tissue expands equally in all directions, i.e. $dL/L = dR/R$, where R is the radius of the section. We now have $Z = f(L, R)$ where L and R are both functions of V.

Thus,

$$\frac{dZ}{dV} = \left[\frac{\delta Z}{\delta L}\right]_R \frac{dL}{dR} + \left[\frac{\delta Z}{\delta R}\right]_L \frac{dR}{dV}.$$

From $Z = \rho L/A$ we find

$$\left[\frac{\delta Z}{\delta L}\right]_R = \frac{\rho}{\pi R^2} \quad \text{and} \quad \left[\frac{\delta Z}{\delta R}\right]_L = \frac{2\rho L}{\pi R^3}.$$

From $V = \pi R^2 L$ and $dL/L = dR/R$ we can obtain

$$dL/dV = 1/3\pi R^2 \quad \text{and} \quad dR/dV = 1/3\pi RL.$$

Substituting these values into the equation for dZ/dV gives

$$dZ/dV = -\rho/3A^2 = -Z/3V. \tag{19.16}$$

The three starting assumptions (radial expansion, axial expansion or isotropic expansion) thus give three different solutions for the change in impedance with volume.

In general, it is not possible to relate impedance changes directly to volume changes. However, impedance plethysmography has been used to measure blood flow, respiration and cardiac output, with varying success. As an example: if current is passed down the leg and two electrodes are used to measure the impedance of the calf and thigh then a baseline impedance of about 200 Ω is found. The impedance will be found to decrease during systole by about 0.1 Ω (see Brown *et al* 1975). If the venous return is occluded then the impedance of the limb falls progressively by about 0.1 Ω on each cardiac cycle.

19.6.3. *Light plethysmography*

The light plethysmograph consists of a light source and a photodetector, which measure the reflection from or the transmission through a capillary bed (see figure 19.14). Common locations are the end of the finger or an ear lobe. Pulsatile changes in the volume of blood in the capillary bed give rise to a signal which looks very similar to that of figure 19.11. The signal is a useful indicator of the pulse, but has proved to be impossible to quantify directly in terms of blood flow. However, the technique is the basis of the technique of pulse oximetry for the measurement of oxygen saturation in the blood. This technique is widely used. It is described in section 17.5.1 in some detail.

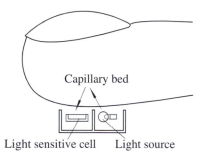

Figure 19.14. *Light plethysmograph probe.*

19.7. BLOOD VELOCITY MEASUREMENT USING ULTRASOUND

All the techniques we have considered so far, with the exception of the electromagnetic flowmeter, have measured volume flow. The technique we are now about to consider in some detail measures blood velocity. You will remember that we said that typical arterial blood velocities are about 1 m s^{-1} during systole and that venous velocities are a few cm s^{-1}. Doppler ultrasound is able to measure these velocities and is the most widely used method of measuring blood flow. Chapter 7 dealt with the generation of ultrasound and its transmission in tissue. We begin now by looking at the Doppler effect for ultrasound (Evans *et al* 1991).

19.7.1. *The Doppler effect*

The change in apparent frequency of a sound source as a result of relative movement of the source and observer was first described by Christian Doppler in 1842. The pitch change from a passing train is usually given as a common example of the Doppler effect. In that case only the source of the sound is moving. However, the effect is also observed if sound is reflected from a moving target. This is how the effect is used to measure blood velocity. Moving blood does not spontaneously emit sound, so ultrasound must be propagated through the tissue to the blood vessel. The red cells are much smaller than the wavelength of the ultrasound, and behave as elastic spheres in the ultrasonic pressure field. Each red cell is caused to alter its size by the pressure field, and acts as a point source of transmitted ultrasound. The sound transmitted by the red cell is detected by the receiver.

Consideration of the Doppler effect for red cells is complicated by the fact that the red cell first acts as a moving receiver, and then as a moving source. We will start by considering these two situations.

Consider first the case of a stationary source and a receiver moving towards the source (figure 19.15). The wavelength of the sound emitted by the source λ_s is given by

$$\lambda_s = c/f_s$$

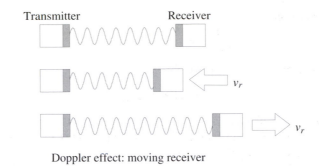

Doppler effect: moving receiver

Figure 19.15. *A static ultrasound transmitter. v_r is the velocity of the receiver.*

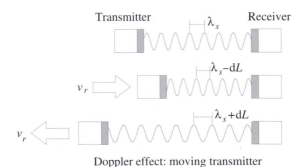

Doppler effect: moving transmitter

Figure 19.16. *A static ultrasound receiver.*

where c is the velocity of sound and f_s is the frequency. In unit time, the receiver moves a distance v_r, and therefore detects an additional number of peaks v_r/λ_s. The total number of peaks received per unit time is f_r:

$$f_r = f_s + v_r/\lambda_s$$
$$= f_s + (v_r/c)f_s$$

i.e. the Doppler frequency

$$f_D = (v_r/c)f_s. \qquad (19.17)$$

For the case of a moving source and a stationary receiver (figure 19.16), the time interval between peaks is $1/f_s$. In this time, the source moves a distance dL given by

$$dL = v_s(1/f_s) \quad \text{where } v_s \text{ is the source velocity.}$$

The wavelength λ_r at the receiver is

$$\lambda_r = \lambda_s - dL$$
$$= \frac{c}{f_s} - \frac{v_s}{f_s} = \frac{c}{f_r}$$

i.e.

$$f_r = \frac{c}{c - v_s}f_s = \frac{1}{1 - v_s/c}f_s.$$

Using the Taylor series expansion,

$$\frac{1}{1-x} = 1 + x + \frac{x^2}{2} + \cdots$$

and ignoring x^2 and higher terms, as $v/c \ll 1$, we obtain

$$f_r = (1 + v_s/c)f_s$$

i.e. the Doppler frequency $f_D = (v_s/c)f_s$.

Not surprisingly this is the same as equation (19.17). Movement of the receiver and the transmitter give the same result. It is the relative velocity that matters.

We now need to apply this to the case of a red cell which is moving at an angle to the ultrasound beam. We will take a general case where the transmitter makes an angle of α with the particle velocity and the receiver makes an angle β. We will also use the angular frequency ω instead of f (figure 19.17).

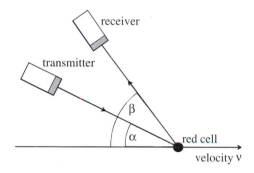

Figure 19.17. *General case of a red cell moving at an angle to the transmitter and receiver.*

The velocity v_p of the particle in the direction of the transmitted beam is $v_p \cos \alpha$, so the frequency seen by the moving particle is

$$\omega_p = \omega_c + (v_p \cos \alpha / c)\omega_c$$

where ω_c is the transmitted (carrier) frequency, ω_p the frequency at the particle and ω_r the received frequency. The frequency ω_p is radiated by the moving particle, and the frequency at the receiver is

$$\omega_r = (1 + v_p \cos \beta / c)\omega_p$$
$$= (1 + v_p \cos \beta / c)(1 + v_p \cos \alpha / c)\omega_c$$
$$= (1 + v_p \cos \beta / c + v_p \cos \alpha / c + (v_p/c)^2 \cos \alpha \cos \beta)\omega_c.$$

Neglecting the $(v_p/c)^2$ term, we obtain

$$\omega_r = \omega_c + (v_p/c)(\cos \alpha + \cos \beta)\omega_c.$$

The Doppler frequency is thus given by

$$\omega_r = (v_p/c)(\cos \alpha + \cos \beta)\omega_c.$$

Now $(\cos \alpha + \cos \beta) = 2 \cos \frac{1}{2}(\alpha + \beta) \cos \frac{1}{2}(\alpha - \beta)$. If we define the angle between the transmitter and receiver to be $\phi = \alpha - \beta$, and the angle between the particle velocity and the bisector of the angle between the transmitter and receiver to be $\theta = \frac{1}{2}(\alpha + \beta)$, we obtain

$$\omega_D = 2(v_p/c) \cos \theta \cos(\phi/2)\omega_c.$$

If the angle ϕ between the transmitter and receiver is zero, this reduces to the familiar equation for the Doppler frequency from a moving red cell:

$$\omega_D = 2(v_p/c)\omega_c \cos\theta. \tag{19.18}$$

It can be shown that having the transmitter and receiver in the same position (i.e. $\phi = 0$) is the optimum arrangement, as this maximizes the received power.

So we have concluded that the Doppler frequency will be directly proportional to the velocity of the red blood cells multiplied by the cosine of the angle between the ultrasound beam and the blood flow. The interesting result is that the Doppler frequency can be either positive or negative depending upon the direction of the blood flow. What does a negative frequency mean? We will consider this question in the next section.

Equation (19.18) is the basic Doppler equation and shows that we can directly relate the Doppler frequency to blood velocity. However, the equation also makes clear the major disadvantage of the technique as a method of calculating blood flow. We need to know the angle between the blood flow and the beam of ultrasound. In practice we do not usually know this angle and it can be difficult to determine unless we have an image of both the blood vessel and the ultrasound beam.

19.7.2. *Demodulation of the Doppler signal*

The Doppler signal is a frequency shift—it increases or reduces received frequency. To measure it we need to be able to find the difference between the transmitted and received frequencies.

What is the magnitude of the Doppler frequency?

We can insert typical values into the Doppler equation (19.18). If $c = 1500$ m s^{-1}, $v = 1$ m s^{-1}, $f_c = 5$ MHz and $\theta = 30°$, we find that $f_D = 5.8$ kHz. In general, Doppler frequencies for ultrasonic detection of blood velocity lie in the audio range. This is an important observation as it means that the ear can be used for interpretation. Most Doppler systems have an audio output so that the blood flow signal can be heard.

The next important question is how we might extract the Doppler signal using either electronics or software. Figure 19.18 shows a block diagram of the simplest Doppler demodulator.

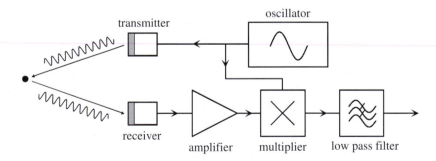

Figure 19.18. *Block diagram of a Doppler demodulator.*

If the transmitted signal is $E\cos(\omega_c t)$, the received signal V_i will be

$$V_i = A\cos(\omega_c t + \phi) + B\cos(\omega_c + \omega_d)t$$

where ϕ is the phase shift in the carrier component relative to the transmitted signal. The first term is the result of reflection from static structures within the body, and imperfect acoustic insulation between the transmit and receive crystals. The Doppler frequency is separated from the component at the transmitted frequency by a demodulator (figure 19.18). The demodulator is a linear multiplier that forms the product of the transmitted frequency and the received signal, giving

$$V_A = AE \cos(\omega_c t + \phi) \cos(\omega_c t) + BE \cos(\omega_c + \omega_D)t \cos(\omega_c t)$$
$$= (AE/2)[\cos(2\omega_c t + \phi) + \cos(\phi)] + (BE/2)[\cos(2\omega_c t + \omega_D t) + \cos(\omega_D t)].$$

The components of this are

$$\cos(2\omega_c t + \phi) \qquad \text{(twice the transmitted frequency)}$$
$$\cos(2\omega_c t + \omega_D t) \qquad \text{(twice the transmitted frequency)}$$
$$\cos(\phi) \qquad \text{(DC level)}$$
$$\cos(\omega_D t) \qquad \text{(Doppler signal)}.$$

The DC level is removed by the high-pass filtering in the system. Remembering that the Doppler frequency is typically three decades below the transmitted frequency, it is clear that the components at twice the transmitted frequency can be removed by a simple low-pass filter.

Note that, as $\cos(\omega_D t) = \cos(-\omega_D t)$, this simple demodulation cannot distinguish forward from reverse flow. This is a major disadvantage. In many cases arteries and veins run side by side in the body so that blood flows both towards and away from the ultrasound probe will exist. We need to be able to separate venous and arterial flows. There is another problem in that both positive and negative flow can occur in a single artery. Because vessels are elastic the positive flow on systole is often followed by a brief period of negative flow. As a rule of thumb, arteries below the heart have reverse flow at some point in the cardiac cycle, so we need to be able to distinguish forward from reverse flow. We need a directional demodulator.

19.7.3. *Directional demodulation techniques*

The problem in directional demodulation is to translate the directional information which is given by the sign of the frequency shift (i.e. is ω_D positive or negative) into some other indicator of direction which is not removed by the demodulation process. In the example we quoted at the beginning of the last section the transmitted frequency was 5 MHz and the Doppler frequency was 5.8 kHz. In a directional system we need to be able to separate the positive flow giving a received frequency of 5.0058 MHz from the negative flow giving a received frequency of 4.9942 MHz.

In spectral terms, the signal received by the transducer is a double-sideband (DSB) signal, and the problem is to distinguish the information carried by each sideband. This is identical to the problem of generating and detecting single-sideband (SSB) radio signals. This problem was thoroughly understood in the 1950s (see, for instance, 1956 *Proc. IRE* **44**). Although not acknowledged, this work forms the basis of all the demodulation techniques used for the ultrasonic detection of blood velocity.

There are four techniques that can be used to separate the upper and lower sideband frequencies from the transmitted signal:

- Filter the signal to separate the sidebands before demodulation (*frequency filtering*).
- Use the timing of the demodulated signal to distinguish forward or reverse flow (*time domain processing*).
- Use the phase information in the demodulated signals to generate demodulated separate upper and lower sidebands (*phase domain processing*).
- Shift the carrier frequency to a non-zero reference frequency such that the upper and lower sidebands both fall in the audio range (*frequency domain processing*).

These techniques generate a demodulated Doppler signal in which the directional information is retained. The resulting signals are then processed to give a mean frequency or a frequency spectrum. The frequency spectrum stage is very important. The blood in an artery does not all have the same velocity. In general, the blood in the middle of a vessel will travel more rapidly than that at the edge. There will therefore be a spread of Doppler frequencies representing the spread of blood velocity. A frequency analysis of the Doppler signal will give us a spectrum of blood velocity.

We will consider all four of the above methods although the first two do not warrant much attention. The last three methods all depend upon a common directional demodulation system so we will consider this first. Figure 19.19 shows the block diagram of the system.

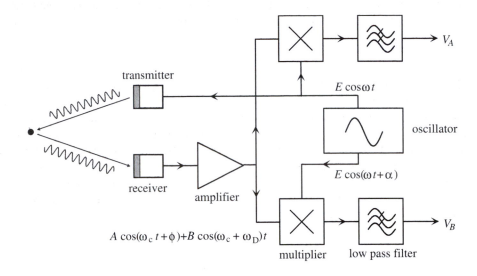

Figure 19.19. *Basic directional demodulator.*

All directional flowmeters start by generating two signals, which are processed to yield directional information (figure 19.19). One signal is derived by multiplying the received signal $A \cos(\omega_c t + \phi) + B \cos(\omega_c + \omega_D)t$ by $E \cos(\omega_c t)$, and the other by multiplying the received signal by $E \cos(\omega_c t + \alpha)$. This gives

$$V_A = \tfrac{1}{2} B E \cos(\omega_D t)$$
$$V_B = \tfrac{1}{2} B E \cos(\omega_D t - \alpha)$$

(19.19)

after removing DC and high-frequency components as before, i.e. the two signals are identical apart from a phase shift α. This phase shift is normally $\pi/2$, so the multiplying signals are sin and cos of the carrier frequency. In this case, the resulting signals are called the quadrature signals.

For $\omega_D > 0$, the phase difference between V_B and V_A is α. For $\omega_D < 0$, V_A and V_B are given by

$$V_A = \tfrac{1}{2} B E \cos[(-\omega_D)t] = \tfrac{1}{2} B E \cos(\omega_D t)$$

$$V_B = \tfrac{1}{2} B E \cos[(-\omega_D)t - \alpha] = \tfrac{1}{2} B E \cos(\omega_D t + \alpha)$$

and the phase difference between V_B and V_A is $-\alpha$. The sign of the flow direction is thus encoded as the sign of the phase difference between V_A and V_B. We will now consider the four methods of preserving this difference to give a blood flow signal.

19.7.4. *Filtering and time domain processing*

Frequency filtering

In principle, as the forward and reverse flow signals are at different frequencies before demodulation, it should be possible to separate them by using band-pass filters. In practice, the roll-off of the filters has to be very steep, as the separation of the sidebands is small (≈ 100 Hz) compared to the centre frequency (a few MHz).

An even more serious problem is the residual carrier frequency. This results from all the static structure echoes and the direct signal from the transmitter and could be 120 dB (10^6) greater than the Doppler signals. It is simply not possible to construct either analogue or digital filters with the performance required.

Time domain processing

This method, also called zero-crossing detection, was used in early flowmeters, but is now only of historical interest as it gives spurious results when there is simultaneous forward and reverse flow. The zero-crossing detector is a device which has a constant positive output when the signal is greater than zero and a constant negative output when the signal is less than zero. The principle of time domain processing is to use the phase difference α in equation (19.19) to switch the Doppler signal between two channels, one for forward flow and one for reverse flow. The two channels are identical, and give a measure of blood velocity by low-pass filtering the output of a zero-crossing detector. It is easy to see that, for a single frequency, the output of a zero-crossing detector is directly proportional to frequency. For an input amplitude spectrum $F(\omega)$, it can be shown that the average number of zero crossings is

$$\bar{N}_0 = \frac{1}{\pi} \left[\frac{\int_0^\infty \omega^2 F(\omega)\, d\omega}{\int_0^\infty F(\omega)\, d\omega} \right]^{1/2}.$$

This is the rms value of the spectrum, i.e. it is related to the rms velocity of the blood, and not to the average velocity. If the range of frequencies is small then the rms and the mean are similar. Therefore, the zero-crossing detector will produce a good approximation to the mean velocity for narrow-band signals, such as those from a pulsed system (see section 19.8) with a small sample volume.

19.7.5. *Phase domain processing*

Equation (19.19) gave us the following signals as the output from figure 19.19:

$$V_A = \tfrac{1}{2}BE \cos(\omega_D t)$$

$$V_B = \tfrac{1}{2}BE \cos(\omega_D t - \alpha).$$

If $\alpha = \pi/2$, then (ignoring the amplitude term) we get the quadrature signals

$$V_A = \cos \omega_D t$$

$$V_B = - \sin \omega_D t.$$

A standard method of separating the sideband signals is to use wideband phase-shifting networks to generate versions of the quadrature signals that are shifted by 0 and $\pi/2$, and sum appropriate pairs of signals as shown in figure 19.20.

To demonstrate how the forward and reverse flow components are separated, we will define forward flow as flow away from the probe, i.e. the forward flow frequency $\omega_f = -\omega_D$, and the reverse flow frequency

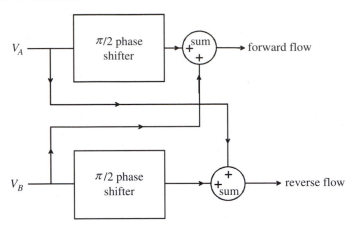

Figure 19.20. *Phase processing of the quadrature signals V_A and V_B.*

$\omega_r = \omega_D$. Inserting these frequencies into the quadrature terms (and remembering that $\sin(-\theta) = -\sin(\theta)$, and $\cos(-\theta) = \cos(\theta)$), we have

$$V_A = \cos \omega_f t + \cos \omega_r t$$
$$V_B = \sin \omega_f t - \sin \omega_r t.$$

Phase-shifting these signals by 0 and $\pi/2$ gives four signals (figure 19.20):

$$\cos \omega_f t + \cos \omega_r t \quad\quad 0 \;\rightarrow\; \cos \omega_f t + \cos \omega_r t \quad\quad (a)$$
$$\pi/2 \;\rightarrow\; \sin \omega_f t + \sin \omega_r t \quad\quad (b)$$
$$\sin \omega_f t - \sin \omega_r t \quad\quad 0 \;\rightarrow\; \sin \omega_f t - \sin \omega_r t \quad\quad (c)$$
$$\pi/2 \;\rightarrow\; -\cos \omega_f t + \cos \omega_r t \quad\quad (d)$$

Adding these signals gives

$$(b) + (c) = \sin(\omega_f t)$$
$$(a) + (d) = \cos(\omega_r t)$$

i.e. the output signals have separated the forward and reverse flow components.

This treatment has assumed that there are no amplitude or phase errors in the system. In principle, amplitude errors can always be made as small as required, but phase errors are a complex function of frequency, and will determine the rejection of the unwanted sideband in each channel. The construction of the $\pi/2$ phase-shifters is difficult and the actual phase shift is not completely independent of frequency.

19.7.6.   *Frequency domain processing*

Our fourth and last method for handling the quadrature signals produced by figure 19.19 is frequency domain processing. Frequency domain processing is the most important method of single-sideband demodulation, because it does not require wideband phase-shifters, which are difficult to construct. The processor has a second pair of multipliers and associated oscillator, and shifts the zero frequency (i.e. zero-velocity) term to

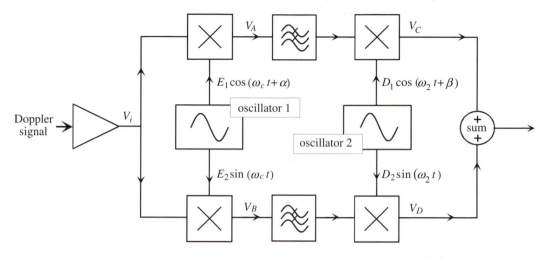

Figure 19.21. *Complete system to separate forward and reverse flows in the frequency domain.*

some secondary frequency ω_2. Positive Doppler frequencies are added to ω_2, and negative Doppler frequencies are subtracted from ω_2. The complete system is shown in figure 19.21.

Figure 19.21 includes both pairs of multipliers, so that the effect of phase errors in both oscillators can be studied. The outputs of the oscillators should have a phase difference of $\pi/2$, i.e. the outputs should be sin and cos waves. α and β represent the *error* in the phase difference. Referring to figure 19.21,

$$V_i = A\cos(\omega_c t + \phi) + B\cos(\omega_c + \omega_D)t.$$
$$\text{(carrier)} \qquad\qquad \text{(Doppler)}$$

Dealing with the signals in the same way as for figure 19.19 and removing the high-frequency and DC terms (by filtering) gives

$$V_A = \tfrac{1}{2}BE_1\cos(\omega_D t - \alpha)$$
$$V_B = -\tfrac{1}{2}BE_2\sin(\omega_D t).$$

The signals following the second pair of multipliers are given by

$$V_C = (BE_1 D_1/2)\cos(\omega_D t - \alpha)\cos(\omega_2 t + \beta)$$

i.e.

$$V_C = \tfrac{1}{4}BE_1 D_1[\cos(\omega_D t + \omega_2 t - \alpha + \beta) + \cos(\omega_D t - \omega_2 t - \alpha - \beta)] \tag{19.20}$$

$$V_D = -(BE_2 D_2/2)\sin(\omega_D t)\sin(\omega_2 t)$$

i.e.

$$V_D = \tfrac{1}{4}BE_2 D_2[\cos(\omega_D - \omega_2)t - \cos(\omega_D + \omega_2)t]. \tag{19.21}$$

Summing (19.20) and (19.21) gives

$$V_C + V_D = \tfrac{1}{4}B[E_1 D_1\cos(\omega_D t + \omega_2 t - \alpha + \beta) + E_2 D_2\cos(\omega_D + \omega_2)t]$$
$$+ \tfrac{1}{4}B[E_1 D_1\cos(\omega_D t - \omega_2 t - \alpha - \beta) - E_2 D_2\cos(\omega_D - \omega_2)t].$$

The first term in this equation is the upper sideband (USB), and the second term is the lower sideband (LSB). Reducing each term to the form $A \sin(\omega t + \phi)$ gives

$$\text{USB} = \tfrac{1}{4} B \left[E_1^2 D_1^2 + E_2^2 D_2^2 + 2 E_1 E_2 D_1 D_2 \cos(\beta - \alpha) \right]^{1/2}$$
$$\times \sin \left[(\omega_D + \omega_2)t - \tan^{-1} \left(\frac{E_1 D_1 + E_2 D_2 \cos(\beta - \alpha)}{E_2 D_2 \sin(\beta - \alpha)} \right) \right] \qquad (19.22)$$

$$\text{LSB} = \tfrac{1}{4} B \left[E_1^2 D_1^2 + E_2^2 D_2^2 - 2 E_1 E_2 D_1 D_2 \cos(-\beta - \alpha) \right]^{1/2}$$
$$\times \sin \left[(\omega_D - \omega_2)t + \tan^{-1} \left(\frac{E_1 D_1 - E_2 D_2 \cos(-\beta - \alpha)}{E_2 D_2 \sin(-\beta - \alpha)} \right) \right]. \qquad (19.23)$$

The lower sideband term vanishes when both

$$E_1 D_1 = E_2 D_2 \qquad \text{and} \qquad (\alpha - \beta) = 0 \quad \text{or} \quad n\pi \quad \text{(excluding trivial cases).}$$

- We reach the conclusion therefore that equation (19.22) gives us the output we require. It can be seen that this is a single sine wave of frequency $(\omega_D + \omega_2)$. We have achieved what we wanted. For zero flow velocity we get frequency ω_2. For forward flow we get a higher frequency than this and for reverse flow a lower frequency. We have removed the ambiguity of negative frequencies that we identified at the end of section 19.7.1.

An example of the output achieved by performing a frequency analysis on the signal as given by equation (19.22) is shown in figure 19.24.

19.7.7. FFT demodulation and blood velocity spectra

The advent of fast processors has made possible the generation of frequency spectra directly from the quadrature signals (equation (19.19)), without an intermediate stage of phase or frequency domain processing. The FFT (fast Fourier transform) algorithm can be used as a combined frequency domain processor and frequency analyser. The frequency analyser stage is very important. The blood in an artery does not all have the same velocity. In general, the blood in the middle of a vessel will travel more rapidly than at the edge. There will therefore be a spread of Doppler frequencies representing the spread of blood velocities. By frequency analysis of the Doppler signal we obtain the spectrum of the blood velocity.

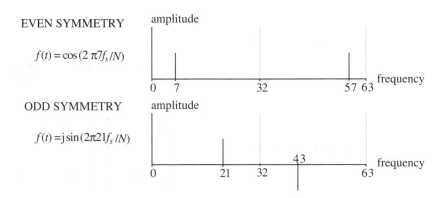

Figure 19.22. *Symmetry properties of FFT.*

The method makes use of the symmetry properties of the FFT. Referring to figure 19.22:

$$
\begin{array}{ccl}
\textit{function} & & \textit{Fourier transform} \\
f(t) & \mapsto & F(s) \\
\text{real and even} & \mapsto & \text{real and even} \\
\text{imaginary and odd} & \mapsto & \text{real and odd.}
\end{array}
$$

The effect on a real function $[\cos(2\pi 7 f_s/N)]$ and on an imaginary function $[\mathrm{j}\sin(2\pi 21 f_s/N)]$ is shown in the figure. The number of points N in the transform is 64. Remember that for an even function, $f(-x) = f(x)$, i.e. the function is reflected about the y-axis (and also, about the Nyquist frequency). For an odd function, $f(-x) = -f(x)$, so that the function is reflected about the origin, i.e. reflected and inverted. Remember also that for an N-point FFT, point 0 is the zero-frequency term, and therefore the Nyquist frequency is point $N/2$.

The general form of the quadrature signals is

$$
\begin{aligned}
V_A &= \cos(\omega_f t) + \cos(\omega_r t) \\
V_B &= \sin(\omega_f t) - \sin(\omega_r t).
\end{aligned} \tag{19.24}
$$

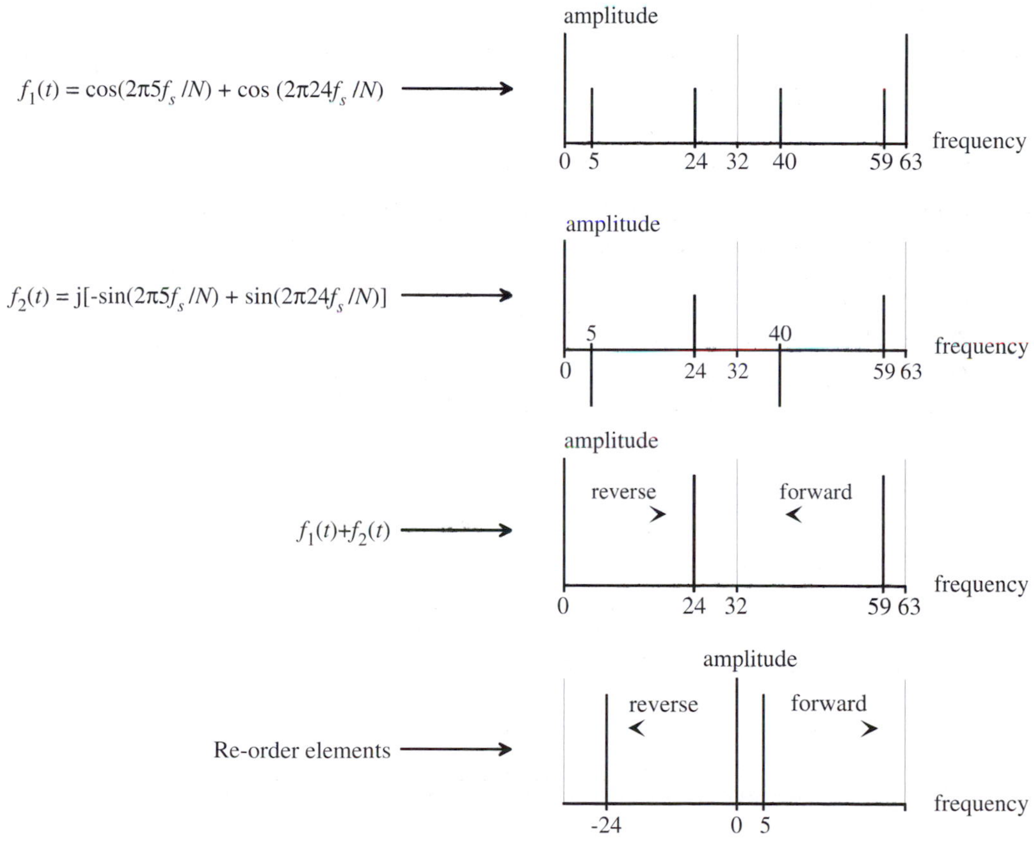

$f_1(t) = \cos(2\pi 5 f_s /N) + \cos(2\pi 24 f_s /N)$

$f_2(t) = \mathrm{j}[-\sin(2\pi 5 f_s /N) + \sin(2\pi 24 f_s /N)]$

$f_1(t)+f_2(t)$

Re-order elements

Figure 19.23. *FFT demodulation.*

Figure 19.23 shows the effect of taking V_A as the real part of the function to be transformed and $-V_B$ as the imaginary part. The forward and reverse signals are

$$\omega_{\mathrm{f}} t = 2\pi 5 f_{\mathrm{s}}/N \qquad \text{and} \qquad \omega_{\mathrm{r}} t = 2\pi 24 f_{\mathrm{s}}/N. \tag{19.25}$$

The figure shows separately the effect of transforming the real and imaginary parts of the composite spectrum generated from the quadrature signals, together with the combined output. The reverse flow signal appears in the first half of the FFT array, and the forward flow in the second half. Re-ordering the values gives the forward and reverse flow spectrum (reflect the reverse flow in the origin and the forward flow about the Nyquist frequency).

Figure 19.24. *Doppler frequency spectrum display of signals from the posterior tibial artery of a normal subject. Velocity (frequency) is plotted against time with the amplitude of the frequency components used to intensity modulate the display. The maximum frequency change is 8 kHz. The flow is zero in late diastole, rises rapidly on systole and is then successively negative then positive before returning to zero flow.*

Mean frequency calculation

The display given in figure 19.24 contains information on the blood velocity spectrum as a function of time. However, there are cases where the only information we want is that of the average velocity. If we want to determine the volume blood flow then we simply want the blood vessel cross-sectional area and the mean velocity. The average velocity is related to the mean frequency by the Doppler equation, and the mean frequency is given by

$$\bar{\omega}_D = \frac{\int_0^\infty \omega P(\omega)\,\mathrm{d}\omega}{\int_0^\infty P(\omega)\,\mathrm{d}\omega}. \tag{19.26}$$

where $P(\omega)$ is the power spectrum. This assumes that the complete cross-section of the vessel is uniformly insonated (exposed to ultrasound), and that the blood is a homogeneous scatterer of ultrasound. The mean frequency can clearly be computed directly from the frequency spectrum.

It is worth noting that the zero-crossing detector and filter, described in section 19.7.4, will work satisfactorily on the outputs of the phase domain processor (section 19.7.5) to give mean flow velocity. This gives a simple method of providing a signal proportional to flow, bearing in mind that the zero-crossing signal is proportional to rms frequency, and that there are additional limitations imposed by noise on the signal.

Considerable effort has been expended to try to develop simple probe systems that will give a volume flow signal directly. For example, devices exist to direct a beam of ultrasound at the aorta and give cardiac output directly. Some of these systems use two beams of different widths and process the signal to identify both average velocity of the blood and vessel cross-sectional area. Unfortunately, none of these systems are very reliable unless used with very great care and consequently results are often treated with suspicion.

19.7.8. Pulsed Doppler systems

One of the most exciting developments in ultrasound over the past two decades has been the combination of ultrasonic imaging (see section 12.3) and ultrasonic Doppler. By combining the two techniques it is possible to image the cardiovascular system and to show the blood flow on the same image. It is a combination of anatomical and functional imaging. In many systems blood flow is 'colour coded' in the image by making flow in one direction red and in the opposite direction blue. The strength of the colour can be related to blood velocity.

So far in this chapter concerned with blood flow we have assumed that we are dealing with a continuous wave (CW) system in which the ultrasound is transmitted continuously. Imaging systems usually use a pulsed system so that the timing of echoes can be used to give spatial information. In pulsed Doppler systems, short bursts of ultrasound are transmitted at regular intervals. Range discrimination is then possible. Pulsed systems can be used to give Doppler information. If the signal from the red cells is only accepted during a fixed time interval after transmission then this determines the sample volume from which the Doppler information is received. The technique is sometimes referred to as 'range-gating'.

Range limitation

The phase ϕ of the received signal (measured with reference to the transmitted frequency f_c) is related to the time of flight T_D by

$$\phi = f_c T_D.$$

Differentiating with respect to time gives

$$\frac{d\phi}{dt} = f_c \frac{dT_D}{dt}.$$

If the target is a distance z from the transducer, the time of flight is related to the total distance $2z$ travelled by the pulse and the velocity of sound c by

$$T_D = 2z/c.$$

The rate of change of z is the particle velocity, therefore

$$\frac{dT_D}{dt} = \frac{2}{c}\frac{dz}{dt} = \frac{2v}{c}.$$

The rate of change of ϕ is the Doppler frequency, therefore

$$f_D = \frac{d\phi}{dt} = f_c \frac{dT_D}{dt} = f_c \frac{2v}{c}$$

which is the usual Doppler equation. However, the maximum Doppler frequency that can be measured is limited by the repetition rate f_p of the pulses. This is a very important constraint. To avoid ambiguity, the maximum time of flight is T_p, where $T_p = 1/f_p$. The maximum Doppler frequency (by Nyquist, section 13.4) is half the repetition frequency, i.e.

$$f_{D_{max}} = \frac{1}{2T_p} = \frac{2v_{max}}{c} f_c. \tag{19.27}$$

The maximum distance $z_{max} = (c/2)T_p$, therefore

$$T_p = \frac{c}{4v_{max}f_c} = \frac{2z_{max}}{c} \quad \text{and} \quad v_{max}z_{max} = \frac{c^2}{8f_c}. \tag{19.28}$$

We can easily assess the importance of these results by taking a typical example. For $f_c = 2$ MHz and $c = 1500$ m s^{-1}, $v_{max}z_{max} = 0.14$ m^2 s^{-1}. At a distance of 10 cm the maximum velocity we can measure will be 1.4 m s^{-1}. Peak velocities in arteries can be considerably above this figure, particularly if the flow is turbulent, so the constraint imposed by equation (19.28) is important.

The range–velocity limit of equation (19.28) is plotted in figure 19.25 for 1, 2, 4 and 8 MHz. The lines at 0.4 and 1.6 m s^{-1} are the maximum velocities usually encountered for venous and arterial flows, respectively.

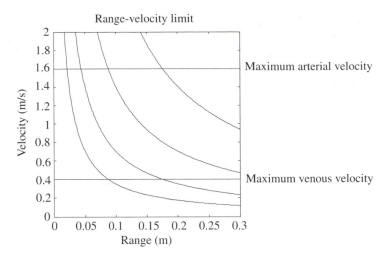

Figure 19.25. *The relation between the maximum range and maximum blood velocity is plotted.*

Bandwidth limitation

There is another limitation on the velocities that we can measure when using a pulsed Doppler system. Ultrasound pulses are typically 3–5 cycles in length. The frequency spectrum of the pulse is the product of the spectra of a continuous sine wave, the gating function and the transducer response. As an approximation, the width of the spectrum is inversely proportional to the number of cycles of the waveform. With 5 cycles, the width of the spectrum will be 20% of the centre frequency. As this spread of frequencies is present in the transmitted frequency ω_c, the same spread will also be present in the Doppler frequency ω_D, and the velocity can only be measured to within 20%.

This problem can also be approached by examining the effect of the size of the sample volume (which is determined by the pulse length). If δz is the axial length of the sample volume (we take the axial length because we can only measure the Doppler shift in the axial direction), then the transit time δt for a particle with velocity v will be

$$\delta t = \delta z / v.$$

The Doppler shift bandwidth is inversely proportional to the length of the Doppler signal, i.e. to δt, therefore

$$\delta f_D = 1 / \delta t = v / \delta z.$$

We also know, from the Doppler expression

$$\delta f_D = 2(\delta v / c) f_c.$$

Eliminating f_D between these two equations gives

$$\frac{\delta v}{v} \delta z = \frac{c}{2 f_c}. \tag{19.29}$$

The fractional accuracy with which we can measure the axial velocity of the target multiplied by the axial length of the target volume is a constant, which decreases with increasing frequency. For a given frequency, an increase in the precision with which we measure the position of the particle is at the expense of the precision with which we know its velocity.

The argument has been developed here for a pulsed Doppler system. A CW Doppler system also has a sample volume, in this case determined by the width of the ultrasound beam. The same argument can therefore be applied to the measurement of red cell velocity with CW ultrasound, and there will be a broadening of the spectrum due to the finite transit time through the beam. Changing the beam size will alter the insonated volume and thus determine the accuracy with which the velocity can be measured.

19.7.9. Clinical applications

The clinical applications of Doppler and Doppler imaging systems are very wide and we cannot hope to describe them here in any detail. References to several excellent books are given in the bibliography at the end of the chapter. All we will do in this brief section is to list some of the more common areas of application.

A long-standing application of Doppler ultrasound is in foetal monitoring. Using a frequency of 1–2 MHz a Doppler signal can be obtained from the foetal heart and used to assess the health of the foetus. The Doppler signal can be used to drive a counter, and so give a foetal heart-rate measurement. A wide range of similar instruments is available to measure blood flow in peripheral blood vessels. Because these vessels are superficial a higher frequency of ultrasound can be used. Remember that the attenuation of ultrasound increases in proportion to frequency so that a low frequency must be used if a long range is required. However, use of a low frequency requires a large transducer (see section 7.3.2) and the beam width will be proportionally larger. It is usually best to use the highest frequency possible. Doppler instruments for peripheral vessels usually operate in the range 5–10 MHz. Simple instruments of this type can be used for pulse detection during blood pressure measurement.

The largest application of Doppler systems and combined Doppler imaging systems is in the assessment of vascular disease. Doppler systems measure blood flow but it is possible to obtain information on the vascular system itself by looking at flow patterns. This is well illustrated in figure 19.26 which shows Doppler blood flow signals obtained from a normal and a diseased artery. In the normal vessel the large positive flow on systole is followed by a small negative flow caused by the elasticity of the blood vessel and then almost zero flow in diastole. In the diseased vessel the systolic pulse is damped and there is no negative flow component as the vessel walls are less elastic as a result of atheroma. Also the flow does not fall to zero in diastole. A large amount of research work has been carried out to determine the association between flow patterns and the state of a blood vessel.

In many vascular laboratories Doppler imaging systems are used to look at vascular disease. Disease can be particularly important where arterial vessels bifurcate. Atheroma can occur at these bifurcations.

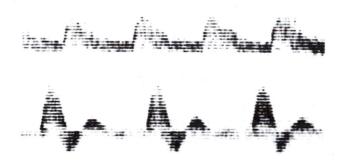

Figure 19.26. *The lower trace is from a normal femoral artery. The upper trace is from a patient with vascular disease. There is a very clear difference in the flow waveforms as explained in the text.*

For example, the common carotid artery in the neck divides into the internal and external carotid arteries and disease at this point can give rise to atheromatous plaque that may be dislodged and cause vascular obstructions in the brain or may completely occlude the flow. Doppler imaging systems can be used to image the flow at the bifurcation and detect phenomena such as the turbulence that arises around obstructions to flow. An image of this situation was given in chapter 12 (section 12.3.6).

Another major area of application for Doppler systems is in cardiac investigations. Cardiac performance can be assessed in a quantitative manner. For example, by looking at the flow patterns around the cardiac valves and detecting features such as turbulence, valvular reverse leakage can be seen. Doppler imaging systems are widely used as a screening test prior to more invasive radiological procedures.

19.8.   PROBLEMS

19.8.1.   *Short questions*

a   In an ultrasonic Doppler velocimeter what is the relationship between Doppler frequency, red cell velocity and the angle between the ultrasound beam and the red cell movement?

b   A red cell has a diameter of about 10 μm. What frequency of ultrasound would have a wavelength of this distance?

c   Would a peak velocity of 1 m s^{-1} for blood in the femoral artery be considered normal?

d   Does blood ever flow backwards in normal arteries?

e   Can we normally consider venous flow as steady?

f   In the Stewart–Hamilton equation for the measurement of flow using an injected bolus of dye what has to be measured and where?

g   Where in a blood vessel is the fastest flow normally found?

h   Is there almost zero blood flow anywhere in a normal artery?

i   What is sometimes referred to as coolth?

j   About how long does it take blood to circulate completely round the body?

k   What is a thin-film flowmeter?

l   Why is the partition coefficient very important when using an indicator clearance technique to measure organ blood flow?

m   Does tissue thermal conductivity appear to increase or decrease with blood flow?

n   Can an electromagnetic flowmeter be used to measure limb blood flow?

o   Does an electromagnetic flowmeter measure flow volume or flow velocity?

p   Why is the output from a light plethysmograph placed on the tip of a finger approximately a sawtooth shape?

q   About what fraction of our cardiac output goes to the brain?

r   Does the weight of an arm change during the cardiac cycle?

s   Is the size of the ultrasonic echo from blood smaller or larger than the echo from surrounding tissues?

t   Can a Doppler frequency be both negative and positive in value?

u   What is the key component of a Doppler directional demodulator?

v   Is 100 kHz a typical value for a Doppler frequency?

w   What is the major limitation on pulsed Doppler systems used to image blood flow?

19.8.2.   Longer questions (answers are given to some of the questions)

Question 19.8.2.1

In a venous occlusion measurement of the blood flow to an arm the change in resistance of the arm is measured during 10 cardiac cycles lasting 10 s. Would the change in resistance be an increase or a decrease?

If the initial resistance of the arm was 50 Ω and the change was 2 Ω then calculate the blood flow averaged over the 10 cardiac cycles. The length of the segment of arm over which the measurement was made was 10 cm and you may assume the resistivity of blood to be 1.5 Ω m. State any further assumptions which you make.

Answer

The resistance will fall. Assume that the blood can be modelled as a parallel resistance to the 50 Ω. 1.25 ml s^{-1}.

Question 19.8.2.2

A bolus injection of a dye has been given to measure the cardiac output. Table 19.1 shows the measured dye concentration at intervals of 0.5 s after the injection. The calculated flow rate was 6.6 l min^{-1}.

(a)   Develop a form of the Stewart–Hamilton equation for sampled data. Use a dimensional analysis to show that the units are correct.
         Correct the data for the effect of recirculation.
(b)   What quantity of dye was injected?
(c)   What effect would taking samples at intervals of 1, 2, 4 and 8 s have on the calculated flow rate?

(Hint: a transformation of the data might help with the recirculation problem.)

Table 19.1. *c(t) measurements for bolus injection of dye.*

Time (s)	Conc. (mg l^{-1})	Time (s)	Conc. (mg l^{-1})	Time (s)	Conc. (mg l^{-1})
0	0	8.5	22.50	17.0	4.792
0.5	0	9.0	24.46	17.5	4.276
1.0	0.010	9.5	25.31	18.0	3.952
1.5	0.015	10.0	25.02	18.5	3.897
2.0	0.020	10.5	23.73	19.0	4.178
2.5	0.022	110	21.65	19.5	4.627
3.0	0.025	11.5	19.10	20.0	4.833
3.5	0.030	12.0	16.43	20.5	4.647
4.0	0.256	12.5	14.00	21.0	4.258
4.5	0.791	13.0	12.08	21.5	3.877
5.0	1.736	13.5	10.61	22.0	3.713
5.5	3.230	14.0	9.479		
6.0	5.410	14.5	8.541		
6.5	8.401	15.0	7.682		
7.0	12.05	15.5	6.874		
7.5	15.91	16.0	6.119		
8.0	19.55	16.5	5.423		

Answer

(a)   The integral Stewart–Hamilton equation is $X = F \int_0^\infty c(t)\, dt$. It is straightforward to develop a sampled data form: $X = F\Delta T \sum_1^N c(n)$ where ΔT is the sampling interval.
Figure 19.2 shows the data and an exponential correction fitted.
(b)   20 mg was injected.
(c)   Surprisingly, using these data, the calculated flow rate is essentially unchanged by sampling at 2 or 4 s, and decreases to 5.7 l min^{-1} at 8 s sampling.

Question 19.8.2.3

A drug is thought to change the perfusing blood supply to the skin and subcutaneous tissue. Assess the feasibility of using a thermal clearance probe to measure the blood flow, paying particular attention to necessary assumptions and problems of the technique.

100 mW of power is supplied to a small heating source on the skin and the radial temperature gradient is measured between 5 and 10 mm from the centre of the probe. What change in temperature gradient would you expect to see if a blood flow change causes the effective thermal conductivity of the tissue to change from 0.4 to 0.41 J m^{-1} K^{-1} s^{-1}? Assume that there is no thermal shunt in the probe and that there is perfect thermal contact between the probe and the skin.

Answer

The principle of the technique is given in section 19.4.4. The assumption is that blood flow will cause the apparent thermal conductivity to change proportionally. Thermal conductivity can be measured by applying a measured quantity of heat and measuring the thermal gradients which result. The problems of the technique relate to the fact that it is impossible to separate the heat loss from blood flow and the heat loss caused by the underlying thermal conductivity of the tissue.

　　　If we assume hemispherical geometry for a probe then the temperature gradient ΔT across a hemispherical element thickness Δr at radius r will be given by

$$\Delta T = \frac{p\,\Delta r}{2\pi r^2 k}$$

where p is the power input and k is the thermal conductivity.
　　　We can integrate this between the two radii of 5 and 10 mm and obtain the temperature drop as 3.98 °C. If k changes from 0.4 to 0.41 the change in temperature gradient can be found as 0.1 °C.

Question 19.8.2.4

We wish to measure the blood flow velocity in a distant artery where there are several intervening smaller blood vessels. If a range-gated Doppler system is considered then what will be the maximum velocity of blood which can be measured, without range ambiguity, if the distance between the probe and the vessel is 80 mm and the frequency of operation is 2 MHz? (Assume the velocity of sound in tissue to be 1500 m s^{-1}.) What pulse repetition frequency will be required?

Answer

The theory of range gating was covered in section 19.7.8. The basic Doppler equation was derived as equation (19.18). Now the time delay between transmission and receipt of a packet of ultrasound dictates the maximum repetition frequency as $c/2R$ where R is the range. As at least two samples per cycle are required to define a sine wave frequency the maximum Doppler frequency is $c/4R$. This corresponds to a maximum velocity given by equation (19.18) as 1.76 m s^{-1}.

It has been assumed that the angle between the probe and the blood vessel is 0°. In practice the angle would not be zero so that the maximum velocity that could be observed in the vessel would be somewhat greater than 1.76 m s^{-1}.

The pulse repetition frequency is given by $c/2R$ as 9375 s^{-1}.

Question 19.8.2.5

Large plates of chlorided silver are placed 200 m apart on opposite banks of the river Thames and wires are run across Tower Bridge to an A–D converter and computer. A voltage of 10 mV is recorded. By observing the surface of the water it is concluded that the water is flowing at a rate of 1 m s^{-1}. What can you deduce from these measurements about the strength of the Earth's magnetic field in London? Note any assumptions you make and comment on the feasibility of the measurement.

Answer

The background to this question is the theory of the electromagnetic flowmeter given in section 19.5. The equation for the potential E derived from Faraday's principle of electromagnetic induction tells us that $E = DBV$ where D is the dimension of the conductor, B the magnetic field and V the velocity of the conductor. The field required to produce a potential of 10 mV is found to be 50 μT.

We can conclude that the vertical component of the Earth's magnetic field in London is 50 μT.

In making this calculation we have assumed that the river Thames is electrically conducting. This is a reasonable assumption as it will contain a number of electrolytes. We have also assumed that the surface velocity of flow reflects the deeper velocity. The measurement is probably feasible although it would not be easy. A signal of 10 mV should be measurable when using AgCl electrodes (see section 9.2), but the interference problems (see section 10.4) in central London could be considerable.

Answers to short questions

a   Doppler frequency is proportional to velocity multiplied by the cosine of the included angle, i.e.

$$f_D = \frac{2vf_c}{c}\cos(\theta).$$

b   150 MHz corresponds to a wavelength of 10 μm.

c   No, a maximum velocity of about 1 m s^{-1} would be expected in the femoral artery.

d   Yes, blood flows backwards at some point during the cardiac cycle in many arteries.

e   Yes, venous flow is usually considered as steady, although there are small variations during the cardiac cycle and also with respiration.

f   The concentration of the dye downstream of the point of injection has to be measured in order to calculate flow from the Stewart–Hamilton equation.

g   The fastest flow is normally in the middle of a blood vessel.

h   Yes, the blood in contact with the vessel walls is at almost zero flow velocity.

i   The thermal energy of the cool saline injected for thermal dilution measurement is sometimes referred to as coolth.

j   It usually takes between 20 and 60 s for blood to circulate completely round the body.

k   A thin-film flowmeter is a small transducer that includes a small heated resistance element. The heat loss from the probe when placed in moving blood is proportional to the blood velocity.

l   The partition coefficient is the ratio of the indicator in the organ to that in the venous blood that is clearing the indicator from the organ. This coefficient can change greatly from tissue to tissue and directly affects the calculation of organ blood flow.

m   Thermal conductivity will appear to increase with increasing blood flow to the tissue.

n   No, an electromagnetic flowmeter can only be used directly on a blood vessel.

o   The electromagnetic flowmeter measures flow velocity.

p   The light plethysmograph gives an output approximately proportional to blood volume. On systole the volume of blood in the finger will increase rapidly but during diastole it will slowly decrease.

q   About 10–15% of cardiac output goes to the brain.

r   Yes, an arm will get heavier during cardiac systole because arterial inflow is greater than venous outflow.

s   The echoes from blood are very much smaller than those from the surrounding tissues.

t   Yes, a Doppler frequency can be both positive or negative depending upon the direction of the blood flow.

u   The key component of a Doppler directional demodulator is a multiplier that multiplies the transmitted and received signals.

v   No, the maximum Doppler frequencies are usually well within the audio bandwidth, i.e. up to 10 kHz.

w   The maximum velocity of blood flow that can be measured is limited by the range and the ultrasound frequency in pulsed Doppler imaging systems.

BIBLIOGRAPHY

Atkinson P and Woodcock J P 1982 *Doppler Ultrasound and its Use in Clinical Measurement* (New York: Academic)

Bellhouse B, Clark C and Schultz D 1972 Velocity measurements with thin film gauges *Blood Flow Measurement* ed V C Roberts (London: Sector)

Brown B H, Pryce W I J, Baumber D and Clarke R G 1975 Impedance plethysmography: can it measure changes in limb blood flow *Med. Biol. Eng.* **13** 674–82

Evans D H, McDicken W N, Skidmore R and Woodcock J P 1991 *Doppler Ultrasound, Physics, Instrumentation and Clinical Application* (New York: Wiley)

Forrester J S, Ganz W, Diamond G, McHugh T, Chonette D W and Swan H J C 1972 Thermodilution cardiac output determination with a single flow-directed catheter *Am. Heart J.* **83** 306–11

Goldberg B B, Merton D A and Pearce C R 1997 *Atlas of Colour Flow Imaging* (London: Martin Dunitz)

Hill D W 1979 The role of electrical impedance methods for the monitoring of central and peripheral blood flow changes *Non-Invasive Physiological Measurement* vol I, ed P Rolfe (New York: Academic)

Holti G and Mitchell K W 1979 Estimation of the nutrient skin blood flow using a non-invasive segmented thermal clearance probe *Non-Invasive Physiological Measurement* vol I, ed P Rolfe (New York: Academic)

Kasai C, Namekawa K, Koyano A and Omoto R 1985 Real-time two-dimensional blood flow imaging using an autocorrelation technique *IEEE Trans. Sonics Ultrason.* **32** 458–63

McDonald D A 1974 *Blood Flow in Arteries* (London: Arnold)

Meier P and Zierler K L 1954 On the theory of the indicator-dilution method for measurement of blood flow and volume *J. Appl. Physiol.* **6** 731–44

Merritt C R 1992 *Basic Doppler Colour Flow Imaging* (Edinburgh: Churchill Livingstone)

Roberts C (ed) 1972 *Blood Flow Measurement* (London: Sector)

Shung K K, Sigelmann R A and Reid J M 1976 Scattering of ultrasound by blood *IEEE Trans. Biomed. Eng.* **23** 460–7

Single Sideband Communication 1956 *Proc. IRE 44* December

Taylor K J W, Burns P and Wells P N T 1995 *Clinical Applications of Doppler Ultrasound* 2nd edn (London: Raven)

Woodcock J P 1975 *Theory and Practice of Blood Flow Measurement* (London: Butterworths)

Wyatt D 1972 Velocity profile effects in electromagnetic flowmeters *Blood Flow Measurement* ed V C Roberts (London: Sector)

CHAPTER 20

BIOMECHANICAL MEASUREMENTS

20.1. INTRODUCTION AND OBJECTIVES

In Chapter 1 we covered the theoretical background to the mechanics of the human body. We saw how relatively simple mechanical models can be used to develop an insight into human performance. In this chapter we look at some of the problems involved in actually making *in vivo* measurements of forces and movements. The human body is a very complex structure and this, combined with the problems of gaining access to specific parts of the structure, can make it very difficult to make reproducible and meaningful measurements. This chapter should only be seen as the introduction to a very complex area of physiological measurement. These are some of the questions we will address:

- Can we describe human gait in terms of a few measurements?
- Is there such a thing as normal gait?
- Does standing cause ischaemia?
- What are normal static and dynamic loads during walking?

When you have finished this chapter, you should be aware of:

- How to measure joint angles.
- How a load cell works.
- How to measure limb positions during gait.
- How to measure foot pressure distribution.

It will be useful if you have read Chapter 1 before reading this chapter. It will also be helpful to have read the section on transducers in Chapter 9 (section 9.5).

20.2. STATIC MEASUREMENTS

We will start by looking at the vertical forces involved in standing. A typical adult male might weigh (have a mass of) 70 kg and this will obviously give rise to a vertical force between the feet and the ground when standing. The force will be $9.81 \times 70 = 686.7$ N. A value of 9.81 has been assumed for g, the acceleration due to gravity. There is actually about a 0.6% difference in g between the equator and the poles so the force would be slightly less at the equator.

The force between the feet and the ground will give rise to a pressure or stress, which we can calculate if we know the area of contact between the feet and the ground. A typical area of contact for each foot is

0.015 m^2 so we can calculate the average pressure as:

$$\text{pressure between feet and ground} = 686.7/0.03 = 22\,890 \text{ N m}^{-2} = 22.9 \text{ kPa}.$$

This is the average pressure distributed over the surface of the feet. There will, of course, be spatial variations of the pressure as the contact between the feet and the ground is not uniform. The maximum pressure may well be a factor of five or more greater than the average pressure. We will return to the subject of the pressure distribution under the sole of the foot in section 20.2.2.

In order to set the average pressure between the feet and ground in context we can compare it with a normal arterial pressure of 16 kPa (120 mmHg) during cardiac systole and 10.6 kPa (80 mmHg) during diastole. The average pressure under the feet is higher than the arterial pressures so that we can expect standing to cause ischaemia in the tissues of the feet. This is indeed the case and explains why people move around when standing and change the weight distribution from foot to foot. This allows the blood to reach the tissues from time to time.

Many problems arise as a result of the pressure between the feet and the ground. However, many more problems arise in the joints between bones and in many cases these problems are caused by the high stresses involved. These stresses arise from the weight of the body itself, from the acceleration of body segments and from the musculature which holds a joint in place. In order to estimate the pressures which will arise in joints we need to know the areas of contact involved. Unfortunately, this is not easy to estimate and there are no satisfactory methods of measuring the stresses in joints *in vivo*. Brown and Shaw (1984) quote stress values measured *in vitro* for the acetabulum of the hip joint. The hip is a ball and socket joint; the ball is the head of the femur and the socket is a region on the edge of the hipbone called the acetabulum ('vinegar cup'). They quote maximum forces of 2100–2700 N and stresses of 5.3 MPa corresponding to an area of about 5 cm^2. The stresses are very high and explain the important role of the articular surfaces and the surrounding soft tissues, which in normal circumstances adapt and regenerate to carry the high stresses. In osteoarthritis the tissues are unable to adapt and regenerate as required so that the joint degenerates. Artificial joints have been developed to treat osteoarthritis and other pathological conditions of the joints. These joints have of course to be designed to withstand the same very high stresses as are encountered in a normal joint.

All of the forces we have considered so far are the static forces involved as a result of body weight. In section 20.3 we will consider the additional forces which arise when we accelerate parts of the body.

20.2.1. *Load cells*

Bathroom scales measure the vertical force exerted by the body when standing. They either contain a direct mechanical action involving a spring, or a transducer which converts force into an electrical signal which can be displayed digitally. The transducer can be referred to as a *load cell* and it contains a *strain gauge*. Load cells can be incorporated into a *force plate* and used to measure the forces involved in both standing and walking (see figure 20.1). By using three load cells as shown in figure 20.1 it is possible to measure the lateral as well as the vertical forces involved when a person walks across the force plate. Force plates or platforms typically have sensors at the four corners of a rectangular plate. If each of the sensors contains transducers to measure force in three orthogonal directions, then it is possible to determine the vertical and two shear forces of stress applied to the surface of the platform. It is also possible to determine the point of application of the force and the moment around the axis perpendicular to the surface of the plate.

The lower part of figure 20.1 shows some typical forces during normal walking. These are expressed as a fractional percentage of body weight and show that the vertical force can exceed body weight during heel strike and kick-off. During normal walking the peak vertical forces are up to 1.5 times body weight but during athletic events such as high jumping they can be 10 times body weight.

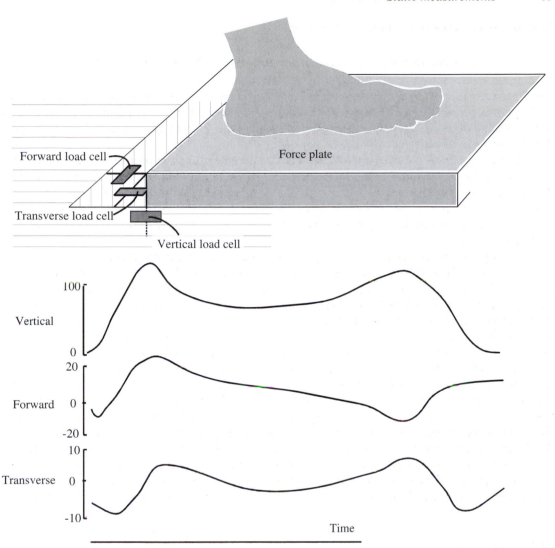

Figure 20.1. *A force plate able to measure vertical, transverse and forward forces during walking. Load cells are shown at only one corner of the plate, but in practice they would usually be placed at all four corners. The lower part of the figure shows typical values for the three components of force obtained during walking. The results are expressed as a percentage of body weight.*

20.2.2. Strain gauges

Strain gauges are a type of transducer (see section 9.5). They are almost always used to convert a mechanical change into an electrical signal. The range of strain gauges is very wide and indeed you may often not think that a particular technique is a strain gauge. For example there are now many laser ranging systems that can be used for measuring biological deformations. They form a type of strain gauge. We will consider just two types of very commonly used gauges: piezoelectric and foil gauges.

Piezoelectric strain gauges

The load cell attached to a force plate contains a strain gauge. Now, of course, we actually want to measure the stress on the plate so the input to the load cell is a stress and the output is an electrical signal proportional to the strain produced within the load cell.

Load cells commonly use a piezoelectric element as the sensor. The expression piezoelectric derives from the Greek *piezein* meaning to press. It was discovered in the 1880s that quartz and some other crystals produce electrical signals in response to mechanical pressure and much more recently various ceramic materials have been developed which exhibit the same effect, but to a much more marked degree. Lead titanate zirconate is a widely used piezoceramic. We can describe the properties of a piezoelectric material as follows.

The mechanical strain s that an elastic material experiences when subject to a stress σ is, according to Hooke's law, given by

$$\sigma = Ys \tag{20.1}$$

where Y is Young's modulus. If an isotropic dielectric material with permittivity ε is placed in an electric field E then a dielectric displacement D occurs, which is given by

$$D = \varepsilon E. \tag{20.2}$$

However, piezoelectric materials are anisotropic and their permittivity and Young's modulus depend upon the direction and magnitude of the stress σ. This interaction can be shown to be described by the following four equations:

$$D = p\sigma \tag{20.3}$$

$$s = pE \tag{20.4}$$

$$s = kD \tag{20.5}$$

$$E = -k\sigma \tag{20.6}$$

where p and k are constants for the piezoelectric material.

From equations (20.3) and (20.4) we can see that the constant p can be seen either as the quotient of the electrical charge (at constant field) produced by a mechanical stress σ, or as the quotient of the strain s produced by an electric field. We can see that these equations describe the fact that piezoelectric materials produce a potential when subject to stress but will also do the reverse and produce a strain when a potential is applied to them. The units of p are either C N^{-1} or m V^{-1}.

From equations (20.5) and (20.6) we can see that the constant k can be seen either as the quotient of the electric field developed when stress σ is applied, or as the quotient of the strain s produced by the application of a dielectric displacement. The constants p and k are important parameters which describe the performance of a piezoelectric transducer. High values are usually best as they correspond to a high sensitivity.

Piezoelectric strain gauges can be used either to produce an output charge in proportion to an applied force as described by equation (20.3), or to produce an output voltage as described by equation (20.6). The former method is normally used because the sensitivity of a voltage amplifier is affected by the capacitance of the input cables and the low-frequency response of the system is limited by leakage of the charge through the piezoelectric material. The type of circuit usually employed is illustrated in figure 20.2, which shows a charge amplifier configuration.

Using the terminology illustrated in figure 20.2 we can use Kirchhoff's law to sum the currents at the inverting input of the operational amplifier to give:

$$i = i_t + i_m - i_f$$

therefore

$$\int i \, dt = \int (i_t + i_m - i_f) \, dt.$$

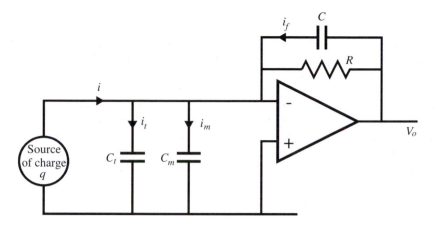

Figure 20.2. *Circuit diagram of a charge amplifier. C_t is the capacitance of the transducer and C_m that of the measurement system.*

Now as the operational amplifier has infinite gain there is no significant input voltage and hence no potential can accumulate across C_t and C_m. Therefore,

$$\int (i_t + i_m)\, dt = 0$$

and

$$V_0 = \frac{\int i\, dt}{C} = -\frac{q}{C}. \tag{20.7}$$

The output is thus proportional to the accumulated input charge q and the feedback capacitance C and independent of C_t and C_m. We have ignored the resistance R in this calculation.

The circuit shown works very well, but when used for measuring slowly changing forces drifts can become very significant. It is left to the reader to calculate the drift which will arise from the bias current flowing into the inverting input of the operational amplifier. This current will be integrated by the feedback capacitance to give a steady drift in the output voltage. This drift can be reduced by including a feedback resistance R as shown, but this will reduce the low-frequency response of the circuit. Transient charges, such as those which will occur when a subject walks over a force plate, will be recorded but very slowly changing forces will not appear.

Foil strain gauges

It was noticed by Lord Kelvin in about 1850 that the resistance of a metal wire increases with increasing strain and that different materials have different sensitivities to strain. What was being observed is simply that the resistance of a given volume of metal will depend upon its shape. If we consider a wire of circular cross-section area a as shown in figure 20.3(a) then its resistance (R) is given by $\rho l/a$ and its volume by al. However, if we keep the volume of the wire constant but double the length to $2l$ (figure 20.3(b)) then the resistance will become $(\rho 2l)/(a/2) = 4R$. The resistance has quadrupled as the length had doubled. If we use the same volume of conductor to form two sections of wire with cross-sections $a/2$ and $3a/2$ (20.3(c)) but each of length $l/2$, then the total resistance of the two sections will be $1.33R$.

Electrical resistance or foil strain gauges are made by laying down the metallic conductor onto a backing material. The sensitivity of the gauge depends upon the configuration of the conductor and the material used

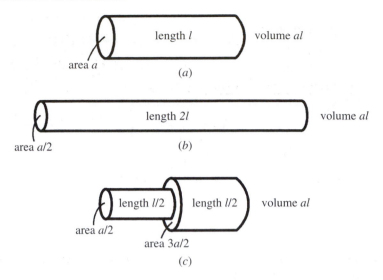

Figure 20.3. *Three sections of wire all with the same volume.*

for the backing. Sensitivity is defined in terms of a gauge factor (GF) which is given by

$$GF = (\Delta R/R)/\text{strain}. \tag{20.8}$$

The gauge factor in the example we gave above was 2 and this is the maximum which is usually obtained.

Electrical resistance strain gauges are affected by temperature simply because metals have a resistivity which increases with increasing temperature. The temperature coefficient of copper is about $0.5\% \,^{\circ}\text{C}^{-1}$. This is a very high temperature coefficient and makes copper unsuitable for use in strain gauges. However, various alloys are available with very much lower temperature coefficients. Another way to avoid temperature effects is to place two conductors on the foil in positions where the resistance will increase with strain in one case but decrease in the other. This can often be achieved by placing the conductors on opposite sides of the backing material. If the two resistances are then placed in opposite arms of a bridge configuration then changes with temperature will tend to cancel, whereas the changes with strain will sum together. The principle is the same as that used to provide temperature compensation in pressure transducers (see section 18.4).

20.2.3. *Pedobarograph*

Measurement of the pressure distribution under the feet is a vital component of the assessment of standing and gait. One way in which to make the measurements is to construct a two-dimensional matrix of force transducers on which the subject stands. This technique certainly works, but it is relatively expensive and complex because a large number of transducers are required. If a spatial resolution of 2 mm is required over a force plate area of 30×30 mm then 225 transducers are required and the associated electronics to interrogate them. An elegant solution to this problem was suggested by Chodera in 1957.

Chodera suggested the system illustrated in figure 20.4. A thick glass plate is edge illuminated as shown such that the light will be totally internally reflected within the glass. Total internal reflection takes place because the refractive index of the glass is higher than that of the surrounding air (see Snell's law in section 3.5.1). On top of the glass plate is a sheet of plastic material whose refractive index is higher than that of the glass. If this sheet of material makes optical contact with the glass then the total internal reflection

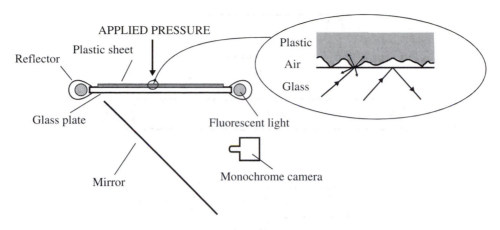

APPLIED PRESSURE

Reflector
Plastic sheet
Glass plate
Mirror
Fluorescent light
Monochrome camera

Plastic
Air
Glass

Figure 20.4. *The principle of a pedobarograph which uses an edge-illuminated glass sheet and a covering plastic sheet to measure pressure distribution.*

will be destroyed and hence light will escape and be scattered by the surface of the sheet of material. If the plate of glass is viewed from the underside then the light scattered by the plastic sheet will be seen. Now if the surface texture of the plastic sheet is such that increasing pressure will give a proportional increase in the area of optical contact with the glass then the intensity of light seen from underneath will be proportional to pressure. The performance of such a system depends very much on the plastic sheet on which the subject places their feet. The material must deform in a uniform fashion and must behave in a reversible fashion when pressure is released.

Calculation of how the area of contact between the glass plate and the plastic sheet might change with applied pressure is not a trivial problem. It was shown in Chapter 1 that many materials exhibit a stress/strain relationship as shown in figure 20.5. The relationship is linear until the yield point Y is reached, but beyond this point the slope of the curve drops off very quickly.

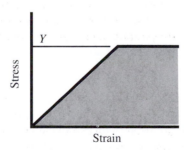

Figure 20.5. *General form of the stress versus strain relationship.*

If the surface of the material is considered to be made up of pyramids as shown in figure 20.6 then we can consider what might happen when pressure P is applied over unit area. A force P will have to be supported by the pyramid of deformable plastic material.

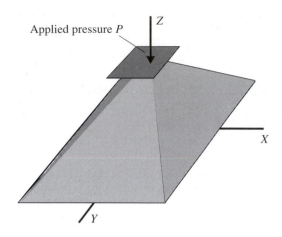

Figure 20.6. *Pressure P applied to a pyramid of deformable material.*

Now the apex of the pyramid will have zero area and hence the applied stress will be infinite. The yield point of the material will be exceeded and so the strain will increase rapidly such that the apex of the pyramid will collapse. It will collapse to the point where the stress no longer exceeds the yield point Y. This point will be given by

$$P = Y \times \text{area of contact.}$$

We can therefore expect the area of contact to be proportional to the applied pressure. This is what is found to be approximately true in practice. However, a more detailed investigation (see Betts *et al* 1980) shows that as pressure increases the number of points of contact between the plastic sheet and the glass increases. This is not unreasonable because the surface of the plastic sheet will be irregular with protrusions above the surface showing a distribution of heights. One of the problems with this technique is that the response time is quite slow and hysteresis also occurs. However, Betts *et al* state that by careful selection of the plastic material a response to an increasing applied pressure occurs within less than 5 ms and to a decreasing pressure within about 20 ms. The longer response time to a decreasing pressure is to be expected if the material is being deformed beyond its yield point.

Figure 20.7 shows measurements of mean pressure over seven regions of interest under a normal foot during walking. It is easy to identify the times of heel-strike and toe-off. It can also be seen that the pressure distribution is very non-uniform and changes during gait. The type of measurement shown has been widely used to investigate gait and to observe the effects of abnormal gait following surgery to the lower limbs.

20.3. DYNAMIC MEASUREMENTS

We now turn from static forces to the forces that result from accelerations of the body. Our musculoskeletal system maintains body posture, but of course posture is a function of time t. We can represent the position vector $\boldsymbol{x}$ of a part of the body and then say that the forces $F(t)$ as a function of time are given by:

$$a_2(\boldsymbol{x})\,\ddot{\boldsymbol{x}} + a_1(\boldsymbol{x})\,\dot{\boldsymbol{x}} + a_0(\boldsymbol{x})\boldsymbol{x} = F(t)$$

where a_2 represents inertial forces, a_1 represents energy losses such as viscous losses and a_0 represents stiffness. In reality a_0, a_1 and a_2 will be functions of $\boldsymbol{x}$ so that the system is nonlinear. For small changes we can often assume linearity and if we are to understand the system then we need to be able to measure a_0, a_1 and a_2. We gave stress values in section 20.1 which showed that the static forces are often exceeded by the dynamic ones, but to quantify these we must be able to measure displacements, velocities and accelerations.

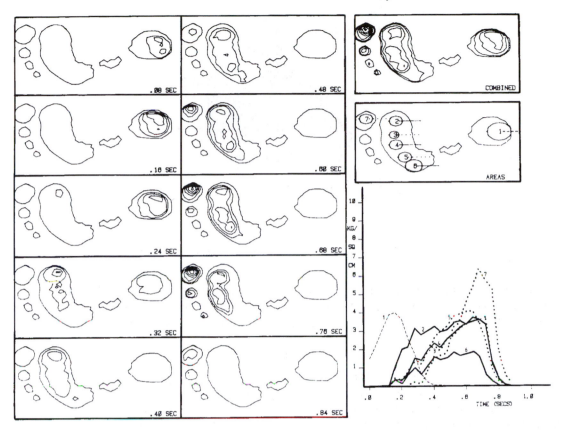

Figure 20.7. *A typical dynamic isopressure contour plot output obtained when a normal subject walks across the plate of a pedobarograph. On the left are shown 10 selected frames equally distributed from throughout the footstep sequence. At the top right is the combined frames image followed by the position of the seven selected areas of interest, and finally the pressure/time curves (1 kg cm² = 98.1 kPa). (Image kindly provided by Dr R P Betts, Royal Hallamshire Hospital, Sheffield.)*

20.3.1. Measurement of velocity and acceleration

It can be argued that it is necessary to measure only one of the three variables, displacement, velocity and acceleration, because the others can be found by differentiation or integration. However, integration has the problem that constants of integration are unknown and differentiation can produce a large amount of high-frequency noise. It is best to measure the variable required as directly as possible.

Relative velocity (v) can be measured by moving a coil through a steady magnetic field (H) which will give rise to an induced voltage u where $u \propto Hv$. The problem with this approach is in producing a steady field H over the volume in which the movement takes place. The technique can be used to measure relative velocity between two limb segments by attaching a permanent magnet to one segment and a coil to the other. However, the output is very nonlinear as the field is not constant around the permanent magnet. If absolute velocity is required then it is often best to integrate the output from an accelerometer.

Accelerometers usually consist of a small mass m attached to a force transducer as shown in figure 20.8. Acceleration a will give rise to a force am on the piezoelectric transducer. It is worth noting that the

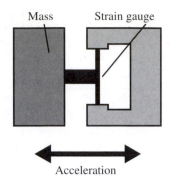

Figure 20.8. *Basis of an accelerometer. Acceleration in the direction shown will give rise to a proportional force on the piezoelectric strain gauge transducer.*

accelerometer cannot distinguish between the force mg due to gravity and the force caused by acceleration of the mass m. If the accelerometer is rotated within the Earth's gravitational field then an output will be obtained. This can be a very significant source of error if such a device is used to measure the acceleration of parts of the body.

20.3.2.   Gait

People are able to move around with relative ease, but exactly how do we walk? How do we put one foot in front of the other and why does it become very difficult for some people? The science of human walking is often called gait analysis. The basic mechanics of walking and running were described in section 1.7.2 of Chapter 1. However, in order to study anything it is necessary to be able to make measurements and in gait analysis these measurements are quite difficult. They fall into three main categories: techniques to measure the forces exerted on the ground during walking, instruments to measure muscle activity and instruments to measure the relative positions of the limbs and the movement of the body in space. We have already described some of the techniques for measuring the forces between the feet and the ground and some of the techniques for measuring muscle activity were covered in Chapter 16. We will now give some thought as to how can we describe body position and then how we can measure these descriptors. Table 20.1 gives some of the factors which have been used to describe normal walking. These parameters need careful definition if comparisons are to be made between measurements made in different laboratories. The stance and swing phases are defined for one foot and correspond to the time when the foot is either in contact with the ground or off the ground, respectively.

It is a struggle to separate the males and females using the data given in table 20.1. In the last column of the table the difference between the mean measurement for males and females has been divided by the mean standard deviation on the measurements. This gives an indication of the overlap between the distributions. A value of 2 would mean that there was an overlap of about 15% if the distributions were normally distributed. Step length and hip flexion would appear to be the best discriminants between male and female gait—not a surprising conclusion.

A considerable amount of the variation in the measurements given in table 20.1 may be accounted for by variations with age. Older people will certainly walk more slowly than younger ones. The variation with age is well illustrated by the data in table 20.2, which gives the variations in six different gait parameters over the first 7 years of life.

Table 20.1. *Gait characteristics of normal men and women at free walking speed. (Data adapted from M P Murray and D R Gore, Gait of patients with hip pain or loss of hip joint motion, Clinical Biomechanics: a Case History Approach, ed J Black and J H Dumbleton (London: Churchill Livingstone).*

Component	Men (mean ± 1 SD)	Women (mean ± 1 SD)	(Diff. in means)/ (mean SD)
Velocity (m min^{-1})	91 ± 12	74 ± 9	1.6
Gait cycle (s)	1.06 ± 0.09	1.03 ± 0.08	0.35
Cadence (steps/min)	113 ± 9	117 ± 9	0.45
Step length (m)	0.78 ± 0.06	0.62 ± 0.05	2.9
Stride length (m)	1.6	1.37	
Single-limb support time (s); (% of gait cycle)	0.44(40)	0.39(38)	
Swing (s)	0.41 ± 0.04	0.39 ± 0.03	0.57
Stance (s)	0.65 ± 0.07	0.64 ± 0.06	0.15
Lateral motion of the head (m)	0.059 ± 0.017	0.040 ± 0.011	1.4
Vertical motion of the head (m)	0.048 ± 0.011	0.041 ± 0.009	0.7
Hip flexion (deg)	48 ± 5	40 ± 4	1.8
A-P pelvic tilting (deg)	7.1 ± 2.4	5.5 ± 1.3	0.86
Transverse pelvic rotation (deg)	12 ± 4	10 ± 3	0.57

Table 20.2. *Gait characteristics of normal children at free walking speeds. (Data from D H Sutherland, R Olshen, L Cooper and S L-Y Woo 1980, The development of mature gait, J. Bone Jt. Surg. 62 336.)*

Component	Age (yrs) 1	2	3	5	7
Velocity (m min^{-1})	38.2	43.1	51.3	64.8	68.6
Gait cycle (s)	0.68	0.78	0.77	0.77	0.83
Cadence (steps/min)	175.7	155.8	153.5	153.4	143.5
Step length (m)	0.216	0.275	0.329	0.423	0.479
Stride length (m)	0.43	0.549	0.668	0.843	0.965
Single-limb support (% gait cycle)	32.1	33.5	34.8	36.5	37.6

It would appear that we are able to characterize normal gait by measuring a number of parameters such as those given in tables 20.1 and 20.2. The next section outlines some of the methods by which we can measure such parameters.

20.3.3. *Measurement of limb position*

A large number of methods have been developed to measure parameters of gait such as those given in table 20.1. Shoes which incorporate switches can be used to record the timing of ground contact. Walkways can incorporate a large array of pressure sensors or have a matrix of exposed wires which sense contact with a shoe with a conducting sole. Light beams can be placed across a walkway and the sequence of interruptions recorded when a subject breaks the beams when walking.

The relative position of limbs can be recorded using goniometers. A *goniometer* is a transducer which measures the angle between two arms of the device. If the arms are attached to limb segments above and below a joint then variations in joint angle during gait can be recorded. Goniometers used in this way have some disadvantages, the main one being that human joints are not simple hinges so that their movement cannot be described in terms of a single angle. Joint movement involves sliding and rolling of the joint surfaces. Another disadvantage of any transducer attached to a limb is that it is invasive. The presence of the transducer and its associated wires and electronics may well affect the way in which the subject walks. The use of telemetry might remove the wires but the weight and size of the device may still be sufficient to change normal gait.

     Video methods are probably the best source of information on gait. By taking a video sequence as the subject walks along a set path it should be possible to determine limb and joint positions as a function of time in a relatively non-invasive way. Unfortunately walking is too rapid for the human eye to be able to make a detailed assessment of gait so that computer methods of image analysis are required. At the moment computer image analysis has not reached the point where it can be used to model the human body as an assembly of components and match a video image to this model. Systems used at the moment require markers to be placed over key anatomical positions on the skeleton.

     Anatomical markers can either be placed onto the video images after image capture or they can be actual markers placed on the subject. Placement of the markers after image capture is done by a human observer who identifies characteristic positions on the computer image using a mouse. This is obviously a time consuming procedure as it has to be done for several positions and for a large number of image frames. It is faster to place actual markers on the subject which can then be recognized by computer analysis of the image frames. These markers may be passive markers with a characteristic colour, high reflection or shape or they may be active markers which emit visible or infrared light. In either case the presence of the markers on the body may be seen as being invasive. Yet another problem is the identification of the markers. If only a few markers are used then there may be only a small chance of ambiguity, but if many are used then they will need to be identified in some way. Active markers may be sequenced by turning the light emission on and off rapidly in a predetermined way, but of course this may involve running wires between all the markers.

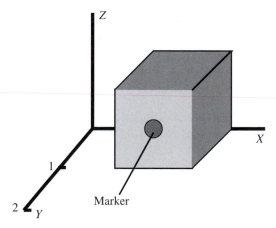

Figure 20.9. *A solid cube with a marker placed in the centre of the face lying in the y = 1 plane.*

     Video systems using markers must use at least two cameras if the fact that the body is three-dimensional is to be taken into account. In practice several cameras are often used so that several markers can always be seen by at least two of the cameras. The problem of markers is illustrated in figure 20.9 where we can consider the problem of localizing a marker placed in the middle of one face of a cube placed such that its faces are in the x, y and z planes. We need six cameras facing in the positive and negative x, y and z directions in order to ensure that at least one of the cameras sees the marker. If we are to localize the marker and cover the possibility that the face of the cube might be concave then we will need more than six cameras.

This illustration shows the disadvantage of a visual marker system. We do not need six or more views of an object in order to make a 3D image of the object because we place constraints on the shape of the object. We know that there will not be sudden discontinuities and that neighbouring points will fall on a surface.

20.4.   PROBLEMS

20.4.1.   *Short questions*

a   What is a goniometer used for?
b   Would you weigh more or less at the equator than at the poles and by how much?
c   Are the peak pressures in a hip joint of the order of Pa, kPa or MPa?
d   What is a load cell?
e   Why do most force plates incorporate at least four load cells?
f   Does walking produce shear forces?
g   What property has quartz that makes it useful in gait measurement?
h   When used in the charge mode in what units can the sensitivity of a piezoelectric strain gauge be expressed?
i   What is a 'charge amplifier'?
j   Will the resistance of a metal wire increase or decrease if you bend it?
k   Why is copper unsuitable as the material in a foil strain gauge?
l   Is the specification of a foil strain gauge likely to specify the gauge factor as 30?
m   What is a pedobarograph used to measure?
n   Are viscous forces associated with displacement, velocity or acceleration?
o   What is the stance phase of walking?
p   Is a normal step length during walking 0.5, 0.7 or 0.9 m?
q   What is a normal male walking speed in mph according to table 20.1?
r   At least six cameras are needed to follow a body marker in all positions. Why do our eyes not require six views to produce a 3D image of a person?
s   Apart from step length what is the most significant difference between male and female gait parameters?

20.4.2.   *Longer questions (answers are given to some of the questions)*

Question 20.4.2.1

The circuit of figure 20.2 is used to produce an output from a piezoelectric force transducer. If the constant p as used in equations (20.3) and (20.4) is 10^{-9} C N^{-1} then calculate the feedback capacitance C required if an input of 1000 N is to give an output voltage of 10 V. What will the drift in the output be over 100 s if the bias current of the operational amplifier is 1 nA?

Answer

100 nF; 1 V.

Question 20.4.2.2

Pressure sores are a serious and costly problem, requiring months of nursing to resolve. If a 70 kg man sits on a hard surface, how large must the contact area be to ensure that the blood supply to the skin was not cut off? For an average size man, would a pressure sore be the result of a localized high pressure, or is the pressure over the whole contact area sufficient to cause pressure sores? Are we in danger of developing pressure sores from sitting, and if so, why do normal individuals not get sores? If the same man sat on a water-filled bag, would you expect pressure sores to develop?

Question 20.4.2.3

We want to select an accelerometer to investigate the movement of the legs during walking. If the normal swing phase of walking takes 0.4 s then estimate the linear acceleration of the middle of the leg on the basis of the following assumptions:

- leg length is 0.5 m;
- total swing angle is 60°;
- the leg accelerates uniformly for the first half of the swing phase and then decelerates uniformly for the second half.

Answer

The distance moved, d, by the mid point of the leg during the first half of the swing phase is given by

$$d = 0.5(\text{m})\pi \frac{30}{360} = \frac{0.5\pi}{6}.$$

Under uniform acceleration Newton's law applies. Therefore,

$$d = 0.5at^2$$

where a is the acceleration and t is the time. This gives the acceleration as 13.1 m s^{-1}.
    We have neglected the forward motion of the whole body but if this is uniform it should not affect our result.

Answers to short questions

a   A goniometer is a transducer used to measure the angle between the two arms of the device. It can be used to measure joint angles.

b   You weigh less at the equator than at the poles by about 0.6%.

c   The peak pressures in the hip joint can be several MPa.

d   A load cell is a device used to measure forces. It incorporates one or more strain gauges.

e   Four load cells are used in a force plate at the four corners. This enables the point of application of the force to be determined.

f   Yes, shear forces arise during walking from the transverse and torsional movements.

g   Quartz has piezoelectric properties that enable it to be used in strain gauges.

h   The sensitivity of a piezoelectric strain gauge can be expressed in $C \text{ N}^{-1}$.

i   A 'charge amplifier' gives an output voltage in proportion to the input charge. They are used to interface to a piezoelectric strain gauge.

j   The resistance of a wire will increase if you bend it.

k   Copper has too high a temperature coefficient of resistance to be useful in a foil strain gauge.

l No, the maximum gauge factor normally encountered is 2.

m A pedobarograph can be used to measure the pressure distribution under the feet.

n Viscous forces are associated with the velocity component.

o During the stance phase of walking the foot in question is in contact with the ground.

p A normal step length during walking is about 0.7 m. It is slightly longer in men than in women.

q 91 m min^{-1} is equivalent to 3.4 mph. This is the normal walking speed for men.

r Our eyes allow the brain to produce a 3D image of a person because we make assumptions about the shape of the person and how they move.

s After step length hip flexion is the most significant difference between male and female gait.

BIBLIOGRAPHY

Betts R P, Duckworth T, Austin I G, Crocker S P and Moore S 1980 Critical light reflection at a plastic/glass interface and its application to foot pressure measurement *J. Med. Eng. Tech.* **4** 136–42

Brown T D and Shaw D J 1984 *In vitro* contact stress distributions in the natural human hip joint. *J. Biomech.* **17** 409–24

Chodera J 1957 Examination methods of standing in man *FU CSAV Praha.* vol 1–3

Hay J G and Reid J G 1988 *Anatomy, Mechanics and Human Motion* (Englewood Cliffs, NJ: Prentice-Hall)

Miles A W and Tanner K E (eds) 1992 *Strain Measurement in Biomechanics* (London: Chapman and Hall)

Ozkaya N and Nordin M 1991 *Fundamentals of Biomechanics: Equilibrium, Motion and Deformation* (London: Van Nostrand Reinhold)

Skinner S R, Skinner H B and Wyatt M P 1988 Gait analysis *Encyclopaedia of Medical Devices and Instrumentation* ed J G Webster (New York: Wiley) pp 1353–63

CHAPTER 21

IONIZING RADIATION: RADIOTHERAPY

21.1. RADIOTHERAPY: INTRODUCTION AND OBJECTIVES

Before reading this chapter you should have read Chapter 5, in particular the sections on the detection of ionizing radiation, absorption and scattering of γ-rays.

X-rays were discovered by Roentgen in 1895, and γ-rays by Becquerel in 1896. Roentgen described most of the fundamental properties of x-rays, including their ability to penetrate tissue and provide a diagnostic image of bones. This was the first tool (other than the knife!) which clinicians were able to use to visualize structures beneath the skin, and it was very soon in widespread use. The damaging effects of radiation were not at first appreciated, and many early radiation workers developed severe radiation damage, particularly to the hands.

In 1902, Marie Curie succeeded in isolating 0.1 g of radium. Pierre Curie exposed his arm to the radiation from the radium, and studied the healing of the resulting burn. Becquerel was burnt by a phial of radium he was carrying in his pocket. As a result of the papers published by Curie and Becquerel, doctors in France started to use radium in the treatment of cancer and, after 1903, factories were set up in France and America to manufacture radium. In 1921, the United States presented Marie Curie with 1 g of radium—one-fiftieth of the separated radium in America.

In 1934, Frederic Joliot and Irene Curie produced radioactive phosphorus by bombarding aluminium with α-particles, and thus produced the first isotope which does not occur in nature. Shortly afterwards, Fermi prepared artificial radioactive isotopes using neutron bombardment, which is now the most important method of producing isotopes. Fermi and a large team of scientists produced the first self-sustaining nuclear chain reaction (i.e. a nuclear reactor) in 1942. The nuclear reactor, with its plentiful supply of neutrons, made possible the production of artificial radioactive isotopes for diagnosis and therapy on a commercial scale.

Radiotherapy is the use of ionizing radiation to treat disease. We hope to address the following questions in this chapter:

- Why can cancer be treated with ionizing radiation?
- Can ionizing radiation be focused onto a tumour?
- How accurate does the dose of radiation have to be?
- What is the significance of radiation energy in radiotherapy?

When you have finished this chapter, you should be aware of:

- The magnitude of radiation doses used in radiotherapy.
- How hazards to staff might arise during radiotherapy.
- The objectives of conformal radiotherapy.
- What factors determine the accuracy of radiotherapy treatments.

650

Research into the nature of fundamental particles has resulted in megavoltage x-ray generators that can be used in therapy. The use of ionizing radiation in therapy depends on the fact that tumour cells are more susceptible to radiation damage than normal cells. The radiation used is usually either x-rays (produced by decelerating electrons) or γ-rays from radioactive materials. The effects of x- and γ-ray photons on tissue are the same, and no distinction will be made between them.

The radiation dose which can be delivered to the tumour depends on the source-to-skin distance, how well the radiation penetrates the tissues, and how much radiation is scattered into the treatment area from the tissues outside this area.

If a beam of x-rays diverges from a point source, and the medium through which the beam passes does not absorb any energy, the intensity of the beam at any given distance from the source will be governed by the inverse square law. Energy will be removed from the beam when an absorbing medium, such as a person, is placed in the beam. Scattering will cause photons to be diverted from their path and absorption will transfer energy from the photons to the electrons of the absorbing medium. The electrons lose their energy by producing ionization and excitation along their tracks, thus causing radiation damage. The excitation process raises the energy level of an electron in an atom, and the ionization process causes electrons to be ejected from atoms. For incident photons of high energy their energy is dissipated over a distance that may be several millimetres in tissue and all electrons lose their energy and come to rest.

Several processes contribute to the absorption, and were considered in detail in section 5.2. In the photoelectric process, all the energy of an incident photon is transferred to an electron which is ejected from the atom. This is the predominant process for low-energy photons in materials of high atomic number, i.e. for diagnostic x-rays in bone or metals. Photoelectric absorption produces the high contrast between bone and soft tissue in diagnostic radiographs.

In the Compton effect, the photon is scattered from a loosely bound electron in the atom, and loses part of its energy to the electron. The recoil electron is ejected from the atom. The Compton process is the most important process for the attenuation of high-energy photons in soft tissue.

If the photon has sufficiently high energy (greater than 1.02 MeV), it can interact directly with the nucleus. The absorbed energy is converted into the mass of a positron and an electron (pair production). In soft tissue, this is only important for photon energies above 20 MeV.

The mass attenuation coefficient is given by the linear absorption coefficient (see section 21.7.1) divided by the density of the material. The total mass absorption coefficient is the sum of the mass absorption coefficients for the photoelectric, Compton and pair-production processes, whose relative importance changes with energy as was illustrated in figure 5.5(*b*).

21.2.   THE GENERATION OF IONIZING RADIATION: TREATMENT MACHINES

There are three principal methods of producing radiation for teletherapy (i.e. therapy in which the radiation source is located at some distance from the body). These are high-voltage x-ray machines, linear accelerators and isotope units. The use of sealed sources of radiation within the body will be dealt with in section 21.6.

21.2.1.   *The production of x-rays*

In Roentgen's original experiments, the x-rays were an accidental by-product of his experiments on the electrical discharge produced when a high voltage is applied between a pair of electrodes in a gas at a low pressure. The gas and electrodes were contained in a glass envelope. The gas in the tube was ionized by the high voltage, and the electrons produced were accelerated towards the anode. Some of the electrons were stopped by the glass envelope, thus producing the x-rays. X-rays are produced when rapidly moving electrons are suddenly stopped. This is an inefficient method of producing x-rays. The efficiency of x-ray production increases with the atomic number of the target, so that tungsten ($Z = 74$) is often used as a target. It is not easy to regulate the number of electrons produced within a gas discharge tube, so that the electrons are now produced by thermionic emission from a heated tungsten filament whose temperature can be controlled. The number of electrons (i.e. the current between the anode and cathode) is controlled by varying the heating current through the filament. There is no longer any need to have any gas in the tube, in fact, it is a positive disadvantage, because ions bombarding the filament will shorten its life—so the tube is highly evacuated. This is the basis of the modern x-ray tube: the Coolidge tube (figure 21.1). The filament is surrounded by a cylindrical focusing cup at the same potential as the filament, and the electrons are accelerated towards the tungsten target embedded in the copper anode. If the potential between the anode and cathode is V volts, the electrons will acquire an energy, E, given by

$$E = eV$$

where e is the charge on the electron. The unit of energy is the electron volt (eV). The quality of the electron beam (see below) may be altered by changing the anode voltage, V.

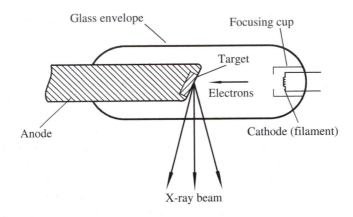

Figure 21.1. *Schematic diagram of an x-ray tube.*

In a conventional x-ray tube, only about 1% of the electron energy is converted into x-rays. The remaining 99% of the energy appears as heat. In a therapy machine with a potential of 250 kV and a tube current of 20 mA there will be 5 kW of heat energy given to the anode by the electrons. It is obviously necessary to find some efficient method of cooling the anode.

Copper is used for the anode, as it is an excellent conductor of heat, and the tungsten target is embedded in the face of the anode. In a low-voltage x-ray set (less than 50 kV), the anode can be at earth potential, and water can be circulated through the anode to cool it. In high-voltage sets, the anode and filament are maintained at equal but opposite voltages, i.e. for a 250 kV set the anode would be at +125 kV with respect to earth, and the filament would be at −125 kV. This halves the maximum voltage to earth within the equipment and reduces the amount of insulation that is required. The x-ray tube is mounted in a housing (figure 21.2) filled with oil, which insulates the tube. The oil is either pumped to a water-cooled heat exchanger, or water-cooling coils are contained within the housing. The oil is pumped through the anode or circulates by convection. A thermal cutout is included, to prevent the tube operating at too high a temperature. In the simple housing shown in figure 21.2 this takes the form of a bellows and cut-out switch, which detects the thermal expansion of the oil.

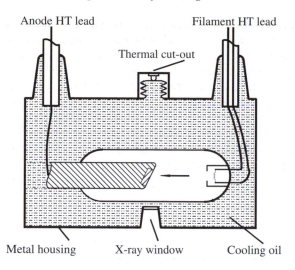

Figure 21.2. *Schematic diagram of an x-ray tube housing. The oil is cooled by water circulating in a copper tube immersed in the oil; this has been omitted for clarity.*

Some means must be provided for supplying the high voltage to the tube, and controls for varying the voltage and the tube current are needed. Figure 21.3 shows a simple x-ray generator circuit. The HT (high-tension) transformer has a fixed turns ratio, and the voltage applied to the tube is altered by varying the input voltage to the transformer primary, using an autotransformer. The voltage applied to the tube is measured in the lower voltage primary circuit, which is safer than measuring the HT voltage directly. The secondary of the HT transformer is in two halves, with the junction of the two halves connected to earth. The tube current is measured at the earthed junction. This gives equal but opposite voltages on the anode and filament. Some means of rectification has to be included, as the tube operates from a DC supply voltage. The current to the filament is supplied by a separate filament transformer, and the heating current to the filament (and hence the tube current) is controlled by adjusting the current flowing in the primary winding.

Several methods of rectification are used, but only full-wave rectification will be described (figure 21.3).

In the full-wave rectifier, two of the four diodes conduct during each half-cycle of the applied voltage. During each half-cycle, the tube current rises until all the electrons emitted by the filament reach the anode. The tube current will then remain constant as the tube voltage changes, until the voltage falls below the level needed to maintain the tube current. It can be seen that the output from the tube is pulsed. The tube current is constant for most of the pulse, so the quantity of x-rays will be constant throughout most of the pulse, but the quality of the x-rays will change continuously as the voltage varies. If a capacitor is added across the tube (shown dotted in figure 21.3), the output voltage will be smoothed. The size of the capacitor is chosen so that the current through the tube will not reduce the voltage across the capacitor by more than perhaps 5%, so that the tube voltage is essentially constant. Consequently, the x-ray quality will be practically constant.

The x-ray machine will be fitted with a timer to control the treatment time, and will normally have an ionization chamber (see section 5.6.1) mounted in the beam, so that the intensity of the x-ray beam can be monitored. The head will be fitted with mountings for filters, and interlocks will usually be fitted so that the machine will not operate unless the correct combination of filter, voltage and current has been selected.

Conventional x-ray tubes can be used for a range of energies from 10 to 500 kV. These energies are often referred to as Grenz rays (10–50 kV), superficial x-rays (50–150 kV) and orthovoltage x-rays (150–500 kV).

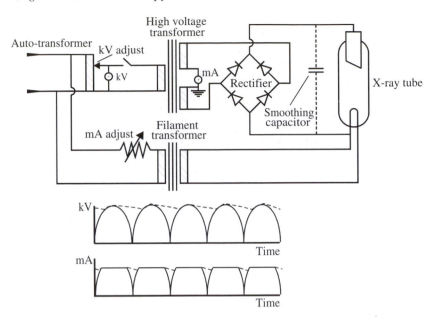

Figure 21.3. *The power supply circuit for an x-ray tube. Note that the tube voltage is measured at the primary (i.e. low-voltage) side of the HT transformer, and the tube current at the earthed centre tap of the secondary. The tube current waveform is the same as the voltage waveform until the tube saturates. Above the saturation voltage the current is independent of the voltage. The dotted lines show the waveform with a smoothing capacitor connected across the full wave rectifier.*

A 250 kV machine, with a tube current of 15 mA, would have an output of 4×10^{-3} Gy s^{-1} at 1 m (see Chapter 5, section 5.4, for definition of the gray).

Electrons striking the target produce x-rays by two processes. In the first process, an electron may collide with the nuclear field around an atom and lose part of its energy, which appears as the energy of an x-ray photon. Interactions of this type will produce a continuous spectrum of x-ray energies at all values up to the maximum energy of the incident electrons. This spectrum is shown in figure 21.4 for three different tube voltages. It can be seen that the maximum x-ray energy and the maximum intensity both increase with increasing tube voltage.

As high-energy x-rays are more penetrating, the x-ray beam becomes more penetrating with increasing tube voltage. In the second process, an electron may be ejected from the target atom, leaving a vacancy. Energy will be released when this vacancy is filled, and the amount of energy will be determined by the atomic structure of the target atom. The resulting x-ray photons will therefore have energies which are characteristic of the target atoms. This characteristic radiation appears as a line spectrum superimposed on the continuous spectrum (figure 21.5).

The *quality* of the x-ray beam is a description of the penetrating power of the beam. For energies of less than 1 MeV, the beam quality is described by the peak tube voltage (which specifies the maximum photon energy) and the half-value thickness (HVT). The half-value thickness is the thickness of a specified metal which will halve the intensity of the beam.

The half-value thickness is not a description of any filters that have been placed in the beam—it is a description of the penetrating power of the beam. For instance, a 300 kV$_p$ (peak energy) x-ray set might have a beam hardening filter (see below) placed in the beam. This filter could be made up of 0.8 mm of tin,

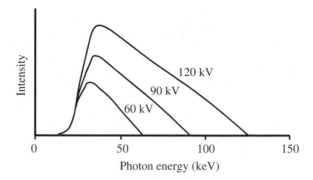

Figure 21.4. *The continuous x-ray spectrum for three tube voltages.*

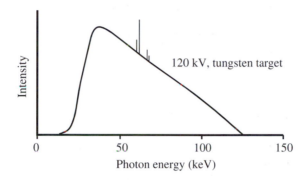

Figure 21.5. *The continuous x-ray spectrum together with the line spectrum which is characteristic of a tungsten target.*

0.25 mm of copper and 1 mm of aluminium. A 3.8 mm thick sheet of copper would reduce the intensity of the resulting beam by half, so that the quality of the beam filtered by 0.8 mm Sn + 0.25 mm Cu + 1 mm Al would be 300 kV$_p$, 3.8 mm Cu. The x-ray spectrum will be altered by any materials, such as the walls of the tube and filters, that are placed in the path of the beam. The effect of placing 0.3 and 0.6 mm copper filters in the beam is shown in figure 21.6. The lower-energy x-rays are less penetrating than the higher-energy ones, and will therefore be preferentially removed from the beam. Whilst filtering reduces the beam intensity it actually increases the average energy of the beam, and therefore makes it more penetrating. This is called a beam hardening filter (see section 12.5.4).

21.2.2. The linear accelerator

Linear accelerators can be used to produce beams of electrons or x-rays at energies between 4 and 25 MV. This is done using radio-frequency (RF) electromagnetic waves at a frequency of, typically 3 GHz, i.e. a wavelength of 10 cm. In free space the velocity of electromagnetic waves is 3×10^8 m s^{-1}. However, if the waves are confined in a waveguide (a hollow metal tube) which has suitably spaced metal diaphragms in it and a hole down the middle, the speed of propagation of the waves can be reduced. If the diaphragms are initially close together and get further apart as the waves travel down the waveguide, the waves will be accelerated (typically from 0.4 to 0.99 times the velocity of light). This principle is used in the linear

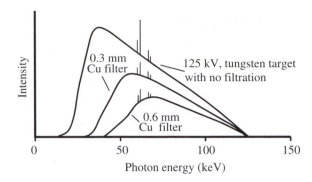

Figure 21.6. *The effect of differing degrees of filtration on the x-ray spectrum from a tungsten target with an applied voltage of 140 kV.*

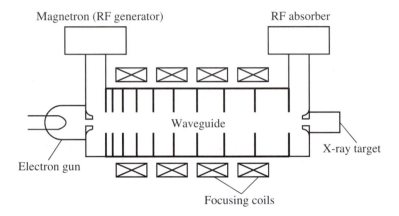

Figure 21.7. *A greatly simplified diagram of a linear accelerator. The interior of the waveguide is maintained at a high vacuum by ion pumps. Electrons from the electron gun are accelerated down the waveguide either to strike the target and thus produce x-rays, or to pass through a thin window to give an external electron beam. The RF energy may be recirculated, instead of being dissipated as heat in the RF absorber.*

accelerator (figure 21.7). Electrons, produced by a heated tungsten filament, are injected into the end of the waveguide. The electrons are carried by the RF wave, and accelerate to a velocity equivalent to 4 MeV in about one metre. At the end of the waveguide, the RF energy is diverted into an absorber, and the electrons are either brought to rest in a transmission target, to produce x-rays, or pass through a thin window to be used as an electron beam.

The waveguide is pumped to a high vacuum, and the beam of electrons is focused by coils surrounding the waveguide. The beam is usually bent through 90° by a magnet before striking the target (figure 21.8) so that the linear accelerator can be conveniently mounted on a gantry. The maximum size of the x-ray beam is defined by the primary collimator. At these high energies, most of the x-rays are emitted along the line of travel of the electrons, so that the beam is more intense on the axis than to either side. The beam profile is made more uniform by correction with a beam flattening filter, which absorbs more energy from the centre of the beam than from the edges. The flattening filter is designed to make the beam intensity constant, to within 3%, from the centre of the beam to within 3 cm of the edge. The beam is pulsed at 100–500 pulses

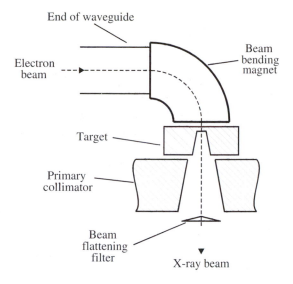

Figure 21.8. *The treatment end of the linear accelerator. The electron beam is deflected magnetically through a right angle (this arrangement makes the whole machine more compact) and strikes the target. The resulting x-ray beam is collimated and passes through a beam flattening filter to adjust the beam uniformity.*

per second, and has an average intensity of about 200 cGy min^{-1} at 1 m from the tungsten–copper target with 200 pulses per second. Each pulse is about 2 μs long.

21.2.3. Tele-isotope units

In a tele-isotope unit an isotope source is sealed in the unit and arrangements are made to expose the patient to the source as required. The only γ-ray-emitting isotope that used to be available for teletherapy units was radium, which had the advantage of a very long half-life (1620 years), so that the treatment time was essentially constant. Radium units had three disadvantages: the danger of a leakage of radioactive radon gas, the low specific activity and the high photon energy which made the construction of the source housing difficult. The availability of caesium-137 and cobalt-60 has made radium units obsolete. Caesium-137 has a useful γ-ray energy of 0.66 MeV (figure 21.9) and a long half-life (30 years), but has a low specific activity and is rarely found in teletherapy units. It is used as a substitute for radium in sealed sources.

Cobalt-60 emits γ-rays at 1.17 MeV and 1.33 MeV (figure 21.9) and has a high specific activity. The half-life is relatively short (5.26 years), so that treatment times have to be increased by about 1% per month to correct for the decay. The useful life of the source is about 3.5 years. The radioactive cobalt is encapsulated inside a pair of stainless steel containers. The cobalt metal is in the form of discs 17 mm in diameter and 2 mm thick; 10 or 12 discs are stacked in the inner container, with any space taken up by brass discs. The capsule lid is screwed on, and then brazed to seal it. The capsule is then placed into another similar container, and the lid of the second container is screwed and brazed in place. This doubly encapsulated source will be fitted into a special source holder before being despatched from the isotope laboratory in a special container.

The transit container is usually designed to fit accurately onto the tele-isotope unit so that source replacements can be implemented on site with the minimum of interruption to the treatment programme. It is good practice regularly to wipe the accessible surfaces of the treatment head with a damp swab, and then count the swab to check for radioactive dust that might have leaked from the source.

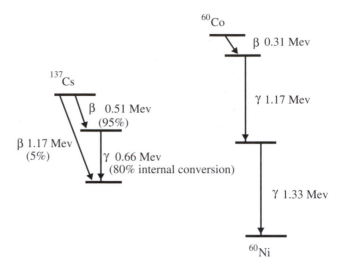

Figure 21.9. *The decay schemes for caesium-137 (left) and cobalt-60 (right). Note that the total energy for the right-hand decay path for caesium-137 (0.51 MeV + 0.66 MeV) is the same as for the left-hand decay path (1.17 MeV).*

It is, of course, impossible to switch off the gamma emission from the cobalt-60, so that some means of interrupting the beam must be provided. The source is mounted within a massive head which is made from lead and depleted uranium; this head will weigh about 1 tonne for a 200 TBq (approximately 5000 Ci) source of cobalt-60. There are two methods of cutting off the beam. In the moving source system, the source is mounted on a turntable, and is rotated to align with the beam exit aperture. In the fixed source system, a rotating shutter interrupts the beam. The source and shutter are moved by an electric motor, and are also spring-loaded so that the beam will be cut off automatically if the motor or the electricity supply fails. A manual method of controlling the beam will also be provided. A 200 TBq source would give an output of 0.013 Gy s^{-1} at 1 m.

21.2.4. Multi-source units

It is obviously possible to control the spatial distribution of dose more precisely if the beam from a linear accelerator can be moved around the patient in a controlled manner. By combining fields from several directions the dose distribution can be matched as closely as possible to the shape of a tumour. Several beam directions are routinely used in linear accelerator treatment and in the technique of conformal radiotherapy (see section 21.4.2) many beams are used.

One problem with the use of multiple beam directions is that it may take a long time to place the beam in many different positions and the patient may move within this time. One solution to this problem is the use of multiple sources to give simultaneous beams. This technique is illustrated in figure 21.10 where 201 cobalt-60 sources, of 1 mm diameter, are used to give a well controlled treatment volume. This method is usually referred to as radiosurgery and is used for treating small volumes within the head. Small tumours and arteriovenous malformations can be treated in this way. A treatment volume as small as 4 mm diameter can be produced. Obviously very good control of patient position is required if the irradiated volume is to be coincident with the treatment volume.

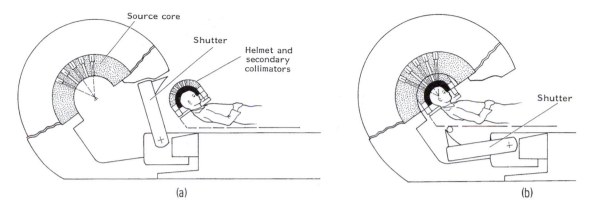

(a)   (b)

Figure 21.10. *The Sheffield stereotactic radiosurgery unit. When the shutter is dropped the patient is moved such that the secondary collimators in the helmet line up with the primary collimators in the source core. Image kindly provided by L Walton.*

21.2.5. Beam collimators

The collimator confines the radiation beam to the appropriate size and direction. The primary collimator sets the maximum field size, and the secondary collimator adjusts the field size to that required for each individual treatment. The primary collimator is a thick metal block with a conical hole through the centre. In an x-ray unit, the primary collimator will be part of the tube housing. In a linear accelerator, the primary collimator will be close to or incorporated in the x-ray target (see figure 21.8), and in a cobalt unit it will be part of the shielding or the shutter mechanism.

    Two types of secondary collimator are used. For low photon energies (less than 500 kV) and a source-to-skin distance (SSD) of less than 50 cm, an applicator will be used. At higher energies, or if the head of

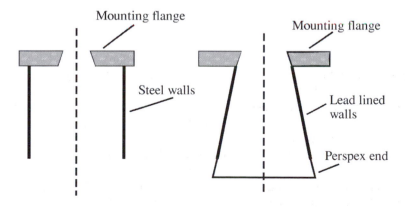

Figure 21.11. *Parallel-sided (left) and 'Fulfield' (right) beam defining applicators. The end of the applicator defines the source-to-skin distance.*

the treatment machine is to be rotated, a diaphragm is used. The applicator (figure 21.11) has a thick base (flange) which is attached to the treatment head, and which reduces the intensity outside the useful beam to less than 2%. The end of the applicator rests on the skin, and therefore defines the source-to-skin distance. The area in contact with the skin defines the useful area of the beam. The walls of a parallel-sided applicator are not irradiated (figure 21.11), and may therefore be made of steel. The *Fulfield* applicator has lead-lined walls parallel to the edges of the beam and a Perspex end-plate to absorb the secondary electrons from the lead. The Perspex can be marked to show the position of the beam axis.

A large thickness of heavy metal is needed to reduce megavoltage beams to less than 2% outside the useful beam, so that an applicator would be excessively heavy. The diaphragm must be placed well away from the skin, as the energetic secondary electrons have a considerable range in air and would give a large dose to the skin. The diaphragm is constructed from a number of lead sheets which can be moved in and out to define the edges of the beam. Several different arrangements are used but they all give a rectangular beam. In the absence of an applicator, some means of visually defining the beam must be provided. This is done by placing a mirror in the beam, and shining a light beam through the diaphragm onto the patient (figure 21.12).

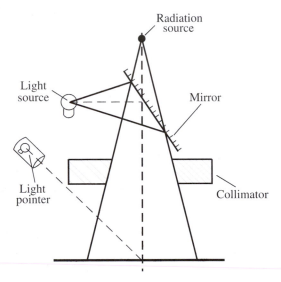

Figure 21.12. *The optical beam-defining system. The mirror, of course, is transparent to x- and γ-radiation. The combination of projected cross-wires (to define the beam centre), and a separate light pointer, can be used to define the source-to-skin distance.*

21.2.6.   Treatment rooms

The treatment set is always located in a treatment room which is designed to reduce the radiation dose to members of staff and the general public to less than that permitted. The floor and ceiling of the room, as well as the walls, may need to include shielding. The shielding must be designed to attenuate both the primary beam and any scattered radiation. The primary beam direction is limited by the available rotation of the treatment head, so that primary beam shielding will only be needed for certain parts of the room. The remainder of the room will have shielding designed to attenuate the scattered radiation.

At photon energies above 50 kV, the equipment is operated from a console outside the treatment room, and some means of observing the patient has to be provided. An interlock must be provided on the door, so that the set cannot be operated unless the door is closed. 2 mm of lead will provide sufficient attenuation of the primary beam at energies of up to 100 kV. This thickness of lead sheet can be incorporated in a door without making it excessively heavy, so that the design of treatment rooms for these energies is relatively easy. At 250 kV, 8 mm of lead is required for the primary shield but only 2 mm is needed for secondary screening. The door will therefore be placed within the secondary screen.

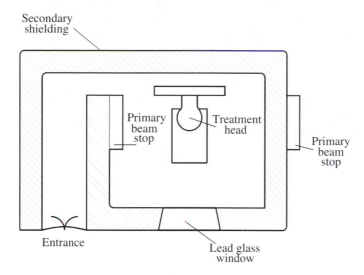

Figure 21.13. *The layout of a megavoltage treatment room with a maze entrance. As the beam cannot be pointed in all directions, the primary shielding (which will attenuate the direct beam) does not have to completely surround the set. Remember that the shielding will also have to cover the ceiling (and the floor if there are rooms beneath the set).*

At higher energies the thickness of the secondary barrier makes the door too heavy, and the room is designed with a maze entrance (figure 21.13). There is no direct path to the door for scattered radiation. The door is a barrier to prevent people entering when the beam is on, and is not intended as a radiation barrier. For a 2 MeV set, the primary barrier would be 108 cm of concrete, and the secondary barrier 51 cm of concrete. This is equivalent to 165 and 87 mm of lead, respectively.

At lower energies the observation window may be lead glass. A 16 mm lead glass window is equivalent to 4 mm of lead. At higher energies, where the shielding is made of concrete, the window may be made of several sheets of plate glass with the same effective thickness as the concrete wall. Closed circuit television is often used as an alternative means of viewing the patient.

21.3. DOSE MEASUREMENT AND QUALITY ASSURANCE

21.3.1. *Dose-rate monitoring*

The dose rate from the x-ray treatment head is monitored by ionization chambers or solid-state detectors placed in the beam after all the filters. Ionization chambers were described in Chapter 5 (section 5.6.1). The whole of the beam may be monitored using a flat chamber which is larger than the beam size defined by the primary collimation. However, most linear accelerators have segmented ionization chambers that allow monitoring of both the beam intensity and the uniformity. Each chamber may have either two or three parallel plates (figure 21.14). The three-plate chamber is said to be safer, as the HT electrode is totally enclosed in the

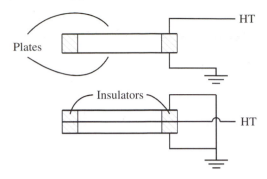

Figure 21.14. *Two- and three-plate ionization chambers for monitoring the beam intensity.*

two earthed plates and the insulation. However, the current which can be drawn from the HT supply is much too low to be a hazard. If the plates are used to measure the actual patient dose, the chamber must be sealed so that the mass of air within the chamber does not change with temperature and atmospheric pressure.

21.3.2. *Isodose measurement*

If we wish to determine the dose given to a particular volume within the patient, we must first know the dose distribution produced by the x-ray beam within the tissue. This is measured by using a tissue equivalent phantom. An ideal soft-tissue equivalent phantom would have the same atomic number as tissue (because photoelectric absorption and pair production depend on the atomic number), the same electron density as tissue (because Compton scattering depends on the electron density) and the same density as the tissue. It must be possible to move the ionization chamber within the phantom, which makes water the material of choice.

The dose distribution is plotted in the form of isodose charts (figure 21.15).

It is conventional to plot the isodose distribution for planes which contain the axis of the beam. Sufficient information can be obtained by plotting the distribution in two planes at right angles. For a rectangular beam, the two planes are parallel to the sides of the rectangle; for a square or circular beam the two planes should be identical. Because the distribution is symmetrical about the axis of the beam, only one-half of the distribution is shown. The isodose lines are lines of constant dose and are expressed as a percentage of the maximum dose on the central axis within the phantom.

Figure 21.15 shows isodose plots for 2 mm Cu HVT x-rays, cobalt teletherapy units and 4 MV x-rays. The dose will fall with depth due to the absorption of the beam by the tissue and the inverse square law fall-off due to the increasing distance from the source. Compton scattering also takes place, and leads to the diffuse edges of the beam at low energies. At higher energies, the recoil electrons are ejected more in the direction of the beam, so the edges of the beam are better defined. This also gives rise to the skin-sparing effect at higher energies. It can be seen that, at higher energies, the 100% dose line is below the surface of the phantom. The depth of the peak in cm is roughly a quarter of the x-ray energy in megavolts, i.e. for 4 MV x-rays, the peak is approximately 1 cm below the surface. For most teletherapy units the surface dose will be 20–50%. This is one of the main advantages of high-energy therapy.

Isodose measurements can be made using an isodose plotter consisting of a water tank about 50 cm in each direction, in which is suspended an ionization chamber or solid-state detector. The beam enters the tank from one side, and the chamber can be traversed across the beam at different distances from the end of the tank. In the simpler plotters, the output from the chamber is plotted on an *X*–*Y* plotter to give a beam profile,

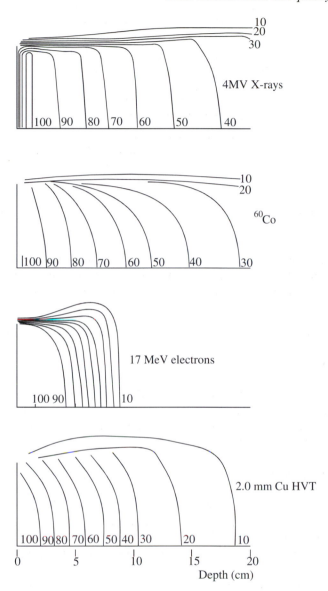

Figure 21.15. *Sketches of isodose charts for megavoltage, tele-isotope, electron and orthovoltage beams. The beam is symmetrical, so only half is shown, the other half being a mirror image. Note the skin-sparing effect of the megavoltage beam and the dose uniformity provided by the electron beam.*

and this is repeated for different distances from the source. The isodose plot is then built up by plotting all the points which have the same dose and joining up the points. It is now more likely that a data set of grid points will be input to a treatment planning system (TPS) and software used to interpolate the isodose curves (see Williams and Thwaites, and Conway, in Cherry and Duxbury (1998), for further information).

21.4. TREATMENT PLANNING AND SIMULATION

The treatment plan describes graphically the dose distribution when one or more radiation beams converge on the tissue volume which is to be treated. We will call the volume which has to be treated the *treatment* volume, and the volume which is actually treated the *treated* volume. ICRU 50 gives definitions of planning target volume (PVT), gross tumour volume (GTV) and clinical tumour volume (CTV). The criteria for judging how good a dose distribution is will vary from centre to centre, but a typical set might be:

- The dose throughout the treatment volume should be uniform to ±5%.
- The treated volume should be as nearly as possible the same as the treatment (PTV) volume.
- The dose to the treated volume should exceed the dose elsewhere by at least 20%.
- The dose to sensitive sites (eyes, spinal cord, etc) should be below their tolerance dose. The integral dose should be minimized.

These criteria can usually be satisfied by using several beams, which are as small as possible and which enter the patient as close to the treatment volume as possible. As an example, a simplified version of the problem of treating a cancer of the bladder using three 4 MV x-ray beams will be given (see Meredith and Massey (1977) for a complete treatment).

21.4.1. Linear accelerator planning

First of all, an outline of the patient at the treatment level is drawn, together with the position of the treatment area (figure 21.16). We will describe this process as though it was being done manually, but in practice the

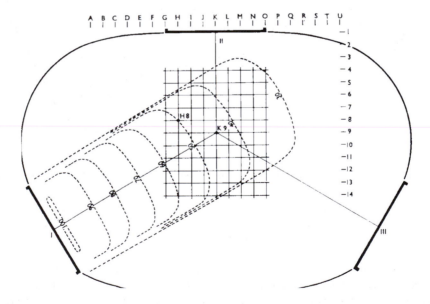

Figure 21.16. *The patient outline showing the three beam positions (I, II and III); a plotting guide placed over the treatment area; and the isodose distribution positioned for beam I. The intensity of the beam is written in each square for each of the beam positions. (From W J Meredith and J B Massey 1977 Fundamental Physics of Radiology (Bristol: Wright). Reprinted by permission of Butterworth-Heinemann Publishers, a division of Reed Educational & Professional Publishing Ltd.)*

manipulation would be done on a computer screen. The axes of the three treatment beams are drawn in, so that they intersect in the centre of the treatment volume. Next, the isodose curve for the machine that is being used is placed along one of the treatment axes.

For each square on the diagram, the percentage dose is read off the isodose curve and written in the square. This is repeated for all the treatment axes. The percentage doses in each square are then summed to give the total dose in each square due to the combination of the three beams. In general, dose uniformity requires that the contributions from each beam are not the same. A typical approach is to adjust the beam intensities so that they all give the same dose at the intersection of the axes. In this example, the beam labelled II is closer to the intersection than the others, and its intensity will have to be reduced. This is done by multiplying the doses in each square by the appropriate factor. If, for instance, the intensities of beams I and III are 50% at the intersection, and that of beam II is 80%, all the values for beam II must be multiplied by 5/8. By examining the dose to each square, the isodose lines can be drawn in (figure 21.17).

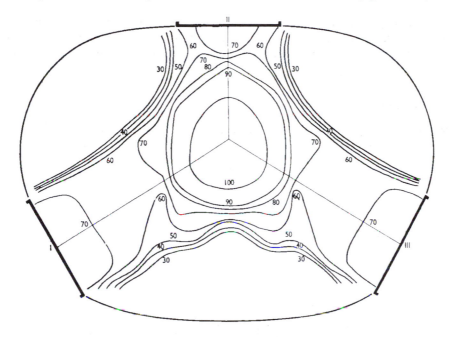

Figure 21.17. *The completed isodose chart for the treatment. The dose delivered by field II has been reduced to 0.615 of the dose delivered by the other two fields, which gives an equal contribution from each field at the intersection of the axes of the beams. (From W J Meredith and J B Massey 1977 Fundamental Physics of Radiology (Bristol: Wright). Reprinted by permission of Butterworth-Heinemann Publishers, a division of Reed Educational & Professional Publishing Ltd.)*

Because the skin is curved, the beam will often enter the skin at an angle other than 90°. This will alter the dose distribution. There are two possible solutions to this problem: either the curvature can be corrected or the distribution can be corrected. At low energies, where there is no skin-sparing effect, the skin can be built up using a tissue-equivalent bolus to give a surface at right angles to the beam (figure 21.18).

The standard isodose curve can then be used to plan the treatment. At higher energies, where the skin-sparing effect is important, a bolus is not used and mathematical corrections are applied to describe the oblique incidence of the beam. It is also possible to correct the isodose distribution by placing a tissue compensator in the beam remote from the skin. This is a suitably shaped attenuator that compensates for the missing attenuation of the tissue. If a large volume close to the surface has to be treated, the isodose

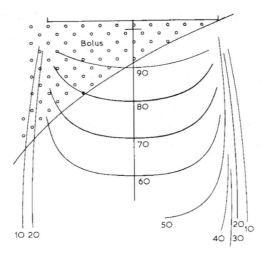

Figure 21.18. *The use of a tissue equivalent bolus to restore the normal incidence of the treatment beam. This maintains the charted isodose distribution, but the skin-sparing effect is lost. (From Bomford et al 1993 Walter and Miller's Textbook of Radiotherapy (Edinburgh: Churchill Livingstone).)*

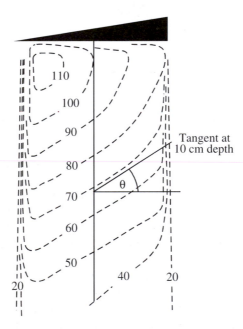

Figure 21.19. *The effect of a wedge on the isodose distribution, and the definition of the wedge angle as the angle of the tangent to the isodose at a depth of 10 cm.*

distribution is often altered by means of wedge-shaped compensators. The wedges will obviously attenuate the beam more at the thick end, and will therefore tilt the isodose curve (figure 21.19). The wedge is usually placed behind the mirror, and the beams are used as 'wedged pairs', with the thick ends of the wedges facing each other.

Nearly all centres now use a computer to do the treatment planning. Details of the isodose distributions for the treatment machines are stored in the computer memory. The outline of the patient is entered using a graphics tablet or direct from a simulator (see below). The size and position of the treatment area is determined from radiographs, and can be entered using a graphics tablet. The computer will then assist the operator to produce an isodose distribution for a selected number of fields. The program will correct for oblique incidence of the beams. The process is fast, so that the operator can alter the details of the treatment and obtain a new plan very rapidly. When the operator is satisfied with the plan, a hard copy is made.

See Conway, in Cherry and Duxbury (1998), and Williams and Thwaites (1993) for much more detail on these techniques.

21.4.2. Conformal techniques

Traditional radiotherapy beams have a simple square or rectangular cross-section. However treatment sites may have an irregular outline and therefore the beam may irradiate tissues outside the area being treated. A simple 'beam's eye view' of the planning target volume (PTV) is given in figure 21.20.

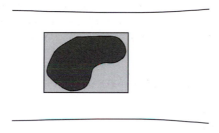

Figure 21.20. *The darker regions are to be treated but the outer regions also receive a similar dose of radiation.*

The central region is the region to be treated, but the other regions receive a similar radiation dose. The radiation dose to these regions can be reduced if the beam is shaped to fit the region being treated. This can be achieved by using adjustable inserts (or multi-leaf collimators) in the beam. These are typically made of tungsten and there may be as many as 40 pairs of them. They are of sufficient thickness to reduce the beam intensity to 1% of the on-axis dose. Given information about the shape of the region to be treated the position of the leaves can be adjusted automatically under computer control to fit the region as closely as possible (see figure 21.21).

The actual number of leaves would be higher than is shown in this image so that the fit to the treatment region might be better than this, although multi-leaf collimators (MLCs) only have a resolution of about 1 cm at the isocentre. The reduction of tissue dose around the object being treated represents a distinct improvement.

Treatment usually consists of irradiation from more than one direction. The projection of the treatment area will be different in different directions so that a different beam shape will be set for each beam. As has been noted before (section 21.2.4), the use of multiple beams is a powerful technique for focusing energy at the treatment site. The more beams that are used the higher the dose that can be delivered to the tumour without damaging surrounding tissue.

In principle the use of a large number of individual beams can be replaced by continuous rotation of the treatment set around the patient. This could be combined with beam shaping, as described above, to shape the treatment volume in three dimensions to match the region to be treated as closely as possible. If accurate

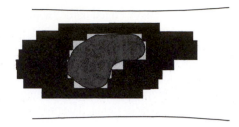

Figure 21.21. *The dose distribution after using tungsten leaves to adjust the beam profile.*

control of the beam position and cross-section is to be possible, detailed information about the position and shape of the treatment region is required. Although this can be obtained from 3D imaging such as CT, or with appropriate corrections, MRI, this information must be accurately mapped into treatment space so that the treatment set 'knows' exactly where the treatment volume is. This is especially critical since an important aim of this approach is to deliver a higher dose to the treatment region than is possible with conventional methods. Irradiation of the wrong area could have serious side effects for the patient. Accurate positioning of the patient and the ability to reposition the patient is important. If the patient cannot be accurately repositioned it is important to know where the treatment region is in relation to the treatment set.

The planning of the treatment is complex. The calculation of the optimal beam shape as a function of position and the dose rate delivered from each direction must determined. Powerful three-dimensional planning systems exist to carry out the required calculations. Because of the increased risk due to the higher radiation doses being delivered, validation of the planning system is critical prior to clinical use.

The above description implies that the treatment set moves in a plane around the patient, but further gains could be obtained if the dose is delivered in three dimensions. An example of this is the 'gamma knife' which is used for treating lesions of various types in the brain (see section 21.2.4). In this system multiple γ-ray emitting ^{60}Co sources are arranged over part of the surface of a sphere and the radiation from these sources collimated so that all beams pass though a focal point. This focal 'hot spot', which can be made as small as a few millimetres in diameter, is then scanned through the treatment volume. In practice the patient is placed on a couch which can be moved in all directions and the patient is then scanned through the focal point. By appropriate motion of the patient the focal point can be scanned through complex volumes. As radiation arrives from three dimensions the surrounding tissue receives a low radiation dose compared to tissue at the focal spot. However, as outlined above, accurate positioning of the patient is essential if the correct area is to be treated. This is currently achieved by fixing a positioning device to the skull of the patient surgically. This device is then attached to a CT or MRI scanner and an image taken of the appropriate region to be treated. The position of this region is known accurately relative to the positioning device. The patient is then positioned in the treatment machine by again fixing the device to the machine. Careful calibration means that the treatment machine 'knows' where the lesion is.

Conformal radiotherapy requires the shape, direction and intensity of the x-ray beam to be continuously changed, e.g. in prostate therapy the beam can be changed to match the outline of the tumour, whether the beam is from the front, the side or at an oblique angle. This spares adjacent organs such as the colon, rectum and bladder. The use of conformal techniques in radiotherapy is likely to increase with the use of volume imaging systems and powerful planning computers.

21.4.3. Simulation

As a check that the treatment beam has actually been placed correctly with respect to the tumour an x-ray film can be exposed during treatment. The film is simply placed close to the patient at the point where the treatment beam emerges. This image is referred to as a portal film and should show the treatment volume and the adjacent anatomy. Unfortunately such images are usually of very poor quality largely because the beam energy is optimized for treatment and not for imaging. The high energies produced by a linear accelerator give very poor contrast between tissues such that the images obtained are very poor. An alternative technique (electronic portal imaging) which has been introduced is to use an array of silicon diodes or ionization chambers to produce an image and this can be used to give rapid images for verifying the position of a treatment beam. However, the need for accurate verification of treatment geometry led to the development of treatment simulators.

The simulator is a diagnostic x-ray set with the same mounting and couch movements as the treatment accelerator. Radiographs taken using the simulator are used to provide information on the treatment area for planning, and for checking the accuracy of the plans and patient positioning. Simulators can also be used for localization of sealed sources used in brachytherapy.

The simulator is a sophisticated diagnostic x-ray unit and is used to localize the target tissues and to verify the treatment plan. It is not intended as a diagnostic aid.

21.5. POSITIONING THE PATIENT

There is little point in producing an accurate plan of the treatment if the position of the patient cannot be guaranteed. The patient must be placed in the correct position in relation to the beam, and must remain stationary during the treatment. A patient shell, which can be fixed to the treatment couch, may be used to position the patient. This is used in about 30% of patients, particularly for radical treatments of the head and neck.

21.5.1. Patient shells

The various stages in making a patient shell are shown in figure 21.22. The patient's skin is first covered with a releasing agent such as petroleum jelly, and a plaster cast (in two sections if necessary) is made of the part of the patient that is to be immobilized, and is then removed from the patient. This negative impression is then used as a mould to produce a positive plaster cast which should have the same dimensions as the patient. The positive plaster cast can then be used as the mould in a vacuum-forming machine. The positive plaster cast of a head, as shown in the diagram, would be cut in half along a plane that avoided the treatment beams, and the two halves of the cast placed in the vacuum-forming machine. A plastic sheet is inserted in the machine and heated to make it pliable. It is then blown into a bubble using compressed air, and the positive cast is placed in the bubble. Finally, the space between the sheet and the cast is evacuated so that the plastic sheet is moulded round the positive cast, to produce a faithful negative impression of the patient. The two halves of this shell are held together by press-studs, and supports are made to attach the shell to the treatment couch.

If the cast is of the head, then provision will have been made for the patient to breathe. If it is necessary to build up the patient's contours using bolus, in order to correct for oblique incidence of the beam, this can be done on the shell. Any local lead shielding that might be necessary can also be attached to the shell.

21.5.2. Beam direction devices

Once the patient has been positioned on the treatment couch, the beam direction must be set accurately. Megavoltage treatment sets are usually mounted on an isocentric gantry (figure 21.23). The beam can be pointed in any direction relative to the patient by means of three rotational movements of the treatment head

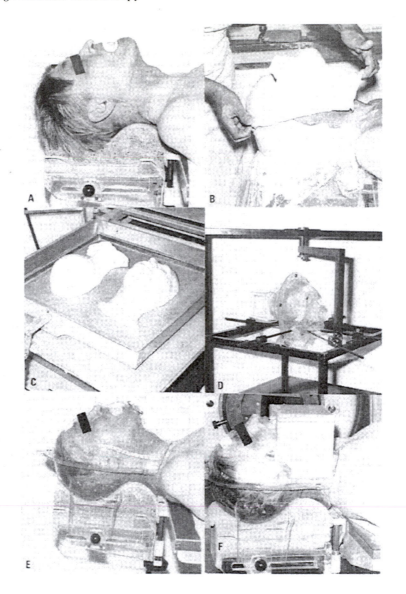

Figure 21.22. *The production and use of patient shells in treatment planning. (a) The patient in the proposed treatment position with a breathing tube. (b) The removal of the front half on completion of a whole-head mould. (c) The two halves in position for vacuum forming. (d) The complete shell in the contour jig. (e) The patient in position for the localization films on the simulator. (f) The shell complete with tissue equivalent wax block in use on the linear accelerator. (From Bomford et al 1993 Walter and Miller's Textbook of Radiotherapy (Edinburgh: Churchill Livingstone).)*

and the couch. The couch can be rotated about a vertical axis, the head can be rotated about a horizontal axis, and the diaphragm system can be rotated about the beam axis. These three axes of rotation intersect at

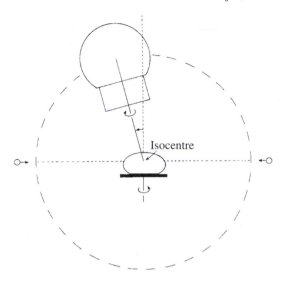

Figure 21.23. *Isocentric mounting of the treatment set. The treatment area is positioned at the isocentre using the linear movements of the table (up/down, left/right, head/foot). The two directions of rotation of the treatment beam (rotation about the beam axis and rotation about the isocentre), and the rotation of the table, will only alter the beam direction. If the treatment area is at the isocentre, and the linear movements of the table are not used, the beam will always pass through the treatment area. (From Bomford et al 1993 Walter and Miller's Textbook of Radiotherapy (Edinburgh: Churchill Livingstone).)*

a fixed point in space called the isocentre. In other words, if the centre of the treatment volume is placed at the isocentre, the direction of the beam can be adjusted by the three rotational movements. The position of the isocentre can be defined by a set of optical pointers fixed to the walls of the treatment room. For instance, if the centre of the treatment volume is 6 cm vertically below a skin mark, the skin mark is first set to the isocentre using the linear movements of the couch (i.e. up/down, left/right, head/foot), and the couch is then raised 6 cm to position the centre of the treatment volume at the isocentre. No further linear movement of the couch is made. The rotational movements of the couch and head are then used to set the beam direction—the appropriate rotations will have been worked out from the plan.

21.6.   THE USE OF SEALED RADIATION SOURCES

A lesion can be irradiated with γ-rays by using a small quantity of a radioisotope sealed in a tube or wire and placed within the tissue to be treated. The sources are precisely arranged around or within the lesion. Radium was used for many years for this purpose. The effective γ-ray photon energy of radium is about 1 MeV, and 0.5 mm platinum filtering (or the equivalent) was used to remove the α- and β-particles. Radium emits γ-rays of up to 2.4 MeV, which necessitated the use of heavy screening around patients with radium implants. A further problem was the build-up of radon gas within the tube, which could not be contained in the event of any damage. ^{60}Co (1.17 and 1.33 MeV, 5.26 year half-life) and then ^{137}Cs (0.66 MeV, 30 year half-life) replaced radium, but the most widely used isotope is now iridium ^{192}Ir (0.61 MeV, 74 days half-life).

The use of small sealed γ-ray sources to irradiate tissue is referred to as *brachytherapy*. It can be used for *interstitial* therapy where the sources are implanted directly into the diseased tissue or for *intracavitary* therapy where an applicator is used to contain the radiation sources which will then irradiate a body cavity from the inside.

21.6.1. Radiation dose from line sources

For a point source of radiation or at a distance greater than three times the maximum dimension of a linear source, the dose rate is given by

$$\text{air kerma dose} = \frac{\text{AKR constant} \times \text{time} \times \text{activity}}{(\text{distance})^2}$$

where activity of the source is in $\text{GBq}\,\text{m}^{-1}$, distance is in m and the air kerma rate constant is in $\mu\text{Gy}\,\text{h}^{-1}\,\text{GBq}^{-1}\,\text{m}^2$.

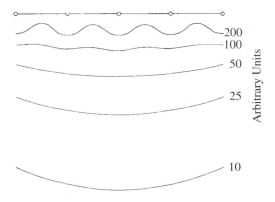

Figure 21.24. *The isodose distribution from five point sources.*

Figure 21.24 shows the dose (in arbitrary units) to be expected from a line of five point sources. It is relatively easy to calculate these isodose curves as the doses from each source will follow an inverse square law and the doses from the sources will be additive. Three useful conclusions can be drawn from a study of this distribution. Firstly, the uniformity of the dose depends on the spacing of the sources and the distance from the sources. Secondly, a line source may be represented by a closely spaced line of point sources. Thirdly, the uniformity of the dose can be increased by increasing the activity of the ends of the lines.

21.6.2. Dosimetry

A volume of tissue can be treated using a geometrical array of line sources. As far as possible a uniform radiation dose must be delivered to the tissue to be treated. The Paterson–Parker rules provided a formalized system by which the approximate layout of the sources could be determined. These have largely been superseded by the Paris system for iridium wire used in interstitial therapy and by the Manchester rules for gynaecological implants. ICRU 38 defines the rules used in intracavitary therapy in gynaecology.

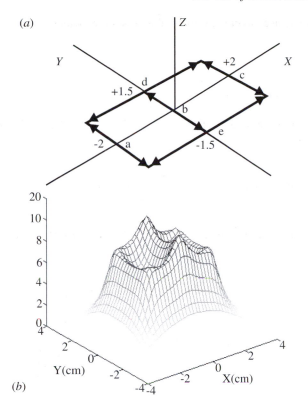

Figure 21.25. *(a) Geometry assumed for the placement of a planar array of five caesium needles (labelled a–e) over a lesion and (b) the spatial distribution of dose (plotted vertically) which results in a plane at a depth of 5 mm.*

We can calculate the uniformity of dose for a given distribution of line sources as follows. Take the geometry of figure 21.25(a) where a treatment area of 3×4 cm is assumed and place sources along the five lines shown. We can calculate the dose rate within the plane given by $z = z_1$ as follows:

$$
\text{dose} \propto \int_{-2}^{2} \frac{a}{(x - x_1)^2 + (-1.5 - y_1)^2 + (-z_1)^2}\, dx + \int_{-2}^{2} \frac{a}{(x - x_1)^2 + (1.5 - y_1)^2 + (-z_1)^2}\, dx
$$
$$
+ \int_{-1.5}^{1.5} \frac{a}{(-2 - x_1)^2 + (y - y_1)^2 + (-z_1)^2}\, dy + \int_{-1.5}^{1.5} \frac{a}{(-x_1)^2 + (y - y_1)^2 + (-z_1)^2}\, dy
$$
$$
+ \int_{-1.5}^{1.5} \frac{a}{(2 - x_1)^2 + (y - y_1)^2 + (-z_1)^2}\, dy
$$

where a is the activity per unit length of the needles and (x_1, y_1) are the coordinates within the plane $z = z_1$.

The distribution which results for $z_1 = 0.5$ cm is given in figure 21.25(b) and shows large inhomogeneities. These inhomogeneities could be reduced by using more needles, modifying the spatial distribution and using different activities in different positions. What the Parker–Paterson rules do is to try to minimize the inhomogeneity within a volume which is 5 mm either side of the plane of the needles.

Comparison between different isotopes

Radium has been superseded by many other isotopes. It is possible to compare isotopes by using the air kerma rate constant. Table 22.1 lists the values for several isotopes.

Table 21.1. *A comparison of several isotopes that have been used as sealed sources in radiotherapy.*

Radionuclide	Maximum energy (MeV)	Half-life	AKR constant (μG h^{-1} GBq^{-1} m^2)
^{137}Cs	0.66	30 years	78
^{60}Co	1.33	5.27 years	307
^{198}Au	0.68	2.7 days	55
^{192}Ir	0.61	74 days	113
^{125}I	0.035	60 days	34
^{226}Ra	2.4	1600 years	195

21.6.3.   *Handling and storing sealed sources*

The gamma and beta radiation from sealed sources is a hazard. However, unless the source is damaged, there is no contamination hazard. They should always be handled using forceps, and should not be picked up by hand.

In most cases sources are now implanted using automated techniques referred to as 'remote after loading'. The sources are kept in a safe and delivered along a tube to an applicator which is placed close to the treatment volume. During nursing procedures the source is returned automatically to the safe and only replaced within the patient afterwards. By this means the radiation dose to staff and visitors can be dramatically reduced.

The radiation hazard can be reduced by spending less time near the sources, increasing the distance from the source, and using screening. Any operation requiring skill, such as threading needles, should be practised with identical, but unloaded, sources. A simple lead bench may be used with higher-activity sources. The sources will be stored in a lockable, shielded safe with several drawers. Each drawer will be fitted out to take a number of sources, and it should be possible to see at a glance whether all the sources are present. All movement of sources to and from the safe should be entered in a record book.

The sources may be sterilized by ethylene oxide, chemically or by boiling. In the case of boiling, the maximum temperature should be limited to 180 °C. Each source should be checked for leaks before sterilizing and at any time when damage may be suspected.

21.7.   PRACTICAL

21.7.1.   *Absorption of gamma radiation*

Objectives

To demonstrate the exponential nature of the absorption of homogeneous γ-rays.

Theoretical basis

In comparison with particulate radiation, γ-rays have a low probability of interaction with matter. In a beam of charged particles, all the particles are gradually slowed down by loss of kinetic energy in collisions. For a beam of γ-rays some of the rays are absorbed or scattered, while the remainder pass on unchanged (see section 5.2). This leads to an exponential absorption law for which no maximum range can be quoted: a small fraction of the γ-rays will always pass through an absorber.

For a collimated beam of γ-rays in which it is assumed that all rays are of the same energy (homogeneous or monochromatic radiation), the intensity of radiation at various depths in the absorbing medium can be calculated from the exponential absorption law:

$$I = I_0 e^{-\mu x}$$

where I_0 is the initial intensity, I is the intensity after passing through an absorber of thickness x and μ is a constant, the 'linear absorption coefficient', which has dimensions of cm^{-1}.

Since γ-ray attenuation is a random process of interaction between photons and atoms, the amount of attenuation will obviously depend upon the number of atoms in any thickness of material. Therefore, the compression of a layer of material to one-half its thickness should not affect its power of attenuation. For this reason the linear attenuation coefficient depends on the density of the material as well as on other features, and is less fundamental than other coefficients which take the density factor into account.

The half-value thickness ($D_{1/2}$) of the absorber is that thickness which reduces the radiation intensity to one-half of its original value, and it can be determined from a plot of count rate against absorber thickness. The absorption coefficient is related to the half-value thickness by the relation

$$D_{1/2} = \frac{0.693}{\mu}.$$

(This follows as $I/I_0 = 0.5 = e^{-\mu D_{1/2}}$, where $D_{1/2}$ represents the half-value thickness. $e^{0.693} = 0.5$, therefore, $\mu D_{1/2} = 0.693$ and $D_{1/2} = 0.693/\mu$.)

Equipment

A collimated source of 1 mCi (37 MBq) ^{137}Cs (γ-ray energy 0.66 MeV).
NaI scintillation detector.
Scaler/timer with high-voltage supply. Range of lead absorbers.
Various clamp stands.

Method

- Using the clamp stands provided, arrange the source absorbing screens and detector so that the source–detector distance is approximately 30 cm and the absorbing screens are placed close to the source.
- Switch on the counting equipment.
- Measure the background count rate for 300 s with the source removed.
- Reposition the source. Measure the count rate for at least 60 s. Close the source.
- Place the thinnest absorber between the source and the detector. Then open the source and measure the count rate.
- Repeat this procedure adding more and more absorbers. Use longer counting periods for the thicker absorbers (to give the same statistical errors for each measurement).

Note. Duplicate readings should be taken for each absorber thickness.

Results

- Correct all counts for the background and then plot (linearly and semi-logarithmically) the relation between the intensity of the radiation at the detector (as a fraction of the unabsorbed intensity) and absorber thickness. This is best carried out in a graphics package but could be done on paper.
- Determine the half-value thickness from the graph and calculate the linear absorption coefficient.

21.8. PROBLEMS

21.8.1.   Short questions

a   Define radiotherapy.
b   Is it possible to focus ionizing radiation onto a tumour?
c   Could an x-ray tube be used to produce a 3 MeV treatment beam?
d   Why are x-ray tubes often surrounded by oil?
e   What is the advantage of using a higher γ-ray energy in radiotherapy?
f   What is a line spectrum in the context of x-ray generation?
g   How accurately does a radiotherapy dose have to be specified?
h   What are the energies of the main peaks in a ^{60}Co spectrum?
i   Why is a smoothing capacitor used in an x-ray generator?
j   In what units are radiotherapy doses usually specified?
k   What is meant by beam hardening?
l   What is the purpose of conformal radiotherapy?
m   Is the output of a linear accelerator a constant intensity beam?
n   What is portal verification?
o   Is the x-ray tube an efficient way of generating radiation?
p   What is meant by 'skin sparing'?
q   Is 200 cGy min^{-1} a typical output for a linear accelerator?
r   What does a collimator do?
s   Why is tungsten usually chosen as the target in an x-ray tube?
t   Does ^{60}Co have a shorter or longer half-life than radium?
u   What is brachytherapy?
v   What does 'remote after loading' mean and why is it used?
w   How is interstitial therapy carried out?

21.8.2.   Longer questions (answers are given to some of the questions)

Question 21.8.2.1

What criteria would you use to select suitable isotope sources for radiotherapy and for radionuclide imaging?

Answer

For radiotherapy high-energy radiation is best when performing external treatment but for internal uses a lower energy will give better localization of the treated volume. The high energy when used for teletherapy gives better penetration and skin sparing.

For radionuclide imaging a pure γ-ray emitter (no β or α emitters) will minimize radiation dose. A medium energy is best as it will not suffer too much internal absorption and yet not be difficult to collimate. A short half-life will reduce patient dose.

Question 21.8.2.2

Explain, with the aid of diagrams, the photon interaction process known as photoelectric absorption and the dependence of its absorption coefficient on the photon energy and the atomic number of the absorber.

Beam hardening filters are frequently used to modify the spectrum of an x-ray beam. Explain briefly:

(a)   how the addition of an aluminium filter 'hardens' a 100 kV beam,
(b)   why hardening filters are used in both diagnostic and therapeutic beams of this energy, and
(c)   why copper and tin are not suitable filter materials for beams of this energy.

Sketch the attenuation curve (% transmission versus thickness of absorber) for an x-ray beam and thus demonstrate that the second HVL (HVT) is greater than the first HVL (HVT).

Question 21.8.2.3

A beam flattening filter is an essential component in the x-ray beam of a linear accelerator. Explain why and describe a method of assessing its effectiveness. What criteria are used to assess beam flatness?

Question 21.8.2.4

In megavoltage radiation treatment, it is often desirable to use a wedge filter. How does the wedge filter affect the isodose distribution and explain what is mean by a wedge angle.

Answers to short questions

a   Radiotherapy is the treatment of disease with ionizing radiation.
b   It is not possible to focus ionizing radiation in the way that a lens can focus light. However, it is possible to direct beams of ionizing radiation to increase the relative dose in a specific region.
c   No, a linear accelerator would be used for 3 MeV γ-ray generation.
d   Oil is used for cooling and electrical insulation of an x-ray tube.
e   A higher γ-ray energy gives greater depth penetration and less relative dose to the skin.
f   The tungsten target and other materials can cause monoenergetic peaks in the x-ray energy spectrum.
g   A radiotherapy dose should be as accurate as possible. Dose uniformity often aims to be accurate to better than 5%.
h ^{60}Co has main γ-ray peaks at 1.17 and 1.33 MeV.
i   A smoothing capacitor is used to smooth out the supply mains half-cycles and thus produce an x-ray beam of stable energy and intensity.
j   The unit of dose is the gray (Gy). Centi-grays (cGy) are also used.
k   In beam hardening a filter is used to increase the average energy of the x-ray beam.
l   The purpose of conformal radiotherapy is to get the best match of treatment and treated volume. The treatment beam conforms to body contours and can be modified in intensity and geometry to optimize the treatment volume.
m   No, a linear accelerator produces very short (μs) pulses of radiation at a few hundred pps.
n   Measurement of the x-ray beam at the output of a linear accelerator.
o   No, an x-ray tube is only about 1% efficient and produces a lot of waste heat.
p   Skin sparing is what can be achieved with high-energy beams where the maximum dose is below the surface of the skin.
q   Yes, 200 cGy min^{-1} is a typical output for a linear accelerator.
r   A collimator is used to narrow the x-ray beam.

s   Tungsten is a suitable target material in an x-ray tube because it has a high atomic number and a high melting point.

t ^{60}Co has a shorter (5.26 years) half-life than radium (1620 years).

u   Brachytherapy is short-range radiotherapy as opposed to long-range teletherapy.

v   In 'remote after loading' interstitial sources are kept shielded and then automatically delivered along tubes to the treatment site. The purpose is to reduce radiation dose to staff and patient visitors.

w   In interstitial radiotherapy the sources are implanted directly into the diseased tissue.

BIBLIOGRAPHY

Bomford C K, Kunkler I H and Sherriff S B 1993 *Walter and Miller's Textbook of Radiotherapy* 5th edn (Edinburgh: Churchill Livingstone)

Cherry P and Duxbury A 1998 *Practical Radiotherapy, Physics and Equipment* (London: Greenwich Medical Media Ltd)

Greene D 1986 *Linear Accelerators for Radiotherapy* (Bristol: Hilger)

ICRU Report 38 1985 *Dose and Volume Specification for Reporting Intracavitary Therapy in Gynaecology* (Bethesda: ICRU)

ICRU Report 50 1993 *Prescribing, Recording and Reporting Photon Beam Therapy* (Bethesda: ICRU)

Klevenhagen S C 1985 *Physics of Electron Beam Therapy* (Bristol: Hilger)

Merideth W J and Massey J B 1977 *Fundamental Physics of Radiology* (Bristol: Wright)

Mould R F 1985 *Radiotherapy Treatment Planning* 2nd edn (Bristol: Hilger)

Williams J R and Thwaites D I (eds) 1993 *Radiotherapy Physics—In Practice* (Oxford: Oxford University Press)

CHAPTER 22

SAFETY-CRITICAL SYSTEMS AND ENGINEERING DESIGN: CARDIAC AND BLOOD-RELATED DEVICES

22.1. INTRODUCTION AND OBJECTIVES

Many of the devices which we use in life can be a source of hazard, but the importance or magnitude of this hazard varies widely between devices. A can opener can certainly be a source of hazard in that we may cut ourselves on it or perhaps be electrocuted by it, but nothing dire happens if the opener fails to work. We will simply find another one or decide to eat something else that is not in a can.

However, when we fly in an aircraft we are in a situation where we are using a device whose failure can be catastrophic. The whole device is a safety-critical system, although not all of the components on board fall into this category. Failure of the public address system is not likely to endanger the aircraft whereas failure of the fuel supply or electrical power system might well do so.

In assessing the safety of a device we need to take into account the importance of the device and what will happen if it does not work. This is true of medical devices as well as household devices or transport systems. The use of a transcutaneous nerve stimulator (TENS) device to give relief from chronic pain may be very important to a patient, but their life is not likely to be threatened by failure of the device. However, if an implanted prosthetic heart valve fails suddenly then the patient is very likely to die, so we describe a prosthetic heart valve as a *safety-critical device*.

Quite obviously more engineering attention needs to be given to a safety-critical device than to other devices. An aircraft will have back-up power supplies and computers to reduce the chance of failure. Building in back-up components into a heart–lung machine may not be very easy, but nonetheless we must consider very carefully the possibilities of failure and how we can minimize these when we design such equipment.

This chapter is about safety-critical systems in health care. Examples of the questions we hope to address are:

- Why can some medical equipment be regarded as 'safety critical'?
- How can failure be predicted?
- How can the effects of equipment failure be minimized?
- Can safety prejudice performance?

When you have finished this chapter, you should be aware of:

- What questions lead to a definition of a safety-critical system?
- How a modern cardiac pacemaker performs.
- The various designs of prosthetic heart valve.

- The method of operation of a cardiac defibrillator.
- Safety-critical features of a haemodialysis system.

The definition of a safety-critical device should relate to the consequences of failure of the device. Failure might be either a failure to function in a therapeutic device, or a failure to give the right answer in a diagnostic device where the consequences can be life threatening.

Here are some examples of safety-critical systems:

Aircraft	Surgical robots
Lifts	Telesurgery
Cars	Pacemakers
Heart valves	Defibrillators
Implants	Hydrocephalus shunts
Haemodialysis equipment	Infusion pumps
Computer software	Lasers
Linear accelerators	

We can list some of the questions we need to ask when considering the safety of a system:

What if it fails to function?
Does it fail safe?
What is the quality of the materials and components?
What are the constructional standards?
What are the maintenance requirements?
What margins of safety are allowed for?
What is the reliability?

22.2. CARDIAC ELECTRICAL SYSTEMS

We will continue our consideration of safety-critical systems by dealing with the design of two very common electromedical devices. The first is the cardiac pacemaker and the second the defibrillator.

22.2.1. Cardiac pacemakers

Electrical stimulators are widely used: physiotherapists use them in order to exercise muscles; anaesthetists test for muscular relaxation during surgery by observing the response to stimulation of a peripheral nerve. A growing use of peripheral nerve stimulators is for the relief of pain, although the mechanism for this effect is poorly understood. Cardiac pacemakers and defibrillators are used to stimulate cardiac muscle directly. The pacemaker corrects for abnormalities in heart rate, whereas cardiac defibrillators are used to restore a fibrillating heart to normal sinus rhythm. The way in which electricity interacts with tissue was covered in Chapter 8 and in section 10.2 of Chapter 10 neural stimulation was described.

Rhythmic contraction of the heart is maintained by action potentials which originate at a natural cardiac pacemaker. There are actually two pacemaker areas in the heart, the sinoatrial (SA) node and the atrioventricular (AV) node, but the SA node normally dominates because it has the higher natural frequency. The normal course of events is that an impulse from the SA node is propagated through the myocardium, spreading over the atria to the AV node where there is a small delay before the impulse is conducted over the ventricles causing depolarization of the musculature. The normal cardiac rhythm is called sinus rhythm.

Any defects in conduction of the cardiac impulse can cause a change in the normal sinus rhythm and this is called an arrhythmia. Heart block occurs when the conduction system between atria and ventricles fails. This will not usually stop the heart because other pacemaking areas of the ventricles will take over or, if the blockage is not complete, some impulses may get through from the atria, but the heart rate will fall. This is called bradycardia (slow heart rate) and it may mean that the heart cannot supply the body's demands and so dizziness or loss of consciousness may occur.

There are three types of heart block:

- In first-degree block the delay at the AV junction is increased from the normal 0.1 to 0.2 s.
- In second-degree block some impulses fail to pass at all but a few get through to the ventricles.
- In complete block no impulses get through and so the ventricles pace themselves, but at a very much reduced heart rate of typically 40 beats per minute.

In all these cases an artificial pacemaker can be used to increase the heart rate to a level where the cardiac output is adequate to meet the body's needs.

The 'His bundle' which connects the atria and ventricles may be damaged by any one of several causes, such as poor blood supply to the bundle, ageing or accidental damage during surgery. Complete heart block can give rise to occasional fainting attacks and this is then described as Stokes–Adams syndrome. If left untreated the prognosis is poor, with 50% mortality after one year.

Brief history of pacemakers

The heart can be stimulated by applying an electrical current for a short time and then repeating this process about once every second. If the current is applied directly to the heart by a pair of wires the current required is only a few milliamperes and can be supplied by a battery. However, the first cardiac pacemakers used in the early 1950s were external devices which applied the current to cutaneous chest electrodes. They were effective but required quite high currents and were painful to the patient and could cause chest burns.

In 1958 the first implanted pacemakers were used. They used stainless steel electrodes sewn onto the myocardium and were implanted at the time of cardiac surgery. The process of implantation required major surgery with associated risks. A major improvement, first tried in about 1965, was to introduce electrodes into the cardiac chambers by passing them along a large vein. By passing the electrodes through the vena cava and right atrium into the right ventricle the need for major surgery was removed.

Very many technical improvements have been made to pacemakers since 1965 and this has improved the reliability and efficacy of the devices, but the basic principles have not changed. There are about 250 000 devices implanted worldwide each year so the financial implications are great.

Output requirements and electrodes

Internal pacemakers typically apply a rectangular pulse of 1 ms duration and amplitude 10 mA to the heart. As the resistance presented by the electrodes and tissue is about 500 Ω we can calculate the power requirement:

$$\text{power} = I^2 R = 5 \times 10^{-2} = 50 \text{ mW}.$$

This is the power required during the pulse. As this pulse of 1 ms duration is repeated at about 1 s^{-1} (60 bpm) we can calculate the average power as 50 μW.

If we supply the pacemaker from a battery then we must know what output voltage we might require. In the above example the pulse amplitude is 5 V (10 mA × 500 Ω) so we could use a 5 V battery. The capacity of a battery is often quoted in ampere hours (A h) and a typical small cell will have a capacity of 1 A h. Now we can calculate the average output current in the above example as 10 μA as the 10 mA flows for one thousandth of each second. Our battery could supply this current for $1/10^{-5}$ h, i.e. about 11 years.

This 'back of the envelope calculation' we have just made shows that we can hope to power a pacemaker for a very long time from a small battery. Our calculation did not take into account the power requirements of the circuitry of the pacemaker, but even if we allow another 10 μA for this the battery life should still be more than 5 years. The output pulses have to be applied to the tissue through an electrode; actually two electrodes as a circuit has to be produced. The cathode is the electrode placed in the ventricle (see section 10.21) as this is where the stimulation is required. Very many types of electrode have been used for pacemakers using metals such as platinum, silver, stainless steel, titanium as well as various alloys. Carbon electrodes have also been used. The subject of electrodes is complex and too difficult to treat here. The electrode can be thought of as a transducer because it has to allow electrons flowing in the pacemaker wires to give rise to ionic flow in the tissue (see section 9.2). This reaction must be stable and non-toxic. Electrodes vary in area between about 10 and 100 mm^2. In all cases a capacitor is inserted in series with the output of the pulse generator in order to prevent any accidental application of DC to the electrodes and also to ensure that in normal operation there is no net current flow into the tissue and so no electrolysis.

Types of pacemaker

A pacemaker must consist of a pulse generator and electrodes. Most pacemakers are implanted or internal types where the entire device is inside the body. However, external pacemakers also exist where the pulse generator is external to the body and the electrodes are located either on or within the myocardium.

External pacemakers can be used on patients with a temporary heart arrhythmia that might occur in critical post-operative periods or during cardiac surgery.

Internal pacemakers are implanted, with the pulse generator put in a surgical pouch often below the left or right clavicle (figure 22.1). The internal leads may then pass into the heart through the cephalic vein.

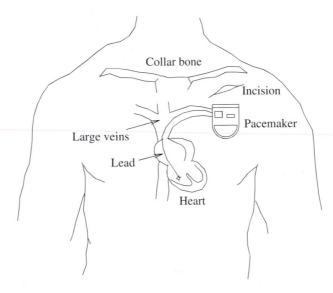

Figure 22.1. *The transvenous lead from the pacemaker enters the subclavian vein and is guided under x-ray control into the heart. The pacemaker is installed in a subcutaneous pouch.*

The simplest type of pacemaker produces a continuous stream of output pulses at about 70 bpm. There are many disadvantages to this simple approach, one being that the heart rate will not vary in response to what the patient is doing. Another disadvantage is that power may be wasted if heart block is not complete because some beats could occur naturally without the pacemaker. In addition this competition between naturally occurring beats and the pacemaker output may not lead to the most effective cardiac performance. These disadvantages to fixed-rate pacing (often referred to as competitive or asynchronous pacing) have led to the development of a range of more complex devices. These are illustrated in figure 22.2.

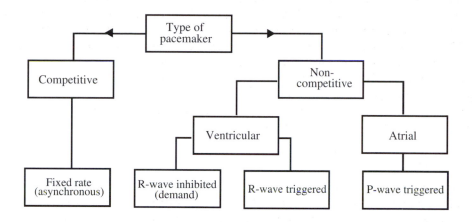

Figure 22.2. *Categories of pacemaker.*

The alternative to fixed-rate or competitive pacing is non-competitive pacing. In this case the pacemaker records the ECG produced by the heart and produces an output in response to this signal. Non-competitive types can be subdivided into ventricular- and atrial-triggered devices (see figure 22.2). Atrial-triggered devices produce an output triggered by the *P-wave* of the ECG (see section 16.2) which is generated by the atria and will not be affected by the heart block. Ventricular-triggered devices use the *R-wave* of the ECG in one of two ways. In a demand-type pacemaker an output pulse is only produced in the absence of a naturally occurring R-wave, i.e. the R-wave is used to inhibit the output of the pacemaker. If the pulse rate falls below the pre-set rate of the pacemaker then output pulses will again be produced. However, in a standby R-wave triggered device an output pulse is produced in response to every R-wave and if one does not occur when expected then the pacemaker will generate one.

For testing purposes at the time of implantation and to enable checks, such as the state of the battery, to be carried out periodically most demand pacemakers can be set into a fixed-rate mode. This is often done by means of a magnet which actuates a reed relay inside the pacemaker. The magnet is placed on the skin above where the pacemaker is implanted.

Ventricular demand-type pacemakers are the most commonly used. However, atrial-triggered devices are used in complete heart block where the normal vagal and hormonal control of the atria allows heart rate to vary in response to demand. In order to record the P-wave electrodes have to be placed in the atria in addition to the pacing electrodes in the ventricles. A block diagram of a typical pacemaker is shown in figure 22.3. The timing circuit, which consists of an *RC* network, reference voltage and comparator, determines the basic pacing rate of the pulse generator. The pulse width circuit determines the output pulse duration and a third *RC* network gives a delay to limit the maximum rate of pulses. The sensing circuit is inhibited during the output pulse in order to stop overload of the circuitry. The sensing circuit detects a spontaneous R-wave and resets the timing capacitor so no output is produced. The voltage monitor has two functions. One is to detect a low battery voltage and use this to reduce the fixed-rate output as a means of signalling a low battery. The other is to increase the output pulse width when the battery voltage falls so that the output pulse energy remains constant.

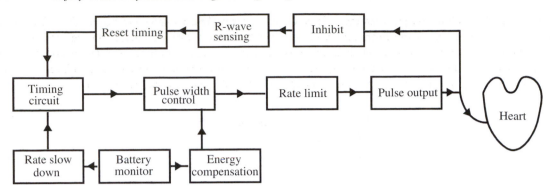

Figure 22.3. *Block diagram of a typical pacemaker.*

There are agreed international codes for the description of pacemakers. These classify devices in terms of the heart chamber which is paced, the heart chamber which is sensed and what type of response is made, i.e. inhibition or triggering. However, the advent of pacemakers with microprocessors and the possibility of very complex algorithms to determine the output of the pacemakers have made it very difficult to describe devices by means of a simple code.

Power sources

It is essential that the pacemaker has a reliable power source which will last as long as possible. It is possible to replace an exhausted pacemaker but surgery is necessary and battery failure if it is not detected at an early stage can give rise to a life-threatening situation. The ideal power source should be very small, have a long life, i.e. a high capacity, be unaffected by body temperature, be easy to test so that exhaustion can be predicted, be cheap, be unaffected by autoclaving and give an output of at least 5 V. No gases must be produced. Types which have been used include mercury cells, nuclear-powered thermoelectric generators and lithium cells.

Many types of power source have been proposed and several used in pacemakers. The most exotic is the nuclear-powered pacemaker which contains a capsule of plutonium (^{238}Pu) which has a half-life of 87 years. This radioactive isotope emits α-particles which generate heat when they are absorbed within the pacemaker; the heat can then be used to generate electricity from a large number of thermocouples which form a thermopile. This type of power source is actually very reliable and has a life of more than 20 years, but there are many disadvantages, so that whilst some remain implanted new ones are not used. The disadvantages are firstly, that plutonium is very toxic so that the encapsulation has to ensure that there is zero possibility of leakage and secondly, that the cost is high. Plutonium is so toxic that cremation of a device would be a significant hazard so that strict control has to be exercised over the disposal of nuclear-powered devices. The radiation dose to the patient is actually very small.

For many years mercury cells were widely used for powering pacemakers but unfortunately their life was only about 2 years. A life of at least 5 years had been anticipated but the deterioration within the body proved to be more rapid than anticipated so that by about 1970 the mercury batteries were the most severe constraint on pacemaker life. In the early days lead failure and problems of encapsulation of the electronics had been the major problems but these had been largely overcome by 1970. People then tried all sorts of alternatives such as piezoelectric devices, self-winding watch-type mechanisms, fuel cells, rechargeable cells and electrochemical generation from body fluids. None of these were very successful.

The battery most commonly used today is the lithium iodine cell. This offers a much longer life than the mercury cell and has the advantage that no gases are given off so that it can be hermetically sealed. It provides a higher voltage output than the 1.35 V of the mercury cell, as shown in figure 22.4. The stable discharge curve is an advantage in that the output of the pacemaker will not change during the life of the battery. However, it is also a disadvantage in that it is quite difficult to assess when a battery is coming to the end of its safe life. It is for this reason that quite complex procedures have to be used to check the batteries, including switching the pacemaker into fixed-rate mode so that the drain on the battery is constant.

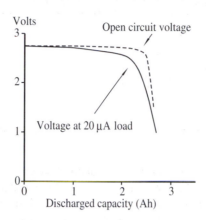

Figure 22.4. *Typical discharge curve of a lithium iodine cell.*

Safety requirements

Many of the requirements for safe operation of a pacemaker have already been mentioned. In the design of the pulse generator attempts are made to ensure that under any single-fault condition the generator will revert to a fixed-rate mode. Attention has also to be given to the effect of electromagnetic interference (see sections 10.4 and 22.2.2) on the performance of demand-type pacemakers. Patients with pacemakers are warned not to expose themselves to the outputs of powerful radio transmitters and devices such as airport security checking devices. They should also avoid magnetic resonance scanners which produce high magnetic and electromagnetic fields. However, inadvertent exposure is possible so that the design of the pacemaker should be such that it will revert to a fixed-rate mode when interference is present.

Hazards do not always concern the electrical output of the pacemaker. The leads from the pacemaker to the heart can also malfunction. There can also be problems of compatibility between the pacemaker encapsulation and body tissues (see Chapter 4) and of course sterilization is necessary before implantation of the pacemaker.

Five methods have been used for sterilization of pacemakers:

- Cold chemicals.
- Radiation.
- Ethylene oxide gas.
- Steam, 120 °C for 15 min is commonly used.
- Dry heat at about 150 °C.

Radiation can affect the electronics of a pacemaker, particularly MOSFET devices. Autoclaving at 120–130 °C and dry heat will damage most cells and some electronic components. This leaves gas sterilization and cold chemicals. Of these gas sterilization is best but it is not always available.

The technology of pacemakers has advanced rapidly over the past 40 years with advances in power sources, biomaterials, sensors and computing. Many pacemakers now incorporate microprocessors and so can be regarded as intelligent implants. These can sense a change from normal cardiac performance and select a suitable therapy which may be a type of pacing, but also perhaps a defibrillation pulse or release of a therapeutic drug. Physiological variables such as temperature, pH, Pco_2, Po_2, respiration, glucose or blood pressure can be sensed and used by the implanted microprocessor to select the appropriate therapeutic action.

22.2.2. Electromagnetic compatibility

Many items of medical equipment now use quite sophisticated electronic control systems. In principle these can both be affected by external electromagnetic fields and also produce such interfering fields. Pacemakers can obviously be a hazard to the patient if they are affected by interference. These problems are covered by national and international agreements on electromagnetic compatibility (EMC). In Chapter 10 we discussed how various types of interference can affect bioelectric measurements. However, a whole range of electromedical devices may be affected by electromagnetic interference.

Examples of equipment which might be affected by external fields:

Infusion pumps	Computers
Patient monitoring equipment	Linear accelerators
Demand pacemakers	EMG/EEG/ECG
Surgical robots	Hearing aids
Defibrillators	Software
Lasers	

Equipment which might produce interference includes:

Computers	Radio communications
Mobile phones	Thermostats
Surgical diathermy/electrosurgery	Electrostatic materials
Linear accelerators	Transformers
Physiotherapy diathermy	

EMC is a growing problem which might be tackled in various ways:

> Design of the equipment
> Earthing/shielding/filtering/separation
> Design of the hospital

In the case of pacemakers there is an intrinsic difference in the susceptibility of unipolar and bipolar devices to interference. The bipolar device is less likely to be affected by interference because the electrodes are close together and hence require a high field gradient to produce an interfering voltage.

22.2.3. Defibrillators

Our second detailed example of a safety-critical system is the defibrillator. Defibrillators are devices that are used to apply a large electric shock to the heart. They are used to restore a normal sinus rhythm to a heart which is still active but not contracting in a co-ordinated fashion. The cause of fibrillation is commonly ischaemia of heart tissue but less common causes are electric shock, drugs, electrolyte disorders, drowning and hypothermia. The use of a defibrillator on a patient following a heart attack is an emergency procedure, as the pumping action of the heart has to be restarted within a few minutes if the patient is to survive. The defibrillator is therefore a 'safety-critical' device; if it fails to work when required then the patient will die.

Defibrillators have a long history in that some animal work was done in 1899, but emergency human defibrillation was not carried out until the 1950s. They have been in widespread use since the 1960s.

Fibrillation

Cardiac muscle is intrinsically active in that it will contract periodically in the absence of any neural connections. If a piece of cardiac muscle is removed from the heart and placed in an organ bath then oscillating electrical potentials can be recorded from the piece of tissue. Now in the intact heart all the pieces of cardiac muscle interact with each other such that they all oscillate at the same frequency and so contract regularly at that frequency. However, if part of the heart muscle is damaged or disrupted then the interaction can be disrupted and fibrillation can occur. All the pieces of cardiac muscle are still oscillating but at different frequencies so that there is no co-ordinated contraction.

Either the ventricles or the atria can fibrillate but the consequences to the patient are very different. Under atrial fibrillation the ventricles still function but with an irregular rhythm. Because atrial filling with blood does not depend upon atrial contraction there is still blood for the ventricles to pump, so that whilst the patient may be aware of the very irregular heart beat blood is still circulating. Ventricular fibrillation is much more dangerous as the ventricles are unable to pump blood so that death will occur within a few minutes. Ventricular fibrillation is not self-correcting so that patients at risk of ventricular fibrillation have to be monitored continuously and defibrillation equipment must be immediately to hand.

Fibrillation is obvious in the resulting ECG as shown. Figure 22.5(*a*) shows a normal ECG and figure 22.5(*b*) the result of cardiac fibrillation. There are still potential changes during fibrillation but they are apparently random and the amplitude is less than occurs during normal sinus rhythm. Care has to be taken in recognizing fibrillation from the ECG as other conditions may change the appearance of the ECG waveform. Figures 22.5(*c*)–(*e*) show the appearance of atrial fibrillation, atrial flutter and ventricular tachycardia.

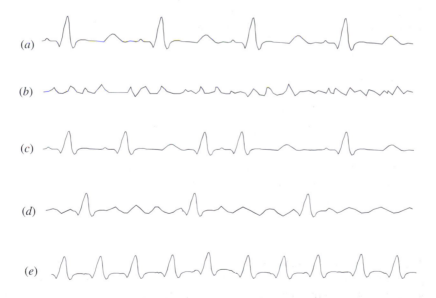

Figure 22.5. *Diagrams of: (a) a normal ECG; (b) fibrillation; (c) atrial fibrillation; (d) atrial flutter; (e) ventricular tachycardia.*

Principles of defibrillation

A defibrillation shock aims to totally stop the heart, or at least to stop enough cells to inhibit fibrillation, in the hope that the heart will restart in an orderly fashion after the shock. If sufficient current is used to stimulate all the musculature of the heart then when the shock stops all the heart muscle fibres enter their refractory period at the same time, after which normal heart rhythm may resume.

Of the two major categories of muscle, striated (skeletal) and smooth, cardiac muscle is most like smooth muscle. Skeletal muscle is composed of long muscle fibres and does not produce activity unless it receives impulses along an associated motor nerve. Smooth muscle is composed of much shorter and smaller cells and it is intrinsically active. Neural impulses and circulating hormones will influence the activity but smooth muscle can contract in isolation. Cardiac muscle is intrinsically active.

Skeletal muscle can be electrically stimulated and will produce an action potential in response to a stimulus of duration 100 μs or less (see figure 10.4). Cardiac muscle is more difficult to stimulate, in part because it is intrinsically active so that our ability to stimulate the muscle depends upon what the muscle is doing at the time we apply our stimulus. Longer duration pulses are required to stimulate cardiac muscle than striated muscle, although high-amplitude stimuli of short duration will have the same effect as lower amplitude impulses of longer duration. This is illustrated in figure 22.6 in the form of a curve often referred to as a strength–duration curve. The 'current' curve shows that the amplitude of current required for defibrillation decreases as duration increases, but that there is a minimum current required whatever the stimulus duration. This is called the rheobase. Take note that the current required is several amps, which is a very high current to apply to the human body.

Figure 22.6 also shows the charge (Q) and energy (E) associated with the current pulse. The charge is the product of the current I and the pulse duration D. The energy is $I^2 R D$ where R is the resistance into which the current is delivered. It can be seen that there is a minimum energy required at a pulse width of about 4 ms.

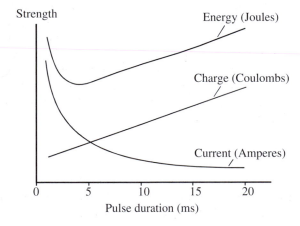

Figure 22.6. *A strength–duration curve for defibrillation applied across the thorax. Curves showing the associated charge and energy are also shown.*

If

$$\text{EOA} = \frac{Q_{\text{rms}}}{51.6\sqrt{\overline{\Delta P}}}$$

where a and b are constants, then

$$I = \frac{a}{D} + b$$

and

$$\text{energy}(E) = \left(\frac{a}{D} + b\right)^2 RD = \left(\frac{a^2}{D^2} + b^2 + \frac{2ab}{D}\right) RD$$

$$\frac{\mathrm{d}E}{\mathrm{d}D} = -\frac{a^2 R}{D^2} + b^2 R.$$

This will be zero at the turning point in E and hence

$$\frac{a^2}{D^2} = b^2 \qquad \text{therefore} \quad D = \pm\frac{a}{b} \quad \text{and} \quad I = 2b.$$

The minimum energy requirement will be for a duration such that the current is double the rheobase.

One reason for the large current which is required for defibrillation is that only a small fraction of the current applied to the chest will actually flow through the heart; if defibrillation is applied directly to the heart during open heart surgery then smaller currents are required.

Another important variable related to the current and pulse duration required for defibrillation is the size of the subject. It has been shown that larger animals require higher currents to defibrillate than smaller ones. It is also true that larger animals are more likely to suffer from fibrillation than smaller ones. Large people require higher defibrillation currents than small people and, in particular, children require low currents.

Pulse shapes. The stimulus waveform which is used in most striated muscle stimulators, such as those used in physiotherapy, is rectangular. However, defibrillators produce more complex waveforms. The reason for this is partly physiological, but it is also related to the way in which the impulses are generated. It is actually quite difficult to generate a very high current rectangular pulse, whereas it is relatively easy to charge up a capacitor to a high voltage and then discharge it through the patient. However, the current waveform then produced has a high peak and a long exponential tail. There is evidence that the long exponential tail can refibrillate the heart and so reverse the defibrillation. For this reason a damped exponential current waveform is used as illustrated in figure 22.7. The pulse width is about 4 ms which corresponds approximately to the minimum energy requirement which was shown in figure 22.6. In the next section we will consider how waveforms of the shape shown in figure 22.7 can be generated.

Design of defibrillators

About 50 A of current is required for defibrillation across the chest. Now if we know the resistance into which this current has to pass we can calculate the power requirement. The resistance will depend upon the size of electrodes used and the size of the patient but a typical figure is 50 Ω:

$$\text{power required for defibrillation} = I^2 R = 125 \text{ kW}.$$

This is a large amount of power. The first defibrillators in clinical use were called AC types and they simply used the mains supply fed through a transformer and a switch for defibrillation. Unfortunately the maximum power which can be drawn from most mains supply outlet sockets is about 15 kW which is just about sufficient to defibrillate a child but not an adult.

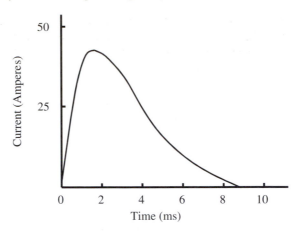

Figure 22.7. *Output waveform from a DC defibrillator.*

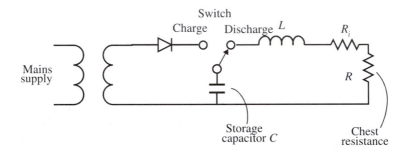

Figure 22.8. *Basic circuit diagram of a DC defibrillator.*

The solution to the problem of limited power availability from a mains supply socket is the DC defibrillator. This uses a large capacitor, which is charged to a high voltage from the mains supply and then discharged rapidly through the patient. A circuit diagram is given in figure 22.8, where the storage capacitor is marked as C. We can determine the necessary size of this capacitor relatively easily.

The voltage V we require is at least 2500 V to produce a current of 50 A into 50 Ω.

The duration of the discharge must be about 4 ms, so the time constant formed by C and R must be about 4 ms. For R equal to 50 Ω this gives C as 80 μF.

The *energy* stored in this capacitor when charged to 2500 V is given by $\frac{1}{2}CV^2$ which is 250 J. In fact, most defibrillators have a maximum stored energy of 400 J, although the maximum output is only required on large patients.

In the previous section on pulse shapes it was argued that a pulse output with a long tail is undesirable because this can cause refibrillation. In order to remove the long tail of the capacitor discharge an inductor L is usually added in series with the defibrillator output as shown in figure 22.8.

We can calculate the shape of the output current as follows. First, by applying Kirchhoff's law we obtain

$$L\frac{\mathrm{d}^2 i}{\mathrm{d}t^2} + (R_{\mathrm{i}} + R)\frac{\mathrm{d}i}{\mathrm{d}t} + \frac{i}{C} = 0 \tag{22.1}$$

where i is the current, L the inductance, C the storage capacitor, R the resistance presented by the patient and R_i is the internal resistance of the defibrillator circuit. The solution of this differential equation has three forms; the first is oscillatory, the second an aperiodic discharge and the third, the fastest aperiodic discharge when the circuit is critically damped. The solution in this case is given by

$$i = \frac{CV(R + R_i)^2}{4L^2} \, t e^{-((R+R_i)/2L)t} \tag{22.2}$$

and

$$L = (R + R_i)^2 \frac{C}{4}. \tag{22.3}$$

This gives a typical value for L of 40 mH.

Stored and delivered energy. Not all the energy stored in the capacitor will be delivered into the chest of the patient because of losses within the output circuit of the defibrillator. The main source of loss is the resistance of the inductor and this is represented as R_i in figure 22.8. Now the *stored energy* in the capacitor charged to voltage V is $\frac{1}{2}CV^2$, but is easy to show that the *delivered energy* to the patient will be $\frac{1}{2}CV^2 \times R/(R + R_i)$. Typically the delivered energy will be about 10% less than the stored energy so that a defibrillator of 400 J maximum stored energy might actually deliver a maximum of 360 J.

Electrodes

Obviously, electrodes have to be placed on the thorax in the best position to ensure that current will flow through the myocardium. If the electrodes have too high an impedance (see section 9.2.3) or are not optimally placed then an unnecessarily high pulse output will be required for defibrillation. The energy output from a defibrillator is sufficient to cause tissue burns so these might also arise if electrode contact impedance is too high.

Chest electrodes and direct heart electrodes usually consist of bare metal electrodes made of a non-corrosive material. Very often output switches are incorporated into the back of the electrodes for easy use by the operator. Typical chest electrode size is 10 cm diameter. Direct heart electrodes are usually smaller and look a bit like spoons on the end of handles so they can be placed in contact with the myocardium. Chest electrodes are usually used with a conductive jelly to reduce contact impedance to about 50–100 Ω. If jelly is not used then burns can occur and defibrillation will be less effective.

When electrodes are placed on the chest, they are usually applied with both electrodes on the anterior chest wall. One electrode is placed over the apex of the heart, which is about the fifth intercostal space in the midclavicular line on the left side of the chest. The other is placed in the second intercostal space adjacent to the sternum.

Use of defibrillators

There are basically three uses for defibrillators:

- The first is direct defibrillation of the heart during surgery. During cardiac surgery the heart may spontaneously fibrillate or the surgeon may intentionally produce fibrillation. The maximum output used to defibrillate directly is about 50 J.
- The second major use of defibrillators is for cardioversion. This is a synchronized shock which is applied across the chest to correct for atrial fibrillation, atrial flutter or ventricular tachycardia. Energies from about 20–200 J are used for cardioversion.

- The third use for defibrillators is emergency defibrillation in cases of ventricular fibrillation. Cardiopulmonary resuscitation is often used to keep the patient alive until the defibrillator is ready for use. Often a first pulse of 200 J will be followed by higher shocks if the first is not successful in restoring a normal sinus rhythm.

Some defibrillators are battery powered and sufficiently small to be portable. Many have ECG/EKG monitoring included and other features, such as impedance detection circuitry to detect poor electrode contact, recording of the actual delivered energy and various alarms to detect dangerous conditions, are also included. Most units have the ECG/EKG monitoring facility and a synchronization circuit, to enable the operator to be certain of the diagnosis. Obviously the outcome to be avoided is the precipitation of fibrillation in a normal heart.

Implanted defibrillators

Totally automatic implantable defibrillators have been developed in the past few years. These are like pacemakers, but treat tachycardias rather than bradycardias and can defibrillate with a pulse of up to about 30 J. Both ventricular fibrillation and ventricular tachycardia are treated, because of the rapid fatality of ventricular fibrillation and the high frequency with which ventricular tachycardia can develop into ventricular fibrillation. These devices have been found to greatly reduce mortality in patients known to be at high risk from sudden cardiac death.

The implanted defibrillator contains sensors to detect cardiac electrical activity in the same way as demand pacemakers and signal processing in order to make the correct decision as to when treatment is required. This is obviously a safety-critical system. Implanted defibrillators contain ECG/EKG sensing leads, a pulse generator, signal processing electronics and electrodes which are placed on the epicardium or in the right ventricle. Obviously such a device makes great demands upon the battery power supply which has to supply large output pulses without impairing its ability to supply the decision making electronics. Current devices are able to supply about 100 shocks of 25 J before the battery is exhausted.

22.3. MECHANICAL AND ELECTROMECHANICAL SYSTEMS

We continue our examples of safety-critical systems by considering some mechanical and electromechanical medical devices.

Disease can be addressed in one of two ways:

- by returning a malfunctioning organ to health using chemical or physical agents, or
- by substituting a functional counterpart, either after removing the sick organ, or ignoring it entirely.

Beyond a certain stage of failure, it is often more effective to replace a malfunctioning organ than seeking in vain to cure it. This has given rise to 'spare parts medicine' where a whole range of implanted artificial organs have been developed. These implants must often be considered as *safety-critical systems*.

The use of implants gives rise to very many complex issues. These concern medical, social, managerial, economic, legal, cultural and political aspects. We will only consider the engineering design aspects. Table 22.1 lists some of the systems which have been developed over the past few decades. We will just consider the design aspects of three of these developments. The first is heart valve substitutes, the second cardiopulmonary bypass and the third haemodialysis systems.

Table 22.1. *Current status of organ replacement technology. (Adapted from The Biomedical Engineering Handbook ed J D Bronzino (Boca Raton, FL: Chemical Rubber Company.)*

Clinical standing	Artificial organ	Transplantation
Generally accepted	Heart–lung machine	Blood transfusion
	Large joint prostheses	Corneal transplants
	Bone fixation systems	Banked bone
	Cardiac pacemakers	Bone marrow
	Large diameter vascular grafts	Kidney, cadaveric donor
	Prosthetic heart valves	Heart
	Intra-aortic balloon pumps	Liver
	Implantable lenses	Heart/lung
	Hydrocephalus shunts	
	Dental implants	
	Skin or tissue expanders	
Accepted with reservations	Maintenance haemodialysis	Kidney, living related donor
	Chronic ambulatory	Whole pancreas
	Peritoneal dialysis	
	Breast implants	
	Sexual prostheses	
	Small joint prostheses	
	Extracorporeal membrane	
	Oxygenation in children	
	Cochlea prostheses	
Limited application	Implantable defibrillator	Pancreatic islets
	ECMO in adults	Liver lobe or segment
	Ventricular-assist devices	Cardiomyoplasty
	Artificial tendons	
	Artificial skin	
	Artificial limbs	
Experimental	Artificial pancreas	Gene transfer
	Artificial blood	Embryonic neural tissue
	Intravenous oxygenation	Bioartificial pancreas
	Nerve guidance channels	Bioartificial liver
	Total artificial heart	
Conceptual stage	Artificial eye	Striated and cardiac muscle
	Neurostimulator	Functional brain implants
	Blood pressure regulator	Bioartificial kidney
	Implantable lung	
	Artificial trachea	
	Artificial oesophagus	
	Artificial gut	
	Artificial fallopian tube	

22.3.1. *Artificial heart valves*

How do natural heart valves work?

In the normal human heart, the valves maintain a unidirectional flow of blood with minimal frictional resistance, whilst almost completely preventing reverse flow. They act passively: the moving parts of the valve, the tissue leaflets, have negligible inertia and open and close in response to pressure changes in the blood generated by the contraction of the surrounding myocardium (see section 2.6). All four valves sit within a flat fibrous supporting structure which is part of the fibrous skeleton of the heart. This separates the ventricles from the atria and the ventricular outflow conduits (the aorta and pulmonary artery), and is perforated by four openings, two larger ones for the atrio-ventricular valves (mitral and tricuspid valves) and two smaller ones for the aortic and pulmonary valves (figure 22.9).

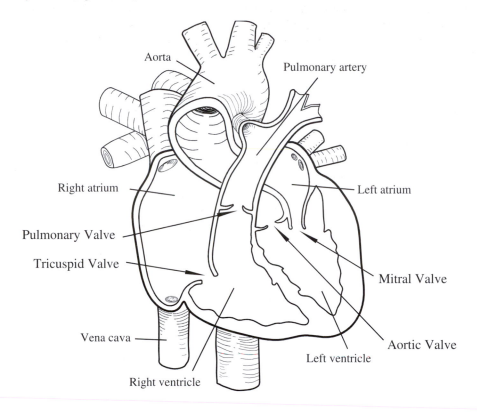

Figure 22.9. *Anatomy and structure of heart valves.*

Why do natural valves sometimes need to be replaced?

Whereas a normal heart valve opens passively with little measurable transvalvular pressure difference, and closes effectively with minimal leakage, diseased valves cause major haemodynamic abnormalities and an increased workload for the heart. Heart valve disease affects primarily the valves on the left side of the heart and is associated with increased leakage, increased resistance to forward flow or a combination of the two. The terms 'insufficiency', 'incompetence' or 'regurgitation' are used to indicate backflow through the closed valve. Excessive regurgitation leads to a progressive enlargement of the cardiac chamber upstream of the valve which now fills from both directions. The term 'stenosis' describes a blocking or narrowing of the valvular orifice and requires an increased amount of energy to drive the blood through. The restriction results in an abnormally high pressure difference across the valve and, with time, leads to hypertrophy of the cardiac chamber upstream of the stenotic valve.

The consequences of valve pathology will depend on the ability of the heart to adapt to the increasing demands. The severity of valve abnormalities can be assessed using ultrasonic imaging which can be used to observe retrograde and turbulent flow around heart valves *in vivo* (see section 19.7).

In cases where cardiac function is severely compromised the damaged natural valve can be replaced with an artificial substitute. This involves the replacement of the moving parts of the diseased valve. Over the years, two distinctly different types of prosthetic valves have been developed: *mechanical valves* which are made of a variety of synthetic materials and *bioprosthetic valves* made of chemically modified biological tissues. Most valve mechanisms have a rigid frame to house the moving flow occluder, whilst anchorage of the device also necessitates suitable accessory structures. The specialized terms used to describe the components of an artificial valve are defined in table 22.2.

Table 22.2. *Nomenclature for prosthetic heart valve components.*

Term	Meaning
Mechanical valve/prosthesis	Heart valve substitute manufactured entirely from man-made materials
Tissue valve/bioprosthesis	Valve substitute manufactured, in part, from chemically treated biological material
Poppet or occluder	Mobile component, typically a ball or disc, which moves to open and close valve
Housing	Assembly which retains the occluder
Frame or stent	Rigid or semi-rigid support for flexible leaflets of a tissue valve
Cusps or leaflets	Flexible components of a tissue valve which open and close in response to flow
Sewing ring	Fabric cuff surrounding stent or housing, used by surgeon to anchor the valve in place by suturing

What does valve replacement involve?

Whilst a detailed description of the operative technique is outside the scope of this book, a typical procedure is described briefly in the section which follows. In order to replace the aortic valve, the surgeon will open the chest, cutting through the sternum to expose the heart. The patient is then placed on cardiopulmonary bypass (see section 22.3.2) to maintain the oxygenation and circulation of the blood. To reduce tissue damage, the body temperature is cooled to 34 °C by passing the blood through a refrigerated heat exchanger. The ascending aorta is clamped, and the heart is arrested by potassium cold cardioplegia. This ensures that the heart is stopped in a relaxed state. The damaged aortic valve tissue is cut away and replaced by a prosthesis which is sutured into the aortic annulus below the coronary arteries. The aorta is closed and the left ventricle is vented to remove any trapped air. The aortic clamp is removed and the heart defibrillated. The patient is rewarmed and removed from cardiopulmonary bypass.

Treatment with anti-coagulants is required during cardiopulmonary bypass to prevent thrombosis and thromboembolism. In addition, patients receiving mechanical valves will require controlled anti-coagulation for the rest of their lives. Bioprosthetic valves are less inherently thrombogenic and may not require long-term anti-coagulants.

A brief history of valve development

The first clinical use of a mechanical heart valve was performed by Dr Charles Hufnagel in 1952, who partially corrected aortic incompetence by implanting an acrylic ball valve into the descending aorta. The introduction of cardiopulmonary bypass in 1953 enabled open heart procedures to be performed and, in 1960, Harken and Starr implanted ball valves enclosed in metal cages, the former in the aortic position, the latter as a mitral valve replacement (see figure 22.10).

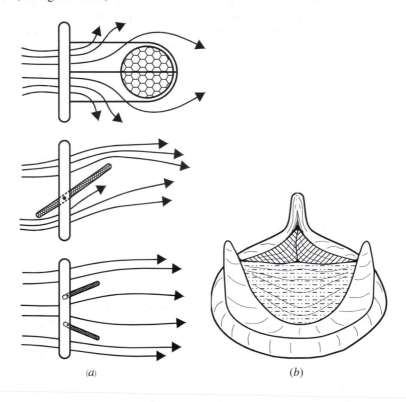

(a)   *(b)*

Figure 22.10. *(a) Three types of mechanical prosthetic valve. A caged ball valve is shown at the top, a tilting disc valve in the middle and a bileaflet valve at the bottom. (b) A porcine bioprosthetic valve.*

It was soon realized that the central occluder of the caged-ball valve presented a degree of obstruction to the blood flow. This was particularly significant in patients with a narrow aortic root. In addition, in the mitral position the high profile of the cage could impinge on the wall of a small ventricle. For these reasons, attempts were made to find less obstructive designs. In consequence, low-profile tilting-disc valves were introduced. Tilting-disc valves presented significantly less resistance to blood flow than the caged-ball valves, and appeared to have lower incidences of thrombosis and thromboembolism. The first tilting disc valves had rigid plastic (Teflon or Delrin) discs. These were replaced by pyrolytic carbon, a material which combines durability with low thrombogenicity, in 1971.

A significant development came in 1978 with the introduction of the all pyrolytic carbon valve by St Jude Medical Inc. This valve comprises two semi-circular discs or leaflets which, in the open position, present minimal flow disturbance. The success of this valve and its relatively low incidence of thrombiotic complications have led to the development of a number of other bileaflet designs which currently account for 75% of valve implants.

Although there have been significant improvements in the design of mechanical valves over the years, all mechanical valves cause flow disturbances which may ultimately led to thrombosis or thromboembolism. For this reason, patients with mechanical valves are required to undergo long-term anti-coagulant therapy.

Many valve designers have sought biological solutions to this problem using natural tissues to fabricate valves. These valves have been developed in parallel with mechanical designs. Pioneers used free-hand constructed valves made of the patient's own tissue. This technique was first reported in 1966 using fascia lata, a sheet of fibrous tissue which covers the muscles of the thigh. The valves gave poor results with premature tearing or calcification and many failed within 5 years of implantation.

The next advance came in 1969 with the introduction of bioprosthetic valves in the form of frame-mounted pig aortic valves. The tissue was chemically modified by glutaraldehyde fixation (tanning). Pig valves had two main disadvantages; they were difficult to obtain in small sizes and, as they were mounted on the inside of a support frame, they presented an obstruction to flow. In an attempt to overcome these problems a valve was introduced in which the leaflets were fashioned from glutaraldehyde-fixed bovine pericardium. In this case, the leaflets were attached to the outside of the frame thus maximizing the area of orifice available for flow. These valves could also be manufactured in a full range of sizes.

An alternative has been the use of human allografts (or homografts). These are human aortic valves which are obtained from cadavers. Prior to use they are cryopreserved or treated with antibiotics. The first allograft procedure was carried out by Ross in 1962. As allografts are normally implanted without a stent, the level of technical skill and operative time required for implantation is much greater than that required for a frame-mounted bioprosthesis. However, as the long-term results have proved encouraging, in recent years, there has been a renewed interest in their use. The use of homografts is restricted by the available supply rather than the demand.

The most recent trend for aortic valve replacement is the use of a stentless bioprosthesis. In acknowledgement of the success of the unmounted allografts, isolated or joined cusps, sometimes surrounded by a conduit of biological tissue, are sutured directly to the patient's aorta. The tissues employed are usually of bovine or porcine origin and are chemically treated. Unstented valves have improved haemodyamics when compared with traditional stented bioprostheses, as they avoid the narrowing of the orifice which is associated with the use of a stent and a sewing ring, thus giving minimal resistance to flow. However, like allografts, they are more demanding in terms of surgical skill and may prove difficult to replace should the need arise.

Research is also being carried out in an attempt to produce a leaflet valve made entirely from man-made materials. To date, attempts to manufacture flexible leaflet valves from synthetic polymers have not been

Table 22.3. *Summary of the material composition of nine key designs of mechanical valves developed over the course of 30 years.*

Year	Name	Type	Poppet	Material
1959	Hufnagel	Ball	Polypropylene	Methacrylate
1964	Starr–Edwards 1000	Ball	Silastic	Stellite
1968	Wada–Cutter	Tilting disc	Teflon	Titanium
1969	Bjork–Shiley	Tilting disc	Delrin	Stellite
1970	Lillehei–Kaster	Tilting disc	Pyrolitic carbon	Titanium
1971	Bjork–Shiley	Tilting disc	Pyrolitic carbon	Stellite
1977	Medtronic–Hall	Tilting disc	Pyrolitic carbon	Titanium
1977	St Jude Medical	Bi-leaflet	Pyrolitic carbon	Pyrolitic carbon
1991	Jyros	Bi-leaflet	Vitreous carbon	Vitreous carbon

successful. No synthetic material yet produced is proven to exhibit the flexural durability of natural valve cusps (see table 22.3).

What are the design criteria?

The aim of the valve designer is to produce a device which:

- Is durable (capable of functioning for 35–40 million cycles per year for up to 30 years).
- Has a large orifice area and presents the lowest possible resistance to blood flow.
- Creates a minimal degree of flow separation and stasis.
- Does not induce regions of high shear stress.
- Has minimal regurgitation.
- Causes minimal damage to blood and plasma constituents.
- Is manufactured from materials which are non-thrombogenic.
- Is easy to implant, is quickly incorporated into, and tolerated by, the patient's tissues.
- Creates no noise.
- Can be simply and consistently manufactured at an acceptable cost.

No current design, other than the native valve, meets all of the above criteria. All prosthetic valves are inherently stenotic when compared with the natural valve. The implantation of any artificial valve unavoidably restricts the tube-like flow paths found in the normal heart and introduces a degree of narrowing. In addition, the shapes of the natural valves change during the different phases of the cardiac cycle and the introduction of a stent or housing imposes a fixed geometry to the valve orifice. Artificial valve design, to date, has been a compromise between an unattainable ideal and the technically feasible.

To simplify matters we should concentrate on a small number of key requirements which we can go some way towards attaining.
Valves should:

- Function efficiently and present the minimal load to the heart.
- Be durable and able to function for the life-time of the patient.
- Not cause thrombus formation or promote the release of emboli.

Current designs

Currently, seven basic configurations of heart valve substitutes are produced commercially: caged-ball valves, tilting-disc valves and bi-leaflet valves, frame-mounted porcine aortic and bovine pericardial bioprosthetic valves and stentless valves. Cryopreserved human valves are also available. Data presented in 1993 showed that the total world market was of the order of 130 000 valves. The USA accounted for 60 000 implants per year and 5000 valve implants were performed every year in the UK.

Table 22.4 summarizes the advantages and disadvantages of mechanical and bioprosthetic valves.

Which valve is best?

There is no simple answer to this question and a number of factors must be considered, including the particular circumstances of the individual patient. The major advantage of mechanical valves is long-term durability. The major advantage of bioprosthetic valves is low thrombogenicity. Enhanced durability coupled with the need for life-long anti-coagulation associated with mechanical valves must be weighed against the better quality of life, freedom from anti-coagulation-related complications such as haemorrhage, but limited durability and risks of reoperation which are associated with a bioprosthetic valve. A recent overview of valve usage suggests

Table 22.4. *A comparison of mechanical and tissue heart valves.*

Valve type	Advantages	Disadvantages
Mechanical	Long-term durability	Unnatural form
	Consistency of manufacture	Patient usually requires long-term anti-coagulant therapy
Tissue	More natural form and function	Uncertain long-term durability
	Less need for long-term anti-coagulant therapy	Consistency of manufacture is more difficult *In vivo* calcification

that bioprosthetic valves are used in 40% of the patients in the USA, in 25% in the UK and in as many as 80% in Brazil.

Many surgeons recommend mechanical valves for patients below the age of 65 and for valve replacements in young patients where durability is of utmost importance. In the case of children and adolescents, calcification of bioprostheses is particularly rapid and severe. Bioprosthetic valves are considered for elderly patients, for patients in developing countries where access to anti-coagulant control may be limited and for patients who are unlikely to comply with the stringent requirements of anti-coagulation therapy. They are essential for patients for whom anti-coagulation therapy is contra-indicated, for example, in women in the early pregnancy.

It must be remembered that, even though currently available valves may not be ideal, the alternative for patients with native valve failure is progressive cardiac failure and death.

Evaluation of valve performance

A true assessment of valve performance can only be obtained from long-term clinical studies. However the initial quantitative information must be obtained by laboratory evaluation using flow simulators. These tests routinely include measurements of pressure difference for a range of simulated cardiac outputs under both steady and pulsatile flow conditions and regurgitation in pulsatile flow. A knowledge of the mean pressure difference $(\overline{\Delta P})$ across the open valve during forward flow enables the effective orifice area (EOA) to be obtained. The EOA gives a measure of the degree of obstruction introduced by the valve and may be calculated in cm^2 from the following formula given by Yoganathan (1984):

$$\text{EOA} = \frac{Q_{\text{rms}}}{51.6\sqrt{\overline{\Delta P}}}$$

where Q_{rms} is the rms flow rate in cm^3 s^{-1} over the cardiac cycle. Note the constant 51.6 that is included. It should be appreciated that this equation is only a guide to the effectiveness of a valve.

Calculations of energy loss enable estimates to be made of the total load the valve presents to the heart. Laser Doppler and flow visualization techniques provide information about flow velocities, shear and shear stress fields (see section 2.6 in Chapter 2). These data allow predications of the likelihood of damage to blood cells to be made.

Pulsatile flow testing is carried out using hydrodynamic test rigs, or pulse duplicators. These model the left side of the heart with varying levels of sophistication and attention to anatomical variation. There is no single universally accepted design and many involve a compromise between accurate simulation and ease of use. It is likely that, in the future, the use of *in vitro* techniques will be superseded by computational fluid-dynamic analyses (CFD).

Wear and durability can also be investigated in the laboratory. Valves are tested at accelerated rates of up to 20 Hz. In this way, the effects of 10 years' mechanical wear can be simulated in a period of 8 months. For mechanical valves, subsequent surface analysis enables the degree of wear to be quantified for individual valve components. Tissue valves are examined for evidence of tissue damage and tears.

22.3.2. Cardiopulmonary bypass

Concept of bypass

'Bypass' is a term employed by surgeons to indicate that fluid normally circulating through an organ is diverted around it, either to reduce the functional work-load and allow the organ to heal, or to isolate the organ for the duration of a surgical procedure.

During cardiopulmonary bypass (heart/lung bypass) the blood is diverted away from the heart and lungs. As this is incompatible with life beyond a few minutes, surgical procedures involving the heart and main blood vessels must be coupled with artificial maintenance of cardiorespiratory function by a heart–lung machine. This is a mechanical system capable of pumping blood around the body and oxygenating it by means of an appropriate gas exchange unit. Such a system is obviously a safety-critical system.

A heart–lung machine was first used for the treatment of pulmonary embolism in 1937 and cardiopulmonary bypass was first used for open-heart surgery in 1953.

Uses of cardiopulmonary bypass

- As a temporary substitute for heart and lung function during surgery.
- As an extracorporeal membrane oxygenation system to assist respiratory exchange.
- To maintain life after severe damage to heart or lung (myocardial infarction (MI), trauma, pulmonary embolism).
- For short-term assist during invasive therapy (lung lavage).
- For the treatment of respiratory imbalance (hypercapnia).

A typical circuit is shown schematically in figure 22.11. The blood is drained from the patient by cannulation of the inferior and superior vena cavae. The heart and lungs are isolated by cross-clamping the aorta downstream of the aortic valve and the venae cavae at the entrance to the right atrium. The venous blood enters the extracorporeal circuit and is transported to the oxygenator where it is oxygenated and carbon dioxide is removed. The example shown incorporates a specific type of oxygenator, a 'bubble' oxygenator. When using this type of device a defoamer and a bubble trap are required to remove gaseous emboli which, if allowed to enter the patient's circulation, may cause brain, lung and kidney damage. A heat exchanger enables the blood to be cooled in a controlled manner inducing systemic hypothermia. Before being returned to the patient the oxygenated blood is filtered. This arterial filter removes microaggregates and any residual gas micro-bubbles. Blood is then returned to the body by means of a cannula in the aorta.

Blood released into the chest during surgery is sucked through a second blood circuit from the surgical field and returned to the system.

Oxygenators

Two types of oxygenator are currently in clinical use. These are direct contact and membrane types.

Direct contact types are usually 'bubble' type oxygenators which allow direct contact between the blood and gas. Oxygen is bubbled through a series of compartments containing venous blood. This process

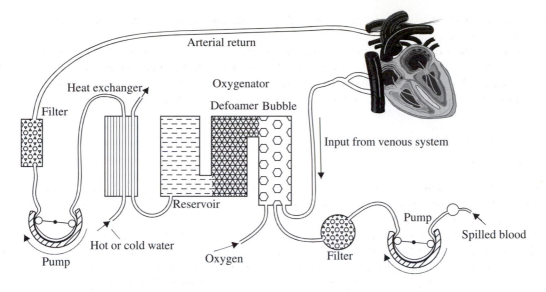

Figure 22.11. *A typical circuit for cardiopulmonary bypass.*

causes foaming of the blood. Defoaming is then carried out by passing oxygenated blood over silicone-coated screens. Contact between the blood and bubbles will result in damage to blood elements and protein denaturation due to the high interfacial energies involved. Foaming precipitates fibrin and increases platelet activation.

In membrane-type oxygenators blood is separated from the gas phase by a permeable polymeric sheet or tube.

Three types of membrane are currently used. Each has advantages and disadvantages.

- *Homogeneous*: the membrane takes the form of a continuous sheet of solution/diffusion membrane.
- *Microporous*: has a high void volume.
- *Composite*: thin film of solution/diffusion polymer on a microporous substrate.

Homogeneous. Gas diffuses into the membrane polymer at the interface and from the polymer into the blood. This process is slow, requiring long perfusion times and a large area for adequate exchange. (Materials used include PTFE, polyurethane and polysiloxane.)

Microporous. Pores are introduced during the manufacturing process (e.g. porous polypropylene).

Composite. A film of homogeneous polymer on a microporous substrate (e.g. 25 μm layer of polysulphone cast onto porous polypropylene.)

The choice of membrane will depend on the balance between gas permeability, strength and blood compatibility. There are two common geometries of membrane oxygenator design. These are multiple flat channels and multiple hollow fibre types. The area of membrane required to obtain the correct level of blood oxygenation will depend on the design and can be calculated if it is assumed that complete saturation of the haemoglobin is required.

Pumps

We can list the design requirements for a suitable pump:

- Be capable of flow rates up to 10 l min^{-1} and be able to achieve this against a pressure of 180 mmHg.
- Cause minimal clotting and thrombus formation.
- Not promote gas emboli.
- Have no hot spots which might damage the blood.
- Be easily sterilized.
- Be capable of being calibrated accurately.
- Be reliable.
- Cause low shear and turbulence.
- Give pulsatile flow.

The last design requirement listed is controversial. As pulsatile flow is more complex to achieve than steady flow, the use of a more sophisticated pump must be fully justified. The benefit of pulsatile flow remains a subject for debate. There is some suggestion that pulsatile flow is associated with an increase in O$_2$ consumption, reduced lactate accumulation and increased capillary blood flow to the brain.

Roller pumps are commonly employed. These have the advantages that the blood is only in contact with the tubing and little priming is required. However, there are disadvantages to the roller pump which does cause shear forces in the blood, will continue to appear to work against a blockage and causes stresses in the tube which may eventually crack.

The purpose of the heat exchanger is to control the blood temperature thus preventing progressive uncontrolled cooling. This is essential as abrupt temperature gradients result in cell damage and the release of gas from solution in the plasma. Filters are placed in the arterial return line and between the cannula used to clear the operative site and the oxygenator in order to remove particulate debris from the blood, thus preventing damage to the lungs, brain or kidney.

There is no ideal design of filter. If the pore size is too small the resistance of the circuit may rise as the filter blocks. In addition, the filter itself may cause blood damage. A typical design is made up of pleated polyester and has a pore size of about 40 μm.

22.3.3. *Haemodialysis, blood purification systems*

Our final example of a safety-critical system is that of haemodialysis. Dialysis is the removal of substances by means of diffusion through a membrane. Dialysis is used to replace the normal function of the kidneys in a patient with kidney failure. The loss of kidney function can be either acute or chronic. In acute renal failure, which can be caused by accident or disease, the kidneys will eventually recover their normal function. In the absence of dialysis the patient would die before the kidneys recovered. In chronic renal failure, the kidneys are permanently damaged and, in the absence of either a kidney transplant or regular dialysis, the patient will die.

Two types of dialysis are used. In peritoneal dialysis, the dialysing fluid is run into, and then out of, the patient's abdomen. This is a relatively simple technique that does not need either expensive equipment or access to the circulation, and it is used for certain patients with acute renal failure. Continuous ambulatory peritoneal dialysis (CAPD) has made peritoneal dialysis suitable for long-term use in chronic renal failure. In haemodialysis, blood is continuously removed from the patient, passed through an artificial kidney machine, and then returned to the patient.

Chronic renal failure patients who have not had a kidney transplant and who are selected as suitable for dialysis will be treated either by haemodialysis or peritoneal dialysis. Alternatively, a kidney can be removed from a live donor (usually a close relative) or from a person who has just died, and can be used to replace the kidneys in the chronic renal failure patient.

Chronic renal failure patients may be trained to use their own haemodialysis machine in their own home. This has many advantages. The risks of cross-infection are much reduced, because all the patients are effectively isolated from one another; the quality of the patient's life is improved; and the cost is reduced. In addition, the patient does not tie up expensive hospital facilities and staff, so that many more patients can be treated. It is worth emphasizing that this is a revolution in patient care as the patient is responsible for his own life-support system, and for doing many things that are usually the province of the doctor or nurse. Obviously the safety-critical aspects of the equipment design are extremely important.

The patient will need two or three dialysis sessions every week, each of several hours duration. The dialysis machine must always be working; and should it fail, it must tell the patient what is wrong. It must, in this situation, be repaired quickly—the patient's life depends on it. Most patients can manage three days without dialysis, but the machine must be repaired by the fourth day.

The function of the normal kidney

The two kidneys are bean-shaped organs, about 12 cm long and 150 g in weight. They are situated on the back of the abdominal wall. The top of the kidneys lies beneath the bottom two or three ribs and each contains about a million nephrons (figure 22.12). The nephron has two parts, the glomerulus and the tubule. The function of the glomeruli is to filter the plasma which is circulating in the capillary loops within Bowman's capsule. This is a passive process—it does not require any energy. The blood pressure in the capillary loops is about 60 mmHg (8 kPa), and about 25% of the cardiac output goes to the kidneys.

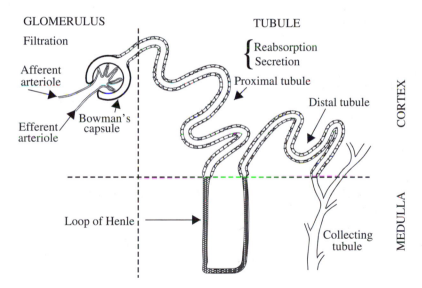

Figure 22.12. *Diagram of a single nephron. (Redrawn from A J Wing and M Magowan 1975 The Renal Unit (London: Macmillan).)*

The output of filtrate is about 1 ml/s/kidney, i.e. about 180 l day^{-1}. The total plasma volume is about 3 l, so that the plasma is filtered through the kidneys about 60 times a day. The filtrate then passes into the tubules. The total length of the tubules in each kidney is about 50 km. The tubules re-absorb electrolytes, glucose and most of the water, giving a total urine output about 1–2 l day^{-1}. This is an active process, which uses energy, and is continuously adjusted to maintain the correct fluid and electrolyte balance in the body.

The composition of the blood is very complex. The most important electrolytes are sodium, potassium, chloride, bicarbonate and calcium. The concentration of these electrolytes in normal plasma is given in table 22.5. The molecular weight of the substance, in grams, dissolved in 1 l of water, gives a concentration of 1 mol l^{-1}. The processes of filtration and re-absorption, together with secretion from the distal tubules, maintain the correct level of the electrolytes. Any departure of the electrolyte levels from normal will have an immediate effect on the health of the patient. If the serum sodium level is elevated, the patient's blood pressure will increase. Potassium changes the excitability of the cells, and an increase of serum potassium above 6 mmol l^{-1} can cause cardiac arrest without warning. An increased calcium level will cause the acid output of the stomach to increase, which can result in bleeding from peptic ulcers. A decrease in the calcium level will cause bone diseases. Most metabolic processes are very sensitive to the pH of the blood, which depends on the bicarbonate concentration.

Table 22.5. *The concentration of the more important electrolytes in normal plasma.*

	mmol l^{-1}
Sodium	132–142
Chloride	100
Bicarbonate	25
Potassium	3.9–5.0
Calcium	2.5

The electrolytes have low molecular weights (e.g. sodium 23, potassium 40). Organic chemicals in the blood have higher molecular weights (e.g. urea 60, bilirubin 600), and proteins have very high molecular weights (e.g. albumin 60 000, fibrinogen 400 000). Diffusion of substances across the dialysis membrane decreases with increasing molecular weight. Failure of the kidneys results in uraemia (retention of urea in the blood) and the concentration of urea in the plasma rises steeply from the normal level of 5 mmol l^{-1}. All the body systems are affected, both by the increasing concentration of waste products in the blood, and by the electrolyte imbalance. Urea is not particularly toxic, and it is thought that one important function of dialysis is to remove unidentified 'middle molecules' with molecular weights between 300 and 1500.

It can be seen that dialysis is much more than just the removal of waste products from the blood. Dialysis has to maintain the correct electrolyte balance within the body, maintain the correct pH of the blood, and control the fluid balance.

History of dialysis

Although the first haemodialysis in an animal was performed more than 90 years ago, the use of the technique in humans had to await the development of suitable membranes and anti-coagulants. Cellophane (cellulose acetate) membranes were developed during the 1920s and 1930s, and the commercial production of heparin started in the middle of the 1930s. The first haemodialysis on a human was performed by Kolff in 1943. He used a cellulose tube wrapped round a cylindrical drum which rotated in a bath of dialysis fluid.

The principle of haemodialysis remains essentially the same. Blood is withdrawn from the body and the patient is heparinized to prevent clotting of the blood in the extracorporeal circuit. The blood passes through an artificial kidney, in which it is separated from the dialysis fluid by a semi-permeable membrane. Electrolytes and small molecules pass freely through the pores in the membrane. The large molecules and the blood cells are unable to pass through the pores. Similarly, any bacteria present in the dialysis fluid are unable to pass into the blood.

Early haemodialysis involved the placing of cannulae into an artery and a vein, using a cutdown technique, in which an incision is made through the skin and the wall of the blood vessel. The blood vessels had to be tied off at the end of the dialysis. The early dialysis techniques altered the blood concentration of the electrolytes too quickly, so that the correction of the intracellular concentrations lagged behind. This could result in 'dialysis disequilibrium', in which the patient appeared to be worse after the dialysis. The development of the arteriovenous shunt in 1961 made repeated long-term dialysis possible, and the introduction of the arteriovenous fistula in 1967 further improved the ease of access to the circulation and minimized the restrictions that access to the circulation cause the patient.

Currently, two types of dialyser are in routine clinical use:

- Parallel plate dialyser, blood flows between parallel sheets of membrane.
- Hollow fibre dialyser, blood flows through hollow fibres which are bathed in dialysate.

Hollow fibre designs have the advantage of low priming volumes with a lower ratio of blood compartment volume to membrane surface area. Typical ratios are 65–86 ml of blood/m^2 of membrane for hollow fibre designs, and 70–100 ml of blood/m^2 of membrane for parallel plate configurations. Hollow fibre designs have the added advantage that the blood volume is unaffected by the transmembrane pressure gradient. Parallel plates bulge under pressure gradients with a consequent increase in the blood volume. Fibre designs have some disadvantages, having a higher residual volume and an increased tendency for clotting. This latter factor is less of a problem in the most recent designs.

Membranes

Membranes are made from a number of different materials including;

- Cellulose (processed cotton), Cuprophan.
- Substituted cellulose, where free OH^- groups on the surface of the cellulose are replaced with acetate groups in order to reduce complement activation.
- Synthetic material, e.g. polyacrylonitrate, polysulfone, polymethylmethacrylate.

It has been estimated that, in the absence of dialysis or transplantation, about 2000 patients between the ages 15 and 50 would die each year from untreated renal failure in a population of 50 million.

Principles of haemodialysis

The purpose of haemodialysis is to remove waste products from the blood, to maintain the correct electrolyte balance and the correct body pH, and to control the body's fluid balance. These functions are controlled by adjusting the composition of the dialysis fluid, and by altering the conditions on each side of the dialysis membrane.

Diffusion

The semi-permeable membranes used in artificial kidneys are typically about 10–20 μm thick (the more common thicknesses are 11, 13 and 15 μm), have a surface area of about 1 m^2 (varying between 0.5 and 2.5 m^2), and have pores which are about 500 nm in diameter. Obviously, substances which are larger than the pore size will not be able to pass through the membrane. This class of substances includes protein molecules and most substances with a molecular weight greater than 40 000. Molecules with a molecular weight less than 5000 will pass fairly easily through the membrane, and molecules of intermediate weights (5000–40 000) will pass slowly. Dialysis will only remove a small quantity of amino acids (molecular weights 75–204) and of some drugs, because they are bound to the plasma protein, and will therefore not pass through the membrane.

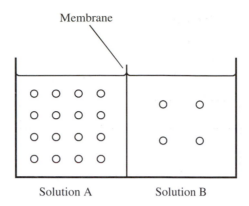

Figure 22.13. *Diffusion. Molecules will diffuse through the semi-permeable membrane from the strong solution A to the weak solution B.*

Figure 22.13 represents two solutions separated by a semi-permeable membrane. Solution A has four times the concentration of solution B. The molecules in the liquid are moving continuously, because of their thermal energy, and because the concentration of solution A is greater than that of solution B.

Four times as many molecules will strike the left-hand side of the membrane. As all molecules of the same size have the same probability of passing through the pores in the membrane, four molecules will move from solution A to solution B for every molecule that moves in the reverse direction. It is easy to see that the end result will be that the two solutions will have the same concentration. Diffusion through the membrane will still continue, but the same number of molecules will pass in each direction and the net result will be that the two solutions will have the same concentration. If each solution contains several different types of molecule, the diffusion of each molecule across the membrane will be proportional to the concentration gradient for that molecule, and the net effect will be to equalize the concentration of each molecule, as if the other molecules were not present.

It is obvious that the concentration of any molecule in the blood which will diffuse through the membrane, can be altered by suitably adjusting the concentration of that molecule in the dialysis fluid. In practice, the dialysis fluid would be stirred, so that the concentration close to the membrane remained the same as that in the rest of the fluid. The best possible concentration gradient would be maintained by continuously replacing the dialysis fluid with fresh fluid, so that the concentration in the dialysis fluid was always correct. If the fluid were flowing past the membrane, there would be a boundary layer, in contact with the membrane, which did not move. This is overcome by making the flow across the membrane turbulent.

Osmosis and ultrafiltration

Water molecules will also diffuse through the membrane. The concentration of water molecules in the stronger solution is less than that in the weaker solution, so that water molecules will diffuse from the weaker solution to the stronger one. The net effect is, again, to make the concentration of the solutions on each side of the membrane the same. This movement of water from the weaker to the stronger solution is called osmosis (figure 22.14).

If pressure is applied to the stronger solution, it is possible to stop the movement of water molecules across the membrane. The pressure needed to stop osmosis is the osmotic pressure. If the pressure is increased further, water molecules will be forced through the membrane. This could also be done by decreasing the pressure of solution B. The movement of water across a membrane as a result of hydrostatic pressure is

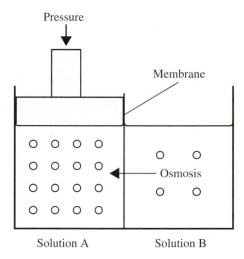

Figure 22.14. *Osmosis. Water will diffuse through the semi-permeable membrane from the weak solution B to the strong solution A. If a hydrostatic pressure is applied to solution A, water molecules can be forced in the reverse direction.*

called ultrafiltration. The amount of water which passes across the membrane is a function of the hydraulic permeability of the membrane, the transmembrane pressure and the surface area of the membrane.

Clearance

The effectiveness of the artificial kidney can be assessed by comparing the concentration, C_A, of a substance in the arterial blood flowing into the machine with the concentration C_V in the venous blood. If the blood flow is F ml min^{-1}, the rate of clearance, Q/t, is given by

$$\frac{Q}{t} = \frac{(C_A - C_V)}{C_A} F \quad \text{(ml min)}^{-1}.$$

The dialysis fluid

The principles of making up a suitable dialysis fluid should now be clear. If the concentration of any dialysable molecule is lower in the dialysis fluid than in the blood, it will be removed from the blood. The rate of removal will depend on the difference in concentration between the blood and the dialysis fluid. Some molecules should be removed as completely as possible, and are therefore left out of the dialysis fluid, so that the concentration gradient is maximized. Examples are urea, uric acid and creatinine. It is important that some molecules are maintained at the correct concentration in the plasma, and these are therefore added to the dialysis fluid at a suitable concentration. For instance, the plasma sodium level should be maintained at 132–142 mmol l^{-1}. Most patients with renal failure tend to retain body sodium, so it is usual to use a slightly lower concentration in the dialysis fluid to correct this. A typical concentration in dialysis fluid would be 130 mmol l^{-1}. The correct pH of the body is maintained by adding lactate or acetate to the dialysis fluid. The concentration of these is lower in the blood, so they diffuse into the blood, and are metabolized by the liver to produce the bicarbonate which regulates the pH. Calcium and magnesium levels in the plasma have to be controlled, and these produce particular problems. Both these ions are present in significant quantities in the water supply and vary both regionally, depending on the hardness of the local water, and seasonally. It is therefore usual to remove these ions from the water by using a water softener or a de-ionizer, so that the concentration in the dialysis fluid is always known.

Some trace elements, such as copper, zinc, manganese and fluoride, are present in low concentrations in the blood. The physiological effect of these is not completely understood, but it is known that, for instance, an excess of copper in the dialysis fluid can cause serious problems. This can happen if copper pipes are used downstream of a water softener.

A serious problem in some areas is caused by the presence of aluminium in the water supply. If the water contains a high proportion of dissolved iron, aluminium (in the form of alum) is added at the treatment works to floculate the iron, so that it can be removed. Aluminium can also occur naturally in the water. If the aluminium level exceeds 0.06 ppm, the dialysis patients will develop a brain disorder known as a dialysis dementia or dialysis encephalopathy. When a dialysis machine is installed in the home, the water supply is analysed; if the aluminium level is less than 0.06 ppm, a water softener will be used in the supply. If the aluminium level is higher than 0.06 ppm, a de-ionizer will be installed instead of the water softener. The ion exchange resin column in the de-ionizer has to be returned to the manufacturer for regeneration, which is expensive, so that they are only used where they are essential.

A further process that is used for purifying water is reverse osmosis. In principle, this involves placing the impure water, under high pressure, on one side of a semi-permeable membrane. The water which is collected on the other side of the membrane is very pure—typically 97% of the organic matter and 90% of the trace elements in the water supply will have been removed. The semi-permeable membranes are expensive to replace, so that the water supply is usually pre-treated. In particular, the membrane is destroyed by chlorine which is often added to water supplies. A large scale reverse osmosis plant might consist of particle filters to remove suspended matter, an activated carbon column to remove chlorine, a water softener, and then the reverse osmosis unit. Final purification would be done by a de-ionizer followed by bacterial filters.

Ultrafiltration is used to remove the excess water during haemodialysis. The dialysis fluid is made up to be isotonic (i.e. the total concentration of dissolved substances in the dialysis fluid is the same as in the blood), and the pressure gradient across the dialyser membrane is adjusted to give the required degree of ultrafiltration. This can be done either by increasing the pressure on the blood side of the membrane, or by decreasing the pressure on the dialysis fluid side.

Access to the circulation

The introduction of shunts and fistulae, which allow repeated access to the circulation, made long-term haemodialysis possible for chronic renal failure patients. Shunts are external to the body and fistulae are internal.

A shunt is formed by permanently placing PTFE and silicone rubber cannulae into an adjacent artery and vein. The cannulae are brought out through the skin, and are normally joined together, thus forming an arteriovenous shunt. The rigid PTFE tip is inserted into the vessel and tied off, and the silicone rubber catheter is brought through the skin and anchored in position. The connection to the dialysis machine is made by clamping off each arm of the shunt, separating the ends and then connecting the blood lines to the machine.

Although the shunt is easy to connect to the dialysis machine, it has many disadvantages. As it is outside the body it is very vulnerable, and restricts the patient considerably. The patient must always carry clamps in case the shunt is damaged, and must be capable of dealing with bleeding from the shunt. The materials used for the shunt are not thrombogenic, but clotting is still possible. The exit through the skin is a direct entry site for infection. Shunts are usually placed at the extremity of an arm or leg. Repair operations are necessary at intervals of a year or less and each repair operation reduces the number of possible sites for the next repair.

A fistula is formed by joining together a peripheral artery and a neighbouring vein. The increased blood flow causes the veins to become enlarged and thickened, so that it becomes easier to insert a large-bore needle into a vein. Blood is removed through this needle and returned through another needle inserted into the same vein. The necessary large amount of arterial blood can be removed without having to puncture an artery and the good flow rate through the fistula prevents clotting.

The patient is much less restricted by a fistula than by a shunt, but has to overcome the problem of repeatedly puncturing the vein. The fistula should last for the life of the patient. In some centres, it is becoming common practice to make a fistula on both sides, for instance, in both the left and right arms. Only one fistula is used for dialysis, and the other fistula is available in case of problems.

Dialysis machines

As there are many different types of dialysis machines in use, the discussion in this section will not deal with any particular machine. Figure 22.15 shows a basic dialysis machine, together with the extracorporeal blood circuit. In a single-pass system, the dialysis fluid would be used at a rate of 500 ml min^{-1}, so that 240 l of dialysis fluid will be used in an 8 h dialysis period.

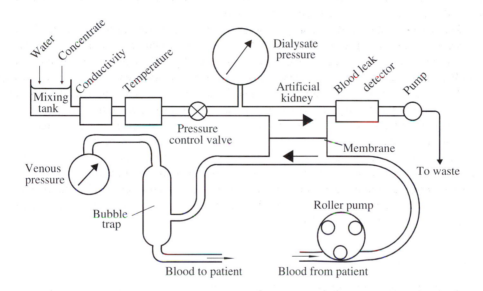

Figure 22.15. *Block diagram of a kidney machine.*

Because such a large volume of fluid is needed, home dialysis machines make up the dialysis fluid continuously from a concentrate, using some form of proportionating pump. The conductivity of the fluid is measured as a check that the concentration is correct; the temperature is also controlled and the pressure measured. The fluid then passes across one side of the dialyser membrane, and a blood leak detector ensures that the patient is not losing blood through the membrane. The dialysis fluid then flows to waste.

Blood is removed from the patient's shunt or fistula, and is pumped across the other side of the membrane. Before the blood enters the artificial kidney, heparin is added continuously to prevent clotting. Because both the flow and return needles are inserted in the same vein, there is no pressure difference to drive the blood through the extracorporeal circuit, and it must therefore be pumped. The heparin is added after the blood pump, where the blood pressure is high. Any leak in the heparin infusing circuit will cause a loss of blood from the system. If the heparin were added before the blood pump, a leak could cause air to be drawn into the extracorporeal circuit.

After leaving the kidney, the blood passes to a bubble trap, where any air trapped in the blood is removed and the venous pressure is measured, before it is returned to the patient. It is obviously undesirable for the patient to have a lengthened blood clotting time after the blood lines are disconnected. The anti-coagulant effects of the heparin can be reversed chemically, but the dosage is difficult to predict. It is therefore more common to stop the heparin infusion 30 min before the end of the dialysis period

Mixing the dialysis fluid

The dialysis fluid is continuously produced by diluting with water a pre-mixed concentrate containing all the necessary electrolytes. The dialysis fluid is not sterile—the sterility of the extracorporeal circuit is maintained by the dialyser membrane. As discussed previously, the water is softened or de-ionized so that the concentrations of calcium and magnesium can be properly controlled. The resulting water is very aggressive (that is, it is a powerful solvent), so copper pipes must not be used after the water softener.

There are several methods used for producing the correct dilution of the dialysis fluid concentrate. One possibility is to use a fixed ratio pump. If a dilution of 10:1 is required, it could be obtained by filling a 1 ml syringe with concentrate and a 10 ml syringe with water, emptying both syringes into a container, and stirring. This is the principle on which the fixed ratio pump works. The two halves of the pump, with their volumes in the appropriate ratio, are mechanically fixed together, so that the correct ratio of concentrate to water is always delivered.

Another alternative is to use variable speed pumps for the concentrate and water, and to alter the pump speed to give the correct dilution. The conductivity of the solution is measured to check the concentration, and the relative speed of the pumps is altered to maintain the correct conductivity.

After the dialysis fluid has been mixed, the conductivity is monitored. The monitoring circuit is completely separate from the conductivity control circuit, and uses a separate conductivity cell. Control and monitoring circuits are always separated, so that a breakdown in a control circuit will not affect the safety circuits. The conductivity is usually displayed on a meter, and an alarm will sound if the conductivity is either too high or too low, the equivalent concentration limits being 145 and 130 mmol l^{-1}, respectively.

Dialysate temperature

The dialysis fluid must be heated to between 36 and 42 °C. If the blood returned to the patient is at a very different temperature to that of the body, the patient will be uncomfortable. Cold blood may cause venous spasm and clotting, while, if the blood temperature rises above 42.6 °C, haemolysis will take place (i.e. the red cells are broken down and haemoglobin is released into the plasma). The temperature is maintained by a sophisticated heater with a three-term controller (integral, derivative and proportional control). Once again, there is a completely separate temperature measuring circuit with high- and low-temperature alarms. There is sometimes an alarm at 42 °C which mechanically latches out the heater.

Dialysate pressure

The pressure difference across the dialyser membrane controls the removal of water from the blood by ultrafiltration. The pressure in the extracorporeal circuit can be increased by using a gate clamp on the outflow from the dialyser, but this does not give very good control of pressure. It is usual to increase the transmembrane pressure by reducing the dialysate pressure below atmospheric pressure. This can be done by using a constant volume pump on the outflow from the dialyser and changing the resistance to flow by a needle valve upstream of the artificial kidney.

If the negative dialysate pressure is too high, the ultrafiltration rate may be excessive and the membrane may rupture, leading to a blood leak. If the dialysate pressure becomes positive with respect to the venous pressure, the reverse ultrafiltration will cause fluid to be added to the blood. An excessive positive dialysate pressure could rupture the membrane, and unsterile dialysing fluid would be added to the blood. The dialysate pressure is therefore controlled so that it is always negative with respect to atmospheric pressure, and alarm limits for the high and low pressure are set. However, what really needs to be controlled is the transmembrane pressure, because this is what controls the ultrafiltration rate. This is done by measuring both the venous and dialysate pressures, and controlling the difference between them. This allows the ultrafiltration rate to be set to an appropriate level for the patient. The maximum pressure difference is about 400 mmHg (50 kPa). The dialysate pressure may be measured either immediately upstream of the artificial kidney (see figure 22.15) or immediately downstream of it. The upstream and downstream pressures will be different, because of the resistance to flow of the kidney.

Blood leak detector

The blood circuit and the dialysate circuits are separated by the membrane. If blood appears in the dialysis fluid, there is a leak and the patient is losing blood. The blood leak detector consists of a light and a photodetector mounted on opposite sides of the outflow tube from the artificial kidney. If blood is present in the dialysate, the amount of light transmitted is reduced and the alarm sounds.

The extracorporeal blood circuit

The alarms on the extracorporeal blood circuit are particularly important and are shown in more detail in figure 22.16. The three main dangers are clotting, massive bleeding and air embolus. The continuous heparin infusion reduces the risk of clotting. The alarms on the blood circuit control the other two problems.

Blood is removed from the patient through either a large bore needle in a fistula, or through a shunt, at a flow rate of about 200 ml min^{-1}.

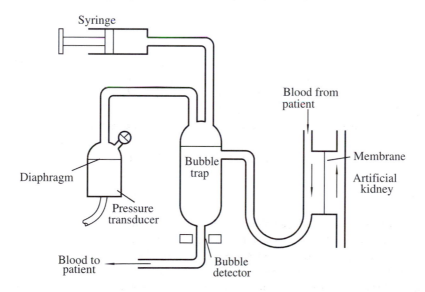

Figure 22.16. *The blood circuit of a kidney machine, showing the positions of the artificial kidney, bubble trap and pressure transducer; the bubble detector; and the syringe for draining off excess air.*

Both of these methods ensure that there is a free flow of blood to the dialysis machine; consequently, if the arterial connection is broken, the patient will suffer massive blood loss. The arterial pressure switch is therefore the first monitoring device in the blood line; it must be placed before the pump. It is only necessary to check either that the pressure in the blood line is greater than atmospheric, or that there is blood in the line. The pressure can be checked by a switch which is held open by the pressure in a distensible sac in the line. If the arterial line is disconnected, the pressure in the sac drops, the switch closes, and the alarm sounds. Alternatively, a photoelectric device (similar to the blood leak detector) can be used to check that there is blood in the line. This alarm will also stop the pump, so that air is not pumped into the patient's bloodstream. The design of the arterial blood lines is now such that a break is very unlikely, and many centres do not use an arterial pressure switch.

Great care must be taken to ensure that air is not introduced into the extracorporeal circuit. If the water used to make up the dialysate contains large amounts of dissolved air, this air will come out of solution in the dialysate. This tendency is increased by reducing the dialysate pressure below atmospheric. The undissolved air will pass across the membrane into the extracorporeal circuit. To remove this danger, the dialysing fluid circuit will contain a de-aerator. The air might be driven out of solution by raising the temperature and lowering the pressure in a chamber before the dialyser.

On the venous side of the dialyser is a bubble trap. The venous pressure is measured at the bubble trap, and some form of bubble detector is also used. The centre of the dialyser, bubble trap and the pressure transducer should be at the same level, so that errors are not introduced in the measured pressure. The pressure transducer is usually separated from the bubble trap by a disposable membrane, so that the transducer cannot be contaminated. As the name suggests, the bubble trap reduces the risk of air embolus by removing any air present in the blood. If the amount of air in the trap increases, it will have to be withdrawn using the syringe, which is normally clamped off. It can also be done automatically using a peristaltic pump controlled by the blood level detector.

The presence of an excessive amount of air in the bubble trap can be detected in two ways. The first is to use a blood level detector (a photoelectric detector) on the trap. This system is not foolproof—it is possible for air to be forced down the venous line without lowering the blood level in the trap, and frothing of the blood can take place without triggering the alarm. A better system is to use a photoelectric detector on the narrow outflow tube from the trap. If air is detected, the blood pump is switched off and the venous line is clamped to prevent air being introduced into the patient's vein.

If there is a break in the venous line, the bubble trap will drain, and the fall of the venous pressure to atmospheric will sound an alarm and stop the blood pump.

Alarms

All the alarms which have been described will have been designed as fail-safe systems, that is, if the alarm itself fails, it should fail in such a way that it indicates that the dialysis system is faulty. It is preferable that a malfunctioning monitoring system should stop the dialysis, rather than the patient having a false sense of security because the alarm does not sound when there is genuinely a fault.

The purpose of the alarms is to indicate to the patient that there is a potentially dangerous malfunction in the dialysis machine. In many cases, the alarms will automatically terminate the dialysis. The dialysate conductivity alarm will automatically divert the dialysis fluid so that it does not flow through the kidney if the concentration is wrong, and the blood line alarms automatically stop the blood pump. There is no justification for over-riding any alarm. Before the dialysis is restarted, the reason for the alarm must be found and corrected—the patient's life can depend on it.

22.3.4. Practical experiments

22.3.4.1. Effect of transmembrane pressure on ultrafiltration

Equipment

Pressure gauge, 0–450 mmHg (0–70 kPa).
250 ml measuring cylinder.
Hollow fibre dialyser.
Retort stand and clamps.

Method

Set up the apparatus as shown in figure 22.17. Establish a flow of 500 ml min^{-1} through the dialyser. With the pressure set to 0 mmHg, record the amount of ultrafiltrate collected in the measuring cylinder over a 30 min period. Repeat the experiment at pressure settings of 100, 200, 300 and 400 mmHg (13.5, 26.6, 40 and 53.2 kPa).

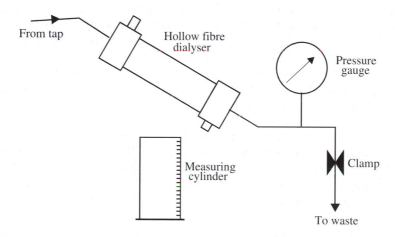

Figure 22.17. *Experimental arrangement for measuring ultrafiltration rate.*

Results

Plot a graph of ultrafiltration rate (ml min^{-1}) versus transmembrane pressure. Transmembrane pressure (TMP) is given by

$$\text{TMP} = P_\text{v} + P_\text{d}$$

where P_v is the pressure inside the hollow fibres and P_d is the negative pressure surrounding the hollow fibres. (In this experiment the outside of the fibres is open to the atmosphere, therefore $P_\text{d} = 0$.)
What would be the effect on ultrafiltration rate if P_d equals:

- A high negative pressure?
- A positive pressure greater than P_v?
- A positive pressure equal to P_v?

22.3.4.2.   Effect of blood flow on clearance rate

Equipment

Hollow fibre artificial kidney.
Retort stand and clamp.
1 set blood and dialysate lines.
4 l blood analogue fluid (isotonic saline).
24 sample bottles.
1 pair clamps.

Method

In this experiment, isotonic saline is used as the blood analogue and water as the dialysate. The experiment measures the sodium clearance rate.

Assemble the apparatus (figure 22.18) with 1 l of blood analogue (retain the other 3 l for use later). Ensure that the blood analogue and dialysate flow in opposite directions through the dialyser.

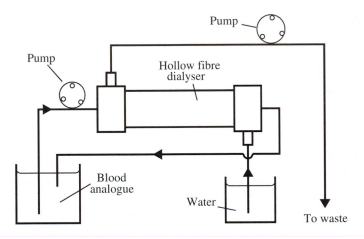

Figure 22.18. *Experimental arrangement for measuring dialyser clearance rate.*

It is also important that the blood analogue is pushed through the dialyser (positive pressure) and the dialysate is pulled through (negative pressure).

Set the pump speed to give a flow rate of 25 ml min^{-1}. Prime the dialysate lines first and then the blood lines. Commence timing the experiment when the blood analogue has filled the kidney. Take the first samples immediately, one before the dialyser and one after. Take the 'before' sample at the tube provided, first allowing some fluid to run to waste (for about 5 s). Take the 'after' sample from the line returning to the blood analogue container. Take further samples at 5, 15 and 30 min after the start.

Repeat the experiment with flow rates of 100, 200 and 300 ml min^{-1}, using a fresh litre of blood analogue each time.

Results

Analyse the samples for sodium content using a flame photometer. Calculate the clearance rates for each flow rate. Plot clearance rate against time for each flow rate on the same graph. Show mean clearance rate for each value of flow. Plot mean clearance rate for each value of flow against the flow. What is the significance of the dialyser clearance rate?

22.4.   DESIGN EXAMPLES

22.4.1.   *Safety-critical aspects of an implanted insulin pump*

Discuss the safety-critical aspects in a feasibility study for an implanted insulin infusion pump. What components are we talking about?

> Syringe
> Motor
> Control circuitry
> Communication system
> Sensors
> Power source
> Encapsulation

What should be the fail-safe condition?

> Fluid contained
> Motor stops
> Output becomes zero
> Any false signals stop the pump
> Rate of change limited
> Rate limited; no leakage

What are the main hazards?

> Motor dumps all the fluid; we can limit the total fluid volume but this then limits the life of the device.
> Leakage of battery power to the tissue; This would cause electrolysis and tissue necrosis.
> Who controls the device? Is it completely automatic or does the patient initiate infusion?
> Is there a problem of output blockage and hence high drive pressure?
> How do we test the system *in vivo*? What will the lifetime be? Can we test the sensors?

22.4.2.   *Safety-critical aspects of haemodialysis*

Consider the safety-critical issues in the design and use of haemodialysis equipment.

The main safety-critical concerns we must apply to a haemodialysis machine and each of its critical components are:

> What happens if it fails to function?
> Does it fail-safe and what margins of safety are allowed for?
> How likely is it to fail—how reliable is it?

Obviously the last question depends upon the quality of the materials used, the construction process and the ease and quality of maintenance.

The functions of haemodialysis for a patient with kidney failure are:

> Regulation of fluid volume.
> Regulation of fluid composition.

The components of a dialysis machine which are essential to its function are:

A mixing pump to mix the water and concentrate in the correct proportions.
A semi-permeable membrane with a large surface area and a higher pressure on the blood side than the dialysate side.
Pumps to move the blood and dialysate round the machine, with counter-current flow across the membrane.
Arterial and venous access to remove and return blood to the patient.

When considering how these components might fail in a critical manner as opposed to a non-critical manner we should probably focus on dialysate concentration, temperature and pressure across the membrane.

22.5.   PROBLEMS

22.5.1.   *Short questions*

a   Define a safety-critical system.
b   List three questions we must ask when considering the safety of a safety-critical system.
c   What should happen to the output of a demand-type cardiac pacemaker under 'fail-safe' conditions?
d   Is 1 mA at 5 V a typical power supply requirement for a cardiac pacemaker?
e   How is a pacemaker electrode normally placed into the heart?
f   In a unipolar pacemaker is the anode or the cathode placed in the heart?
g   What is second-degree heart block?
h   What does the acronym EMC stand for?
i   What type of interference does surgical diathermy/electrosurgery equipment produce?
j   Why are there usually two output buttons on a cardiac defibrillator?
k   Is the typical peak power output from a defibrillator 1, 10 or 100 kW?
l   Why does the design of a defibrillator aim to provide an output that returns rapidly to zero after the peak of the waveform?
m   Why can a defibrillator not be fired several times in quick succession?
n   What are the two major categories of prosthetic heart valve?
o   What was thought to be the main advantage of a tilting-disc heart valve?
p   List four of the design criteria for a prosthetic heart valve.
q   What is the purpose of a defoamer and a bubble trap in a cardiopulmonary bypass system?
r   What does CAPD stand for?
s   What three processes give rise to movement across a haemodialysis membrane?
t   What is the maximum temperature to which blood can be allowed to rise in a dialysis system?
u   Should software be regarded as a safety-critical component of a radiotherapy treatment system?
v   Name three of the more important electrolytes in blood.

22.5.2.   *Longer questions (answers are given to some of the questions)*

Question 22.5.2.1

Suppose 400 J is discharged into a patient through circular electrodes of 10 cm diameter and that 25% of the energy is dissipated underneath each electrode. How likely is it that burns will result? Assume a power density equal to that of the sun (~ 2 kW m^{-2}) is the limit and that we average the power over 10 s.

Answer

400 J over 10 s is 40 W.
25% of this (10 W) is dissipated underneath each electrode.
The electrode area is 78 cm^2, so the power density is 127 mW cm^{-2}.
The sun gives 200 mW cm^{-2} so a burn is unlikely. If the area was much less of course then the outcome would be very different.

Question 22.5.2.2

Discuss why software included in medical equipment should be classified as 'safety critical'.

Question 22.5.2.3

Discuss the question 'Can safety considerations in equipment design prejudice performance/efficacy?'

Question 22.5.2.4

Describe the principles and practice of transcutaneous electrical nerve stimulators (TENS) and give a reasoned argument as to whether or not these should be considered to be safety-critical devices.

In the design of a TENS device the sensation threshold (I_t), in milliamperes, is measured as a function of the duration (d), of the output pulse. The measurements can be fitted to an equation of the form

$$I_t = k + s/d$$

where k and s are constants.
    If k has the value 2 and $s = 200$, where d is measured in microseconds, then calculate the pulse duration that should be used to maximize the battery life of the TENS device.

Answer

The principle of TENS is to introduce impulses that will compete with the transmission of pain. This theory is not proven. However, you can also argue for the release of endorphins or indeed that the whole effect is a placebo effect.
    The practice is to use surface electrodes to apply a signal consisting of pulses of duration up to 0.5 ms and amplitude up to 70 V. Pulse trains are used. TENS is quite widely used during childbirth and for the relief of chronic pain.
    The definition of a 'safety-critical' device should be given. This should ask the question 'What if a failure occurs?'. In this case it should be considered what happens if the output rose to maximum as a result of a failure. My opinion is that these devices need not be considered as safety critical as they can be designed such that the output could not cause cardiac fibrillation.
    Numerical part: 100 μs.

Answers to short questions

a   A safety-critical system is one where the consequences of failure of the system can be life threatening. It may be either a diagnostic or therapeutic system.
b   What if it fails to function? Does it fail safe? What margins of safety have been applied? What are the maintenance requirements? This list is not exhaustive of the questions to be asked when designing a safety-critical system.

c   A demand-type pacemaker should revert to fixed-rate pacing under fault conditions.

d   No, battery life would be too short. 10 μA at 5 V is more likely as the power requirement for a pacemaker.

e   The pacemaker electrode is normally passed up the venous system into the ventricles.

f   The cathode is placed in the heart.

g   In second-degree heart block only some of the atrial impulses pass through to the ventricles.

h   EMC is electromagnetic compatibility.

i   Surgical diathermy/electrosurgery equipment produces radio-frequency interference.

j   Two output buttons are used on a defibrillator to reduce the chance of accidental discharge.

k   100 kW is a typical peak power output from a defibrillator.

l   If a defibrillator has a long tail on the output pulse then refibrillation can occur.

m   Defibrillators store charge on a capacitor before firing and the capacitor takes time to recharge between firings.

n   The main categories of heart valve are mechanical and tissue types.

o   The tilting-disc valve should offer less resistance to flow than a ball and cage valve.

p   A prosthetic heart valve should last for about 35 million cycles, have a large orifice area, be biocompatible, have minimal regurgitation, be easy to implant. These are some of the design criteria for a prosthetic heart valve.

q   Gaseous emboli can be removed from the blood using a bubble trap and defoamer.

r   CAPD stands for continuous ambulatory peritoneal dialysis.

s   Diffusion, osmosis and ultrafiltration take place through a dialysis membrane.

t   42.6 °C. Above this temperature haemolysis (red cell destruction) occurs.

u   Yes, software should be regarded as a safety-critical component of a radiotherapy treatment system. Software errors can cause incorrect calibration and hence the wrong patient dose.

v   Sodium, potassium, chloride, bicarbonate and calcium are the most important electrolytes in blood.

BIBLIOGRAPHY

Bodnar E and Frater R 1991 *Replacement Cardiac Valves* (Oxford: Pergamon)

Bronzino J D 1995 *The Biomedical Engineering Handbook* (Boca Raton, FL: CRC)

Chandran K B Heart valve prostheses—*in vitro* flow dynamics *Encyclopaedia of Medical Devices and Instrumentation* ed J G Webster (New York: Wiley) pp 1475–83

Dorson W J and Lona J B 1988 Heart lung machines *Encyclopaedia of Medical Devices and Instrumentation* ed J G Webster (New York: Wiley) pp 1440–57

Drukker W, Parsons F M and Maher J F (eds) 1979 *Replacement of Renal Function by Dialysis* (Hague: Nijhoff)

Living with Risk The BMA Guide (Penguin)

Pocock S J 1983 *Clinical Trials* (New York: Wiley)

Polyer S and Thomas S A 1992 *Introduction to Research in Health Sciences* (Edinburgh: Churchill Livingstone)

Shim H S and Lenker J A 1988 Heart valve prostheses *Encyclopaedia of Medical Devices and Instrumentation* ed J G Webster (New York: Wiley) pp 1457–74

Unger F (ed) 1984 *Assisted Circulation* (Berlin: Springer)

Yoganathan A P, Chaux A and Gray R 1984 Bileaflet tilting disc and porcine aortic valve substitutes: *in vitro* hydrodynamic characteristics *JACC* **3** 313

GENERAL BIBLIOGRAPHY

This is a general bibliography to the subject of Medical Physics and Biomedical Engineering. Texts which are specific to the subject of a particular chapter in the book are given in the bibliography which follows each chapter.

Ackerman E, Ellis V B and Williams L E 1979 *Biophysical Science* 2nd edn (Englewood Cliffs, NJ: Prentice-Hall)

Aird E G A 1975 *An Introduction to Medical Physics* (London: Heinemann Medical)

Alpen E L 1997 *Radiation Biophysics* (New York: Academic)

Belcher E H and Vetter H 1971 *Radioisotopes in Medical Diagnosis* (London: Butterworths)

Bomford C K, Kunkler I H and Sherriff S B 1993 *Walter and Miller's Textbook of Radiotherapy* (Edinburgh: Churchill Livingstone)

Brown B H and Smallwood R H 1981 *Medical Physics and Physiological Measurement* (Oxford: Blackwell)

Cameron J R and Skofronick J G 1978 *Medical Physics* (New York: Wiley)

Cromwell L, Weibell F J, Pfeiffer E A *et al* 1973 *Biomedical Instrumentation and Measurements* (Englewood Cliffs, NJ: Prentice-Hall)

Curry T S, Dowdey R C and Morray R C 1990 *Christensen's Physics of Diagnostic Radiology* (Lea and Febiger)

Dendy P P and Heaton B 1987 *Physics for Radiologists* (Oxford: Blackwell)

Diem K and Lentner C (eds) 1975 *Documenta Geigy: Scientific Tables* (Basle: Ciba-Geigy Ltd)

Duck F A 1990 *Physical Properties of Tissue* (London: Academic)

Duck F A, Baker A C and Starritt H C 1999 *Ultrasound in Medicine* (Bristol: IOP Publishing)

Geddes L A and Baker L E 1975 *Principles of Applied Biomedical Instrumentation* (New York: Wiley Interscience)

Hobbie R K 1997 *Intermediate Physics for Medicine and Biology* 3rd edn (New York: Springer)

Horowitz P and Hill W 1989 *The Art of Electronics* (Cambridge: Cambridge University Press)

Johns II E and Cunningham J R 1983 *The Physics of Radiology* 4th edn (Illinois: Thomas)

Marieb E N 1991 *Human Anatomy and Physiology* 3rd edn (New York: Benjamin/Cummings)

Memmler R L 1987 *The Human Body in Health and Disease* (London: Lippincott)

Meredith W J and Massey J B 1977 *Fundamental Physics of Radiology* (Bristol: Wright)

Plonsey R and Fleming D G 1969 *Bioelectric Phenomena* (New York: McGraw-Hill)

Polyer S and Thomas S A 1992 *Introduction to Research in Health Sciences* 2nd edn (Edinburgh: Churchill Livingstone)

Stanford A L 1975 *Foundations of Biophysics* (New York: Academic)

Underwood J (ed) 1992 *General Systematic Pathology* (Edinburgh: Churchill Livingstone)

Webb S 1988 *The Physics of Medical Imaging* (Bristol: Hilger)

Webster J G 1988 *Encyclopaedia of Medical Devices and Instrumentation* (New York: Wiley) p 800

Webster J G (ed) 1998 *Medical Instrumentation: Application and Design* 3rd edn (New York: Wiley)

INDEX